The Molecular Biology of Cancer

Dedication

We would like to dedicate this work to our parents, whose support and encouragement throughout our lives, education, and subsequent career development has been a major factor in our success and peace of mind. We know that Stella's father, Andrew, who we recently lost to us, would be especially proud of this. We also thank our daughter, Charlotte, for always loving us and helping us to maintain a balance between work and private life and for being a constant reminder of the joy that family life can bring.

Mike Khan and Stella Pelengaris

The Molecular Biology of Cancer

Edited by

Stella Pelengaris and Michael Khan
University of Warwick

With contributions from

Maria Blasco, Centro Nacional de Biotecnologia
Norbert C.J. de Wit, University of Warwick
Martine Roussel, St Jude Children's Research Hospital
Nicky Rudd, University Hospitals of Leicester and LOROS Hospice
Christiana Ruhrberg, University College of London
David Shima, Cancer Research UK
William Steward, University of Leicester
Charles Streuli, University of Manchester
Anne Thomas, University of Leicester
Esther Waterhouse, University Hospitals of Leicester and LOROS Hospice
Cassian Yee, Fred Hutchinson Cancer Research Centre

Blackwell
Publishing

BLACKWELL PUBLISHING
350 Main Street, Malden, MA 02148-5020, USA
9600 Garsington Road, Oxford OX4 2DQ, UK
550 Swanston Street, Carlton, Victoria 3053, Australia

First published 2006 by Blackwell Publishing Ltd

1 2006

Library of Congress Cataloging-in-Publication Data

The molecular biology of cancer/edited by Stella Pelengaris and Michael Khan; contributors, Maria Blasco . . . [*et al.*].
    p. ; cm.
  Includes bibliographical references and index.
  ISBN-13: 978-1-4051-1814-9 (pbk. : alk. paper)
  ISBN-10: 1-4051-1814-8 (pbk. : alk. paper) 1. Cancer–Molecular aspects.
  [DNLM: 1. Neoplasms–genetics. 2. Molecular Biology. QZ 202 M7182 2006] I. Pelengaris, Stella.
II. Khan, Michael. III. Pelengaris, Stella. IV. Khan, Michael. V. Blasco, Maria.

  RC269.M65 2006
  616.99′4042–dc22

                           2005015464

A catalogue record for this title is available from the British Library.

Set in 9/11pt Trump Mediaeval
by NewGen Imaging Systems (P) Ltd, Chennai, India
Printed and bound in the United Kingdom
by TJ International Ltd, Padstow, Cornwall.

The publisher's policy is to use permanent paper from mills that operate a sustainable forestry policy, and which has been manufactured from pulp processed using acid-free and elementary chlorine-free practices. Furthermore, the publisher ensures that the text paper and cover board used have met acceptable environmental accreditation standards.

For further information on
Blackwell Publishing, visit our website:
www.blackwellpublishing.com

Contents

Contributors

Maria Blasco, PhD, is Director of the Molecular Oncology Program at the Spanish National Cancer Center in Madrid. She has been at the forefront of research into mammalian telomeres and telomerase.

Norbert C.J. de Wit, PhD, is currently completing training in Clinical Chemistry at Jeroen Bosch Hospital in the Netherlands. He previously completed a PhD in proteomics in a joint program through the University of Warwick and Utrecht University.

Mike Khan, PhD, FRCP, is a Consultant Endocrinologist and Physician at the University Hospitals of Coventry and Warwickshire, and is Head of Molecular Medicine at the University of Warwick and Warwick Medical School. He was elected as a fellow of the Royal College of Physicians in 2002 and as a member of the Association of Physicians in 2004. His main research interests have been in the regulation of tissue growth and plasticity during development and in adult tissue homeostasis. Currently, he is collaborating with mathematicians and others in a systems biology approach to define key gene networks involved in regulating cell fate.

Stella Pelengaris, PhD, is a Senior Research Fellow in Molecular Medicine in the Department of Biological Sciences at the University of Warwick Medical School. While working as a PDRA at the Imperial Cancer Research Fund, she established a series of unique model systems for studying the role of c-MYC and apoptosis in cancer initiation and reversal. Since 1988 she and Mike Khan have jointly run the cancer research group at the University of Warwick where, in collaboration with Gerard Evan, they have confirmed the inherent tumor-suppressor activity of c-MYC (apoptosis) as a major barrier to oncogenic deregulation of c-MYC. The group is currently researching the role of c-MYC in both oncogenesis and in the apoptosis of beta cells in diabetes. Dr Khan and Dr Pelengaris jointly run various postgraduate training courses at masters level and beyond in cancer biology and in lipid metabolism/diabetes. They also are coauthors of the textbook *Lipids and Diabetes* (2002).

Martine Roussel, PhD, is Professor in the Department of Tumor Cell Biology at St Jude's Children's Cancer Research Hospital in Memphis, TN, and is also affiliated with the Department of Molecular Sciences at UT Memphis. She is currently researching the role of cell cycle regulators and tumor suppressors in development and cancer.

Nicky Rudd is Consultant and Head of the Department of Palliative Medicine at the University Hospitals Leicester. She also holds an appointment at the Leicestershire Hospice and is Chair of the LNR Palliative Care Network.

Christiana Ruhrberg, PhD, is a Medical Research Council Fellow at University College London. Her current research focuses on the codevelopment of nerves and blood vessels during development.

David T. Shima, PhD, is Senior Vice President of Research and Preclinical Development at the Eyetech Research Center in Lexington, MA. Before joining Eyetech in 2002 to establish and oversee research and pipeline initiatives, Dr Shima was the Director of the Endothelial Cell Biology Laboratory at Cancer Research in the UK.

Will Steward is Professor of Medical Oncology and Head of the University of Leicester Department of Cancer Studies and Molecular Medicine. His main research interests are chemoprevention and new drug development with a focus on targeted therapies.

Charles Streuli is Professor of Cell Biology at the University of Manchester. He has worked at the Imperial Cancer Research Fund in London and has been the recipient of the EMBO Long Term and Fogarty International Fellowships, as well as a Wellcome Senior Research Fellowship in Basic Biomedical Science.

Anne Thomas is Senior Lecturer and Honorary Consultant in Medical Oncology at the University of Leicester. She specializes in the management of patients with gastrointestinal tumors, and her research interest is conducting phase I clinical studies; particularly those involving angiogenesis inhibitors.

Esther Waterhouse is a Consultant in palliative care at University Hospitals Leicester. She has trained in General Practice and Palliative Medicine and is involved in educating students and health professionals in all aspects of care for patients with life-limiting illness.

Cassian Yee is an Associate Member of the Program in Immunology at the Fred Hutchinson Cancer Research Center and an Associate Professor at the University of Washington. His research focuses on the development of immunotherapeutic strategies for the treatment of patients with cancer.

Preface

Based on our experience of teaching undergraduates and postgraduates it became clear that no single current resource covered in detail the cellular and molecular changes that give rise to cancer alongside the basic principles of biology without which these can not be readily understood. We had not intended to write a textbook at this stage in our careers, but realized that there was a real need for such a work for undergraduates, medical students, and even established researchers in the field. Very few cancer molecular biology textbooks were available that started at the beginning, using a format and language easy to digest and included not only a comprehensive description of all aspects of cancer biology but also important chapters on diagnosis, treatment, and care of cancer patients. Students are first introduced to an overview of the cancer cell (Chapter 1), and of selected human cancers (Chapter 2), following which the textbook covers in depth those key cellular processes of greatest relevance to cancer. Chapters 3–13 cover the full range of cancer-relevant biology, including highly topical and important areas not traditionally covered in any depth in such works, such as apoptosis, telomeres, DNA damage and repair, cell adhesion, angiogenesis, immunity, epigenetics and the proteasome, as well as traditionally important areas such as cell cycle control, growth regulation, oncogenes, and tumor suppressors. The book then gives a description of cancer diagnosis, treatment and care of cancer patients, essential not only to medical students and oncologists, but also important for cancer researchers and biology students who need to have a broader view of cancer and its impact. Finally, Chapter 18 describes the application of new technology in cancer research.

The role of textbooks as information repositories is increasingly under threat. Yet, even now that we are well into the new millennium, with students and researchers alike bathed in seemingly limitless available information on the worldwide web, textbooks still exist. Why is this? With the near universal availability of internet access to students and researchers, the most current information is potentially available to any interested party almost instantaneously. No printed source can hope to provide the same immediacy of the latest breakthroughs or experimental findings, although they are free of the distractions of on-line gambling, 24 hour shopping, and less savoury diversions that plague the internet. However, limitless information creates new problems, namely how to evaluate, correlate and place into context this wealth of knowledge. More than a million cancer-related publications are referenced on Medline alone, and even for the initiated it can prove daunting to attempt to construct a balanced overview of the many aspects of cell and molecular biology that impact cancer. Because of these difficulties, one of the key aims of this book is to provide in a single source the necessary framework within which new information can subsequently be aligned and a more comprehensive, but still contextual, understanding of cancer achieved. In particular, we have taken the opportunity to highlight controversial areas and to identify areas of research promise, whilst establishing potential links between often diverse sub-disciplines in a coordinated and accessible way. Hopefully, having read this book, the reader will be suitably equipped to better understand the significance and relevance of new research papers on cancer and be able to place it into an overall picture of the disease. Moreover, the book also provide valuable insights into the important questions that remain to be addressed.

The issue of references, how many and where to cite, is often difficult to judge. One has to balance the flow of the text with the need to give pointers to the reader for further information and to highlight key studies.

This textbook can be used by undergraduates in biology and medical students and can be used in cancer biology courses structured either for a quarter or semester system. Moreover, the book will be of value to those preparing for professional exams in medicine and oncology and for established cancer researchers seeking a single-source overview of all aspects of cancer.

Features

We have included a number of features to facilitate the use of this textbook to teach cancer biology:

- Each chapter builds on concepts learned in previous chapters and is organized in a similar fashion, starting with key points and an introduction and concluding with a summary and future directions section, key experiments in the field, study questions and topical suggestions for further reading. Answers to questions are provided at the end of the book.
- Chapters also include a list of key outstanding questions remaining in the field.
- Each chapter contains text boxes that provide additional and relevant information as it relates to a described concept and are fully illustrated throughout.
- An accompanying website contains a more comprehensive reference list, additional text boxes, all original art in downloadable format, and is accessible through www.blackwellpublishing.com/pelengaris. The website is regularly updated.

Stella Pelengaris

Acknowledgments

An enormous number of talented scientists contributed to the knowledge described in this textbook. We acknowledge the many colleagues, past and present, whose important work could not be referenced in the text proper due to space constraints. Please refer to the accompanying website for the more detailed reference list. In addition, we apologize if we failed to adequately identify contributions in the text and website reference sections. This oversight was not intentional, but rather a reflection of the overwhelming number of contributors to this field.

We thank mentors past and present for their help and encouragement; Martin Raff and Anne Mudge for making cell biology interesting and intelligible, and Gerard Evan for introducing us to the world of cancer research. We thank our friends and colleagues who took time from their hectic research and clinical commitments to contribute to this book. We also thank our dedicated research team, Sylvie Abouna, Linda Cheung, Vicki Ifandi, Goran Mattsson, and Sevasti Zervou for bearing with us while we were writing and editing this book. A special thanks is due to our graduate student Sam Robson for reading and commenting on several of the chapters, to David Epstein FRS, our friend and colleague, for taking on too many tasks while we were occupied with this venture and to our other work colleagues at the University of Warwick for their encouragement and for helping to provide a supportive environment. Finally we thank Nancy Whilton, Elizabeth Frank, Rosie Hayden, and Sarah Edwards, at Blackwell Publishing, for their patience, help and advice without which this book would not have been possible.

We also acknowledge the contributions of our outside reviewers: Stewart Martin of Nottingham University; Brian Keith, University of Pennsylvania; S.J. Assinder of the University of Wales, Bangor; Satya Narayan of the University of Florida; Mary Jane Niles of the University of California, San Francisco; Fiona Yull of Vanderbilt University; Alison Sinclair of the University of Sussex; Amy H. Bouton of the University of Virginia; Elizabeth J. Taparowsky of Purdue University; Gary Gallick, University of Texas, MD Anderson Cancer Center; Michael Sulzinski of the University of Scranton; as well as those who have chosen to remain anonymous.

Reviewing is an enormous and time-consuming activity. We greatly appreciate the time spent by our reviewers, generating insightful and helpful comments.

The following conventions are used in this book

Readers will find various different conventions employed in different sources, we use the following here.

1. Proteins are given in roman and genes in italics. Greek letters and hyphens are generally not used in the gene. Full gene names if used are not in italics.
2. Human proteins and the genes from which they are encoded when abbreviated appear in capitals, except where the first letter is by normal usage lower-case, for example, c-MYC and *c-MYC*; p14^{ARF} and *ARF*, RAS and *RAS*.
3. Mammalian but non-human (e.g. mouse, rat) proteins as below have the first letter only capitalised, and similarly the genes from which they are encoded have the first letter capitalised, for example, c-Myc and *c-Myc*; p19^{Arf} and *Arf*.
4. Yeast, zebrafish, fruitfly proteins as other non-human have the first letter capitalised but genes take all lower-case letters, for example, Cdc25 and *cdc25*.

1
Introduction

Stella Pelengaris and Mike Khan

A new scientific truth does not triumph by convincing its opponents and making them see the light, but rather because its opponents eventually die, and a new generation grows up that is familiar with it.
Max Planck

KEY POINTS

- Cancer is a genetic disease.
- Despite the undoubted biological importance of studying rare familial "monogenic" cancer syndromes, most cancers are "sporadic", with cancer-causing gene mutations occurring in adult somatic cells.
- Most cancers may arise through combinations of avoidable or unavoidable carcinogens on a background of, as yet, largely unknown inherited polymorphic alleles.
- In general, factors that cause mutations and those that increase replication can combine to cause cancer.
- Cancer is a clonal disease arising by the multistep accumulation of genetic or epigenetic changes in tumor-suppressor genes, oncogenes and "caretaker" genes that favor expansion of the new clone over the old – a process akin to Darwinian evolution. Natural selection will favor expansion of a given clone that carries advantageous characteristics.
- These changes allow a normal cell to achieve the "hallmark" features of cancer, which have been neatly summarized by Hanahan and Weinberg as:

  a. Capacity to proliferate irrespective of exogenous mitogens;
  b. refractoriness to growth-inhibitory signals;
  c. resistance to apoptosis;
  d. unrestricted proliferative potential (immortality);
  e. capacity to recruit a vasculature (angiogenesis);
  f. ability to invade surrounding tissue and eventually metastasize.

- There is still intense debate about when and where, in the life history of the cancer, the genetic changes required for metastases occur. It is not clear how natural selection would favor expansion of a clone of cancer cells with metastatic capabilities within the primary tumor, unless such mutations provide a growth advantage. Possibilities include:

  a. Potential metastatic behavior is serendipitously acquired early in tumorigenesis as a by-product of mutations promoting growth of the primary tumor (supported by some gene expression profiling studies of whole tumors); or
  b. mutations in specific metastasis-suppressor genes, that do not confer a growth advantage to the primary tumor, occur at a later stage, possibly once cancer cells have begun circulating.

- Also a matter of debate is the identity of the cell of origin for any given cancer – stem cells that partially differentiate or differentiated cells that dedifferentiate.
- The cancer microenvironment, including the tissue stroma and inflammatory milieu, is a major determinant of the success of tumorigenesis.
- Cancer-contributing mutations that enable some of the hallmark behaviors noted above may occur in cells other than the cancer cell itself. Thus, mutations in stromal cells, that are not themselves cancerous, may allow them to promote a microenvironment that is more supportive for the cancer cells.
- Expression profiling of tumors (as exemplified by the "poor prognosis signature" defined by van't Veer and colleagues) may enable more accurate diagnosis/prognosis of cancers. As this process is ultimately refined to analyses of single cells, such profiling may address numerous unanswered questions in cancer biology.

This chapter is deliberately aimed to be provocative and stimulating. We will introduce and summarize the concepts and topics to be covered in the book, but with continual emphasis on new thinking and key novel models for oncogenesis.

Cancer has been recognized throughout recorded history and was known to the ancient Egyptians (see Appendix 1.1, history of cancer) but it was not until the seventeenth century that the formal study of cancer (oncology) was first documented. As with much of biology, the last 50 years has witnessed spectacular progress in describing the fundamental molecular basis of cancer following the advent of molecular biology and genetics. What is still a matter of some concern is that such exponential progress in describing the biology of cancer has, as yet, not translated into an equally impressive progress in the treatment of most common cancers (see Fig. 1.1). At first glance the biology of cancer appears straightforward: cancer cells stop obeying the restraints imposed on cells within the adult organism and instead multiply uncontrollably and in places they should not. However, in order to achieve such independence cancer cells must overcome numerous intrinsic and extrinsic barriers which seek to prevent such selfish behavior before it can threaten the survival of the entire organism. In this book we will describe the means by which normal cells become cancer cells and the key cellular processes which are disrupted along the way. We will also describe how this basic knowledge has been translated into improved diagnostics and therapeutics for cancer

patients. Along the way we will make some predictions as to where the new breakthroughs may come from and offer our humble opinions as to why knowledge has as yet failed to yield the anticipated more effective anticancer therapies.

Cancer poses a major threat to already overstretched healthcare services and the magnitude of the problem has been elegantly summarized by Dr Gro Harlem Brundtland, Director-General, World Health Organization: "The global burden of cancer continues to increase. In the year 2000, 5.3 million men and 4.7 million women developed a malignant tumor and 6.2 million died from the disease. The number of new cases is expected to grow by 50% over the next 20 years to reach 15 million by 2020."

Cancer is a major health issue of our time and is responsible for more than 10% of deaths worldwide and more than 25% in some countries. Lung cancer remains the commonest cancer worldwide, accounting for 1.2 million new cases per year, followed closely by breast cancer and colorectal with around 1 million new cases.

The high incidence of this disease, its life-threatening nature, and often unsatisfactory management has motivated academic researchers and those from the biotechnology and pharmaceutical industries to focus on the causes and potential treatments of cancer on a scale unparalleled in almost any other disease area. At present, there are almost 500 products in clinical trials of which 100 are in phase III, with breast cancer and non-small-cell lung cancer receiving the most attention.

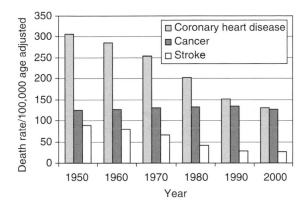

The incidence of cancer is not declining when compared to other major diseases, yet in the US alone more than $4.7 billion per year is spent on cancer research. Leland Hartwell and others at a recent meeting of the American Association of Cancer Research identified the following areas, in addition to developing new therapies, as key targets to address this major public health issue:

1. More coordinated and concerted activity between researchers – this would require establishing the necessary infrastructure for facilitating collaborative working and information exchange.
2. Testing drugs/agents in early stage of the disease rather than as at present largely in the end-stage cancer (we may be underestimating the potential of many drugs/therapies for this reason).
3. Real-time monitoring of treatments in early-stage cancers – though to identify earlier stages will require improved biomarkers.
4. Use RNAi to explore combinations of targets.
5. Understanding chromosomal aberration (occurs very early in mouse tumors).
6. Exploiting genomic instability in therapeutics. Understand more about DNA repair and repair of double strand breaks (latter unusual in mouse unless telomeres shortened).
7. Improved diagnostics from blood and body fluids – proteomics (less than 1% of proteins in blood identified, and less than 20% of these licensed for diagnostics).

Fig. 1.1 Data from Center for Disease Control.

In general cancers begin with a mutational event in a single cell and then develop in multiple stages through the acquisition of further mutations that are inherited by the progeny of that cell when it divides – cancer is a clonal disease (Fig. 1.2). Mutations are not the only way in which a would-be cancer cell acquires inactivation or activation of a key gene/protein, in a way

that can be passed on through successive cell generations, as the same outcome can arise by epigenetic factors that alter chromatin structure, without altering the coding DNA. Thus, sometimes the term "epimutations" is used to encompass these two major routes by which cancer cells acquire aberrant expression/activity of key genes and proteins. The average adult human has been estimated to contain as many as 10^{14} cells (i.e. 100,000,000,000,000 cells), most of which could theoretically become a cancer cell given the right sort of genetic (mutations) and epigenetic changes. In fact, cancer is really quite unique in that epimutations in a single cell can give rise to a devastating disease. Replicating cells may be most vulnerable to cancer-causing mutations and although some cell types, of which adult nerve cells are a good example, may avoid becoming cancer cells because they are essentially non-proliferating in the adult, most cells either regularly do or can at a pinch replicate. Most adult cells survive on average for 4–6 weeks and then have to be replaced; over a thousand billion cells may die each day and are renewed either by replication of existing cells or from stem cell precursors. Given that every cell gets a substantive amount of daily DNA damage and 10^{11} or more of them will replicate each day, that is a lot of potential cancer cells! With this in mind a cancer might be expected to be a frequent occurrence and yet this only happens in 1 in 3 people and usually even then only after 60 or 70 years of potentially mutation-causing events.

So why does a clinically apparent cancer only arise in every third individual when there are somewhere in the region of 10^{14} good potential cellular targets, and moreover we live in a world in which each of those cells is continually exposed to a myriad of avoidable and unavoidable DNA damaging agents? In fact, a fairly bold statement can be made: at a cellular level cancer is surprisingly very, very rare, and this can only be accounted for by the existence of some extraordinarily effective barriers to cancer cell development – even though these do occasionally break down!

Cancers may well originate in a single bad cell, but are usually only clinically detectable by direct observation or conventional investigations like an X-ray, when replication has increased the number of cancer cells to around one billion

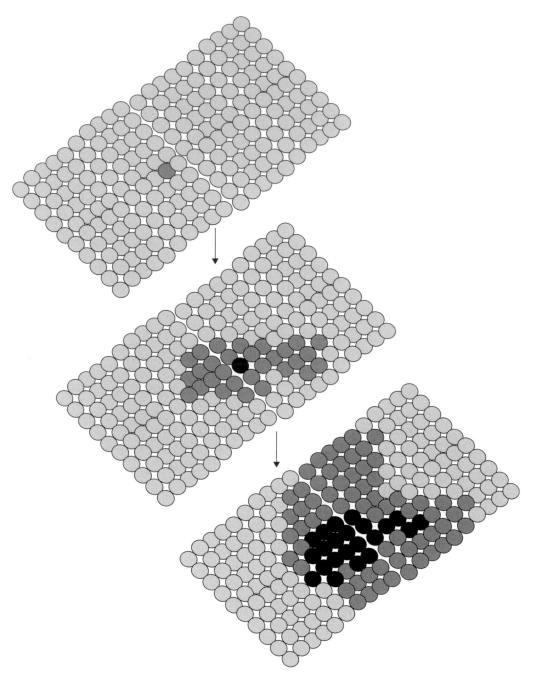

Fig. 1.2 Cancer is a clonal disease. Expansion of the original clone (dark gray) is followed by emergence of a new clone (black), which gradually replaces the original.

(10^9) – a small nodule. In other words, by the time a cancer is discovered the original cancer cell has proceeded through some 30 or more cell divisions, and acquired a host of further epimutations – a situation compounded by the near universal loss of normal DNA repair processes. Not surprisingly, this has complicated studies attempting to unravel the initial causes of cancer in man, which happened some 30 generations in the past. If cancer cells may be compared to human societies then this becomes a challenging problem more akin to archeology than traditional biology – imagine today trying to identify which one of our ancestors was responsible for losing the battle of Hastings in 1066 (activation of William or inactivation of Harold notwithstanding). A cartoon schematic of how we believe cancers arise and progress is provided in Fig. 1.3. This can be used as an overview to be referred to whilst reading the more detailed (and complex) description of the basis of cancer in this book.

CANCER INCIDENCE/EPIDEMIOLOGY

In the United Kingdom and North America, the lifetime risk of developing cancer is more than one in three, and cancer is responsible for around one in four deaths. Yet the undoubtedly justified fear of cancer experienced by many individuals should be balanced by an appreciation that one is still far more likely to die or become disabled due to a heart attack or stroke (see Fig. 1.4), if that knowledge may be regarded in any way as reassuring.

Given that almost every cell type can give rise to cancer, more than 200 different types of cancer are recognized, but four – breast, lung, large bowel (colorectal) and prostate – account for over half of all new cases. Non-melanoma skin cancer (NMSC) is very common, with a recorded 60,000 new cases each year, but is nearly always curable. The national statistics for NMSC are incomplete and are now routinely omitted from the overall total of new cases of cancer.

Different cancers affect people at different ages, but the overall risk of developing a cancer rises significantly with aging, with 65% of cancers in the UK occurring after the age of 65 years. In children, leukemia is the most common

cancer (around 30% of all pediatric cancers); in young men aged 20–39 it is testicular cancer.

The incidence of cancer has changed over the last 20 years, with a decline in lung cancer in the United Kingdom and North America in males, mainly due to changes in smoking habits, but an increase in breast and prostate cancer. In 1981, there were 78 cases of breast cancer per 100,000 women in Great Britain, and 38 cases of prostate cancer per 100,000 men. By 1996, these rates had risen to 104 and 65 respectively.

It has long been appreciated that there is a geographical variation in cancer incidence and deaths. Globally, the most common cancer, affecting women, is breast cancer followed by cervical cancer. However, in North America the most common cancer after breast cancer is lung or colon cancer. Of the estimated 371,000 new cases of cervical cancer in 1990, around 77% were in developing countries.

In this latter case, this likely reflects socioeconomic factors. However, race and gender also influence rates of cancer and this is graphically illustrated by the most recent available data (1999) from the United States Department of Health and Human Services (HHS). Some of the findings are predictable, including lower rates of melanomas in men and women of Afro-Caribbean origin, attributed to inherent protection from UV exposure. However, others are less so. Thus, although prostate cancer is the most frequent cancer in males, rates are 1.5 times higher in Afro-Caribbean men than in white. Similarly, the leading cancer in women, regardless of race, is breast cancer, followed by lung/bronchus and colon/rectal in white women, and colon/rectal and lung/bronchus in Afro-Caribbean women. Breast cancer rates are about 20% higher in white women. Multiple myeloma (cancer that arises in plasma cells) and cancer of the stomach are among the top 15 cancers for Afro-Caribbean women, but not for white women.

TOWARDS A DEFINITION OF CANCER – WHAT IS CANCER?

The terms tumor or neoplasia are used interchangeably to describe a diverse group of conditions associated with uncontrolled cell

Intrinsic factors

Inherited susceptibility:
High-penetrance genes – rare;
Low-penetrance – likely;
polymorphisms at multiple alleles
(100s or 1000s) may all confer a
degree of sensitivity or resistance
to cancer (however slight the
effect)

Initiation:
Spontaneous mutation in an
oncogene, tumor-suppressor or
caretaker gene. (Could be
"Knudson", second hit in rare
familial cancers). DNA repair
genes and p53 pathway will try
and protect if intact.

Promotion:
Selective growth advantage
leads to start of clonal
expansion. Antiapoptotic lesion
probably required before a
"mitogenic" lesion, in order to
block "default" cell death.
Properties acquired: minimal
platform, deregulated cell
proliferation, avoidance of
apoptosis. Genetically
homogeneous clone. May be
"premalignant"

Progression:
Further mutations confer
additional growth advantage to
successive clones.
Genetic instability, aneuploidy.
Properties acquired: deregulated
proliferation, avoidance of
apoptosis (and senescence),
loss of differentiation, loss of cell
adhesion, invasiveness,
angiogenesis. Invasion of
lymphatics/vasculature. Clones
genetically heterogeneous.

Metastatic spread:
Mutations in "metastasis
suppressor"genes (possibly
some already acquired earlier).
Eventually cancer cells entering
lymphatics/vessels are able to
colonize distant organ/tissue.

The would-be cancer cell

Germ cell

Somatic cell

Clonal expansion

Further rounds of clonal expansion.
Progression

Metastases

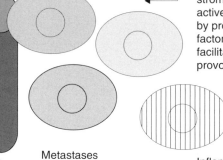

Extrinsic factors

Carcinogens (mutagens) may
increase risk of DNA damage
and mutation.

Carcinogens (mutagens) may
increase risk of mutation.
Important cancer-causing
mutations may also occur in
stromal cells (i.e., not necessarily
in the cancer cell). Crosstalk with
microenvironment also critical.

Carcinogens (mitogens) may
support promotion. Stroma may
actively support tumor growth
by providing survival/growth
factors. Immune surveillance
may try and eliminate cancer
cells.

Carcinogens may support
progression. Important cancer-
causing mutations may occur in
stromal cells also. Stroma may
actively support tumor growth
by providing survival/growth
factors; angiogenesis; MMPs
facilitate invasion and may
provoke DNA damage.

Inflammatory cells may help
"convey" cancer cells. "Seed" and
"soil" may determine where a
given cancer cell can establish
colonies. Gross factors such as
sites of lymph drainage will also
dictate sites of metastases.

Fig. 1.3 A highly stylized potential "life history" of a cancer cell. Cancer cells are shown in grey (different shades, subclones); stromal cells are shown in vertical lines, vessels in dark grey.

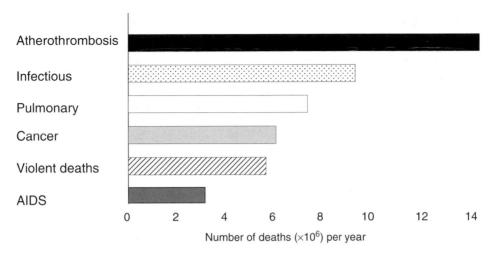

Fig. 1.4 Causes of death. (Leading causes of death worldwide* up to 1997, expressed as millions of deaths per annum).*In eight defined regions of the world, including developed and developing areas. From Murray, C.J.L. and Lopez, A.D. (1997) Mortality by cause for eight regions of the world: global burden of disease study. *Lancet,* **349**: 1269–76.

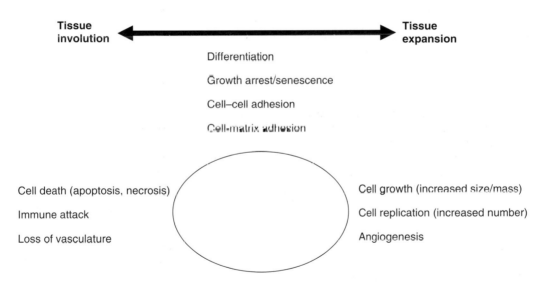

Fig. 1.5 Processes contributing to regulation of tissue mass. Cell mass is determined by the balance of various cellular processes including at two extremes growth/replication and cell death.

replication. Tissue mass is normally tightly controlled to serve the needs of the organism. This control is achieved by the balancing of various cellular processes (see Fig. 1.5). Disturbing the balance of these processes results in diseases; if cell losses exceed renewal this results in

degeneration/involution, whereas the converse results in tissue expansion, hyperplasia, or neoplasia. If the expansion in cell numbers is confined locally then it is described as benign, if however this unscheduled cell replication is accompanied by invasion of surrounding tissues or spread to distant sites (metastasis) then it is unambiguously described as malignant. These terms are relatively straightforward as they are based on gross observations. It should be appreciated that the pathological definitions of benign and malignant do not always translate into similarly benign or malign outcomes for the patient. Thus, a benign brain tumor causing severe neurological disturbance may be inoperable or require potentially life-threatening surgery, whereas a malignant prostatic cancer or microscopic metastases may be discovered accidentally at post mortem. Adenomas are benign tumors originating in glandular or secretory tissues (such as lactotroph adenomas of the pituitary, which secrete prolactin, or parathyroid adenomas, which secrete parathyroid hormone, PTH). Such adenomas can result in substantial morbidity as a result of deregulated secretion of hormones and may also progress to become malignant, when they are termed adenocarcinomas.

Classification of cancer

This is complicated by the variety of human cancers, with hundreds of different tumor types arising from almost every tissue and in every organ. This is further complicated by the ability of a cancer cell to invade surrounding tissues and metastasize to distant organs. Cancer biologists and oncologists have agreed on a classification based on the tissue of origin, regardless of organ location, focusing on the similarities in cellular structure and function among these tumors. Tumors are generally classified as either liquid or solid. The former includes leukemias and lymphomas comprising neoplastic cells whose precursors are usually motile. Solid tumors comprise either epithelial or mesenchymal cells that are usually immobile. Pathologically, cancers are classified as:

1. Carcinoma, originating from epithelial cells in the skin or in tissues that line or cover internal organs, and typically representing over 80% of diagnosed human cancer each year;
2. sarcoma, originating in bone, cartilage, fat, muscle, blood vessels, or other connective or supportive tissue;
3. leukemia, a cancer originating in blood-forming tissues such as the bone marrow, and causing large numbers of abnormal blood cells to be produced and enter the bloodstream; and
4. lymphoma, which originates in the cells of the immune system.

It is worthwhile remembering the purposes behind disease classification – in the clinic to help make the most accurate predictions about prognosis and response to particular therapies, and in the laboratory to ensure that as far as possible like is studied alongside like. Thus, terms including carcinoma *in situ* are used to refer to lesions regarded as cancer that remains localized to the tissue of origin, often constrained by an intact basement membrane. Such tumors often respond well to treatment, resulting in a good prognosis for the patient. In contrast, invasive carcinomas, by disrupting basement membranes and growing into surrounding tissues, are more difficult to treat successfully. Additionally, since invasion is usually a prerequisite for metastasis, which is the ultimate cause of most cancer-related deaths, even when the local lesion is treated, the prognosis is often poor. Importantly, disease classification is not written in stone. As technical advances are made and larger numbers of individuals with a given disease are studied, it is often possible to recognize previously unappreciated "subclasses" of disease that can readily be detected and further improve accuracy of prognosis and prediction of treatment responses. Most recently, advances in post-genome era technologies such as oligonucleotide arrays and proteomics have led to attempts to further classify cancers in terms of global gene/protein expression patterns – molecular tumor "fingerprinting". In the future, it is hoped that such powerful means of measuring all the genes/proteins expressed by a tumor (or different single tumor cells) or even in blood tests, will ultimately allow more accurate determination of prognosis and even "tailored" therapy, whereby each patient can be uniquely classified and treated on the basis of such tests. These aspirations are often referred to as "individualized medicine", reflecting the ideal of being able to treat each individual in a uniquely appropriate way, based on variation in one or more of the following parameters: gene alleles; gene expression/protein expression by tumor cells; proteins in the blood.

It is surprisingly difficult to define cancer in practice

Cancer itself is potentially a much more difficult term to define accurately. Simply, cancer is synonymous with malignancy and refers to a group of conditions which have manifested malignant behavior, namely unscheduled and uncontrolled cell growth leading to invasion and/or metastases. There is no ambiguity in this case as the definition is "retrospective" and based on the readily observable behavior of the "cancer". However, such a narrow definition is of limited practical value in the laboratory and particularly in the clinic – as it precludes true preventative or even early treatment. This seemingly abstract issue is placed in context when it is remembered that for those cancers where rates of death have actually been reduced over the last few decades, this has resulted primarily from improvements leading to earlier diagnosis and earlier administration of treatment.

It is clear that certain features at a gross macroscopic level and at a microscopic level can accurately be employed to identify a tumor as cancer before it manifests overt malignant behavior. In most cases this requires the demonstration of evidence of invasion into surrounding tissue (histological examination) or the presence of "cancer cells", namely cells exhibiting defined changes, which from experience are the same as or similar to those seen in circumstances that are incontrovertibly cancer (cytological examination). *In other words a cancer is a cancer before it necessarily declares itself by behaving as one.* In a clinical setting where the primary purpose is to identify a tumor or lesion that requires surgical excision, or other treatment, it may be sufficient to know that a particular lesion (based on gross appearance or histological examination) poses a risk of proceeding to a cancer, whether that lesion is already regarded as a cancer, or because of its still unrealized potential is called precancerous (or premalignant). This forms the basis for identifying "high risk" lesions such as Barrett's esophagus (a precursor of esophageal cancer), colonic polyps (a precursor of colon cancer), and others. Cytological examination can identify premalignant cells and is employed where such cells can readily be obtained, including cervical screening for the early detection and prevention of cervical cancer (see Box 1.1).

Box 1.1 Cancer screening

In 1968, Wilson and Jungner of the WHO stated 10 principles, which should govern a national screening program. These are:

1. The condition is an important health problem;
2. its natural history is well understood;
3. it is recognizable at an early stage;
4. treatment is better at an early stage;
5. a suitable test exists;
6. an acceptable test exists;
7. adequate facilities exist to cope with abnormalities detected;
8. screening is done at repeated intervals when the onset is insidious;
9. the chance of harm is less than the chance of benefit;
10. the cost is balanced against benefit.

The aim of screening is to identify individuals at risk for whom effective interventions/treatments are available, and it should also be limited to situations where that treatment is more effective if administered early and before the condition to be treated becomes readily apparent. If the above criteria are satisfied, then in general, the ideal screening test for any given condition should be highly sensitive (few false negatives – patients deemed normal who actually have the condition) and highly specific (few false positives – normal patients deemed to have the condition). In many cases increasing sensitivity may result in decreasing specificity, and often health policy decisions have to be made, which take account of the prevalence and severity of the condition to be screened for, economic factors relating to the cost of screening and the subsequent proposed interventions, and also both the efficacy and the safety of the available interventions (risk – benefit ratio). Broadly, two types of screening are applied: (i) population screening, where mechanisms are put in place to ensure that all appropriate individuals are screened at given times/intervals – largely the responsibility of public health organizations; (ii) opportunistic screening, where healthcare workers undertake screening when individuals present to them for whatever reasons – this is largely the responsibility of healthcare professionals. The latter approach is cheaper, but will provide lesser cover of the population.

For a research scientist these distinctions are also of critical importance; to define at what point a premalignant benign lesion ends and a malignant cancer begins is a prerequisite to understanding the initiation and key early events in cancer formation. At least in the laboratory one can study the progressive behavior of transformed cells or tumor progression in animal models, as long as the necessary investigative tools are available, an opportunity that is self-evidently usually lacking in the study of cancer in man. In other words, the cancer researcher can validate predictions made about the future behavior of a given lesion by prospectively tracking the eventual emergence of invasive metastatic cancer.

Cancers may not always be clinically apparent

Difficulties of definition notwithstanding, the clinical situation is further complicated by the increasing awareness that microscopic colonies of cancer cells (*in situ* tumors) can be detected in different tissues (e.g. thyroid, breast, prostate) at autopsy in most older individuals. In fact, such clinically irrelevant *in situ* cancers may be a 100–1000-fold more common than clinically apparent cancers arising in those same tissues during life (e.g. most older individuals have *in situ* thyroid carcinomas at autopsy, whereas only around 0.1% of similarly aged individuals are found to have thyroid cancer during life). Although biologically intriguing and testifying to the potential effectiveness of innate anti-cancer defenses (such as antiangiogenic factors), such findings may increasingly be problematic in the clinic. Until recently, we have generally not detected the vast majority of such *in situ* tumors during life, largely because we do not routinely biopsy tissues in apparently healthy individuals. However, one area in which detection of such *in situ* tumors may pose difficult and as yet unresolved clinical dilemmas is in the increasing use of diagnostic prostatic biopsy in older men, and the discovery of so-called "incidentalomas" during routine imaging procedures such as CT and MRI scanning. Guidelines have had to be developed to assist clinicians in deciding which of those individuals with such findings actually require any form of treatment or if simply reassurance would suffice.

There are many cases where the actual ability to predict the risk of future invasive cancer based on the appearances of a given lesion are not yet sufficiently mature. Moreover, in many cases it is not technically possible to detect the early lesion let alone examine it, which ironically may in some cases be for the best, until our ability to more accurately predict the future behavior of these early lesions improves and/or we greatly increase our current arsenal of sufficiently well-tolerated and nonharmful therapies to exploit early diagnosis. However, it is abundantly clear that in order to prevent or cure cancer effectively it is essential to diagnose disease as early as possible. Failure to do so will inevitably mean that potentially life-saving early treatment for some individuals destined to develop clinically important cancer will be delayed. To resolve this conundrum is theoretically simple – we just need to distinguish early lesions that will never progress to disease from those that will progress to cancer. We just need better tests and tools. Fortunately, the research community has responded to this challenge and much progress is being made in finding new "biomarkers" for various cancers that might give important information about prognosis and treatment response (see Box 1.2, biomarkers).

Box 1.2   Cancer biomarkers

Leland Hartwell, in his keynote address at the 2004 meeting of the AACR, suggested that earlier diagnosis and improved monitoring of cancer progression by noninvasive means could dramatically improve the outcome for many patients. Early detection represents one of the most promising approaches to reducing the growing cancer burden and has been revolutionized with the advent of post-genome era technologies that can identify cellular changes at the level of the genome or proteome and new developments in data analyses and modeling. Gene expression profiling of various human tumor tissues has led to the identification of expression patterns related to disease outcome and drug resistance, as well as to the discovery of new therapeutic targets and insights into disease pathogenesis. However, as these approaches require removal of cancer tissues they are not ideal for achieving earlier diagnosis or for general screening, where a noninvasive test would have numerous advantages. Therefore, considerable efforts are now directed at finding "biomarkers" in blood tests that can be obtained relatively noninvasively and rapidly, and which could much more readily be employed in screening large numbers of individuals. Their role could also be extended into surgical surveillance for potentially operable disease and postoperative follow-up for disease recurrence.

Broadly, three overlapping technologies can be employed to look for cancer biomarkers:

1. Analyses of proteins by
   a. immunoassay of a single known protein predicted to be of interest;
   b. proteomics, including 2D gel-based separation or liquid chromatography followed by mass spectrometry to identify potentially thousands of different proteins;
   c. proteomic pattern analysis or "fingerprinting", which relies on the pattern of proteins observed and does not rely on the identification of individual traceable biomarkers.
2. Analyses of free RNA in the circulation, some of which derives from the cancer.
3. Isolation and study of circulating tumor cells (CTC), which can in turn be profiled for gene expression by microarrays.

As mentioned earlier, in order to improve our predictive/diagnostic abilities, traditional examination of patients in the clinic, application of imaging techniques, and cytology/histology of the tumor are increasingly being supported by newer techniques, such as molecular profiling. Traditionally, genetic analysis looks for single susceptibility genes, which confer a high risk of cancer formation, but in future may include more complex genomic testing (of multiple polymorphic alleles – see below), or direct analyses of gene/protein expression in the tumor by various techniques including gene chip microarrays and proteomics. Considerable enthusiasm has been generated by the possibility of using relatively noninvasive tests to identify cancer "biomarkers" in blood samples from patients with cancer or at risk of cancer. Thus, the analyses of proteins/gene transcripts or even cancer cells (or their DNA) in the patient's blood circulation may reveal important information on prognosis and treatment response.

If we find a cancer what do we do with it?

Not only do we often not know whom to treat, we are often unsure what treatments to use, particularly before the development of an obvious cancer. This situation has not been helped by the majority of therapeutic trials that have focused largely on the end stages of cancer, by definition the point at which these therapies are least likely to successfully cure the disease. Improved ability to predict treatment response is fundamental to avoiding the morbidity and mortality associated with cancer while also restricting potentially harmful or even life-threatening treatments to those individuals most likely to benefit. Most treatments are justifiable when a life-shortening cancer is prevented, but would be very undesirable if employed in an individual never destined to develop cancer, or whose life would have been affected less by the cancer than by the treatment. One thing is, however, clear: early treatment offers the best chance of a successful outcome. This problem is addressed by various screening programs aimed at identifying premalignant or early-stage cancers (see Box. 1.2). Importantly, in these cases suitable treatment strategies have been defined.

As discussed in the previous section, it is hoped that detailed molecular analyses of tumor samples or body fluids will not only improve our understanding of the "road map" to cancer for any given cancer, which might in turn guide us to the application of specific drugs to target particular genes/proteins, but may also improve our ability to predict therapeutic responses. Such detailed analysis of individual tumors starts to realize the potential of post-genome era science and may finally deliver the ultimate goal of "tailored" therapy – where treatment is fitted specifically to an individual.

The best treatment is prevention This requires a combination of activities involving different organizations, including public health strategies aimed at the whole population and exemplified by activities targeting adverse lifestyles including smoking and poor diet. More targeted advice and possibly interventions may be needed for individuals at the highest predicted risk of disease. However, with the rare exceptions of individuals with known familial cancer syndromes this has proved far more difficult a strategy for cancer prevention than it has done for preventing coronary heart disease (CHD). Robust tools have been developed allowing reasonably accurate estimation of future risk of heart attacks or strokes based on using simple information such as age, sex, blood pressure, and level of circulating fats (readily determined in the clinic) in order to calculate a risk score. For cancer the hope is that improved genetic testing (see Box 1.3), measurement of new disease

biomarkers, and improved clinical investigational tools will match these successes in heart disease. In some cases, exemplified by cervical screening, the entire female population of a certain age are deemed at risk and subjected to screening. Screening is discussed in more detail below and in Box 1.1.

What's next best The early detection of cancer or precancer syndromes is self-evidently the next best to prevention – based on the assumption that small numbers of cells of potentially less advanced malignancy will prove easier to treat or cure. This forms the basis of screening for cervical, breast, and colon cancers (see Box 1.1). Improved early detection also involves the speedy selection of patients with appropriate symptoms or signs for early application of diagnostic tests (including X-rays, blood tests, biopsy, etc). The nature of such tests is also under continuing evolution, with great interest growing in the possibility of improved molecular diagnostics employing cancer biomarkers, serum protein signatures (proteomic analyses), and isolating small numbers of cancer cells from the circulation for analyses. Not all such new diagnostic tests will necessarily result from ever more advanced molecular and cellular biology; thus, harking back to ideas of the original pioneers of cancer biology (see Jean Astruc, in Appendix 1.1), even highly creative or eccentric ideas such as training "sniffer dogs" to identify certain cancers such as bladder from the "smell" of urine are being evaluated.

Currently available treatment options These have expanded dramatically in recent years with the emergence of specific therapies targeting individual cancer-relevant molecules or signaling pathways. However, choice of appropriate treatment regimens for any given patient remains challenging. In general, the first decision to be made is whether the cancer may be cured by surgical resection, drugs, or both. A more detailed discussion of cancer therapies is presented in Chapter 16, but a few interesting aspects will be highlighted here.

Achieving lasting remissions in patients suffering from nonlocalized malignancies remains

---

Box 1.3   Genetic testing

The identification of disease-related genes has led to an increase in the number of available genetic tests that detect disease or an individual's risk of disease. Gene tests are available for many disorders, including Tay–Sachs disease and cystic fibrosis, and in cancer testing for the genes *BRCA1* in breast cancer, *MEN1* and *RET* in endocrine tumors. As more disease genes are discovered, more gene tests can be expected.

elusive – we are rarely, if ever, able to kill all the cancer cells in the primary tumor and metastatic lesions. Such failures may be the result of poor access of effective treatments to all tumor locations, varying susceptibility to conventional DNA-damaging anticancer agents, or the rapid evolution of resistance. A particular problem is posed by those cancers where cells spread early via the circulation to establish micrometastases in the bone marrow or elsewhere. While increasing drug dosage can overcome some of these barriers it also increases toxicity to normal cells. Traditional cytotoxic treatments aim to kill all cancer cells, whereas some newer approaches may be directed at disabling cancer cells (inducing growth arrest, differentiation, etc.), without necessarily killing them.

Despite some notable successes, concerns remain about potential adverse effects of traditional radio- and chemotherapy on normal tissues and intriguingly also on the surviving cancer cells themselves. Cancer progression is an evolutionary process driven by acquisition of epimutations, which provide a selective growth advantage to particular cell populations. Therapies which induce irreparable damage to cell DNA may have undesirable consequences on those cancer cells which fail to undergo apoptosis but rather survive this onslaught. In fact, one means of resistance in cancer cells may be the increased mutation rate and selection pressure provided by such drugs. In fact, the net effect of unsuccessful cancer therapies could be to speed the progress of the disease as more mutated cells expand without the competition of their less aggressive predecessors and their offspring. Increasingly, therefore, new combinations of drugs are employed to reduce the likelihood of cancer cells surviving to become resistant to all these agents.

An interesting parallel may be drawn here between evolution of species and evolution of a cancer. Evolution is driven not only by mutations and natural selection, but also by catastrophic extinctions which by removing less-hardy competitors clear the path for the survivors to fill the vacuum. Maybe sub-total cancer cell killing with chemotherapy, radiotherapy or even surgery may be the cancer equivalent of a meteor impact. Given that only some 1 in 10,000 of the estimated 50 billion or more species which have evolved on earth still exist,

and if we are anything to go by these include some of the hardiest and nastiest, then maybe extinctions of some of the less able to survive may be undesirable if you don't cull the lot. Moreover, the situation in cancer therapy is likely a lot worse as cancer cells are repeatedly selected for their ability to not be killed by cancer therapy – species have not necessarily been selected largely on their ability to survive repeated meteor impacts or volcanic activity, but probably somewhat more randomly. The risks inherent to increasing "selection pressure" have been ably demonstrated by the emergence of antibiotic resistance in bacteria. What might be an alternative? Theoretically, arresting replication in cancer cells might be a good alternative or addition to traditional treatments that offer anything other than complete extinction of cancer cells, as this would prevent expansion of an aggressive surviving clone and might instead foster "stagnation" of the cancer cell population. Assuming that no treatment will ever immediately kill all cancer cells – what proportion is effectively total extinction? The 90% extinction of species believed to have occurred at the Permian era was followed by a substantially slower recovery (based on fossil records and therefore only really applicable accurately to "big organisms") than after those in different eras which resulted in 60–70% extinction – but they still eventually recovered. Arguably, we might wish to know what proportion of cancer cells need to be killed in an individual for no symptomatic recurrence of the tumor to take place during that individual's lifespan!

Causes of cancer

Much has been learned about the causes of cancer, including the role of genetic predisposition, gene–environment interactions, and infectious agents. Intriguingly, recent research points to the considerable overlap between the behavior of cancer cells and that of cells during normal physiological wound healing and during embryogenesis. Similarities include replication, less differentiated state, invasion/migration, with the major differences reflecting the lack of control and the unscheduled nature of replication which characterizes cancer. One intriguing question, addressed later, is how the organism is able to distinguish between normal growth and tissue repair (normal cell cycles) on one hand and neoplastic growth (cancer cell cycles) on the other.

The clonal evolution theory

Most cancers derive from an individual somatic cell in the adult organism, with the initiation and progression of tumorigenesis dependent on the accumulation of genetic or epigenetic changes that determine the emerging cancer phenotype. Initiation is believed to be through DNA damage, which renders the cell capable of forming a cancer; initiated precancer cells then multiply during a promotion phase. A "multistage" model of carcinogenesis was first articulated by Armitage and Doll in the 1950s based largely on their epidemiological observations. Following increasing knowledge of the molecular genetic basis of disease, in 1976 Nowell proposed that cancers arise by a process of multistep clonal evolution. He proposed that most neoplasms arise from a single cell, and tumor progression results from acquired genetic variability within the original clone allowing sequential selection of more aggressive sublines. He also stated, rather prophetically, that acquired genetic instability may result in apparently similar advanced tumors being very heterogeneous both at a molecular and at a behavioral level, requiring individual specific therapy. He also predicted that therapy could be thwarted by emergence of genetically variant resistant sublines.

Becoming a cancer cell – multistage carcinogenesis

A wealth of data has supported the view that cancers are multistage diseases progressing via protracted accumulation of multiple genetic changes (lesions) that compromise control of cell proliferation, survival, differentiation, migration, and social interactions with neighboring cells and stroma. Hanahan and Weinberg (2000), in a highly recommended review, construed the axiomatic requirements of cancer cells as the:

- Capacity to proliferate irrespective of exogenous mitogens;
- refractoriness to growth-inhibitory signals;
- resistance to apoptosis;

- unrestricted proliferative potential (immortality);
- capacity to recruit a vasculature (angiogenesis);
- ability to invade surrounding tissue and eventually metastasize.

By implication, tumor progression proceeds by the acquisition of lesions that provide the tumor cell with these attributes and thereby shape its complex phenotype (see Fig. 1.6). Mostly, these lesions are acquired in somatic cells, but in the inherited cancer syndromes (see Chapter 3), one of the lesions is inherited and is present in all somatic cells. It is important to note that seemingly phenotypically similar cancers may arise through differing combinations of lesions: there are likely many different routes to cancer even in the same cell type (Box 1.4). Many key cancer-relevant signaling pathways may be activated or inactivated by mutations at various different points that could result in largely identical cell behaviors.

The multistage theory of cancer formation is illustrated by models proposed by Fearon and Vogelstein to explain the observed behaviors of carcinogenesis in the colon (see Fig. 3.5, Chapter 3). A normal colonic enterocyte acquires a mutation that confers a growth advantage and begins to expand clonally. This stage may be protracted as the progression to full malignancy requires not one mutation, but between 8 and 12 independent mutations. The chances of a single mutation occurring, amongst the billions of gut cells over a 70-year or more lifespan, is substantial. However, the chance of two mutations occurring in one cell is much smaller (the square of the original probability) and for all eight or more mutations occurring in one cell in the lifetime of an individual it is very small. However, if one also assumes that each mutation results in clonal expansion then these probabilities begin to increase rapidly (a second mutation is clearly going to be more likely in a few million proliferating cells than in one cell). An alternative explanation for the infrequency of cancer development is that interlocking combinations of mutations might be required from the outset – in other words for the initial expansion of a clone of cells more than one mutation is needed.

Recently, work by several laboratories has supported this notion because in certain cases the mutational route to cancer may be rather

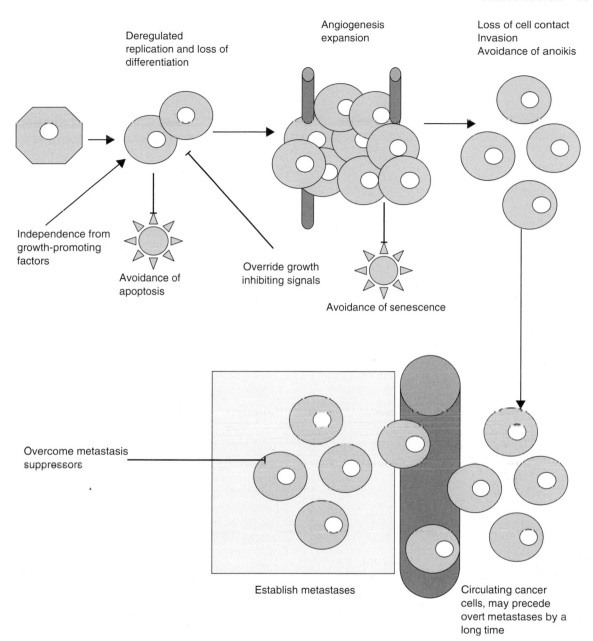

Deregulated
replication and loss of
differentiation

Angiogenesis
expansion

Loss of cell contact
Invasion
Avoidance of anoikis

Independence from
growth-promoting
factors

Avoidance of
apoptosis

Override growth
inhibiting signals

Avoidance of senescence

Overcome metastasis
suppressors

Establish metastases

Circulating cancer
cells, may precede
overt metastases by a
long time

Fig. 1.6 Processes contributing to cancer formation. The "hallmark" features of cancer shown appearing in a potential sequence. It should be noted that this does not imply that this is the actual sequence in which such features are acquired in any particular cancer.

short (in molecular terms) with as few as two interlocking mutations required for initiation and progression of cancers, in animal models, and at least where one of these lesions involves particularly "dangerous" oncogenes such as *c-MYC* also in man (Box 1.4).

Viruses and the beginnings of cancer biology

The identification of the genetic mechanisms of transformation owes much to the study of transforming viruses, in which the transforming effect could be attributed to specific oncogenes.

Box 1.4   Two steps to seven? The road map for cancer

"Pluralitas non est ponenda sine necessitas"

William of Ockham, the most influential philosopher and theologian of the fourteenth century, is best known for applying the medieval rule of parsimony to formulate one of the best-known principles of science, Ockham's razor. This rule translates as "entities should not be multiplied beyond necessity", and as a principle in science may be defined as "favor the simplest model that explains the observations". Even earlier, Aristotle made statements such as "nature operates in the shortest way possible".

It has been widely assumed that (i) as human solid tumors when examined carry a plethora of genetic and epigenetic alterations, and (ii) it is genetically difficult to transform cells under tissue culture conditions, cancer formation can only occur under the influence of multiple (possibly seven or more) genetic lesions. However, in some cases the situation can be much simpler. Namely, as the key requirements for tumorigenesis are deregulated cell proliferation and suppression of cell death, mutations enabling these may constitute the "minimal platform" for the development of a cancer, at least where one of those lesions is deregulated expression of *c-MYC*. It is clear that there are far fewer "pathways" implicated in cancer than genes. Therefore, some cancer cells may indeed "arrive" at this destination via a protracted route involving multiple mutations, as

the way in which a given cell activates/suppresses the requisite pathways needed to complete this "journey" may be very variable. Some of the pathways strongly implicated in cancers include those regulating G_1/S transition in the cell cycle including the Rb protein; the p53 tumor-suppressor pathway and other apoptosis pathways; and the angiogenesis/HIF1 pathway. In fact, there are now numerous examples of only two genetic lesions fulfilling these requirements and promoting neoplastic progression, suggesting that at least in some cases the genetic basis of a given cancer may be remarkably simple. In this model, the genetic complexity of an advanced tumor is more a reflection of evolutionary pressures and natural selection of clones with a growth advantage, rather than an indication of the mutations required to initiate that tumor. The "mission critical" mutations are concealed within the plethora of mutations, many of which are likely irrelevant to tumorigenesis.

This minimal platform model may be reconciled with studies of cell transformation *in vitro* – it may be much harder to establish transformation and immortality in a cultured cell as compared to producing a cancer cell within the organism. The intact organism comprises a network of usually highly effective anticancer barriers, but once these become breached it may instead support the developing tumor. This is not pure conjecture; it is clear that the organism provides the developing tumor with a blood supply as long as it is instructed to do so; in some cases this may require an "angiogenic switch" (an acquired mutation that allows the tumor to "request" to stromal cells for angiogenesis), but might also be an inevitable accompaniment of tissue growth, no matter how inappropriate. In fact, much is now known about the interactions between proangiogenic factors produced by the tumor (such as FGF, VEGF, and PDGF) and antiangiogenic factors produced in the tissues or within the circulation (such as thrombospondin, tumstatin, endostatin, angiostatin, and interferons alpha and beta respectively). The initiation of angiogenesis is likely dictated by the balance of these factors, and in turn by the genes expressed by a given cancer cell on the one hand and by the tumor microenvironment on the other.

DNA viruses express proteins analogs to key pro-liferation factors that substitute for or replace the function of the cellular factors. In contrast, the oncogenes of RNA retroviruses are derived from the hijacking of critical cellular regulatory genes with the addition of gain of function mutations (see Chapters 3 and 6).

Knowledge gained about DNA tumor viruses, and the molecular biology of viral transform-ation, have played a major role in further-ing the understanding of oncogene and tumor-suppressor function and in the development of cancer biology in general, though the actual contribution of viruses to the formation of most human cancers is rather modest by compar-ison. The studies of SV40 large T antigen and HPV E6/E7 proteins, along with the studies of the familial cancers, proved critical in under-standing the importance of the *RB* and *p53* tumor-suppressor genes. This is one example of how several fields of study often converge in science to help illuminate a key process (see Chapter 3).

The cell of origin in cancer

In the 1950s, the histologist Charles Leblond described three main mechanisms by which adult organs are maintained: static, where essen-tially no replication occurs (e.g. nervous system); self-renewal, where stem cells compensate for rapid losses of differentiated cells (e.g. gut and skin epithelia, blood); and simple duplication, where tissues are maintained by proliferation of their own differentiated cells (tissues with slower turnover, such as pancreas, liver, kid-ney, and blood vessels). Interestingly, this early view has been largely discarded over the last decades in favor of the notion that essentially all adult tissues are maintained primarily from a local minority subpopulation of progenitor cells, which retain a strong proliferative capa-city, as well as the ability to differentiate into the required mature cell types after dividing – the so-called stem cells (see Box 5.1 – stem cells). Only recently, with seminal studies employing direct lineage tracking using "pulse-chase" techniques (see Chapter 18), have experimental data actu-ally provided unambiguous support for Leblond's original idea, at least with respect to simple duplication being important in pancreas.

It is a widely held view that cancers origin-ate primarily in stem cells. In fact, the stem cell origin of cancer originates from mid-nineteenth century microscopic observations, which noted the similarity between embryonic tissue and cancer, suggesting that tumors arose from embryo-like cells. The later demonstration of the so-called embryonic remnants in adult tis-sues that could become activated in cancer in the late nineteenth century gave rise to the "embryonal rest" theory of cancer – now understood as the origin of cancer from adult stem cells.

Given their longevity and unique abilities to self-renew and proliferate, it is not surpris-ing that cancers might originate in stem cells. Importantly, the evidence for this is strongest for cancers of the blood and epithelial cells; tis-sues usually maintained by stem cell replication. The "cancer stem cell" model has recently been supported by a study with another tissue where progenitor cells are the major or only source of cell renewal in the adult, the brain. Thus, only a subpopulation of brain cancer cells, expressing a marker indicating their progenitor cell status, were able to generate tumors when implanted into mice.

A major factor often cited in support of the stem cell origin theory of cancer is the observed similarity between many cancer cells and vari-ous embryonic or adult stem cells. However, it is a frequently observed fact that overex-pression of many different oncogenes, such as *c-MYC* or *RAS*, may result in a rapid loss of differentiation and re-entry into the cell cycle for various previously differentiated cell types (Chapter 6). In other words, the initiating muta-tion could equally well occur in a post-mitotic differentiated cell as long as such mutations will confer or capitalize on the potential of that cell to re-enter the cell cycle. In this scenario, the phenotypic similarities between cancer cells and primitive precursors or stem cells arises not necessarily because this reflects the nature of the cell of origin but rather as one of the associated consequences of the initiating oncogenic lesion, whatever the ori-ginal state of differentiation of the cell involved (see Fig. 1.7).

If "dedifferentiation" is an inevitable accom-paniment of cancer-causing mutations, then the preferential role of stem cells in the initiation of

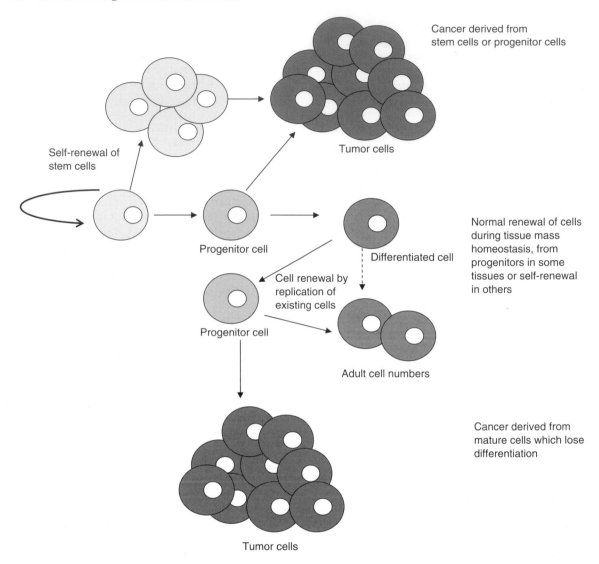

Fig. 1.7 Cell of origin of cancer. Cancers probably originate most frequently in progenitor or stem cells, but could also arise from more differentiated cells that lose differentiation as part of the oncogenic process.

cancer may instead reflect the higher intrinsic rate of replication or their longevity in adult organisms. This is more plausible as it is extremely likely that mutations would occur more frequently during cell division because of the vulnerability during DNA replication. However, this is by no means the only way in which mutations occur (see Chapter 3), and also it is not only the stem cells that may

replicate in the adult organism. The observation that "promotion" of an epidermal cancer may be accomplished months or even years after the initial exposure to a carcinogen ("initiation") is often taken to imply that the original carcinogenic event occurs in a long-lived epithelial stem cell population. While this is highly likely in the skin, where mature cells are continually removed by shedding at the surface, it is equally plausible in other tissues where the original mutation has conferred longevity (particularly likely given the repeated observation in mouse models that an antiapoptotic lesion may be amongst the earliest required mutational events in cancer formation), or that cell turnover of differentiated cells in a given tissue is usually slow (thus, unless the mutation conferred an immediate growth advantage, it would only be passed on to a small number of progeny). It must be remembered that it is now unarguable that differentiated cells can and do replicate in the adult even under normal physiological circumstances and in some tissues may be the sole or major source of new cells. The cellular events during development of liver cancer suggest that cancers may arise from cells at various stages of differentiation in the hepatocyte lineage.

Much experimental data support the view that dysregulation of specific genetic pathways, rather than the cell of origin, dictates the emergence and phenotype of various cancers, including high-grade glioma and others.

Whatever the actual outcome of these scientific debates, it is unarguable that treating cancer by inducing its differentiation (differentiation therapy), whatever that may have been in the cell of origin, offers considerable promise. However, it cannot be assumed that this alone will suffice if the cell of origin was differentiated to begin with. Thus, for example, inducing differentiation in c-MYC-induced ostesarcomas, by transiently inactivating c-MYC, has recently been shown to alter the epigenetic context surrounding c-MYC signaling so as to change this from procancer to proapoptotic (anticancer). Whereas, in the case of a c-MYC-induced tumor arising from a more differentiated cell type, which in consequence loses differentiation as part of c-MYC activation, inducing "redifferentiation" by transient c-MYC inactivation does not change the context and reactivation of c-MYC, thereby resulting in further tumor progression. Once again, a general rule holds true, namely, that most things related to cancer are a matter of timing and are also determined by numerous factors including the cell of origin, the mutations accumulated, and the cancer environment – together referred to as the molecular "road map" of that cancer.

Cancer is a genetic disease

With the availability of the reference genome for man and mouse, the last decade has witnessed an explosion of new knowledge in human genetics. Our understanding of the genetic basis of disease has grown dramatically, with nearly 5000 diseases identified as heritable. Moreover, it is now known that genes contribute to common conditions such as heart disease, diabetes, and many types of cancer.

Currently, more than 1% of all human genes are "cancer genes", of which approximately 90% exhibit somatic mutations in cancer, 20% bear germ-line mutations that predispose to cancer, and 10% show both somatic and germ-line mutations. A recently published "census" of cancer genes (see the Sanger Institute website – http://www.sanger.ac.uk/genetics/CGP/Census/) is dominated by genes that are activated by somatic chromosomal translocations in leukemias, lymphomas, and mesenchymal tumors. Interestingly, the protein kinase domain was the most frequently represented domain encoded by cancer genes, providing support for the development of therapies targeting this domain in cancer, followed by domains involved in DNA binding and transcriptional regulation.

Broadly, cancers arise due to genetic (or epigenetic – see Chapter 11) alterations in three types of genes, oncogenes (Chapter 6), tumor-suppressor genes (Chapter 7) and caretaker genes – such as DNA repair genes (Chapter 10). Combinations of epimutations in these classes produce tumors. Genetic (but most probably not epigenetic) alterations may occur in the germ line, resulting in inherited cancer predisposition, or more commonly occur in somatic cells, giving rise to sporadic tumors. The first somatic epimutation in an oncogene or tumor-suppressor gene that enables clonal expansion may be regarded as the initiating insult. Unfortunately, in the vast majority of human cancers this key early step is not known. Tumors progress through the acquisition of further somatic epimutations, which allow further rounds of clonal expansion. Broadly, therefore, tumor cells evolve, with those cells that have a growth advantage selected for at each mutational event. Individuals with an inherited abnormality in any of these genes are cancer-prone presumably because they are one step ahead of those without such germ-line abnormalities.

Inherited single gene defects and susceptibility to cancer

Single gene inherited predispositions are not often contributors to cancer and predominantly involve inactivation silencing of a "caretaker" gene. Inherited forms of cancer represent perhaps about 5–10% of all cancers and include two rare inherited cancers, studies of which have resulted in disproportionately spectacular insights into cell and cancer biology in general; a childhood eye cancer known as retinoblastoma (caused by loss of the *RB* tumor suppressor) and the Li–Fraumeni syndrome (caused by loss of the *p53* tumor suppressor), in which children and young adults of the family develop an assortment of cancers, including sarcomas, brain tumors, acute leukemia, and breast cancer.

More recently, gene mutations associated with common cancers, including colon cancer and breast cancer, have been identified. The familial adenomatous polyposis coli gene (*APC*) has been identified as a cause of inherited precancerous polyps, and a contributor to colon cancers. Women with an altered copy of the *BRCA1* breast cancer susceptibility gene, in particular, are susceptible to ovarian as well as breast cancer. It is estimated that as many as 1 in 300 women may carry inherited mutations of breast cancer susceptibility genes. People who inherit cancer genes are more likely to develop cancer at a young age, because the predisposing gene damage is present throughout their lives.

Loss of heterozygosity and comparative genome hybridization   Deletion of genetic material is a very common event in human cancer. Indeed, it is the most frequently observed genetic abnormality in solid tumors. There are several mechanisms through which a somatic cell, with an inherited mutated gene allele, can lose the normal gene copy and become vulnerable to cancer (Fig. 1.8, also see Chapter 10). These mechanisms may result in what has been described as loss of

heterozygosity (LOH). LOH can occur by deletion of the normal allele; deletion of part of or the entire chromosome (the latter is referred to as aneuploidy), possibly followed by duplication of the chromosome containing the mutated allele; mitotic recombination or crossing over, with genetic recombination that can occur in mitosis (it is a normal part of meiosis). Thus, a particular chromosomal region might be found in zero, one, two, or many copies; whereas the similar region in normal cells always has two copies. These extreme genetic aberrations in cancer cells (loss or gain of chromosomal regions) may be readily detectable during cytological examination and such abnormalities can form the basis of diagnostic and prognostic decisions.

Haploinsufficiency   Knudson's two-hit model of tumor suppressor genes (Knudson, 1971) proposes that two mutations are required to cause a tumor, one occurring in each of the two alleles of the gene (Chapter 7). Recently, however, tumor suppressors that do not conform to this standard definition have been described, including genes requiring inactivation of only one allele (also referred to as haploinsufficient), and genes inactivated by epigenetic silencing (Chapters 7 and 11).

Complex polygenic mechanisms and inherited susceptibility to cancer

What remains to be uncovered is how low-penetrance genetic variants (polymorphisms) contribute to the risk of developing so-called sporadic cancers. Polymorphism refers to a gene that exists in more than one version (allele), and where the rare allele can be found in more than 2% of the population. They broadly encompass any of the many types of variations in DNA sequence found within a given population. Specific subtypes of polymorphisms include mutations, point mutations, and single nucleotide polymorphisms, SNPs (Chapter 10). Although an oversimplification, polymorphisms may be regarded as having less dramatic or overt functional effects than mutations.

Although we are a long way from describing variations in these multiple potential gene alleles, we know that polymorphisms contribute

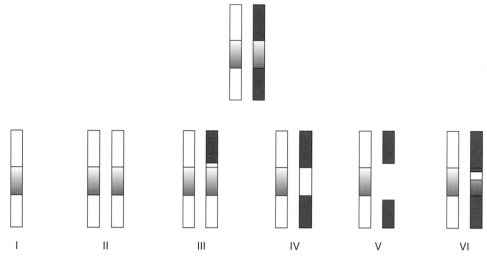

Various somatic events giving rise to loss of heterozygosity for a tumor-suppressor gene.

I, Nondisjunction; II, Nondisjunction and reduplication; III, Mitotic recombination; IV, Gene conversion;

V, Gene deletion; VI, Point mutation.

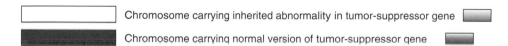

Chromosome carrying inherited abnormality in tumor-suppressor gene

Chromosome carrying normal version of tumor-suppressor gene

Fig. 1.8 Loss of heterozygosity through various genetic events.

to response to carcinogens, variations in drug responses, and undoubtedly to many other aspects of cancer. Recently, much interest has been sparked by the identification of polymorphisms which may contribute to the risk of lung cancer by influencing the susceptibility to carcinogens in tobacco smoke.

Cancer is an epigenetic disease

Epigenetic information is not contained within the DNA sequence itself, but is transmitted from one cell to all its descendants. Such a control is referred to as "epigenetic," as the DNA sequence is not altered. This is a major potential flaw inherent in attempts to understand diseases by sequencing genomes, as these epigenetic factors will be missed. Many key genes may be silenced by epigenetic changes during successive cell differentiation stages during development, and

two epigenetic events in particular that have been associated with transcriptional silencing in cancer cells include methylation of *CpG* islands in gene promoter regions and changes in chromatin conformation involving histone acetylation. Genes known to be epigenetically silenced in cancers include more than half of all known tumor suppressors, with much data in particular available for *p53* and *PTEN*, and the *MLH1* mismatch repair gene, silencing of which can cause genetic instability – thus linking epigenetic and genetic factors.

Loss of imprinting (LOI), the silencing of active imprinted genes or the activation of silent imprinted genes, is frequently observed in human cancers and is responsible for overexpression of the gene encoding insulin-like growth factor (IGF)-2 in the pathogenesis of Wilms tumor in Beckwith–Weidemann syndrome, and in some epithelial cancers, including colonic cancer.

The cancer "roadmap"

It has been estimated that up to seven rate-limiting genetic/epigenetic events are needed for the development of common human epithelial cancers. These may be ordered in multiple different combinations depending on which particular tissue or cell-specific "anticancer" barriers need to be circumvented and also there may be a number of different effective "routes" available for overcoming any given barrier. Importantly, many key molecular contributors to cancer progression may themselves not be deregulated at the gene level. Thus, downstream signaling proteins may become upregulated because of alterations upstream in growth-factor signaling genes, altered catabolism, genes inactivated by epigenetic factors, protein expression altered by enzyme activity, degradation, chaperones, etc. Again, it should be noted that events contributing to cancer are not restricted to the cancer cells. Thus, for example, expression of key cancer-contributing proteins in the cancer cell, such as NF-κB in hepatocytes, may be upregulated through changes in expression of TNF-α in neighboring stromal inflammatory cells.

Cancer is rare

Given the evolutionary nature of cancer, it is maybe surprising that three lifetimes are required to generate a cancer cell. The mutation rate has been estimated as one in 2×10^7 per gene cell division. Given that there are around 10^{14} target cells in the average adult human, with a myriad of potential target genes involved in regulation of cell expansion, and that the chances of further mutations are greatly increased by clonal expansion of those cells carrying the initial lesion, highly effective innate barriers to cancer must exist. Some of these barriers are now well described and include the coupling of oncogenic proliferative signals to those that induce apoptosis, senescence, or differentiation and the tumor-suppressor pathways involving *p53* and *RB*. Large long-lived animals like man, have a large potential somatic mutational load. It has been estimated that point mutations resulting in activation of *RAS* occur in thousands of cells daily in the average human. As the vast majority of these do not result in neoplasia, it is assumed that the usual outcome of such mutations is apoptosis, differentiation, or growth arrest. It should also be remembered that epithelia, such as gut, have the unique advantage of being able to shed potential would-be cancer cells from the surface – such cells removed from the nurturing environment of the body undergo a form of apoptosis (anoikis) before ending life in the waste disposal.

The barriers to cancer

The tumor suppressors   Two key pathways, those involving the tumor suppressors p53 and RB, are amongst the most critical barriers to cancer development. Not surprisingly, the p53 and RB pathways are frequently inactivated in human tumors and may be disrupted at different points. Thus, genetically, the RB pathway (cyclin D, CDK4, p16^{INK4A}, RB), a critical determinate of the G_1/S transition in the replication cell cycle, acts as one "critical target" in cancer cells, but the mechanism of disruption varies by tissue. Thus, for example, cyclin D is overexpressed by amplification in breast cancer and by translocation in parathyroid cancer; CDK4 is mutated or overexpressed in melanoma; p16^{INK4A} is inactivated by deletion or silencing in melanoma and pancreatic cancer; RB expression is lost by mutation or deletion in retinoblastoma and soft-tissue sarcomas. Such patterns may not be random. Specific associations of events are seen within individual tumors, and these presumably reflect the evolution of the tumors along particular pathways.

The p53 tumor-suppressor protein is a major component of the natural defenses against cancer. The p53 protein acts by arresting the cell cycle and promoting apoptosis (programmed cell death) in response to DNA damage, hypoxia, or unscheduled activation of oncogenes such as *c-MYC*. The *p53* gene is altered in more than half of all human cancers and because of its role in mediating growth arrest or apoptosis in response to DNA damage it has been termed the *"guardian of the genome"*. However, given the equally important role of p53 in initiating apoptosis in response to inappropriate oncogene activation, this term is somewhat underrepresentative. Mediators and regulators of p53 activities are also targeted in cancer, and inactivation of p21^{CIP1} or ARF or activation of MDM2 (an inhibitor of p53) are all observed in cancers.

Over the last decade numerous links between the p53 and RB tumor-suppressor pathways have been identified, including regulation of the G_1/S transition and its checkpoints, and this has highlighted the crucial role of the E2F transcription factor family in these pathways. Virtually all human tumors deregulate either the pRB or p53 pathway or both. Many other tumor suppressors are known and are discussed in Chapter 7.

Avoiding suicidal urges   In 1972, John Kerr, Andrew Wyllie, and Alistair Currie published a description of an unusual form of cell death distinctly different from necrosis that they termed apoptosis, which is now amongst the most published areas of biology (Chapter 8). Robert Horvitz, who along with Sir John Sulston was awarded the Nobel Prize for his work on apoptosis, has rather succinctly summarized the three stages of apoptosis as follows: "first, killing the cell, then getting rid of the body and then destroying the evidence."

Probably the single most critical barrier against cancer is the selfless suicide (apoptosis) of a "potential cancer cell", which either because it has been unable to repair damaged DNA or because it is being inappropriately pushed into the cell cycle disassembles and repackages itself as an "energy-giving snack" for its neighbors, rather than pose a threat to the whole organism. Apoptosis offers several distinct advantages to the organism, not least of which is a relative absence of inflammation (that might well result if the body had been required to "murder" the potential cancer cell – necrosis). Such an absence of collateral damage during apoptotic death is largely due to the ability of neighboring cells and phagocytes to swiftly recognize and cannibalize the apoptotic cell (usually before it has actually "died"). Moreover, when operating correctly, this also prevents the release of viruses or harmful cellular contents into the environment and instead seamlessly passes them from the apoptotic cell to another cell where they can be neutralized. Arguably, the ability to undergo apoptosis is one of the major hallmarks of moving from a unicellular to being part of a multicellular organism, where "social responsibility" amongst constituent cells becomes paramount for the survival of the whole organism. Apoptosis and necrosis are not the only forms of cell death described; others include anoikis, endoplasmic reticulum stress. More recently another potential balance against inappropriate replication has been described, namely irreversible exit from the cell cycle (senescence – see Fig. 1.9). This is described in detail in Chapter 9.

Cells are continually receiving and integrating a variety of both positive and negative growth signals. One intriguing result of much research over the last 20 years has been the appreciation that cells seem only too willing to commit suicide. In fact, cells require continuous signals from neighboring cells in order to survive. Loss of these normal "survival" signals or an increase in negative growth signals will tip the balance and a cell will undergo apoptosis. Two major pathways of apoptosis are known: one is intrinsic and is integrated by a variety of signals operating at the mitochondria, and the other extrinsic, triggered by activation of cell surface receptors such as FAS or TNF receptor. Both pathways eventually activate cascades of caspases, expressed as inactive zymogens, which when activated in cells destined to undergo apoptosis execute the necessary steps for apoptosis. However, the initiating caspases (apical caspases) differ – the intrinsic pathway commences with activation of caspase 9, the extrinsic with caspase 8.

In cancer, the intrinsic pathway of apoptosis may be triggered by "sensors" that determine the presence of irreparably damaged DNA or inappropriate attempts to engage the cell cycle machinery, which in turn may be modulated by external signals, which either prevent or provoke apoptosis. In general, these mechanisms are largely integrated at the mitochondria. Although, the body rarely "murders" would-be cancer cells it can certainly drive these cells to suicide. The extrinsic pathway is utilized by the immune system to engage the apoptotic machinery via surface "death receptors". These death receptors respond to some secreted inflammatory cytokines and to some populations of T cells, and include those for TNF and FASL. The pathways activated by these receptors include those able to trigger caspase cascades independent of the mitochondria.

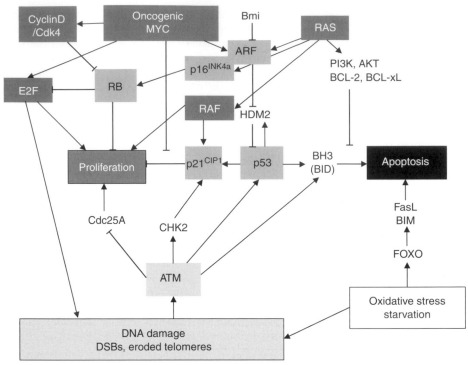

Several links exist between mitogenic signaling and those regulating growth arrest, and apoptosis. Moreover, DNA damage response pathways may be involved in linking oncogenic cell cycles with growth arrest and apoptosis.

Activation of Ras and MYC via growth factor signaling results in potential engagement of both replication and growth but also of apoptosis and possibly growth arrest. If either MYC or RAS levels are excessive (as might occur during oncogenesis) or other proapoptotic signals are received then the balance may be tipped away from replication. Oncogenic RAS can promote senescence through either p16^{INK4a} or ARF, which activate RB or p53 pathways respectively. Intriguingly, c-MYC may activate apoptosis through activation of ARF, possibly at least in part via DNA damage responses. Though, it remains unclear as to how the cell can distinguish between a normal cell cycle and an aberrant "cancer cell cycle", one possibility is that the latter may be more likely to result in DNA damage. Apoptosis may be blocked by RAS activation of PI3K and AKT pathways.

Fig. 1.9 Linkage between signaling regulating replication, DNA damage, apoptosis, and growth arrest.

Apoptosis can also be executed by caspase-independent death effectors, such as apoptosis-inducing factor (AIF), endonuclease G, and a serine protease (Omi/HtrA2) released from mitochondria during permeabilization of the outer membrane (MOMP). It is worth noting that many of these proteins have important or even essential roles in cellular processes unrelated to cell death. AIF and Omi/Htra2 are involved in redox metabolism and/or mitochondrial biogenesis; caspase activation is essential in some cells for terminal differentiation, lipid metabolism, inflammatory responses, and proliferation. This has important ramifications, as it implies that certain key parts of the apoptotic response could not be ablated therapeutically without impeding normal cellular functions, unless drugs can be designed to target only the lethal (and not vital) role of these proteins.

Avoiding senescence In 1961, Leonard Hayflick and Paul Moorhead found that many human cells such as fibroblasts had a limited capacity to replicate themselves in culture. In fact, they observed that cells can undergo between 40 and 60 cell divisions, but then can divide no more, a process described as senescence (or die). This number is often referred to as the Hayflick limit. Cellular senescence is associated with aging and longevity and has also been termed replicative senescence. The Hayflick limit for dividing cells may in part be determined by the length of the cells' telomeres (see Chapter 9). Telomeres are non-coding regions at the tips of chromosomes. Cell division requires the duplication of chromosomes, but each time a chromosome reproduces itself, it loses a part of the telomere (telomere attrition). Once a cell's telomeres reach a critically short length, the cell can no longer replicate its chromosomes and thus will stop dividing. Such cells are termed "senescent." Cells taken from older humans divide fewer times before this occurs.

Oncogene-induced senescence

As if this were not already complex enough, senescence can also be triggered by activation of various signaling pathways (see Chapter 9). Long appreciated as a major restraint to replicate to replicate potential *in vitro*, several recent studies have now confirmed the key role of oncogene induced senescence (OIS) as another potential inherent restraint to tumorigenesis (along with apoptosis and growth arrest) *in vivo*. Although the exact signaling pathways most critical for OIS may vary for different cell types and cancers, there are common features that overlap with activation of DNA damage responses such as those seen with telomere attrition and various engagement of either the ARF-p53-p21^{CIP1} and/or p16^{INK4a}-Rb pathways. What remains unclear is for how long such senescent cells persist before being culled and whther this state is true and always irreversible.

A key feature of cancer cells is that they avoid death and senescence, which is frequently termed immortalization – though clearly cancer cells do die by hypoxia, extensive DNA damage/chromosomal instability etc. Cellular senescence may have evolved as one mechanism to avoid cancer, which clearly increases in frequency with aging. Several studies have shown that the induction of cellular senescence can inhibit particular cancers. Importantly, although the large majority of cancer cells seem able to avoid telomere attrition (shortening) and the telomere-stabilizing enzyme telomerase is induced in tumors and linked to unrestrained replicative potential, this is not as straight-forward as it might at first appear. First, inactivating telomerase in some models of viral oncogene-induced cancers does not impede tumorigenesis or replicative potential, suggesting alternative methods for telomere maintenance are also important. Moreover, in other cancer models, where p53 is inactivated, telomere shortening instead of promoting apoptosis or senescence may instead lead to a more genetically unstable cancer as chromosome rearrangements are favored.

Oncogenes as tumor suppressors

Studies over the last two decades have revealed another crucial antineoplastic mechanism, namely that many signaling networks that can promote cellular replication also possess intrinsic growth-suppressing activities. Under normal growth conditions such as tissue maintenance and repair, signaling networks are activated in a coordinated fashion by appropriate extracellular signals, which can block the growth-suppressing pathways and the cell replicates and survives. By contrast, inappropriate activation of a potentially powerful replicative signal such as c-MYC, for instance by mutation, occurs in the absence of activation of those other key pathways, and instead may result in apoptosis of the mutated cell, thereby eliminating the risk of further mutations and cancer. This "intrinsic tumor suppressor" activity of some mitogenic proteins to induce either apoptosis or growth arrest allows them to function as a critical "fail-safe" mechanism in the avoidance of cancer through unscheduled cell cycling. By implication, therefore, in cancer the inherent growth-suppressing activities of oncogenes such as *c-MYC* must be suppressed in order for cancer to develop or progress – an example of oncogene cooperation discussed in detail in later chapters.

Location, location, location – the cancer environment

Numerous studies now point to the crucial interplay between the cancer cell and its local and systemic microenvironment. It is often assumed that the body is largely a hostile environment for an incipient cancer, with triggering of immune responses and inflammation recognizing and attempting to eliminate the cancer cells, stromal cells resisting cancer spread, and the body denying blood supply and nutrients to the tumor. Thus, cancers are assumed to have to overcome these hostile forces in order to progress. Recent studies have increasingly challenged this view and have thrown light on the importance of environmental interactions in supporting the cancer (Chapter 12). Chronic inflammation has long been known to increase risk of many cancers, possibly by increased mitogenesis (and thereby mutagenesis) or through paracrine effects from inflammatory cells. However, it is clear that even in the absence of preceding inflammation, malignant transformation takes place within the context of a dynamically evolving "microenvironment" and is accompanied by fibroblast proliferation and transdifferentiation, extracellular matrix deposition and remodeling, increased matrix metalloproteinase expression and activity, infiltration of immune cells (Chapter 13), and angiogenesis (Chapter 14). Many such changes may actively support tumor cell invasion, survival, and growth and this is particularly important in epithelial carcinogenesis (Chapter 12).

Interactions between cancer cells and their environment are thus key determinants of tumor progression. In recent years, a considerable interest has developed in "immune privilege" (Chapter 13). Foreign antigens that enter immunologically privileged sites, of which the eye, brain, and testis are examples, can survive for an extended period of time, whereas the same antigens would normally be swiftly eliminated elsewhere. Despite the existence of tumor-specific immune cells, most tumors appear to have acquired a means to avoid immune attack. It has been proposed that the tumor microenvironment may become a site of immune privilege, possibly through factors produced by the tumor, which might impair immune surveillance. Immune privilege could provide a "safe haven" for cancer cells. Recent studies, in ovarian cancer, have suggested that one means of immune privilege is recruitment of regulatory T cells by the tumor. These regulatory T cells can block the activity of those T cells that are reactive to tumor antigens, thereby interfering with tumor-specific T-cell immunity and enabling progression of ovarian cancers *in vivo*. Other possibilities include production by the cancer cells of cytotoxic or inhibitory factors for tumor-reactive T cells, such as galectin-1, TGF-β, or Fas ligand.

It's all about timing

The exact role of any given protein may be largely a matter of timing with respect to the stage of a cancer's evolution and likely also the developmental stage of the cell under consideration. Thus, even individual proteins within the cancer cell can exert widely differing effects on phenotype. Mitogenic proteins like c-MYC may prevent the initiation of cancer through their inherent apoptotic activity, but once the cancer cell has acquired the ability to avoid apoptosis, or the environment provides sufficient survival signals, may instead confer a wide range of cancer-promoting behaviors.

A recent study has shown that brief inactivation of c-MYC was sufficient for the sustained regression of c-MYC-induced invasive osteogenic sarcomas in transgenic mice; subsequent reactivation of c-MYC led to extensive apoptosis rather than restoration of the neoplastic phenotype. Possible explanations for this outcome include changes in epigenetic context that may have occurred within the cell type, that is, between the immature cell in which c-MYC was originally activated and the differentiated cell resulting from subsequent (brief) inactivation of c-MYC. In this tumor model, although c-MYC expression is initiated in immature osteoblasts during embryogenesis, subsequent inactivation of c-MYC in osteogenic sarcoma cells induces differentiation into mature osteocytes. Therefore, reactivation of c-MYC now takes place in a different cellular context and induces apoptosis rather than neoplastic progression.

Initially TGF-β was identified in culture media from transformed cells as part of a factor that could produce a transformed phenotype in a nontransformed cell line. The observations that TGF-β1 inhibited the growth of epithelial cells, and that inactivating mutations within the TGF-β1 signaling pathway occurred in many cancers, supported the view of TGF-β1 signaling as a tumor-suppressor pathway for early stages of cancer. However, many human carcinomas overexpress TGF-β1 and this is associated with a poor prognosis and metastasis. Similar results pertain to tumor cell lines and animal models. Together, this suggests that TGF-β1 switches from tumor suppressor to oncogene, as the context changes probably due to genetic or epigenetic alterations in tumor cells or stromal cues. Thus, the role of TGF-β1 in cancer is stage-specific.

The end of the journey – metastatic spread

As tumors progress, cells within them develop the ability to invade into surrounding normal tissues and through tissue boundaries to form new growths at sites distinct from the primary tumor. The seeding and growth of cancer cells in distant organs is termed metastasis and is the ultimate cause of death in around 90% of cancer patients. Metastasis was first described in 1839 by the French gynecologist Joseph Recamier, and soon thereafter, physicians found that certain cancers were most likely to spread to certain organs. Breast and prostate cancer, for example, move to lymph nodes, bones, lung, and then the liver. Skin cancer tends to spread to the lungs, colon cancer targets the liver, and lung cancer typically moves to the adrenal glands and the brain.

In 1889, Stephen Paget proposed that cancer cells shed from an initial tumor were dispersed randomly throughout the body by the circulatory system. He called these circulating cancer cells "seeds" and proposed that only some seeds fall onto "fertile soil", organs where they can grow. About 30 years later, a researcher named James Ewing proposed an alternative nonrandom model by which circulating cancer cells become trapped in the first small blood vessels, or capillaries, they encounter and then grow in the surrounding organ.

While much is now known about molecular alterations that contribute to tumorigenesis, the genetic and epigenetic alterations that result in metastatic spread of the disease are less well understood. Millions of tumor cells can be shed into the vasculature daily; yet few secondary tumors are formed. The general explanation for this has relied on the assumption that a number of additional genetic events had to occur

in order for a small subclone of cells to arise with the capabilities to enter, navigate, and exit the vasculature and thence to colonize a distant site. However, some recent studies suggest that genes required for metastatic spread may already be expressed in primary tumors and before any metastatic spread, suggesting that metastatic ability might be preprogrammed in tumors by the initiating oncogenic mutations. One problem with such data is that even though multiple genes were aberrantly expressed in such primary tumors, they may not all have been so in any individual cell (gene expression profiles were generated from mushed-up whole tumors), and epigenetic factors were not addressed.

In the past decade, much has been learned about how cancers metastasize. Key findings have included the observation that cancer cells are subject to growth regulation at the secondary site and that the molecular characterization of proteins can suppress the metastatic phenotype. These proteins are encoded by metastasis-suppressor genes, defined as genes that suppress *in vivo* metastasis without inhibiting primary tumor growth when transfected into metastatic cell lines and injected into experimental animals. Key processes required for metastatic spread include migration and invasion of tumor cells. Various proteins have been implicated in these processes including cell adhesion molecules, proteolytic enzymes, and members of the RHO family, including RHO, RAC, and CDC42, that are involved in cytoskeletal organization.

Recent exciting data suggest that invasive and metastatic potential is related to reactivation of general embryonic pathways involved in morphogenesis and might include mutations that deactivate E-cadherin and other cell adhesion molecules, those that activate transcription factors and signaling molecules such as NF-κB and TWIST, which might promote epithelial–mesenchymal transition (EMT). EMT, originally described *in vitro* as dedifferentiation of epithelial cells to fibroblastoid, migratory, and more malignant cells, with an accompanying altered mesenchymal gene expression program, correlates well with late-stage tumor progression. Typical phenotypic features of EMT include loss of E-cadherin and acquisition of vimentin immunoreactivity. EMT also occurs during embryonic development and is regulated by a complex network of signaling pathways, including the RAF/MEK/MAPK pathway, PI3K/AKT pathway, NF-κB, and TGF-β. In various animal models systems, metastatic potential strictly correlates with the ability of epithelial tumor cells to undergo EMT. A question that is currently of tremendous interest is at what time cancer cells acquire the capabilities to undergo metastatic spread – this is addressed in the next section.

The "poor prognosis" metastatic signature

Increasingly, it is apparent that understanding major genetic and epigenetic factors will still provide only a partial picture of the disease. In practical terms, cancer patients with ostensibly identical clinical stages of disease (and probably even those with apparently similar genetic factors) may have markedly different treatment responses and overall outcome. In the same way that genomics offers the possibility of a more complete understanding of disease by describing multiple polymorphisms, so also advances in molecular biology raise the possibility of going a step further. Cell behavior and disease pathogenesis ultimately arise through the differential expression of multiple genes, and thus in turn by their protein products in the diseased cells and also in other cells, neighboring and more distant, and within the affected organism. At best, genomic sequences will have only a partial relationship to gene/protein expression particularly as they will largely overlook epigenetic factors and moreover large-scale identification of polymorphisms may be far more difficult to comprehend than a molecular profile from a given cell/tissue.

Recently, landmark studies indicate that gene-expression profiling can produce relatively unique tissue "molecular signatures" that can be employed to improve ability to predict disease prognosis and response to therapy. Using DNA microarray analysis on primary breast tumors, researchers from the Netherlands Cancer Institute identified a gene-expression signature that was predictive of a short latency of development of metastases. Importantly, this "signature" was able to identify those individuals likely to progress amongst those otherwise generally regarded as low risk. This

"poor prognosis signature" was shown to include genes regulating cell cycle, invasion, metastasis, and angiogenesis. This study provides support for the current "holy grail" of post-genome era medicine, namely disease fingerprinting and individualized medicine. Importantly, these studies examined gene expression in "lumps" of tumor, which contained cancer cells, stromal cells, and others. This does not lessen the clinical utility of such a whole-tumor "poor prognosis" signature, but imagine the "power" of a similar study looking at gene or indeed protein expression in individual cells – and not just different cancer cells, but stromal cells, vascular endothelial cells. Some preliminary steps have been made toward this ultimate goal. Recently, Dennis Sgroi and colleagues (2003) have combined microarray analysis of gene expression on specific bits of tumor isolated from tissue sections by means of laser capture microdissection. These were obtained from premalignant, preinvasive, and invasive stages of human breast cancer. The resultant "*in situ* expression signatures" intriguingly suggested that there were extensive similarities at the transcriptome level for these different stages of cancer progression, supporting a notion that gene expression alterations conferring the potential for invasive growth might already be present in early preinvasive stages. In contrast to tumor stage, different tumor grades were also associated with distinct gene expression signatures, particularly between preinvasive and invasive.

However, it must be remembered that "groups" of cells rather than single cells had been profiled; it is by no means certain that all the genes expressed applied to any individual cell. Tumors are usually genetically heterogeneous and therefore tumor profiling unless supplemented by "single cell" analyses may lead to erroneous conclusions, particularly if the assumption is made that all abnormalities detected apply to all individual tumor cells – clonal expansion does not equate to all cancer cells being identical but simply means that all cells will in some way carry the initiating genetic lesions alongside those additional mutations acquired during "cancer evolution".

With this in mind, very recent studies suggesting that single cancer cells from primary tumors may indeed carry the "poor prognosis" signature

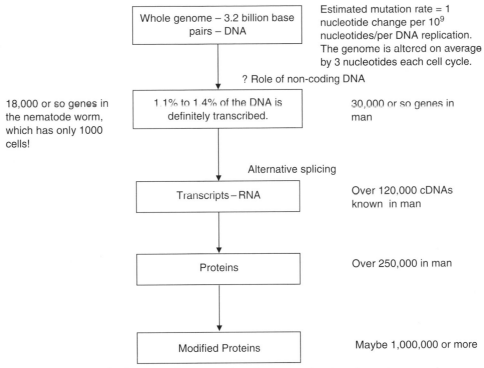

Although the issue of information flow seems hopelessly complex there is much reason for hope. First, the availability of the reference genome for man and many experimental models alongside new technologies for analyzing the expression of multiple genes and proteins will with the appropriate techniques for analyzing and distributing experimental data hopefully result in major progress in "discovery science", and second, as many key genes/proteins have homologs in more primitive and experimentally amenable organisms (Friedrich Nietzsche, "You have made your way from worm to man, and much within you is still worm"), we should have a much greater scope for functional studies.

Fig. 1.10 The complexity of cellular information flow in cancer.

for metastases are very exciting, but will need confirmation.

A cancer protein expression profile Ultimately, it is proteins that determine phenotype. Not all genes are expressed in any given cell, and even of those genes expressed (for which mRNA is formed) alternative regulatory events may still take place after transcription, which determine protein levels. Although considerable correlation exists between gene and protein expression, there are far more proteins than genes (see Fig. 1.10). Thus, alternative splicing of RNA, post-translational modifications, and enzyme activities can all contribute to the generation of a multitude of different proteins. Importantly,

not all of these different proteins can therefore be directly inferred from examination of either the genome or even the transcriptome of a given cell at any given time. This has been the major impetus behind efforts to describe the cell proteome using mass spectrometry and other techniques (Chapter 18).

CONCLUSIONS AND FUTURE DIRECTIONS

Cancers arise by the stepwise accumulation of mutations and epigenetic factors that alter gene expression to confer cancer properties on the cell. The presence of inherited cancer-causing

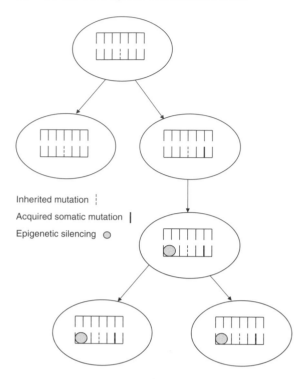

Inherited mutation ┊

Acquired somatic mutation ▮

Epigenetic silencing ◯

Fig. 1.11 Tumorigenesis ultimately results from dis-ordered gene expression. Tumor cells arise through aberrant expression of genes and the proteins they encode. This may result from: mutations in the cod-ing or noncoding regulatory regions of genes, which can be either inherited or acquired in somatic cells or even by major rearrangements of the chromosomes; epigenetic factors such as altered patterns of methyl-ation and acetylation, which control the "accessib-ility" of genes for transcription. These events may in turn affect the stability and processing of RNA or proteins.

mutations will give a would-be cancer cell a head start, but somatic mutations and epi-genetic alterations are still needed for can-cer development (see Fig. 1.11). It is likely, given the increasing susceptibility of progress-ing cancer cells to mutations, that not all such mutations are actually cancer-relevant. It is anticipated that improved knowledge about these various processes regulating aberrant gene expression in cancer will lead to new treatments aimed at specifically targeting the expression of genes/proteins – "mission critical" for the initi-ation and progression of cancer.

Appendix 1.1 History of cancer (see also: http://press2.nci.nih.gov/sciencebehind/cioc)

Early Egyptian papyri from around 1600 BC, such as the "Edwin Smith" and "George Elbers" papyri, already include descriptions of benign and malig-nant tumors and treatments based on castor oil and various animal parts, including pigs' ears. But much of the early written descriptions of can-cer originate from the classical Greek and Roman physicians Hippocrates and Galen, who laid the foundations for modern medicine by emphasiz-ing that diseases were natural physical processes. In fact, we owe the names for cancer to Hippo-crates, who first employed the terms *karkinos* and *karkinoma* (ancient Greek for "crab") to various diseases including cancers of the breast, uterus, stomach, and skin. Cancer is the Latin equival-ent. Interestingly, although Galen performed some early surgical interventions for cancer, he main-tained that cancer was generally best left untreated. However, Galen also believed that diseases resul-ted from imbalances in the four bodily "humors" (blood, phlegm, yellow bile, and black bile), which were also responsible for differing temperaments such as melancholy!

Unfortunately, relatively little progress was recor-ded during the so-called middle ages (from the fall of the Roman empire till the renaissance). Though, clearly in the Arab world, Moorish Spain, Constantinople, and in the West in monastic com-munities, much classical learning was preserved and recorded for the future benefit of renaissance scholars. This generally negative view of human progress in the middle ages as being largely the copying and preservation of classical texts for the future benefit of renaissance scholars is rather over-stated, as illustrated by an intriguing quotation from Theodoric, Bishop of Cervia (1267) – "The older a cancer is, the worse it is. And the more it is involved with muscles, veins and nutrifying arteries, the worse it is, and the more difficult to treat. For in such places incisions, cauteries and sharp medications are to be feared." Much import-ant scholarship was undertaken in the Arab world, not the least of which was laying the foundations for modern mathematics. With respect to cancer, the insightful writings of two prominent Arab schol-ars have been recorded. Thus, to quote Avicenna (981–1037):

The difference between cancerous swelling and induration. The latter is a slumbering silent mass, which ... is painless, and stationary .... A cancerous swelling progressively increases in size, is destructive, and spreads roots which insinuate themselves amongst the tissue-elements;

and Albucasis (1050):

The Ancients said that when a cancer is in a site where total eradication is possible, such as a cancer of the breasts or of the thigh, and in similar parts where complete removal is possible, and especially when in the early stage and small, then surgery was to be tried. But when it is of long standing and large you should leave it alone. For I myself have never been able to cure any such, nor have I seen anyone else succeed before me.

From classical times until the late renaissance when Vesalius and artists such as Michelangelo and Leonardo da Vinci developed an interest in anatomy, cancer was still believed to be caused by various acts of god or still, in deference to Galen, by an excess of black bile. Although still believed to be incurable, a wide variety of arsenic-containing preparations were employed to treat it. Based on his observations in Austrian mines, Theophratus Bombastus von Hohenheim, better known as Paracelsus, described the "wasting disease of miners" in 1567. He proposed that the exposure to natural ores such as realgar (arsenic sulfide) and others might have been causing this condition. Paracelsus was actually among the first to consider a chemical compound as an occupational carcinogen. Paracelsus was probably the first prominent objector to Galen's humoral doctrine, and instead proposed that mineral salts, when concentrated in a particular part of the body and unable to find an outlet, were the real cause of cancer.

The beginnings of recognizably modern science took place in the seventeenth century; William Harvey described the continuous circulation of the blood, finally resulting in the rejection of the humoral theory of disease, and cancer was no longer attributed to bile. A contemporary of Harvey, Gaspare Aselli identified the lymphatic system, which he suggested as a primary cause of cancer. However, on the basis of this discovery, Descartes developed a new theory, termed the "sour lump" theory, in 1652, whereby it was suggested that lymph became hard through some congealing process and formed a scirrhus. If this fermented (i.e. became acid or sour) then a cancer would develop. Surgery for cancer now began to include removal of the lymph nodes, when enlarged and near the tumor site. A renowned German surgeon, Fabricius Hildanus, removed enlarged lymph nodes in breast cancer operations, but in the absence of either septic techniques or anesthetics it was an extremely hazardous procedure.

In the eighteenth century oncology became a recognized discipline, with early experiments conducted. The French physician Claude Gendron (1663–1750) concluded after 8 years of research that cancer arises locally as a hard, growing mass, untreatable with drugs, that must be removed with all its "filaments." The Dutch professor Hermann Boerhaave believed inflammation could result in a scirrhus or tumor, capable of evolving into cancer. John Hunter, one of the earliest modern surgeons, taught that if a tumor were moveable, it could be surgically removed, as could resulting cancers in proper reach. If enlarged glands were involved, he advised against surgery.

Two eighteenth century French scientists, physician Jean Astruc and chemist Bernard Peyrilhe, conducted experiments to confirm or disprove hypotheses related to cancer. Their efforts may appear amusing to us now, but they helped establish the discipline of experimental oncology. For example, in 1740 Jean Astruc, a professor of medicine at Montpellier and Paris, sought to test the validity of the humoral theory by comparing the taste of boiled beef steak with that of boiled breast tumor; he found no black-bile-like taste in the tumor – he may also have had a lasting influence on French culinary practices! Peyrilhe attempted to demonstrate an infective cause for cancer by injecting human cancer tissue into a dog. The resultant infected abscess (no cancer!) resulted in a housemaid drowning the poor dog to end its misery.

Later in the same century, two English physicians, John Hill and Percivall Pott, described the occurrence of cancerous alterations in the nasal mucosa and at the skin of the scrotum in a few patients, and traced it to the local long-term exposure to snuff

Appendix 1.1 (continued)

and to repetitive local contamination by soot, respectively.

The nineteenth century heralded the beginnings of modern biology. Virchow focused pathology on the cell; and anesthesia and antisepsis improved surgery. Oncology progressed as Röntgen described X-rays, the Curies isolated radium, and Müller observed abnormalities of cancer cells. By the mid nineteenth century, French and Italian researchers had found that women died from cancer much more frequently than men, and that the cancer death rate for both sexes was rising. Domenico Rigoni-Stern concluded that incidences of cancer increase with age.

Throughout the early decades of the last century, researchers pursued different theories of the origin of cancer; the early part of the twentieth century saw the proposal by Theodor Boveri, professor of zoology at Wurzberg, that cancer was due to abnormal chromosomes. This was remarkably prescient given that it was more than 40 years before the discovery of the structure of DNA. A viral cause of cancer in chickens was documented in 1911, and both chemical and physical carcinogens were conclusively identified. Radium and X-rays were employed against cancer early in the century, and it was found that X-rays selectively damaged cancer cells, causing less harm to other tissues. As safe levels of dosage were determined, the therapy became standard. Chemical and radiation-induced cancers were first reliably confirmed. While the smoking – cancer link was noted in the 1930s, causality was only proven following extensive epidemiological studies in 1950.

Molecular biology has revolutionized both medicine and cancer research; following the identification of the structure of DNA by Francis Crick and James Watson in 1953, the genetic code was soon broken, and the foundations were laid for much of what is discussed in this book.

We conclude with two quotes, illustrating how far we have progressed in cancer therapy:

Paul of Aegina [625–690]:

Cancer is an uneven swelling, rough, unseemly, darkish, painful, and sometimes without ulceration ... *and if operated upon, it becomes worse ... and spreads by erosion; forming in most parts of the body, but more especially in the female uterus and breasts. It has the veins stretched on all sides as the animal the crab (cancer) has its feet, whence it derives its name.*

Aulus Aurelius Cornelius Celsus, [25 BC – AD 50]:

A carcinoma does not give rise to the same danger [as a carbuncle] *unless it is irritated by imprudent treatment.* This disease occurs mostly in the upper parts of the body, in the region of the face, nose, ears, lips, and in the breasts of women, but it may also arise in an ulceration, or in the spleen .... At times the part becomes harder or softer than natural .... After excision, even when a scar has formed, none the less the disease has returned, and caused death.

BIBLIOGRAPHY

General biology

Alberts, B., Johnson, A., Lewis, J., Raff, M., Roberts, K., and Walter, P. (2002). *Molecular Biology of the Cell*, 4th edn. New York, Garland Publishing.

Kimball, J.W. Kimball's Biology Pages. users.rcn.com/jkimball.ma.ultranet/BiologyPages/

Kufe, D.W., Pollock, R.E., Weichselbaum, R.R., Bast, R.C., Jr., Gansler, T.S., Holland, J.F., and Frei, E. (ed.) (2003). *Cancer Medicine*, 6th edn. Hamilton (Canada), BC Decker Inc.

Genes and cancer

Balmain, A., Gray, J., and Ponder, B. (2003). The genetics and genomics of cancer. *Nature Genetics*, **33**(Suppl): 238–44.

Futreal, P.A., Coin, L., Marshall, M. *et al.* (2004). A census of human cancer genes. *Nature Reviews: Cancer*, **4**: 177–83.

Knudson, A.G. (2002). Cancer genetics. *American Journal of Medical Genetics*, **111**: 96–102.

Ma, X.J., Salunga, R., Tuggle, J.T., *et al.* (2003). Gene expression profiles of human breast cancer progression. *Proceedings of the*

National Academy of Sciences of USA, **100**(10): 5974–95.

van 't Veer, L.J., Dai, H., van de Vijver, M.J., *et al.* (2002). Gene expression profiling predicts clinical outcome of breast cancer. *Nature*, **415**(6871): 530–6.

Multistage carcinogenesis

Hanahan, D. and Weinberg, R.A. (2000). The hallmarks of cancer. *Cell*, **100**, 57–70.

Nowell, P.C. (1976). The clonal evolution of tumor cell populations. *Science*, **194**(4260): 23–8.

Rangarajan, A., Hong, S.J., Gifford, A., and Weinberg, R.A. (2004). Species- and cell type-specific requirements for cellular transformation. *Cancer Cell*, **6**(2): 171–83.

Vogelstein, B. and Kinzler, K.W. (2004). Cancer genes and the pathways they control. *Nature Medicine*, **10**(8): 789–99.

Tumor suppressors

Lowe, S.W. and Sherr, C.J. (2003). Tumor suppression by Ink4a-Arf: progress and puzzles. *Current Opinion in Genetics and Development*, **13**(1): 77 83.

Vousden, K.H. and Prives, C. (2005). P53 and prognosis: new insights and further complexity. *Cell*, **120**(1): 7–10.

Whyte, P., Buchkovich, K.J., Horowitz, J.M., *et al.* (1988). Association between an oncogene and an anti-oncogene: the adenovirus E1A proteins bind to the retinoblastoma gene product. *Nature*, **334**(6178): 124–9.

Christophorou, M.A., Martin-Zanca, D., Soucek, L., *et al.* (2005). Temporal dissection of p53 function in vitro and in vivo. *Nature Genetics*, **37**(7): 718–26.

Bourdon, J.C., Fernandes, K., Murray-Zmijewski, F., *et al.* (2005). p53 isoforms can regulate p53 transcriptional activity. *Genes and Development*, **19**: 2122–37.

Robanus-Maandag, E., Giovannini, M., van der Valk, M., *et al.* (2004). Synergy of Nf2 and p53 mutations in development of malignant tumors of neural crest origin. *Oncogene*, **23**(39): 6541–7.

Telomeres, instability, and chromosomal abnormalities

DePinho, R.A. and Polyak, K. (2004). Cancer chromosomes in crisis. *Nature Genetics*, **36**(9): 932–4.

Feldser, D.M., Hackett, J.A., and Greider, C.W. (2003). Telomere dysfunction and the initiation of genome instability. *Nature Reviews Cancer*, **3**: 623–627.

Oncogene-induced senescence

Braig, M., Lee, S., Loddenkemper, C., *et al.* (2005). Oncogene-induced senescence as an initial barrier in lymphoma development. *Nature*, **436**(7051): 660–5.

Chen, Z., Trotman, L.C., Shaffer, D., *et al.* (2005). Crucial role of p53-dependent cellular senescence in suppression of Pten-deficient tumorigenesis. *Nature*, **436**(7051): 725–30.

Michaloglou, C., Vredeveld, L.C., Soengas, M.S., *et al.* (2005). BRAFE600-associated senescence-like cell cycle arrest of human naevi. *Nature*, **436**(7051): 720–4.

Apoptosis

Lowe, S.W., Cepero, E., and Evan, G. (2004). Intrinsic tumor suppression. *Nature*, **432**(7015): 307–15.

Pelengaris, S., Khan, M., and Evan, G.I. (2002). Suppression of MYC-induced apoptosis in beta cells exposes multiple oncogenic properties of Myc and triggers carcinogenic progression. *Cell*, **109**(3): 321–34.

Microenvironment, inflammation, immunity, and cancer

Curiel, T.J., Coukos, G., Zou, L., *et al.* (2004). Specific recruitment of regulatory T cells in ovarian carcinoma fosters immune privilege and predicts reduced survival. *Nature Medicine*, **10**(9): 942–9.

Zhu, Y., Ghosh, P., Charnay, P., Burns, D.K., and Parada, L.F. (2002). Neurofibromas in NF1: Schwann cell origin and role of tumor environment. *Science*, **296**(5569): 920–2.

Metastases

Crnic, I. and Christofori, G. (2004). Novel technologies and recent advances in metastasis research. *International Journal of Developmental Biology*, **48**(5–6): 573–81.

Gupta, P.B., Kuperwasser, C., Brunet, J.P., *et al.* (2005). The melanocyte differentiation program predisposes to metastasis after neoplastic transformation. *Nature Genetics*, **37**(10): 1047–54.

Steeg, P.S. (2003). Metastasis suppressors alter the signal transduction of cancer cells. *Nature Reviews Cancer*, **3**(1): 55–63.

Yang, J., Mani, S.A., Donaher, J.L., *et al.* (2004). Twist, a master regulator of morphogenesis, plays an essential role in tumor metastasis. *Cell*, **117**(7): 927–39.

Cancer profiling and "tailored" medicine

Carr, K.M., Rosenblatt, K., Petricoin, E.F., and Liotta, L.A. (2004). Genomic and proteomic approaches for studying human cancer: prospects for true patient-tailored therapy. *Human Genomics*, **1**(2): 134–40.

Screening and prevention

Miller, A.B. (1992). *Cervical Cancer Screening Programmes*. Geneva, WHO.

Stewart, B.W. and Coates, A.S. (2005). Cancer prevention: a global perspective. *Journal of Clinical Oncology*, **23**(2): 392–403.

2

The Burden of Cancer

William Steward and Anne Thomas

All interest in disease and death is only another expression of interest in life.   Thomas Mann

---

KEY POINTS

- The incidence of cancer in developed countries is rising due to increased aging of the population and increasing exposure to carcinogens.
- Cancers can cause symptoms from local mass effects leading to pain or organ dysfunction, or may cause systemic symptoms that are nonspecific and include weight loss, lethargy, muscle wasting, and debility.
- The suspected diagnosis of cancer should be confirmed by histological assessment of biopsy specimens taken from areas of abnormal tissue and may be supported by results of imaging and serum markers.
- Lung cancer is the most common malignancy causing death in Western nations and comprises small- and non-small-cell variants, which each have different biological behaviors, treatments, and outcomes.
- Breast cancer is the most common malignancy in females and has widely differing incidences in different nations, suggesting the importance of environmental factors in its etiology.
- Prostate cancer has a 10-fold difference in incidence between Japanese and Afro-Caribbean men and carries a worse prognosis in the Afro-Caribbean population.

KEY CLINICAL STUDY

Doll and Hill's study (1964) documented smoking habits among medical practitioners after following this group over many years. They detected a significant increase in incidence of lung cancer in the smokers and correlated lung cancer risk with increasing exposure to cigarettes. This was a landmark epidemiological study that had profound implications for society and defined a major carcinogen.

INTRODUCTION

The global cancer burden is estimated at approximately nine million new cases per year, and this figure will increase steadily in the future. More than 50% of patients with cancer are in developing nations, where the estimated cancer incidence is about 100 per 100,000 population. The incidence is, however, three- to four-fold higher in Western nations. Cancer as a global cause of death ranks third in developing countries after infectious and cardiovascular diseases, where it is second in developed countries after cardiovascular disease.

The five most common sites of cancer in 1985 were estimated for males and females. Combining both sexes, lung cancer was the most frequent type of cancer globally, contributing 11.6%, followed by stomach (9.9%), breast (9.4%), colon/rectum (8.9%), and uterine cervix (5.7%). The global incidence of lung cancer was increasing by about 0.5% per year at that time.

35

In 1997, the WHO estimated the global incidence of the major cancers in men and women. Carcinoma of the lung was again the most common cancer in men, followed by stomach, colon/rectum, prostate, mouth/pharynx, and liver. Cancer of the breast was the leading malignancy in women, followed by cervix, colon/rectum, stomach, lung, and mouth/pharynx. There has been a global increase of malignancies ascribed to HIV/AIDS over the last decade, with marked increases of lymphomas and hepatocellular carcinomas.

The incidence of cancer in Western societies has risen steadily over the last century, predominantly because of the increasing median age of the population but also because of rising exposure to carcinogens. It has been estimated that two-thirds of all malignancies in Europe and North America are related to dietary factors and cigarette smoking. Approximately 40% of all individuals can expect to develop some form of malignancy during their lifetime and this figure is expected to rise further. Lung cancer is by far the most common cause of death from malignancy in both males and females, followed by colorectal cancer in men and breast cancer in women.

This chapter will focus on the clinical features of the most important cancers and will include information on clinical manifestations, diagnosis, and treatment (summarized in Box 2.1). More detailed information on diagnostic procedures is given in Chapter 15.

All malignancies may cause symptoms, which result from different biological effects on the host. The earliest manifestations may result from the local effect of the tumor and include the detection of a mass, discomfort from compression of local organs or nerves, hemorrhage from the involvement of blood vessels, or obstruction of airways, ureter, bile duct, and other key structures. Tumors may also present with nonspecific effects such as cachexia (the loss of body mass frequently seen in patients with chronic debilitating diseases), lethargy, weight loss, fever, and a variety of neuromuscular syndromes. Some malignancies may produce hormones which act on distant tissues via the circulation. Such hormones include ADH, PTH, erythropoeitin, ACTH, and calcitonin, though this is less common than the production of

---

Box 2.1   Key concepts of chapter

Selected malignancies

- Lung
- Breast
- Colorectal
- Prostate
- Skin
- Hematological
  - Lymphomas
  - Leukemias
- Incidence
- Pathology
- Etiology

Presentation

- Local symptoms from primary
- Symptoms from metastases
- Systemic symptoms of malignancy
- Paraneoplastic syndromes
- Symptoms from biochemical abnormalities

Therapy

- Surgery
- Chemotherapy
  - Adjuvant
  - Neoadjuvant
  - Advanced
- Hormones
- Radiotherapy
- Biological therapies
- Modulators of cell signaling pathways

---

locally acting growth factors. Ultimately, most cancer deaths are the result of invasion and spread of the primary cancer to other more distant tissues (metastases), which may in some cases become clinically apparent before the primary tumor – in fact in some cases the primary tumor may never be discovered. The diagnosis of malignancy can therefore be difficult with a wide variety of presentations, which may not necessarily indicate an obvious site of the primary tumor.

For those patients who complain of specific symptoms that raise the possibility of a particular malignancy (e.g. altered bowel habit and carcinoma of the colon) the necessary investigations and diagnostic procedures are usually self-evident. For those patients who present with nonspecific symptoms (e.g. cachexia and weight loss) the choice of investigations is difficult. The majority of such individuals will probably not have an underlying malignancy and it is essential to balance the potential morbidity and occasional mortality of investigations (together with the cost) against the likelihood of detecting a malignancy and the potential value of making an early diagnosis. The gold standard of diagnosis is obtaining tissue that on histological examination confirms the presence of neoplastic cells. In the majority of cases this is straightforward with symptoms guiding the choice of a site for biopsies. Support for a diagnosis can be provided by tumor markers (usually proteins specifically expressed by or in response to a tumor) and abnormalities of hematological and biochemical blood results, but in a small percentage of patients no obvious site of neoplastic focus is detected and a diagnosis has to be made on the basis of probabilities. A further occasional clinical problem arises when a metastatic site is detected (most commonly an isolated area of lymphadenopathy) and histological examination reveals malignant cells (frequently adenocarcinoma or squamous carcinoma). Investigation of common primary sites (e.g. breast, gastrointestinal tract, head, neck, and lung) reveals no obvious origin and in this instance a diagnosis of "carcinoma of unknown primary" is made, which can lead to a dilemma when trying to choose appropriate therapies.

Key features, particularly the prevalence, pathogenesis, and clinical manifestations of selected cancers are outlined below. Current approaches to diagnostic techniques and treatment are summarized. It must be stressed that oncological practice is changing rapidly with the availability of new imaging technologies (e.g. positron emission tomography (PET) scanning) and targeted, noncytotoxic, approaches to therapies.

LUNG CANCER

Incidence

The incidence of carcinoma of the bronchus rose steadily during the twentieth century throughout the Western world. The rate of rise mirrored the increase in cigarette smoking. It has one of the highest mortality rates of all malignancies with the death rate approaching that of incidence. As a result it has been the highest cause of death from malignancy among males in the West for several decades and in females since the late 1980s. The overall annual incidence in most Western nations is approximately 100 per 100,000 males and 40 per 100,000 females. There has been a recent decline in the incidence among males, but as smoking has become more popular in females there has been an increasing incidence in women such that the male : female ratio for lung cancer incidence has risen steadily since the 1950s from approximately 13 : 1 to 2 : 1.

Of all lung cancers 80–90% are attributable to cigarette smoking. The first study to clearly demonstrate this association was published by Doll and Hill in 1964 and was based on the incidence of lung cancer among doctors in the United Kingdom. Several subsequent studies have confirmed this link and have also shown a relationship between the number of cigarettes smoked and the risk of lung cancer. Importantly, there is also clear evidence that the cessation of smoking is associated with a subsequent fall in lung cancer risk such that approximately 12 years after discontinuation, the risk of lung cancer almost reaches that of a lifelong nonsmoker. There has also been an observation that lung cancer risk rises among nonsmokers who live or work in an environment with smokers leading to an approximately two-fold rise in the risk of lung cancer (passive smoking). Young females who smoke appear to be at a particularly high risk of subsequent lung cancer, an extremely worrying observation with the current vogue for cigarette smoking among young peripubertal females in most Western nations. There is increasing information becoming available on individuals who may be at increased risk of susceptibility to lung cancer following inhalation of tobacco smoke. In particular, genetic polymorphisms for phase I/II metabolizing enzymes have been described, which may alter the production and clearance of carcinogens (see Chapter 3).

Other factors that have been attributed to increased risk of lung cancer include exposure to asbestos, radon, industrial air pollution, chromium, nickel, and inorganic arsenic compounds. Of these, asbestos exposure appears to have the greatest impact for both lung cancer and mesothelioma. There is an additive effect with cigarette smoking such that for an asbestos worker who also smokes, the risk of lung cancer is approximately 45 times that of the normal population.

Numerous studies have investigated the possible link between genetic abnormalities and the risk of developing lung cancer. In non-small-cell lung cancer (NSCLC) the presence of *K-ras* abnormalities, upregulation of the expression of the epidermal growth factor receptor and abnormalities of tumor-suppressor genes, particularly *p53*, have been demonstrated. To date, although numerous genetic changes have been reported, no clear links between specific genetic abnormalities and the risk of lung cancer have been proven.

Pathology

Lung cancer is divided into two major groups – small-cell and non-small-cell. These differ markedly in their biology, therapeutic approaches, and outcomes. NSCLC comprises several different subgroups. Adenocarcinoma is the most common and accounts for approximately 40% of all the cases of lung cancer. Its incidence has steadily risen over the past decade for reasons that are still unclear. These tumors are frequently peripheral in site of origin and will often invade the pleura. There is some evidence that it is less related to cigarette smoking than other tumor types. Squamous cell carcinoma was the most frequent subtype until the early 1990s, since when its incidence has been reducing. Squamous cell carcinomas are more likely to be centrally placed and may be more easily detected by cytology. As these tumors grow they may obstruct major airways, with resulting distal pneumonia. They appear to begin as *in situ* carcinomas with subsequent development over 3 to 5 years before they become clinically apparent. The final major group of NSCLCs is large-cell carcinoma. This comprises approximately 15% of all non-small-cell tumors and frequently appears undifferentiated on light microscopy. It frequently arises in more distal bronchi.

Small-cell lung cancer is identified by the presence of diffuse small cells, which contain fine granular nuclei. Granules are frequently seen and contain a variety of hormones including ACTH and ADH. Markers of neural differentiation are often expressed by these tumors and overexpression of one or more of the oncogenes from the *myc* family is frequently observed (Chapter 6). Twenty to twenty-five percent of all lung cancers show a mixed picture, frequently with a small and non-small-cell component. The exact cell of origin of all lung malignancies remains a subject of debate.

Clinical features

Although lung cancer is the most common cause of paraneoplastic syndromes, most patients present with symptoms related to the primary tumor. Initially there may be a cough, which is often persistent and may be associated with a wheeze. This is frequently confused with similar symptoms induced by smoking and may simply be perceived as an exacerbation of existing chronic obstructive airways disease. Consequently, seeking medical advice is often delayed. Hemoptysis may follow with frequent blood-streaked sputum but rarely massive hemorrhage. Dyspnea can result from obstruction of an airway or the development of a pleural effusion. This may be exacerbated by segmental collapse of one of the lungs. Pain is a frequent associated feature and may indicate mediastinal involvement or invasion of ribs or pleura. Hoarseness of the voice is seen when tumor invasion of the recurrent laryngeal nerve occurs and dysphagia may occur with mediastinal lymphadenopathy (enlarged lymph nodes in the chest), which compresses the esophagus. Apical tumors (arising in the uppermost area of lung) may compress the superior vena cava causing the syndrome of superior vena cava obstruction (SVCO). This is associated with a reduced return of venous blood from the head and neck resulting in swelling of the face, arms, and neck (Fig. 2.1).

Numerous symptoms follow the development of metastases, and may relate directly to the site of the secondary tumors, and include bone pain, liver capsule distension from hepatomegaly, headaches, epileptic fits from cerebral metastases (Fig. 2.2), and discomfort at the sites of lymphadenopathy. Some patients may actually

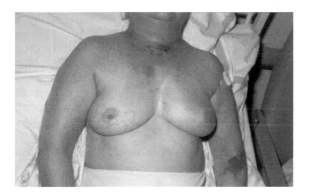

Fig. 2.1 Typical picture of SVCO in patient with lung cancer, showing dilated veins in neck and on chest wall.

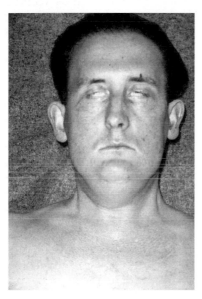

Fig. 2.2 Patient with cerebral metastases from lung cancer, showing features of cranial nerve damage with bilateral ptosis.

present with a mass in lymph nodes (commonly in the neck) from a metastasis.

Several paraneoplastic syndromes are associated with lung cancer, predominantly SCLC. Some of these may result from the abnormal production and release of hormones by the SCLC:

- The syndrome of inappropriate ADH secretion (SIADH) is associated with profound hyponatremia (due to water retention, secondary to the action of ADH on the kidney) causing lethargy and somnolence.

- The production of a protein related to parathyroid hormone, PTHrp, may cause hypercalcemia leading to nausea, polyuria and, if untreated, coma.
- The elevated and unregulated secretion of ACTH ("ectopic ACTH" – because it is not coming from its usual source in the pituitary gland) may result in dramatic increases in secretion of the steroid hormone cortisol from the adrenal gland, causing Cushing's syndrome.

In fact, ectopic production of protein hormones by cancer cells is often far more problematic than the excess production of the same or related hormones from their usual cell type. Primarily, this is because mutations or altered differentiation of the cancer cells that have conferred the ability to produce these hormones rarely if ever is matched by production of proteins needed for feedback control – there are no brakes on this system. Other paraneoplastic effects can include cerebellar syndromes, muscle weakness (myasthenia), and a variety of other neurological abnormalities have been described.

Small-cell lung cancer has a more aggressive rate of development, with one of the highest tumor doubling times of all solid tumors. As a result it is rarely localized at the time of diagnosis and, almost invariably, metastases can be detected. Bone marrow involvement is frequent and up to 90% of cases can be found to have bone marrow metastases if highly sensitive tests such as PCR are used to detect cancer cell genes. If patients with SCLC are untreated, symptoms progress rapidly over 2 to 3 months whereas the rate of progression with NSCLC is much slower with a proportion of patients having truly localized disease. This difference in the biological behavior of the disease has a significant impact on management.

Diagnostic and staging investigations

The possibility of a lung cancer may first be raised when an opacity ("shadow") is seen on a chest X-ray, which may be a mass often associated with an area of pneumonia. Confirmation of a diagnosis relies on obtaining neoplastic cells and in almost all cases this is undertaken by fiberoptic bronchoscopy. This is a rapid outpatient procedure allowing direct visualization and biopsy of a majority of tumors (Fig. 2.3). Brushings may also be taken

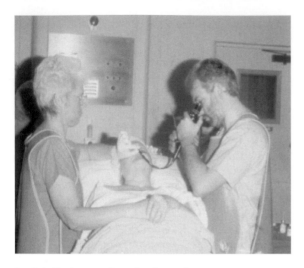

Fig. 2.3 Patient undergoing bronchoscopy.

from the bronchi and can reveal the presence of cancer cells (cytological evidence) in up to 80% of cases. In some cases metastases, usually lymph nodes, may be biopsied to confirm a suspicious X-ray.

Following diagnosis, staging is undertaken to determine extent of the disease. For SCLC, surgical resection is unlikely to be performed and staging is therefore relatively simple, the aim being to obtain a rough estimate of the total tumor burden and divide patients into those who will receive more or less aggressive therapy (Chapter 16). Staging for SCLC is divided into "limited" disease, in which tumor is confined to one side of the chest (hemi-thorax), or "extensive" disease where cancer has spread beyond this. Staging involves blood tests for routine biochemistry, liver function and full blood count and, often, isotope bone scans and imaging of the liver either by ultrasound or CT scanning. If blood counts are abnormal, it is a practice in many centers to undertake a bone marrow examination. If there is any clinical evidence of cerebral metastases, imaging of the brain is undertaken.

For NSCLC, staging is more important as surgical resection may be possible such that it is essential to exclude metastatic disease for those who might undergo surgery. A widely accepted tumor, node, metastasis (TNM) staging system is utilized and divides patients into stage I (a), (b), stage II (a), (b), stage III (a), (b), or stage IV disease.

Survival is closely linked to the stage. Staging investigations include rigid bronchoscopy in many centers for those patients considered potential candidates for resection, CT scanning of the thorax and abdomen, mediastinoscopy to directly visualize the mediastinum and, increasingly, PET imaging. These investigations complement each other to minimize the risk of undertaking a futile major surgical procedure (thoracotomy) on a patient with metastatic disease.

Treatment

Surgery Surgical resection is the only form of therapy that provides the possibility of cure in NSCLC. This reflects the potential for this disease to be detected at a time prior to dissemination. Unfortunately, for SCLC, metastases are almost invariable (albeit microscopic) in almost all patients and so surgery is rarely offered. An exception, in some centers, is for patients with small peripheral lesions that have been detected coincidentally.

For all patients with NSCLC, the potential for resectability is considered in the first instance. Such surgery is only undertaken in patients with localized disease (stage I/II). Approximately 30% of patients will be offered resection (many will be excluded because the cancer is deemed inoperable or the patient's general condition is poor). Surgery depends on the site of disease, and may include local segmental or wedge resections, lobectomy, or even pneumonectomy (removal of an entire lung). Success of surgery depends on the stage of cancer, with 5-year survival rates of approximately 54% in stage I, but only 24% for stage IIb. For those centers who operate on stage IIIa disease, neoadjuvant chemotherapy is often utilized, but 5-year survival is only 15–18%.

Radiotherapy Radiotherapy is used to good effect in NSCLC to manage immediate symptoms or to delay the time until symptomatic progression rather than in an attempt to cure (palliation) – see also Chapters 16 and 17. SVCO also often responds to radiation treatment. The timing of palliative radiotherapy remains controversial with no clear evidence that early initiation prior to the development of symptoms provides better quality of life or improved survival compared with waiting for symptoms to develop and treating at that point. Nevertheless, in most centers, early radiotherapy is initiated. Radical radiotherapy may occasionally be offered to patients with the intent of producing long-term survival. In particular, this option is favored for localized disease unsuitable for surgery, but the 5-year survival is only 10%. Continuous hyperfractionated accelerated radiotherapy (CHART) has been demonstrated to improve survival (by ~10%) compared with standard techniques. Improved survival has also been demonstrated with the addition of cisplatin to radiotherapy. Both approaches are now widely adopted.

Chemotherapy is frequently used in SCLC as this is one of the most radiosensitive of all tumors. Rapid reduction in tumor volume occurs in 80–90% of patients but recurrence is almost inevitable. The addition of radiotherapy to chemotherapy regimens appears to improve median and 2-year survival rates by approximately 5%. For patients with limited-stage disease, thoracic radiotherapy is widely used early during the administration of chemotherapy and prophylactic cranial radiation is also routinely offered with chemotherapy, reducing the frequency of brain metastases from approximately 40% to 8%. Local radiotherapy for the palliative treatment of painful metastases is of frequent value during the course of disease.

Chemotherapy Chemotherapy is widely utilized in the treatment of all forms of lung cancer. A wide variety of combinations of agents are used in NSCLC as single-agent response rates are only around 15–20%. The most commonly used agents are cisplatin, gemcitabine, paclitaxel, docetaxel, and ifosfamide (Chapter 16). Many combinations have been compared and it appears that, so long as cisplatin is included, activity for two or more agents is similar with response rates of 40–50%. For inoperable disease, the early introduction of chemotherapy appears to have significant, though modest, benefit in terms of median and 1-year survival and is associated with an improvement in the quality of life. Surprisingly, the early use of chemotherapy also appears to be cost-effective when compared

with palliative care alone. The use of preoperative chemotherapy for those deemed surgically resectable is gaining increasing support. Large randomized trials comparing surgery alone with preoperative chemotherapy are ongoing as the results of published studies to date remain conflicting. One meta-analysis has, however, suggested an improvement of 1-year survival of approximately 5%.

Chemotherapy has a more established role in the management of SCLC. This disease is highly chemosensitive with single-agent response rates of approximately 60% for carboplatin and etoposide and a slightly lower response rate of 30–40% for a wide variety of other agents. Limited-stage disease is usually treated aggressively with combinations of agents (including cisplatin or carboplatin for best results) and a median survival of 14–16 months is now achieved (compared with just 4 months without treatment). In fact, up to 10% may survive beyond 5 years. For those with extensive-stage disease, chemotherapy is offered for palliation and, as a result, regimens are chosen with less toxicity. Median survival rates of 8–10 months are usually achieved, with approximately 2% of patients alive after 2 years.

There has been enormous interest in the use of inhibitors of the epidermal growth factor (EGF) receptor in lung cancer in recent years (see Chapter 5). This receptor is expressed in the majority of patients with NSCLC and an early study using a single agent, Iressa, an inhibitor of the EGF receptor, in patients who have relapsed after chemotherapy, showed response rates of approximately 20%. This was associated with an improvement in quality of life and was so encouraging that it led to two large randomized trials comparing chemotherapy with or without Iressa. Disappointingly, survival rates were almost identical in the two arms of both trials. A recent review of patients receiving Iressa has suggested that only those with the mutated EGF receptor have a likelihood of response and clearly future trials will need to focus on this subgroup of patients. Several other inhibitors of the EGF pathway are under investigation in clinical trials in NSCLC and there is also early encouragement for the use of vascular endothelial growth factor receptor inhibitors in this disease (Chapter 14).

BREAST CANCER

Breast cancer is the most common malignancy in women in the Western nations. The highest frequency is in the United States with an annual incidence of 85 per 100,000. In Europe it is commonest in the Netherlands with an incidence of 70 per 100,000 per year and lowest in Spain with an incidence of 43 per 100,000 per year. It is much less common in Asia, and Japan has an incidence of just 25 per 100,000. The incidence increased steadily during the twentieth century but there is evidence that it may have plateaued over the past 5 years. On average, approximately one in nine women develop breast cancer in Western nations and one in three of these will die of the disease.

Many potential risk factors for the development of breast cancer are known. A family history of carcinoma of the breast increases the risk, with one first-degree relative conferring a three-fold increase and an even greater risk if the relative was premenopausal when diagnosed. Lifetime exposure to estrogen is an important risk factor, with breast cancer risk increased by early menarche or late menopause and by the use of hormone replacement therapy after menopause. Women who have their first child over the age of 30 years experience a three-fold increased risk. Previous benign breast disease, a prior exposure to radiation, and high dietary fat intake all appear to increase the risk of breast cancer.

The key role of genetic factors in breast cancer was illustrated by the identification of mutations in the *BRCA1* in patients with familial breast and ovarian cancer (Chapter 10). Mutations of this gene are associated with a 50–85% risk of developing breast cancer during the lifetime of a woman.

Pathology

The majority of breast cancers are adenocarcinomas, which may be either infiltrating lobular or ductal. With the increasing uptake of mammographic screening, preinvasive tumors are being detected – ductal or lobular carcinoma *in situ*. Rarely, primary lymphomas, sarcomas, and squamous carcinomas may be found in the breast.

Clinical features

Breast cancers vary considerably between individuals in their rate of growth and pattern of spread. Some may remain predominantly localized, infiltrating local structures, some may spread via lymphatics to draining lymph node areas while others may disseminate widely in the blood stream. The majority of women will present with a lump in the breast and some will have pain in this region, discharge, or bleeding from the nipples. Increasing numbers of women are being diagnosed following screening and will have no symptoms (Fig. 2.4). In some cases, local infiltration can produce a fungating tumor with erosion of the skin (Fig. 2.5). If

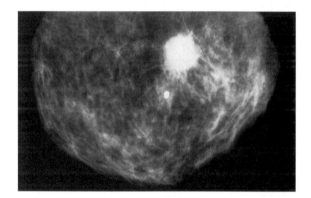

Fig. 2.4 Mammographic image of primary breast tumor.

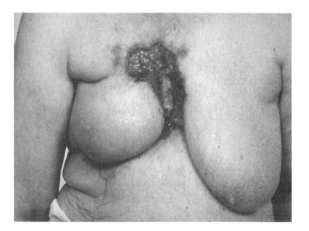

Fig. 2.5 Primary breast tumor showing extensive local skin involvement.

fungation does not occur, there may be widespread skin involvement leading to thickening and a typical pattern of irregularity termed *"peau d'orange"* (Fig. 2.6). In this instance, there is local lymphatic infiltration and obstruction leading to edema. Presentations may occur, less commonly, from sites of metastases and include the detection of an enlarged lymph node or symptoms from distant metastases to bone, brain, or liver. These may cause pain and neurological symptoms and signs. As with all malignancies, nonspecific symptoms including anorexia and lethargy may predominate.

Diagnostic procedures and investigations

Physical signs associated with breast cancer include the presence of a firm irregular mass and fixation to skin or deep structures such as the chest wall. Mammography is widely used in the initial assessment of lumps thought to be possible breast cancer; the typical X-ray appearances are fine calcification with areas of irregularity. Ultimately diagnosis depends on a histological or cytological assessment of a specimen from the mass. Fine-needle aspiration usually provides sufficient cells for cytological assessment but a core biopsy is often obtained using a percutaneous biopsy needle, which will yield sufficient tissue for histological assessment. Occasionally the diagnosis remains uncertain and an excision biopsy may be required. It is clinically valuable to obtain

cells not only for confirmation of the diagnosis of malignancy but also to measure the presence of hormone receptors – both progesterone and estrogen. Positive receptor status correlates with the likelihood of benefit using hormone manipulation therapy and assessment is therefore associated with considerable clinical utility.

Treatment decisions are based on the histological grade and size of the primary tumor together with the presence or absence of metastases. The vast majority of patients will undergo resection of the primary tumor and preoperative assessment of the extent of disease varies between centers. Basic investigations including chest X-ray, liver function tests, and full blood count would be considered standard and provide an indication of possible metastases. Most centers would also perform abdominal imaging to exclude liver metastases. This can be with CT or ultrasound scanning. The presence of bone pain with elevated serum calcium or alkaline phosphatase indicates the possibility of bone metastases that can be detected with isotope bone scans. Imaging of the brain should only be performed routinely in patients who complain of symptoms suggesting possible cerebral metastases. The TNM staging system is widely used alongside the histological grade of tumor and receptor status when deciding on appropriate therapy (Table 2.1).

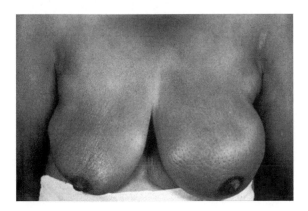

Fig. 2.6 Breast cancer showing *"peau d'orange"* in left breast.

Table 2.1 TNM staging system for breast cancer

Stage	Tumor description
T1	Tumor less than 2 cm in diameter
T2	Tumor 2–5 cm in diameter
T3	Tumor more than 5 cm
T4	Tumor of any size with direct extension to chest wall or skin
N0	No palpable node involvement
N1	Mobile ipsilateral node(s)
N2	Fixed ipsilateral nodes
N3	Supraclavicular or infraclavicular nodes or edema of arm
M0	No distant metastases
M1	Distant metastases

Treatment

Surgery Surgical resection of primary breast tumors aims to control local disease and prevent recurrence in regional draining lymph nodes. The understanding of the biology of breast cancer is a major example of how clinical and fundamental research have modified a therapeutic approach. Until relatively recently, the surgical approach to breast cancer was to undertake a radical operation with removal of the breast, block dissection of the axillary nodes, and often removal of underlying muscle. Radical mastectomy is a mutilating and disfiguring operation that often led to marked long-term morbidity. However, once it was recognized that blood-borne metastases and not lymphatic and local spread were the main causes of dissemination, the role of surgery was adapted. Particularly, as most patients die from metastases that may have been "seeded" prior to the opportunity for surgery, one might expect only a limited benefit from aggressively removing the primary tumor, surrounding tissues, and lymph nodes. Surgery is now less radical with the widespread adoption of breast-conserving surgical approaches, such as lumpectomy (removal of the local tumor alone) when possible. The move to more conservative surgery has been slower in the United States than in Europe and it is estimated that 50% of women with early-stage breast carcinoma are still being treated with modified radical mastectomies in the United States. The choice of surgery is affected by the size of the primary tumor and of the breast (large tumors in small breasts are often removed with better cosmetic results using a simple mastectomy than with lumpectomy) and the presence of multiple tumors within the same breast or widespread intraduct carcinoma (both providing indications for mastectomy rather than lumpectomy).

Radiotherapy Conservative surgery alone is followed by local recurrence in approximately 30% of cases, but can be reduced to just 5–10% by employing postoperative radiotherapy. Recurrence is more common with tumors greater than 5 cm in diameter or in patients with axillary node disease. The results from the United States suggest local recurrence rates following radical surgery alone are 4–14%, indicating the equivalence with breast-conserving surgery followed by radiotherapy. Likewise survival appears to be identical in the two groups and these results strongly support the use of breast-conserving surgery with postoperative radiotherapy whenever technically possible. Interestingly (and perhaps not surprising when considering the biology of the disease) the vast majority of randomized studies have shown no effect on survival of post-operative radiotherapy despite a reduction in local recurrence rates. One recent meta-analysis (pooled statistical analysis of multiple single studies) that included 8000 patients did suggest a slight, but significant, improvement in survival for the radiotherapy arm but these results must be interpreted with some caution, particularly as no single prospective study has demonstrated this. Meta-analyses are particularly prone to bias, because negative studies may not have been reported in the literature (and thus self-evidently are not available for the meta-analysis) and because there may be considerable variation between groups of patients in the different studies included.

For patients with large primary tumors (particularly those >5 cm or with direct extension to the chest wall or skin), and for a large proportion of patients with clear nodal involvement, surgery appears to have little role other than occasionally to debulk the tumor for symptomatic benefits. For such patients, radiotherapy has an important role, producing high local control rates (80–90%). It is usually given in combination with hormones and/or chemotherapy.

As with all malignancies, radiotherapy can be particularly useful for patients who have developed painful areas of metastasis. This is particularly the case with bone lesions, where symptomatic benefit is achieved in the majority of patients (Fig. 2.7). About 15–20% of patients will develop cerebral metastases and radiotherapy can reduce tumor volume and improve symptoms in the majority of these patients. Likewise skin, lymph node, and other areas of metastasis can benefit from local radiotherapy. One final role for radiotherapy is the induction of menopause with pelvic irradiation. This appears to be as valuable as oophorectomy for the relief of symptoms from metastatic disease if patients are premenopausal.

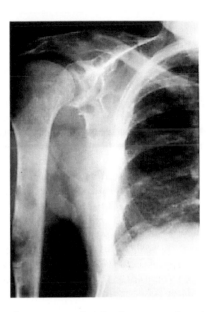

Fig. 2.7 Plain X-ray of right humerus showing lytic bone metastases.

Chemotherapy Breast cancer is a relatively chemosensitive disease with numerous single agents producing responses in 30–50% of patients. The anthracyllines, taxanes, alkylating agents, and vinca alkaloids are most widely utilized, generally in combinations. A large proportion of patients will receive chemotherapy following a diagnosis of breast cancer, either in the adjuvant (immediate postoperative) or metastatic setting. There are increasing numbers of trials currently under way which also utilize chemotherapy in the preoperative, neoadjuvant, setting with the hope of reducing tumor volume, treating any micrometastatic disease early, and potentially reducing the need for radical surgery. While the role of neoadjuvant therapy remains to be proven, there is clear evidence of the value of chemotherapy in the adjuvant and metastatic settings.

The role of adjuvant therapy is based on the observation that even at relatively early stages of development, a proportion of patients with breast cancer already have micrometastases. In fact, when taken together with the demonstration of large number of circulating tumor cells (CTCs) in patients' blood, and the demonstration of metastasis-related gene expression in early primary tumors, which are present before overt metastatic disease, these observations have a profound impact on our understanding of metastases (Chapter 12).

Even with stage I disease, metastatic breast cancer cells can be detected within the bone marrow by a variety of techniques including gene expression analyses, and this figure rises with advancing stage. Although it is likely that many of these cells would not have long-term viability, clearly in a significant proportion of patients they are responsible for local systemic relapse and death. The use of systemic therapy at an early stage is now widely adopted with the aim of eradicating these cells whenever possible and prolonging relapse-free and overall survival. Since the introduction of adjuvant therapy, there has been a steady decrease in the death rate from breast cancer, suggesting a significant impact on public health. Numerous trials have been performed in different countries and by many cooperative groups. These have all shown a reduction in the risk both of recurrence and of death. The effect of chemotherapy is more marked in younger patients, particularly those below the age of 50. In fact, for premenopausal women, the annual odds of recurrence are decreased by approximately 35% and of death by 25–30%. The benefits of adjuvant chemotherapy appear to be largely confined to patients with node-positive disease and high-grade tumors. Adjuvant chemotherapy is now routinely recommended for premenopausal women with stage II disease.

Combination chemotherapy is offered to most women with metastatic disease, particularly premenopausal node-negative patients and patients who have failed hormone therapy. Although response rates tend to be high (50–60%) with improved symptoms and quality of life, survival is not prolonged. Thus, potential benefits must be carefully weighed against toxicity. The recent use of herceptin for tumors expressing the Her-2 receptor (Chapter 5) does appear to prolong survival in this subgroup and it is hoped more targeted drugs will soon be available.

Hormone therapy It has been recognized for over 100 years that hormone manipulation can reduce tumor volume in patients with metastatic or locally advanced breast cancers.

More recently, it has been found that estrogen-receptor-positive cancers are most likely to respond to hormone manipulation. Knowledge of progesterone receptor status adds to the ability to predict response to endocrine therapy – 10% of those who have negative status of both recept-ors will respond whereas 70% of those who have estrogen and progesterone receptor positiv-ity will respond. About 30–50% of patients who have either estrogen or progesterone positivity obtain a response to hormone manipulation. For many years, the antiestrogen, tamoxifen, has been the most widely used means of altering hor-mone activity in the tumor cells. It may also have a direct cytotoxic effect. More recently, aromatase inhibitors (e.g. anastrozole) have been widely used, particularly as they are not associ-ated with the development of uterine malignan-cies, which are a rare complication of tamoxifen therapy.

Several studies, and a key overview of all pro-spective randomized trials published in 1992, demonstrated the clear benefit of tamoxifen given in the adjuvant setting. The annual odds of death appear to be reduced by approximately 30% with 40–50% reductions in the annual odds of recurrence. Both node-negative and node-positive patients appear to benefit as do pre-menopausal and postmenopausal women. The optimum duration of treatment remains unclear but studies have shown that at least 5 years of therapy appears superior to shorter durations and current trials are exploring the role of longer durations.

For patients with metastatic disease, hor-mone therapy is appropriate if receptor status is positive. For patients who are premenopausal, radiation-induced menopause or the use of LHRH antagonists (e.g. goserelin) are often util-ized, whereas tamoxifen may be the preferred first approach in postmenopausal women. 30–50% of patients will respond to the first-line hormone manipulation and there is increasing evidence that the early use of aromatase inhib-itors may be superior to tamoxifen, providing longer disease-free survival and reduced toxicity. Hormone therapy is preferred prior to chemo-therapy for patients with receptor-positive dis-ease even though the response rate is lower, because toxicity is also less and response more durable. Chemotherapy is introduced when hor-mone therapy fails or first-line if the cancer may cause early complications and where a more rapid response is required.

COLORECTAL CANCER

Malignancies arising in the large bowel cause the second largest number of deaths from can-cer in the Western world. The incidence rate varies markedly between nations, being lower in Africa and Asia than in Western Europe and North America. The annual incidence var-ies between 30–60 per 100,000 across Western nations with overall 5-year survival rates being approximately 50%. The age-adjusted incidence rates have remained relatively constant over the past 30 years although the number of cases has risen because of an increasing population and age demographics. Sixty percent of these tumors arise in the colon and 40% in the rectum. Tumors arise more frequently in distal sites of the large bowel, providing evidence for the sug-gested association of dietary intake factors and risk of colorectal cancer. As bowel contents move distally, they become more solid and their transit time is lengthened. They thus have a greater likelihood of prolonged intimate contact with the bowel epithelium. It is therefore sug-gested that if potential carcinogens are present, they are more likely to have an effect in the distal bowel, explaining the greater propensity of this region to develop malignancies.

More is known about the development of colorectal cancer than any other malignancy – in fact colon cancer has proved a paradigm for the development of "multistage" theories of carcinogenesis. Several clear stages of pro-gression from normal epithelium through early adenomas, late adenomas, early carcinoma, and ultimately to metastatic disease have been described and many of the key genetic lesions responsible have been identified – also see Chapter 3. There is a clear association between family history of colon cancer and the likelihood of developing this disease, par-ticularly for an individual with more than one first-degree relative diagnosed prior to the age of 40 years. Several familial syndromes are known to be associated with an increased risk of developing colorectal cancer and these include Peutz–Jeghers, Lynch, and Gardner's syndromes. Familial adenomatous polyposis coli (FAP) has been widely studied and is a clear example of an autosomal dominant inherited malignancy. Multiple polyps develop throughout the bowel, predominantly in the distal colon. There is an inevitable conversion to malignancy and a panprocto-colectomy (removal of the entire colon and rectum) is recommended prophylactically when an individual is found to have inherited FAP. Sporadic polyps may also occur in individuals and will convert to malignant lesions in approximately 15% of cases. Larger lesions are associated with a greater risk and prophylactic removal at the time of colonoscopy is generally undertaken. There is increasing information becoming available on inherited risks of nonpolyposis malignancies of the colon (HNPCC)–Chapters 3, 10, and 11.

Several chromosomal changes have been described as being associated with the risk of colorectal cancer. The familial polyposis gene is located on chromosome 21 and the loss of an allele at this site can be found in up to 40% of cases of sporadic carcinoma of the colon. Loss of an allele of the *p53* gene on chromosome 7 and of the *DCC* gene (deleted in colorectal carcinoma) on chromosome 18 have also been implicated in colorectal cancer risk. Mismatch repair genes form the genetic basis for hereditary nonpolyposis colon cancer and families with this predisposition appear to have an excess risk of many other adenocarcinomas apart from colorectal cancer. To date, at least four separate mismatch repair genes that contain mutations have been implicated in the etiology of hereditary nonpolyposis colon cancer. These include *hMSH2*, *hMLH1*, *hPMS1*, and *hPMS2* (Chapter 10).

Several other factors have been linked with the risk of developing colorectal cancer. Dietary intake of fat and fiber appear to play a role and ulcerative colitis, particularly when the onset is at a young age, is associated with a significant risk of the development of malignancies.

Diagnostic and surgical procedures

The diagnosis of colorectal cancer is usually relatively straightforward once the suspicion is raised and the patient has been referred to a specialist center. Simple rectal examination will detect approximately 75% of all tumors within the rectum (i.e. almost 30% of all large-bowel tumors). With the advent of high-quality flexible endoscopes, flexible sigmoidoscopy or colonoscopy are generally the first investigation of choice. Sigmoidoscopy is rapid and will examine the distal 25–30 cm of large bowel, the site of approximately 60% of all tumors. Colonoscopy allows visualization of the entire large bowel through to the cecum but is a more lengthy and costly investigation. However, the advantages are considerable ease of use and speed together with the high sensitivity and ability to biopsy any abnormal lesions. Colonoscopy also allows the removal of polyps and the detection of other sites of disease. There is a small risk (~0.01%) of serious complications – usually hemorrhage or perforation. Barium enemas are also widely used and allow visualization of the entire large bowel. A mass can be identified or a stricture that has typical features in a malignancy (Fig. 2.8). The disadvantage of barium investigations is the discomfort they produce and the fact that they will often miss lesions on the right side of the colon. In addition, histological confirmation is not obtained.

As with all tumors, histological confirmation of malignancy is essential as abnormalities such as diverticular disease or ulcerative colitis can mimic malignancy both in their clinical presentation and, occasionally, on their appearance at investigation. The vast majority of tumors are adenocarcinomas but, rarely, other tumors can present in the large bowel and include carcinoids, sarcomas, and lymphomas.

Staging

Given that the gold standard is to surgically resect the primary tumor in order to cure the patient, it is important to first determine the extent of local disease and the presence of any possible metastases. A relatively simple staging system is widely utilized (Dukes' system–Table 2.2). While the Dukes' system has considerable clinical value in determining prognosis and indicating optimal therapy, it is widely recognized that it provides only a crude assessment of local involvement, particularly with Dukes' B and D disease. A more refined

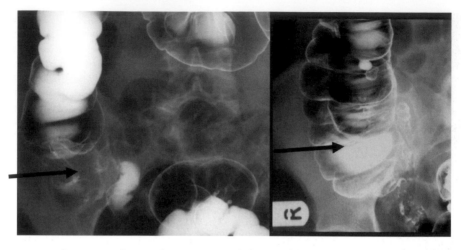

Fig. 2.8 Barium enema showing malignant lesion at site of abnormal barium filling (arrowed).

Table 2.2   Dukes' staging system for colorectal cancer

Stage	Description of tumor extent
A	Tumor confined to bowel wall (mucosa and submucosa
B1	Tumor penetrating the bowel wall involving muscularis propria (but not penetrating through it)
B2	Tumor penetrating through the bowel wall
C	Tumor extending into local lymph nodes
D	Distant metastases

staging system has been developed by the American Joint Committee on Cancer (AJCC staging system) and this provides information on the degree of local involvement in the bowel wall and the number of regional lymph nodes involved.

Colorectal cancer usually metastasizes to the liver but may also disseminate to lung and para-aortic lymph nodes. Distant metastasis is more common with colon cancer than rectal malignancies, whereas the latter will often invade locally into adjacent structures. For rectal tumors, high-quality imaging of the local extent of disease is beneficial when making a decision

about surgical techniques and MRI scans of the pelvis are usually performed (Fig. 2.9). CT imaging of para-aortic nodes and ultrasound or CT of the liver is usually recommended. Imaging of the thorax is often confined to plain radiology, although some centers will also perform CT scans. Routine biochemistry and blood count together with measures of the tumor biomarker carcinoembryonic antigen (CEA), can provide useful information, particularly by giving an indication of possible metastases at the time of presentation or in subsequent follow-up to help screen for recurrent disease. It was widely hoped that CEA would prove to be a specific marker of colorectal malignancy, but it is elevated in many other conditions and its specificity is therefore low. It is, however, useful to obtain a preoperative measure of CEA as, if it is elevated, a subsequent fall can be useful in monitoring response to treatment.

Clinical manifestations

Colorectal cancer produces a variety of symptoms, the predominant nature of which depends on the site of the tumor. Right-sided tumors arise where the bowel contents are fluid and therefore tend not to produce obstruction early in the course of their growth. They are often associated with chronic hemorrhage (bleeding), which leads to symptoms of anemia but may also cause pain or a detectable abdominal mass.

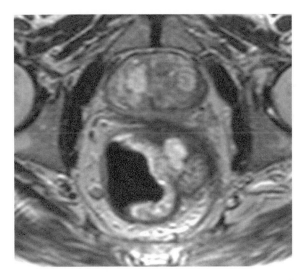

Fig. 2.9 MRI image of pelvis showing large rectal tumor.

A change in bowel habit occurs in approximately 50% of patients and nonspecific symptoms such as anorexia and weight loss are common. In the left side of the colon, the bowel contents are more solid with a slower transit time. As a result, the tumors more readily cause obstruction with subsequent colicky abdominal pain and vomiting. Altered bowel habit occurs in 60–70% of patients and blood may be noticed with the stools. Abdominal pain is common. Rectal tumors are more likely to be associated with obvious blood loss (in over 60% of patients) and altered bowel habit is almost universal. All tumors may be associated with nonspecific symptoms such as lethargy, anorexia, and weight loss and all may cause symptoms from metastases, if present.

Many groups have attempted to develop scoring systems for risk factors of malignancy. Altered bowel habit, rectal bleeding, and anemia (particularly in women) are not uncommon and none are specific for a bowel malignancy. It is clearly important to balance the costs and risks of investigation against the risk of missing the diagnosis of a malignancy, when a patient complains of symptoms that could indicate colorectal cancer. Many programs have investigated the role of screening for colorectal cancer and there has been a demonstration of a reduction in colorectal cancer death rates of up to 40%

where widespread screening programs have been performed. These have usually included measurement of fecal occult blood (a chemical test is performed on a stool sample to reveal the presence of blood), followed by endoscopy in patients who have positive results. Unfortunately, this is a highly labor-intensive and costly program and more information on its utility will be necessary before it is widely adopted in the Western nations.

Management of colorectal cancer

High-quality surgery by experienced surgeons is essential to optimize the chance of cure in colorectal cancer. Excision of the primary tumor together with the involved segment and draining lymph nodes should be undertaken and the exact extent of bowel removed depends on the site of the primary lesion. For some patients who have obstruction but in whom there is disease which is either too locally extensive for resection or who have widespread metastases, a defunctioning colostomy may be performed to control symptoms. Rectal tumors are technically more difficult to remove and require great expertize to minimize the risk of nerve damage and attempt sphincter preservation. Local recurrence is much more common with rectal cancers than for tumors in the colon and it is clearly essential to remove as wide an area of normal tissue as is possible around the primary rectal cancer to reduce this risk.

Surgery is increasingly employed for patients with liver metastases. The mortality from this procedure has fallen dramatically and encouraging long-term survival rates (up to 30%) have been demonstrated in specialist centers.

Radiotherapy

The role of radiotherapy in the management of large-bowel tumors is relatively limited. Local recurrence is very unusual following surgery for colon tumors, so there is no role for perioperative radiotherapy. However, radiotherapy has been shown to reduce local recurrence in rectal tumors. Survival rates have been higher in several series where radiotherapy has been added to surgery but these have not usually reached the levels of significance. A study from Scandinavia has suggested the superiority of preoperative radiotherapy compared with postoperative treatment and a large ongoing study in the United Kingdom is comparing preoperative radiotherapy with selective postoperative radiation treatment (for patients with "inadequate" resection margins). Although survival benefits for preoperative radiotherapy in rectal cancer are equivocal, it is routinely used in the United States and much of Europe.

Chemotherapy

Unfortunately, colorectal cancer is one of the more chemoresistant malignancies and until the late 1990s only 5-fluorouracil (5-FU) was widely used in the treatment of this disease. Its response rate when modulated by folinic acid is modest at 20–30%. Over the past 5 years, oxaliplatin and irinotecan have shown significant activity and have been widely incorporated into combination regimens. The oral fluoropyrimidine, capecitabine, has demonstrated a greater response rate than 5-FU and is more acceptable to patients. As a result, it is increasingly replacing 5-FU in the treatment of this disease. Chemotherapy is administered postoperatively in the adjuvant setting for patients with Dukes' C disease (i.e. disease involving lymph nodes) and in many centers is often administered to patients with Dukes' B disease. Randomized prospective trials have shown an approximately 30% relative risk reduction in the death rate from colorectal cancer with the use of adjuvant 5-FU-based therapy. Early data from the X-ACT international trial has shown at least equivalence for the substitution of oral capecitabine as adjuvant therapy and the preliminary analysis from the MOSAIC trial has suggested that the combination of oxaliplatin with 5-FU and folinic acid may produce superior relapse-free and overall survival rates compared with 5-FU and folinic acid alone. More mature data will be necessary to confirm these promising early results.

The treatment of metastatic colorectal cancer has improved markedly over the past 5 years. The use of 5-FU and folinic acid alone resulted in a median survival of approximately 8 months but the addition of oxaliplatin and irinotecan combined with 5-FU and folinic acid (with either combination used in the first- or second-line setting) have improved median survival to 18–20 months. The use of chemotherapy at the time of establishment of metastases confers a survival advantage compared with waiting until the development of symptoms before commencing treatment. Oral capecitabine as a single agent is widely used for metastatic disease, particularly for patients who are unable to tolerate more aggressive regimens. This agent is also increasingly combined with oxaliplatin and irinotecan. Several new approaches including use of VEGF inhibitors appear particularly promising and a randomized trial combining bevacizumab (a monoclonal antibody to VEGF-A) improved median survival by 5 months when combined with irinotecan and 5-FU as compared with chemotherapy alone. A monoclonal antibody directed against the EGF receptor (erbitux) has also shown promise in combination with chemotherapy and improves responsiveness to irinotecan and increases disease-free survival when combined with chemotherapy in the relapse setting. Oxaliplatin appears to have a role combined with 5-FU in the preoperative treatment of patients with liver metastases who are potentially able to undergo resection.

CARCINOMA OF THE PROSTATE

There is a wide range of incidence of prostate cancer in different countries from a level of as high as 80 per 100,000 per year in the Afro-Caribbean population in the United States to only 8 per 100,000 per year in Japan. In most Western countries, it is the third most common cause of death from cancer in males. The incidence of prostate cancer increased markedly during the 1970s and the 1980s but, since 1992, the incidence rate has steadily declined. In parallel 5-year cancer survival rates have increased from approximately 70% in the early 1980s to more than 90% in the 1990s. It is likely that a significant contribution to these figures relates to increased diagnosis with the use of screening, with earlier-stage disease being detected.

Little is known about the etiology of prostate cancer, although there is some evidence that testosterone levels and androgen receptor gene activity may be related to the risk of developing this disease. There is, however, likely to be an environmental contribution in that migrant studies have demonstrated that men who move from a low-risk country to an area of high risk increase their incidence and mortality from prostate cancer within a generation. No specific gene abnormalities have consistently been related to prostate cancer risk.

Pathology

The commonest form of malignancy to arise in the prostate is adenocarcinoma although, more rarely, squamous and small-cell carcinomas may arise, as may transitional-cell carcinomas. The histological grade of the tumor has an impact on outcome. Patients with low-grade tumors survive significantly longer than those with high-grade disease. The Gleason system is widely used to describe the degree of differentiation and pattern of glandular histology. Grades I–V are described and correlate well with outcome. Additional information can be obtained from TNM staging. Diagnosis requires histological assessment of the tumor and increasingly this is obtained using transrectal biopsy. Prostate-specific antigen (PSA) is a useful biomarker that is usually elevated and is an excellent indicator of the presence of disease. There is a good correlation between extent of disease and the level of PSA found in the serum.

Clinical features

A large proportion of patients with prostate cancer are asymptomatic until late in the course of their disease. Interestingly, autopsies performed for other reasons (e.g. following road traffic accidents) on individuals over the age of 50 years have revealed carcinoma in the prostate in up to 30% of men. This indicates the prolonged period of development of this disease in many individuals. Once the prostate has enlarged to the point of reducing bladder outflow, typical symptoms of hesitancy of micturition, nocturnal frequency, and occasionally

hematuria may be experienced. Metastatic disease is often the cause of presentation and the most common site is in the bones where metastases are often sclerotic in nature. Bone pain, most frequently in the back and pelvis, may occur.

Diagnostic and staging procedures

Serum PSA levels are an excellent early indicator of the presence of prostate cancer and this should be confirmed by histological assessment of a biopsy. Isotope bone scans can detect the extent of metastatic bony involvement and plain skeletal X-rays of abnormal areas may be performed. Renal function should be assessed and this should include intravenous pyelography to determine whether there is hydronephrosis (enlargement of the kidneys from outlet blockage). The extent of local disease, especially the presence of nodal involvement, can be determined by CT or MRI scanning. Transrectal ultrasound may also provide useful information on the extent of disease.

Treatment

If disease is localized to the prostate gland, a curative approach to therapy is taken. Radical prostatectomy or radical radiotherapy may be offered and different centers will often predominantly utilize one or other of these approaches. In general, surgical approaches are more widely utilized in North America than Europe. Radiotherapy generally involves external beam treatment using high doses. An alternative used increasingly is the use of I^{125} seeds placed interstitially in the prostate. Cryosurgery and brachytherapy are also being utilized in several specialized centers. For larger tumors, some centers may offer local radical therapy after tumor volume reduction with hormone intervention.

There has been enormous debate as to the optimal management of localized disease but several studies have suggested similar long-term survival rates from both approaches. Both approaches have significant complication rates and for surgery this involves impotence and urinary incontinence in approximately 10% of patients. Radiotherapy can result in urethral stricture and reduction in bladder capacity. There is a lower rate of impotence.

Controversy remains as to the optimal management of screen-detected early prostate cancer with some centers recommending radical intervention and others following a "watch and wait" policy (see Chapter 1 – screening).

Hormonal therapy is of value for patients whose disease is inoperable and for those with metastatic disease. A variety of estrogens have been offered for several decades and orchidectomy can produce similar clinical benefits. Gonadotropin-releasing hormone analogs are now widely utilized as an alternative to systemic estrogens or orchidectomy and are generally much better tolerated. These interfere with gonadotropin release, leading to a fall in circulating testosterone. Goserelin is a depot form of gonadotropin-releasing hormone analog and is given on a monthly basis. A more recent alternative is flutamide, which is a pure antiandrogen. Flutamide does not lead to central inhibition of luteinizing hormone release and appears to have less toxicity. Many centers increasingly use combinations of hormonal agents in an attempt to provide blockade of both testicular and adrenal output of androgens. Some studies have suggested superior outcome compared with single-agent therapies.

Chemotherapy has a limited role in the management of prostatic cancer although a combination of mitozantrone and steroids appears to provide clinical benefit in a small proportion of patients. For those individuals with widespread bone pain, the use of strontium-89 has been reported to improve pain, as has the use of bisphosphonates.

SKIN CANCER

Malignancies arising in the skin are common but the vast majority are curable and therefore mortality is very low. A variety of industrial carcinogens implicated in causing skin cancers have been described and include arsenic, chimney soot, sunlight, and ultraviolet and ionizing radiation. Patients who have received kidney transplants have a markedly increased risk of all forms of malignancy and, in particular, of squamous carcinomas of the skin, basal cell carcinoma, and melanomas. There are a variety of inherited disorders associated with skin malignancies including xeroderma pigmentosum, which involves a deficiency in the DNA excision repair mechanism (Chapter 10). Such individuals have an increased sensitivity to UV light leading to the development of malignant skin tumors on exposed areas. Ultraviolet light has been shown to be linked with causing mutations in the $p53$ gene and to cause defects in nucleotide excision repair genes. Basal cell carcinomas have been linked to specific gene defects including inactivation of the patched gene. Approximately 90% of patients with squamous cell carcinoma of the skin have $p53$ mutations, and $BRaf$ abnormalities have been detected in 80–90% of patients with malignant melanoma. $p16$ has been identified as a familial melanoma susceptibility locus (Chapters 6 and 7 discuss the concepts of oncogenes and tumor suppressors in more detail). Squamous carcinomas of the skin are more common in individuals with albinism, and familial traits for malignant melanoma have been described with gene defects being defined (Chapter 3).

The major malignancies involving the skin have all increased in frequency significantly over the past two decades, presumably related to the increase in intensity of sunlight exposure in many Western nations as a result of reduction of the ozone layer and also because of increased travel to warm climates for holidays with rapid changes in sunlight exposure.

Malignant melanoma

Malignant melanoma varies in frequency and is generally more common in nations with high levels of sunlight exposure among the Caucasian population. Its incidence is 10 times greater in Australasia than Europe but the frequency is increasing at up to 7% per year in some parts of Europe, with Scotland in the United Kingdom having the highest rise in incidence in the world. The annual incidence is 5–7 per 100,000 of the population in Europe and the United States. It is more common in Caucasians and, in particular, in those with fair skin, red hair, and pigmented naevi. The majority of malignant melanomas occur in sites that are exposed to sunlight.

Clinical features

Most malignant melanomas arise at sites of previous naevi and are associated with enlargement of the naevus, ulceration, or bleeding in the majority of patients (Fig. 2.10). Any of these observations merits consideration of an excision biopsy. Familial melanomas have been described and alterations of pigmented lesions in first-degree relatives of patients who have had a melanoma should be treated with suspicion.

There are three well-described patterns of presentation of malignant melanoma – superficial spreading melanomas, nodular melanomas, and lentigo maligna.

Patients may present with a cutaneous lesion but occasionally present with symptoms of metastasis. Melanoma can spread to any organ, lymph nodes, or bones.

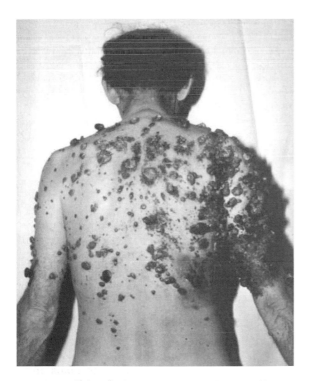

Fig. 2.10 Skin lesions in extensive malignant melanoma.

Treatment

The prognosis for malignant melanomas is particularly affected by the degree of invasion. The Breslow stage describes the thickness in millimeters of vertical tumor invasion into the dermis, with tumors less than 0.75 mm rarely metastasizing and those invading more than 1.5 mm having a high risk of subsequent metastasis. The presence of lymph node or distant metastasis at the time of diagnosis confers a very poor prognosis, lymph node involvement being associated with a 5-year survival rate of less than 20% and distant metastasis being rarely curable. Treatment is aimed at excision of the tumor together with a margin of surrounding skin. This will often involve skin grafting. Removal of the regional lymph nodes can provide prognostic information as survival is reduced if there is microscopic involvement but the role of lymphadenectomy in improving outcome is unclear. If lymph nodes are clinically involved, a block dissection of regional nodes is usually performed although 5-year survival rates are poor at only approximately 10%.

For patients with localized disease with invasion less than 0.75 mm, 5-year survival rates of 70% can be achieved.

Malignant melanoma is one of the more resistant tumors to both radiotherapy and chemotherapy. Radiotherapy may be beneficial for symptomatic metastases and chemotherapy can be used for disseminated disease although response rates are poor at only 10–20% for most agents. Dacarbazine is the most widely used drug for this disease, but median survival is only 6–8 months once metastases have been demonstrated, regardless of therapies employed. When disease is apparently localized to an extremity, some centers utilize regional perfusion of chemotherapy agents in the hope that obtaining high concentrations will produce greater tumor-cell kill but avoid the systemic toxicity that would otherwise follow. Cutaneous lesions can be dramatically reduced with this approach but survival seems to be little affected.

There has been considerable interest in the use of immunotherapy with interferons, interleukin-2, and a variety of vaccines but unfortunately early promise has not been followed by significant improvements in survival rates (Chapter 13).

Basal cell carcinoma

Basal cell carcinomas, also termed "rodent ulcers," are the most common skin malignancies, comprising approximately 75% of all cases. The incidence rises steadily with age such that at the age of 80, there are approximately 450 per 100,000 of the male population who develop a basal cell carcinoma (BCC). These tumors are presumed to originate from stem cells in the basal layers of skin, which have the potential to produce a variety of structures including sweat glands and hair follicles. These tumors predominantly develop in areas that are exposed to sunlight and are most frequent on the face. There is usually a characteristic clinical presentation with a firm pink nodule that has a distinct raised edge. There may be ulceration in the center of the nodule, which frequently bleeds.

Histologically, these tumors appear malignant with local infiltration and pleomorphic cells. Surprisingly, however, metastasis is very uncommon but there is local invasion that may penetrate to cartilage and occasionally to bone.

Management involves initial biopsy to exclude other forms of skin lesion, and when a diagnosis is confirmed a variety of treatment options are available. These predominantly involve surgical excision but cryosurgery, radiotherapy, and topical chemotherapy are all effective. Radiotherapy has the advantage of avoiding the need for a general anesthetic and often providing a better cosmetic result than surgery, which frequently involves skin grafting. Cure rates are high and approach 100%. There does appear to be a higher recurrence rate with radiotherapy than surgery (in one series 7.5% for radiotherapy and 0.7% for surgery) but patients treated with radiotherapy are often those with larger or more inaccessible tumors or individuals with high risk factors for an anesthetic.

Squamous cell carcinoma

Squamous cell carcinomas of the skin comprise approximately 20% of skin malignancies. They are also related to sunlight exposure but in addition they may occur at the periphery of previous areas exposed to radiation and have also been linked to exposure to arsenic.

These tumors have the histological appearance of keratinizing squamous carcinomas and vary in their degree of differentiation. The typical clinical presentation is with a crusted ulcer or nodular lesion, most commonly on the face, neck, and some exposed areas of the limbs. More rarely, they may arise at mucocutaneous junctions, particularly in the anus and vulva. As with BCCs, the initial procedure is a diagnostic biopsy and, if it confirms squamous carcinoma, it should be followed by surgical excision, or if this is not technically feasible radiotherapy can be undertaken. Squamous cell carcinomas of the mucocutaneous junctions are usually treated by chemoradiotherapy initially, often followed by surgical resection. Cryosurgery and topical chemotherapy may be used but are followed by higher rates of recurrence. For squamous carcinomas of the head, neck, and limbs, cure rates exceed 90%.

HEMATOLOGICAL MALIGNANCIES

The major diseases of the hemopoietic systems are lymphomas and leukemias. The lymphomas include Hodgkin's disease and non-Hodgkin's lymphomas. Leukemias are acute or chronic and predominantly involve the lymphatic or myeloid series.

Information on genetic causes for the majority of hematological malignancies is becoming increasingly available. One of the earliest abnormalities to be described was the translocation between chromosomes 9 and 22, involving the *bcr* and *abl* genes, resulting in a fusion protein with deregulated growth-promoting properties (Chapters 5 and 6).

Hodgkin's disease

The incidence of Hodgkin's disease follows an unusual bimodal distribution with an early peak in individuals between the ages of 10 and 20 years and a second peak after the age of 50 years. There is a predominance in males (~50% greater frequency than females). Links between Hodgkin's disease and infection with Epstein–Barr virus have been made and some clusters of disease have been seen in small geographic areas, suggesting environmental causes. No other definite etiological factors have been identified.

Diagnosis Biopsy of abnormal lymph nodes reveals typical changes in Hodgkin's disease. The diagnosis is confirmed by the presence of the Reed–Sternberg cell, a binucleate cell with a typical appearance (Fig. 2.11). Immunophenotyping reveals B-cell characteristics for the Reed–Sternberg cell and a variety of chromosomal changes have been detected. Typical histological patterns are seen in different patients such that the disease is divided into *lymphocyte rich, nodular lymphocyte predominant, nodular sclerosis, mixed cellularity*, and *lymphocyte depleted*. Eighty percent of all patients have either nodular sclerosing or mixed cellularity histology.

Clinical features Almost all patients present with a painless enlarged lymph node, usually in the neck, but it may involve other sites including the inguinal region (10%) and axilla (25%). There may be a long history of slow growth of these nodes and they may fluctuate in size. Approximately one-quarter of all patients will have systemic symptoms such as fever, sweats (may be drenching and predominantly during the night), and weight loss. Patients may also complain of alcohol-induced pain at the site of lymphadenopathy, pruritus, and symptoms suggestive of anemia. The enlarged nodes may compress airways, blood vessels, and bile duct leading to symptoms including cough, dyspnea, edema,

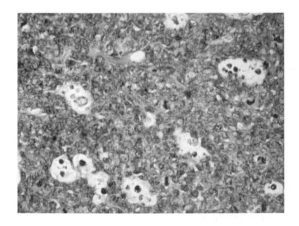

Fig. 2.11 Histological section of lymph node from patient with Hodgkin's disease showing typical large Reed–Sternberg cell.

and jaundice. Occasionally, skin lesions may occur and central nervous system involvement can lead to spinal cord compression.

Staging Investigations are undertaken to determine the volume of disease and its anatomical spread. Chest X-ray, CT scan of the thorax, abdomen, and pelvis, full blood count, and biochemical screen are routinely performed and provide sufficient information on staging for the majority of patients. Disease stage is described according to the Ann Arbor system with stage I disease defining a single lymph node area of involvement, stage II being two lymph node areas on the same side of the diaphragm, stage III describing two or more lymph node areas on opposite sides of the diaphragm, and stage IV disease being dissemination to non-lymph node sites including liver, lungs, and bone marrow.

Treatment The treatment of Hodgkin's disease represents a major therapeutic advance in oncology of the past 40 years and most patients are now cured. Both chemotherapy and radiotherapy are used; localized disease is usually managed with involved field radiotherapy whereas more widespread disease, and all patients with significant symptoms, will receive combination chemotherapy. Several combinations are used in different centers with the most frequent regimen including adriamycin, bleomycin, vinblastine, and dacarbazine (ABVD). Radiotherapy is often given after chemotherapy to sites of previous large-volume disease. Late toxicity is important given the high cure rate and a large number of second malignancies may follow protracted chemotherapy (particularly if radiotherapy is also administered). Approximately 8% of patients may develop second malignancies after 10 years and a major research focus is now to adapt therapies to risk and to minimize treatment without compromising cure rates.

Non-Hodgkin's lymphomas

Non-Hodgkin's lymphomas (NHLs) form a large and heterogeneous group of diseases likely arising from different types of lymphocytes at different stages of development (the loss of cellular differentiation that almost inevitably accompanies tumorigenesis further complicates the determination of the cell of origin – still a fundamental question to be answered in cancer biology). Their pathological classification has been the subject of several revisions over the years such that a large number of systems of classification exist. The incidence of NHL has risen rapidly over the past decade with increasing numbers of cases being related to AIDS. This is predominately a disease of older people with a steady rise in frequency throughout the decades following the thirtieth year. There is a slight male preponderance. Several etiological factors have been described including viral infections (HTLV-1, EBV, human herpes virus type 8). NHL is also markedly more common in patients who have undergone organ transplantation with prolonged immunosuppression and in patients who have celiac disease. Increasing information is becoming available on genetic factors relating to the etiology of non-Hodgkin's lymphomas. More than 80% of patients with Burkitt's lymphoma (Fig. 2.12) have a reciprocal translocation between chromosomes t(8;14) (q24;q32). For the other 20% of patients with this disease, translocations between chromosomes 8 and 22, and 2 and 8 are demonstrated. The break on the 8 chromosome occurs at the site of the *c-myc* proto-oncogene (Chapter 6). Over 80% of cases of follicular lymphomas have a chromosome translocation between 14q32 and 18q21. The breakpoint on chromosome 18 is the *bcl-2* gene, encoding a protein that inhibits apoptosis (Chapter 8). Increasing numbers of chromosome abnormalities are being described in association with different types of lymphoma and can be used to support diagnosis.

Pathology   The pathological classification of NHLs is complex and evolving. The most recent classifications have been developed by the World Health Organization, but both the Kiel and REAL systems are widely used. These divide the NHLs into B-cell or T-cell origins and subdivide these into histological patterns, which have different prognostic significance. T-cell lymphomas comprise approximately 10% of all NHLs.

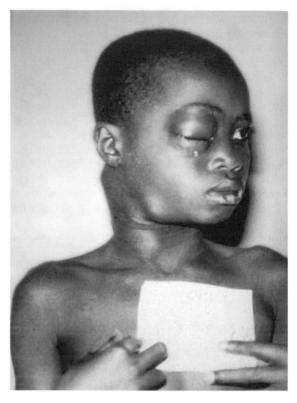

Fig. 2.12 Enlarged lymph node in a patient with Burkitt's lymphoma.

The diagnosis of NHL is made from a biopsy that is usually derived from an enlarged lymph node. NHL may involve a wide variety of sites and tissue may be obtained from these (e.g. liver, bowel, brain). The diagnosis can be difficult and relies on obtaining enough tissue to undertake immunohistochemistry and to obtain a large enough field of cells to comment on the histological pattern. Needle biopsies should be avoided wherever possible as a result.

Clinical presentation   In the majority of patients, NHL presents as enlargement of one or more lymph node areas. The lymphadenopathy is usually painless and, as with Hodgkin's disease, is most frequently found in the neck, but any lymph node area may be involved. Symptoms may thus arise from local compression of nerves, airways, or blood vessels. NHL tends to spread widely at a much earlier stage than Hodgkin's disease, so several sites of disease may be present at diagnosis. Examination may reveal hepatosplenomegaly in addition to lymphadenopathy.

Staging is undertaken to determine the extent of disease and the majority of patients will have stages III or IV disease at presentation, if careful investigation and examination are undertaken. Full CT scanning, biochemical assays, and blood count together with bone marrow aspirate and trephine should be undertaken. Several centers are increasingly using PET scanning to provide further information on stage.

Treatment The different pathological classifications can broadly divide patients into those with "low-grade" and those with "high-grade" disease. The approach to treatment differs in that low-grade lymphomas tend to have a chronic history with median survivals of approximately 10 years. These are not curable and will inevitably recur after treatment. High-grade lymphomas behave in a much more aggressive fashion with median survivals, if untreated, of only 4–6 months. They are, however, potentially curable. For low-grade lymphomas there does not appear to be a survival advantage to immediate treatment for patients who have no serious symptoms. As a result, localized radiotherapy is usually offered for stage I and II disease with the expectation of 8–10 years of freedom from recurrence. For those with stages III and IV low-grade lymphoma, most centers would carefully watch patients in clinic without intervening if they are free of symptoms or have no evidence of hematological or biochemical compromise from the sites of disease. If symptoms, abnormal blood counts or biochemistry develop, relatively simple outpatient chemotherapy can be used and is effective in the majority of patients. Chlorambucil remains the drug of choice for first-line therapy. Once this disease becomes refractory to chemotherapy, rituximab (a monoclonal antibody directed at CD20) can produce durable responses.

For patients with a high-grade lymphoma, combination chemotherapy is used for all stages of disease. Subsequent radiotherapy is usually offered to previous sites of bulky disease and has been shown to reduce the risk of relapse and prolong overall survival. The most commonly used chemotherapy involves the combination of cyclophosphamide, doxorubicin, vincristine, and prednisolone (CHOP). More intensive regimens have been investigated and there is conflicting evidence as to whether they are superior to CHOP. Response rates are high (~80%) with 60–65% of patients being cured of their disease.

For those patients who relapse following initial chemotherapy, high-dose chemotherapy and autologous stem cell transplantation is usually offered. Unfortunately, there is little data on benefit from randomized trials although results from several uncontrolled trials suggest that the majority of patients who are transplanted will achieve a response that improves quality of life and approximately 40% will have 3 or more years of survival, which is significantly greater than historical data for those patients who were not transplanted. There is one small randomized study that seems to suggest a survival advantage for patients being transplanted.

Leukemias

Leukemias are malignancies arising from the white blood cell lineage. This is one of the most common malignancies in children and, in the elderly, chronic lymphatic leukemia (CLL) is a frequent coincidental finding on routine blood counts. Acute lymphoblastic leukemia makes up approximately 80% of all pediatric cases, whereas acute myeloid leukemia (AML) is commoner in adults.

Etiological factors that have been identified include ionizing radiation, chemotherapy for previous malignancies, the HTLV virus, smoking, and exposure to benzene. There is also a variety of inherited syndromes, including Down's, which are associated with an increased risk of acute leukemia.

Chromosomal translocations have been described in several leukemias. For chronic myelogenous leukemia, a translocation between chromosomes 9 and 22 (t9;22) (q34;q11) has been identified and shown to bring the *bcr* gene and the *abl* gene together resulting in the production of a tyrosine kinase, which is central to growth regulation. This abnormality is seen in the vast majority of such cases and is used to help make the diagnosis. Translocations contribute to over 50% of all leukemias.

Diagnosis and pathology

The diagnosis of leukemia is usually suspected from an abnormal peripheral blood count. This frequently shows an elevated white count and is usually accompanied by anemia. There are often leukemic blasts visible in the blood. A bone marrow aspirate and trephine will usually provide the diagnosis. The marrow is almost completely replaced by blast cells, which have typical appearances and staining patterns of lymphoid or myeloid cells. Several subtypes of myeloid and lymphoid leukemia are described according to the histological appearance and staining of the malignant cells. Lymphoid leukemias may be of B- or T-cell origins.

Clinical history and management

Patients may present with a variety of features including lethargy from anemia, hemorrhage or bruising from thrombocytopenia, and infection from immunocompromise. Extensive involvement of bone marrow may cause bone pain. Some patients may have central nervous system involvement with associated signs and symptoms suggestive of meningeal involvement (stiff neck, headache, photophobia).

Treatment involves the use of chemotherapy for all cases of acute leukemia. This is usually divided into an "induction" phase that is intense and requires prolonged admission during periods of pancytopenia. Subsequent "consolidation" therapy is given after induction of remission and additional therapy with central nervous system prophylaxis to prevent relapse may be offered. High-dose treatment with transplantation (allogeneic or autologous) may be offered to high-risk individuals following remission induction if they are fit enough to tolerate this.

The chronic leukemias involve older patients who may be relatively asymptomatic, being detected from a routine blood count. Chronic lymphatic leukemia (CLL) is incurable and treatment can be deferred if the patient is asymptomatic. Increasing symptoms or bone marrow failure are indications for considering therapy and simple outpatient oral chemotherapy (usually with chlorambucil) may be offered. Combination chemotherapy is occasionally used and fludarabine is increasingly offered as it has a high level of activity and may improve survival.

Chronic myelogenous leukemia may be a coincidental finding from a routine blood count but may also cause fatigue from anemia, lethargy, and weight loss together with pain over the spleen from splenomegaly. Considerable interest exists in this disease with the advent of new treatments. Traditionally, patients were offered hydroxyurea or busulfan, which could reduce blood counts and improve symptoms but were associated with an inevitable relapse and progression to an acute phase, which was rapidly fatal. Alpha-interferon is now widely used and produces a remission in approximately 80% of patients. Several remissions can be obtained with sequential courses of interferon but toxicity is a problem and resistance inevitably develops. Gleevec (that is targeted to the *bcr–abl* oncoprotein) became available in 2001, and produces remissions in approximately 90% of patients (Chapter 5). It is well tolerated and remissions appear to be durable. The use of gleevec has largely replaced chemotherapy. For younger patients who are fit, the only potentially curative treatment is with an allogeneic stem cell transplant following high-dose chemotherapy and total body irradiation. With transplantation the prognosis appears to be markedly improved with 5-year relapse-free survival rates in excess of 75%.

QUESTIONS REMAINING

1 Will preoperative (neoadjuvant) chemotherapy improve the resectability rate and outcome in the common solid tumors?
2 How can we optimally develop and utilize signal transduction inhibitors to manage malignancies? What will be their impact on outcome?
3 Can we develop pretreatment markers that can better guide choices of optimal individualized therapy? This could allow focusing of expensive potentially toxic treatment on those who are most likely to benefit and avoid unnecessary toxicity for those who are unlikely to benefit.
4 What is the optimal method of following patients after treatment is completed? Can we develop noninvasive methods to detect recurrence?

5 Can we develop methods for effective screening of populations to detect tumors at an early stage with greater potential for cure? Do these techniques improve survival rates in the population?

CONCLUSIONS AND FUTURE DIRECTIONS

The rapid diagnosis and appropriate management of the common malignancies is essential to optimize the chance of cure. Unfortunately, many tumors are diagnosed at a time when they have already metastasized and are likely to be incurable. It is essential that patients are well educated and able to recognize symptoms that may indicate an underlying malignancy and that primary care clinicians are also aware of patterns of presentation so that investigations and referral for a specialist opinion are organized without delay. Most malignancies present with typical symptoms and signs that should automatically trigger an appropriate series of investigations, culminating in a histological diagnosis, if an abnormal area of tissue is detected. Surgery is usually the most appropriate form of curative treatment for solid tumors, but is increasingly preceded (neoadjuvant) or followed (adjuvant) by chemotherapy and/or radiotherapy in an attempt to improve resection rates and reduce the risk of recurrence. A large number of new cytotoxic agents have been developed in recent years and often have greater activity and reduced toxicity compared with previous anticancer drugs. Of great promise is the increasing availability of targeted therapies directed at specific components of cell signaling pathways or circulating biologically active molecules. The effectiveness of therapies for malignancies should markedly improve in the next decade and the side effects will be minimized. Unfortunately, these benefits will require huge increases in funding for healthcare providers – a challenge only just becoming apparent in many countries.

QUESTIONS

1 The pattern of cancer incidence
   a. Has remained relatively stable in Western nations over the last 50 years.
   b. Is related to the average age of the population in a nation.
   c. Is related to environmental factors.
   d. Has shown increases in all tumor types over the last century.
2 Lung cancer
   a. Is steadily increasing in incidence in males and females.
   b. Risk reduces after discontinuing cigarette smoking.
   c. Usually presents as a result of symptoms from metastases.
   d. Has a worse prognosis than breast cancer.
3 Breast cancer
   a. Is more common in Japanese women than in women living in the United States.
   b. Is fatal in 75% of all cases.
   c. Is more common in postmenopausal women.
   d. Is managed differently in pre- and postmenopausal women.

BIBLIOGRAPHY

Epidemiology of cancer

Doll, R. and Hill, A.B. (1964). Mortality in relation to smoking: ten years' observation on British Doctors. *British Medical Journal*, **1**. 1399.

Parkin, D.M., Pisani, P., and Ferlay, J. (1999). Global cancer statistics. *CA: Cancer Journal for Clinicians*, **49**: 33.

Lung cancer

Mountain, C.F. (1997). Revision in the international system for staging lung cancer. *Chest*, **111**: 1710.

Herbst, R.S., Giaccone, G., Schiller, J.H., *et al.* (2004). Gefitinib in combination with paclitaxel and carboplatin in advanced non-small-cell lung cancer: a phase III trial – INTACT 2. *Journal of Clinical Oncology*, **22**: 785.

Breast cancer

Bluman, L.G., Rimer, B.K., Berry, D.A., *et al.* (1999). Attitudes, knowledge and risk perceptions of women with breast and/or ovarian cancer considering testing for *BRCA1* and *BRCA2*. *Journal of Clinical Oncology*, **17**: 1040.

Early Breast Cancer Trialists' Collaborative Group. (1998). Tamoxifen for breast cancer. An overview of the randomised trials. *Lancet*, **352**: 930.

Fisher, B., Bauer, M., Margolese, R., *et al.* (1985). Five years' results of a randomised clinical trial comparing total mastectomy and segmental mastectomy with or without radiation in the treatment of breast cancer. *New England Journal of Medicine*, **312**: 665.

Colorectal cancer

Hurwitz, H., Fehrenbacher, L., Novotny, W., *et al.* (2004). Bevacizumab plus irinotecan, fluorouracil, and leucovorin for metastatic colorectal cancer. *New England Journal of Medicine*, **350**: 2335–42.

Vogelstein, B., Fearon, E.R., Hamilton, S.R., *et al.* (1988). Genetic alterations during colorectal tumor development. *New England Journal of Medicine*, **319**: 525.

Skin tumors

Dennis, L.K. (1999). Analysis of the melanoma epidemic, both apparent and real: data from the 1973 through 1994 surveillance, epidemiology, and end results program registry. *Archives of Dermatology*, **135**: 275.

Spratt, J.S., Jr. (1999). Cancer mortality after non-melanoma skin cancer. *JAMA*, **281**(4): 325.

Prostate cancer

Eisenberger, M.A., Blumenstein, B.A., Crawford, E.D., *et al.* (1998). Bilateral orchiectomy with or without flutamide for metastatic prostate cancer. *New England Journal of Medicine*, **339**(15): 1036.

Gleason, D.F. and Mellinger, G.T. (1974). Prediction of prognosis for prostatic adenocarcinoma by combined histological grading and clinical staging. *Journal of Urology*, **111**: 58.

Hematological malignancies

Fisher, R.I., Gaynor, E.R., Dahlberg, S., *et al.* (1993). Comparison of a standard regimen (CHOP) with three intensive chemotherapy regimens for advanced non-Hodgkin's lymphoma. *New England Journal of Medicine*, **328**: 1002.

Harris, N.L., Jaffe, E.S., Diebold, J., *et al.* (1999). World Health Organization classification of neoplastic diseases of the hematopoietic and lymphoid tissues: report of the clinical advisory committee meeting–Airlie House, Virginia, November, 1997. *Journal of Clinical Oncology*, **17**: 3835.

3

Nature and Nurture in Oncogenesis

Stella Pelengaris and Mike Khan

Every politician, clergyman, educator, or physician, in short, anyone dealing with human individuals, is bound to make grave mistakes if he ignores these two great truths of population zoology: (1) no two individuals are alike, and (2) both environment and genetic endowment make a contribution to nearly every trait.
Ernst Mayr

KEY POINTS

- Rare highly penetrant genes cannot explain much of the familial risk for most common cancers.
- A high proportion of cancers may arise in a susceptible minority who carry low-penetrance genes or gene combinations.
- Environmental carcinogens (and somatic mutations) have been widely held to be the major causes of most common cancers.
- Geographical differences in cancer incidence, and the "migration effect" suggest the importance of environmental factors.
- Smoking, diet, sex hormones, and increasing age influence risk of cancer.
- More recently, mechanisms have been identified whereby such risk factors and gene polymorphisms may interact to cause common cancers.
- Some cancer-causing viruses carry oncogenes or promote cancer by insertional mutagenesis.

INTRODUCTION

A risk factor is anything that increases the likelihood (probability) of an individual developing a particular disease, such as cancer. A risk factor for cancer is just that – its presence does not invariably guarantee that an individual will go on to develop a particular cancer nor does its absence imply that they will not. Obviously, some risk factors are much more predictive than others, but those that are known to predict a very high risk of developing cancer are relatively few and can only rarely be applied to predicting individuals most likely to develop common noninherited (sporadic) cancers.

By implication, identifying risk factors that predict the future development of cancer might enable us to identify particular individuals who have a particularly high risk of developing cancer, opening up the possibility of preventative medicine. Such an approach, involving random routine screening of unselected individuals for cardiovascular risk factors in order to most effectively target preventative drug treatments, has been remarkably effective in preventing heart attacks. However, this has been achieved largely because cardiovascular risk markers can readily be determined by combining simple clinical consultation (blood pressure, age, sex, history of smoking, and family history of heart

61

disease) with a routine blood test (to measure cholesterol, other blood fats, and glucose). Moreover, extensive epidemiological and prospective studies of huge numbers of patients have allowed us to create reasonably accurate and effective risk charts that relate all these parameters to future risk of heart attack. Preventative treatments, which are themselves remarkably free of potentially harmful side-effects, can then be recommended to at-risk individuals, above a certain percentage threshold for disease development (for coronary disease prevention a risk of events above 15% over 10 years has been chosen – many feel this is too high).

In cancer prevention, although there have been some notable successes, including screening for breast and cervical cancer, readily determined risk factors for future development of cancer are generally not available for the common cancers. Most tests for potential "markers" of risk or of early disease are either complex, invasive, or require costly and not universally applicable imaging (cervical cytology, biopsy, X-ray, and scanning). Moreover, these largely pick up established disease, whereas the ideal screening test would identify individuals who have as yet not developed cancer at all. Even where noninvasive markers that precede cancer onset have been described, such as the *BRCA* genes in breast cancer, the predictive strength is not robust enough, particularly as the preventative treatment based on these is not a simple well-tolerated drug, but may be "mutilating" surgery. One approach to improving the accuracy of predicting the future risk of cancer in individuals is to look at more than one risk factor concurrently (as is done for heart disease), many of which have been and continue to be described.

Risk factors may be environmental (lifestyle-related or unavoidable) or genetic (inherited) and are generally specific for particular cancers. Inherited factors are clearly of major importance and a historical overview of theories of heredity is given in Box 3.1 and on the website. However, most cancers arise through a combination of genetic and environmental factors and less commonly by extreme exposure to radiation, viruses, cancer-causing chemicals (carcinogens), or combinations of genetic factors only (Fig. 3.1). Having a risk factor, or even several, does not guarantee that an individual will get

---

Box 3.1 Development of the theories of heredity

Gre gor Mendel's now famous early research into heredity in the mid nineteenth century, using hybrid common garden peas, is among the earliest descriptions of what has in recognition been termed Mendelian inheritance. Importantly, he also proposed a simple hypothesis for the operation of these laws, namely that observed traits are determined by discrete "factors," now called genes. Ironically, Mendel's work was largely ignored for 40 years, after which its "rediscovery" helped launch the new field of genetics. At the same time as Mendel was breeding peas, Charles Darwin published his work *On the Origin of Species*, describing his theory of evolution namely that species evolved under the pressure of environmental factors through natural selection. With Darwin's theory and Mendel's laws, the foundations for genetics had been laid.

The chromosome theory of heredity (that hypothetical entities called genes are parts of chromosomes) was originally proposed by the German biologist Theodor Boveri and an American student Walter Sutton. In 1902, they observed that the behavior of chromosomes at meiosis was compatible with the observations and hypotheses of Mendel; genes are in pairs (so are chromosomes); the alleles of a gene segregate equally into gametes (so do the members of a pair of homologous chromosomes); different genes act independently (so do different chromosome pairs). However, it was a further 50 years before the nature of the genetic material was discovered.

Eventually, the X-ray crystallographic work of Rosalind Franklin and Maurice Wilkins would lead to the model proposed by James Watson and Francis Crick in 1953. This model postulated that the structure of a gene lay in the sequences of the bases (A, G, C, and T), replication that distributed these genes to offspring, and the ability of changes in the base sequence to result in mutations to the genes. Finally, a molecular explanation was beginning to emerge for the earlier work of Darwin, namely that it was the DNA upon which natural selection acted. Thus, changes in the sequence of bases in the DNA (mutations) would lead to different genetic traits that could be selected for, or against, by changes in the environment.

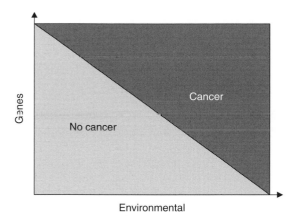

Fig. 3.1 Cancer susceptibility is due to a combination of genetic and environmental factors. The cumulative effects of both genetic/epigenetic changes and exposure to environmental factors determine the likelihood of developing cancer. In this model, only in extreme cases will cancers arise entirely due to either genetic or environmental factors alone.

the disease, but such guarantees rarely exist in disease prevention (where hindsight is not yet available!) and are unlikely to exist in the foreseeable future. What is practically required is a means of calculating the risk of cancer (in the future) accurately enough to be able to safely apply preventative treatments – in other words, if that treatment is via an entirely nontoxic and cheap drug then treating a few individuals who would never have developed the disease is fine, whereas if the treatment is mastectomy it is not.

Importantly, many risk factors for cancer are avoidable and there is no need to screen individuals before advocating smoking cessation, healthy diet, and sunscreen use – such recommendations are a public-health issue and should apply to all. It was 50 years ago that Richard Doll and A. Bradford Hill reported findings on a cohort of British doctors showing a strong link between cigarette smoking and lung cancer.

Risk factors and cancer can broadly be divided into two categories.

1. Environmental, including
   a. Lifestyle-related risk factors, such as smoking, unhealthy diet, or unprotected exposure to strong sunlight.
   b. Factors pertaining to the environment not under the direct control of the individual, such as high levels of radiation, cancer-causing chemicals, and certain infections.

   c. Unavoidable natural processes, such as free radical generation, endogenous hormones, cosmic rays.
2. Genetic, including
   a. Single gene (monogenic) disorders that are inherited from a parent and are direct causes of a given disease. These are generally rare and include inheriting a defective copy of the *p53* gene (Li Fraumeni syndrome).
   b. Polygenic disorders that are caused by variations in multiple genes and often occur in combination with environmental influences to which an individual with a particular pattern of genes may be particularly vulnerable.

RISK FACTORS ACT IN COMBINATION

It is often difficult to attribute tumorigenesis to any single causative agent, but exposure to specific environmental factors is strongly predictive of the future development of particular cancers. Particularly notable are the strong links between cigarette smoking and lung cancer and exposure to ultraviolet (UV) light and skin cancer. Even in these cases, not all individuals who smoke or experience repeated sun exposure develop cancer. Likewise, cancers rarely develop solely because of genetic factors – inherited cancers are uncommon, but their study has contributed greatly to our understanding of the molecular basis of common forms of cancer. By implication, and as is the case for many chronic diseases, both genetic and environmental factors contribute to disease pathogenesis (Lichtenstein *et al.* 2000) – see Figure 3.2. Possible ways in which mutations in various genes might directly interact with exposure to environmental factors are shown in Figure 3.3. Traditionally, the genetics of inherited cancer predisposition have been considered in terms of single, high-risk genes with high penetrance, such as in retinoblastoma (*RB*), Li Fraumeni (*p53*), familial breast cancer (*BRCA1/2*), familial colorectal cancer (*HNPCC*), and Wilms' tumor (*WT1*). In fact, Knudson's hypothesis was first confirmed by the study of retinoblastoma – see Chapter 7 for a detailed account. A list of the more common single-gene inherited cancer syndromes is given in Table 3.1.

It is worth emphasizing here that the absence of an overtly inherited monogenic disorder does

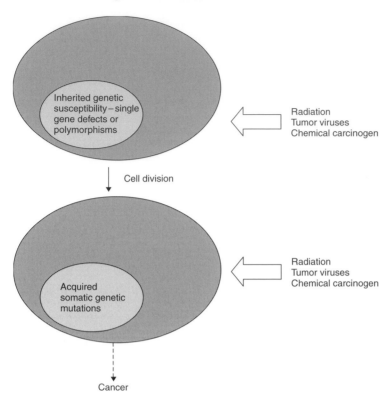

Fig. 3.2 The combination of genes and environment in the genesis of a cancer cell. This diagram describes a potential example of how inherited and acquired genetic alterations together with environmental factors may lead to tumorigenesis and cancer.

not imply that inherited genetic factors have not played a role in formation/progression of that cancer; subtle variations in multiple alleles (polymorphisms) may increase disease susceptibility through numerous mechanisms including alteration of metabolism of carcinogens, efficiency of DNA repair, free-radical handling, angiogenesis, etc. Such low-penetrance genes frequently vary by single nucleotide polymorphisms (SNPs) – see Table 3.2. All of these inherited factors may in turn influence the risk of cancer after exposure to environmental carcinogens or after acquisition of various somatic mutations (Fig. 3.2).

Three classes of environmental agents – ionizing radiation, tumor viruses (viruses that cause cancer in animals), and chemical carcinogens (cancer-causing chemicals) – have been shown to increase the risk of cancer in both laboratory animals and people. Each of these agents may produce mutations in genes that contribute to abnormal growth.

Given that only some 5% of cancers arise through inherited genes, environmental factors, such as tobacco, alcohol, diet, infections, and occupational exposures, are strongly implicated as contributors to the pathogenesis of as many as 80–90% of cancers, though these may not invariably act through promoting gene mutations. In at least some of these cases, the "environmental" agents (such as background radiation) may not be avoidable, but this notwithstanding, it has been estimated that about 75% (smokers) or 50% (nonsmokers) of all deaths due to cancer in the United States could be avoided by elimination of avoidable risk factors, making the prevention of many cancers a realistic prospect. For example, smoking cigarettes while young puts a person at 10–20 times higher risk of developing cancer later in life than persons who do not smoke. A high-fat diet and obesity may contribute to the development of certain cancers.

Epidemiological studies including studies of migrant populations have provided the strongest support for the role of environmental factors in cancer. In particular, the fact that migrants often acquire the disease profiles of their adopted country is often taken as support for the notion

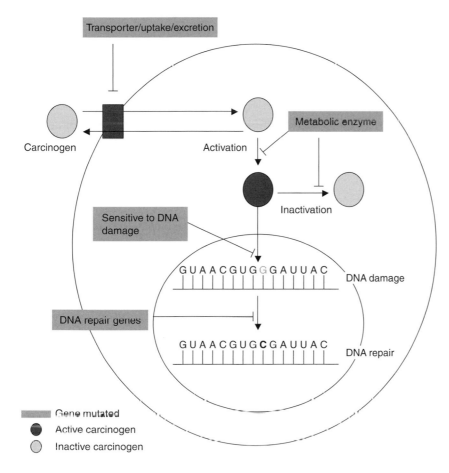

Fig. 3.3 Genetic variation may influence susceptibility to carcinogens. Variation in gene alleles encoding proteins responsible for uptake, transport, and metabolism of a carcinogen may all affect the likelihood that the carcinogen will cause DNA damage. Moreover, variation in DNA repair processes will also play a role.

that such diseases are thus more likely to be caused by environment and lifestyle rather than genetic factors. It must, however, also be appreciated that genetic factors may still play a major role. One notable example is provided by the fate of migrant South Asians in the United Kingdom and North America. In the Indian subcontinent both diabetes and coronary heart disease (CHD) have been notably less frequent than in the United Kingdom (although this situation may be changing somewhat, particularly in urban and more affluent areas). Migrant Asian populations in the United Kingdom were then found to acquire a risk profile for these diseases thought to be more typical of the United Kingdom. However, over the last decade it has become apparent that, in fact, Asians in the United Kingdom may experience a far greater risk of diabetes and CHD than the native population – in large part due to presumed genetic factors that actually increase susceptibility to dietary and other factors to which they are exposed in the new environment.

ENVIRONMENTAL RISK FACTORS

Carcinogens are agents that may damage genes involved in cell proliferation and migration, and may selectively enhance growth of tumor cells or their precursors. Free radicals may result in nonspecific DNA damage; some viruses can accelerate the rate of cell division. Importantly, genetic factors and inherited predisposition may play a major role in determining how a given individual responds to these extrinsic factors. Thus, at the most obvious level, fair skin strongly predisposes to melanoma and nonmelanoma skin cancers; whereas more subtle variation at individual gene

Table 3.1 Hereditary cancer syndromes (based on OMIM data)

Syndrome	Cloned gene	Function	Chromosomal location	Tumor types
Li Fraumeni syndrome	*p53*; tumor suppressor	Cell cycle regulation, apoptosis	17p13	Brain tumors, sarcomas, leukemia, breast cancer
Retinoblastoma	*RB1*; tumor suppressor	Cell cycle regulation	13q14	Retinoblastoma, osteogenic sarcoma
Wilms' tumor	*WT1*; tumor suppressor	Transcriptional regulation	11p13	Childhood kidney cancer
Neurofibromatosis, type 1	*NF1*; protein = neurofibromin 1; tumor suppressor	Catalysis of RAS inactivation	17q11.2	Neurofibromas, sarcomas, gliomas
Neurofibromatosis, type 2	*NF2*; protein = merlin or neurofibromin 2; tumor suppressor	Linkage of cell membrane to cytoskeleton	22q12.2	Acoustic neuromas, Schwann cell tumors, astrocytomas, meningiomas, ependynomas
Familial adenomatous polyposis coli	*APC*; tumor suppressor	Signaling through adhesion of molecules to nucleus	5q21	Colon cancer
Tuberous sclerosis 1	*TSC1*; protein = hamartin; tumor suppressor		9q34	Facial angiofibromas
Tuberous sclerosis 2	*TSC2*; protein = tuberin; tumor suppressor	GTPase activation	16	Benign growths (hamartomas) in many tissues, astrocytomas, rhabdomyo-sarcomas
Deleted in pancreatic carcinoma 4	*DPC4*; also known as *Smad4*; tumor suppressor	Regulation of TGF-b/BMP signal transduction	18q21.1	Pancreatic carcinoma, colon cancer
Deleted in colorectal carcinoma	*DCC*; tumor suppressor	Transmembrane receptor involved in axonal guidance via netrins	18q21.3	Colorectal cancer
Familial breast cancer	*BRCA1*; tumor suppressor	Repair of double strand breaks by association with Rad51 protein	17q21	Breast and ovarian cancer
Familial breast cancer	*BRCA2*; tumor suppressor	DNA damage repair	13q12.3	Breast and ovarian cancer
Peutz–Jeghers syndrome	*STK11*; tumor suppressor; protein = serine-threonine kinase 11	Potential regulation of vascular endothelial growth factor (VEGF) pathway	19p13.3	Hyperpigmentation, multiple hamartomatous polyps, colorectal, breast, and ovarian cancers

Table 3.1   *(continued)*

Syndrome	Cloned gene	Function	Chromosomal location	Tumor types
Hereditary nonpolyposis colorectal cancer type 1 HNPCC1	*MSH2*; tumor suppressor	DNA mismatch repair	2p22-p21	Colorectal cancer
Hereditary nonpolyposis colorectal cancer type 2 HNPCC2	*MLH1*; tumor suppressor	DNA mismatch repair	3p21.3	Colorectal cancer
von Hippel–Lindau syndrome	*VHL*; tumor suppressor	Regulation of transcription elongation	3p26-p25	Renal cancers, hemangioblastomas, pheochromocytoma
Familial melanoma	*CDKN2A*; protein = cyclin-dependent kinase inhibitor 2A tumor suppressor	Inhibits cell cycle kinases, CDK4 and CDK6	9p21	Melanoma, pancreatic cancer, others
Gorlin syndrome: nevoid basal cell carcinoma syndrome (NBCCS)	*PTCH*; protein = patched tumor suppressor	Transmembrane receptor for *hedgehog* signaling protein	9q22.3	Basal cell skin cancer
Multiple endocrine neoplasia type 1	*MEN1*; tumor suppressor	Unknown	11q13	Parathyroid and pituitary adenomas, islet cell tumors, carcinoid
Multiple endocrine neoplasia type 2	*RET, MEN2*	Transmembrane receptor tyrosine kinase for glial-derived neurotrophic factor (GDNF)	10q11.2	Medullary thyroid cancer, type 2A pheochromocytoma, mucosal hamartoma
Beckwith–Wiedmann syndrome	*p57, KIP2*	Cell cycle regulator	11p15.5	Wilms' tumor, adrenocortical cancer, hepatoblastoma
Hereditary papillary renal cancer (HPRC)	*MET*	Transmembrane receptor for hepatocyte growth factor (HGF)	7q31	Renal papillary cancer
Cowden syndrome	*PTEN*; tumor suppressor	Phosphoinositide 3-phosphatase protein tyrosine phosphatase	10q23.3	Breast cancer, thyroid cancer, head and neck squamous carcinomas
Hereditary prostate cancer numerous loci: HPC1(PRCA1), HPCX, MXI1, KAI1, PCAP	*HPC1* and *PRCA1* are same designation ribonuclease L (RNaseL) maps to this locus	RNaseL involved in mRNA degradation	1q24-q25	Prostate cancer
Ataxia telangiectasia (AT)	*ATM*	DNA repair	11q22.3	Lymphoma, cerebellar ataxia, immunodeficiency

Table 3.1 (*continued*)

Syndrome	Cloned gene	Function	Chromosomal location	Tumor types
Bloom syndrome	*BLM*	DNA helicase?	15q26.1	Solid tumors, immunodeficiency
Xeroderma pigmentosum (XP) seven complementation groups for XPA, XPC, XPD	*XPA – XPG*	DNA repair helicases, nucleotide excision repair	XPA = 9q22.3 XPC = 3p25 XPD = 19q13.2-q13.3 XPE = 11p12-p11 XPF = 16p13.3-p13.13	Skin cancer
Fanconi's anemia	*FANCA – FANCH*	Components of DNA repair machinery	FANCA = 16q24.3 FANCC = 9q22.3 FANCD = 3p25.3 FANCE = 11p15	Acute myeloid leukemia (AML), pancytopenia, chromosomal instability

loci may, for example, increase the risk of lung cancer in smokers.

Socioeconomic differences are major contributors to patterns of disease, with poorer individuals tending to experience several specific adverse lifestyle factors including higher rates of smoking, alcohol consumption, poor nutrition (variously malnutrition in developing countries and ironically obesity in developed countries), and exposure to certain infectious agents including *Helicobacter pylori*. Conversely, more affluent individuals are more likely to develop cancers of the breast or prostate, at least in part through reduced parity, and older age at time of first pregnancy and probably earlier diagnosis, respectively.

Recent strong support for the role of carcinogenic environmental factors in causing sporadic cancers has come from two large Scandinavian studies – (i) twin cohort from Sweden, Denmark, and Finland; (ii) the Swedish family-cancer database, which includes around 10 million individuals. Although still controversial, both studies suggest that the influence of factors not shared among siblings or relatives (including inherited genetic factors and shared exposure) predominates. Such "nonshared environmental factors" include smoking, radiation, infections, and occupational exposures, as well as sporadic mutations.

The risk of developing some cancers has declined dramatically in developed nations in this century, but unfortunately the risk of the most prominent forms has increased. This may be largely a manifestation of increasing longevity and age within populations coupled with improved recording of disease incidence and causes of death, rather than any increases in environmental cancer-causing agents. However, the possibility of depletion in the ozone layer contributing to increased rates of UV-exposure-induced cancers is a prominent example of one such explanation.

Individual environmental risk factors

Carcinogenic chemicals Carcinogenic chemicals such as azo dyes, asbestos, benzene, formaldehyde, and diesel exhaust are dangerous in high concentrations. This level of concentration often used to exist in some workplaces. Strict control over the past 50 years of such occupational carcinogens, however, has greatly reduced cancers caused by these substances.

One of the earliest recorded associations between chemical exposure and cancer was noted by the famous Renaissance intellectual and father of pharmacology Paracelsus. In 1567, Paracelsus proposed that the "wasting disease"

Table 3.2 Polymorphisms and cancer

Gene	Location	Cancer
ALDH2	12q24.2	Aldehyde dehydrogenase 2
APC	5q21-q22	Adenomatous polyposis coli – I1307K polymorphism in colorectal cancer
CCND1 (Cyclin D, PRAD1)	11q13	Cyclin D – head and neck cancers
CDKN2A (P16, INK4A)	9p21	P16 tumor suppressor – melanoma and mole density
CYP17	10q24.3	Cytochrome P450 (*CYP17*) – breast and prostate cancers
CYP19	15q21.1	Aromatase cytochrome P450 gene (*CYP19*) and breast cancer risk
CYP1A1 (CYP1)	15q22-q24	Cytochrome P450, subfamily I (aromatic compound-inducible), polypeptide 1 breast, lung
CYP1B1	2p22-p21	Cytochrome P450, subfamily I (dioxin-inducible), polypeptide 1 (glaucoma 3, primary infantile) – breast cancer
CYP2A6	19q13.2	CYP2A6; cytochrome P450, subfamily IIA (phenobarbital-inducible), polypeptide 6 – nicotine metabolism and lung cancer
CYP2E	10q24.3-qter	CYP2E; cytochrome P450, subfamily IIE (ethanol-inducible) – lung cancer
GSTM1 (GST1)	1p13.3	GSTM1. Metabolism of tobacco carcinogens, lung cancer
GSTP1 (GST3)	11q13	GST pi – breast
HRAS	11p15.5	Harvey RAS – minisatellite alleles and cancer susceptibility. *v-Ha-ras* Harvey rat sarcoma viral oncogene homolog – breast, ovary, and lung
LTA	6p21.3	Lyphotoxin alpha (TNF superfamily, member 1) – risk of MGUS and myeloma
MC1R (MSH-R)	16q24.3	Melanocortin 1 receptor (alpha melanocyte stimulating hormone receptor) – melanoma
MTHFR	1p36.3	5,10-Methylenetetrahydrofolate reductase (NADPH) – colon cancer, acute lymphoblastic leukemia
NAT1	8p23.1-p21.3	*N*-acetyltransferase 1 (arylamine *N*-acetyltransferase) – bladder cancer, lung cancer
SRD5A2	2p23	Steroid-5-α-reductase, α polypeptide 2 – prostate cancer, breast cancer
TNF	6p21.3	Tumor necrosis factor α – myeloma.

of Austrian miners might be caused by their exposure to natural ores, such as realgar (arsenic sulfide). The eighteenth century witnessed one of the earliest descriptions of a carcinogenic chemical with the link identified between soot and cancer of the scrotum in chimney sweeps, by the rather aptly named Percival Pott. Another English physician, John Hill, described the occurrence of cancer in the nose in long-term snuff users. By the end of the nineteenth century occupational exposure of workers in the paraffin industry had been linked to certain cancers, and the development of certain skin cancers had been attributed to use of arsenicals.

By the start of the twentieth century further examples of "occupational cancers" had been described – including bladder tumors in workers in the aniline dyestuff industry – "aniline cancer." At around the same time, tar or tarry compounds were recognized as causes of skin cancers, which was further supported by the induction of rodent skin cancer by exposure to coal tar. Over the next few decades higher molecular weight polycyclic aromatic hydrocarbons (PAHs), such as dibenz-(*a*, *h*)anthracene (DBA), were described and found experimentally to be the cancer-causing constituents of coal tar. By the middle of the last century, chemical carcinogenesis in humans was already well established with observational data supported by experimental studies on defined compounds in model systems. Numerous carcinogenic compounds are recognized now and will be discussed later. Importantly, many cancer chemotherapeutic agents are themselves carcinogenic. Alkylating agents are associated with risk of several cancers; cyclophosphamide is associated with risk of bladder cancer.

Chemical carcinogens taken up by cells are usually metabolized, and the resulting metabolites are then excreted, but may on occasion be retained by the cell. Such internalized carcinogenic compounds can then directly or indirectly alter gene expression. Some carcinogens are genotoxic, forming DNA adducts or inducing various chromosomal abnormalities. For example, carcinogenic ions or compounds of nickel, arsenic, and cadmium can induce aneuploidy. Other carcinogens may act by nongenotoxic mechanisms, including promoting inflammation, suppressing immunity, forming damaging reactive oxygen species (ROS), or by activation of signaling pathways, such as receptors for aryl hydrocarbon receptor (AhR), estrogen receptor, PKC, and epigenetic silencing. Together, these genotoxic and nongenotoxic mechanisms may provoke the "hallmark" features of cancer.

Two terms are frequently used for classifying chemical carcinogenesis:

Initiation: the result of exposure of a cell or cells to a carcinogen that permanently alters its genetic material but does not immediately influence phenotype. Carcinogens in this category are described as mutagens or genotoxic.

Promotion: these factors cause tumors from cells that have been initiated already. Promoters are nongenotoxic carcinogens. The best-known promoters are the phorbol esters, which activate the PKC signaling pathway and promote mitogenesis and survival.

The classification of carcinogens in this way has limitations (Box 3.2) in its application to many cancers and is increasingly falling out

---

Box 3.2 Initiation and promotion of cancer

It can reasonably be assumed when modeling the development of cancer, that all cancers are the result of the initiation, promotion, and progression phases of carcinogenesis (see Fig. 3.4). By implication, there are genes that can:

1. protect or predispose proto-oncogenes and tumor-suppressor genes from activation or inactivation.
2. bring about or suppress the growth and expansion of initiated cells.
3. prevent or enhance the acquisition of genetic/epigenetic instability by the initiated cells in order for them to become malignant.

Various genetic syndromes in humans support this notion.

---

of favor. However, the basic notion of cancers arising in a stepwise fashion through acquisition of successive mutations likely does apply to many human cancers, and forms a useful model for understanding how oncogenes and tumor suppressors may cooperate in carcinogenesis. A view of multistage carcinogenesis incorporating these concepts is shown in Figure 3.4.

Activation of carcinogens Some carcinogens are direct-acting and activation-independent. However, the majority are actually procarcinogens that require metabolic activation to produce carcinogens. Well-known direct-acting carcinogens include the alkylating agents. Polycyclic hydrocarbons (smoke), aromatic amines, amides, azo dyes, and nitrosamines all require activation by the hepatic cytochrome P450 mixed function oxidase system.

Many carcinogenic compounds are mutagenic – that is, they can induce mutations in DNA. Only sufficiently reactive, electrophilic compounds can interact directly with DNA, but such "direct carcinogens" form a minority of known human carcinogens, including ethylene oxide, *bis*(chloromethyl)ether, and some aziridine or nitrogen-mustard derivatives used in anticancer chemotherapy. Most mutagenic chemicals are actually formed by metabolic activity after exposure from initially inert nucleophilic compounds, such as aromatic and heterocyclic amines, aminoazo dyes,

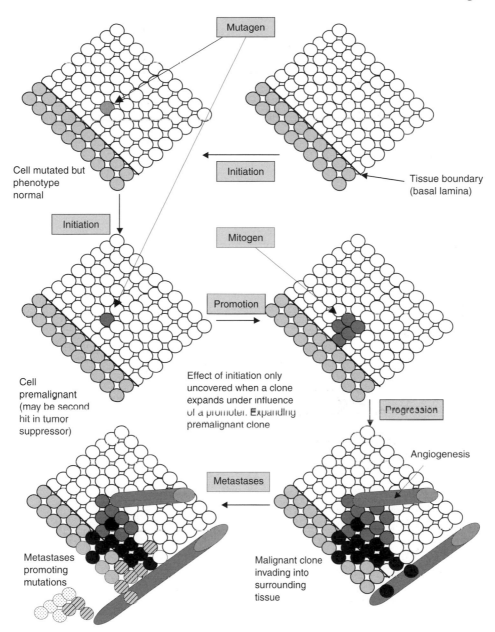

Fig. 3.4 Multistage carcinogenesis, concept of initiation and promotion. Genetic changes induced by mutagens are irreversible but may be phenotypically occult until further events such as proliferation or loss of differentiation unmask them. The mutagen/carcinogen is the "tumor initiator," but other factors (tumor promoters) affect whether mutated cells proliferate and form tumors. Promoters can contribute to cancer formation but do not alter DNA. Promoters increase the frequency of tumor formation in tissues previously exposed to mutagen/tumor initiator. For example, skin papillomas form after exposure to carcinogens, but there may be considerable latency between the mutation in the stem cell pool and exposure to a tumor promoter that promotes proliferation, resulting in a visible lesion. This expanded clone will then itself be vulnerable for subsequent mutational events (hits) for tumor progression. Some evidence suggests that angiogenesis may accompany the growth of the tumor, or may arise through a distinct mutational event (angiogenic switch). Finally an invasive cancer forms with cells entering the circulation. However, it is generally believed that a further mutational event (such as inactivation of a "metastasis-suppressor" gene) is needed for colonies of cancer cells to establish in a distant location.

PAHs, and *N*-nitrosamines – they are termed "procarcinogens."

Metabolic conversion of procarcinogens involves in particular the action of microsomal enzymes such as cytochrome-P450-dependent monooxygenases (CYPs). The initial step during conversion of organic xenobiotics into hydrophilic and excretable derivatives is mainly catalyzed by CYP enzymes, of which more than 50 types are currently known. The types of CYP known to be involved in activation of carcinogens include CYP1A1 in lung and CYP1A2, CYP2A6, and CYP2E1 in the liver. Carcinogens may also be metabolized to intermediates that can undergo transferase-catalyzed conjugation to polar molecules, such as glutathione by glutathione *S*-transferases (GSTs), to glucuronic acid by glucuronosyltransferases, or to small residues such as sulfate by sulfotransferases (SULTs) and acetic acid by *N*-acetyltransferases (NATs). These enzymes may also contribute to activation of procarcinogens into compounds that interact with DNA. Thus, *N*-hydroxy derivatives of procarcinogenic arylamines/amides, aminoazo dyes, or heterocyclic amines are converted by NAT or SULT enzymes into highly reactive ester intermediates that bind to DNA. Some compounds do not bind to DNA and are not mutagenic, yet they are carcinogenic in animal models.

Dioxins The nongenotoxic carcinogen 2,3,7,8-tetrachlorodibenzo-*p*-dioxin (TCDD) is produced as a by-product during the manufacture of polychlorinated phenols. TCDD is generated during incineration of waste and has been strongly implicated in a range of human cancers by epidemiological studies. One famous case took place in 1976 in Seveso, Italy, when following an accident at a chemical plant, several thousand kilograms of chemicals were released into the air – including 20 kg of dioxins. Over 30,000 people were exposed to these chemicals, and decades later cancer rates among those exposed were significantly increased compared with control populations. TCDD is also carcinogenic in animal models, promoting cancer in skin, lung, and liver, among others. TCDD is an agonist of the AhR, a cytosolic protein involved in induction of microsomal aryl hydrocarbon hydroxylase (AHH) activity. The AhR belongs to the basic helix–loop–helix/PAS family of transcription factors, which upon ligand binding translocates to the nucleus and forms heterodimers with AhR nuclear translocator to form an active transcriptional activator. Binding to so-called xenobiotic responsive elements (XREs) regulates transcription of a variety of genes encoding among others protein enzymes involved in xenobiotic metabolism, such as CYP1A1, CYP1B1, CYP1A2, and GST. Recent studies suggest that TCDD binding to AhR also influences expression of factors that regulate cell growth, proliferation, differentiation, and apoptosis. Recent studies using gene microarray analysis of gene expression in human hepatoma cells exposed to TCDD demonstrated that a few hundred genes were differentially expressed, including *CYP1A1* and *KRAS*. AhR–ligand complexes also affect pathways involving SRC.

PAHs These are higher molecular weight polycyclic aromatic hydrocarbons such as DBA, and require metabolic activation to become carcinogens. Present in coal tar, PAHs can be formed by incomplete combustion of most organic materials, including tobacco leaf. Following activation of PAHs, diol–epoxide metabolites have been implicated in mediating the DNA-binding activity, and they are also highly carcinogenic. Disruption of genes that encode enzymes and factors that are involved in this activation route, such as CYP1B1, microsomal epoxide hydrolase (mEH), or AhR, renders mice resistant to PAH-induced cancers. PAHs can induce both somatic mutations in crucial genes (tumor "initiation") and subsequent growth of transformed cells (tumor "promotion") in mice – such agents are described as "complete carcinogens." Among genes known to be mutated by PAHs, *HRAS* is of particular interest.

Arylamines/amides These are aromatic amines, such as 4-aminobiphenyl (ABP) or the arylamide AAF, and induce bladder tumors in dogs, and occasionally in rodents. The primary amines are known to be human bladder carcinogens. In rodent tumor models, these compounds primarily induce tumors in the liver, lung, or mammary gland because of a high rate of *N*-hydroxylation in rodent hepatocytes, whereas humans and dogs primarily produce *N*-glucuronides in the liver, which are then transported to the kidney. These amines are then released in the acid urine and converted into genotoxic derivatives in bladder epithelial cells.

Tobacco smoke   It causes 30% of all cancer deaths in developed countries, particularly lung, but also upper respiratory tract, esophagus, bladder, pancreas, and probably others. Despite this, over one billion people continue to smoke, making tobacco smoke the single most lethal carcinogen. Carcinogens in tobacco smoke link nicotine addiction and lung cancer. People smoke because of addiction to the nicotine in tobacco smoke, but nicotine is not itself a prominent carcinogen. However, the cigarette poses numerous problems as a nicotine delivery device, because carcinogens are inhaled alongside nicotine in every "satisfying" puff. The cumulative exposure to carcinogens in a lifetime of smoking is substantial even though the actual carcinogen content of each cigarette is tiny. The risk may be determined by the number of "pack years" (number of cigarettes smoked per year and duration of habit) and debatably by the tar content. However, this latter perception may falsely reassure smokers of so-called light cigarettes that they may, as a result, experience only equally "light" heart attacks or cancers, though the risk of some chronic respiratory conditions (chronic airflow limitation, emphysema) may be reduced by lowered tar content. Polycyclic hydrocarbons, including 3-methylcholanthrene, benz(a)anthracene, and benzo(a)pyrene are believed to be the key carcinogens in tobacco smoke. Tobacco carcinogens in most cases are processed by enzymes to form reactive carcinogens (metabolic activation) that bind to DNA, forming covalent binding products called DNA adducts. Individuals vary in the way they metabolically activate tobacco carcinogens and in the ways that they process DNA adducts. These and other factors may determine cancer susceptibility. See also polymorphisms and cancer.

A tremendous amount of publicity has surrounded the ongoing debate about passive inhalation of second-hand smoke and cancer. The balance of evidence would seem to strongly support avoiding such "passive smoking" as a means of reducing the risk of cancer, although this risk is considerably lower than that experienced by active smokers. Importantly, here we must make a distinction between individuals who have chosen to smoke (no matter how much we may sympathize with them as victims of tobacco advertising and peer pressures) and those who have chosen not to smoke and yet may be put at risk by the actions of other individuals. Yet it is not just individuals who should be encouraged to act responsibly – tobacco advertising, tax revenues, and air pollution may be deemed more a matter for governments and hence all play a role in health.

Ionizing radiation
X-rays and nuclear radiation can damage DNA. Importantly, the risk of DNA damage from ionizing radiation depends on lifetime exposure. Marie Curie, who discovered radium and paved the way for radiotherapy and diagnostic X-rays, died of cancer as a result of radiation exposure in her research. Nuclear radiation is clearly linked to risk of various cancers and in the cases where radiation exposure has been concentrated to specific tissues (in diagnostic or more usually therapeutic situations, or inadvertently following nuclear disasters), the risk of cancer is highest in these tissues. Fortunately, however, relatively few individuals receive such exposure, with workers in power plants, laboratories, hospitals, etc. closely monitored for the cumulative exposure to ionizing radiation. Radon is a radioactive gas emitted from the earth in some locations, and is now actually one factor in determining property values and features in surveys undertaken by prospective purchasers in several countries. Much controversy has surrounded the role of low-frequency electric and magnetic fields and radio frequencies, particularly those associated with overhead power lines, mobile telephone masts, and mobile phones. No empirical evidence supports the view that mobile phones may increase the risk of cancers: though individual studies have suggested a possible increase in risk of some brain tumors, other studies have failed to show this.

Ultraviolet radiation is primarily part of the sun's radiation and is also generated artificially in sunlamps and tanning facilities. The most harmful of this type of radiation are the high-frequency, DNA-damaging UV B rays. These are the rays that cause 90% of all skin cancers.

Dietary factors   Diet may play a role in up to 20% of fatal cancers. A diet high in saturated fat and deficient in fiber may correlate with risk of cancers of the colon and rectum, but remains contentious. Obesity per se, however, is a major risk factor for both diabetes and some cancers, including breast and colon. Diets lacking in antioxidant vitamins correlate with risk of heart disease and cancer, yet are difficult to disassociate from the commonly associated problems of low socioeconomic class in general. Carcinogenic foods remain contentious. Salted, pickled, and smoked foods, such as pickles or smoked fish, and meats treated with nitrites have been implicated. Hot drinks have also been shown to be associated with higher esophageal cancer risk. Aflatoxin is contained within grain and peanuts that have become moldy with an *Aspergillus* species endemic in Africa, and is associated with hepatocellular carcinoma. Pickled fish and vegetables have been associated with risk of nasopharyngeal and esophageal cancers in China. Alcohol is estimated to contribute to about 3% of deaths from cancer. People who drink alcohol heavily have a higher risk of mouth, throat, esophagus, stomach, and liver cancer.

Medical treatments   Medical treatments may increase the risk of cancers. Radiation and chemotherapy may result in secondary cancers and intriguingly even in increased aggressiveness of cancers not fully cured by such treatments. Some immunosuppressive drugs used for chronic inflammatory diseases or after transplantation can cause lymphoma. Women exposed to excess estrogen are placed at increased risk for some gynecological cancers (e.g. breast, uterus), because estrogens can stimulate cell proliferation in these tissues. The level of exposure to estrogen is determined by several factors, including age at menarche (first period), pregnancy and age at pregnancy, age at menopause, weight, physical activity, and diet. Thus, a woman with an early age at menarche and late age at menopause would have a greater exposure to estrogen. Obviously, taking postmenopausal hormone replacement therapy (HRT) will increase overall exposure, whereas conditions associated with early menopause or low estrogen levels will reduce exposure.

Carcinogenesis and oxidative stress   The process of aging is believed to be in part due to ROS produced as by-products of normal metabolism. These ROS, such as superoxide and hydrogen peroxide, are the same mutagens produced by radiation, and can cause damage to DNA, proteins, and lipids. In rodent models, the DNA in each cell accumulates around 100,000 oxidative lesions per day, which are usually repaired by DNA repair enzymes. However, this system is not foolproof: a young rat has about one million oxidative lesions in the DNA of each cell, which increases to about 2 million in an old rat. Luckily, human cells receive about 10 times less damage than a rat cell, compatible with the higher incidence of cancer and shorter lifespan in a rat.

The degenerative diseases of aging, such as cancer and CHD, all share an oxidative origin. It has often been suggested that intake of dietary antioxidants, such as vitamins C and E, might reduce the incidence of degenerative conditions, but to date clinical trials using antioxidant supplementation have been disappointing, possibly in part due to the relatively short duration of such trials. Epidemiological studies show that the incidence of most types of cancer as well as CHD is much higher in populations where intake of antioxidant-containing vegetables and fruits is low. Smokers may be particularly at risk as their antioxidant pools are depleted (cigarette smoke is high in oxidants).

The three main potentially avoidable causes of cancer are smoking, poor diet (excess fat and calories; inadequate intake of fruits, vegetables, and fiber), and chronic infections leading to chronic inflammation (hepatitis B and C viruses, *Helicobacter pylori* infection, etc.). All of these may in some way result in increased ROS and oxidative stress.

Testing for carcinogens   For obvious reasons such tests are not conducted in humans but in animal models, primarily the rat. Animal cancer tests are generally conducted at the maximum tolerated dose (MTD) of carcinogens, which creates a certain amount of difficulties when interpreting results for human cancer. Formation of cancer in rat at MTD does not necessarily imply that a low dose of the chemicals tested will cause cancer in man. In fact, this issue is highlighted by the fact that in animal cancer tests around 50% of all naturally occurring chemicals are carcinogens for rats. In an illuminating example from Bruce Ames, there are over 1000 chemicals in a cup of coffee. Only 26 have been tested in animal cancer tests and more than half are rodent carcinogens. One explanation for such results may be that high-dose animal cancer tests are largely measuring increased rates of cell division in some cell types induced in response to nonspecific toxic cell killing by the high doses used.

However, low doses of rat carcinogens (those we are usually exposed to) may not have these same effects in man, because innate defense mechanisms such as DNA repair and oxidant defenses such as glutathione transferases may cope at these levels of exposure and may be induced to cope with increasing exposure.

It is important to put the results of animal tests in context. Again to quote Bruce Ames, "The effort to eliminate synthetic pesticides because of unsubstantiated fears about residues in food will make fruits and vegetables more expensive, decrease consumption, and thus increase cancer rates. The levels of synthetic pesticide residues are trivial in comparison to natural chemicals, and thus their potential for cancer causation is extremely low."

MUTATIONS

The term mutation can refer to any type of change in DNA. Exposure to genotoxic carcinogens can result in various differing forms of mutation. However, it is important to remember that mutations can also occur under the influence of unavoidable DNA damage such as that induced by oxidative stress, background or cosmic radiation, and also simple errors arising during DNA replication. The process by which proteins are made – translation – is based on the "reading" of mRNA that was produced via the process of transcription (see website). Any changes to the DNA that encodes a gene will lead to an alteration of the mRNA produced. In turn, the altered mRNA may lead to the production of a protein that no longer functions properly. Even changing a single nucleotide along the DNA of a gene may lead

KEY EXPERIMENT – THE AMES TEST

A widely used test for determining if a chemical is a mutagen is named after its developer, Bruce Ames. The test was developed in 1975 by Ames and his colleagues at the University of California at Berkeley. The Ames test is based on the assumption that substances that are mutagenic for a strain of bacterium (*Salmonella typhimurium*) might also be carcinogenic in man. Some substances that cause cancer in laboratory animals (e.g. dioxin) do not give a positive Ames test (and vice versa). However, the test is cheap and easily performed, and has proved extremely useful in rapidly screening substances in our environment for possible carcinogenicity. The bacterium used in the Ames test carries a gene mutation making it unable to synthesize the essential amino acid histidine from the culture medium. However, this mutation can be reversed, a back mutation, with the gene regaining its function. These revertants are able to grow on a medium lacking histidine. If cultures of these bacteria are exposed to mutagenic chemicals then increased numbers of bacteria might regain the ability to grow without histidine, forming visible colonies (above those spontaneously reverting in control dishes). Many chemicals are not mutagenic (or carcinogenic) by themselves, but become converted into mutagens (and carcinogens) as they are metabolized by the body. This is the reason the Ames test includes a mixture of liver enzymes. The Ames test yields a number, specifically the number of growing bacterial colonies, which is a measure of the mutagenic activity (potency) of a treatment chemical. This value is often expressed as the number of revertants per microgram of a pure chemical (mutagen) or per gram of food containing that mutagen.

to a completely nonfunctional protein. Mutations in one or more genes can therefore lead to disease. The genetic changes that lead to unregulated cell growth may be acquired in two different ways – they can be inherited or they can develop in somatic cells. The phenotype of cancer cells result from mutations in key regulatory genes. The cells become progressively more abnormal as more genes become damaged, particularly when the genes that regulate DNA repair and checkpoints become damaged (see Chapters 7 and 10). Most cancers are thought to arise from a single mutant precursor cell (in other words they are "clonal"), with further clones originating by accumulation of further mutations; those clones that gain a growth advantage will tend to take over the population (clonal expansion). One aspect of this view of cancer is that the transition from a normal, healthy cell to a cancer cell occurs via the stepwise accumulation of mutations in multiple different oncogenes, tumor-suppressor genes, or caretaker genes (Fig. 3.4). This model also accounts for the prevalence of cancer particularly in older individuals. Although the number, identity, and order in which mutations occur will vary enormously between individuals and different cancer types, attempts have been made to quantify the likely number of mutations required to generate a transformed human cell in culture. Intriguing studies from the laboratory of Robert Weinberg support the view of cancer formation initially voiced by Armitage and Doll in the 1950s by demonstrating that at least 4–6 interlocking mutations may be needed to transform cultured human primary cells. However, if this also equates to the requirements for formation of all cancers in the context of the intact organism remains controversial. Controversy remains because as few as two interlocking mutations may suffice to generate cancers in rodent models, a hypothesis that is difficult to test in man, as the very earliest stages of cancer are generally not available to the researcher. It is possible that intrinsic differences between human and mouse may be important and these might include the relatively longer telomeres found in mouse cells, for example. However, it is also possible that the larger number of mutations required to generate a transformed cell *in vitro*, than required for a cancer *in vivo*, might be explained by the absence of the potentially supportive effects of environmental factors extrinsic to the cancer cell, such as the tissue location, stroma, and vasculature, from transformation assays *in vitro*. Defining the minimal platform for oncogenesis *in vivo* is likely to be an important and interesting area for study over the next decade.

The actual number required notwithstanding, mutations can occur gradually in somatic cells over a number of years, leading to the development of a "sporadic" case of cancer. Alternatively, it is possible to inherit dysfunctional genes leading to the development of a familial form of a particular cancer. A model of cancer development in the colon is shown in Figure 3.5 – this scheme is based largely on observation and genetic analyses of tissues obtained from patient colon at various stages of disease.

Genetic alterations can be placed into two large categories. The first category comprises changes that alter only one or a few nucleotides along a DNA strand, termed point mutations; the second comprises various major rearrangements in genes or entire (or parts of)

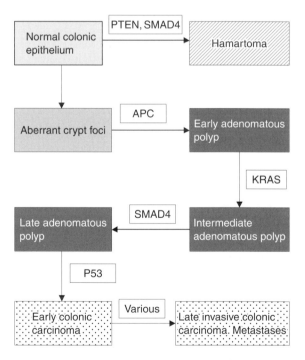

Fig. 3.5 Colorectal cancer as a model of multistage carcinogenesis – the adenoma–carcinoma sequence. In this cancer model sequential acquisition of mutations in various genes, shown in hatched boxes, is associated with initiation and progression of cancer. However, this does not mean that all colonic cancers arise in this sequence – activation of oncogene and inactivation of tumor-suppressor pathways is the key factor (alternative gene mutations could achieve the same net effect).

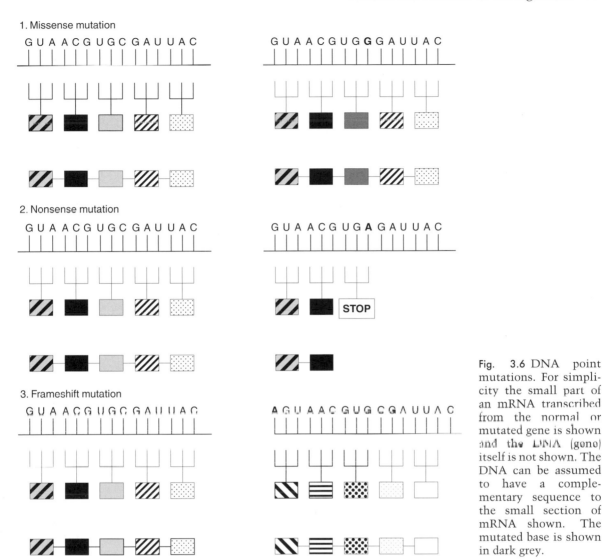

1. Missense mutation

G U A A C G U G C G A U U A C

G U A A C G U G G **G** A U U A C

2. Nonsense mutation

G U A A C G U G C G A U U A C

G U A A C G U G **A** G A U U A C

STOP

3. Frameshift mutation

G U A A C G U G C G A U U A C

A G U A A C G U G C G A U U A C

Fig. 3.6 DNA point mutations. For simplicity the small part of an mRNA transcribed from the normal or mutated gene is shown and the DNA (gene) itself is not shown. The DNA can be assumed to have a complementary sequence to the small section of mRNA shown. The mutated base is shown in dark grey.

chromosomes (this is discussed in more detail in Chapter 10).

The structure of proteins is encoded in the nucleotide sequence of DNA. A particular sequence of nucleotides gives rise to a particular sequence of amino acids, and that in turn determines the way that protein will function. Many changes in the nucleotide sequence will alter the amino acid sequence of the protein, and perhaps will change its function. There are several different forms of mutations, with different causes. Mutations can be classified under various headings and are briefly described below and in Figure 3.6.

Single-base substitutions

A single base is substituted by another, otherwise termed a point mutation; if a purine (adenosine or guanine) or a pyrimidine (cytosine or thymine) is replaced by the other member of the same class, the substitution is called a transition; whereas if a purine is replaced by a pyrimidine or vice versa, the substitution is called a transversion. This can happen if DNA polymerase mismatches two bases during replication. There are "proofreading" functions that correct most such errors, but about one in a million is not detected and becomes incorporated permanently in the DNA. Some chemicals can interact with DNA and greatly increase the chance of such mutations, particularly those that alter base structure. Such alterations can result in DNA polymerase inserting an inappropriate base on the opposite strand.

Missense mutations   The new nucleotide alters the codon thus producing an altered amino acid in the protein product; one of the three nucleotides making up the codon is replaced. This change in the nucleotide sequence results in an altered amino acid in the protein product after translation.

Nonsense mutations   The new nucleotide changes a codon that specified an amino acid to one of the STOP codons (TAA, TAG, or TGA), resulting in the premature arrest of RNA translation and a truncated protein product.

Insertions and deletions   Extra base pairs may be added (insertions) or removed (deletions) from a gene. These may have major consequences if only one or two base are involved, as translation of the gene is "frameshifted." Altering the reading frame, one nucleotide to the right or left, will result in multiple alterations in the amino acid sequence as multiple codons will now be altered and the mRNA translated in new groups of three nucleotides. Frameshifts may also create new STOP codons and thus generate nonsense mutations. Alterations in three nucleotides or multiples of three may be less serious because they preserve the reading frame.

Silent mutations   Most amino acids are encoded by several different codons. Thus TCT, TCG, TCA, and TCC all code for the amino acid serine. Any mutation altering the base at position 3 will have no effect on the resultant protein. Such mutations are silent and are detected only by gene sequencing.

Splice-site mutations   Intronic sequences are removed during the processing of pre-mRNA to mature mRNA under the influence of various proteins acting at the splice site. If a mutation alters one of these signals, then the intron is not removed and remains as part of the final RNA molecule. The translation of its sequence alters the sequence of the protein product.

More substantive DNA mutations

Chromosomal translocations   In some cases, diseases result not from changes in individual genes but from changes in the number or arrangement of chromosomes. Inherited abnormalities in chromosomes are unusual because they generally are incompatible with normal *in utero* development, but one notable exception is Down's syndrome, where individuals have three copies of chromosome 21, instead of the usual two. However, the most common scenario involving altered numbers of chromosomes is within cancer cells. In fact, this is one of the hallmark features of cancer (see also Chapter 10 – genomic instability) – examples are shown in Box 3.3.

Box 3.3   Chromosomal rearrangements

Karyotyping (examination of chromosome numbers) of tumor cells highlighted the presence of abnormal chromosomes, such as chromosomal translocations, in which a part of one chromosome becomes joined to another. For example, in almost all patients with chronic myelogenous leukemia (CML), the leukemic cells show the same chromosomal translocation event between chromosomes 9 and 22. The *BCR* gene on chromosome 22 becomes joined to the *c-abl* gene on chromosome 9, generating two abnormal chromosomes: a longer chromosome 9, and a small chromosome 22 called the "Philadelphia chromosome" after the city where the abnormality was first recorded. The resulting BCR/ABL fusion protein has the amino terminus of the BCR protein joined to the carboxyl terminus of the ABL tyrosine protein kinase. In consequence, the ABL tyrosine protein kinase becomes inappropriately active in hematopoietic cells, driving their excessive proliferation by activating multiple pathways normally regulated by extrinsic growth factors.

In Burkitt's lymphoma, the proto-oncogene c-MYC on chromosome 8 is translocated to one of the three chromosomes containing the genes that encode antibody molecules: immunoglobulin heavy chain locus (chromosome 14) or one of the light chain loci (chromosome 2 or 22). In every case, c-MYC now finds itself in a region of vigorous gene transcription, and it may simply be the overproduction of the c-MYC protein that turns the lymphocyte cancerous. The risk of translocations involving the heavy chain gene locus is probably especially high because breaks in its DNA occur naturally during the synthesis of antibodies.

Fusion of the promyelocytic leukemia (PML) protein to the retinoic acid receptor-alpha (RAR-α) generates the transforming protein of acute PMLs. PML appears to be involved in multiple functions, including apoptosis and transcriptional activation by RAR, whereas PML–RAR-α blocks these functions of PML. PML interacts with multiple corepressors (c-SKI, N-CoR, and mSin3A) and histone deacetylase 1, and this interaction is required for transcriptional repression mediated by the tumor-suppressor MAD.

The *BCL2* gene is located on chromosome 18 and encodes the antiapoptotic BCL-2 protein, which protects cells from cell death (see Chapter 8). The *BCL2* gene was discovered as the translocated locus in a B-cell leukemia (BCL). In B-cell cancers, the portion of chromosome 18 containing the *BCL2* locus has undergone a reciprocal translocation with the antibody heavy chain locus on chromosome 14. The heavy chain enhancer can thus regulate the *BCL2* gene, resulting in excessive BCL-2 protein in these t(14;18) cells.

Gene amplification Instead of making a single copy of a region of a chromosome, many copies are produced, resulting in the production of multiple copies of genes on that region of the chromosome. In extreme cases, these copies may form their own small pseudo-chromosomes called double-minute chromosomes. This is often observed in cancer cells and can result in deregulated expression of oncogenes. Examples of this include amplification of c-MYC in several tumors and amplification of NEU in breast cancers. Gene amplification in the *MDR* gene encoding the protein multiple drug resistance

(MDR) contributes to drug resistance in cancer. The MDR protein is a membrane pump capable of eliminating chemotherapeutic agents from the cancer cell, rendering them ineffective.

Inversions DNA fragments are released from a chromosome and then reinserted in the opposite orientation, which can either activate an oncogene or deactivate a tumor-suppressor gene.

Duplications/deletions Through replication errors, a gene or group of genes may be copied more than one time within a chromosome. Duplications are a doubling of a section of the genome. During meiosis, crossing over between sister chromatids that are out of alignment can produce one chromatid with a duplicated gene and the other having two genes with deletions. However, unlike gene amplification, genes are not replicated outside the chromosome and only single copies are produced. Similarly, genes may become lost. Gene duplication has occurred repeatedly during the evolution of eukaryotes. Genome analysis reveals many genes with similar sequences in a single organism. If two or more such paralogous genes are still similar in sequence and function, their existence provides redundancy. This may be a major reason why knocking out certain genes in yeast or mice may have little or no effect on phenotype.

Aneuploidy Entire chromosomes may be lost or replicated during cell division if the replicated chromosomes fail to separate into the daughter cells accurately.

INHERITED SUSCEPTIBILITY TO CANCER

In inherited diseases, the disease-causing mutation is present at birth and all cells in the body have a mutation. Although most cancers are believed to arise from mutations occurring in single somatic cells in the adult, several inherited cancer syndromes have been described. In these cases, all somatic cells carry a mutation, which on its own does not cause cancer, as additional somatic mutations are also needed. Importantly, inherited mutations usually predispose to the development of specific cancers, suggesting that certain cell types in certain tissues exhibit different sensitivity to cancer-susceptibility genes. Because the DNA is already damaged, cells have a head start when accumulating subsequent mutations and individuals with an inherited mutation consequently have a much higher risk of getting cancer compared with the general population, and develop cancer at a younger age. An example of this for breast cancer is shown in Figure 3.7.

Knowledge of a hereditary predisposition to cancer dates back at least to the sixteenth century, when physicians and families were clearly aware that certain overtly visible phenotypic features tended to cluster in families. Notably, the cutaneous findings in von Recklinghausen's neurofibromatosis, now known to be caused by germ-line inheritance of one defective copy of the *NF1* tumor-suppressor gene (the other allele being eventually lost in some somatic cells).

Over the last few decades remarkable progress has been made, leading to the identification of numerous genes that contribute to germ-line inheritance of cancer susceptibility, including

- The Li Fraumeni syndrome, due to inherited inactivation of one copy of *p53* in the germ line.
- Retinoblastoma, caused by inactivation of a copy of the retinoblastoma *RB* gene.
- Familial adenomatous polyposis, *APC*, and cancer of the colon.

- Cowden syndrome and *PTEN*.
- *VHL* in von Hippel–Lindau syndrome.
- Lynch syndrome, due to mutations in mismatch repair genes, *MSH2*, *MLH1*, and *MSH*.
- The hereditary breast–ovarian cancer syndrome with *BRCA1* and *BRCA2*.
- The familial atypical multiple mole melanoma in association with pancreatic cancer, due to the *CDKN2A (p16)* germ-line mutation.

A more comprehensive list is shown in Table 3.1.

Polymorphisms and cancer

Some of the inherited susceptibility to common cancers, which aggregate in families, may thus result from highly penetrant germ-line mutations in individual known genes, but this still leaves much unaccounted for. Although in a few cases such unexplained familial risk may reflect the presence of high-penetrance mutations in as yet unidentified genes, polygenic mechanisms are increasingly being appreciated as likely contributors to inherited susceptibility to cancer (and many other common chronic diseases). Most diseases that appear to be inherited are not caused by mutations in a single gene. Rather, combinations of alleles of various genes contribute to the phenotype (thus the prefix "poly",

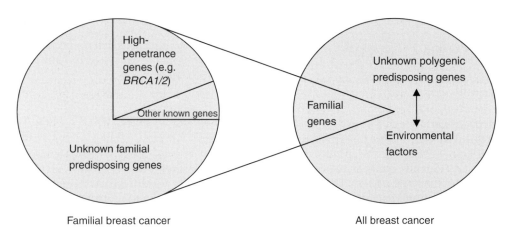

Fig. 3.7 Breast cancer susceptibility (adapted from Balmain *et al.* 2003). Familial breast cancer accounts for around 5–10% of all breast cancers. However, known genes, such as *BRCA1* and *BRCA2* account for only 20% of the familial risk, so that most genetic factors contributing to breast cancer are unknown. These unknown genetic variants (probably at multiple different loci) interact with environmental factors in the pathoetiology of around 80% or more of breast cancers.

meaning "many"). Each of the contributing alleles on their own is not disease-causing, so the method of identifying and studying the genetics of these diseases is quite different from the study of mutations in single genes. Not surprisingly, therefore, and as is the case in many chronic diseases, the detection of low-penetrance susceptibility genes has proved difficult – but technical advances and the availability of the reference human genome will hopefully accelerate the progress in the coming years, with the major challenge probably being the availability of appropriately large well-characterized patient cohorts. Some cancer-relevant polymorphic alleles are shown in Table 3.2.

Polymorphisms are defined as the regular and simultaneous occurrence in a single interbreeding population of two or more alleles of a gene, where the frequency of the rarer alleles is greater than can be explained by recurrent mutation alone (typically greater than 1%). Thus, these gene variants are inherited and may be regarded as "normal" variants of genes occurring in the population. There may be as few as two or three such commonly occurring variants for individual genes in a typical population and in general none of these are strong predictors of pathology or disease. However, they may increase the risk of disease developing if both alleles are of the same polymorphic type (or occur in a particular combination), under particular environmental conditions, or if combined with several other specific polymorphisms in other genes.

The most common polymorphisms studied currently are SNPs. The difficulties in determining the role of these in cancer susceptibility should not be understated. Even in the case of smoking-related lung cancers polymorphisms could influence carcinogen metabolism, DNA repair, drug responses, or even the likelihood of smoking in the first place! Separating these factors will require vast patient cohorts who have been rigorously characterized, and importantly by the use of standard protocols across all centers involved in these studies. This notwithstanding, a considerable body of evidence suggests that gene polymorphisms may contribute to carcinogen metabolism (e.g. of tobacco smoke).

Various polymorphisms have been associated with cancer susceptibility (see Table 3.2).

Genetic polymorphisms in enzymes involved either in detoxification of procarcinogens or DNA repair may affect cancer risk. Human cytochrome P450 (CYP) enzymes play a key role in the metabolism of drugs and environmental chemicals. Several CYP enzymes metabolically activate procarcinogens to genotoxic intermediates, and not surprisingly, associations have been identified among SNP polymorphisms, gene duplications and deletions, CYP enzyme activity, and the risk of several forms of cancer. Strong support for an association between *CYP* polymorphisms and lung, head and neck, and liver cancer now exists. Polymorphism in the *TPMT* gene can result in slower than usual metabolism of the chemotherapeutic agent mercaptopurine, resulting in greatly increased toxicity to the drug. The lack of glutathione S-transferase M1 (GSTM1-null genotype) is associated with increased sensitivity to genotoxicity of tobacco smoke, and GSTM1-null smokers also show an increased frequency of chromosomal aberrations.

GENE–ENVIRONMENT INTERACTIONS

The actions of environmental factors are not independent of cancer genes. Sunlight may induce aberrant tumor-suppressor genes in skin cells, and cigarette smoke may make lung cells more vulnerable to carcinogenic compounds in smoke. These factors probably act directly or indirectly on the genes that are already known to be involved in cancer. Individual genetic differences also affect the susceptibility of an individual to the carcinogenic effects of environmental agents. About 10% of the population has an alteration in a gene causing them to produce excessive amounts of an enzyme that breaks down hydrocarbons present in smoke and various air pollutants. The excess enzyme reacts with these chemicals, turning them into carcinogens. These individuals are about 25 times more likely to develop cancer from hydrocarbons in the air than others are.

How hereditary factors may contribute to the development of sporadic cancer is still unclear. A recent landmark study involving 44,788 pairs of twins listed in the Swedish, Danish, and Finnish twin registries provides estimates of the overall contribution of inherited genes to the development of various cancers (Leichtenstein *et al.* 2000). Although, 28 cancer types were looked at, an increased risk was found among the twins of affected persons only for stomach, colorectal, lung, breast, and prostate cancer. Statistically significant effects of heritable factors were observed for prostate cancer (42%), colorectal cancer (35%), and breast cancer (27%). These results suggested overall that inherited genetic factors make a minor contribution to susceptibility to most types of cancer, and therefore by implication that the environment takes the lead role in causing sporadic cancer. However, this notwithstanding, there were relatively large effects of heritability in some common cancers, though the genetic bases underlying these observations remain to be defined.

MUTATIONS AND TREATMENT

Various tyrosine kinase inhibitors, including imatinib and gefitinib, can cause tumor regression in certain patients, but it has proved difficult to ascertain which patients would be responsive to the drugs. Imatinib (gleevec) is most effective against certain leukemias and gastrointestinal stromal tumors. Importantly, the presence of specific mutations (such as the BCR/ABL fusion protein) can identify patients who are likely to respond. More recently, it has been shown that the presence of a specific mutation in a common adult cancer, non-small-cell lung cancer (NSCLC), may identify patients most likely to respond to gefitinib. Mutation in the epidermal growth factor receptor (*EGFR*) gene is involved in NSCLC, and the presence of this particular mutation can identify patients most likely to respond successfully to gefitinib. However, the situation is more complex than this would imply, because marked differences in response were observed between Japanese and American patients. In fact, Japanese women with adenocarcinoma showed the highest percentage of *EGFR* mutations and also showed the best clinical response to gefitinib. This suggests that genetic variation in different ethnic, cultural, and geographic groups might also contribute to sensitivity of NSCLC to particular drugs. Thus, combinations of

inherited polygenic factors and single somatic mutations may all contribute to determine treatment responsiveness.

MICROORGANISMS AND CANCER

In 1911, Peyton Rous discovered that sarcomas in chickens could be transmitted between animals using a "cell-free" filtrate. The active agent was subsequently identified as a virus now known as the Rous sarcoma virus; and a single gene, *src*, was identified as responsible for these cancers. In 1981, Michael Bishop and Harold Varmus discovered that normal cells contain a gene homologous to the viral *src* gene, termed *c-SRC*. These were key steps in defining the genetic basis of cancer (Chapter 6).

Tumorigenesis can be driven by the actions of specific tumor viruses. Viral causation of cancer is well documented in the lab, and is important in some (but probably not most) human cancers. Tumor viruses comprise two distinct types, viruses with DNA genomes (e.g. papilloma and adenoviruses) and those with RNA genomes (retroviruses).

DNA viruses

Cellular transformation by DNA tumor viruses, in most cases, has been shown to be the result of protein–protein interaction. Proteins encoded by the DNA tumor viruses, the tumor antigens, or T antigens, can interact with cellular proteins. This interaction effectively sequesters the cellular proteins away from their normal functional locations within the cell. The predominant types of proteins that are sequestered by viral T antigens have been shown to be of the tumor-suppressor type (see Chapter 7). It is the loss of their normal suppressor functions that results in cellular transformation.

Hepadnaviruses Hepatitis B virus (HBV) is associated with some cancers in man. In areas of Asia and Africa with high rates of HBV infection there are also high rates of hepatocellular carcinoma, though no HBV genes have been clearly shown to be transforming. Chronic irritation or cirrhosis of the liver may in fact be the major explanation. In animal models, such as the woodchuck, virtually 100% of the infected animals with chronic hepatitis develop liver cancer.

Papillomavirus

Human papilloma viruses (HPV) cause common warts, and have been strongly implicated in a variety of genital cancers, particularly cervical carcinoma. Cervical carcinoma has a classic epidemiology pattern suggestive of an infectious sexually transmitted disease; it is rare in the sexually inactive and conversely is common in women with multiple sexual partners. The HPV 16 and 18 strains are most frequently identified in cervical carcinoma and also display the greatest capacity to transform cells *in vitro*. The HPV proteins E7 and E6 can bind and inactivate the *p53* and *RB* tumor suppressors. In fact, binding of cell tumor-suppressor genes by viral proteins is a common property of many DNA tumor viruses.

Herpes viruses

Epstein–Barr virus (EBV), the etiological agent of infectious mononucleosis, has been associated with the genesis of Burkitt's lymphoma and nasopharyngeal carcinoma (viral DNA and sometimes virus are present in these cancer cells). Furthermore, EBV has been detected in the Reed–Sternberg cells in a high percentage of Hodgkin's lymphomas. EBV transformation is multigenic and at least five different viral genes appear to have some involvement. Recent studies have suggested that, like other DNA viruses one of the EBV gene products can bind with tumor-suppressor proteins. One of the first viruses shown to be involved in cancer was an avian herpes virus known as Marek's disease virus. In chicken this virus causes a fatal T-cell lymphoma in virtually all infected birds.

Adenoviruses

These viruses have never been associated with cancer in humans, but they transform cultured cells and cause cancer in animals. At least two gene products have been implicated in transformation studies (E1A and E1B). These two genes alone are sufficient to transform cells in culture. Studies have shown that the E1A protein binds *RB* and that the E1B protein binds *p53*.

Poxviruses

Disease of rabbits causing cancerous myxomas.

Polyomaviruses

Oncogenic in laboratory animals and can transform human cells in culture. The human JC virus has been detected in numerous brain tumors of various pathologies, but no clear correlation has been established. Polyoma virus and SV40 are linked to a variety of animal tumors. The latter is known to inactivate both the *p53* and RB tumor suppressors. So, not surprisingly, it can strongly predispose to cancer formation; in fact, many early animal models of cancer were founded on the platform of first overexpressing the SV40 large T antigen responsible for this activity.

Retroviruses

RNA tumor viruses are common in chickens, mice, and cats but are rare in humans. Currently, the only known human retroviruses are the human T-cell leukemia viruses (HTLVs) and the related retrovirus, human immunodeficiency virus (HIV).

Tumorigenic retroviruses (oncoviruses) are grouped into three categories based on their mechanism of oncogenicity: (i) transducing retroviruses; (ii) *cis*-activating retroviruses; and (iii) *trans*-activating retroviruses.

Two features of the retrovirus life cycle permit the acquisition and activation of oncogenes. Namely, integration into the cell chromosome provides the opportunity to hijack cellular genes and allows genetic permanence once an oncogene has been transduced; and furthermore, most retroviruses do not kill the cells they infect so that genetic alterations are transmitted to daughter cells. When a retrovirus infects a cell, its RNA genome is converted into DNA by the viral encoded RNA-dependent DNA polymerase (reverse transcriptase). The DNA can then integrate stably into the genome of the host cell and be copied as the host genome is duplicated during normal cell division. Contained within the sequences at the ends of the retroviral genome are powerful transcriptional promoter sequences termed long terminal repeats (LTRs). The LTRs promote the transcription of the viral DNA, leading to the production of new virus particles. The

Table 3.3   Examples of viral oncogenes

Oncogene	Nature of proto-oncogene	Virus-induced tumor
src	Tyrosine kinase	Chicken sarcoma
erb-b	EGF-receptor/tyrosine kinase	Chicken fibrosarcoma
abl	Tyrosine kinase	Mouse leukemia
myc	Transcription factor	Chicken myelocytoma
fos	Transcription factor	Mouse osteosarcoma
jun	AP-1 transcription factor	Chicken fibrosarcoma
bcl-2	Antiapoptotic factor	Lymphoma
h-ras	GTP protein	Rat sarcoma
sis	Platelet-derived growth factor	Monkey sarcoma

process of integration may result in rearrangement of the viral genome and the consequent incorporation of a portion of the host genome into the viral genome, termed transduction.

Transducing retroviruses   Occasionally, transduction results in the virus acquiring a host gene normally involved in cellular growth control. The transduced gene is always altered in some way by either point mutation or deletion of protein sequences. In many instances the gene is fused to a viral gene and during transduction all introns are lost from the cellular gene. These alterations serve to activate the transduced gene and the end result of this process is unrestricted cellular proliferation. The transduced genes are termed oncogenes (see Chapter 6). The expression of the transduced gene is driven by the virus promoter/enhancer region (LTR) and its expression is no longer under cell control. Transducing retroviruses cause tumors at high efficiency (100% of animals) and with short latency periods (days). In addition, they readily transform cells in culture, with almost 100% of the cells being transformed. In very rare instances the virus transduces the cellular oncogene without the concomitant deletion of viral coding sequences. In these unusual cases, the virus is replication-competent. Numerous oncogenes have been discovered in the genomes of transforming retroviruses. Typically, viral oncogenes are mutated, have lost regulatory sequences, or are amplified, and are capable of causing cancer by themselves. Aside from the key role that retroviruses

carrying oncogenes have played in the discovery of cancer-causing genes, they probably have no role in the great majority of human cancers (Chapter 6).

Cis-activating retroviruses   The second mechanism for retroviral transformation of cells is through the transcription-promoting action of the LTRs. Retroviral genomes integrate randomly into the host genome, which can occasionally result in the viral LTRs being in proximity to a gene that encodes a growth-regulating protein. Thus, if the growth-regulatory gene is now overexpressed it can result in cellular transformation. This is termed retroviral integration-induced transformation and is exemplified by the induction of certain cancers by HIV infection. These viruses activate a cellular proto-oncogene by integrating adjacent to it and increasing or altering its expression (promoter or enhancer insertion model). Tumors formed by these viruses take longer to occur and not all animals form tumors. These viruses are replication-competent.

Trans-acting retroviruses   These viruses may upregulate cell oncogenes through the action of a viral transactivator protein. The latency period is long (years) and the efficiency of tumor induction is very low (1%). They are replication-competent and do not transform cells in culture. HTLV-1, Human T-cell leukemia virus, was first isolated by Gallo and coworkers in 1980 and is linked to adult T-cell leukemia in Japan. All cases of ATL show evidence of HTLV-1 infection, but the lifetime risk of developing ATL is only 1% for infected individuals. The viral *tax* gene can activate transcription of important T-cell growth factors (e.g. IL-2), suggesting that an autocrine mechanism of transformation may be operative, but verification of this theory awaits further studies.

QUESTIONS REMAINING

1 To define which individual genes/proteins involved in tumorigenesis are "mission critical" for cancer development and moreover which are most critical for maintaining the cancer, as these will be the optimum targets for new drug development.
2 To understand how combinations of various low-penetrance genetic variants (polymorphisms) interact with, protect from, or predispose to different types of cancer, determine prognosis once cancer develops, and dictate the likely success of particular therapies.
3 To improve the predictive value of traditional toxicity studies of carcinogens, involving relatively short-term exposure to high doses of single compounds, to more accurately define the risk of exposure to lower levels of carcinogens and combinations of carcinogens over more protracted periods in man

CONCLUSIONS AND FUTURE DIRECTIONS

Over the last decades environmental factors have been accepted as predominant causes of human cancer, largely on the strength of epidemiological studies showing that cancer incidence varied widely depending on geographical factors, and, moreover, that immigrants tended to acquire the cancer risk of their new adopted location (Doll & Peto 1983; Peto 2001). Inherited factors have been largely explored in rare familial cancer syndromes, and genes identified from the study of such families, notably *RB* (Friend *et al.* 1986) and *p53* (Malkin *et al.* 1990), have made major contributions to the understanding of cancer and to biology in general. Yet such high-penetrance genes contribute little to inherited susceptibility to the common cancers, though they are frequently the subject of cancer-causing somatic mutations. However, advances

in molecular biology over the last two decades have resulted in increasing experimental support for the role of low-penetrance genetic variants (polymorphisms) in determining the risk of sporadic cancer development, and have provided a molecular explanation for gene–environment interactions in the pathogenesis of cancer. Notably, the demonstration of genetic linkages in breast cancer families (Hall *et al.* 1990) strongly supports the role of inherited genetic factors in a common cancer. Together with the study of inherited cancer syndromes, the study of viruses in tumorigenesis has also had a disproportionately large impact on cancer biology. Thus, although few human cancers are caused by viruses, the study of virus-induced cancers in animal models and of those few human examples has contributed substantially to the discovery of several important oncogenes and also tumor suppressors.

Major concerns still exist about how much weight can be placed on animal models of carcinogen-exposure-induced cancer. In general, the levels in these studies greatly exceed those to which most humans will ever be exposed and moreover, such studies almost invariably address single carcinogens, whereas humans will be exposed often to complex mixtures. Even tobacco smoke is believed to contain upwards of 50 different carcinogenic compounds – the effects or potential synergistic effects of such mixtures are poorly understood. Improved understanding of the mechanisms underlying the genotoxic effects of carcinogens and how such lesions are recognized and repaired is also often incomplete. This information would increase our ability to determine the minimal levels of carcinogen exposure needed to cause cancer, and thereby to set safe thresholds for these factors in the environment.

Interindividual variability in susceptibility to carcinogens and other cancer-contributing factors is also an important issue. With this in mind, we must continue to explore the interactions between susceptibility and resistance genes targeted by carcinogens or influencing the consequences of exposure to carcinogens. The ability to incorporate large numbers of genes and lifestyle factors into the accurate calculation of risk will result from our increasing ability to understand the clinical implications of human DNA sequence variability. At present, this goal is still some way ahead for cancer, though one should take heart from the spectacular progress made in the area of CHD. Advances in cancer biology, bioinformatics, systems biology, and mathematics may finally allow us to exploit the results of the Human Genome Project in order to improve the accuracy of cancer risk evaluation and benefit the patient by the implementation of tailored chemopreventative, screening, and lifestyle strategies.

QUESTIONS

1 The following processes contribute to the majority of cancers
   a. Inherited mutations in a single major cancer-causing gene.
   b. Polymorphic alleles.
   c. Viral infection.
   d. Inactivation of a tumor-suppressor gene.
   e. Exposure to mutagenic factors.
2 The following statements are true of breast cancer
   a. Most breast cancers arise through inherited abnormalities in the *BRCA1* gene.
   b. Most breast cancers are not familial.
   c. Breast cancers only affect females.
   d. Most breast cancers at diagnosis are genetically unstable.
   e. Most familial breast cancers arise through mutations in the *ATM* gene.
3 Cancer incidence is influenced by
   a. Geographical factors.
   b. Genetic predisposition.
   c. Diet.
   d. Gender.
   e. Age.
4 The following may significantly influence the cancer risk associated with cigarette smoking
   a. Polymorphisms in genes involved in metabolizing components of tobacco smoke.
   b. Numbers of cigarettes smoked.
   c. Age.
   d. Gender.
   e. Time following smoking cessation.
5 Viruses may contribute to tumorigenesis in the following ways
   a. By introducing oncogenes.
   b. By introducing functional tumor-suppressor genes.
   c. By inactivating tumor-suppressor genes.
   d. By adversely affecting aspects of immune function.
   e. By promoting apoptosis in the nascent cancer cell.

BIBLIOGRAPHY

Mutltistage oncogenesis and gene-environment interactions

Armitage, P. and Doll, R. (1954). The age distribution of cancer and a multi-stage theory of carcinogenesis. *British Journal of Cancer*, **8**: 1–12.

Lichtenstein, P., Holm, N.V., Verkasalo, P.K., *et al*. (2000). Environmental and heritable factors in the causation of cancer – analyses of cohorts of twins from Sweden, Denmark, and Finland. *New England Journal of Medicine*, **343**(2): 78–85.

Peto, J. (2001). Cancer epidemiology in the last century and the next decade. *Nature*, **411**(6835): 390–5.

Vineis, P. (2004). Individual susceptibility to carcinogens. *Oncogene*, **23**: 6477–83.

Carcinogens and the Ames test

Ames, B.N., Durston, W.E., Yamasaki, E., and Lee, F.D. (1973). Carcinogens are mutagens: a simple test system combining liver homogenates for activation and bacteria for detection. *Proceedings of the National Academy of Sciences of USA*, **70**: 2281–5.

Luch, A. (2005). Nature and nurture – lessons from chemical carcinogenesis. *Nature Reviews: Cancer*, **5**(2): 113–25.

McCann, J., Choi, E., Yamasaki, E., and Ames, B.N. (1975). Detection of carcinogens as mutagens in the Salmonella/microsome test: assay of 300 chemicals. *Proceedings of the National Academy of Sciences of USA*, **72**: 5135–9.

Genetics (see Chapter 7 references for tumor suppressors and inherited cancers)

Balmain, A., Gray, J., and Ponder, B. (2003). The genetics and genomics of cancer. *Nature Genetics*, **33**(Suppl): 238–44.

Hall, J.M., Lee, M.K., Newman, B., *et al.* (1990). Linkage of early-onset familial breast cancer to chromosome 17q21. *Science*, **250**(4988): 1684–9.

Houlston, R.S. and Peto, J. (2004). The search for low-penetrance cancer susceptibility alleles. *Oncogene*, **23**(38): 6471–6.

Park, J.Y. *et al.* (2002). Polymorphisms of the DNA repair gene xeroderma pigmentosum group A and risk of primary lung cancer. *Cancer Epidemiology, Biomarkers and Prevention*, **11**: 993–7.

Viruses and cancer (see also Chapters 6 and 7)

Baltimore, D. (1970). RNA-dependent DNA polymerase in virions of RNA tumor viruses. *Nature*, **226**: 1209–11.

Bishop, J.M. (1995). Cancer: the rise of the genetic paradigm. *Genes and Development*, **9**: 1309–15.

Lane, D.P. and Crawford, L.V. (1979). T antigen is bound to a host protein in SV40-transformed cells. *Nature*, **278**: 261–3.

Varmus, H.E. (1990). Retroviruses and oncogenes. I. *Bioscience Reports*, **10**: 413–30.

Whyte, P., Buchkovich, K.J., Horowitz, J.M., *et al.* (1988). Association between an oncogene and an anti-oncogene: the adenovirus E1A proteins bind to the retinoblastoma gene product. *Nature*, **334**: 124–9.

4

DNA Replication and the Cell Cycle

Stella Pelengaris and Mike Khan

It has not escaped our notice that the specific pairing we have postulated immediately suggests a copying mechanism for the genetic material.
                                                                    Francis Crick and James Watson

---

KEY POINTS

- The cell cycle is a highly regulated process during which cells replicate themselves to produce two genetically identical daughter cells as a result of faithful DNA duplication, and is conserved across species.
- The cell cycle is comprised of four phases, which can vary in length depending on cell type and the signals the cell receives: a gap period G_1 (growth phase); S (DNA synthesis phase); a gap period G_2 (growth phase); M (nuclear and cell division). Cells can enter a "resting" non-proliferative state, which can be very prolonged or even permanent – termed G_0, which equates with terminal differentiation.
- Each phase of the cell cycle is driven by sequential activation of cyclin-dependent kinases (CDKs) in association with regulatory subunits – the cyclins. CDKs are also negatively regulated by phosphorylation events and in particular by a key group of proteins – the cyclin-dependent kinase inhibitors (CKIs), including $p16^{INK4A}$, $p21^{CIP1}$, and $p27^{KIP1}$.
- The G_1/S transition is a key point in the cell cycle as in general mitogenic stimuli are no longer required once the cycle has passed this point. Not surprisingly, all cancers find a way to circumvent the normal regulation of this stage. The RB tumor suppressor is a key negative regulator of G_1/S transition and normally prevents synthesis of S-phase genes required for DNA synthesis and cell cycle progression by sequestering members of the E2F transcription factor family, required for expression of S-phase genes.
- Mitogenic stimuli activate signaling pathways that allow progression through G_1 and then G_1/S transition by sequentially activating complexes of CDK4 (and CDK6) in association with cyclin D, and then of CDK2 in association with cyclin E. Together these complexes result in hyperphosphorylation of the tumor suppressor RB, thus releasing E2F, which is now available for activating S-phase genes. Conversely, various negative signals prevent G_1/S transition by preventing RB phosphorylation. In particular the $p16^{INK4A}$ CKI, which interferes with activity of CDK4 (and CDK6), and $p27^{KIP1}$, which interferes with CDK2, can both prevent RB phosphorylation and S-phase entry.
- In fact, all key stages in cell cycle progression are subject to stringent "checkpoints" that determine if various steps in the cell cycle, such as DNA replication, duplication, and partitioning of chromosomes have been successfully completed. If a problem is detected then the cell cycle is arrested until it is corrected, or failing that the cell may die or irreversibly exit from cell cycling (senescence). These checkpoints are monitored by, amongst others, important tumor suppressors, including RB, p53, and various CKIs, and two damage-sensing pathways – the ATM and the ATR-CHK1 signaling pathways.
- An important checkpoint involves the tumor suppressor p53, which ensures that the S phase is blocked after DNA damage. This key checkpoint prevents replication of mutated DNA until the damage has been repaired or the cell is eliminated by its own death (apoptosis).

- Importantly, p53 is also activated in response to "oncogenic stress" (cancer cell cycles driven by deregulated expression of oncogenes such as c-MYC) in this case via the ARF (p14^{ARF} in man and p19^{ARF} in mouse) protein.
- Failure of such checkpoints can prevent cell cycle arrest or cell death (apoptosis) normally induced by deregulated oncogenes or damaged DNA. This results in aberrant proliferation and in the propagation of mutated DNA to daughter cells that are likely to pose a cancer risk.
- Mutations in genes that promote the cell cycle (oncogenes) or checkpoints that arrest the cycle or promote apoptosis (tumor suppressors) are ubiquitous and central to cancer development.

INTRODUCTION

The highly regulated processes by which cells make exact replicas of themselves are central to life and development of complex organisms. In multicellular organisms, many rounds of cell division are required, both during the development of a new individual and also in the adult, where cell division serves to maintain tissue homeostasis – new cells to replace damaged or dying cells or to adapt tissue mass to varying demands. An adult human being consists of around 100,000 billion (10^{14}) cells, all originating from a single cell, the fertilized egg cell. Furthermore, in an adult human, millions of new cells are made every day to replace dying ones and to adapt tissue mass to new demands. In fact, it has been estimated that humans will have undergone around 10^{16} or more cell divisions in a lifetime. Cells replicate themselves by duplicating their contents and then dividing in two – a process known as the mitotic cell cycle (Fig. 4.1). It is crucial that during each cell cycle, two genetically identical daughter cells are produced. Thus, the DNA must be faithfully copied and then the replicated chromosomes segregated into two separate cells. Apart from DNA replication, all the cytoplasmic organelles are duplicated during each cell cycle, and most cells will double their mass. Not surprisingly, there are key regulatory mechanisms in place to ensure that mistakes are not made during cell replication, which might otherwise lead to daughter cells receiving abnormal DNA, or either too many or too few chromosomes (see Chapter 10). Cells containing such damaged DNA may pose a neoplastic risk to the host, for example if a mutation allows cells to proliferate excessively (see Chapters 6 and 7) or to avoid cell death

(apoptosis) – Chapter 8. The degree of precision with which DNA synthesis and mitosis are carried out is remarkable, but occasional errors are unavoidable – potentially resulting in mutations in the genome or aneuploidy (lack or excess of chromosomes). Such mutations will in general be harmless if involving only a single cell, as it is hard to imagine how one out of 10^{14} cells could threaten the organism. However, if a cell with a mutation replicates and so do the progeny, thus propagating the mutation, then that situation becomes very different and the foundations have been laid for the development of cancer. For this reason, cells have developed very complex and usually effective means of arresting replication and initiating DNA repair when mutations are detected. However, it should be remembered that a certain minimal capacity for genetic information to change is an essential requirement for evolution.

THE CELL CYCLE – OVERVIEW

Many students have found the cell cycle conceptually difficult and for this reason numerous helpful analogies have been employed to illustrate some of the cardinal points of cell cycle regulation (Box 4.1). Confusion is often generated by the use of different terminology when describing events during cell replication. This becomes less confusing if one remembers that historically different tools have been available to experimenters studying cell replication and current terminologies still reflect this. During the late 19th century, using microscopy, Walther Flemming observed structures he called threads (now called chromosomes) and how these threads change

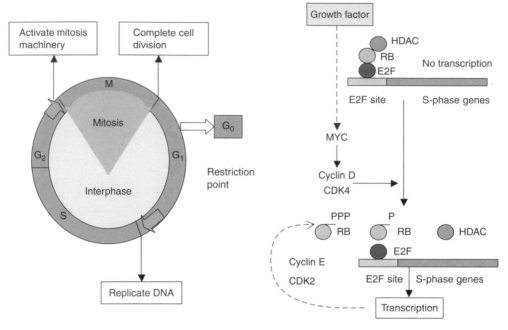

The cell cycle consists of the following phases:

- G_1 = gap phase for growth and preparation of the chromosomes for replication
- S = synthesis of DNA (and centrosomes)
- G_2 = gap phase for growth and preparation for mitosis
- M = mitosis (nuclear division) and cytokinesis (cell division).

After going through mitosis, a normal cell that re-enters G_1 can either start cycling again or exit the cell cycle and enter a resting state. It may stay in this quiescent state termed G_0 for days, weeks, or years, until the balance of growth-stimulatory and inhibitory signals from outside the cell indicates that it is time to divide again; only then will the cell become active again, re-entering the cell cycle in G_1.

The point at which cells commit to replicate their DNA and enter the cell division cycle is controlled by the RB protein, the product of the tumor suppressor gene *RB*. RB acts as an "off switch" for entry into S-phase by binding the transcription factor E2F required for expression of S-phase genes. After appropriate growth factor stimulation, the inhibitory effects of RB are removed by phosphorylation, which allows E2F to activate S-phase genes, during early G_1, required for cell cycle progression. One such S-phase gene is cyclin E, which together with CDK2 then acts to further phosphorylate RB and amplify the process, in part by fully activating transcription factors normally held inactive by RB.

Fig. 4.1 The phases of the mammalian cell cycle.

during cell replication, a process he called mitosis. For many years the cell cycle was studied primarily by observing chromosomal behavior by light microscopy and other techniques (Box 4.2 and Fig. 4.2), which has generated a very specific terminology still employed in cytogenetics today. More recently the unraveling of DNA structure and replication (Box 4.3) and the subsequent revolution in molecular biology has enabled us to describe the molecular machinery

that underlies and regulates these observed behaviors and identify the key role of changes in various genes/proteins in regulating the cell cycle.

Not surprisingly, key events observed microscopically during mitosis are now known to equate with specific molecular events during cell cycle transitions, each regulated by complex signaling networks. The cell cycle is normally very strictly regulated, with cell division only taking place under appropriate circumstances

---

Box 4.1   An analogy for the cell cycle – *"It will all come out in the wash!"*

It is often helpful when thinking about important but complex processes such as the cell cycle to have a more easily appreciated analogy based on common experience to hand, which proves reasonably representative of the process being considered. Such helpful analogies have been widely used and variously the cell cycle has been likened to an automobile or an automatic washing machine cycle (or a relay race). In all these cases the common theme is that the various operations can be switched on or off and then follow a set pattern that directs the various processes occurring in a predetermined sequence hopefully to a successful outcome (driving the car out of the garage then along the road; clothes washing then drying). Thus the processes, be it driving or washing, may be taken to represent the active processes of the cell cycle, namely DNA replication followed by chromosome segregation. Moreover, because all these processes involve a complicated machinery that is prone to intrinsic errors as well as adverse environmental influences (low engine oil, improperly closed doors, locked brakes, pets and small children – DNA copying errors, radiation, improper chromosome separation) occurring at various stages, there will be disastrous consequences (engine damage, a crash, flooding – replication of damaged DNA, aneuploidy, cancer) unless a means of stopping the process exists. Therefore, all these processes have inbuilt "failsafe" mechanisms that aim to stop the process until the process is resolved (lights flash, engine won't start – cell cycle checkpoints). Equally, as driving a car inappropriately or with dangerous faults may damage other neighboring cars and drivers, various external controls also operate (codes of behavior, various traffic signals, traffic jams, and ultimately a policing system – body plan, homeostasis, local negative and positive growth signals, contact inhibition by neighboring cells, circulating regulators of growth) all designed to protect society (the organism) from selfish drivers or dangerous cars. It should also be remembered that this society usually has a very strong honor code – with very poor drivers (cells) encouraged to commit suicide (apoptosis) rather than risking damaging others. This is crucial because the

offspring will inevitably inherit the same poor driving skills as the parent, thus amplifying the problem until society is overrun with bad drivers. At this point it would seem in "bad taste" to extend the driving analogy further downstream from suicide – as this society also embraces cannibalism.

Box 4.2   Chromosomes

In eukaryotes, chromosomes consist of a single molecule of DNA together with associated positively charged proteins, the histones, tightly bound to negatively charged phosphate groups in the DNA. The molecule of DNA in a single human chromosome ranges in size from 50 to 250×10^6 nucleotide pairs. Stretched end-to-end, the DNA in a single human diploid cell would extend over 2 meters. In the intact chromosome, however, this molecule is packed into a much more compact structure, so that during mitosis a typical chromosome is condensed into a structure about 5 μm long (a 10,000-fold reduction in length).

Five main types of histone are known, all of which are rich in lysine and arginine residues and associate with DNA. Before mitosis, each chromosome is duplicated (during S phase). Then as mitosis begins, the duplicated chromosomes, termed dyads, condense into short (~5μm) structures. The duplicates are held together at the centromere, which in man comprises over 3 million base pairs of mostly tandem arrays of repeated short sequences of DNA. While they are still attached, it is common to call the duplicated chromosomes sister chromatids. The kinetochore is a complex of proteins (11 in budding yeast) that forms at the centromere and helps to separate the sister chromatids as mitosis proceeds into anaphase. The shorter of the two arms extending from the centromere is called the p arm; the longer is the q arm. Dyads occur in homologous pairs, one member of each pair having been acquired from one of the two parents of the individual whose cells are being examined. All species have a characteristic number of homologous pairs of chromosomes in their cells, called the diploid number. The complete set of chromosomes in the cells of an organism is its karyotype, which in human females contains 23 pairs of homologous

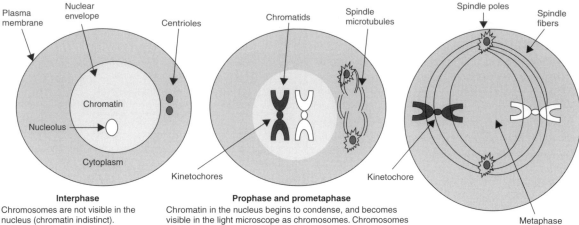

Interphase

Chromosomes are not visible in the nucleus (chromatin indistinct). Nucleus intact. Nucleolus may be visible. Centrioles may be detected. DNA synthesis (S phase), and G_1 and G_2, all encompassed by interphase.

Prophase and prometaphase

Chromatin in the nucleus begins to condense, and becomes visible in the light microscope as chromosomes. Chromosomes appear as sister chromatids. The nucleolus disappears. Centrioles begin moving to opposite ends of the cell and fibers extend from the centromeres. Some fibers cross the cell to form the mitotic spindle.
Nuclear envelope disappears at the beginning of prometaphase. Proteins attach to the centromeres creating the kinetochores. Microtubules attach at the kinetochores and the chromosomes begin moving.

Metaphase

Spindle fibers align the chromosomes along the middle of the cell nucleus. This line is referred to as the metaphase plate. Mitotic spindle complete. Chromatid sets move to spindle equator.

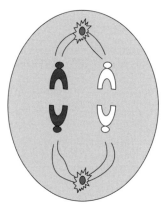

Anaphase

The paired chromosomes separate at the kinetochores. Kinetochores divide, freeing sister chromatids as individual chromosomes, which move to opposite sides of the cell (poles). Motion results from kinetochore movement along the spindle microtubules and through the physical interaction of polar microtubules.

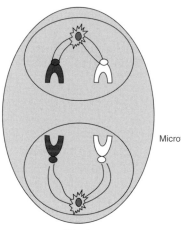

Telophase

Chromatids at opposite poles, and new membranes, form around new daughter nuclei. Chromosomes disperse and are no longer visible under the light microscope. The spindle fibers disperse, and cytokinesis or the partitioning of the cell may also begin during this stage.

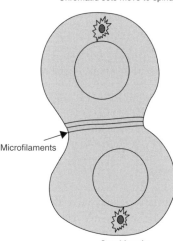

Cytokinesis

Cytokinesis results when the actin ring pinches the cell into two daughter cells, each with one nucleus.

Fig. 4.2 During the majority of the cell cycle (interphase), chromosomes are not visible under light microscopy. However, this is not the case during the nuclear division phase of mitosis, which is traditionally categorized, on the basis of early microscopy-based studies, into distinct observable stages that constitute the "chromosome cycle". Research over the last two decades has shown that the chromosome cycle is inextricably linked with the "cell cycle" (also referred to as the cyclin-dependent kinase – CDK cycle).

During mitosis the centrosomes duplicate, separate, and generate a microtubular spindle between them, thus providing a bipolar framework for the rest of mitosis. The chromosomes comprising two sister chromatids condense and become attached to microtubules arising from opposing spindle poles. Sister chromatids then separate and move apart to form two nuclei, which are ultimately separated by cytokinesis. Briefly, mitosis is divided into various phases based on early microscopy studies that correlate with distinct observable events.

During prophase the two centrosomes, each with a pair of centrioles, move to opposite poles of the cell, and the mitotic spindle forms and the chromosomes shorten and compact.

chromosomes (22 pairs of autosomes and 1 pair of X chromosomes, and in human males contains the same 22 pairs of autosomes, one X, and one Y chromosome. Karyotype analyses can reveal abnormal copy numbers of chromosomes (ploidy). Karyotype analysis can also reveal translocations between chromosomes. A number of these cause cancer, for example the Philadelphia chromosome in chronic myelogenous leukemia (CML).

Box 4.3 DNA replication in the cell cycle

Briefly, the coordinate activity of several enzymes, working together in replication complexes, is required to unravel the DNA strands and for new copies of each strand to be synthesized. These replication complexes initiate DNA replication in association with regulatory proteins required for recognition of the site of origin and which ensure that DNA is replicated only once during S phase (licensing).

Prior to cell division, the DNA is duplicated during S phase. First a portion of the double helix is unwound by a helicase enzyme, following which a DNA polymerase (delta) binds to one DNA strand and begins moving along it. This strand then acts as a template for the assembly of a leading strand of nucleotides and the regeneration of a double helix. Because DNA synthesis only occurs in 5' to 3', a molecule of a second type of DNA polymerase (epsilon) binds to the other template strand as the double helix opens. This molecule synthesizes discontinuous polynucleotide segments termed Okazaki fragments, which are then joined together by the enzyme DNA ligase to form the lagging strand (Fig. 4.10). DNA

replication is semiconservative, as strands of the DNA are retained and serve as templates for the newly synthesized complementary strands.

The average human chromosome contains 150×10^6 nucleotide pairs copied at a rate of almost 50 base pairs per second. To enable this process to be completed within a short timespan (6–10 hours in mammalian cells), there are multiple sites of origin on each chromosome. Replication begins at some replication origins earlier in S phase than at others, but the process is completed for all by the end of S phase. Given these multiple origins, several control steps are required to ensure that all original DNA is replicated and that freshly synthesized DNA is not replicated again. The integrity of the genome is maintained by licensing; DNA is only replicated if each origin of replication is bound by an origin recognition complex of proteins (ORC). Also required are various licensing factors, such as CDC-6 and CDT-1, which accumulate during G_1 and coat DNA with MCM proteins, without which DNA is not replicated. The protein geminin is also an important negative regulator of replication that can prevent the assembly of MCM proteins on freshly synthesized DNA; when mitosis has been completed, geminin is degraded in order that DNA of the two progeny will be able to replicate their DNA at the next S phase by reacting to licensing factors.

Endoreplication

Endoreplication is the replication of DNA during S phase without subsequently completing mitosis and/or cytokinesis. Endoreplication is a feature of certain types of cells in both animals and plants and takes distinct forms including: replication of

---

Fig. 4.2 (continued) During prometaphase the nuclear envelope disappears and the kinetochore forms at each centromere. Spindle fibers attach to the kinetochores as well as to the arms of the chromosomes. For each dyad, one of the kinetochores is attached to one pole, the second (or sister) chromatid to the opposite pole.

At metaphase all dyads are arranged midway between the poles, called the metaphase plate. The chromosomes are at their most compact at this time.

At anaphase sister kinetochores separate (cohesions are broken down by separase, normally kept inactive by the chaperone securin; anaphase is initiated when the APC targets securin for proteasomal degradation). Each kinetochore moves to its respective pole along with its attached chromatid (chromosome).

During telophase a nuclear envelope reforms around each group of chromosomes, which re-adopt less compacted form.

Finally, during cytokinesis, the cell divides into two, through the action of actin filaments that constrict at the "waist" of the cell.

DNA with completion of mitosis without cytokinesis; repeated DNA replication but no new nuclei form in telophase, resulting in either polyploidy (multiple chromosome copies – cells are polyploid if they contain more than two haploid sets of chromosomes) or polyteny (multiple chromosomes remain aligned as giant chromosomes); mitosis without cytokinesis, producing multiple nuclei. Polyploidy is observed normally in liver cells, placental trophoblasts, and some hematopoietic lineages.

such as to replace dying cells or allow tissue growth. Self-evidently, most cancer-causing gene or epigenetic alterations act in some way to deregulate normal cell cycle control and result in inappropriate cellular replication. In fact, given the central role in cancer of processes that drive or stop the cell cycle on the one hand and those that monitor the accuracy of the process on the other, it is not surprising that the cell cycle features in most chapters in this book.

The molecular mechanisms regulating the cell cycle are highly conserved across species, a fact which has enabled seminal discoveries using various model organisms, such as yeasts, sea urchins, and frogs to be readily applied to understanding cell cycle control in man, and are exemplified by the recent awards of the Nobel prize for physiology or medicine to Leland Hartwell, Tim Hunt, and Paul Nurse (Nobel lectures published in: Hartwell 2002; Hunt 2002; Nurse 2002) – Box 4.4.

The cell cycle is divided into four distinct phases (Fig. 4.1): the replication of chromosomal DNA during the synthesis phase (S phase), the partitioning of replicated chromosomes during mitosis (M phase), and two gaps, one before and one after S phase, that are referred to as G_1 and G_2 respectively. Broadly, M phase in the cell cycle includes the various microscopically observed stages of nuclear division and cytokinesis (mitosis) and is itself divided into phases termed prophase, prometaphase, metaphase, anaphase, and telophase (Fig. 4.2). Interphase is the term that encompasses stages G_1, S, and G_2 of the cell cycle.

Box 4.4 Key studies in the cell cycle field. Cell cycle regulation "from yeast to beast"

The appreciation that essentially all cells arise by the binary division of preexisting cells was eloquently stated by the pathologist Rudolf Virchow: "Omnis cellula e cellula." Cell division is now thought of as a process that has been endlessly perpetuated ever since the first cell came into existence – the beginning of life on this planet.

Originating from a single cell, the fertilized ovum, an adult human being consists of approximately 10^{14} cells and needs to produce about three million new cells every second simply to maintain normal tissue homeostasis.

The cell cycle includes key cellular processes including the growth of the cell, replication of the genome, and the distribution of the chromosomes during mitosis. Not surprisingly, cell cycle progression is subject to extensive "checkpoint" monitoring and regulation by various factors that ensure that cell replication is appropriate. Even single-cell organisms such as yeast must be able to link replication to the availability of nutrients in the environment. In multi-cellular organisms these controls are even more important since cell division must be adapted to meet the needs of the whole organism. Importantly, it is clear that the regulatory processes underlying cell division are extremely similar in all animals and plants. Particularly in the last two decades it has been recognized that certain "pacemakers" of the cell cycle have been conserved through several hundred million years of evolution.

It was for the characterization of these cell cycle pacemakers that the 2001 Nobel Prize in Physiology or Medicine was awarded to Leland Hartwell, Paul Nurse, and Tim Hunt. They discovered protein kinases that are periodically switched on and off during the cell cycle, and control practically all important stages, particularly the replication of DNA and mitosis. Since all of them depend on special regulatory subunits, the cyclins, they are known as "cyclin-dependent kinases" (or CDKs).

These landmark studies on cell cycle regulation started in the late 1960s. Hartwell used budding yeast to identify mutants that blocked specific stages of cell cycle progression. Nurse, working in fission yeast in the 1970s, went on to isolate mutants that could also speed up the cell cycle, thus focusing his attention on the original CDK kinase, cdc2. In the 1980s, Hunt identified proteins in sea urchin extracts, the levels of which varied through the cell cycle, hence "cyclins". Cyclins were first identified by work on sea urchin eggs. Noting that the concentration of cyclins went up rapidly through most of the cell cycle, then suddenly dropped to zero halfway through the M phase, it was suggested that cyclins might act as a sort of switch, turning on mitosis whenever their concentration reaches a certain level. Similar cyclins were later found in budding yeast. It soon became clear that cyclins bound to, and activated, a type of protein kinase called the cyclin-dependent kinases (CDKs). This activation was required for cells to move from one stage of the cell cycle to the next.

A large part of contemporary cell cycle research has been conducted on mutant strains of the fission yeast (*Schizosaccharomyces pombe*) and the budding yeast (*Saccharomyces cerevisae*) that have genetic lesions in some phase of the cell cycle. The cell division cycle (cdc) mutant strains have been a major contributor to our current understanding of how the cell cycle is regulated. The yeast cell cycle has two points at which it commits to progressing into the next phase in the cycle: start near the end of G_1 at which the cell commits to DNA synthesis (S phase); and at the beginning of M phase when the cell commits to chromosome condensation and mitosis. Seminal studies have shown that key protein complexes are formed at each of the two committal points, comprising a cyclin and a protein kinase called p34. The existence of such a complex was described biochemically when a factor called maturation-promoting factor (MPF) was isolated that could initiate mitosis in certain mutant yeast strains whose cell cycle was arrested at this stage. It was the coupling of this type of biochemical research with genetics that defined and elucidated many of the steps in the cycle.

The duration of the cell cycle can vary remarkably between cell types. For example, cells in early embryos can proceed through continuous cycles with each cell cycle completed in a mere half hour. This is in contrast with cells of the adult, where a fairly rapidly dividing mammalian cell would have a cycle time of 12 to 24 hours, whereas the cell cycle of a human liver cell can last longer than a year! The much longer duration of cell cycle transit that occurs in adult tissues compared to early embryonic cells is due to the presence of the gap phases, G_1 and G_2, which allow for growth and importantly the repair of DNA damage and replication errors (see below).

Cell cycle events are tightly regulated and subject to rigorous quality-control steps. First, commitment to DNA synthesis is dictated by exposure of the cell to a variety of growth-promoting and inhibitory factors; G_1/S transition in normal cells only occurs if the balance favors growth promotion. Second, at different steps in the cell cycle the cell determines if an earlier event has occurred correctly before proceeding to a further step. These are referred to as checkpoints (Fig. 4.3).

The point at which cells commit to replicate their DNA and enter the cell division cycle is controlled by the RB protein, the product of the tumor suppressor gene *RB*. RB acts as an "off switch" for entry into S phase by binding the transcription factor E2F, required for expression of S phase genes (Fig. 4.1). After appropriate growth factor stimulation, the inhibitory effects of RB are removed by phosphorylation, which allows E2F to activate S-phase genes, during early G_1, required for cell cycle progression (Fig. 4.4). In fact after this point the cell cycle can proceed even if growth factor stimuli are no longer present. RB or its regulatory pathways are frequently inactivated in many types of cancer, contributing to inappropriate and growth-factor-independent cellular replication. The engine which drives the cell cycle machinery consists of various proteins acting together and in a predetermined sequence – in particular changes in levels of two classes of proteins, the cyclins and the cyclin-dependent kinases (CDKs), during different phases are central to cell cycling (Fig. 4.5). Conversely, various inhibitory inputs into cell cycle progression, including inhibitory growth factors and triggering of various checkpoints, operate in part via the activity of a key group of proteins – the cyclin-dependent kinase inhibitors (CKIs).

A very important checkpoint involving the p53 tumor suppressor ensures that S phase is blocked after DNA damage, thus preventing the replication of mutated DNA until the damage

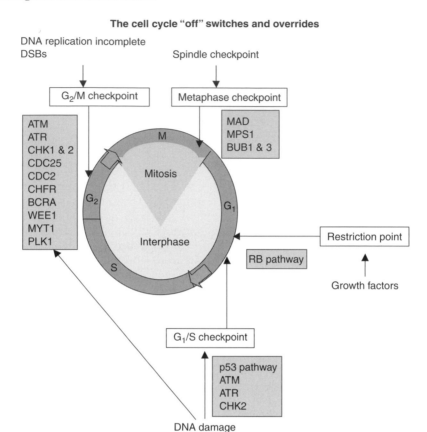

ATM = ataxia telangiectasia mutated, detects among others double-strand breaks (DSBs) and together with p53 can arrest the cell cycle; ATR= ataxia telangiectasia related; MAD= mitotic arrest deficient, detects attachment of microtubules to kinetochores and prevents aneuploidy; if fails then blocks anaphase entry (other important proteins acting at the spindle checkpoint include MPS1, BUB1, and BUB3). CHFR is a ubiquitin ligase that can target polo-like kinase for proteosome degradation thus arresting cells at the G2/M.

Fig. 4.3 Major checkpoints activated by DNA damage and mitotic defects in the cell cycle.

has been repaired, or the cell eliminated through its own death (apoptosis) (Fig. 4.6). A further checkpoint involving p53 can block entry into mitosis (Fig. 4.7) – the G2/M checkpoint. One key mediator of these actions of p53 is the CKI, p21. Lastly a checkpoint arrests mitotic progression and prevents anaphase if the spindle is not assembled or chromosomal orientation is disordered – the spindle checkpoint (Fig. 4.8). Failure of these checkpoints promotes genetic instability and can allow cells with damaged DNA or incorrectly partitioned chromosomes to divide and thereby propagate cells with potentially cancer-causing mutations.

This chapter sets out to describe the phases of the cell cycle, the key regulatory proteins that ensure cell replication proceeds without mistakes, and some examples of how this complex process becomes derailed in cancer cells.

PHASES OF THE CELL CYCLE

The passage of a cell through the cell cycle is regulated by multiple cytoplasmic proteins, described below, which are remarkably well conserved amongst species. In fact, seminal work

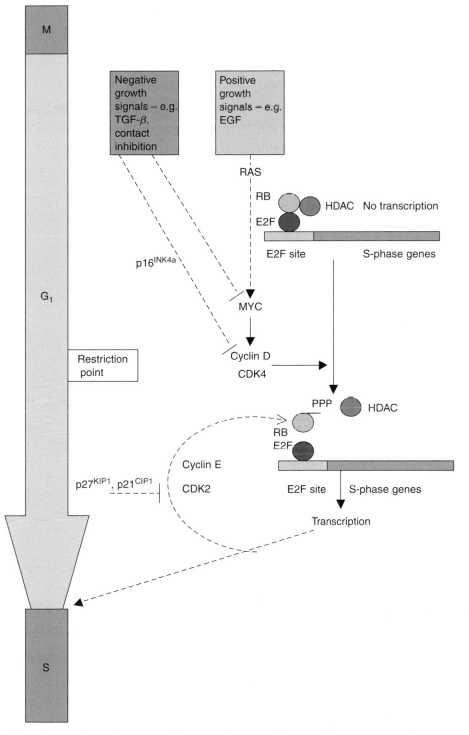

Fig. 4.4 Balancing of growth-regulating signals received in G_1, determines entry into S phase. Growth factor mitogens primarily act on the cell during the G_1 phase of the replication cycle.

The engines of the cell cycle

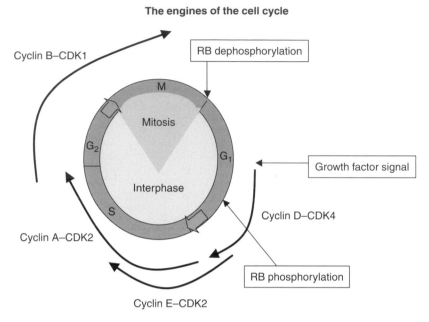

Fig. 4.5 The cyclins and CDKs during the cell cycle – running the cell cycle "relay." The cell cycle resembles a relay race with a cyclin/CDK "runner" responsible for each of the four legs of the race. There are four classes of cyclins, each defined by the stage of the cell cycle at which they bind CDKs and function.

1. G_1/S-cyclins bind CDKs at the end of G_1 and commit the cell to DNA replication.
2. S-cyclins bind CDKs during S phase and are required for the initiation of DNA replication.
3. M-cyclins promote the events of mitosis.
4. G_1-cyclins help promote passage through Start or the restriction point in late G_1.

In G_1, the runners of the first leg (cyclin D–CDK4/6) are in the blocks waiting for the start gun (in G_0 they are stretching or milling around watching the long jump, some may have retired). The race starts in G_1 when the appropriate signal is received and the first baton is handed over by cyclin D – CDK4/6 to the S-phase cyclins after the restriction point. Once the race is on, each of the four runners drops out in sequence and another takes over for each of the four legs. Like in a race, the runners of the next leg sometimes start running before they get the baton and the previous runners take a while to stop running after the baton is handed over. If something minor goes wrong (a runner trips over) the race stalls but can restart once the problem is resolved. However, if something irretrievable happens (a baton is dropped) then the race is over.

involving the study of budding and fission yeasts, sea urchins, and frogs has substantially contributed to our understanding of the cell cycle in mammals. In fact, the apparent complexity of terminology used to describe genes/proteins involved in cell cycle regulation reflects the organisms in which they were first described or studied (Table 4.1).

It has also been appreciated since the 1920s that cell cycle progression is linked to cell growth, with early theories speculating that cells divided if they reached a particular cytoplasmic size, which in some way rendered them unstable (see Box 5.2).

The cell cycle consists of the following phases (see Fig. 4.1):

- G_1 = gap phase for growth and preparation of the chromosomes for replication
- S = synthesis of DNA (and centrosomes)
- G_2 = gap phase for growth and preparation for mitosis
- M = mitosis (nuclear division) and cytokinesis (cell division)

The M phase (division phase) is relatively short, lasting about 1 hour for a cell cycle time of 24 hours compared to the rest of the cycle (interphase) during which time much cell growth

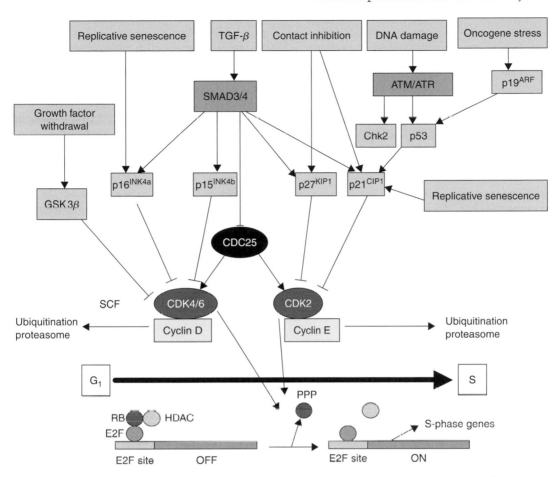

Fig. 4.6 The G_1/S checkpoint. The G_1/S cell cycle checkpoint controls the passage of eukaryotic cells from the first "gap" phase (G_1) into the DNA synthesis phase (S). Two cell cycle kinases, CDK4/6-cyclin D and CDK2-cyclin E, and the transcription complex that includes RB and E2F, are key regulators of this checkpoint. In G_1, the RB–HDAC repressor complex binds to the E2F transcription factors, inhibiting transcription of S-phase genes. Phosphorylation of RB by CDK4/6 and CDK2 dissociates the RB–repressor complex, permitting transcription of S-phase genes (encoding proteins that amplify the G_1 to S-phase switch and that are required for DNA replication). Different stimuli exert checkpoint control including (i) TGF-β, (ii) DNA damage, (iii) contact inhibition, (iv) replicative senescence, (v) oncogenic stress, and (vi) growth factor withdrawal. The first five [(i)–(v)] act by inducing INK4 or KIP/CIP families of CKIs. TGF-β additionally inhibits the transcription of CDC25. Growth factor withdrawal activates GSK3β, which phosphorylates cyclin D, leading to its rapid ubiquitination and proteosomal degradation. Ubiquitination, nuclear export, and degradation are mechanisms often employed to rapidly reduce the concentration of cell cycle control proteins.

takes place. Although lengths of each phase vary from cell to cell, the greatest variation is in G_1: cells that have not committed to DNA synthesis can pause and enter a resting state termed G_0, where they can remain for days, weeks, or years before resuming proliferation. This state is often used to refer to cells which are growth-arrested or terminally differentiated, but

does not imply that these cells cannot re-enter the cell cycle at some later stage. Compared to DNA replication (S phase) and mitosis (M phase), which follow canonical steps that vary little from cell to cell, the regulation of entry and progression through G_1 largely depends on cell type and context. For instance, different signals are received by a stem cell of the intestinal

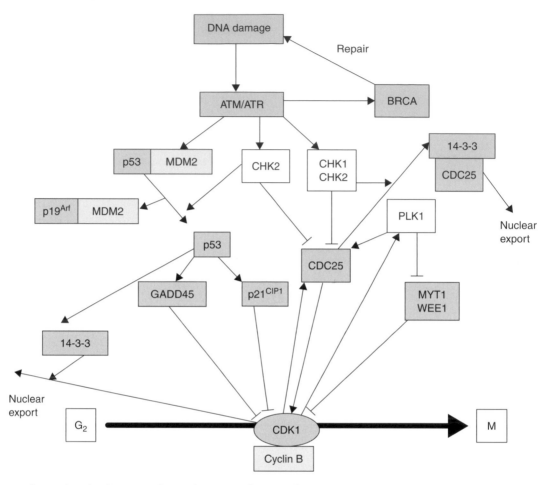

Fig. 4.7 The G_2/M checkpoint. The G_2/M DNA damage checkpoint prevents the cell from entering mitosis (M phase) if the genome is damaged. The CDK1 (CDC2)-cyclin B kinase is a key regulator of this transition. During G_2, CDK1 is maintained in an inactive state by the kinases WEE1 and MYT1. As M phase approaches, the phosphatase CDC25 is activated by the polo-kinase PLK1. CDC25 then activates CDK1, establishing a feedback amplification loop that efficiently drives the cell into mitosis. DNA damage activates the ATM/ATR kinases, initiating two parallel cascades that inactivate CDK1-cyclin B. The first cascade rapidly inhibits progression into mitosis: the CHK kinases phosphorylate and inactivate CDC25, which can no longer activate CDK1. Phosphorylation of p53 dissociates it from MDM2, activating its DNA binding activity. Acetylation by p300/PCAF further activates its transcriptional activity. The genes that are activated by p53 encode proteins including 14-3-3s, which bind phosphorylated CDK1-cyclin B promoting nuclear export; GADD45, which apparently binds to and dissociates the CDK1-cyclin B kinase; and p21^{CIP1}, an inhibitor of a subset of the CDKs including CDK1.

lining compared to a lymphocyte stimulated by an antigen, or an angioblast responding to vascular injury, all of which ultimately proceed through the G_1 phase.

To proceed or not to proceed The G_1 phase is an important gap phase, when the cell may

receive many signals that can influence cell division (Fig. 4.3). Many of these are related to growth factors (discussed in Chapter 5) as well as cell–cell contact (discussed in Chapter 12). In other words, on the basis of signals received, the cell will decide whether to enter the S phase or to pause or arrest in G_0, to differentiate or to die. Diverse metabolic, stress, and environmental

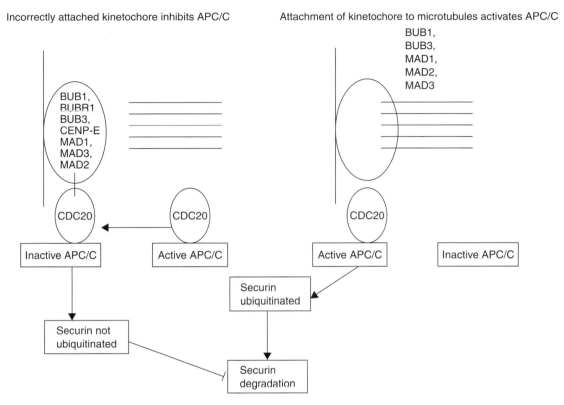

Fig. 4.8 The spindle checkpoint. A schematic of the spindle checkpoint pathway – in the left panel incorrectly attached kinetochores recruit BUB and MAD proteins, which form an inhibitory complex that catalyzes inactivation of the anaphase-promoting complex (APC/C) through MAD2 binding to CDC20. In the right panel, correct attachment and "tension" favors release of the checkpoint proteins from the kinetochore, releasing inhibition of APC/C allowing anaphase. This allows APC/C to ubiquitinate securin, which is then targeted for degradation by the proteasome. The APC/C also regulates degradation of B cyclins for mitotic exit. In mammalian cells securin at sites other than the centromere is removed by action of polo-like kinase during prophase.

signals are integrated and interpreted during this phase and influence the activity of various cell cycle regulatory proteins (described below and in Chapter 7) and consequently cell cycle progression. Signals could be external – from other cells or from the circulation – or arise within the cell, as a result of DNA damage caused by genotoxic agents, physical or chemical stresses, or oncogenic stimuli. Given the task of interpreting a flood of signals in order for the correct outcome – cell cycle progression, arrest, differentiation, or death – it is perhaps not surprising that mistakes in this process can lead to cancer. In fact, it is hard to imagine any cancer developing at all without deregulation of some aspects of cell cycle control.

The cell cycle engine: cyclins and kinases

Once the cell has received appropriate signals for entry into the G₁ phase of the cell cycle, activation of various regulatory proteins called cyclin-dependent kinases (CDKs) must occur if the cell is to proceed and enter S phase. CDKs are serine/threonine protein kinases that require binding to their regulatory subunits, called cyclins, in order to become catalytically active. Active CDKs are universal regulators of the cell cycle that alter the biological functions of regulatory proteins through phosphorylation, and have been described as a "cell cycle engine" (see below).

Different members of the CDK family are activated by their appropriate cyclins during

Table 4.1 Nomenclature of proteins involved in cell cycle in budding yeast, fission yeast, and mammals.

Factor	Function	Saccharomyces cerevisiae	Schizosaccharomyces pombe	Mammalian
CDK	Cyclin-dependent kinase	Cdc28	Cdc2	Multiple CDKs: CDK1 – 6
G_1-cyclin	Regulatory subunit of CDK for cell cycle entry	Cln1, 2, and 3		CDK4-cyclin D
S-phase cyclin	Regulatory subunit of CDK for S-phase entry	Clb5, 6	Cig2	CDK2-cyclin E
Late S-phase cyclin	Regulatory subunit of CDK for S-phase progression	Clb3, 4		CDK2-cyclin A
M-phase cyclin	Regulatory subunit of CDK for mitosis	Clb1, 2	Cdc13	CDC2 < CDK1)-cyclin B
APC	Multicomponent E3 ubiquitin ligase required for degradation of substrates in mitosis and G_1	Many proteins	Many proteins	Many proteins
APC specificity factors	Target the APC towards different substrates	Cdc20; Hct1	Slp1; Srw1	CDC20, fizzy HCT1, FZR
Securin	An APC target, inhibits sister chromatid separation	Pds1	Cut2	Securin
Separase	The securin target, a protease that degrades cohesin	Esp1	Cut1	Separase
Cohesin	A complex of proteins that holds sister chromatids together	Scc1	Rad21; Psc3; Psm1; Psm3	RAD21 SCC1; SCC3; SMC1; SMC3
SCF	Multi-component ubiquitin ligase required for degradation of phosphorylated substrates in G_1	Skp1; Cdc53; Cdc4	Skp1, Pop1, 2	S is SKP1; C is cullin; F is F-box protein

Table 4.1 *(continued)*

Factor	Function	*Saccharomyces cerevisiae*	*Schizosaccharomyces pombe*	Mammalian
CKIs	CDK inhibitors	Sic1	Rum1	p16^{INK4a}, p15^{INK4b}, p18^{INK4c}, ARF, p21^{CIP1}, p27^{KIP1} and p57^{KIP2}
ATM/ATR	Master kinase regulators of checkpoint pathways	Mec1; Tel1	Rad3; Tel1	ATR; ATM
Checkpoint sensor	Complex of proteins consisting of a clamp loader and a clamp that binds DNA and monitors damage	Rad24, Mec3, Rad17; Ddc1	Rad17; Hus1; Rad1; Rad9	RAD17; HUS1; RAD1; RAD9 (9-1-1 complex). (also in mammals the MRN complex MRE11, RAD50 and NBS1, Chapter 10)
Effector kinases	Downstream of sensor kinase, respond to different challenges	Chk1 (damage); Rad53 (hydroxy urea)	Chk1; Cds1	CHK1; CHK2; BRCA1
Targets	Downstream of kinases regulators of the cell cycle	Cdc5; Pds1	Wee1; Cdc25	CDC25; p53; MDM2
Transcription factors	Regulated transcription; the ones here are active for synthesis of S-phase genes	Swi6; Swi4, Mbp1	Cdc10; Res1	E2F
preRC	Prereplication complex, which marks a replication origin as ready to fire	Orc1-6; Cdc6; Mcm2-7	Orp1-6; Cdc18; Mcm2-7	ORC1-6; CDC6; MCM2-7
CDC7	Origin-activating kinase, which may play other roles in maintaining genome integrity. Requires a subunit (DBF4) which does not look like, but acts like, a cyclin	Cdc7; Dbf4	Hsk1; Dfp1	CDC7; DBF4/ASK1

Source: After S.L. Forsburg, UCSD.

Table 4.2   A summary of metazoan cyclins and CDKs

Cyclin	CDK	Role
Cell cycle		
A1,2	CDC2, CDK2	DNA replication and mitotic entry
B1,2	CDC2	Mitosis
B3		Unlike B1 and B2, enters the nucleus during interphase
D1,2,3	CDK4, CDK6	Required for G_1/S transition
E1,2	CDK2, CDK3	DNA replication and centrosome duplication
F	?	Binds to cyclin B
Transcription		
C	CDK8	Phosphorylates C-terminal of RNA polymerase II large subunit. Target of vitamin D
H	CDK7	May phosphorylate and activate other CDK/cyclin complexes, modifies transcription machinery
T1,2a,2b	Cdk9	Transcription elongation factor (Tef-2b). Binding of cyclin T1 to HIV Tat is required for efficient binding to the Tar element of HIV RNA
Other/unknown		
p35	CDK5	Found in postmitotic neurons. p35 proteolysis associated with neurodegenerative diseases
?	CDK10	Unknown
?	PCTAIRE	Unknown
?	PFTAIRE	Unknown
?	PITSLRE	Unknown, possible role in inducing apoptosis
G1,2	?	Induced by proapoptotic stimuli. Reported association with CDK5
I		Unknown, related to cyclin G
L1,2		Unknown. Homology to transcriptional cyclins. Induced by dopaminergic stimulation
M	?	Homology to transcriptional cyclins
O	?	Described as uracil-DNA glycosylase. Homology to cyclins
P	?	Distant relative of A and B type cyclins. Function unknown. Restricted to vertebrates?

Source: Murray, A.W and Marks, D. (2001). Can sequencing shed light on cell cycling? *Nature*, **409** (6822): 844–6.

certain phases of the cell cycle (see Table 4.2). While levels of CDKs, albeit inactive, remain fairly stable throughout the cell cycle, levels of cyclins rise and fall (with the exception of cyclin D, which rises early in G_1 and remains constant thereafter), thus determining at what stage of the cycle the CDK partner is activated.

Levels of cyclins rise and fall during the progression of the cell cycle, and each cyclin is categorized with respect to the stage at which it is elevated. In general levels of cyclins are determined by rate of synthesis and also by degradation by the ubiquin–protreasome pathway described later and in Chapter 11. Table 4.2 gives a fairly comprehensive list of currently known metazoan cyclins and CDKs

- G_1 cyclin (cyclin D1, D2, and D3)
- S-phase cyclins (cyclins E1, E2, and A)
- mitotic cyclins (cyclins B and A)

Cyclin-dependent kinases (CDKs):

The progression of a cell through the cell cycle is promoted by CDKs, whose periodic activation is driven by cyclins and negatively regulated by CDK inhibitors (CKI). CDKs are categorized numerically. CDK protein levels remain relatively stable during the cycle, but activity is dependent on each CDK forming an active holoenzyme complex by binding to its relevant cyclin. Cyclin-CDK complexes in turn promote phosphorylation of target substrates needed for cell cycle progression. Substrate specificity is likely determined by both subcellular localization and structural factors.

- G_1 CDK (CDK4&6) : cyclin D-CDK4/6 complex: key target primarily RB.
- an S-phase CDK (CDK2) : cyclin E or A-CDK2 complex: key target RB and others such as Cdc6.
- an M-phase CDK (CDK1 or Cdc2) : cyclin B or A-CDK1: key target not well defined but include WARTS and PAK1.

Not surprisingly, many of the cyclins and CDKs are involved in cancer (Table 4.3).

As mentioned earlier, CDK activity during the cell cycle is either high or low; the change in state underlies temporal control of DNA replication and links it to mitosis. Initiation of DNA replication in S phase requires the formation of a prereplication complex at chromosomal sites known as replication origins, and the activation of DNA-unwinding and polymerase functions (Fig. 4.10). The formation of a prereplication complex occurs when CDK activity is low whereas recruitment of DNA helicases (to unwind DNA) and polymerases occur when CDK activity is high. The transition from low to high CDK activity is essential for correct DNA replication. Moreover, the high CDK activity prevents the formation of further prereplication complexes until the completion of mitosis, when CDK activity is reduced.

Table 4.3 Cyclins, CDKs, and CKIs in cancer

Cyclin/CDK/CKI	Cancer
Cyclin A	Liver, breast
Cyclin B	Colon, breast
Cyclin C	Melanoma, leukemia
Cyclin D1	Lymphoma, breast
Cyclin D2	Colon, testicular, leukemia
Cyclin D3	Lymphoma, leukemia
Cyclin E	Colon, breast, prostate, ovary, lung, stomach, pancreas, leukemia
Cyclin E2	Breast, lung, cervix
Cyclin K	Kaposi's sarcoma
CDK2	Colon
CDK4	Melanoma, colon, breast
CDK6	Glioma
$p16^{INK4a}$	Melanoma, leukemia, glioma, lung, breast, esophagus, pancreas
$p15^{INK4b}$	Melanoma, leukemia, lung
$p18^{INK4c}$	B-cell lymphoma, testicular cancer
$p19^{INK4d}$	Testicular tumors
$p21^{WAF1}$	Brain, colon, leukemia, melanoma
$p27^{KIP1}$	Breast, colon, melanoma
$P57^{KIP2}$	Lung, breast, mesothelioma

These events ensure that DNA will be replicated once and only once per cell cycle.

Inhibitors of the cell cycle engine

Cyclin–CDK complexes are themselves negatively regulated by two classes of CDK inhibitors (CKI) listed below. These proteins – many of which are known to be tumor suppressors in man – are able to inhibit cell cycle entry and progression and prevent replication of abnormal DNA, by allowing cells to stall at appropriate points during the cell cycle so that DNA damage can be repaired. These so-called "checkpoints" are discussed below, but briefly are activated by various forms of "cell stress", including "genotoxic stress" (DNA damage) and "oncogenic stress" (inappropriate and excessive activation of oncogenes). It should be noted here that CKIs are key participants in the signaling pathways of both the RB and p53 tumor suppressors and, therefore, it is not surprising that loss of function of certain CKIs is associated with many cancers. This will be more thoroughly discussed in Chapter 7:

- The INK4 (inhibitor of cyclin-dependent kinase 4) family of proteins: $p15^{INK4b}$, $p16^{INK4a}$, $p18^{INK4c}$, and $p19^{INK4d}$ (not to be confused with $p19^{ARF}$, the mouse homolog of human $p14^{ARF}$). This is a highly conserved and critical group of cell cycle regulatory tumor suppressors that all inhibit the activity of cyclin D–CDK4/6 complexes and prevent RB phosphorylation. However, at least one member is also involved in activating another very important tumor suppressor, p53. ARF ($p19^{ARF}$ in mouse and $p14^{ARF}$ in man) is produced from an alternative reading frame of the same gene locus as $p16^{INK4a}$, hence the name ARF. ARF can sequester MDM2 (a protein which otherwise binds and targets p53 for degradation), thus stabilising the p53 protein in response to "oncogenic stress" (see below) – this results in G_1/S or G_2/M cycle arrest at least in part by activating the $p21^{CIP1}$ CKI discussed below. Whereas $p16^{INK4a}$ directly induces G_1 arrest by inhibiting the cyclin D-dependent kinases, CDK4 and CDK6, and thus preventing phosphorylation of RB.
- The CIP/KIP family of CKIs, $p21^{CIP1}$, $p27^{KIP1}$, and $p57^{KIP2}$, are negative regulators of cyclin E– and cyclin A–CDK2 and of cyclin B–CDK1 and are discussed in Chapter 7. $p27^{KIP1}$ is a key inhibitor of the G_1/S transition and acts downstream of various inhibitory growth factors such as TGF-β. $p21^{CIP1}$ is one of the major transcriptional targets of p53. High levels of $p21^{CIP1}$ and $p27^{KIP1}$ arrest the cell cycle by inhibiting cyclin E–CDK2 activity, thereby preventing RB hyperphosphorylation, whereas lower levels may instead stabilize cyclin D–CDK complexes. As cyclin D–CDK4/6 complexes form in mid-G_1, $p27^{KIP1}$ and $p21^{CIP1}$ redistribute from cyclin E–CDK2 to cyclin D–CDK4/6, thereby allowing active cyclin E–CDK2 complexes to complete phosphorylation of RB. Phosphorylation of $p27^{KIP1}$ by cyclin E–CDK2 then targets the protein for degradation, a prerequisite for S-phase entry.

REGULATION BY DEGRADATION

Two enzyme complexes, SCF (after its three main protein subunits – SKP1/CUL1/F-box protein) and APC/C (the *anaphase-promoting complex/cyclosome*), are also crucial components of the cell cycle control system. SCF and APC/C are ubiquitin ligases that promote the ubiquitination of several cell cycle regulators, including cyclins, which target them to the proteasome for destruction (Chapter 11). This degradation is critical for many key events in the cell cycle, with SCF and APC/C acting at distinct stages – SCF mainly regulates the G_1/S transition whereas the APC/C mainly targets proteins involved in mitosis.

SCF is a multi-unit ubiquitin ligase comprising several subunits; an F-box protein which acts as a receptor for the protein to be ubiquitinated, an adaptor protein SKP1, and a catalytic core of cullin (CUL1) and ROC1. Some of the SCF components are tumor suppressors inactivated in a wide range of cancers (see Chapter 11). The F-box FBW7 protein was first recognized by its ability to bind cyclin E, but is now known to also recognize c-MYC, c-JUN, and NOTCH as substrates for degradation. As will be seen in Chapter 6, FBW7 thus negatively regulates several key oncoproteins, explaining the key role in tumor suppression; in fact *Fbw7* is a haploinsufficient tumor suppressor in mice and inactivated in many different human cancers. Inactivation of SCF thus contributes to cancer by preventing degradation of key positive cell cycle regulators. In G_1 and S phase, SCF is responsible for destruction of G_1/S-cyclins and certain CKI proteins that control S-phase initiation. The importance of this SCF pathway is further highlighted by the fact that it is activated by genotoxic stress and p53 and that targets of SCF include cyclin E and c-MYC (Chapter 6), which are frequently involved in cancer – inactivation of SCF would result in increased levels of cyclin E or c-MYC by reducing degradation in the proteasome, and might impair some p53 responses.

Activation of SCF activity can paradoxically also contribute to cancer. $p27^{KIP1}$ protein downregulation is usually achieved by proteasomal degradation and is often correlated to a worse prognosis in several types of human cancers, resulting in the reduction of disease-free and overall survival. SCF targets $p27^{KIP1}$ CKI (an inhibitor of CDK2 and CDK1 activity), and other inhibitory proteins, for degradation a key step in S-phase entry and cell cycle progression. In fact, one way in which c-MYC may promote replication is by activating CUL1. Importantly, overexpression of SKP2 (another SCF component), observed in many human cancers, results in aberrantly increased degradation of $p27^{KIP1}$ and increased tumor aggressiveness. Another ubiquitin ligase often overexpressed in cancer is MDM2, which targets p53 for degradation. It is highly likely that the specific phenotypes of mutations in SCF are due to deregulated activity of specific protein substrate recognition components (such as the F-box proteins) within the SCF complex.

The APC/C comprises at least 13 subunits in yeast and regulates mitosis probably in large part by degradation of cyclin B and securin. In M phase, APC/C is responsible for initiation of anaphase and exit from mitosis. APC/C is activated by various factors including the CDC20 and CDH1/HCT1 proteins, resulting in destruction of M-cyclins (cyclin A and B and associated CDK1) and other regulators of mitosis including securin. Ubiquitination of cyclin B and securin depends on a destruction box (D box) sequence in these proteins, which is recognized by APC/C bound to CDC20 or CDH1. APC/C may also promote synthesis of G_1-cyclin for subsequent cycles. Interestingly, the APC/C also promotes degradation of geminin, a key protein involved in preventing re-replication of newly synthesized DNA in S phase.

These complexes are regulated in different ways. SCF activity is constant during the cycle and ubiquitinylation is regulated by changes in phosphorylation of its target proteins: only specifically phosphorylated proteins are recognized, ubiquitinated, and destroyed. APC/C activity, by contrast, changes at different stages of the cell cycle. APC/C is turned on mainly by the addition of activating subunits to the complex.

With inhibitors of the proteasome pathway already in clinical use, it is predicted that future therapeutic developments will target specific ubiquitin ligases and thus more specifically interfere with this pathway, for instance by preventing degradation of tumor suppressors but increasing that of oncogenes.

DNA REPLICATION AND MITOSIS

A license to replicate at the G_1/S transition

Once a cell progresses past a certain point in late G_1 termed the restriction point (analogs to start in yeast) it becomes irreversibly committed to entering S phase and replicating DNA. Moreover, once the cell has passed the restriction point cell cycle progression no longer depends on stimulation by external growth factors (Chapter 5). Cyclins, CDKs, and RB are all important regulators of the restriction point. The transit of cells from G_1 to S phase is regulated by both proto-oncogenes and tumor suppressors and will therefore also feature strongly in Chapters 6 and 7.

Prior to the restriction point, accumulation of p16[INK4a] and p27[KIP1] can bind and inhibit cyclin D–CDK and cyclin E–CDK activity, arresting cells in G_1 – the G_1/S checkpoint. The G_1/S transition is promoted by mitogenic signals through several mechanisms. Sequentially, activation of key growth regulators such as the transcription factor c-MYC first induces expression of G_1-cyclins (cyclin D) and at least one of its partners CDK4, (remember c-MYC also increases the degradation of p27[KIP1]). Cyclin D–CDK4/6 consitutes an active kinase complex and phosphorylates distinct substrates, such as the tumor suppressor, retinoblastoma (RB) pocket proteins (see Fig. 4.5). When RB is unphosphorylated it binds to E2F transcription factors, preventing the expression/regulation of several genes/proteins required for DNA replication, such as DNA polymerase. First low-level phosphorylation of RB by cyclin D–CDK4 (or cyclin D–CDK6) disrupts interactions between RB and E2F proteins, thus releasing RB-mediated repression of E2F target genes. Early transcription of cyclin E by E2F results in formation of cyclin E–CDK2 complexes, which functions in a positive-feedback loop to promote RB hyperphosphorylation – importantly this is antagonized by the CKI p27[KIP1]. This releases E2F, which results in the full activation of E2F target genes that promote entry into S phase. RB hyperphosphorylation is a key event in cell replication as it inhibits the ability of RB to sequester E2F transcription factors; E2F can then in turn promote transcription of genes required for S-phase DNA synthesis, such as cyclin A and E. In normal cells, RB phosphorylation is mediated by cyclin D1–CDK4 in response to growth factor stimulation. However, mutations inactivating RB or the RB pathway are ubiquitous in cancer cells, resulting in inappropriate cell cycling. Thus, the *CDKN2A* gene encoding the p16[INK4a] CKI, is inactivated in many human cancers, thus removing an inhibitor of cyclin D–CDK4. Similarly, genes encoding CDK4 or cyclin D1 are amplified in some glioblastomas and breast cancers, all of which may inappropriately promote replication.

Leaving G_0

In contrast to G_1/S transition, less is known about regulation of the G_0/G_1 transition. Both transitions involve RB, as inactivation is sufficient for cells growth-arrested in G_0 to re-enter the cell cycle. A recent study suggests that RB phsophorylation can be mediated by cyclin C–CDK3 complexes during the G_0/G_1 transition, in similar fashion to the action of cyclin D–CDK4 for S-phase entry.

Replicating DNA in S phase

Under the influence of E2F transcriptional activity, rising levels of S-phase proteins (also referred to as S-phase promoting factor – SPF), including cyclins E and A bound to CDK2, promote duplication of DNA and centrosomes. In fact, on CDK2 activation, DNA helicases and polymerases are recruited to unwind the double helix and to replicate DNA. As DNA replication proceeds, cyclin E is destroyed, and the level of mitotic cyclins rise in G_2. Importantly, CDK2 levels remain elevated from the G_1/S transition through to M phase when chromosomes have segregated. Together with another protein, geminin, elevated CDK2 levels play absolutely critical roles in preventing formation of new pre-replication complexes. CDKs are therefore key factors that restrict DNA replication origin firing to once per cell cycle by preventing the assembly of prereplicative complexes outside of G_1 phase. Thus, DNA gets replicated once only in each cell cycle. If this process is impeded then one can get re-licensing of already fired replication origins and re-replication of parts of the genome that have already been replicated – resulting in increased gene copy number. Such abnormalities will usually be detected in the cell inducing DNA damage responses (Chapter 10), including arrest and apoptosis (see below). Paradoxically, however, cyclin E–CDK2 may also be necessary to promote licensing. In a very recent study, Mailland and Diffley show that cyclin E–CDK2 can phosphorylate and stabilize the essential licensing factor Cdc6, by preventing APC/C ubiquitination. Intriguingly, this enables Cdc6

to accumulate before the licensing inhibitors geminin and cyclin A (also APC/C substrates), thus establishing a "window of time" prior to S phase when pre-RCs can assemble. Faults can be detected by monitoring of various products or damage that will trigger the activation of checkpoints. Thus, for example the presence of the Okazaki fragments is monitored on the lagging strand during DNA replication and cycling arrested until these have disappeared. However, failure to activate these responses will result in propagation of cells with altered gene copy numbers – a frequent cause of increased oncogene activity in cancer.

Mitosis, the spindle, and chromosome segregation: The so-called M-phase promoting factor (comprising mitotic cyclin–CDK complexes: cyclin B or A/CDK1) initiates assembly of the mitotic spindle, breakdown of the nuclear envelope, and condensation of chromosomes. These events take the cell to metaphase, at which point the M-phase promoting factor and proteins such as CDC20 and CDH1/HCT1 activate the APC/C discussed earlier. Degradation of the protein securin then leads to separation and movement of sister chromatids to the poles (anaphase), completing mitosis.

Accuracy of mitosis is essential for cell survival and is thus subject to tight control and is monitored by important checkpoints that can arrest mitosis if errors are detected. Errors in mitosis can lead to genomic instability, a major factor in progression and even initiation of cancers (Chapter 10).

The mitotic spindle is a highly dynamic structure comprising microtubules originating from the centrosomes, which provides the structural framework for chromosome segregation in cell division. Microtubules of the mitotic spindle are the key drivers of chromosome segregation and during metaphase contribute to the "search and capture" of chromosomes for bipolar alignment on the spindle. The taxanes and vinca alkaloids, which target tubulin, perturb mitotic spindle microtubule dynamics and thus interfere with cell cycling and have proved successful in the treatment of a number of human cancers (Chapter 16). Accurate chromosome segregation depends on proper assembly and function of the kinetochore and the mitotic spindle. Mitotic kinases are important in this process and also play a role in many cancers (Chapter 10).

The centrosome is attached to the outside of the nucleus and duplicates just prior to mitosis. The two centrosomes then move apart to opposite sides of the nucleus (Fig. 4.2). As mitosis proceeds, clusters of microtubules, called spindle fibers, grow out from each centrosome with plus ends growing toward the metaphase plate and play a key role in the assembly of the chromosomes at metaphase. Spindle fibers growing from opposing centrosomes attach to one of the two kinetochores (protein complex assemblies on the centromere or "waist" of condensed metaphase chromosomes in higher eukaryotes) of each sister chromatid pair (dyad) and some bind to the chromosome arms. The two kinetochores thus capture microtubules emanating from opposite spindle poles to produce the so-called metaphase plate in which all the chromosomes are bilaterally attached and aligned at the equator of the spindle. Fibers also extend from the two centrosomes in a region of overlap. Microtubule attached to opposite sides of the dyad shrink or grow until they are of equal length. Microtubule motors attached to the kinetochores move them toward the minus end of shrinking microtubules (dynein); toward the plus end of lengthening microtubules (kinesin). The chromosome arms use a different kinesin to move to the metaphase plate.

In early mitosis, the two sister chromatids of the chromosome are held together along their entire length by multi-subunit complexes of cohesin proteins. Anaphase is marked by acute loss of cohesion between sister chromatids, under the influence of the protease separin; sister kinetochores then separate and, with an attached chromatid, move along the microtubules powered by dynein minus-end motors, while the microtubules shorten. The overlapping spindle fibers move past each other (pushing the poles farther apart) powered by kinesin plus-end motors. Sister chromatids thereby end up at opposite poles of the spindle. Until anaphase, separin is part of an inactive complex with the protein securin. Anaphase starts once the APC/C covalently ubiquitinates securin, targeting it for proteasome degradation. Separin is then free to cleave its cohesin target and sister chromatids can separate. In mammalian cells securin is targeted for APC/C ubiquitination during prophase through phosphorylation by polo-like kinase (PLK1) – see below. However, cohesin complexes remain at the centromere until anaphase.

Cancer cells often have more than the normal number (1 or 2 depending on the stage of the cell cycle) of centrosomes as well as chromosomes. Mutations in *p53* (also referred to as the *TP53* gene) predispose the cell to excess replication of the centrosomes. Each centrosome contains a pair of centrioles, made up of a cylindrical array of 9 triplets of microtubules. Each centriole is duplicated during G_1/S transition. Centrioles appear to be needed to organize the centrosome in which they are embedded.

All chromosomes must be correctly attached to the spindle before anaphase can begin to prevent potentially dangerous formation of cells with altered chromosome content (ploidy) – this is the role of the spindle checkpoint (see below).

Mitotic kinases

To summarize, during mitosis, cells undergo centrosome maturation, chromosome condensation, nuclear-envelope breakdown, centrosome separation, bipolar-spindle assembly, chromosome segregation, and cytokinesis. Mitosis is brought about by a highly dynamic bipolar array of microtubules, the mitotic spindle. The formation and function of the mitotic spindle during M phase of the cell cycle is regulated by protein phosphorylation, involving multiple phosphatases and a group of serine/threonine protein kinases, known as mitotic kinases. Mitotic kinases include CDK1, PLK, NIMA-related kinases, WARTS/LATS1-related kinases and Aurora/IPL1-related kinases, all of which are highly conserved through evolution.

Three classes of Aurora kinases have been identified in man, Aurora A, B and C. Aurora A and B are expressed in most normal cell types and are involved in cell-cycle progression from G_2 through to cytokinesis, though with different location and timing of activation during the cell cycle. Aurora A is a serine/threonine-specific protein kinase important for spindle assembly. In several common human tumors, Aurora A is overexpressed, and deregulation of this kinase was shown to result in mitotic defects and tetraploidy and aneuploidy. In mouse models overexpression of Aurora A results in defects in the G_2 checkpoint and in spindle assembly. Moreover, recent genetic evidence directly links the human Aurora A gene to cancer susceptibility. Recently, potent and selective small-molecule inhibitors of Aurora kinases have been described that block cell cycle progression and induce apoptosis in a diverse range of human tumor cell types.

Polo-like kinases (PLKs) are key contributors to cell cycle regulation. PLK1 pushes cells through mitosis and is overexpressed in many human tumors, in which it may also correlate with prognosis. Although all PLKs share two conserved elements, the N-terminal Ser/Thr kinase domain and a highly homologous C-terminal region termed the polo-box motif, their functions differ markedly. PLK1 contributes to anaphase entry by phosphorylating target proteins such as securin that are then recognized by the APC/C. PLK1 is normally inhibited by various checkpoints in the cell cycle whereas PLK2 and PLK3 are activated. Deregulation of PLK1 activity contributes to genetic instability, which in turn leads to oncogenic transformation. In contrast, PLK2 and PLK3 are involved in checkpoint-mediated cell cycle arrest to ensure genetic stability, thereby inhibiting the accumulation of genetic defects. Several interacting partners of PLK1 have been identified that are tumor-suppressor gene products.

CHECKPOINTS – PUTTING BREAKS ON THE CELL CYCLE ENGINE

The cell has several options for arresting the cell cycle if something goes wrong – the so-called checkpoints (Fig. 4.3). For instance, if a cell sustains DNA damage following exposure to radiation or chemical agents, or replication errors have occurred during the cell cycle, then the duplication of that cell might pose a potential cancer risk to the organism. Far better for the organism is to eliminate the cell by death (apoptosis) or for it to permanently exit from the cell cycle (senescence), unless the DNA damage can be efficiently repaired. For these reasons, there exist various checkpoints throughout the cell cycle: DNA damage checkpoints operate before S phase (G_1/S checkpoint), during S phase, and after DNA replication (a G_2/M checkpoint) – Figures 4.3, 4.6, and 4.7. These checkpoints serve to detect DNA damage and stall G_1 or G_2 until

Fig. 4.9 The two families of CKI proteins. The INK4 family of CKI (p15[INK4b], p16[INK4a], p18[INK4c], and p19[ARF]) inhibit cyclin–CDK action by binding to the CDK, whereas CIP/KIP family members (p21[CIP1], p27[KIP1], and p57[KIP2]) bind to the cyclin component of the cyclin–CDK complex.

the damage is repaired, or else trigger the elimination of the cell by apoptosis. In addition, mitotic (or spindle) checkpoints arrest the cell in metaphase if chromosomes are not properly aligned on the spindle prior to cell division (cytokinesis). As mentioned above, various CKIs are instrumental in allowing the cell to stall at a particular checkpoint, and these are highlighted in Figures 4.3, 4.6, 4.7, 4.8, 4.9, and Plate 4.1.

Defects in a number of cell cycle regulators are implicated in cancer, including key proteins such as p53, pRB (and related proteins, p107 and pRB2/p130), their pathways and various CKI such as p15[INK4b], p16[INK4a], p18[INK4c], p19[INK4d], p21[CIP1], p27[KIP1], all of which can arrest the cycle in response to DNA damage until this is repaired. Virtually all human tumors deregulate either the pRB or p53 pathways and likely in most cases both.

G_1/S checkpoint

One of the key players in the G_1/S checkpoint is the p53 tumor-suppressor protein (Fig. 4.6), also discussed in Chapter 7. A key player in stress responses and in the DNA damage response, p53 indirectly senses DNA damage (Chapter 10) and can either arrest the cell cycle in G_1 by inducing expression of p21[CIP1], until the damage is repaired, or if repair is not possible can trigger apoptosis, via induction of various proapoptotic factors (PUMA, BAX, NOXA) (see Chapters 7 and 8). Two checkpoint pathways involved in DNA damage have been described. In the first, double-stranded breaks (DSBs) activate the ATM (ataxia telangiectasia mutated) pathway, which can arrest the cell cycle in G_1 in a p53-dependent manner (discussed in Chapter 10

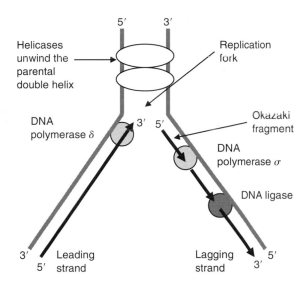

Fig. 4.10 DNA synthesis during S phase. DNA is synthesized from 5′ to 3′ by the action of DNA polymerases.

and outlined in Fig. 4.3). The second involves the ATR (ATM- and RAD3-related) and CHK1 proteins, discussed in the next section, and responds to "less extreme" DNA damage by inducing G_2 arrest, independently of p53.

Importantly, p53 is the most frequently mutated tumor-suppressor gene in human cancer – cells lacking p53 can survive levels of DNA damage and mitogenic stimulation that would otherwise kill the cell and can enter and progress through the cell cycle thus propagating DNA damage. The p53 protein is also activated via a complex system that responds to "oncogenic stress". Discussed in more detail in later chapters, this process broadly involves a series of processes activated in a cell in response to deregulated expression of oncogenes such as *c-MYC* or *RAS*, but intriguingly not to any significant extent during the activity of these proteins in normal growth. This also raises the important concept that cells can distinguish between "normal" cell cycles and "cancer" cell cycles. For example, in response to oncogenic c-MYC, the cell activates the p19[ARF] tumor suppressor that in turn can both indirectly, via MDM2 and p53, and probably directly promote apoptosis or cell cycle arrest. Interestingly, very recent work suggests that even early in the evolution of a neoplasm, oncogenic stress and aberrant cell cycles may often also result in DNA damage and activation of the ATR-CHK1 DNA damage

response pathway (Chapter 10). The explanation for this is not clear but at least in the case of c-MYC might arise through oxidative stress. Another key tumor-suppressor pathway involves the RB pathway. The protein p16[INK4a] inhibits CDK4 activity in G_1 and thereby induces cell cycle arrest and may also contribute to replicative senescence in response to oncogenic RAS or other oncogenes. Once again, as seen for p53, the Ink4 locus is frequently altered by mutations that inactivate one or both of its protein products, p16[INK4a] and p14[ARF] (p19[ARF] in the mouse) – see Chapter 7.

G_2/M checkpoint

The G_2/M DNA damage checkpoint prevents the cell from entering mitosis (M phase) if the genome is damaged (Fig. 4.7). The cyclin B-CDK1(CDC2) complex acts as a regulator of this transition. During G_2, the kinases CHK1, WEE1, and MYT1 act directly or indirectly to phosphorylate CDK1 (CDC2), thereby inhibiting activity. As M phase approaches, the CDC25 phosphatases are activated by the polo-like kinase PLK1. CDC25 then activates CDK1, in part by reversing inhibitory phosphorylation, thus establishing a feedback amplification loop that efficiently drives the cell into mitosis. In other words, at G_2 the activity of

WEE1-family kinases exceeds that of CDC25 phosphatase. However, at M-phase entry the situation is reversed, such that the activity of CDC25 exceeds that of WEE1.

As mentioned previously, DNA damage activates the ATM/ATR kinases, initiating two parallel cascades (the ATM–CHK2 and ATR–CHK1). There is redundancy and overlap between these pathways but broadly, amongst other actions discussed in Chapter 10, they negatively regulate cyclin B–CDK1 thus arresting the cycle; CHK kinases phosphorylate and inactivate CDC25, preventing both progression of S phase and entry into M phase. The ATM pathway is largely responsible for detecting major damage such as DSBs and arrests cells in G_1, in a large part through activation of the p53–p21^{CIP1} pathway. The ATR–CHK1 pathway seems largely responsible for detecting less pronounced DNA damage, possibly triggered by stalling of the DNA polymerase during DNA replication. This process is complex and for ATR to activate CHK1 likely requires several additional proteins including those involved in detection of single-stranded DNA, RAD protein containing complexes (9-1-1 complexes, comprising RAD9-RAD1-HUS1) and others discussed in detail in Chapter 10. M-phase entry can also be regulated by a second pathway involving p53 either through the damage response or via stabilization of p53 by p19^{ARF}. The genes that are then activated by p53 include 14-3-3sigma, which bind phosphorylated CDK1–cyclin B promoting nuclear export; GADD45, which apparently binds to and dissociates the CDK1–cyclin B kinase; and p21^{CIP1}, an inhibitor of a subset of the cyclin-dependent kinases including CDK1.

Intriguingly, the ATR–CHK1 pathway may be important in maintaining the survival of a cell with potentially repairable DNA damage. It is likely that this will in some way balance signals involving p53 and/or p19^{ARF} which in addition to promoting cell cycle arrest also promote apoptosis. CHK1 can also help restart stalled replication forks and may interact with DNA repair pathways. Faults can be detected by monitoring of various products or damage that will trigger the activation of checkpoints. Thus, for example during S phase, the presence of the Okazaki fragments is monitored on the lagging strand during DNA replication and cycling arrested until these have disappeared.

Much is therefore known about how DNA damage triggers multiple checkpoint pathways to arrest cell cycle progression. However, comparatively little is known about the mechanisms that allow resumption of the cell cycle once checkpoint signaling is silenced. In a recent study from Medema's group in Amsterdam, PLK1 was shown to be essential for mitotic entry following recovery from DNA damage. At least in part this effect was mediated by targeting of WEE1 for degradation at the onset of mitosis, thus favoring CDC25 mediated activation of CDK1.

The spindle checkpoint, centromeres, and kinetochores

The metaphase-to-anaphase transition is regulated by multiple proteins that together comprise the spindle (mitotic) checkpoint, essential for ensuring the fidelity of chromosome segregation. Activated by failure of kinetochore occupancy by microtubules or through lack of tension, this checkpoint delays anaphase by inhibiting the APC/C, and also the mitotic exit network required for exit from mitosis (Fig. 4.8). Conversely, failure of the spindle checkpoint is a major contributor to cancer as it can result in genomic instability and aneuploidy (Chapter 10).

The centromere is a chromosomal locus that ensures delivery of one copy of each chromosome to each daughter at cell division. The kinetochore, the protein complex assembled at each centromere, serves as the attachment site for spindle microtubules and the site at which motors generate forces to power chromosome movement. Unattached kinetochores are also the signal generators for the mitotic checkpoint, which arrests mitosis until all kinetochores have correctly attached to spindle microtubules, thereby representing the major cell cycle control mechanism protecting against loss of a chromosome. The spindle checkpoint prevents the onset of anaphase before all chromosomes are attached to spindle microtubules. Centrosomes play important roles in cell polarity, regulation of cell cycle, and chromosomal stability. Centrosome abnormalities are frequently found in cancers and contribute to chromosomal instability (including aneuploidy, tetraploidy, and/or micronuclei) in daughter cells through the assembly of multipolar or monopolar spindles during mitosis. Many cancer-causing mutations may result in hyperamplification of centrosomes.

Key proteins essential for metaphase arrest in the presence of a disrupted mitotic spindle were initially identified in the budding yeast, *S. cerevisiae*. The MAD1-3 (mitotic arrest defective) and the BUB1-3 (budding uninhibited in benzimidazole) gene products are essential for this process. Subsequently, MAD1, MAD2, BUBR1 (MAD3), BUB1, and BUB3 homologs have been identified in vertebrates. A further protein, the kinesin CENP-E, is also an essential component of kinetochore acting regulators of this checkpoint.

The checkpoint is generally accepted as operating by the generation of a diffusible stop signal in response to unattached kinetochores. This stop signal then prevents the onset of anaphase. Unattached kinetochores recruit a complex of BUB and MAD proteins (Fig. 4.8). The proteins BUB1, MAD1, and a portion of MAD2 comprise a catalytic platform that recruits, activates, and releases the stop signal, which may itself be part of the MAD2 protein. The MAD2 component can block entry into anaphase by inhibiting the APC/C (MAD2 binds to the CDC20 subunit required for APC/C activation). Conversely, the release of MAD1 and MAD2, but not BUB1, from kinetochores upon attachment separates the elements of the catalytic platform, thereby avoiding generation of the anaphase inhibitor despite continued rapid cycling of MAD2 at spindle poles.

Mutations in MAD do not prevent mitosis but damage this checkpoint, resulting in daughter cells with aberrant chromosome numbers (aneuploidy) and tumorigenesis arising from chromosome missegregation (see Figure 4.8). Loss of one copy of MAD2 results in aneuploidy, but intriguingly near-complete elimination of MAD2 protein in tumor cells results in p53-independent cell death. However, mutations in the checkpoint genes BUB1, BUBR1, and MAD2 are rarely found in cancers. But alternative mechanisms may be able to deregulate the checkpoint. Thus, amplification of Aurora A overexpressed in some epithelial cancers might override the checkpoint – Aurora A is able to induce defects in chromosome-spindle attachment if overexpressed and can also prevent MAD2 from effectively preventing APC/Cactivation.

Approximately one in twenty patients infected with the human T-cell leukemia virus 1 (HTLV-1) develops adult T-cell leukemia. HTLV-1 encodes a protein, TAX, that can sequester MAD proteins, thereby damaging the spindle checkpoint resulting in aneuploidy. The leukemic cells in these patients show many chromosome abnormalities including aneuploidy.

CELL CYCLE CONTROL

What are the extracellular signals that determine whether the cell will grow and divide?

So far, we have discussed the engine that drives the cell cycle (cyclin/CDKs), and inhibitory proteins (CKIs) that play a key role in stalling the cell cycle at various checkpoints to prevent the propagation of cells with damaged or abnormal DNA. Now we turn to some of the external signals a cell might receive that will subsequently determine whether it will grow and divide, or indeed arrest, differentiate, or die. These are discussed in more detail in the following chapter and in Chapter 12.

Growth factors stimulate cell division

As mammalian cells are not easily accessible to detailed analyses in the intact animal, most studies on mammalian proliferation derive from cells grown in culture. When cells are grown in culture, they are bathed in medium usually containing fetal calf serum, which contains a cocktail of growth factors such as PDGF, EGF, IGF, essential for cell growth and division. Without serum, cells stop dividing and become arrested usually in G_0. Thus, in the intact animal, the availability of growth factors, either from the circulation or from neighboring cells, is likely to play an important role in regulating cell proliferation. Once a growth factor has bound to its appropriate cell surface receptor, signaling pathways are transduced within the cell that will ultimately drive the cell cycle. Some of these pathways are described in the next chapter and include induction or activation of proto-oncogene products, such as c-MYC, FOS,

JUN, and RAS. In contrast, in the absence of appropriate growth factors, the cell is unable to progress past the G_1 checkpoint, arresting in G_0 until it is awakened by further signals (Chapter 5).

In contrast to growth factor mitogens, which are positive regulators of the cell cycle, negative regulators such as TGF-β are also important in cancer. Importantly, the exact role of TGF-β may depend on the timing during tumorigenesis as variously TGF-β has been shown to inhibit cancer, but less intuitively can also support tumor growth (though this may largely be a product of the interactions with stromal cells rather than the cancer cells themselves). TGF-β acts through surface receptors, which in turn activate SMAD2 or SMAD3 proteins by serine phosphorylation. Activated SMAD2 and 3 can then bind to SMAD4, with the resultant complexes entering the nucleus and activating transcription of genes including p15^{INK4b}, a G_1 CKI, which can mediate G_1 arrest.

Cell adhesion to neighboring cells and the extracellular matrix regulates cell division

Cells cultured *in vitro* require adherence to a substratum in order to proliferate. When normal fibroblasts or epithelial cells are grown in suspension unattached to any solid surface, they almost never divide. This contrasts with many cancer cells that no longer require these focal contacts for cell division. Focal contacts are sites of adhesion for intracellular actin filaments and extracellular matrix (ECM) molecules – binding of ECM molecules, such as fibronectin or laminin, to transmembrane integrin proteins can generate signals inside the cell to control cell proliferation (Chapter 12).

Nutrients and oxygen supply

Particularly in single-cell organisms such as yeasts, the major environmental regulators of cell division are the presence or absence of nutrients, which ensure that cells replicate under suitable environmental conditions. Although such signals are less relevant to cell cycle regulation in metazoans, where growth factors are more prominent, one must remember one important thing. Namely, that replication for a single-cell organism in a pond or other environment over which it has no control must be determined by environmental factors; the emphasis is on the word organism rather than cell. On a metazoan scale this scenario is closer to reproduction (replication of the whole organism): recent advances in understanding of hormones such as leptin that reflect energy homeostasis (and thereby nutrient availability in the environment), show that they have a profound impact on fertility (i.e. ability to reproduce), thus linking environmental nutrients to replication of the whole organism, as it is in yeasts – the difference being the nature of the regulatory signal. Although the cell cycle machinery may be similar, there must be major differences in the controls required over replication of somatic cells within a metazoan (in a regulated environment at least in part divorced from immediate changes in nutrient availability in the environment) and replication of an entire organism, no matter how few cells it may comprise, that is much more directly dependent on environmental support.

CHANGES IN GLOBAL GENE EXPRESSION DURING THE CELL CYCLE

Most attention has been directed to regulation of cell cycle protein activity by phosphorylation and protein levels by degradation. However, complex transcriptional networks also help maintain the cyclic behavior of dividing cells and progress has been made in describing these in budding and fission yeast. Maybe surprisingly, little is understood about the biological significance – with most attention directed at regulation of expression of a handful of genes by c-MYC and E2F family members during G_1/S transition and S phase. Recently, global gene expression analyses of proliferating cells, using gene microarrays and other high-throughput techniques (Chapter 18) have identified hundreds of periodically expressed genes occurring in three major transcriptional waves approximating to three main cell cycle transitions (initiation of DNA replication, entry into mitosis, and exit from mitosis). In fact, nearly 1000 genes demonstrate significant oscillations in expression and as many as 2000 genes undergo slight oscillations during the cell cycle. The functional role of most of these is unknown and may in fact be irrelevant, driven as a by-product of chromatin condensation. This not-withstanding, many of these oscillating genes are concentrated in early/mid G_2 phase and near the G_2/M transition and some can be grouped by function into those involved in ribosome bio-genesis and protein synthesis and those involved in mitosis, mitotic exit, and cell separation respectively. It is predicted that much import-ant progress will result from work in this area in mammalian cells in the coming years. In fact, such studies involving the elucidation of genes regulated by c-MYC have already provided intriguing results (Chapter 6).

CELL CYCLE AND CANCER

In cancer, alterations in the genetic control of cell division result in unrestrained proliferation. In general genetic or epigenetic alterations occur in two classes of genes: proto-oncogenes and tumor-suppressor genes, discussed in detail in Chapters 6 and 7 respectively. In normal cells, the proteins encoded by proto-oncogenes act in signaling pathways that stimulate cell prolif-eration. Mutated versions of proto-oncogenes or oncogenes promote tumorigenesis. Inactiva-tion of tumor-suppressor genes like RB and p53 results in loss of regulatory pathways that can inhibit cell cycle. In cancer, mutations have been observed in genes encoding CDK, cyclins, CDK activating enzymes, CKI, CDK substrates, and checkpoint proteins (Table 4.3).

G_1 is the phase of the cell cycle wherein the cell is responsive to growth-factor-dependent signals. As already discussed in general, unless some-thing goes awry, once the cell has passed the restriction point and makes the transition from G_1 into S phase the cell cycle will proceed under its own steam. Not surprisingly, control of the G_1/S transition is often disrupted in cancer cells, which can thus potentially replicate autonom-ously and independently of mitogenic stimuli. As mentioned earlier, the key event in the G_1/S transition is the phosphorylation of RB and the release of E2F for the transcription of S-phase genes. In human cancer, the RB pathway is fre-quently non-functional, though this is rarely due to a mutation of the RB gene itself, with the notable exception of some childhood cancers. In

fact, loss of RB function may usually result from inappropriate activation of CDK/cyclins or loss of CKIs, discussed below and in Chapter 7.

The CDKs and cancer

Occasionally, though infrequently, the CDKs themselves may be mutated in cancers. Amp-lification of CDK4 may occur in melanoma, sarcoma, and glioma. Mutations in CDK4 and CDK6, rendering these unresponsive to CKIs, can also occur. CDK1 and CDK2 may be overex-pressed in some colonic tumors. Some activating proteins of CDKs are implicated in cancers. As mentioned already, CDK activation is in part determined by dephosphorylation by the CDC25 phosphatase family, important regulators of the G_1/S transition. Other family members are important during S phase and entry into mitosis. Deregulation of CDC25 can result in inappro-priate activation of CDK–cyclins and has been shown in some cancers. CDC25 is overexpressed in 32% of primary breast cancers. Interestingly, CDC25 is a MYC target gene (Chapter 6).

Cyclins and cancer

Cyclin D1 binds to CDK4 and CDK6 in early G_1, and is a key mediator of growth-factor activation of cell cycling. Since cyclin D1 was first linked to tumors of the parathyroid gland in 1991, dereg-ulated expression of cyclin D1 has been reported in many other human cancers. A cyclin D1 gene translocation is now known to occur in B-cell tumors, including a type of lymphoma. In mantle cell lymphoma a characteristic t(11;14) trans-location juxtaposes the cyclin D1 gene to the immunoglobulin heavy chain gene, leading to cyclin D1 overexpression. Amplification of cyc-lin D1 is found in cancers of the breast, bladder, and lung.

The S-phase CDK2 interacting cyclins are key regulators of DNA replication and cellu-lar proliferation. The human genome encodes two E-type cyclins (E1 and E2) and two A-type cyclins (A1 and A2). Dysregulation of the CDK2-bound cyclins plays an important role in the pathogenesis of cancer.

The E-type cyclins (cyclin E1 and cyclin E2) are expressed during the late G_1 phase of the cell cycle till the end of S-phase. Cyclin E is limiting for passage through the restriction point, the point of no return for cells progressing from G_0 or G_1 into S phase. Expression of cyclin E is not only regulated by E2F transcription factors but also by proteasome-mediated degradation. Cyclin E binds and activates the kinase CDK2 and by hyperphosphorylating RB may further enhance the E2F-dependent transcription of S-phase genes. High levels of cyclin E are associated with the initiation or progression of breast cancer, leukemia, and others, and is associated with aggressive disease and poor prognosis. In addition to potential deregulation of G_1/S transition, over-expressed/deregulated cyclin E may also provoke chromosome instability and loss of heterozygosity at tumor-suppressor loci (Chapter 7). In part, therefore, the tumorigenic activity of cyclin E may arise from the aberrant assembly of prereplication complexes and incomplete DNA synthesis and also from resultant mitotic defects – increased risk of non-disjunction, impaired chromosome segregation, and ploidy.

Cyclin A2 is associated with cellular proliferation and can be used for molecular diagnostics as a proliferation marker. In addition, cyclin A2 expression is associated with a poor prognosis in several types of cancer, including some lung cancers. Cyclin A1 is a tissue-specific cyclin that is highly expressed in acute myeloid leukemia and in testicular cancer.

CKIs, checkpoints, and cancer

CKIs largely suppress cell cycle and growth by activation of RB and prevent G_1/S transition. Approximately 90% of human cancers have abnormalities in some component of the RB pathway (including RB itself, p16^{INK4a}, or cyclin D–CDK4/6). Cancer-promoting mutations of the E2F family of transcription factors have not yet been described.

Loss of cell cycle checkpoints is extremely common in human cancer. The *p53* gene is the most frequently mutated gene in human cancers (see Chapter 7). However, mutations in other genes in the p53 pathway may also result in avoidance of normal regulation. The INK4a-ARF locus is often the target of inactivating mutations in human tumors and can affect either or both protein products – p16^{INK4a} and p14^{ARF}.

Loss of expression of another key regulator of the cell cycle, p27^{KIP1}, has been reported for a number of human tumor types (lung, breast, bladder) and has been correlated with poor prognosis and tumor aggressiveness. It has been shown in colorectal carcinomas that increased proteasome-dependent proteolysis, rather than gene deletion, is responsible for p27^{KIP1} downregulation. This may result from activating mutations in SCF components required for degradation of p27^{KIP1} or by inactivating mutations required for degradation of proteins, such as c-MYC, that can inhibit p27^{KIP1}.

In normal cells, DNA damage if detected results in the induction of p53 and one of its downstream targets, p21^{CIP1}, which results in G_1 and/or G_2 growth arrest. Cancer cells that are deficient in p53 or p21^{CIP1} will fail to arrest and continue through the cell cycle into mitosis. p21^{CIP1}, originally described as a universal inhibitor of CDKs, is now known to have other key roles. In fact, p21^{CIP1} may also help maintain cell survival following DNA damage and subsequent p53 induction, in order for the cell to effect repairs. Thus, the increase in p21^{CIP1} seen in some cancers may impart these cells with a survival advantage. Another way in which cancer cells may avoid p53 regulation is by overexpression of MDM2, the negative regulator of p53, and this has been shown in leukemia, breast carcinoma, and glioma.

Checkpoint proteins involved in DNA damage responses are discussed in more details in Chapter 10. Ataxia telangiectasia (AT), a rare human disease caused by mutations in the AT-mutated gene (ATM), is characterized by sensitivity to radiation and predisposition to cancer, a phenotype also shared with many other diseases. At least in part because of the numerous proteins involved in the ATM signaling pathway, activation of the ATM kinase is a critical factor in triggering of all three cell cycle checkpoints following DSBs. AT-like diseases are due to inactivating mutations in genes encoding other key proteins in the DNA double-strand break (DSB) repair, such as those in the MRN protein complex (MRE11, RAD50, and NBS1), the DNA-dependent protein kinase complex consisting of the heterodimer KU70/KU80 and its catalytic subunit DNA-PKcs, H2AX or p53. H2AX functions primarily as a downstream mediator of ATM, whereas NBS1 operates as both an activator of ATM and as a co-activator of ATM downstream processes (but may not be essential for the former). NBS1, H2AX, and p53 play synergistic roles in ATM-dependent DNA damage responses and tumor suppression. The number of genes involved in DNA damage responses implicated in cancer is increasing at a steady pace. In addition to the ATM pathway described above, the ATR–CHK1 pathway is also important in activating checkpoints after genotoxic stress. CHK1 deficiency results in stabilization of the CDC25A protein and overriding of G_1/S and G_2/M arrest, which can lead to genomic instability and accumulation of DSBs. In a recent study, RAD9 (part of the 9–1–1 complex), involved in activation of ATR, was shown to be a potential oncoprotein in breast cancer. Increased levels may arise by gene amplification or differential methylation of Sp1/3 binding sites within the RAD9 gene, and correlate with tumor size and local recurrence.

Abrogation of checkpoints through epigenetic factors is well described (Chapters 7 and 11) and may be a key means by which tumor suppressors are inactivated in cancer cells. Loss of heterozygosity may occur through epigenetic, as well as genetic, loss of the wild-type allele of *RB*, *p53*, *MLH1*, and others. In fact, a wider role for chromatin remodeling (Chapter 11) in checkpoint failure is underlined by several recent observations. The hSNF5 subunit of human SWI/SNF ATP-dependent chromatin remodeling complexes is a tumor suppressor which, when inactivated in malignant rhabdoid tumors, leads to chromosomal instability. hSNF5 normally operates through the RB growth arrest pathway, possibly to maintain silencing of E2F-responsive genes, thus blocking cell cycle progression.

QUESTIONS REMAINING

1. To translate increasing knowledge about cell cycle regulation into improved new therapeutics, earlier cancer diagnosis and improved accuracy of prognosis.
2. To define at what point and in what respects signaling pathways regulating cell division and those regulating apoptosis diverge.
3. To increase knowledge about the mechanisms that regulate and monitor DNA synthesis and

the accurate distribution of chromosomes during mitosis.
4. To exploit the human genome reference database and new high-throughput techniques for examining gene expression to identify and functionally model new potential cell cycle regulatory genes/proteins.
5. To develop specific therapeutics to target cell cycle regulatory proteins.

CONCLUSIONS AND FUTURE DIRECTIONS

During the following chapters we will continually return to regulation of the cell cycle as a central process in cancer development and treatment. Many cell cycle proteins are now known to be directly involved in cancer and will be discussed in Chapters 6 and 7 in particular. Though most is known about proteins regulating the G_1/S transition and various cell cycle checkpoints, essentially all cell cycle proteins will likely play a role in cancer, and some may have either therapeutic or diagnostic potential.

Until recently most developments in cancer prevention and therapy have resulted from trial and error. But this is changing – through basic research, scientists are acquiring a detailed understanding of how cancer cells originate and grow. Armed with this new knowledge, scientists can now design therapies that are directly targeted at key molecules critical to the behavior of cancer cells. Given that abnormal replication is one of the "hallmark" features of cancer, recent approaches are increasingly being directed towards key cell cycle regulators. Thus, much work is now ongoing to examine the role of genes/proteins that regulate the G_1/S transition of the cell cycle and either promote or inhibit RB phosphorylation, and thereby control S-phase entry.

Many human cancers express increased amounts of cyclin D1, including cancers of the breast, prostate, esophagus, stomach, and colon. Recent studies have also begun to focus on other cell cycle proteins. Cyclin E plays a critical role for G_1/S transition and high levels of cyclin E are found in many types of cancer. Overexpression of cyclin E may arise by gene amplification and transcriptional mechanisms, but may arise largely through impaired degradation by the ubiquitin/proteasome pathway. In addition, proteolytically cleaved forms of cyclin E that show oncogenic functions have been described. Overexpression of cyclin E is now linked both to aberrant proliferation and also to a more malignant phenotype associated with chromosomal instability.

Increasingly, advances in the basic understanding of cell cycle regulation will be translated into novel therapeutic approaches, and will likely include drugs that reconstitute aspects of regulation lost in cancers.

Several new drugs targeting cell cycle proteins are currently being studied, including non-selective inhibitors of multiple cyclin–CDK holenzymes or those more specifically targeting cyclin D–CDK4/6 activity.

QUESTIONS

1. The cell cycle is
   a. Driven by the action of CKIs.
   b. Subject to various checkpoints.
   c. Driven by Cyclin–CDK complexes.
   d. Linked to cell growth.
   e. Not invariably deregulated in cancer.
2. In the cell cycle the G_2/M checkpoint
   a. Prevents DNA synthesis.
   b. Prevents mitosis.
   c. Is triggered by DNA damage.
   d. Involves the activity of the ATM and ATR protein kinases.
   e. Primarily involves the RB tumor-suppressor pathway.
3. The spindle checkpoint
   a. Responds to aberrant chromosomal segregation.
   b. Blocks cell cycle by activating the APC/C.
   c. May be triggered by abnormal tension on kinetochores during mitosis.
   d. If defective can result in chromosomal instability.
   e. Is rarely involved in human cancers.
4. The following have little influence on cell division
   a. Interactions with extracellular matrix.
   b. Cell size.
   c. Telomeres.
   d. Secreted growth factors.
   e. The p53 signaling pathway.
5. In the cell division cycle G_1/S transition is promoted by

   a. Hypophosphorylayion of the RB protein.
   b. E2F transcriptional activation.
   c. Mitogenic growth factors.
   d. Cyclin D–CDK4 complexes.
   e. The product of the *p16* gene.

BIBLIOGRAPHY

We apologize to all our colleagues past and present whose important work could not be referenced here due to space constraints. Please refer to website for more detailed reference list.

Cell cycle- general

Nurse, P. (2000). A long twentieth century of the cell cycle and beyond [review]. *Cell*, **100**: 71–8.

Kittler, R., *et al.* (2004). An endoribonuclease-prepared siRNA screen in human cells identifies genes essential for cell division. *Nature*, **432**(7020): 1036–40.

Vogelstein, B. and Kinzler, K.W. (2004). Cancer genes and the pathways they control. *Nature Medicine*, **10**: 789–99.

Cell cycle entry at G_1

Aleem, E., Kiyokawa, H., and Kaldis, P. (2005). CDC2-cyclin E complexes regulate the G1/S phase transition. *Nat Cell Biol*, **7**(8): 831–6.

Buchkovich, K., Duffy, L.A., and Harlow, E. (1989). The retinoblastoma protein is phosphorylated during specific phases of the cell cycle. *Cell*, **58**: 1097.

Massague, J. (2004). G1 cell-cycle control and cancer. *Nature*, **432**(7015): 298–306.

Sherr, C.J. and Roberts, J.M. (2004). Living with or without cyclins and cyclin-dependent kinases. *Genes and Development*, **18**(22): 2699–711.

Cyclins, CDK, CKI and checkpoints.

Castro, A., Bernis, C., Vigneron, S., Labbe, J.C., and Lorca, T. (2005). The anaphase-promoting complex: a key factor in the regulation of cell cycle. *Oncogene*, **24**(3): 314–25.

Drayton, S., Rowe, J., Jones, R., Vatcheva, R., Cuthbert-Heavens, D., Marshall, J., Fried, M., and Peters, G. (2003). Tumor suppressor p16^{INK4a} determines sensitivity of human cells to transformation by cooperating cellular oncogenes. *Cancer Cell*, **4**(4): 301–10.

Evans, T., Rosenthal, E.T., Youngblom, J., Distel, D., and Hunt, T. (1983). Cyclin: a protein specified by maternal mRNA in sea urchin eggs that is destroyed at each cleavage division. *Cell*, **33**: 389.

Kastan, M.B. and Bartek, J. (2004). Cell-cycle checkpoints and cancer. *Nature*, **432**(7015): 316–23.

Lygerou, Z. and Nurse, P. (2000). Cell cycle. License withheld–Geminin blocks DNA replication. *Science*, **290**: 2271–73.

Loog, M. and Morgan, D.O. (2005). Cyclin specificity in the phosphorylation of cyclin-dependent kinase substrates. *Nature*, **434**(7029): 104–8.

Mailand, N. and Diffley, J.F. (2005). CDKs Promote DNA Replication Origin Licensing in Human Cells by Protecting CDC6 from APC/C-Dependent Proteolysis. *Cell*, **122**: 915-26.

Moroy, T. and Geisen, C. (2004). Cyclin E. *Int J Biochem Cell Biol*, **36**(8): 1424–39.

Paulovich, A.G., Toczyski, D.P., and Hartwell, L.H. (1997). When checkpoints fail. *Cell*, **88**(3): 315–21.

Mitosis

Cleveland, D.W., Mao, Y., and Sullivan, K.F. (2003). Centromeres and kinetochores: from epigenetics to mitotic checkpoint signaling. *Cell*, **112**: 407–21.

Hoyt, M.A. and Geiser, J.R. (1996). Genetic analysis of the mitotic spindle. *Annual Review of Genetics*, **30**: 7–33.

Mitchison, T.J. and Salmon, E.D. (2001). Mitosis: a history of division. *Nature Cell Biology*, **3**: E17–21.

Rieder, C.L. and Maiato, H. (2004). Stuck in division or passing through: what happens when cells cannot satisfy the spindle assembly checkpoint. *Developmental Cell*, **7**(5): 637–51.

Shah, J.V., Botvinick, E., Bonday, Z., Furnari, F., Berns, M., and Cleveland, D.W. (2004). Dynamics of centromere and kinetochore proteins; implications for checkpoint signaling and silencing. *Current Biology*, **14**(11): 942–52.

Xie, S., Xie, B., Lee, M.Y., and Dai, W. (2005). Regulation of cell cycle checkpoints by polo-like kinases. *Oncogene*, **24**(2): 277–86.

5

Regulation of Growth: Growth Factors, Receptors, and Signaling Pathways

Stella Pelengaris and Mike Khan

What is the most rigorous law of our being? Growth. No smallest atom of our moral, mental, or physical structure can stand still a year. It grows – it must grow; nothing can prevent it.

Mark Twain

KEY POINTS

- All cells in a multicellular organism must coordinate their behavior for the overall good of the organism. This is only possible because cells have a variety of ways to communicate with one another and an elaborate machinery that allows them to interpret and integrate often diverse messages into responses (e.g. grow–don't grow; live–die).
- Cell numbers are determined by balancing often-opposing processes such as: cell division and senescence, cell survival and cell death.
- The actual size of an organism and constituent tissues is normally tightly controlled by regulation of cell proliferation, growth, and death. Size is largely determined by genetic factors, but a certain degree of plasticity in response to environmental factors is allowed in some tissues more than others. However, the actual means by which "appropriate organ size" is determined by the organism and if deemed necessary cell number and size adjusted are poorly understood, but will involve coordination of diverse processes, self-evidently including cellular growth and survival, but likely also cell adhesion, pattern formation and polarity. The whole can usefully be thought of as an "organ-size checkpoint", analogs to the cell cycle checkpoints described previously.
- Regulatory signals controlling growth can be both local (paracrine) and operating at a distance via the circulation (endocrine) and can be either positive or negative. Much of the integration of growth-regulatory signaling occurs in the cell during the G_1 phase of the cell cycle.
- Cellular replication is regulated by complex pathways activated via cell surface or intracellular receptors, to which growth factors bind, and a cascade of intracellular enzymes activated by the growth factor/receptor complex, many of which can remove or add phosphate groups (a common mechanism for activating or deactivating signaling molecules); and transcription factors which engage the growth/cell cycle machinery.
- Although some tumor cells produce and secrete growth factors, many growth factors are synthesized by stromal cells, e.g. fibroblasts, which can in turn influence behavior of other cell types including tumor cells. In order to activate signaling pathways, growth factors usually require integrin-mediated cell adhesion. This ensures activation only occurs when cells are in their correct environment.

- Intracellular signaling networks promoting cell proliferation also have intrinsic growth-suppressive properties, which will negate the cancer-causing potential of any single mutation occurring in these networks.
- Cellular replication requires the coordinated action of extracellular factors that activate multiple pathways in order to circumvent growth suppression. Thus, for example, mitogenic signals must in general be accompanied by signals which promote survival (prevent apoptosis), though these may occasionally derive from the same growth factor. Failure to provide such survival signals will result in mitogenic stimuli inducing apoptosis, a potential failsafe mechanism to guard against unscheduled deregulation/mutation in mitogen encoding genes.
- Amongst the most important pathways for transducing mitogenic signals is the RAS-MAPK, and for survival signals the PI3K-AKT.

INTRODUCTION

The size of an organ or organism is determined by the total cell mass – in other words the total number of cells and their size. Three overlapping processes control cell mass, namely cell growth, cell division, and cell death, and all are subject to specific regulatory factors and signaling pathways. The extracellular signals regulating cell size and cell number are either soluble secreted proteins, proteins bound to the surface of cells, or components of the extracellular matrix (ECM). As can be seen in Table 5.1, various regulatory processes impact on cell replication and cell number, many by means of extracellular growth-regulatory factors. The factors regulating cell mass are conveniently divided into three classes based on their predominant cellular action, though individual factors may have all these properties.

- *Mitogens:* promote cell division largely by allowing G_1/S transition in the cell cycle.
- *Growth factors:* promote an increase in cell mass (cell growth), by enhancing protein synthesis.
- *Survival factors:* promote cell survival by suppressing apoptosis.

Students should be aware that the term *growth factor* (GF) is frequently employed to describe a protein that has any of these properties – as evidenced by the names by which almost all of them are known.

Platelet-derived growth factor (PDGF) was amongst the first proteins found to have a mitogenic action. Originally described as a serum mitogenic factor, PDGF was subsequently identified as the product of platelets released during blood clotting. Now more than 50 different protein mitogens have been identified with different degrees of selectivity for certain cell types. In addition, there are factors, exemplified by members of the transforming growth factor-β (TGF-β) family, which also inhibit cell replication. Many factors have actions in addition to that on cell division and can promote cell growth, survival, or motility (Table 5.2 gives a description of some known GFs).

In this chapter we will commence by identifying the links between GFs and the cell cycle before discussing general concepts of growth and differentiation. We will then outline the normal regulation of cell proliferation by GFs, the receptors through which they act, the downstream signaling pathways activated by these receptors, and finally the activation of nuclear transcription factors, which regulate genes involved in cell cycling (see Fig. 5.6 for a simple schematic of GF signaling). As all of these steps involve proteins whose deregulated activity has been shown to contribute to cancer (in particular the oncogenes), we will conclude by giving a flavor of how such processes become derailed during cancer formation. A more detailed discussion of key proteins and signaling pathways can be found in other chapters, for instance: oncogenes (Chapter 6), tumor suppressors (Chapter 7), senescence (Chapter 9), and cell adhesion in cancer (Chapter 12).

GROWTH FACTOR REGULATION OF THE CELL CYCLE

Cells in early embryos replicate very rapidly (a mere half hour) and can start DNA replication as soon as mitosis is completed. This is due to the absence of gap phases G_1 and G_2 (normally

Table 5.1   Regulators of cellular replication

Various regulatory processes may act to constrain cell proliferation and include:

GF dependence: cell proliferation requires tissue-type-specific mitogenic and survival factors, the absence of which may also trigger apoptosis. Cells are also exposed to a variety of growth-inhibitory factors such as TGF-β.

Anchorage dependence: cell proliferation requires transmembrane proteins called integrins to interact with components of the ECM – again such contacts may be required to prevent apoptosis.

Contact inhibition: contact with like cell types inhibits cell movement and proliferation.

Senescence: observation of vertebrate somatic cells in culture have identified a limitation on the number of times a cell can divide (around 50–70 divisions for human cells) before the cells cease further division – the Hayflick limit. However, this is difficult to examine in the intact organism, though loss of telomeres, which is a major factor in determining the Hayflick limit, is also clearly a mediator of replicative arrest *in vivo*. What is clear is that senescence can be coupled to signals which regulate replication and together with apoptosis (cell death) may act as a counterbalance particularly to inappropriate or excessive activation of mitogenic signals within the cell.

Apoptosis: many mitogenic signals intracellularly are also coupled to pathways which can induce apoptosis.

Nutrient and oxygen supply: expansion of cell numbers may result in a tissue out-growing its blood supply and thereby could result in deprivation of oxygen (hypoxia) or nutrients or accumulation of damaging waste products, all of which might in turn limit further tissue growth. Moreover, the endothelial cells, which line the inside of the vessels, may secrete survival factors, also contributing to regulation of tissue growth. Thus, cells may undergo apoptosis or necrosis unless tissue growth is accompanied by new vessel development/expansion (angiogenesis).

All of these processes are also inherent barriers to neoplasia and must be overcome in order for tumors to form and progress.

present in cycling cells of the adult – Chapter 4), and likely serves the needs for rapid expansion of cell numbers following fertilization. However, later in development and particularly in the adult, this replication potential is restrained and subject to tighter controls. As discussed in the previous chapter, the adult cell cycle contains two gap periods, G_1 phase between M (nuclear division) and S (DNA synthesis) and a G_2 phase between S and M. Not only do these gaps allow for a period of monitoring and correction of any DNA damage or errors that may have arisen during DNA replication, but critically G_1 also provides the opportunity for many diverse signals to be integrated into a decision as to whether the cell will actually replicate, pause or indeed remain alive, and to a degree a similar scenario applies to terminally differentiated cells in G_0, seeking to re-enter the cell cycle (Fig. 5.1). Importantly, it is increasingly apparent that any given cell is probably exposed simultaneously to a variety of stimulatory and inhibitory signals, arising locally and via the circulation (Fig. 5.2), all of which have to be integrated by that cell before it decides whether to replicate. This integration is performed by complex interacting networks of intracellular signals generated in response to these extrinsic regulating factors – it is the net balance of all these responses that determines whether a cell enters the cell cycle (Fig. 5.3). Growth factor signaling systems are now believed to operate as complex networks rather than the traditional view of signaling as linear pathways. Complex signaling networks enable integration of multiple inhibitory and stimulatory inputs into binary choices by the cell including: replicate or arrest; survive or die. Moreover, there are numerous sites of overlap between signals regulating progression through the cell cycle and those that can permanently prevent a cell from replicating by promoting either senescence or even cell death.

Although there is considerable overlap and crosstalk, broadly two of the most important signaling pathways activated by various growth factors are:

1. The mitogenic RAS/MAP Kinase cascade and a key transcription factor, c-MYC (Fig. 5.5), and
2. The key PI3K–AKT survival pathway.

Together these pathways may act synergistically to allow cell cycle entry and progression

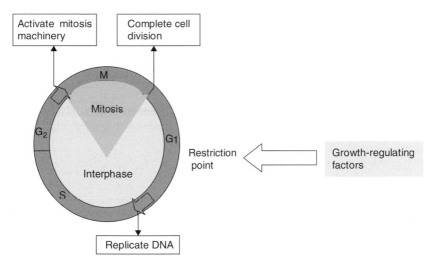

The point at which cells commit to replicate their DNA and enter the cell division cycle is controlled by the RB protein, the product of the tumor-suppressor gene *RB*. RB acts as an "off switch" for entry into S phase by binding the transcription factor E2F required for expression of S-phase genes. After appropriate GF stimulation, the inhibitory effects of RB are removed by phosphorylation, which allows E2F to activate S-phase genes, during early G_1, required for cell cycle progression

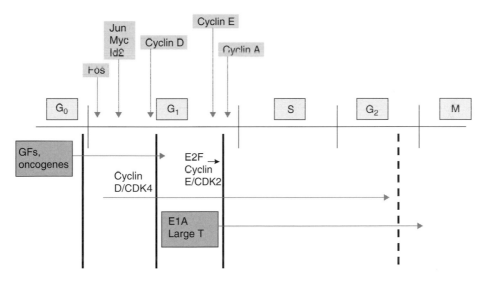

A special case for differentiated cells re-entering cell cycle. Blocks to replication in "terminally differentiated" quiescent cells are somewhat different to those generally described for the regulation of cell cycle in other cells, probably because less is known, but clearly show some overlap. Three blocks to cell cycle progression in "quiescent cells" are shown in thick black lines, which can be overcome by different mitogenic stimuli. The mid-G_1 block is overcome by cyclin D/CDK4 activity induced by GF-regulated genes such as *c-MYC*. Viral proteins such as E1A and T antigen are not susceptible to these blocks and are capable of pushing previously quiescent cells through one or more cell cycles, but then result in apoptosis, which restrains oncogenic potential. E2F or cyclin E/CDK2 expression alone cannot overcome the late G_1 block, while cyclin D1 and CDK4 alone cannot take the cycle beyond G_2.

Fig. 5.1 GFs act particularly at the G_1 phase of the cell cycle.

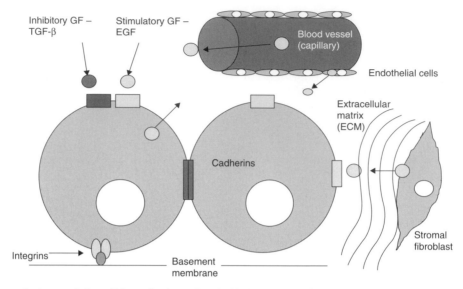

Factors regulating cellular replication and survival include secreted GFs as well as cell–cell and cell–matrix interactions. GFs may be produced locally by the cancer cell or cells of a different type and act by an autocrine (on the same cell) or paracrine (on a neighboring cell) route. GFs can be delivered via the circulation (endocrine) or be secreted from the vascular endothelial cells or by stromal cells. Stromal cells can also produce MMPs that can act to liberate GFs from matrix or can directly affect production of ROS and thus DNA damage in the cancer cell. Recruited immune cells, including activated T lymphocytes and macrophages, may also produce growth-regulating cytokines.

Fig. 5.2 Growth-regulatory factors derive from local and circulating sources.

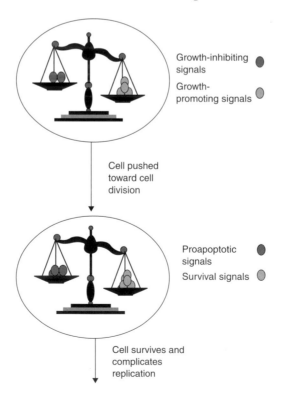

Fig. 5.3 Cell division depends on the balance of different growth-regulating signals. Ultimately, the decision whether a cell lives or dies, replicates or does not, is dependent on the balance of positive and negative signals received.

whilst also avoiding various failsafe mechanisms seeking to trigger cell death and growth arrest. Not surprisingly, RAS, c-MYC (Chapter 6), and various activators of the PI3K–AKT survival pathway are frequently involved in cancer and will often feature in subsequent chapters.

Mammalian cells must pass through a restriction point, or "R" point (equivalent of the start point in yeast cells), before committing themselves to entering the S phase and duplicating their entire DNA. Importantly, progressed beyond this point (G_1/S transition) the cell cycle will complete even without further mitogenic stimuli. The net result of many positive and negative growth-regulating signals normally determines whether a cell makes the G_1/S transition and proceeds to DNA replication, and unsurprisingly such controls are usually lost in cancer cells. Diverse growth-regulating factors operate largely at the level of activation of cyclin-dependent kinases (CDKs), which drive the "cell cycle engine". Prominent amongst these regulatory factors are the polypeptide growth factors (GFs), which either promote or inhibit the G_1/S transition (Fig. 5.6). Factors that promote the cell cycle (mitogens) generate signals resulting in increased activity of cyclin D–CDK holenzyme complexes during the G_1 phase, which as described in Chapter 4 (Fig. 4.2) promote phosphorylation of the tumor suppressor, retinoblastoma (RB) protein required for expression of S-phase genes. Conversely, inhibitory "cytostatic" signals, in particular the transforming growth factor-β (TGF-β), activate inhibitors of cyclin–CDK complexes such as p15^{Ink4b}, p21^{Cip1}, and p57^{Kip2}, thus preventing RB phosphorylation.

It is also important to reiterate here that apart from controlling cell growth and cell cycle, some growth factors also activate signaling pathways that affect other cellular processes including cell survival, migration, and differentiation – cellular processes that if disrupted inappropriately are important in the behavior of the cancer cell (Table 5.1).

GROWTH HOMEOSTASIS

Multicellular organisms, like human societies, rely on the cooperative behavior of their

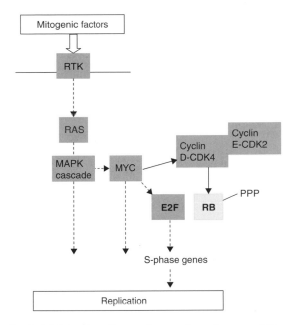

Fig. 5.4 Main signaling pathways for mitogens. Mitogenic factors largely, but not exclusively, act via RTK to activate in particular the MAPK cascade.

constituents. This is particularly important with respect to ensuring that cell numbers, functions, and localization are appropriately maintained. In general, such control over cell behaviors becomes increasingly stringent as the organism completes its development – the majority of cells sacrificing a large part of their potential to replicate and becoming more specialized in their functions. Thus, cells stop cycling and differentiate to give rise to populations of cells that exhibit specialized functions, such as skin, nervous system, etc. Importantly, within the adult organism, cells repress their innate potential to grow and divide beyond their normal boundaries and contrary to patterns dictated by the overall developmental plan of the organism. This can be a particularly tricky business given that the rules that normally restrain these behaviors can be relaxed when required for wound healing or regeneration of damaged tissue, or to allow adaptive growth of a particular tissue or cell population to changing demands (e.g. growth and modulation of breast tissue during lactation; growth of uterine tissue during the menstrual cycle) – see Fig. 5.7. Under such circumstances, many cells may be given a degree of freedom, to

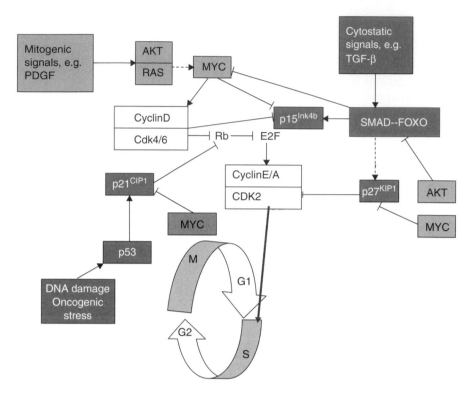

SMAD and FOXO together mediate inhibitory action of TGF-β on cell replication by activating CDK inhibitors (CKI). In contrast growth factors and c-MYC act to promote cell replication by upregulating cyclin D and inhibiting activation of CKIs such as p15^{Ink4b} and p27^{KIP1}.

Fig. 5.5 Positive and negative growth regulators determine G_1/S transition and cell cycling.

expand numbers and to migrate into what is for them an unusual location, which they would not normally be allowed. The major challenge, given the obvious similarities in behavior to those of cancer cells, is to ensure that enhanced cell replication and motility are temporary. Therefore, when the tissue in question has returned to its normal state (e.g. the wound has healed, or lactation has ended), the normally strict controls are re-applied successfully. A simple, clear, example of such a case is given in Chapter 12 (Box 12.4).

Thus, in multicellular organisms, cellular cooperation is normally ensured by numerous extrinsic and intrinsic factors, which together regulate cell growth, proliferation, motility, and survival in order to best serve the requirements of the whole organism (Fig. 5.2). It is not

surprising then that disruption of the tight controls normally imposed on these processes in the adult can have drastic effects on cells and tissues of the body, leading to diseases such as cancer. In fact, one of the common defining features of cancer is that of uncontrolled and unscheduled cellular proliferation; cancer cells divide relentlessly and are able to resist the usual mechanisms regulating replication in normal cells. Normal cells require external GFs (mitogens) to divide – at least in order to make the G_1/S transition. When synthesis of these GFs is inhibited or access restricted by some other means, then cells stop dividing. Normal cells also respond to growth-inhibiting signals, such as TGF-β. One of the "hallmark" features of cancer is autonomous cell division, such that cancer cells are no longer

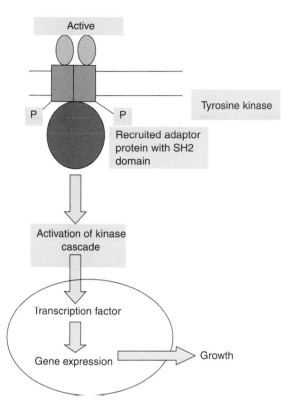

Fig. 5.6 Simple schematic of RTK signal transduction. Dimerization of RTK is triggered by ligand binding and activates the receptor. Activation involves tyrosine autophosphorylation and recruitment of various adapter proteins which then activate downstream signaling.

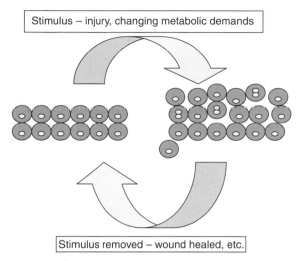

Fig. 5.7 Restraints on cell replication and motility may be temporarily released. During injury to tissues, cells may be allowed to replicate and migrate, but this is transient.

dependent on such positive or negative GFs. Cell behavior is not just regulated by secreted factors but also by molecules on the surface of neighboring cells, such as integrins and adhesion molecules (Chapter 12). Normal cells show contact inhibition *in vitro*; that is, they respond to contact with other cells by ceasing cell division. This characteristic is lost in cancer cells, as demonstrated in cell culture where cancer cells continue to divide and "pile up" to form transformed colonies *in vitro* (see Chapter 6, Appendix 6.1, and Fig. 5.3). Normal cells also age and die, and are then replaced in a controlled and orderly manner by new cells when possible. It is widely accepted that there are limits on the number of times that normal cells can divide (at most fifty times in cell culture) before they exit permanently from cell division either by

becoming senescent or by dying. This intrinsic cell division "counter" is operated at least in part by the progressive shortening of structures at the ends of chromosomes called telomeres, a process referred to as telomere attrition. In cancer cells, the enzyme involved in maintaining telomeres, telomerase, is reactivated and this contributes to continued cell divisions. Cellular senescence may also be promoted by mechanisms involving tumor suppressors such as RB (Chapters 7 and 9). Lastly, normal cells will cease to divide and undergo apoptosis when there is DNA damage too extensive to be repaired (Chapters 8 and 10), or when cell division is abnormal. In contrast, cancer cells continue to divide under these conditions (e.g. in cells that have lost p53), enabling the propagation of cells with abnormal DNA. The signaling pathways activated by these various factors show considerable overlap (Fig. 5.8).

An organ-size checkpoint

One area of great biological interest involves exactly how an organism determines that a given tissue is of the correct size (or for that matter how an individual cell determines how big it is).

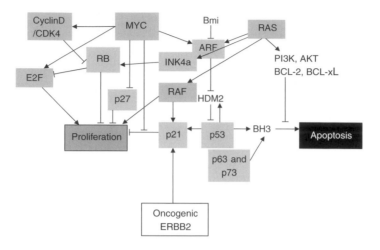

Activation of RAS and MYC results in potential engagement of both replication and growth but also of apoptosis and possibly growth arrest. MYC triggers G$_1$/S transition and cell cycle entry by regulating expression of cyclin D and CDK4 (upregulated) and p21^{CIP1} and p27^{KIP1} (downregulated). Together with cyclin E–CDK complexes (not shown), the net effect is hyperphosphorylation of RB and release of E2F for transcription of S-phase genes. If either MYC or RAS levels are excessive (as might occur during oncogenesis) or other proapoptotic signals are received then the balance may be tipped away from replication. RAS can promote senescence through either p16^{INK4a} or ARF, which activate RB or p53 pathways respectively, while MYC can inhibit growth arrest/senescence by inhibiting p21^{CIP1} and inducing TERT (telomerse reverse transcriptase). RAS may also activate p21^{CIP1} via RAF activation. Although MYC may activate the apoptotic pathway (eg. via ARF), RAS is able to suppress apoptosis by activating the PI3K pathway and subsequently, AKT. It can readily be appreciated how oncogenic MYC and RAS may conspire in oncogenesis. The combination of RAS and MYC acting together provides a potential means of avoiding apoptotic and growth arrest mechanisms activated by either acting individually. Moreover, it can also be appreciated how inactivating mutations in RB or p53 may collaborate in tumorigenesis by enabling the cancer cell to avoid either growth arrest or apoptosis or both.

Fig. 5.8 Cooperation between oncogenic signaling pathways that regulate cell replication, and apoptosis.

A checkpoint, analogs to those described in Chapter 4, has been proposed; this "organ-size checkpoint" in some way determines the size of an organ and if required triggers corrective processes that will adjust cell size and number. Much remains to be learned about this fundamental process but studies in fruit flies and nematode worms have revealed some basic principles. First, nutrient signals are important players, and possibly via hormones such as insulin and insulin-like growth factors (discussed later) activate intracellular signaling pathways that regulate both cell size and replication/survival. One protein activated by such signaling, mTOR, is of particular interest and is discussed later. Very important regulatory processes also involve cell death control. The two most important in this context are: (1) the activity of proteins such as c-MYC which can activate both cell replication/growth on the one hand and also apoptosis on the other; and (2) a particular form of apoptosis, termed anoikis, which is triggered in cells which have inadvertently left their usual tissue location, have breached a tissue boundary, and wandered into an inappropriate location (Chapter 8). Finally, recent studies suggest that the organization of a tissue plays an important role in controlling cell growth. Thus, cellular polarity (many cells particularly in epithelia usually face in a particular direction – "up or down") and interactions between neighboring cells, basement membranes, and extracellular matrix all influence cellular behavior, including differentiation, replication, and survival (Chapter 12). Not surprisingly, these processes will be referred to frequently later in the book as they must be overcome during initiation and progression of cancer.

DEREGULATED GROWTH

In the living intact organism changes in cell number reflect the balance of cell proliferation and cell death, so that an increase in tissue mass can result from either suppression of cell death (apoptosis) or increase in proliferation. However, processes regulating cell volume or size are increasingly appreciated as independent contributors to cell or tissue mass (see Box 5.2).

Self-evidently, in order for oncogenic (cancer-causing) mutations (Chapter 6) to give rise to cancer, the mutated cells must divide or else the mutation will not be propagated. It is widely accepted that cancers are predominantly formed from the most replicatively active cells, such as stem cells or progenitor cells (see Box 5.1). This is partly due to the increased likelihood of acquiring mutations during DNA replication in the cell cycle. In general, the turnover of cells in most adult organs is slow with small numbers of cells lost being replaced by replication of surviving cells of the same type or by neogenesis from stem cell precursors (Fig. 5.9) or even from cells of another type (transdifferentiation).

Box 5.1 Stem cells – additional information

Stem cells generate the differentiated cell types within many organs throughout the lifespan of an organism. Stem cells have been defined as clonogenic cells that undergo both self-renewal and differentiation to more committed progenitors and functionally specialized mature cells. Stem cells characteristically self-renew and differentiate into a variety of cell types. It is believed that of the two progeny of stem cell division, one normally undergoes differentiation to the required mature cell and the other to replenish the stem cell compartment, this is also described as asymmetrical cell division. Some stem cells, described as totipotent cells, have tremendous capacity to self-renew and may differentiate into all cell types. Embryonic stem cells have pluripotent capacity, able to form tissues of all 3 germ layers, but cannot produce a complete organism. Embryonic stem cell research has opened up many exciting possibilities, including cell or tissue replacement therapy for diseases such as diabetes, reproductive cloning, and the improved understanding of many aspects of developmental biology. However, these remarkable opportunities are matched by difficult ethical issues primarily relating to the harvesting of stem cells from human embryos. Adult stem cells are therefore a very attractive option as their acquisition and application is not associated with the same degree of ethical objections. However, such adult stem cells may have a restricted capacity to differentiate, though this view is currently being challenged. The existence of multipotent somatic stem cells in bone marrow has been reported. Under appropriate experimental conditions some bone marrow stem cells may differentiate (or transdifferentiate) into a variety of non-hemopoietic cells of ectodermal, mesodermal, and endodermal origins. However, this area is controversial with detractors suggesting that promising *in vivo* data using adult stem cells to repair various tissues may not be transdifferentiation, but instead fusion of adult stem cells with tissue cells and subsequent adoption of their phenotype.

There are important links between stem cells and cancer cells; many cancers may originate in stem cells, a not surprising fact given their greater replicative potential at the outset. Mammalian aging occurs in part because of a decline in the restorative capacity of tissue stem cells. Replicative senescence is under the control of the p53 and retinoblastoma (RB) tumor-suppressor pathways. Stem cells may be protected from malignant transformation by tumor-suppressor mechanisms including p16^{INK4a}–RB, ARF–p53, and the telomere, which may also limit the life span of the stem cell and contribute to aging. Moreover, cellular senescence by limiting the proliferation of damaged cells may act as a natural barrier to cancer progression. One mechanism for senescence may be the recruitment of heterochromatin proteins and the retinoblastoma (RB) tumor suppressor to E2F-responsive promoters and repression of E2F target genes. Interestingly, recent studies suggest that acute loss of RB in senescent cells can lead to a reversal of cellular senescence. Given the importance of stem cell replication throughout life, processes may exist to limit senescence. Bmi1 is required for the maintenance of adult stem cells in some tissues partly because it represses genes that induce cellular senescence and cell death.

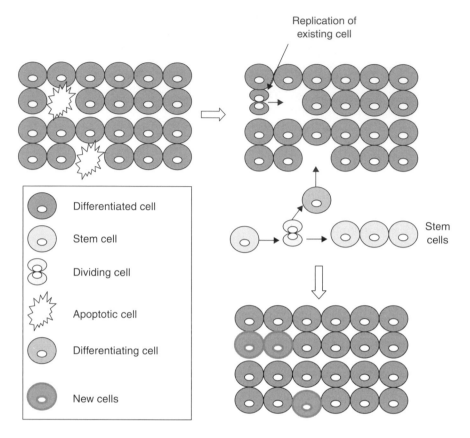

Tissue mass in the adult is maintained by balancing cell losses with cell renewal. Cells lost through apoptosis (necrosis or other cell death) or by shedding are normally replaced by replication of existing differentiated cells of the same, or occasionally even of another, cell type, or by replication and differentiation of adult stem cells. Self-evidently if cell losses in any tissue exceed the ability for renewal then the tissue will shrink (good examples being loss of insulin-producing beta cells in human type 1 and type 2 diabetes), conversely if cell gains exceed losses then the tissue will expand (as exemplified in the initial response of beta cells to conditions demanding more insulin production). Many tissues are capable of switching between these conditions, thus enabling adaptation to demands – tissue plasticity. Some tissues are regarded as static with effectively no turnover in the adult. Any losses in such tissues will result in progressive involution/atrophy of that tissue with time.

Fig. 5.9 Tissue mass homeostasis.

Despite rapid progress in this area, it may be surprising to learn that there is still considerable debate in many situations as to the relative contributions of these various processes of cell renewal to normal tissue homeostasis and, moreover, to situations where more substantive cell renewal is required such as after injury or during regeneration. During development, in general cell replication exceeds cell death and the organism grows, whereas in adult life the processes are balanced and tissue size is maintained.

When mutations occur in non-dividing cells such as neurons, they rarely induce cancer, which accounts for the rarity of such tumors in adults. However, some tissues with usually low rates of turnover exhibit considerable plasticity during adult life, for example many endocrine glands can undergo adaptive increases in mass in response to increased demands for their hormone output. Importantly, such adaptive growth is not inevitably a response to a pathological process, but is a normal part of physiology, for example during pregnancy, where demands for various hormones such as insulin increase. Particularly, in this latter scenario adaptive growth of insulin-secreting pancreatic β cells is usually followed by β-cell apoptosis and involution of the previously expanded islet tissue postpartum when the demand for insulin is reduced. Intriguingly, a recent study employing a lineage-tracking system demonstrates that β-cell numbers may normally be maintained primarily be replication of preexisting β cells themselves, rather than from a progenitor cell, and even following surgery to remove a part of the pancreas, including some of the islets that contain the β cells, new β cells formed during recovery arise from replication. Adaptive growth may also occur in response to nutrient deprivation or disorders in metabolism. Thus, the parathyroid glands may undergo hyperplasia (increase beyond normal cell mass) in response to reducing serum calcium (secondary hyperparathyroidism). Importantly, under some circumstances, such as that following a prolonged or frequently repeated period of adaptive growth, hyperplasia of a tissue may become autonomous and continue even after the stimulus has been removed. A well-known example of this occurs in the parathyroid glands – a condition known as tertiary hyperparathyroidism. In other circumstances, hyperplasia may indeed progress to formation of a benign (adenoma) or malignant (adenocarcinoma) neoplasia. Presumably in these cases, the growth-promoting environment may also provide a fertile one for the accumulation of oncogenic mutations. In this scenario, such oncogenic mutations (e.g. activated *c-MYC* or *RAS*, loss of *p53* or R*b*) would drive excessive cell replication independently of the normal regulatory signals (Chapters 6 and 7).

Cancers occur in many tissues populated predominantly by non-dividing differentiated cells. Stem cells are the major or only usual source of new cells in such tissues, which include epidermis and gut epithelium where cell losses are extensive (frequent mechanical damage, normal regular sloughing of cells from the surface), or when the normally minimal cell losses in a given tissue are increased such as after injury or inflammation. It is generally believed that stem cells may be the major origin of cancers in these cases, though many recent studies have shown that terminally differentiated cells can readily become less differentiated and re-enter the cell cycle under certain circumstances, suggesting that not all cancers in these tissues necessarily arise in the stem cell compartment. In general, differentiation of a cell greatly reduces its replication potential. Under some circumstances such as during inflammatory conditions even differentiated epidermal keratinocytes may replicate, though this is associated with some loss of or delay in acquiring the normal fully differentiated phenotype. The association between differentiation and replication is clearly seen in human tumors, with the most poorly differentiated cells (for example in the metastases) exhibiting the shortest cell division times.

CELLULAR DIFFERENTIATION

The cells comprising the various tissues of a multicellular organism often exhibit dramatic differences in both structure and function. In fact, if you consider the differences between a mammalian nerve cell and a lymphocyte it becomes difficult to imagine that the two cells have identical genomes. In fact, for many years differentiation was held as evidence against the central dogma of genetics and indirectly contributed to the beginnings of what we know as epigenetics (Chapter 11) – namely, information that can be passed on to the progeny during cell division not encoded in the DNA. Originally, it had been suggested that differentiation might actually involve the irreversible loss of some genes (DNA). This is now known not to be the case; rather differentiation involves the progressive alteration of gene expression. Various factors may produce the unique patterns of gene expression characteristic of any given cell type, including silencing of some genes by methylation and chromatin modification (epigenetics) and selective activation of others.

Importantly, it is increasingly apparent that many cell types do not irreversibly alter gene expression during differentiation, rather they may actually retain the ability to adopt a more primitive developmental state if required and stimulated by the appropriate signals or potentially under the influence of cancer-causing mutations. In fact, it may be a normal feature of cells which retain the potential to replicate in the adult (and this does not just apply to stem cells) to "de-differentiate" somewhat in order to be able to replicate. This is well illustrated in cancer cells, which are highly replicative and almost invariably show some loss of the normal differentiated features of the cell of origin. Thus, in several tumor models activation of c-MYC may cause loss of differentiation alongside cell cycle entry.

One important developmental regulator, the hedgehog protein (Hh) controls cell proliferation and differentiation in normal limb formation and bone differentiation and is also implicated in the self-renewal of stem cells in adult tissues. Persistent Hh pathway activation has been shown to contribute to various cancers including some forms of skin cancer. In an intriguing study, it was recently shown that loss of imprinting (LOI) of *Igf2* (see also Chapter 11) in a mouse model doubles the risk of developing intestinal tumors and that this is associated with a less differentiated phenotype of cells even in the normal colonic mucosa. Similar observations have also been made in humans with LOI of *IGF2*. The authors suggest that altered maturation of non-neoplastic tissue may be one mechanism by which epigenetic changes affect future cancer risk.

Differentiation therapy

The rationale behind differentiation therapy is the assumption that cancer cells have been fixed in an immature or less differentiated state and that this contributes to accelerated replication and avoidance of apoptosis. Differentiation therapy aims to drive the cancer cell back into a more differentiated state and resume the process of maturation. Differentiation therapy is therefore not in itself cytotoxic but instead restrains growth and may sensitize the cells to potential apoptotic triggers or chemotherapy. The first successful differentiation agent was all-trans-retinoic acid (ATRA) used to treat acute promyelocytic leukemia (APL). APL is the result of a translocation (an exchange of chromosome material) between chromosomes 15 and 17, which results in formation of a chimeric protein comprising the promyelocytic leukemia protein (PML) and the retinoic acid receptor alpha (RARa), which regulates myeloid

differentiation. PML/RAR causes an arrest of maturation in myeloid cells at the promyelocytic stage, contributing to expansion of promyelocytes. Treatment with ATRA causes the promyeloctes to differentiate and may result in remission rates of up to 70%.

The concept of terminal differentiation

Traditionally, cells that have irreversibly lost the ability to proliferate during the process of developing specialized functions (differentiation) have been described as **terminally differentiated**. However, this term can be misleading because it implies that a given adult cell type cannot proliferate. In fact, many cell types, previously believed terminally differentiated, have subsequently been shown to be capable of proliferation; in the previous chapter the concept was introduced that some cells may remain quiescent (in G_0) for extremely long periods of time waiting for an appropriate stimulus to re-enter the cell cycle. So in general it may be preferable to think about differentiated cells that are not replicating as quiescent unless one is absolutely certain. Thus, in some cases, of which skin keratinocytes are a notable example, cells lose their nuclei during terminal differentiation and are as a consequence unambiguously irreversibly growth-arrested. However, such an obvious example aside, one must be cautious in adopting the definition of terminal differentiation, as has been done in the past, to all those cell types that are perceived to be unable to proliferate to any useful extent and therefore would not usually contribute to tissue growth, maintenance, or repair. One illuminating example would be that of the insulin-secreting β cells of the pancreatic islets; in this case it was long accepted that the β-cell mass was maintained and underwent adaptive increases in growth through the replication of a progenitor cell/cells, but recent studies using "lineage tracking" have clearly shown that β cells can themselves re-enter the cell cycle and are surprisingly the predominant usual source of new β cells in the adult. Crucial to resolving the seeming paradox, namely, that differentiation and replication are generally incompatible, has been the repeated demonstration that differentiated cells can shake off some of their specialized functions (and change gene and protein expression to more closely resemble that of an earlier developmental stage) – termed cell "plasticity" – in order to replicate. Figure 5.1 shows a schematic representation of how cell cycle entry and progression may be regulated in a well-studied usually quiescent cell type, the myotube.

In fact, in most cancer cells, at least some of the "markers" of the presumed mature cell type from which the cancer cell may have originated are absent or reduced (compared to the normal cell counterpart). This also reflects directly on an important question in cancer biology – what is the cell of origin for various cancers? This is addressed in other chapters, but is worth noting here – *because cancer cells are in general less differentiated, does this imply that they have originated in stem cells that may have "differentiated" a bit, or conversely in mature cells that may have "de-differentiated" a bit?*

These intriguing questions aside, what is clear is that increasing or restoring differentiation (specialized behaviors) in cancer cells may be one means of reducing or even removing their replicative or invasive potential. Thus, understanding the mechanisms that cause permanent loss of proliferative capacity or at least prolonged quiescence in differentiated cells is of major interest and may lead to the design of new therapeutic strategies aimed at targeting differentiation of cancer cells. Differentiation is not the only means by which cells may lose their ability to proliferate, however, and replicative senescence, which may be induced by activation of tumor suppressor pathways, is discussed in more detail in Chapter 9.

ANGIOGENESIS AND TISSUE GROWTH

During development and under some circumstances in the adult (e.g. wounding), tissue adaptation and regeneration are accompanied by the growth of new blood vessels, a process known as angiogenesis (Chapter 14). The resultant supply of blood to the growing tissue – aside from delivering nutrients and oxygen and removing waste products and carbon dioxide – establishes an access route for circulating regulatory factors (endocrine route – Figs 5.2, 5.10) and immune cells (Chapter 13). Importantly, it

now seems increasingly likely that the vascular endothelial cells also provide GF signals that have a direct local influence on cell survival and growth of the tissue.

The notion that tissue expansion per se can be associated with the concurrent provision of an appropriate vasculature is not a novel one. For angiogenesis to take place an appropriate increase in angiogenesis-promoting factors and or a decrease in angiogenesis-suppressing factors must occur. However, for many years based on landmark studies by Judah Folkman, Doug Hannahan, and others the prevailing view has been that during tissue expansion in cancer an "angiogenic switch" was required and probably conferred by a specific and distinct mutagenic event which enabled a would-be cancer to become angiogenic and thereby for cancer to progress. By implication, such a switch would activate angiogenesis-promoting factors or deactivate suppressors, thus enabling a tumor to become angiogenic and expand. However, although this may be the case in some cancers, recent results challenge the idea that this is a general requirement. Thus, tissue growth under the influence of some oncogenes, notably *c-MYC* and *RAS*, may automatically be associated with an angiogenic response without a further "oncogene-driven switch"; in other words the activity of these proteins includes activation of angiogenesis. In some cases this may relate to the development of a deficiency in oxygen supply (hypoxia) as a growing tissue begins to outstrip its existing blood supply or the direct secretion of angiogenic factors such as VEGF. A more detailed discussion of angiogenesis is given in Chapter 14.

KEY EXPERIMENTS

In 1956, Rita Levi-Montalcini and Viktor Hamburger performed the seminal studies that culminated in the first identification of a growth factor. NGF was the first growth factor identified, for its action on the morphological differentiation of neural-crest-derived nerve cells. Later, its effect on neuronal cells of the peripheral and central nervous systems, and on several non-neuronal cells, was also determined.

In these studies a particular mouse tumor known as a sarcoma was implanted into developing chick embryos. These tumors became heavily innervated, with sympathetic and some sensory neurones, and it was reasoned that a factor was being produced by the tumor which promoted nerve growth. Extracts from these tumors were next found to promote neurone outgrowth in explant cultures. Further studies demonstrated that a similar activity could be observed in some snake venom, and finally in submandibular salivary glands, allowing the factor to be isolated and purified. The factor was later named Nerve Growth Factor (NGF). Rita Levi-Montalcini was awarded the Nobel prize for this work in 1986.

Aloe L. Rita Levi-Montalcini: the discovery of nerve growth factor and modern neurobiology. Trends Cell Biol. 2004 Jul;14(7):395–9.

GROWTH FACTOR SIGNALING PATHWAYS

The transmission and transduction of extracellular growth signals into the expression of genes involved in cell replication is fundamental to our understanding of cancer biology. Not surprisingly, most of the major cancer-causing mutations (but by no means all) relate in some way to these signaling pathways; with mutations that either activate signaling molecules which promote growth (oncogenes) or disable molecules required for arresting cell replication or inducing apoptosis (tumor suppressors – Chapter 7). Broadly speaking, proteins encoded by oncogenes can be ordered or classified according to the level they occupy within a growth factor signaling pathway and also by the nature of the key functional domains of the oncoprotein. One such scheme is outlined here and will be followed in Chapter 6, where oncogenes are discussed in more detail.

Growth factors: Including platelet-derived growth factor (PDGF), epidermal growth factor (EGF).
Receptor tyrosine kinases (RTK): Including the receptors for various growth factors, such as EGF.

Membrane-associated non-receptor tyrosine kinases: Including SRC.
G-protein coupled receptors: Including the angiotensin receptor.
Membrane-associated G-proteins: Including three different homologs of the *c-RAS* gene, each of which was identified in a different type of tumor cell.
Signaling pathways: Primarily involving a series of enzymes that regulate activity of proteins by either adding or removing phosphate groups.
Serine/threonine kinases: Including the oncoprotein RAF, involved in the signaling pathway of most RTK, and responsible for threonine phosphorylation of MAP kinase following receptor activation.
Lipid kinases: Including PI3 kinase, involved in activating AKT.
Nuclear DNA-binding/transcription factors: Including c-MYC and FOS.
Nuclear receptors: Not all growth factors act via receptors on the cell surface, but some can enter the cell and bind to nuclear receptors. These are ligand-dependent transcription factors that regulate cell growth and differentiation in many target tissues and include receptors for steroid hormones such as estrogen.
Specific regulators of survival: Including the pro- and antiapoptotic members of the BCL-2 family of proteins (discussed in Chapters 6 and 8).

GROWTH FACTORS

GFs, as with oncogenes, are generally named after experimentally observed activities associated with their original identification (e.g. fibroblast growth factor, FGF, platelet-derived growth factor, PDGF, transforming growth factor, TGF-α and TGF-β, insulin-like growth factor, IGF, and so on).

GFs are polypeptides that bind to cell surface receptors, with the major action of promoting cellular proliferation, but may also have major effects on differentiation, migration, and survival (when they are sometimes referred to as "survival factors"). Many GFs are very versatile and act on multiple cell

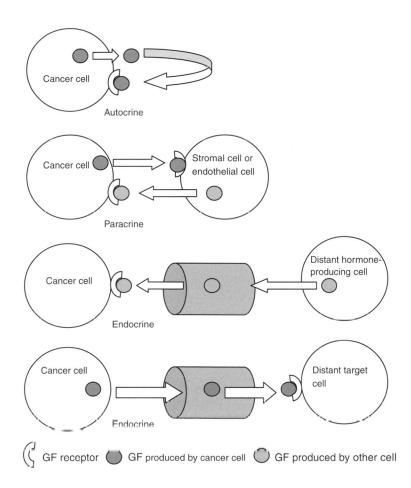

Cancer cell

Autocrine

Cancer cell — Stromal cell or endothelial cell

Paracrine

Cancer cell

Distant hormone-producing cell

Endocrine

Cancer cell

Distant target cell

Endocrine

(GF receptor   GF produced by cancer cell   GF produced by other cell

Fig. 5.10 Growth factors may act via different routes. Secreted growth factors may act locally on the same cell (autocrine), or on neighboring cells (paracrine), or, at a distance (endocrine).

types, while others are specific to a particular cell-type. As can be seen from Table 5.1, which contains a partial list of GFs implicated in cancer, these factors are produced by various cell types including stromal cells, such as fibroblasts and adipocytes, and even more distant tissues that can secrete GFs, which act via the circulation (see Fig. 5.10 for a summary of the various sources of GFs). Although many tumor cells become a source of GF production, tumor cells can also influence stromal cells to secrete GFs that, in turn, can promote tumor cell behavior (see Fig. 12.7, Chapter 12). Many GFs are synthesized by stromal cells and are normally sequestered by proteins within the extracellular matrix (ECM). In this way, tight control over their activity can be maintained within normal tissues. However, enzymes known as matrix metalloproteinases (MMPs) are often secreted by tumor cells, and can

release the sequestered GFs (e.g. IGFs, TGF-α, EGF, TGF-β) – see Fig. 5.18. Such GF activity can promote tumor behavior as described in Chapter 12. It is also worth remembering that GFs usually require integrin-mediated cell adhesion in order to activate signaling pathways, ensuring that activation only occurs when cells are in their correct environment. This homeostatic function becomes disrupted in cancer.

Cytokines are a distinct category of GFs that are secreted by leukocytes, but also by other cells, and are involved in directing a wide range of processes including immune responses and activation of phagocytic cells. The interleukins are cytokines secreted by multiple cell types and act as growth factors for cells of hematopoietic origin, as well as other functions. The list of identified interleukins grows continuously with at least 22 recognized at the time of

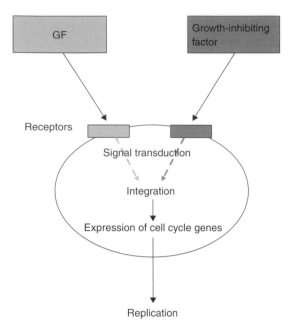

Fig. 5.11 Signaling through membrane-bound GF receptors. Positive and negative signals are integrated at various levels before a decision on whether to respond or not is completed.

going to print. A more detailed discussion of cytokines in the immune system is provided in Chapter 13.

Normal cells growing in culture will not divide unless they are stimulated by one or more GFs present in the culture medium. Extracellular polypeptides and proteins act via cell membrane receptors whose signals are transduced to the nucleus via intermediary signals including protein–protein interactions, phosphorylation, and cytoskeletal changes (Fig. 5.11). Examples of such signaling transduction pathways are described later. Lipid-soluble regulators such as steroid hormones and retinoic acid readily traverse cell membranes and bind to intracellular receptors to form transcription factor complexes (Fig. 5.12). The source of these extracellular signals can be endocrine (hormones which act on distant targets reached via the circulation), paracrine (produced by nearby cells of a different type and acting locally), or autocrine (produced by the same cells on which they act) – see Fig. 5.10. In cancer each of these is important, and such signals may be acting directly on the cancer cell or on other cells in its environment, such as blood vessel endothelium (Chapter 14), or the extracellular matrix, which

can potentially be induced to support aspects of neoplastic growth.

Growth factors and cancer

Although most GFs are not oncogenes, the excess production of (or increased sensitivity to) existing GFs may also have a major role in supporting cancer growth (Table 5.2). Such examples include IGF-1, levels of which are increased in the disease acromegaly, characterized by excess production of growth hormone (GH) by pituitary tumors and generalized expansion of soft tissues. GH increases secretion of IGF-1 into the circulation primarily from the liver, which in turn promotes expansion of connective tissue cells amongst others, and may increase the likelihood of some cancers. IGFs are important activators of the IRS–PI3K–AKT pathway. Small-cell lung cancers can produce GFs such as bombesin; some tumors produce IGF-2; and circulating estrogens can support the growth of estrogen receptor-positive breast cancer cells. Autocrine transforming interactions have been identified in a number of human malignancies (Fig. 5.2). At least one PDGF chain and one of its receptors

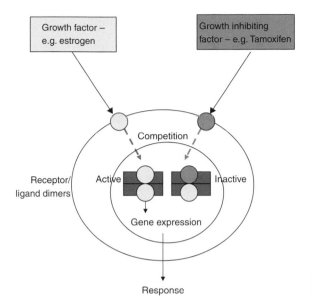

Fig. 5.12 Signaling through nuclear receptors. Some hormones may act as ligands for nuclear receptors that may then act as transcription factors.

have been detected in a high fraction of sarcomas and in glial-derived neoplasms. Tumor cells cultured from some of these cancers demonstrate a functional autocrine loop, in which persistent PDGF receptor activation is shown.

TGF-α is often detected in carcinomas that express high levels of EGF receptor. However, the role of acidic or basic FGF (bFGF) in tumors is still unclear as neither molecule possesses a secretory signal peptide sequence. However, recent studies have implicated bFGF in melanoma; human melanoma cell lines express bFGF, which is not observed with normal melanocytes. Moreover, bFGF antagonists can inhibit growth of melanoma cells. Although several growth factors have been shown to induce transformation by an autocrine mode, it is also worth considering the possible role that growth factors might have in predisposing to cancer. It can be hypothesized that overexpression of growth factors by a paracrine mode might increase the proliferation of a polyclonal target cell population (Fig. 5.2). By such a model, increased growth factor production could act in a manner analogs to that of a tumor promoter. For example, continuous stimulation by growth factors in paracrine as well as autocrine modes, during chronic tissue damage and repair associated with cirrhosis and inflammatory bowel disease, may predispose to tumors. Finally, paracrine-acting angiogenic growth factors such as the VEGFs (vascular endothelial growth factors) are produced by some tumor cells. Such growth factors cause the paracrine stimulation of endothelial cells, inducing new blood vessel growth (angiogenesis) and lymphangiogenesis, which contribute to tumor progression.

Growth factors may also act via endocrine routes (Fig. 5.2). With respect to human cancers this particularly includes estrogens produced from the ovary acting on breast cancers and IGF-1 from the liver acting on diverse tumor types. It is worth reiterating here that these GFs may exert equally important effects on the tumor's environment as on the tumor directly (for detailed discussion, see Chapter 12).

Binding proteins, which are not receptors, exist in the circulation and may bind and inactivate a ligand or serve as a pool of available ligand, because binding is reversible. Protein binding acts as an important reservoir in the circulation for IGFs and also for hormones such as thyroxine.

Growth factor oncogenes

The *v-sis* gene was the first viral oncogene found to encode a growth factor and was subsequently identified as having homology to a known

Table 5.2   A short list of some known GFs

GF	Principal source	Main physiological role	Role in cancer
PDGF	Platelets, endothelial cells	Proliferation of connective tissue cells, glia, and smooth muscle cells	Two different protein chains form three distinct dimer forms: AA, AB, and BB. The *c-Sis* proto-oncogene has been shown to be homologous to the PDGF A chain.
EGF	In large quantities in salivary glands; produced by a variety of other cell types	Proliferation of mesenchymal, glial, and epithelial cells	The EGF receptor is a member of a family of four receptors [EGFR (HER1 or ERBB1), ERBB2 (HER2/neu), ERBB3 (HER3), and ERBB4 (HER4)]. EGFR promotes deregulated growth in many types of cancer, including breast and lung, and activating mutations are known. Two new drugs, which target the receptor or its downstream kinase activity, are now in use in cancer: cetuximab and gefitinib. The *Neu* gene was identified as an EGFR-related gene in an ethylnitrosourea-induced neuroblastoma and is implicated in breast cancer and others. This is also targeted by a monoclonal anticancer agent, trastuzumab.
TGF-α	Transformed cells	Normal wound healing	Often produced by cancers. May activate similar receptors and signaling pathways as EGF.
FGF	Multiple cell types, protein is associated with the ECM	Promotes proliferation of many cells; inhibits some stem cells; induces mesoderm to form in early embryos	At least 19 family members, 4 distinct receptors. Kaposi's sarcoma cells (prevalent in patients with AIDS) secrete a homolog of FGF, encoded by the *K-FGF* proto-oncogene. The *Flg* gene (*Fms-like gene*) named because it has homology to the *Fms* gene, encodes a form of the FGF receptor
M-CSF	Monocytes, granulocytes, endothelial cells, and fibroblasts	Proliferation and differentiation of hematopoietic cells	The *c-Fms* gene encodes the M-CSF receptor (CD115), first identified as a retroviral oncogene.
NGF	Schwann cells, fibroblasts, sites of sensory and sympathetic innervation	Promotes neurite outgrowth, neural cell survival	NGF identified as secreted by adrenal medullary tumors. The *Trk* genes encode the NGF receptor-like proteins. The first *Trk* gene was found in a pancreatic cancer. Now known to comprise several related genes: *trkA*, *trkB*, and *trkC*. Numerous other "neurotrophic" factors are now known, including BDNF, CNTF, GDNF.
HGF	Many cell types including stromal cells	Potent mitogen for mature parenchymal hepatocyte cells; involved in cell–matrix interactions and cellular migration in a wide range of tissues and cell types	HGF and its receptor c-Met are involved in cancer invasion and metastasis. Paracrine activation of c-Met by stromal-derived HGF facilitate invasion and metastasis. Autocrine or mutational activation of *c-Met* associated with the progression of malignant tumors.

Table 5.2 (*continued*)

GF	Principal source	Main physiological role	Role in cancer
Erythropoietin	Kidney	Promotes proliferation and differentiation of erythrocytes	Used therapeutically in some cases of anemia.
TGF-β	Inflammatory cells, stroma	Anti-inflammatory, promotes wound healing, inhibits proliferation of many adult cell types	May either be anti- or procancer depending on timing.
IGF-1	Liver endothelial cells	Promotes survival of many cell types and proliferation of some	Related to IGF-2 and proinsulin, also called somatomedin C
IGF-2	Variety of cells	Promotes proliferation of many cell types primarily of fetal origin	Related to IGF-1 and proinsulin. Secreted by some cancers as an autocrine/paracrine survival factor.

cellular gene, *c-sis*. *Sis* is the oncogene incorporated into the genome of the simian sarcoma virus, and it encodes the platelet-derived growth factor PDGF-B chain. This virus is able to form tumors in its host as a result of overproduction of PDGF-B under the control of viral regulatory elements rather than its own promoter.

Box 5.2 Regulation of cell size – additional information

Tissue mass may increase by increases in cell volume as well as increases in cell number.

Although clearly linked, it has become increasingly apparent that the processes of cell replication and cell growth (increase in cell volume) may be independently regulated, though growth is under normal circumstances a usual precursor to replication. Increased cell proliferation requires a general increase in protein synthesis and a specific increase in the synthesis of replication-promoting proteins. Transient increase in the general protein synthesis rate, as well as preferential translation of specific mRNAs coding for growth-promoting proteins (e.g. cyclin D1), occur during normal mitogenic responses. Several signal transduction pathways involved in growth such as RAS–MAPK, PI3K, and mTOR-dependent pathways activate the translational machinery. Cell growth (an increase in cell mass and size through macromolecular biosynthesis) and cell cycle progression are generally tightly coupled, allowing cells to proliferate continuously while maintaining their size. The target of rapamycin (TOR) is an evolutionarily conserved kinase that integrates signals from nutrients (amino acids and energy) and growth factors (in higher eukaryotes) to regulate cell growth and cell cycle progression coordinately. In mammals, mTOR is best known to regulate translation through the ribosomal protein S6 kinases (S6Ks) and the eukaryotic translation initiation factor 4E-binding proteins.

In cancer, oncogenic activation of RAS or MYC may lead to continuous upregulation of key elements of translational machinery. On the other hand, tumor-suppressor genes (p53, pRB) downregulate ribosomal and tRNA synthesis. The RAS and Akt signal transduction pathways play a critical role in regulating mRNA translation through activation of the initiation factor eIF4E, which binds the 5′cap of mRNAs. Not surprisingly, therefore, IF4E is overexpressed in many human cancers and controls the translation of various malignancy-associated mRNAs which are involved in polyamine synthesis, cell cycle progression, cell survival, angiogenesis, and invasion.

As mentioned earlier, viruses can also integrate in the vicinity of a proto-oncogene and activate transcription of the host gene, leading to formation of the tumor. The *Int-2* gene is a common site of integration of the MMTV, and encodes an FGF-related growth factor. MMTV has also been shown to integrate in the vicinity of various members of the FGF family, such as FGFs 3, 4, and 8, and consequently can activate transcription of the particular FGF in the mammary gland, as the host gene is now under the control of the viral promoter.

Potentially, viral proteins may also activate GF receptors. For example, the human papilloma virus (HPV) protein E5 can result in aggregation of PDGF receptors, mimicking dimerization and resulting in sustained receptor autophosphorylation and activation.

TGF-β – "playing both sides" in cancer

The role of transforming growth factor beta type I (TGF-β) in growth regulation and cancer is sometimes seen as confusing. TGF-β is a ubiquitous cytokine that is known to inhibit proliferation of numerous cell types, particularly those of epithelial origin. Moreover, mutations inactivating various components of the TGF-β signaling pathway are found in many cancers, particularly of gut origin, supporting the role of TGF-β as a tumor suppressor. In complete contrast, other cancers manifest enhanced growth in response to TGF-β and TGF-β may contribute to promoting epithelial-to-mesenchymal transition (EMT) and metastatic behavior. Recent studies have shed light on the potential "switch" of TGF-β from growth inhibitor to promoter during oncogenesis for some epithelial cancers. Thus, the TGF-β "switch" may be thrown by impaired CKI activity (e.g. loss of p27[KIP1]), which unmasks the growth-promoting actions of this cytokine. This is one of the best-described examples of how important the timing (as well as the nature) of cancer-causing events is in the life history of a tumor.

As discussed in Chapter 12, epithelial tissues require interactions with the stroma both in normal tissue homeostasis and particularly during tumorigenesis. Cancer-related changes in the epithelium are inevitably accompanied by changes in the tissue stroma that in turn contribute to the cancer phenotype. Stromal changes may foster invasion, of which a "transdifferentiation" of fibroblasts into myofibroblasts may be an important example. Some of these cancer-promoting effects of the stroma are mediated by release of factors such as TGF-β by the tumor, which act in a paracrine fashion to contribute to angiogenesis, "immune privilege" (avoidance of immunosurveillance), and recruitment of myofibroblasts. TGF-β may also have an autocrine effect on the cancer cells themselves, particularly if CKIs are inactivated (TGF-β switch) or the cancer cell has activated β-catenin or RAS mutations. A model for TGF-β signaling is shown in Fig. 5.5.

TYROSINE KINASES

Results from the human genome project suggest that there may be around 1000 protein tyrosine kinases (TK) in the genome. Deregulated activation of TK is involved in most cancers and in fact over 70% of the known oncogenes and proto-oncogenes involved in cancer encode TK. There are two main classes of TK: receptor TK (RTK) and cellular, or non-receptor, TK. Of the known TK around two-thirds are RTK. Not surprisingly, receptor and non-receptor TK have emerged as clinically useful drug target molecules for treating certain types of cancer.

Receptor TK share an extracellular ligand binding domain, a transmembrane domain, and an intracellular catalytic domain. Different receptor TK have different extracellular domains that confer ligand specificity, including one or more recognized structural motifs, such as cysteine-rich regions, fibronectin III-like domains, immunoglobulin-like domains, EGF-like domains, cadherin-like domains, kringle-like domains, Factor VIII-like domains, glycine-rich regions, leucine-rich regions, acidic regions, and discoidin-like domains.

Activation of the TK is achieved by ligand binding to the extracellular domain, which induces dimerization of the receptors. Activated receptors autophosphorylate tyrosine residues outside the catalytic domain, stabilising the receptor in the active conformation and recruiting various adapter proteins required for signaling.

Unlike receptor TK, non-receptor TK are located in the cytoplasm or nucleus, or anchored to the inner leaflet of the plasma membrane. They are divided into various families, each comprising several members: SRC, JAK, ABL, FAK, FPS, CSK, SYK, and BTK. Aside from carrying homologous kinase domains (SRC Homology 1, or SH1 domains), and some protein–protein interaction domains (SH2 and SH3 domains), these families are structurally unrelated. Biological activities are also variable; many, exemplified by SRC, are involved in cell growth (Chapter 6). In contrast, FPS is involved in differentiation, ABL in growth inhibition (Chapter 6), and FAK with cell adhesion. Some members of the cytokine receptor pathway interact with JAKs, which phosphorylate the transcription factors, STATs (see below).

GROWTH FACTOR RECEPTOR TYROSINE KINASES

GF ligands bind to membrane receptors on the external surface of the cell membrane and subsequently generate a signal on the inside, which is then ultimately relayed to the nucleus via a signal transduction pathway involving multiple proteins and a series of phosphorylation events.

The receptors for many GFs either contain intrinsic TK within their cytoplasmic domain (RTK), or like the cytokine receptors can recruit TK to the active receptor. Although GF receptors can be constitutively activated by autocrine loops as discussed above, various other mechanisms are known. Oncogenic activation of GF receptors invariably results in forced dimerization of the receptor. The outcome is activation of signal transduction pathways without the requirement for GF stimulation. This uncontrolled activation may result in cell proliferation, survival, or indeed other cell behaviors, such as increased motility.

Here we will describe a model RTK signal transduction pathway, the basics of which are illustrated in Fig. 5.6. For many GF receptors, autophosphorylation is mediated by dimerization following ligand binding (mutual phosphorylation of receptors) – see Fig. 5.13.

Autophosphorylation generates key phosphotyrosine sites, which enable binding of various signaling adapter proteins, many of which contain the SRC Homology domain 2 (SH2 domain) – see Fig. 5.14. Multiple signaling pathways are regulated by RTK – see Fig. 5.15 and below. The SH2 domain is a sequence originally described in several tyrosine kinases such as c-SRC and ABL and is distinct from the kinase catalytic domain (SH1), but is required for interaction with targets. In addition to the primary phosphorylation sites in receptors, there are secondary sites that bind other SH2 domain proteins. Several SH2-containing proteins have been show to interact with GF receptors, such as PLC-γ (phospholipase C), SRC, and GRB (growth-factor receptor bound). GRB2 consists of a single SH2 domain flanked by two SH3 domains. SH3 domains act as binding sites for proline-rich domains in effector molecules acting downstream in the signaling pathway. Many of these aspects are illustrated in the RAS–MAPK and ABL signaling pathways shown in Fig. 5.16.

GRB2 also recruits SOS (from the *sos* gene son of sevenless – a *Drosophila* gene responding to the sevenless RTK required to specify the seventh receptor cell of the fly eye). This localizes SOS to the membrane, where it acts as a guanine nucleotide exchange factor (GEF) for activation of RAS (exchange of GDP for GTP). Conversely, GTPase activating proteins (GAP) negatively regulate RAS by stimulating GTPase activity (GTP is converted to GDP). Key signaling components such as RAS and c-SRC are anchored to the membrane, by isoprenyl and myristoyl chains, respectively (Chapter 6). Recruitment of the signaling adapter proteins close to the membrane increases their local concentration, allowing the signal to pass effectively. RAS in turn activates RAF, a serine/threonine protein kinase, which is the starting point for the MAP kinase pathway (Fig. 5.16) controlling cell proliferation, and is discussed in more detail in Chapter 6.

RTK and cancer As is the case for other proto-oncogenes, viral oncogenes encoding mutant growth factor receptors have been identified, the best known of which is *v-erbB*, the viral oncogene of avian erythroblastosis virus. *v-erbB* is the oncogenic counterpart of the epidermal

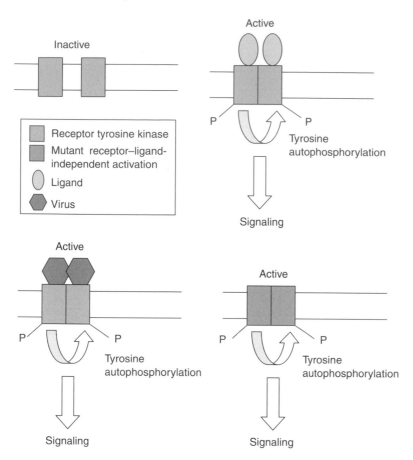

Fig. 5.13 Dimerization of ligand-bound RTK activates intrinsic TK activity. RTK are activated by ligand binding, which induces conformational changes, and receptor dimerization that activates intrinsic TK activity (panel 1). RTK can also become activated by viruses (panel 2) or by mutations (panel 3)

growth factor receptor (EGFR). The *c-Fms* gene encoding the colony stimulating factor-1 (CSF-1) receptor was also first identified as a retroviral oncogene.

RTK can also be activated as a result of mutations. The *Neu* gene was originally isolated from a chemically induced (ethylnitrosourea) neuroblastoma in the rat, and it encodes the HER-2 receptor, encoded by an epidermal growth factor (EGF) receptor-related gene (Box 5.3). Conversion of proto-oncogenic *Neu* to oncogenic *Neu* requires only a single amino acid change in its transmembrane domain – this subsequently leads to the spontaneous dimerization (and thus activation) of the HER-2 receptor. In man, oncogenic *NEU* is now known to be a very important contributor to the pathogenesis of many breast cancers and NEU is the target of the drug herceptin, used successfully in the treatment of appropriately selected cancers. Deletions within

the external domain of the EGFR can result in ligand-independent activation of RTK activity. Several other receptors, including MET (*the hepatocyte growth factor(HGF)/scatter factor (SF) receptor*) PDGF-β, RET, c-KIT, ALK, and the neurotrophin receptors (NGF receptor-like proteins) TRKA (encoded by the *NTRK1* gene) and TRKC, have been shown to be oncogenically activated in human malignancies by gene rearrangements that lead to fusion products containing the activated TK domain (Table 5.3).

RTK which have not been structurally altered by mutation also play a role in many cancers. Thus, normal RTK may be upregulated or amplified in human cancers, examples of which include EGFR, ERBB2, and c-MET. *ERBB2* was initially identified as an amplified gene in a primary human breast carcinoma and a salivary gland tumor and was shown to transform NIH/3T3 fibroblasts when

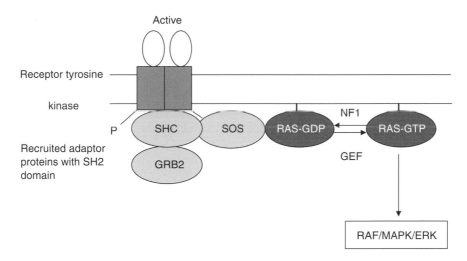

Binding of GF ligand to the receptor results in dimerization and activation of intrinsic tyrosine kinase activity and tyrosine autophosphorylation of the receptor. The resultant activated receptor recruits the GRB2/SOS complex to the receptor via SH2 domains. These adaptor proteins interact with the inactive GDP-bound RAS to catalyze exchange for GTP, a process facilitated by the guanine nucleotide exchange factor (GEF), thus activating RAS and downstream signaling. RAS is subsequently inactivated by reversion to the GDP-bound form by intrinsic GTPase activity, which is in turn facilitated by the NF1 protein.

Fig. 5.14 Activation of RAS by RTK.

overexpressed *in vitro*. The normal *ERBB2* gene is now known to be often amplified or overexpressed in human breast carcinomas and in ovarian carcinomas, and high *ERBB2* levels correlate with poor survival. Overexpression of normal EGFR has been found in squamous cell carcinomas and glioblastomas, where it is often activated by autocrine stimulation by TGF-α.

ERBB/HER2 The ERBB/HER protein-tyrosine kinases, which include the EGF receptor (EGFR) are implicated in many cancers (See Box 5.3).

OTHER RECEPTOR TYROSINE KINASES

RET

Glial cell line-derived neurotrophic factor (GDNF), a ligand for the RET tyrosine kinase, promotes the survival and differentiation of a variety of neurons. GDNF–RET signaling is crucial for the normal development of the kidney and various populations of neurons in

the peripheral nervous system, including sympathetic, parasympathetic, and enteric neurons. RET can activate several signaling pathways including the RAS/extracellular signal-regulated kinase (ERK), phosphatidylinositol 3-kinase (PI3K)/AKT (see below), p38 mitogen-activated protein kinase (MAPK), and c-Jun N-terminal kinase (JNK) pathways. These signaling pathways are activated via binding of adaptor proteins to intracellular tyrosine residues of autophosphorylated RET, as for other RTK.

THE MEN SYNDROMES

Multiple endocrine neoplasia (MEN) types 1 and 2 syndromes are rare hereditary cancer syndromes expressing a variety of endocrine and non-endocrine tumors. MEN1 results from inherited inactivating mutations in a putative tumor-suppressor gene encoding the protein menin and is discussed elsewhere. MEN2 arises through various activating mutations in the RET RTK. Constitutive activation of RET by somatic rearrangement with other partner

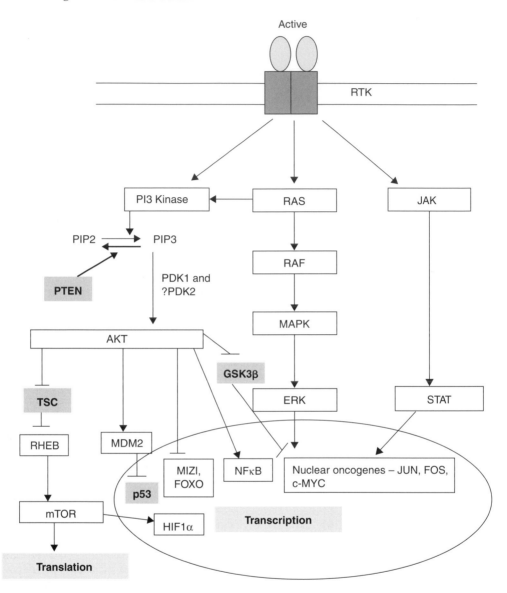

Fig. 5.15 Major growth-regulating signaling pathways activated by RTK. Amongst others, RTK can activate three major signaling pathways, RAS-RAF-MAPK; JAK-STAT and PI3K–AKT, all of which eventually signal to the nucleus and regulate gene expression. The latter is shown in greater detail as the other pathways are well covered in chapter 6. Some key tumor suppressors regulated or regulating this pathway are shown in bold.

genes or germ-line mutations results in human papillary thyroid carcinomas or multiple endocrine neoplasia – MEN type 2A and 2B, respectively – whereas the dysfunction of RET by germ-line missense and/or nonsense mutations causes Hirschsprung's disease, a disorder of nervous development in the gut resulting in motility

problems. The disease phenotype is affected by the nature of the mutant RET protein – for example, the MEN2B activating mutation also alters the substrate specificity of the RET tyrosine kinase, and thus induces a different set of gene targets to RET carrying the MEN2A activating mutation, which conversely primarily

Box 5.3 ERBB family of receptor tyrosine kinases-additional information

Growth factors and their transmembrane RTK regulate cellular processes, particularly proliferation and migration, during embryogenesis, during tissue mass regulation in the adult, and in cancer. The ERBB/HER family of receptors play essential roles in the development of epithelial cell lineages and serve as a useful model of RTK in growth regulation and deregulation in cancer.

Four major forms of ERBB RTK are known and together with a variety of ligands comprise a complex signaling network in mammals. This network is present, though in a more simplified form, in primitive organisms such as *C. elegans* but has likely evolved through gene duplication and diversification to the more complex system found in higher organisms. This greater level of organization may allow for more highly tuned responses and integration of growth-regulatory inputs. For instance, ERBB-2/HER2 can only signal effectively in the presence of active ERBB-3. Moreover, ERBB-2/ERBB-3 heterodimers result in amplified responses to growth factors largely through their ability to engage both the RAS–MAPK cascade and the PI3K–Akt pathway, thus stimulating cell proliferation and blocking apoptosis. Heterodimers may also be more oncogenic because they can avoid receptor endocytosis, a major means of inactivating RTK.

ERBB proteins and various EGF-like ligands are major contributors to several human cancers. Three major mechanisms have been identified whereby ERBB is involved in cancer:

1. Through autocrine loops, whereby coexpression of a receptor and the respective ligand (e.g. ERBB-1/EGFR and the transforming growth factor alpha) are found in the same cells.
2. Genetic aberrations, which affect primarily ERBB-1/EGFR, and result in loss (deletion) of the regulatory domains. In fact, such mutant forms of ERBB-1 are found in both brain and lung tumors.
3. Overexpression of ERBB-2/HER2. Once overexpressed, excessive ERBB-2/HER2 may favor the formation of receptor heterodimers – that is to say, form dimers with other growth factor receptors. Dimerization normally only happens after growth factor stimulation (ligand binding). Thus, in cancer ERBB-2 can become a pre-activated receptor, serving to amplify a variety of different growth signals, without the need for binding of its own ligand.

Overexpression of ERBB-2 identifies aggressive subsets of mammary and other tumors, testifying to the potentially wide-ranging effects of inappropriate activation of growth-signaling pathways.

Several novel treatments involve the targeting of ERBB signaling, in particular the humanized monoclonal antibodies to ERBB-2 and to ERBB-1 (herceptin). These antibodies accelerate receptor internalization, thereby reducing availability of receptors for heterodimerization. Alternatives include the use of low molecular weight antagonists of the tyrosine kinase domain shared by all four ERBB proteins (e.g. gleevec). These analogs of ATP bind to the nucleotide-binding site of the tyrosine kinase and block all signaling events that depend on tyrosine phosphorylation. A more novel approach entails promoting degradation of ERBB in the proteasome.

results in deregulated activity of the usual repertoire of RET targets. The different downstream activities of the mutant RET account for the phenotypic differences between MEN2A and MEN2B and thus produce a different portfolio of endocrine tumors.

MEN1 is associated with hyperplasia or tumors of several endocrine glands, including the parathyroids, the pancreas, and the pituitary. The most frequent problem in MEN1 is hyperparathyroidism. Hyperplasia of the parathyroid glands and excess production of parathyroid hormone result in elevated blood calcium levels, kidney stones, constipation, weakened bones, and depression. Almost all MEN1 patients show parathyroid symptoms by age 40. Various gut and pancreatic endocrine tumors are also common in MEN1 (including gastrinomas that can result in excess secretion of gastrin and gastrointestinal ulcers and insulinomas that cause inappropriate secretion of insulin resulting in lowered blood glucose levels). The anterior pituitary and the adrenal glands can also be involved.

MEN2A and MEN2B have some features in common, likely due to overlapping downstream activities of the two different RET mutant proteins. Namely, medullary thyroid cancer (MTC), in which there is usually excess secretion of the

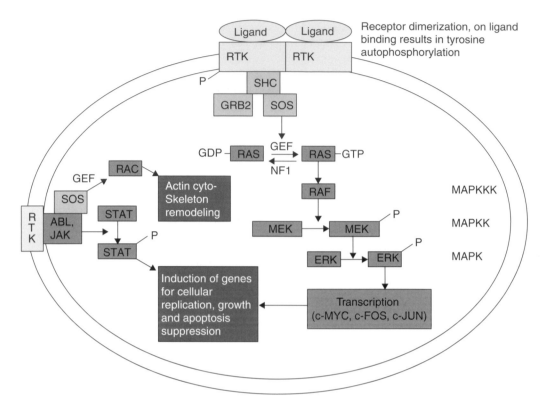

Receptor dimerization on ligand binding induces tyrosine autophosphorylation, binding of GRB2 to the receptor via SH2 domain and translocation of SOS to the cell membrane. SOS in turn promotes RAS activation by enhancing GDP–GTP exchange. A third adaptor protein SHC (also carrying an SH2 domain) may be phosphorylated by growth factor binding and can recruit the GRB2/SOS complex. Activated RAS by either of these two overlapping pathways in turn phosphorylates the serine/threonine kinase RAF. RAF phosphorylates and activates MAP kinase/ERK kinase (MEK). MEK then activates ERK, which can then translocate to the nucleus and activate transcription factors such as c-MYC, c-FOS, and c-JUN. This is a prototypic kinase cascade generically referred to as MAPKKK–MAPKK–MAPK. Growth factors (particularly cytokines) can also activate the Janus cytoplasmic protein tyrosine kinases (JAKs). These kinases lack SH2 or SH3 domains and phosphorylate the receptors themselves after ligand binding. JAK phosphorylated receptors then bind STAT transcription factors via SH2 domains, which become activated oligomers. STAT dimers can then translocate to the nucleus and activate transcription. The ABL cytoplasmic tyrosine kinase is also important in mediating growth factor signals. In particular ABL may link to RAS family members such as RAC involved in cell motility and it also plays a role in regulation of the cell cycle through interactions with RB.

Fig. 5.16 RAS–MAPK, a model growth factor-receptor-activated signaling pathway.

hormone calcitonin, and a tumor of the adrenal gland medulla known as pheochromocytoma, in which various adrenal catecholamine hormones are overproduced resulting in high blood pressure. However, the two forms of MEN2 can be distinguished by additional features, suggesting the activation of unique downstream targets specific to the different mutations in RET. MEN2A is associated with parathyroid gland hyperplasia or tumors of the parathyroid gland in around 20% of individuals. MEN2B is associated with more aggressive tumors and several morphological abnormalities often detectable by 5 years of age including a characteristic facial appearance with swollen lips; tumors of the mucous membranes of the eye, mouth, tongue, and nasal cavity; enlarged colon; and skeletal abnormalities.

Currently, DNA testing can facilitate early diagnosis of MEN syndromes by identification of germ-line mutation in asymptomatic mutant gene carriers.

Children who are identified as carriers of the RET gene can be offered total thyroidectomy on a preventative (prophylactic) basis to prevent the development of MTC. Prevention is the key goal as although MTC can be treated by surgical removal of the thyroid, metastatic spread of the cancer may already have occurred and neither chemotherapy nor radiotherapy is effective in controlling spread. Pheochromocytoma in both types of MEN2 can be cured by surgery.

PDGFr and KIT

KIT and platelet-derived growth factor receptors (PDGFrs) are oncogenic in a variety of hematologic and solid tumors. These RTK, as well as the non-receptor TK, ABL, and BCR–ABL, are inhibited by imatinib (see below).

The platelet-derived growth factors (PDGF) are a pleiotropic family of peptide growth factors that signal through cell-surface RTK (PDGFr) and stimulate various cellular functions including growth, proliferation, and differentiation. To date, PDGF expression has been demonstrated in a number of different solid tumors, from glioblastomas to prostate carcinomas. In these various tumor types, the biologic role of PDGF signaling can vary from autocrine stimulation of cancer cell growth to subtler paracrine interactions involving adjacent stroma and vasculature.

The tyrosine kinase inhibitor imatinib mesylate (gleevec) blocks activity of the BCR–ABL oncoprotein (Chapter 6) and the cell surface tyrosine kinase receptor c-KIT, and has recently been approved for treatment of chronic myelogenous leukemia and gastrointestinal stromal tumors. Imatinib mesylate is also a potent inhibitor of PDGFr kinase and is currently being evaluated for the treatment of chronic myelomonocytic leukemia and glioblastoma multiforme, based upon evidence in these diseases of activating mutations in PDGFr. Given the numerous mechanisms by which increased GF may arise during tumorigenesis, the PDGF pathway may represent a therapeutic target in tumors even when it is not in itself the target of an activating mutation.

G-protein coupled receptors and their ligands

In some cell types, such as Schwann cells of the peripheral nervous system, activation of G-protein coupled receptors (GPCRs), which couple to activation of adenylate cyclase and elevate intracellular cAMP, may stimulate replication in synergy with traditional growth or survival factors. Conversely, in these types of cells activation of GPCRs, such as somatostatin receptors that negatively regulate cAMP production, will reduce growth. In some cell types, such as those involved in growth hormone (GH) production in the pituitary gland, elevation of cAMP via GPCRs (e.g. GHrh) is pro secretory. Human pituitary adenomas express multiple somatostatin receptor subtypes, the pattern of which may in part determine the response to medical treatments with the somatostatin agonists octreotide and lanreotide, which preferentially activate the type 2 receptor. Human growth hormone-secreting (somatotroph) pituitary adenomas express both type 2 and type 5 receptors that are involved in the regulation of GH secretion – this adenoma is most responsive to octreotide treatment. Prolactinomas (secreting the pituitary hormone prolactin) rarely express the type 2 receptor, explaining the lack of efficacy of octreotide in lowering the elevated prolactin in patients. However, these tumors frequently express dopamine receptors and may be sensitive to treatment with dopamine agonists that can suppress growth and abnormal hormone secretion.

Amidated and non-amidated gastrins (gastrin precursors) may contribute to growth of gastrointestinal and pancreatic cancers. Progastrin and gastrin are antiapoptotic and amidated gastrins may promote migration of epithelial cells. In support of this, targeting gastrins therapeutically via a vaccine has had some limited success in treating GI and pancreatic cancers.

It has long been appreciated that the GPCRs can activate the MAPK signaling pathway and thereby contribute to cell replication. However, the mechanisms underlying this crosstalk have remained elusive. Work from the group of Axel Ullrich and others has identified crosstalk between GPCRs and epidermal growth factor receptor (EGFR) signaling systems in a variety of normal and transformed cell types. GPCR agonists can promote activation of the RAS–MAPK pathway and cell proliferation via the EGFR in fibroblasts and some cancer cells. The metalloprotease-disintegrin tumor necrosis factor-alpha-converting enzyme (TACE/ADAM17) may be important in determining this interaction. In part activation of stromal cell EGFR signaling by TACE may be mediated by proteolysis-induced release of the ligand amphiregulin.

Regulation occurs at multiple levels in these pathways, including at the level of the G-proteins themselves. G-protein alpha-subunits (G_α) appear to be capable of interacting with different effectors leading to engagement of distinct signaling pathways. Expression of activated G_α has been shown to cause transformation of cells, and G_α subunits can activate MAPK signaling. Moreover, $G_\alpha o$ and $G_\alpha i$ subunits can signal through both SRC and STAT3 pathways to contribute to cell transformation.

Effective new cancer therapies targeting key signaling molecules in the RTK signal transduction pathway are now available, whereas GPCR-mediated disorders are still largely treated with receptor-specific agonists or antagonists. Progress in understanding GPCR signaling may, however, in the future yield new therapies based on targeting GPCR activity at different molecular levels. Thus, not only could we make the targeting of specific receptor subtypes more selective, we may in the future also target various regulatory factors in G-protein signaling pathways. These include receptor-associated proteins such as receptor-activity-modifying proteins (RAMPs) and receptor-activated solely by synthetic ligands (RASSLs); and also proteins which modify G-protein activity such as activators of G-protein signaling (AGS) and regulators of G-protein signaling (RGS).

Non-receptor tyrosine kinases

Non-receptor tyrosine kinases, including SRC, ABL, and JAK, are major components of signaling networks implicated in diverse processes including cell growth, replication, survival, differentiation, migration, and genome maintenance. JAK is discussed below under signaling and SRC and ABL are discussed in more detail in Chapter 6 – oncogenes.

SIGNALING PATHWAYS

RAS–RAF–MAP kinase pathway

This is one of the earliest to be identified and best described of all the cancer-causing signaling pathways (see Fig. 5.16) and is covered in depth in the section on RAS in Chapter 6 – oncogenes.

RAF is the first member of a three-kinase modular sequence, with the generic designations MAPKKK, MAPKK, and MAPK, standing for mitogen activated protein kinases. At least six distinct kinase pathways follow this general pattern in mammalian cells, including the JUN kinase/stress activated kinase (SAPK).

The mitogenic pathway involves RAF, MAPK/ERK kinase (MEK 1 and 2), and extracellular receptor kinase (ERK 1 and 2). Sequential kinase cascades may function to amplify the input signal. However, the linking of some of these signaling molecules to scaffolding proteins such as **MP1** (MEK partner) and **JIP1** (JUN N-terminal kinase interacting protein) could prevent such amplification and might also act to prevent undesired crosstalk. In fact, scaffolding proteins may be a general mechanism whereby the potentially undesirable spread of signaling across multiple pathways can be prevented, as only the appropriate targets of signaling components are brought together. The end result of the MAPK sequence is relocation of active ERK to the nucleus, and phosphorylation of regulated transcription factors of the ETS family. Ultimately this leads to enhanced expression of other key transcription factors such as FOS and c-MYC, which is key to cell cycle control (see Chapter 6).

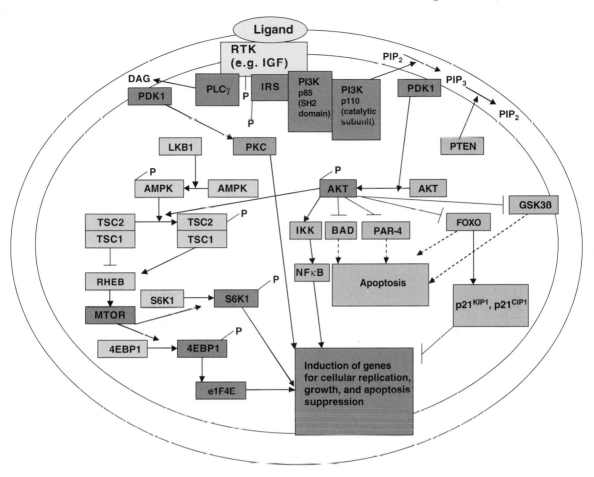

Fig. 5.17 Lipid signaling from receptor tyrosine kinases (such as IGF-1R, HER2/Neu, VEGF-R, PDGF-R): the PI3K and DAG–PKC pathways. RTK become activated by tyrosine autophosphorylation. Activated RTK associate with the p85 SH2 domain containing subunit, which recruits and activates the catalytic domain of PI3K. Activated PI3K in turn phosphorylates membrane inositol phospholipids in the 3′ position, which can then act as docking sites for proteins with pleckstrin homology (PH) domains, such as phospholipase cγ (PLCγ), PDK1, and AKT. PLCγ activates protein kinase C (PKC), important for transducing mitogenic signals, via diacylglycerol (DAG). The serine/threonine kinase AKT is activated by phosphorylation by PDK1/PIP3. AKT regulates phosphorylation of several downstream effectors, such as NF-κ B (NF-κB), mTOR, FOX01, BAD, glycogen synthase kinase (GSK-3), and MDM-2, which in turn influence cell growth, proliferation, protection from proapoptotic stimuli, and stimulation of neo-angiogenesis. AKT activates mTOR by phosphorylating the tuberous sclerosis 2 protein (TSC2); the resultant complex of TSC1/TSC2 is inactivated (loss of GTPase activity for RHEB), thus increasing active RHEB–GTP levels, which activate mTOR, leading to cell growth and replication. GSK-3 is a constitutively active proline-directed serine/threonine kinase that may induce apoptosis through multiple pathways. AKT can inactivate GSK-3 via phosphorylation thus reducing apoptosis.

Phosphatidylinositol-3 kinase signaling pathway

The phosphatidylinositol-3 kinase (PI3K) signaling pathway is as important as the RAS–MAP kinase pathway in cell survival and proliferation, and hence its potential role in cancer is of great interest (Plate 5.1 shows an integrated view of these signaling pathways). PI3Ks constitute a lipid kinase family characterized by

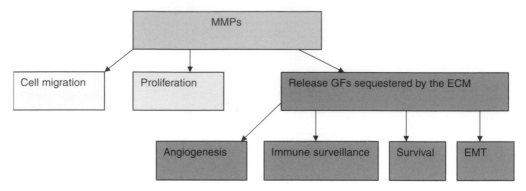

Fig. 5.18 Growth factor activation by matrix metalloproteinases. Some growth factors are inactive whilst bound to extracellular matrix, but can be released by activity of matrix-degrading enzymes such as the matrix metalloproteinases.

their ability to phosphorylate inositol ring 3′-OH group in inositol phospholipids to generate the second messenger phosphatidylinositol-3,4,5-trisphosphate (PIP-3). PIP-3 may be important in mediating the survival effects of various growth factors in part by regulating activation of downstream kinases such as AKT/PKB and S6K. PIP-3 activates the serine/threonine kinase AKT by promoting translocation to the inner membrane, where it is phosphorylated and activated by PDK1 and PDK2. In turn activated AKT modulates the phosphorylation and activity of numerous substrates involved in the regulation of cell survival, cell cycle progression and cellular growth (Fig. 5.18), including NF-kappa B, mTOR (and thence S6K), Forkhead, **BAD**, GSK3β, and MDM2 (Fig. 5.18). GFs such as insulin and IGFs can activate the PI3K–AKT pathway via the IRS-2 adaptor protein recruited to the activated RTK. This is an important pathway in normal regulation of beta-cell growth regulation and defects in this pathway may favor loss of beta cells in diabetes.

Deregulation of the PI3K pathway is a feature of many common cancers, either by loss of the tumor-suppressor protein PTEN (phosphatase and tensin homolog deleted on chromosome 10) discussed further in Chapter 7, or by constitutive activation of PI3K isoforms or downstream elements such as AKT and the mammalian target of rapamycin (mTOR). This pathway potentiates not only cell survival and proliferation, but also cytoskeletal deformability and motility – key elements in tumor invasion (Chapter 12). AKT is frequently hyperactivated

in human cancers, as is one key downstream effector, mTOR. AKT activation is regulated by two tumor suppressors, PTEN (a lipid phosphatase involved in breaking down PIP-3 to PIP-2), which inhibits PI3K upstream of AKT and the TSC1/TSC2 heterodimer, which acts downstream of AKT and upstream of mTOR.

The *PTEN* and *p53* tumor suppressors are among the most commonly inactivated or mutated genes in human cancer. Importantly, acute *PTEN* inactivation induces growth arrest through the p53-dependent cellular senescence pathway both *in vitro* and *in vivo*, suggesting that p53 is an essential failsafe protein preventing tumorigenesis in PTEN-deficient tumors. Hereditary mutations of *PTEN*, in Cowden disease, are associated with increased tumor-susceptibility and defects in angiogenesis. The implication of the PI3K pathway in angiogenesis is relatively novel and recent studies have explained the phenotype observed in Cowden disease, as the PI3K pathway is important for angiogenesis both in normal development and in tumorigenesis.

Because AKT and its upstream regulators are involved in key cancer behaviors and are deregulated in a wide range of cancers the AKT pathway is being explored as a potential prognostic marker and as a target for novel anticancer therapies. Activation of the PI3K–AKT pathway renders cancer cells resistant to apoptotic signals and promotes tumor growth. However, exactly how AKT prevents apoptosis is not fully understood. Inhibition of GSK3β is one important contributor as is phosphorylation of

BAD, but recent studies suggest that AKT may prevent apoptosis in some cases by phosphorylation of another proapoptotic protein, PAR-4. In a study by Goswami and colleagues, suppressing AKT activation by PTEN or other means caused apoptosis in cancer cells. Importantly, apoptosis resulting from inhibition of AKT was blocked by inhibition of PAR-4 expression, but not by inhibition of other apoptosis agonists that are AKT substrates.

It is increasingly likely that cellular proliferation requires the coordinated activation of signaling pathways: mitogenic, involving RAS–MAPK, as well as pro-survival, involving PI3K–AKT, and various downstream targets of these signaling cascades. This is illustrated by a recent publication from the Rosen laboratory. In this study, both pathways combine to prevent apoptosis by phosphorylation of the proapoptotic protein BAD. EGFR/MEK/MAPK may phopshorylate BAD on serine 112, whereas PI3K–AKT phosphorylates BAD on serine 136 phosphorylation; either is sufficient to sequester BAD to 14-3-3, blocking proapoptotic activity. Interestingly, apoptosis is only triggered if BAD is dephosphorylated on both serines in response to inhibition of both pathways.

mTOR

The mammalian target of rapamycin (mTOR) protein, a downstream effector of the PI3K/AKT signaling, is increasingly of interest to cancer researchers. Not only because it is a target of a known chemotherapeutic agent (rapamycin!), but mTOR is also known to be involved in key cellular processes such as cell size, survival, and proliferation. mTOR is normally involved in regulating many aspects of cell growth, including membrane traffic, protein degradation, protein kinase C signaling, ribosome biogenesis, and transcription, in part by activating both the 40S ribosomal protein S6 kinase (p70s6k) and the eukaryotic initiation factor 4E-binding protein 1 (EIF4E). Survival signals generated by PI3K and phospholipase D target mTOR, which in turn contributes to suppression of apoptotic pathways in cancer cells. Conversely, inhibitors of mTOR prevent G_1/S transition by nonspecific inhibition of cell growth and also by preventing CDK activation.

Tuberous sclerosis (TSC) is an autosomal dominant human genetic syndrome, caused by germline mutations in either *TSC1* or *TSC2* genes, and is characterized by the development of specific tumors called hamartomas. The *Tsc1* and *Tsc2* genes, encoding the proteins hamartin and tuberin respectively, have been shown to contribute to regulation of cell and organ size in several organisms including the fruit fly. The *TSC* genes are in the PI3K–AKT–mTOR–S6K pathway. Activated AKT phosphorylates TSC2 in the TSC1/TSC2 protein complex, inactivating it; while TSC1/TSC2 has GAP activity for the RHEB GTPase (a member of the RAS family), and activated RHEB–GTP activates mTOR. Thus, in cells lacking TSC1 or TSC2 there are increased levels of RHEB–GTP, which leads to activation of mTOR, leading to cell size increase and growth. Malignancies with PTEN mutations, which are associated with constitutive activation of the PI3K–AKT pathway, are relatively resistant to apoptosis and may be particularly sensitive to mTOR inhibitors (Chapter 7).

Translational control and growth

Increased cell proliferation requires a general increase in protein synthesis and a specific increase in the synthesis of replication-promoting proteins. The RAS–MAPK and PI3K–AKT–mTOR pathways are particularly well known to activate the function and expression of various components of the translational machinery. Protein synthesis in eukaryotic cells following binding of mRNA to small ribosomal subunits is catalyzed by a family of eukaryotic translation initiation factors (eIF). One such family the eIF4, comprises at least four (but with completion of genome sequencing projects many more are likely) members, eIF4E, eIF4G, eIF4A, and eIF4B. Polypeptide chain initiation involves the assembly of the 43S initiation complex catalyzed by polypeptide chain initiation factor eIF2 and the binding of eIF4E to eIF4G during the recruitment of mRNA to the ribosome. eIF2 activity is controlled by phosphorylation of the alpha subunit of this factor by various kinases (including GCN2, and eIF2alpha kinase 4), whereas eIF4E is regulated by phosphorylation of a small family of binding proteins (the 4E-BPs). Importantly, these phosphorylation steps are regulated by numerous oncogenes and tumor suppressors in cancer. The situation is complex: different eIF4 proteins can promote or inhibit translation of specific mRNAs, and may be active in different tissues and developmental stages. mTOR, discussed already, regulates ribosome biogenesis, protein synthesis and cell growth in large part by controlling the translation machinery, through activation of S6K and eIF4E (by inactivating the binding protein- 4E-BP1). There is increasing evidence that deregulation of gene expression at the level of mRNA translation can contribute to cell transformation and the malignant phenotype. Interestingly, eIF4E cooperates with c-MYC in B-cell lymphomagenesis; in this case c-MYC could overcome growth arrest triggered by deregulated eIF4E and eIF4E could prevent c-MYC-dependent apoptosis, identifying another protein in the PI3K–AKT pathway able to suppress apoptosis. eIF4E overexpression has been demonstrated in various different human tumors, including breast, head and neck, colon, prostate, bladder, cervix, and lung. Although eIF4E regulates the recruitment of mRNA to ribosomes, and thereby globally regulates cap-dependent protein synthesis, eIF4E also selectively enables translation of a select number of directly cancer-relevant mRNAs, such as those encoding cyclin D1, c-MYC, MMP-9, and VEGF. However, upregulating eIF4E is not the only means by which cancer cells increase general protein synthesis. In fact, in many even aggressive cancers eIF4E may be barely detectable. Eukaryotic cells can respond to diverse stresses by activating remedial and damage-limitation responses. eIF2 and various regulatory kinases (eIF2 kinases) play a key role in the regulation of protein synthesis in response to such stresses. Phosphorylation of eIF2 reduces general protein synthesis, but increases translation of a defined number of specific mRNAs that encode transcription factors.

A greater appreciation of regulatory processes for protein translation, particularly how specific mRNAs are selectively targeted for enhanced or reduced translation, will certainly be a very important area in cancer research and drug design in the next few years.

Cytokine signaling

Cytokines and growth factors activate the MAP kinase pathways resulting in the stimulation of ERK1/2, c-Jun N-terminal kinases, and p38 kinases, which in turn activate transcription factors like AP-1 and ATF-2. Other pro-inflammatory agents such as TNF-α and IL-1 can activate the transcription factor NF-κB that in turn regulates the expression of immediate early genes involved in immune, acute phase, and inflammatory responses. Besides the transcription factors NF-κB and AP-1, which are immediate-early transcriptional activators, components of the JAK/STAT pathway play an important role in the transcriptional activation of many inflammatory genes.

JAK/STAT pathways

The JAK–STAT signaling pathway comprises three families of proteins, Janus kinases (JAKs), signal transducers and activators of transcription (STATs) and their endogenous inhibitors- the SOCS family. JAK–STAT pathways are utilized by a wide variety of cytokines and growth factors, whereas inappropriate function of this pathway is implicated in many human cancers. Defective JAK–STAT–SOCS pathways may impair tumor responses to immunotherapy. In particular, processes controlled by this pathway include cell growth, differentiation, senescence, and apoptosis. STAT-family proteins are latent cytoplasmic transcription factors that convey signals from cytokine and growth factor receptors to the nucleus (see Fig. 5.16). They were originally discovered and characterized through the study of interferon-induced responses. Binding of cytokines to cell surface receptors results in two categories of signaling, either activation of cytoplasmic tyrosine kinases (particularly JAK or SRC kinase families) or activation of receptor-intrinsic tyrosine kinase activity (such as discussed for PDGF and EGF above). Tyrosine phosphorylation activates STAT monomers, which dimerize through interaction of SH-2 domains. The resultant STAT dimers translocate to the nucleus and bind to STAT-specific sites known as gamma-activated sites (GAS) of target genes to induce transcription. There are seven known STAT proteins each encoded in a distinct gene but subject to alternative RNA splicing or post-translational proteolytic processing, which may result in additional STATs found in malignant cells. Tumor cells acquire the ability to proliferate uncontrollably, resist apoptosis, sustain angiogenesis and evade immune surveillance. STAT proteins, especially STAT3 and STAT5, regulate all of these processes and are persistently activated in a number of human cancers. Epigenetic silencing of the SOCS-1 gene, an inhibitor of the JAK/STAT pathway, has been described in liver cancer.

Constitutive activation of the JAK–STAT pathway is known to occur in HTLV-I-transformed T cells, Sezary's syndrome, and v-*abl* or v-*src* transformation. STAT3 and STAT5 are the most frequently observed in cancers and active forms have been detected in cancers of the head/neck, lung, kidney, prostate, breast, ovaries, and blood. Aberrant signaling through a number of upstream pathways can result in constitutively activated STATs in tumor cells. Cellular transformation with v-src or BCR–ABL activates STAT3 and STAT5 and constitutively active STAT3 will transform immortalized fibroblasts. STAT5 can control transcription of various factors implicated in cancer including BCL-X$_L$, PIM-1, and cyclin D1, suggesting a role both in resistance to apoptosis and in cell replication.

JAK tyrosine kinases are activated by interleukins and other growth factors, and promote survival and proliferation of cells in multiple tissues and are constitutively active in many hematopoietic malignancies and certain carcinomas. JAK can activate both STAT and PI3K signaling. JAK activity is essential for lymphoma invasion and metastasis, independent of its role in survival and proliferation, and independent of STAT and PI3K signaling. Currently potential therapeutic agents targeting the JAK–STAT pathway are under investigation, including tyrphostin (AG 490), a JAK protein kinase inhibitor, and sant 7, an IL-6 receptor antagonist.

Growth signaling via cell adhesion – the Wnt–β-catenin pathway

Cell–cell adhesion determines cell polarity and is critical for tissue homeostasis. Cell–cell adhesiveness is decreased in cancer, enabling

deregulated proliferation, migration, invasion, and metastases, but in general only if cells can survive loss of cell–cell contacts (Chapter 12). E-cadherin, and interacting catenins, which connect the cadherins to the cytoskeleton, are located at adherens junctions of epithelial cells and establish firm cell cell adhesion. Silencing of the E-cadherin gene by DNA hypermethylation around the promoter region occurs frequently, even in precancerous conditions. In diffuse infiltrating cancers, mutations are found in the genes for E-cadherin and alpha- and beta-catenins. At the invading front of cancers, the E-cadherin cell adhesion system is inactivated by tyrosine phosphorylation of β-catenin. β-catenin also connects the E-cadherin system to the Wingless/Wnt signaling pathway.

The Wnt–β-catenin pathway is involved in specification and localization of new cell types during tissue differentiation, primarily by regulating interactions between neighboring cells. Wnt ligands bind to their target "frizzled" membrane receptor and interfere with the multi-protein destruction complex, resulting in downstream activation of gene transcription by β-catenin. The multi-protein destruction complex requires the presence of Axin and the adenomatosis polyposis coli (APC) protein, which act as scaffolds, to facilitate phosphorylation of β-catenin by the enzyme GSK3β. Phosphorylated β-catenin is degraded in proteasomes, whereas unphosphorylated β-catenin accumulates and associates with nuclear transcription factors such as TCF/LEF, leading to expression of growth-regulating target genes such as c-*MYC*, c-*Jun*, *Fra*, and cyclin D1 (Fig. 5.19). Mutations resulting in active β-catenin are found in at least 80% of colon cancers and include activating mutations in the β-catenin gene, increased protein stability, or loss of the APC tumor suppressor. Activation of β-catenin/TCF in turn results in deregulated expression of key target genes including c-*MYC* and cyclin D.

Recent appreciation of similarities between developmental morphogenetic processes and cancer has had a major impact on cancer biology. Wnt signaling regulates morphogenetic processes such as gastrulation, which requires cells to undergo epithelial to mesenchymal transitions – EMT – (but these are usually transient), enabling the cells to dissociate and migrate. Cell dissociation and migration are also

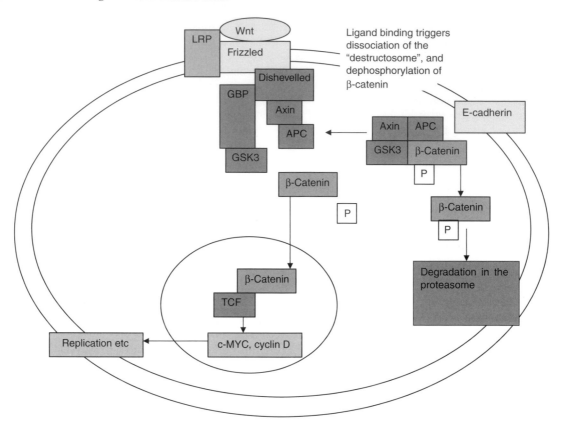

Fig. 5.19 The Wnt signaling pathway. Binding of Wnt ligands to Frizzled allows the disassembly of the "destructo-some" (comprising Axin, APC, and GSK3β). Specifically, Wnt signaling activates dishevelled, dishevelled binds to axin, excluding GSK3; thus β-catenin is able to avoid GSK3-mediated phosphorylation and thereby avoid proteasomal degradation; β-catenin accumulates in the nucleus and activates TCF/LEF transcription factors, which in turn activate growth-regulatory genes, such as c-MYC and cyclin D.

essential for tumor cell invasion and metastases and recent data suggest that EMT is a major determinant of the latter.

Notch signaling

The Notch signaling pathway has been intimated for nearly a century since the identification of a mutant strain of *Drosophila* fruit flies with "notched" wings in Thomas Hunt Morgan's lab in 1910. Later studies revealed that the *Notch* gene encodes a receptor protein capable of interacting with a ligand protein Delta, presented on the surface of a neighboring cell. This juxtracrine interaction results in release of an intracellular region of the Notch protein resulting in transcriptional activation

of various genes. In mammals, there are now several known core components of the Notch signaling cascade, including ligands (Delta-like 1, -3, -4, and Jagged-1 and -2) and a family of four receptors (Notch1, -2, -3, and -4), which all share an extracellular ligand-binding domain comprising a series of EGF-like repeats and a cytoplasmic domain containing several functionally important motifs, including a transcriptional activation domain and a PEST sequence which regulates protein stability. Following ligand binding the receptor is subjected to proteolytic cleavage: (1) removal of the ligand-binding domain by the metalloproteinase enzyme TACE and (2) release of the cytoplasmic domain (intracellular or icNotch) by another enzyme, a complex of ?-secretase and presenillin. Gene expression is in turn activated by icNotch in part by binding to the ubiquitous CSL transcriptional repressor; together icNotch and CSL recruit other proteins, including mastermind-like proteins, resulting in expression of numerous genes. Genes activated in this way by icNotch include various regulators of differentiation, including other transcription factors such as members of the Hairy/enhancer of split (HES) family, and also some cell cycle and apoptosis regulators.

Cell–cell communication is a key factor in determining cell fate, and Notch signaling plays a key role in the normal development of many tissues and cell types, through diverse effects on differentiation, survival, and/or proliferation. Notch can either inhibit or delay differentiation. The Notch pathway is important in embryonic development and is usually inactivated in adult cells. However, active Notch signaling has been found in several human cancers including a small subset of human pre-T-cell acute lymphoblastic leukemias (T-ALL), which show a recurrent chromosomal translocation involving the Notch receptor gene and the T-cell receptor β); the result is that icNotch is activated independently of ligand binding in a manner regulated by the TCR-β. Deregulated icNotch contributes to T-ALL by increasing numbers of cells proceeding down the T-cell lineage during lymphoid development, at the expense of the B-cell lineage, and also by preventing full T-cell differentiation. Notch is also a causative agent in mouse mammary tumor virus (MMTV) induced breast cancer in mice. Notch may also be activated by loss of the inhibitors of this pathway – NUMB or Deltex. Inactivation of Numb is seen in a large number of human breast cancers and is associated with increased activity of Notch1. Interestingly, at least in pancreatic cancers, the majority of which show activation of the EGFR–RAS–MAPK system, crosstalk results in activation of Notch signaling also, which might in turn contribute to loss of differentiation of pancreatic cells.

However, under some circumstances, conversely, loss of Notch might also contribute to tumorigenesis. In some tissue Notch, rather than blocking differentiation, may do the opposite; in rodent models, inactivating Notch in skin results in keratinocyte overgrowth and carcinoma, in part by blocking differentiation.

A number of stages in Notch signaling are subject to ubiquitinylation and degradation by the proteasome pathway (see Chapter 11). In fact, ligand-bound notch receptor signaling may in part be terminated by this means. Interestingly, recent studies have now shown that ubiquitinylation may also be a means of maintaining inactivity of the unbound (inactive) Notch receptor. The ubiquitin ligase Nedd4 can mimic the nicked wing phenotype characteristic of Notch loss in *Drosophila* and inhibition of Nedd4 results in ligand-independent activation of the Notch pathway. A further putative ubiquitin ligase, Deltex, can bind to and activate the Notch receptor, suggesting that Nedd4 and Deltex compete in regulating Notch activity. This also raises the possibility that Notch signaling could be manipulated therapeutically in cancer by targeting of the ubiquitin–proteasome pathway.

NUCLEAR PROTO-ONCOGENES AND TRANSCRIPTION FACTORS

Ultimately signals from growth factor receptor-signaling pathways control growth by regulating gene expression in the nucleus. Thus, ERK and STAT can enter the nucleus, where they in turn activate the proto-oncogenes c-MYC, c-FOS, and c-JUN and others involved in cell cycle regulation, protein synthesis, cell survival, etc. described in Chapter 6.

JUN protein, the target of the JNK signaling pathway, is a constitutive transcription factor that regulates survival but is only a weak regulator of proliferation as a homodimer. JUN combines with FOS (Finkel osteosarcoma), which is a product of c-MYC-induced transcription, to give a heterodimer, which activates genes with tumor response element enhancers. Long-lived or overexpressed JUN homodimers can themselves activate transcription of *FOS*. JUN/FOS heterodimers regulate genes involved in cell growth and may result in loss of differentiation.

QUESTIONS REMAINING

1. To understand how the size/proportion of various tissues/organs (cell size and number) is determined during development and maintained during adult life.
2. To further understand how GF signaling is regulated, from receptor numbers and desensitization

down to the nuclear transcription factors that ultimately control gene expression.

3. To progress beyond the current view of signaling pathways as largely linear cascades, towards a more integrated view, and in particular to begin to unravel the complex interacting functional protein networks within which GF signaling pathways participate.

4. To develop further small-molecule drugs specifically targeting various GF signaling pathways involved in cancer.

5. To further understand the means by which TGF-β may be both a growth inhibitor and also a promoter of cancer behaviors.

CONCLUSIONS AND FUTURE DIRECTIONS

Deregulation of GF signaling pathways is involved in essentially all cancers and most but not all known oncogenes activate some parts of these signaling pathways. GF signaling may trigger many cancer-relevant behaviors in addition to cell replication, such as prevention of apoptosis and increased motility. Aberrant activation of GF signaling pathways can arise through a variety of mechanisms (mutations, epigenetic alterations, upregulation) and act at different levels including the GF itself, the RTK, signaling pathways, or downstream transcription factors. Given the central role of GF signaling in cancer, it is not surprising that a new generation of small-molecule drugs and monoclonal antibodies that focus on elements of these pathways has emerged as the "new frontier" in cancer therapy.

QUESTIONS

1. Cell numbers in adult metazoans are
   a. Tightly controlled by regulatory processes.
   b. Determined solely by rates of cell division.
   c. Normally fixed in most adult tissues.
   d. Regulated by circulating as well as local factors.
   e. Maintained only by balancing losses through replication of stem cells.
2. Mitogens
   a. Act primarily by promoting G_1/S transition in the cell cycle.
   b. Only act via cell surface receptors.
   c. Often act via the RAS/MAP kinase cascade.

   d. Do not influence differentiated cells.
   e. Can be produced by the same cells on which they act.
3. Signaling pathways activated by growth factor mitogens
   a. Involve a series of steps regulated by phosphorylation of various proteins.
   b. May be initiated by autophosphorylation of growth factor receptors on tyrosine residues.
   c. Do not include serine/threonine kinases.
   d. Include the PI3K signaling pathway.
   e. Are rarely mutated in cancers.
4. The PI3K signaling pathway
   a. Is far less frequently involved in cancer than the RAS/MAPK pathway.
   b. May be important in regulating cell death.
   c. Is not involved in regulating cell replication.
   d. May be activated by inactivation of *PTEN*.
   e. Can result in activation of AKT and mTOR.
5. The RAS/MAPK pathway
   a. Eventually results in activation of nuclear factors such as c-MYC.
   b. In cancer, is only known to be activated by mutations in *RAS*.
   c. Is not involved in cell death regulation.
   d. Is often activated via receptor tyrosine kinases.
   e. Is not involved in replicative senescence.

BIBLIOGRAPHY

Growth factors

Hermansson, M., Nistér, M., Betsholtz, C., Heldin, C.-H., Westermark, B., and Funa, K. (1988). Endothelial cell hyperplasia in human glioblastoma: coexpression of mRNA for platelet-derived growth factor (PDGF) B chain and PDGF receptor suggests autocrine growth stimulation. *Proc Natl Acad Sci U S A.* **85**: 7748–52.

Foulstone, E., Prince, S., Zaccheo, O., *et al.* (2005). Insulin-like growth factor ligands, receptors, and binding proteins in cancer. *J Pathol.* **205**(2): 145–53.

Growth factors and cancer

Sporn, M.B. and Roberts, A.B. (1985). Autocrine growth factors and cancer. *Nature.* **313**: 745–7.

Inoki, K., Corradetti, M.N., and Guan, K.L. (2005). Dysregulation of the TSC-mTOR pathway in human disease. *Nat Genet.* **37**(1): 19–24.

Benitah, S.A., Valeron, P.F., van Aelst, L., Marshall, C.J., and Lacal, J.C. (2004). Rho

GTPases in human cancer: an unresolved link to upstream and downstream transcriptional regulation. *Biochim Biophys Acta.* **1705**(2): 121–32.

Roberts, A.B. and Wakefield, L.M. (2003). The two faces of transforming growth factor beta in carcinogenesis. *Proc Natl Acad Sci U S A.* **100**(15): 8621–3.

GF and cell cycle

Nasmyth, K. (1996). Viewpoint: Putting the cell cycle in order. *Science.* **274**: 1643–5.

Growth regulation and differentiation

Owens, D.M. and Watt, F.M. (2003). Contribution of stem cells and differentiated cells to epidermal tumors. *Nat Rev Cancer.* **3**: 444–51.

Reya, T., Morrison, S.J., Clarke, M.F., and Weissman, I.L. (2001) Stem cells, cancer and cancer stem cells. *Nature.* **414**: 105–11.

Sakatani, T., Kaneda, A., Iacobuzio-Donahue, C.A. et al. (2005). Loss of imprinting of Igf2 alters intestinal maturation and tumorigenesis in mice. *Science.* **307**(5717): 1976–8.

Teitelman, G., Alpert, S., and Hanahan, D. (1988). Proliferation, senescence, and neoplastic progression of beta cells in hyperplasic pancreatic islets. *Cell.* **52**(1): 97–105.

Scaglia, L., Smith, F.E., and Bonner-Weir, S. (1995). Apoptosis contributes to the involution of beta cell mass in the post partum rat pancreas. *Endocrinology.* **136**(12): 5461–8.

Signaling

Chen, Z., Trotman, L.C., Shaffer, D., et al. (2005). Crucial role of p53-dependent cellular senescence in suppression of Pten-deficient tumorigenesis. *Nature.* **436**(7051): 725–30.

Guertin, D.A. and Sabatini, D.M. (2005). An expanding role for mTOR in cancer. *Trends Mol Med.* **11**(8): 353–61.

Vanhaesebroeck, B., Ali, K., Bilancio, A., Geering, B., and Foukas, L.C. (2005). Signalling by PI3K isoforms: insights from gene-targeted mice. *Trends Biochem Sci.* **30**(4): 194–204.

Fang, J.Y. and Richardson, B.C. (2005). The MAPK signalling pathways and colorectal cancer. *Lancet Oncol.* **6**(5): 322–7.

Teitell, M.A. (2005). The TCL1 family of oncoproteins: co-activators of transformation. *Nat Rev Cancer.* **5**(8): 640–8.

WNT signaling

Nelson, W.J. and Nusse, R. (2004). Convergence of Wnt, beta-catenin, and cadherin pathways. *Science.* **303**(5663): 1483–7.

van Amerongen, R. and Berns, A. (2005). Re-evaluating the role of Frat in Wnt-signal transduction. *Cell Cycle.* **4**(8).

Targeted therapies

Adams, G.P. and Weiner, L.M. (2005). Monoclonal antibody therapy of cancer. *Nat Biotechnol.* **23**(9): 1147–57.

Demetri, G.D., von Mehren, M., Blanke, C.D., et al. (2002). Efficacy and safety of imatinib mesylate in advanced gastrointestinal stromal tumors. *New England Journal of Medicine.* **347**(7): 472–80.

Kantarjian, H., Sawyers, C., Hochhaus, A., et al. (2002). Hematologic and cytogenetic responses to imatinib mesylate in CML. *New England Journal of Medicine.* **346**(9): 645–52.

Richardson, P.G., Barlogie, B., Berenson, J., et al. (2003). A phase 2 study of bortezomib in relapsed, refractory myeloma. *New England Journal of Medicine.* **348**(26): 2609–17.

Wagner, R.W. (1994). Gene inhibition using antisense oligodeoxynucleotides. *Nature.* **372**: 333–5.

6

Oncogenes

Stella Pelengaris and Mike Khan

To retain respect for sausages and laws, one must not watch them in the making.

Otto von Bismarck

We have seen that cells are formed and grow in accordance with essentially the same laws: hence that these processes must everywhere result from the operation of the same forces.

Theodore Schwann, 1839

KEY POINTS

- Oncogenes (cancer-causing genes) are activated versions of normal cellular genes involved in regulating cell replication, growth, survival, differentiation, or motility – these normal versions are referred to as proto-oncogenes and their products as proto-oncoproteins.
- Oncogenes were first described in association with a class of viruses known as the retroviruses. A classic example of a retrovirus causing growth of a tumor on infecting a host animal is the Rous sarcoma virus, which induces sarcomas in chickens and contains the *v-src* oncogene.
- Oncogenes become activated by genetic mutations, chromosomal translocations, or gene amplification, but increasingly it is appreciated that upregulation of expression per se (as is often observed for *c-MYC*) is a major contributory factor in many human cancers. As inappropriate expression of an oncoprotein may result from factors such as increased stability of the oncoprotein or deregulated upstream signaling, this will not necessarily be apparent by looking for mutations in that oncogene.
- The majority of oncogenes fall into four familiar functional classes: transcription factors, growth factors, receptors, and signal transducers. In addition, some oncogenes also encode proteins with diverse functions such as prevention of the mitochondrial (intrinsic) pathway of cell death – apoptosis (e.g. BCL-2 and BCL-X_L).
- In general individual oncogenes are themselves insufficient to cause cancer but rather collaboration with other oncogenes or with loss of tumor suppressors is required – termed "oncogene collaboration" or "oncogene cooperation".
- Various mouse models of cancer have enabled issues of oncogene cooperation to be addressed *in vivo*. More recently, the generation of regulatable expression systems has allowed the role of various oncogenes in tumor maintenance to be explored and "proof of hypothesis" studies for oncogene-targeted therapies to be tested.

INTRODUCTION

As discussed in Chapters 4 and 5, somatic cells depend on mitogenic signals in order to promote entry into the cell cycle. Growth factor binding can trigger cascades of intracellular signals that enable cells to progress through the G_1/S transition. However, these complex signaling networks are normally tightly regulated to ensure that cells proliferate only when this is useful to the whole organism, such as during development, after tissue injury, or in normal homeostasis. Cancers arise when mutations in key regulatory genes disrupt these signaling networks, resulting in unscheduled cell division, growth, and avoidance of apoptosis in defiance of normal controls. Although the specific genes that are affected and the order in which mutations occur will differ from one cancer patient to another (the cancer roadmap), most mutations impede the function of a finite number of key regulatory pathways. The identification and characterization of the genes involved – the oncogenes, tumor suppressors, and most recently caretaker genes, has been one of the great triumphs of molecular biology.

Oncogenes (from Greek *onkos*, a tumor) are activated versions of normal genes (proto-oncogenes), and in general share the property of accelerating cell division and growth, but increasingly are being shown to play key roles in loss of differentiation, cell motility, avoidance of apoptosis, and invasion. In healthy cells the proto-oncogenes are under tight regulation to avoid excessive proliferation, or indeed the inappropriate survival of a cell that has sustained DNA damage and would otherwise have been got rid of. It is deregulated expression or hyperactivity that often distinguishes oncogenes from their normal counterparts. As only one of the two gene copies in a cell need undergo a change in order to become hyperactivated this type of mutation is considered "dominant". Conversely, avoidance of cell death (apoptosis) or unrestrained proliferation can result from mutations that inactivate both copies of a tumor-suppressor gene, thus releasing normal survival or proliferative restraints. In contrast to oncogenes, both copies of the gene must be lost or inactivated to bring about these effects, and so the mutant phenotype is considered "recessive". As often happens, however, the recent appreciation that some tumor suppressors may be "haploinsufficient" (and in the case of cancer can contribute to tumorigenesis when only a single copy is inactivated) has blurred this distinction. It must also be borne in mind that seemingly identical cellular behaviors can result from inactivation of tumor suppressors as from activation of oncogenes – most of the control mechanisms in the cell involve both inhibitory (tumor suppressor) and stimulatory (proto-oncogene) components. Put simply, the behavior of a cancer cell will reflect which regulatory or signaling pathways have become derailed rather than the means by which this has been achieved.

Moreover, to complicate matters further, intriguing studies over the last 15 years have forced a major reassessment of traditional views of the role of oncogenes in cancer. Thus, despite their undisputed importance in tumorigenesis, it is now clear that oncogenes (such as *c-MYC*, *E2F*, and *RAS*) capable of driving uncontrolled cell-cycle progression also possess *intrinsic tumor suppressive mechanisms* that can trigger cell death (apoptosis) or cellular senescence.

In other words oncogenes may act as their own tumor suppressors! The ability of oncogenes to promote such contrasting cell fates in a given cell – proliferation and death – may at first seem somewhat contradictory and confusing to the reader. In fact, the discovery of this phenomenon in the early 1990s (see Chapter 8) was a great surprise to the scientific community but has since had an enormous impact on understanding the evolution as well as the treatment of cancers. To this end, we will discuss the importance of such intrinsic tumor suppression later in this chapter using animal models, and how putative cancer cells have found ways to block tumor-suppressing activities of oncogenes, for example through oncogene collaboration. With this in mind we will also discuss members of the BCL-2 family of proteins that specifically act as oncogenes by suppressing apoptosis.

The previous chapters have described the basic cell cycle machinery (Chapter 4) and how cell cycle entry and progression are regulated under the influence of mitogens and survival signals (Chapter 5); this chapter will describe how these processes become derailed by the activation of oncogenes, by mutations or otherwise, in cancer. Oncogenes include a diverse range of genes encoding proteins involved in the regulation of cell division, differentiation, and death/survival. The majority of oncogenes fall into four familiar functional classes: transcription factors, growth factors, receptors, and signal transducers (Tables 6.1–6.4) and some key examples of representative oncogene protein products are shown in Figure 6.1. Following a general overview, a select sample of key proto-oncogenes will be described in detail to illustrate the role of their protein products in normal tissue growth and how their normally tight

Table 6.1   List of oncogenes. Vogelstein and Kinzler (1998)

Some known oncogenes classified by site of action in growth-regulating pathways

Oncogene	Chromosome	Discovery	Cancer	Means of activation	Protein
Growth factors					
HST	11q13.3	Transfection	Stomach carcinoma	Constitutive Synthesis	FGF family
INT2	11q13	Proviral insertion	Mammary carcinoma	Constitutive synthesis	FGF family
KS3	11q13.3	Transfection	Kaposi sarcoma	Constitutive synthesis	FGF family
v-sis	22q12.3-13.1	Viral homolog	Glioma/ fibrosarcoma	Constitutive synthesis	B-chain PDGF
Growth factor tyrosine kinases					
HER1/EGFR	7p1.1-1.3	DNA amplification	Squamous cell carcinoma	Gene amplification/ increased protein	EGF receptor
HER2/NEU	17q11.2-12	Point mutation amplification	Neuroblastoma/ breast carcinoma	Gene amplification	Human EGFR-2; preferred heterodimer-ization partner for other EGFR family members.
HER3		Amplification	Squamous carcinoma	Gene amplification	
FMS	5q33-34 (FMS)	Viral homolog	Sarcoma	Constitutive activation	CSF1 receptor
KIT	4q11-21 (KIT)	Viral homolog	Sarcoma	Constitutive activation	Scf receptor
ROS	6q22 (ROS)	Viral homolog	Sarcoma	Constitutive activation	
MET	7p31	DNA transfection	Mnng-treated human osteo-carcinoma cell line	DNA rearrangement/ constitutive activation (fusion proteins)	HGF/SF receptor
ALK		Chromosome	Lymphoma	Fusion proteins	
RET	10q11.2	DNA transfection	Carcinomas of thyroid; men2a, men2b	DNA rearrangement/ point mutation constitutive active	GDNF receptor
TRK	1q32-41	DNA transfection	Colon/thyroid carcinomas	DNA rearrangement/ constitutive activation (fusion proteins)	NGF receptor
G protein coupled receptor					
MAS	6q24-27	DNA transfection	Epidermoid carcinoma	Rearrangement of 5" noncoding region	Angiotensin receptor

Table 6.1 *(continued)*

Some known oncogenes classified by site of action in growth-regulating pathways

Oncogene	Chromosome	Discovery	Cancer	Means of activation	Protein
Signal transduction pathways					
Non-receptor tyrosine kinases					
ABL	9q34.1	Chromosome	CML	DNA rearrangement translocation (fusion proteins)	Tyrosine kinase
FGR	1p36.1 - 36.2 (FGR)	Viral homolog	Sarcoma	Constitutive activation	Tyrosine kinase
BCR-ABL		Chromosome	CML	Fusion protein	Tyrosine kinase
FES	15q25-26 (FES)	Viral homolog	Sarcoma	Constitutive activation	Tyrosine kinase
SRC	20p12-13	Viral homolog	Colon carcinoma	Constitutive activation	Tyrosine kinase
YES	18q21-3 (YES)	Viral homolog	Sarcoma	Constitutive activation	Tyrosine kinase
Membrane-associated G proteins					
H-RAS	11p15.5	Viral homolog/ transfection	Colon, lung, pancreas carcinomas	Point mutation	GTPase
RAS	12p11.1- 12.1	Viral homolog/ transfection	AML, thyroid carcinoma, melanoma	Point mutation	GTPase
N-RAS	1p11-13	Transfection	Carcinoma, melanoma	Point mutation	GTPase
GSP	20	DNA sequencing	Adenomas of thyroid	Point mutation	G_s alpha subunit
GIP	3	DNA sequencing	Ovary, adrenal carcinoma	Point mutation	G_i alpha subunit
GTPase exchange factor (gef)					
DBL	Xq27	DNA transfection	Diffuse b-cell lymphoma	DNA rearrangement	Gef for rho and cdc42
VAV	19p13.2	DNA transfection	Hematopoietic cells	DNA rearrangement	Gef for ras?
Serine/threonine kinases: cytoplasmic					
MOS	8q11 (MOS)	Viral homolog	Sarcoma	Constitutive activation	Protein kinase (Ser/Thr)
RAF	3p25 (RAF-1)	Viral homolog	Sarcoma	Constitutive activation	Protein kinase (Ser/Thr)
PIM-1	6p21 (PIM-1).	Insertional mutagenesis	T-cell lymphoma	Constitutive activation	Protein kinase (Ser/Thr)
Cytoplasmic regulators					
CRK	17p13 (CRK)	Viral homolog		Constitutive tyrosine kinase activity	SH-2/SH-3 adaptor
PRAD-1			Breast		Cyclin D1

Table 6.1 (*continued*)

Some known oncogenes classified by site of action in growth-regulating pathways

Oncogene	Chromosome	Discovery	Cancer	Means of activation	Protein
Trancription factors					
ETS-2	21q24.3	Viral homolog	Erythroblastosis	Deregulated activity	Transcription factor
AML1		Chromosome	AML	Deregulated activity	Transcription factor
ERBA1	17p11-21	Viral homolog	Erythroblastosis	Deregulated activity	T3 transcription factor
ERBA2	3p22-24.1	Viral homolog	Erythroblastosis	Deregulated activity	T3 transcription factor
FOS	14q21-22	Viral homolog	Osteosarcoma	Deregulated activity	Transcription factor api
JUN	p31-32	Viral homolog	Sarcoma	Deregulated activity	Transcription factor api
c-MYC	8q24.1 (MYC)	Viral homolog	Multiple cancers, breast, stomach, lung, cervix, colon, neuro-blastomas and glio-blastomas. Burkitt's lymphoma.	Amplification, Translocation Deregulation	Transcription factor
N-MYC	2p24	DNA amplification	Neuroblastoma; lung carcinoma	Deregulated activity	Transcription factor
L-MYC	1p32	DNA amplification	Carcinoma of lung	Deregulated activity	Transcription factor
MYB	6q22-24	Viral homolog	Myeloblastosis	Deregulated activity	Transcription factor
REL	2p12-14	Viral homolog	Lymphatic leukemia	Deregulated activity	Mutant nfkb
SKI	1q22-24	Viral homolog	Carcinoma	Deregulated activity	Transcription factor
TS-1	11p23-Q24	Viral homolog	Erythroblastosis	Deregulated activity	Transcription factor
Apoptosis regulators and others					
BCL2	18q21.3	Chromosomal translocation	B-cell lymphomas	Constitutive activity	Antiapoptotic protein
MDM2	12q14	DNA amplification	Sarcomas	Gene amplification/ increased protein	Complexes with p53

AML = Acute myeloid leukemia; CML = Chronic myelogenous leukemia; CSF = Colony stimulating factor; EGF = Epidermal growth factor; FGF = fibroblast growth factor; HGF = Hepatocyte growth factor; NGF = Nerve growth factor; PDGF = Platelet-derived growth factor.

Table 6.2   List of oncogenes activated by amplification in cancer

Cellular oncogene	Location	Protein function	Type of cancer
ABL	9q34.1	Protein tyrosine kinase	Chronic myelogenous leukemia
BCL1	11q13.3	G1/S-specific cyclin D1	Breast cancer, squamous cell carcinoma of the head and neck, bladder cancer
CDK4	12q14	Cyclin-dependent kinase	Sarcomas
EGFR/ERBB-1	7p12	Epidermal growth factor receptor	Glioblastoma multiforme, epidermoid carcinoma, bladder cancer, breast cancer
ERBB2(NEU)	17q12-q21	Growth factor receptor	Breast cancer, ovarian cancer, stomach cancer, renal adenocarcinoma, adenocarcinoma of salivary gland, colon carcinoma
HSTF1	11q13.3	Fibroblast growth factor	Breast cancer, esophageal carcinoma
INT1/WNT1	12q13	Probably growth factor	Retinoblastoma
INT2	11q13.3	Fibroblast growth factor	Breast cancer, esophageal carcinoma, melanoma, squamous cell carcinoma of the head and neck
MDM2	12q14.3-q15	p53-binding protein	Sarcomas
MET	7q31	Hepatocyte growth factor receptor	Amplified in cell lines from human tumors of nonhematopoietic origin, particularly gastric tumors
MYB	6q22-q23	DNA-binding protein (essential for normal hematopoiesis)	Leukemias, colon carcinoma, melanoma
MYC (c-MYC)	8q24.12-q24.13	DNA-binding protein	Small-cell lung cancer, giant cell carcinoma of lung, breast cancer, colon carcinoma, acute promyelocytic leukemia, cervical cancer, gastric adenocarcinoma, chronic granulocytic leukemia
MYCN (N-MYC)	2p24.3	DNA-binding protein	Neuroblastoma, small-cell lung cancer, retinoblastoma, medulloblastoma, glioblastoma, rhabdomyosarcoma, adenocarcinoma of lung, astrocytoma
MYCL1 (L-MYC) MYCLK1	1p32 7p15	DNA-binding protein	Small-cell lung cancer
RAF1	3p25	Serine/threonine protein kinase	Non-small-cell lung cancer
HRAS1	11p15.5	GTPase	Bladder cancer
KRAS2	12p12.1	GTPase	Adrenocortical tumor, giant cell carcinoma of lung
NRAS	1p13	GTPase	Breast cancer
REL	2p12-p13	DNA-binding protein	Non-Hodgkin's lymphomas

Table 6.3   Growth factor receptors and cancer

Growth factor Receptor	Activation	Cancer
ALK	Gene rearrangement: NPM-ALK t(2;5)	Anaplastic large-cell lymphomas
CSF-1R/fms	Extracellular domain mutations	Acute myeloid leukemia (AML)
CSFR/KIT	Point mutations and deletions	AML, mastocytomas, gastrointestinal stromal tumors
EGFR	Autocrine activation, amplification/overexpression	Squamous cell carcinoma, glioblastoma
EGFR	Extracellular domain deletions	Glioblastoma
HER2/NEU	Point mutation, amplification/overexpression	Breast and ovarian carcinoma
FGFR1	Chromosomal translocations	Myeloid malignancies
FGFR3	Translocation	Multiple myelomas
FLK2/FLT3	Internal tandem duplication	AML, PML
c-MET	Gene rearrangement: Tpr-met t(1;7)	Stomach
c-MET	Amplification/overexpression	Thyroid, ovarian, and colorectal cancers
c-MET	Point mutations	Hereditary and sporadic papillary renal carcinoma
PDGFR	Autocrine activation	Osteosarcoma, melanoma, glioblastoma
PDGFR	Tel-PDGFR-β t(5;12) translocation	Chronic myelomonocytic leukemia (CML)
RET	Germ-line point mutations	Men2A, Men2B
TIE1, TIE2	Paracrine activation	Tumor angiogenesis and lymphangiogenesis
TRKA	Gene rearrangement: TPM-TRKA t(1;1)	Papillary thyroid carcinoma
TRKC	Gene rearrangement: Tel-TRKC t(12;15)	Congenital fibrosarcoma, AML, breast carcinoma
VEGFR1, -2, -3	Paracrine activation	Tumor angiogenesis and lymphangiogenesis

Table 6.4   Oncogenes in signaling pathways

Signaling molecule	Activation	Cancer
H-RAS	Point mutations	Thyroid and bladder carcinoma
K-RAS	Point mutations	Pancreatic, lung, and colon adenocarcinomas, non-small-cell lung carcinoma, myeloid leukemia, thyroid carcinomas
N-RAS	Point mutations	Melanoma, myeloid leukemia, thyroid carcinomas
B-RAF	Point mutations	Melanoma, colon carcinoma, small-cell lung cancer
Cyclin D_1	Amplification/overexpression	Breast cancer, head, and neck squamous carcinoma, esophageal cancer, lymphoma.
PI3K p110	Amplification/overexpression	Ovarian and cervical cancer

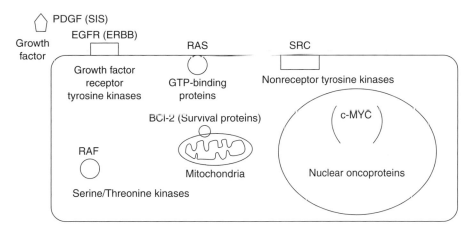

Fig. 6.1 Representative oncogene products. In healthy cells, the normal nonmutated counterpart, proto-oncogenes, encode proteins that in general promote cell proliferation or cell survival but which are under tight regulation to avoid excessive proliferation, or indeed the survival of a cell that has sustained DNA damage and would normally be got rid of. Oncogenes include a diverse range of genes encoding proteins involved in the regulation of cell division, differentiation, and death/survival. The majority of oncogenes fall into five functional classes: transcription factors, growth factors, receptors, signal transducers, and survival proteins. Some key examples of representative oncogene protein products are shown here.

regulation by mitogenic or survival signaling may be circumvented in oncogenesis.

ONCOGENES – WHERE IT ALL BEGAN

The concept of an oncogene (cancer-causing gene) was established by the observation of viral genes that could promote cancers in animal cells. Although viruses have no role in the majority of common human cancers, they are more prominent causes of cancer in some animal species. However, the analysis of these animal viruses has played a remarkable part in identifying genes associated with human cancer [see Appendix 6.1. Historical Perspectives – The Discovery of Oncogenes]. In this section, we will mention various oncogenes that were identified initially as viral oncogenes, and have highlighted some well-known viral and cellular oncogenes in Table 6.5.

As mentioned above, oncogenes are mutant genes that have a dominant effect on cell growth: enhanced cellular replication and avoidance of cell death are common examples, and will be discussed later. Historically, oncogenes were defined using cell culture experiments as genes able to "transform" cells into tumorigenic ones. Such transformed cells would manifest various abnormal cell behaviors in the culture dish, or be able to form tumors if implanted into a mouse. Some abnormal features of transformed cells include uncontrolled cell proliferation in low or absent concentrations of serum, and the "piling up" of cells on top of each other to form characteristic foci. The appearance of these multi-layered foci formed an experimentally convenient method of identifying transformed cells from their normal counterparts, which due to a strict requirement for attachment to a solid substrate only form a single layer of cells. However, it is worth remembering that cell behavior in a culture dish – removed from normal contacts with neighboring cells and extracellular matrix (see Chapter 12), as well as lacking signals derived from a vascular and lymphatic supply (Chapter 14) – may differ to that of a cell in the intact organism, and the ramifications of this will be highlighted later in the chapter.

Mechanisms of converting proto-oncogenes to oncogenes

A proto-oncogene can be converted to an oncogene in a number of ways. In general, such mutations either affect the coding region of the

Table 6.5   Viral and cellular oncogenes

Viral disease	Viral	Cellular	Function
Simian sarcoma	*v-sis*	*PDGFB*	Platelet-derived growth factor B subunit
Chicken erythroleukemia	*v-erb-b*	*EGFR*	Epidermal growth factor receptor
McDonough feline sarcoma	*v-fms*	*CSF1R*	Macrophage colony-stimulating factor receptor
Harvey rat sarcoma	*v-ras*	*HRAS1*	Cell signaling, activation of MAPK cascade
Abelson mouse leukemia	*v-abl*	*ABL*	Protein tyrosine kinase
Avian sarcoma 17	*v-jun*	*JUN*	Transcription factor
Avian myelocytomatosis	*v-myc*	*MYC*	Transcription factor
Mouse osteosarcoma	*v-fos*	*FOS*	Transcription factor

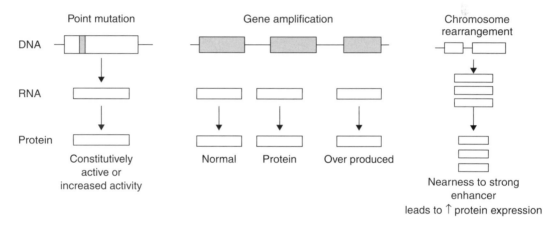

Fig. 6.2 Conversion of proto-oncogenes to oncogenes. A proto-oncogene can be converted to an oncogene in a number of ways. These include point mutations in a single gene (e.g. *RAS*) that can either affect the coding region of the gene resulting in the formation of an abnormal oncoprotein with enhanced stability or activity, or may affect regulatory elements resulting in enhanced or deregulated expression. Chromosomal translocations or rearrangements can lead to overexpression of an oncoprotein. For example, in Burkitt's lymphoma, the proto-oncogene *c-MYC* on chromosome 8 is translocated to one of the three chromosomes containing the genes that encode antibody molecules: immunoglobulin heavy chain locus (chromosome 14) or one of the light chain loci (chromosome 2 or 22). *c-MYC* now finds itself in a region of vigorous gene transcription, leading to over-production of the c-MYC protein. Gene amplification can lead to overexpression of the oncogene. Lastly, the fusion of one protein to another might lead to its constitutive activity. For example, fusion of the promyelocytic leukemia (PML) protein to the retinoic acid receptor-alpha (RARα) generates the transforming protein of acute promyelocytic leukemias.

gene resulting in the formation of an abnormal oncoprotein with enhanced stability or activity, or may affect regulatory elements resulting in enhanced or deregulated expression of the protein (Fig. 6.2). Oncogene mutations span a spectrum of genetic alterations from simple point mutations altering only single amino acids in the protein product, to major chromosomal rearrangements that completely alter gene regulation (Box 3.3, Chapter 3). The major mechanisms involved in activating oncogenes can be classified as: (i) structural alterations – from point mutations in a single gene to chromosomal translocations; and (ii) gene amplification, leading to overexpression of the oncogene. Table 6.2 shows some well-known examples of oncogenes activated by amplification in different cancers.

TYPES OF ONCOGENES

In this section we will give a brief introduction to a selection of well-known oncogenes. Although all these oncogenes are likely to play an important role in various cancers, due to space constraints we have chosen a small number of oncogenes, *c-MYC*, *RAS*, and *SRC*, to discuss in more detail, because they are present in a large number of human cancers and give important historical and mechanistic insights into oncogenesis in general.

A fairly comprehensive list of known oncogenes is given in Table 6.1, and what immediately becomes obvious is that the vast majority of these are involved in pathways that normally serve to regulate cell division and survival.

A simple classification of oncogenes

The vast majority of oncogenes are involved in pathways that normally serve to regulate cell division and survival (and other processes including cell motility and invasiveness – though it is likely that mutations influencing these behaviors would also need to confer some growth advantage). One way of subdividing oncogenes is, therefore, through the site in growth-regulating signal transduction pathways at which they act. Thus, oncogenes may be divided into five groups based on the functional and biochemical properties of protein products of their normal proto-oncogene counterparts. These groups are (1) growth factors, (2) growth factor receptors, (3) signal transducers, (4) transcription factors, and (5) regulators of cell death and others.

Growth factors   The reader is referred to Chapter 5, where this subject is covered in more depth. Oncogenic activation of growth factors (GF) invariably results from transcriptional activation of the gene, leading to overproduction of the growth factor. The best known GF oncogene is *SIS*, the oncogene incorporated into the genome of the simian sarcoma virus, and it encodes the platelet-derived growth factor PDGF-B chain.

Growth factor receptors (receptor tyrosine kinases, RTK)   Several mutations resulting in constitutively active receptors have been identified. For example, a single point mutation converts the normal HER2 receptor into the NEU oncoprotein. Other RTK oncogenes have been identified and are listed in Table 6.3. The reader is again referred to Chapter 5 for a more detailed discussion.

G-Protein coupled receptors   There are few known examples of G-protein coupled receptors (Chapter 5) acting as oncogenes, the most notable being the *MAS* gene, first identified in a mammary carcinoma. *MAS* is also believed to be involved in human epidermoid tumors and has been shown to encode an angiotensin receptor.

Signal transducers

(i) *Nonreceptor membrane-associated tyrosine kinases*   The v-src gene is historically of great importance and was the first oncogene to be identified (see Appendix 6.1 and later this chapter). Moreover, the normal cellular gene *c-SRC* (SRC) is the archetypal protein tyrosine kinase (Fig. 6.18), and the functional domains of other subsequently identified tyrosine kinases are still referred to on the basis of homology to those in SRC (SH-domains for SRC homology). The role of oncogenic SRC in cancer is experiencing something of a renaissance in recent years and is discussed in detail later.

(ii) *The RAS family*   Amongst the first human oncogenes to be identified and frequently mutated in cancers, *RAS* genes encode members of the RAS family of membrane-associated small G proteins. Given its importance in human cancer, we will discuss RAS in more detail later in the chapter. There are three different oncogenic homologs of the c-RAS gene (Box 6.1), each of which was identified in a different type of tumor cell. In all cases, oncogenic *RAS* genes have undergone point mutations that eliminate the intrinsic GTPase activity of the RAS protein. The result is that RAS is permanently "switched on".

Serine/threonine kinases   The *RAF* gene was originally identified as a transforming oncogene from the rat fibrosarcoma virus, from where its name derives. Incorporation of the *RAF* gene into the viral genome resulted in truncation of the protein, with the resulting protein possessing constitutive kinase activity.

Box 6.1   Nomenclature of oncogenes

The nomenclature of oncogenes is straightforward. They are described by a three-letter code relating to the cancers in which they were first identified. Thus, *RAS* oncogenes were first described in *rat sarcomas*. They may carry a prefix distinguishing between cellular (**c-**) and viral (**v-**) homologs, or between differing members (mutant forms) of the same oncogene family. Members of the *RAS* family include *H-RAS, K-RAS,* and *N-RAS,* which refer to **H**arvey and **K**iersten murine sarcoma virus and **N**euroblastoma, respectively, indicating where these variants were first described (in all cases the italicized forms refer to the genes). *RAS* genes have been incorporated into the genomes of such retroviruses and ultimately have contributed to tumor development in certain mouse tissues.

The *c-MYC* proto-oncogene encodes the transcription factor, c-MYC, originally identified as the cellular homolog to the viral oncogene (*v-myc*) of the avian **my**elo**c**ytomatosis retrovirus. There are now several members of the myc family: *c-MYC, N-MYC,* and *L-MYC,* where the prefix denotes the normal cellular (c-) gene, and the dominant tissues in which a particular homolog is expressed: neuronal (N-), lung (L-).

Since the normal *RAF* gene product, RAF, is responsible for threonine phosphorylation of MAP kinase (MAPK) following receptor activation, oncogenic RAF leads to constitutive activation of the downstream MAPK pathway (Fig. 6.14). The *BRAF* gene has been found to be mutated in some human cancers. Table 6.4 summarizes some of the known oncogenes in signaling pathways.

Transcription factors   A considerable number of transcription factors have been shown to possess oncogenic activity when deregulated. Not surprisingly, given the potentially large number of genes whose expression could be deregulated as a consequence. Examples of well-known oncogenic transcription factors include *FOS, JUN, MYC, NFKB, GLI, MYB, ETS,* and others (see Tables 6.1 and 6.2).

As discussed in Chapter 3, various mechanisms of oncogene activation have been demonstrated and depend on the host and the oncogene in question. Thus, oncogenes may be activated by integration of a virus in the vicinity of the host gene or through direct incorporation of the oncogene into a transforming retroviral genome (Chapter 3). Oncogenes may also be activated by point mutations and by chromosomal translocation (Box 3.3, Chapter 3). In some cases, there is evidence for mutations that lead to an increased stability of the oncoprotein (e.g. c-MYC and FOS).

The *FOS* gene was identified as a transforming oncogene of the feline osteosarcoma virus. FOS interacts with a second proto-oncoprotein, JUN, to form a transcriptional regulatory complex. The mechanism of activation of the *FOS* gene results from truncations of the coding and noncoding sequences of the gene following incorporation into the viral genome. This results in the loss of "destability" sequences as well as sequences involved in transcriptional repression of *FOS.* As a consequence a stable FOS protein is produced and, unlike the normal situation in replicating cells where FOS expression is transient, high levels of FOS may persist throughout the cell cycle.

ETS family (ETS) transcription factors, characterized by a conserved ETS domain, are downstream nuclear targets of RAS–MAP kinase signaling (Fig. 6.13), and the deregulation of *ETS* genes results in the malignant transformation of cells. Several *ETS* genes are rearranged in human leukemia and Ewing tumors to produce chimeric oncoproteins (fusion of elements of at least two different proteins). ETS transcription factors are involved in malignant transformation and tumor progression, including invasion, metastasis, and neo-angiogenesis.

The *c-MYC* gene was originally identified in the avian myelocytomatosis virus. The oncogenic functions of the transcription factor c-MYC play a key role in well-known human cancers, such as Burkitt's lymphoma as a result of chromosomal translocation (Box 3.3, Chapter 3). In addition, this oncogene appears to be activated in the majority of human cancers at some stage during tumor development. For this reason, we have selected c-MYC for further discussion in the next section, and will highlight both its normal biological functions as well as those during oncogenic activation *in vitro* and in genetically altered mouse models. Although the mechanism for c-MYC activation remains unclear in many human cancers, there is evidence of mutations that confer increased stability of the c-MYC protein. In all cases, though, the transforming effect of oncogenic c-MYC (described in the next section) is the result of its elevated or deregulated levels of expression.

Regulators of cell survival and death Normal tissues are maintained by a regulated balance between cell proliferation and cell death (apoptosis). In fact, apoptosis is a normal process during embryonic development and tissue homeostasis in the adult – otherwise referred to as programmed cell death (Chapter 8 covers this in depth). Studies of cancer cells have shown that both uncontrolled cell proliferation and failure to undergo programmed cell death can contribute to neoplasia and insensitivity to anticancer treatments. The only proto-oncogenes specifically regulating programmed cell death described to date are members of the *BCL2* family (Chapter 8), first identified by the study of chromosomal translocations in human lymphoma (Box 3.3, Chapter 3). The *BCL2* gene encodes a protein, BCL-2, localized to the inner mitochondrial membrane, endoplasmic reticulum, and nuclear membrane. This BCL-2 family is now known to include a large number of proteins with either anti- or proapoptotic activities. If expression of antiapoptotic BCL-2 family members is increased, or that of proapoptotic members lost, then cells are vulnerable to other cancer-causing mutations, in particular those which increase proliferation. Such cells will now no longer be able to activate key innate tumor-suppressive mechanisms and instead of dying, as they would normally do, they are now able to survive and give rise to cancer. We highlight the importance of BCL-2 family members in tumor progression later in the chapter.

Oncogene collaboration – from cell culture to animal models

A mutation in a single oncogene is not sufficient to cause cancer – the act of collaboration (or cooperation) between two oncogenes is the minimum that is required. Oncogene collaboration can be demonstrated *in vitro* using cultured cells in which various oncogenes are introduced, and also *in vivo* using genetically altered mice.

In different cell types, different combinations of oncogenes are required for such collaborative events to transform normal cells into cancerous ones (transformation).

If normal rat embryo fibroblasts cultured in serum are transfected with the *c-MYC* oncogene alone or *RAS* oncogene alone, then they behave normally. However, when both oncogenes are transfected, the cells become transformed, that is, show various features or behavior associated with cancer cells (Fig. 6.3). These include enhanced proliferation and motility, and the ability to divide without adherence to a substratum, resulting in cells piling up on top of each other to form colonies. Although much important data has been generated, it is worth emphasizing again that the behavior of cells studied *in vitro* – in this case, to identify oncogenes that cause transformation – might not always be the same for cells within the intact organism *in vivo*. For this reason, genetically altered mice are studied in which expression of particular oncogenes is targeted to certain tissues and can be examined in the context of the whole animal. The oncogene is linked to a suitable promoter and then injected into a mouse egg nucleus. The oncogenic construct becomes integrated into the mouse genome leading to the generation of a strain of genetically altered mice that carry the oncogene in all cells (Chapter 18). Depending on the promoter selected, expression of the oncogene will ultimately be driven in a specific cell type or tissue in the animal. Historical studies show that mice carrying either a *c-MYC* or a *RAS* oncogene linked to the MMTV promoter (which ultimately drives MYC or RAS overexpression in the mammary and salivary gland) develop tumors much more frequently than normal. However, when these two strains of mice are cross-bred to generate mice carrying both *c-MYC* and *RAS* oncogenes, tumors develop at a far higher rate – the result of oncogene collaboration. Nevertheless, tumors arise only after a delay and in a small proportion of cells within the tissue. This implies that a further genetic alteration is needed for tumors to develop in this particular tissue – thus the delay while this is "acquired" in one or a few cells through chance. We will see later in this chapter, however, that there are examples of cell types within the mouse that require only two oncogenes for tumors to arise.

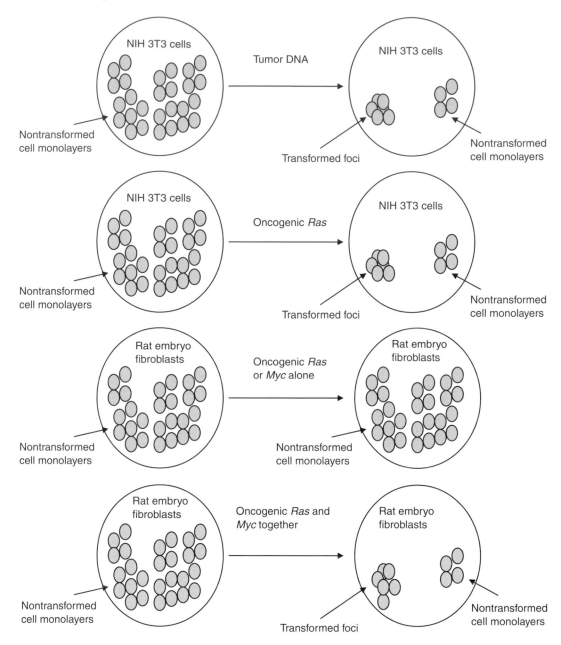

Fig. 6.3 Oncogene collaboration *in vitro*. Transfection of NIH 3T3 fibroblasts with DNA from tumors or tumor cell lines results in some cells becoming transferred as demonstrated by loss of normal contact inhibition and formation of foci (colonies) of cells piled up rather than in a monolayer. This same result could subsequently be achieved by transfecting NIH 3T3 cells with oncogenic *RAS*. In contrast, transfection of a "more normal" rat embryo fibroblast with any single oncogene (either *RAS* or *MYC* alone) did not result in transfomation. However, if both oncogenic *RAS* and *MYC* were transfected into rat embryo fibroblasts then foci did form – oncogene cooperation. By implication NIH 3T3 fibroblasts have already undergone mutations such as those likely to occur in multistep tumorigenesis, thus enabling them to be more readily transformed in cell culture.

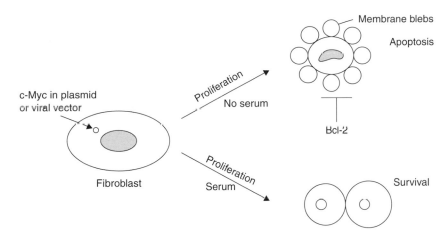

Fig. 6.4 c-MYC promotes cell death by apoptosis (*in vitro*). In the early 1990s, several laboratories made an intriguing discovery: oncoproteins such as c-MYC and the adenovirus E1A – both potent inducers of cell proliferation – were shown to possess apoptotic activity. Ectopic expression of c-MYC in fibroblasts that were cultured in the absence (or limited supply) of survival factors [e.g. factors that are present in fetal calf serum, such as the extracellular molecule insulin growth factor 1 (IGF-1), which mediates cell survival via its receptor] led to apoptosis, with the eventual loss of the entire cell population. Although interpreted by some as a conflict of growth signals, oncogenes activate apoptosis if the proliferative pathway is blocked in some way. The most widely held view of oncoprotein-induced apoptosis is that the induction of cell cycle entry sensitizes the cell to apoptosis, in other words, cell proliferative and apoptotic pathways are coupled. However, the apoptotic pathway is suppressed as long as appropriate survival factors deliver antiapoptotic signals. In this scenario, the predominant outcome of these contradictory processes will depend on the availability of survival factors.

Further *in vitro* work in the early 1990s showed that if rat fibroblasts transfected with *c-MYC* oncogene alone are cultured in very low amounts of serum, then instead of cells arresting (as they would normally do), they were driven to proliferate uncontrollably. Paradoxically, cells also started to die, with the end result being overwhelming cell death by apoptosis (Fig. 6.4). However, if the oncogene *BCL2* was also overexpressed, then cells were rescued from death and allowed to proliferate excessively: *BCL2* acts as an oncogene because the encoded BCL-2 protein inhibits apoptosis. These important findings gave birth to the concept that oncogenic mutations that drive uncontrolled proliferation (as is the case with c-MYC) also possess intrinsic tumor-suppressing activity. In other words, cells that acquire such oncogenic mutations have a built-in "failsafe" mechanism ensuring that potential cancer cells will die by apoptosis rather than survive to accumulate further genetic mutations that may otherwise lead to cancer. Such a cell population would be unable to outgrow its environment unless apoptosis was inhibited. There is now compelling evidence

in vivo to support this concept, as shown by the dramatic synergy between oncoproteins such as c-MYC and mechanisms that suppress apoptosis. An early example is the acceleration of lymphoma development in genetically altered mice that overexpress both c-MYC and the antiapoptotic protein BCL-2 in B lymphocytes, compared with mice expressing c-MYC alone. More recent mouse models include the overexpression of c-MYC together with an antiapoptotic protein, such as BCL-2 or BCL-x_L, or loss of p19^{ARF} or p53 tumor suppressors, some of which will be highlighted later in the chapter. Later we will discuss another critical inherent tumor-suppressing activity associated with oncogenic *RAS*, namely growth arrest (senescence).

THE c-MYC ONCOGENE

Activated c-MYC expression in human tumors

The normal cellular proto-oncogene, c-*MYC*, encodes the protein c-MYC, whose key biological function is to promote cell cycle progression. However, as you will see later, this enigmatic protein appears to be a key player in various other biological processes, such as differentiation, cell death, and angiogenesis. In fact, increasingly c-MYC is being viewed as a "master regulator" of large numbers of genes and processes that prepare the cell to grow, replicate, or die. c-MYC is a transcription factor that requires dimerization with another protein, MAX, in order to become transcriptionally active (Fig. 6.5). Most effects of c-MYC on cell behavior are likely to be the result of expression (and in some cases repression) of various target genes.

Before launching into a description of the mechanisms by which c-MYC activates various biological processes, whether it be within a normal healthy cell or in a tumor cell, this section highlights what we know about c-MYC in human cancers and why we have had to look towards mouse tumor models in order to gain further insight into MYC's oncogenic role *in vivo*.

In the intact organism, c-MYC is expressed ubiquitously during development in growing tissues. There also exist other homologs of c-MYC (L-MYC, N-MYC – Box 6.1), and the expression of each is normally restricted to certain

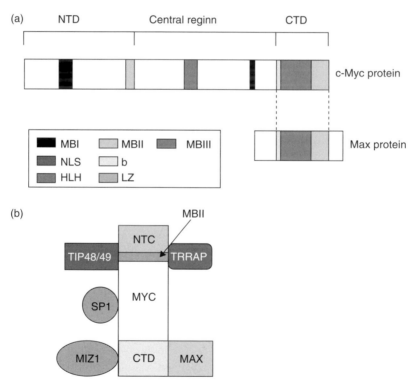

Fig. 6.5 Functional domains of human c-MYC protein. (a) The C-terminal domain (CTD) of human c-MYC protein harbors the basic (b) helix–loop–helix leucine zipper (bHLH-LZ) motif for dimerization with its partner, Max, and subsequent DNA binding of c-MYC–Max heterodimers. The N-terminal domain (NTD) harbors conserved "c-MYC boxes" I and II (MBI and MBII) essential for transactivation of *c-MYC* target genes. Recently the MBIII situated in the central region has been found to be important for negatively regulating the apoptotic response. (b) Some major c-MYC-interacting proteins that may or may not bind simultaneously to c-MYC. These include coactivator TRRAP, part of a complex possessing histone acetyltransferase (HAT) activity, which interacts with MBII region and mediates chromatin remodeling. TIP48 and TIP49 proteins interact with the NTD of c-MYC and are implicated in chromatin remodeling due to their ATP-hydrolyzing and helicase activities. Proteins involved in transcriptional regulation, such as Miz-1, interact with the CTD of c-MYC, whereas SP1 interacts with the central region of c-MYC. NLS; nuclear localization signal.

tissues. Likewise, in the adult body, c-MYC is expressed in tissue compartments possessing high proliferative capacity (e.g. skin epidermis and gut), whereas it is undetected in cells that have exited the cell cycle. With respect to human cancer, oncogenic c-MYC, resulting from a translocational event, is instrumental in the progression of Burkitt's lymphoma (Box 3, Chapter 3). In addition, elevated or deregulated expression of c-MYC has been detected in a wide range of human cancers, and is often associated with aggressive, poorly differentiated tumors. Such cancers include breast, colon, cervical, and small-cell lung carcinomas, osteosarcomas, glioblastomas, melanoma, and myeloid leukemias. In most cases the causes of c-MYC overexpression remain to be specifically described, but include gene amplification (Table 6.2), whereby multiple copies of the c-MYC gene arise under the influence of genome instability (Chapter 10), stabilization of c-MYC mRNA transcripts, or through enhanced initiation of translation due to mutation of the internal ribosomal entry site, or indirectly via activation of various upstream signaling pathways, such as β-catenin/LEF. Recent data suggest that the stability of the c-MYC protein may itself be regulated. Cells have evolved a number of mechanisms to limit the activity and accumulation of c-MYC. One of the most striking of these mechanisms is through the ubiquitin–proteasome pathway (Chapter 11), which typically destroys c-MYC within minutes of its synthesis.

Activation of c-MYC, as is the case in many human tumor cells, is defined as deregulation of the normal, highly controlled expression pattern of the c-MYC proto-oncogene. Until very recently, however, distinguishing between cells that harbor normal or activated c-MYC has proved difficult. In normal dividing cells, c-MYC expression is maintained at a relatively constant intermediate level throughout the cell cycle, whereas in its oncogenic form c-MYC might be constitutively expressed at levels ranging from moderate to very high, and is non-responsive to external signals. Alternatively, the regulated pattern of c-MYC expression can remain intact, but exceed normal levels of expression for the given cell type.

Historically, molecular pathologists have relied on the presence of gross chromosomal abnormalities, such as translocation or amplification, of the c-MYC locus to define activation of this oncogene in tumor cells (as is the case in Burkitt's lymphoma and neuroblastoma, respectively). However, this restrictive diagnostic criterion has resulted in an underestimate of the numbers of tumors with deregulated c-MYC. Therefore, the detection of c-MYC activation in tumor cells now relies on both elevated expression of c-MYC mRNA and genetic alterations of the c-MYC locus. Advances in technology allow us to evaluate expression levels of c-MYC mRNA in tumor cells (precisely excised using laser-capture microdissection) with real-time reverse transcription polymerase chain reaction (RT-PCR) and expression profiling using microarray technology (Chapter 18). These approaches allow rapid, quantifiable, high-throughput screening of c-MYC activation in tumor cells. However, these techniques do not measure levels of c-MYC protein expression, which are largely assayed by immunohistochemical methods.

The fact that c-MYC activation is present in a broad range of human cancers, and is often associated with a poor prognosis, suggests that analysis of c-MYC deregulation in certain tumor types may be used as a diagnostic marker. Moreover, inactivating c-MYC, or downstream targets of c-MYC, may provide important therapeutic targets. To this end, much important information has derived from regulatable mouse tumor models (see later) – these models have exposed various oncogenic properties of c-MYC during tumor progression as well as determining whether inactivating c-MYC in tumors can lead to their regression. Although still in its infancy, various approaches to target c-MYC in human tumors are presently under investigation. Not surprisingly, a major challenge is achieving specific delivery of the c-MYC inhibitor only to the nucleus of tumor cells *in vivo*, an issue that is relevant for many therapeutic agents that target specific tumorigenic proteins.

c-MYC – cell growth and proliferation

When a normal cell *in vitro* receives a signal to proliferate, such as that following binding of a growth factor to its transmembrane receptor (Chapter 5), the result is activation of signaling pathways inside the cell that will ultimately lead to changes in gene expression. Consequently, the induction of specific genes will activate the cell cycle engine, allowing the cell to leave its arrested state (G_0) and re-enter the cell cycle (described in Chapter 4). The *c-MYC* proto-oncogene is one of these induced genes, known as an early-response gene (induced within 15 minutes of growth factor treatment *in vitro*) and plays a crucial role in allowing cells to exit G_0 and proliferate. Thus, when cells are not proliferating, the *c-MYC* gene is silent. It is not difficult, then, to imagine the potential effect a gene like *c-MYC* might have on a cell, if it were continuously expressed or activated – that is, oncogenic *c-MYC*.

c-MYC is crucial for normal cell proliferation. Without it, cells would have a remarkably difficult time replicating themselves: normal development does not proceed in mice in which both *c-MYC* alleles have been inactivated, and embryos die *in utero*. More recently, the cell cycle effects of inactivating c-MYC have been investigated using a rat fibroblast cell line in which both alleles of *c-MYC* were ablated. Cells show greatly reduced rates of cell proliferation mainly due to a major lengthening of G_1 phase (four- to fivefold) and the significantly delayed phosphorylation of the retinoblastoma protein (RB) (see Chapter 4). Thus, progression from mitosis to the G_1 restriction point and the subsequent progression from the restriction point into S phase are both drastically delayed. Moreover, loss of c-MYC also leads to marked deficiency in cell growth (accumulation of mass), as demonstrated by a decreased global mRNA and protein synthesis. Very recent studies in mouse knockdown models suggest that at least under some circumstances cells may be able to replicate without normal levels of c-MYC, but the significance is still not clear. Normal healthy cells need to grow in size during the cell cycle before they divide. The first notion that c-MYC influenced cell growth came from the correlation between c-MYC and the expression of the rate-limiting translation initiation factors eIF4E and eIF2α, now known to be directly transcribed by c-MYC. Later studies showed that c-MYC has a direct role in the growth of invertebrate and mammalian cells: while diminished expression of the *Drosophila* ortholog of vertebrate *MYC* (*DMYC*) resulted in smaller but developmentally normal flies, *DMYC* overexpression resulted

in larger cells, with no significant change in division rate. A different outcome occurs in genetically altered mice when c-MYC is overexpressed in B lymphocytes: c-MYC induces cell growth but not cell cycle progression in this cell type. Having said this, it is most likely that c-MYC induces both cell growth and proliferation in most cell types. It is indeed plausible that c-MYC's role in regulating cell proliferation could at least in part be mediated through its effects on cell growth.

How does c-MYC mediate these effects on cell growth? Although the picture is not completely clear, RNA polymerase III (pol III) which is involved in the generation of transfer RNA and 5S ribosomal RNA (required for protein synthesis in growing cells), is activated by c-MYC via binding to TFIIIB, a pol III-specific general transcription factor. In fact, recent studies suggest that c-MYC may regulate the activity of all three known nuclear RNA polymerases, indicating a potentially central role in mediating ribosomal biogenesis and cell growth.

Insights into how c-MYC might promote cell proliferation have resulted from a number of important studies revealing c-MYC's ability to activate or repress target genes involved in cell cycle progression (Fig. 6.6).

If you cast your mind back to Chapter 4, progression of the cell cycle from G_1 phase to S phase (DNA synthesis) is controlled by the activities of the cyclin-dependent kinase (CDK) complexes: cyclin D–CDK4 and cyclin E–CDK2. c-MYC induces cyclin E–CDK2 activity early in the G_1 phase of the cell cycle, which is regarded as an essential event in MYC-induced G_1–S progression. But how does c-MYC activate cyclin E–CDK2? It was recently shown that *CCND2* gene (which encodes cyclin D2) and *CDK4* are direct target genes of c-MYC. Expression of *CCND2* and *CDK4* leads to sequestration of the CKI p27^{KIP1} by cyclin D2–CDK4 complexes. The subsequent degradation of p27^{KIP1} has recently been shown to involve two other c-MYC target genes, *CUL-1* and *CKS*. Therefore, by preventing the binding of p27^{KIP1} to cyclin E–CDK2 complexes, c-MYC allows inhibitor-free cyclin E–CDK2 complexes to become accessible to phosphorylation by cyclin-activating kinase (CAK) – see Figure 6.6. As a consequence, increased CDK2 and CDK4 activities would result in RB hyperphosphorylation

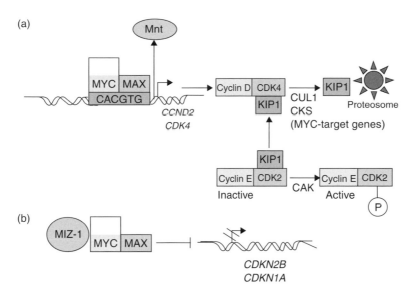

Fig. 6.6 c-MYC induces cell cycle entry through activation and repression of target genes. (a) MYC–MAX heterodimers activate target genes, *CCND2* (cyclin D2) and cyclin-dependent kinase 4 (*Cdk4*), which leads to sequestration of CDK inhibitor KIP1 p27^{KIP1} in cyclin D2–Cdk4 complexes. Subsequent degradation of KIP1 involves two further MYC target genes, *CUL-1* and *CKS*. In so doing, KIP1 is not available to bind to and inhibit cyclin E–CDK2 complexes, thereby allowing cyclin E–Cdk2 to be phosphorylated by cyclin-activating kinase (CAK). Activation of some genes by MYC–MAX involves displacement of the putative tumor suppressor MNT from target genes. (b) MYC–MAX heterodimers repress CDK inhibitors, p15^{INK4B} and WAF1 (p21), which are involved in cell cycle arrest. By interacting with transcription factors, MIZ-1 (and/or Sp1), MYC–MAX prevents transactivation of INK4B (CDKN2B) and WAF1 (CDKN1A).

and subsequent release of E2F from RB (see Chapters 4 and 7).

Recent studies support the idea that c-MYC may also exert important influences on the cell cycle by repressing genes, such as those encoding the CKIs p15^{INK4b} and p21^{CIP1}, that are involved in cell cycle arrest (Chapter 4 and 7), through the c-MYC–MAX heterodimer interacting with positively acting transcription factors such as MIZ-1 and SP1 (Fig. 6.6). Consequently, the interaction of c-MYC–MAX with MIZ-1 blocks the association of MIZ-1 with its own coactivator (P300 protein), with the subsequent downregulation of p15^{INK4b} and p21^{CIP1}.

Finally, we now know that c-MYC (with its partner MAX) activates target genes through its ability to remodel chromatin, and this is described in Figure 6.7. The mechanisms by which c-MYC silences gene expression are not well understood but likely include functional interference with transcriptional activators. Recent studies have implicated altered methylation as one means (Chapter 11): thus c-MYC binds the corepressor DNMT3a and associates with DNA methyltransferase activity. Moreover, reducing DNMT3a levels results in reactivation of c-MYC-repressed genes such as p21^{CIP1}, whereas the expression of c-MYC-activated E-box genes is unaffected. Interestingly, DNMT3a and c-MYC form a complex with MIZ-1, which appears to be required for methylation and repression of the p21^{CIP1} promoter.

Regulating c-MYC activity

Cells have evolved a number of mechanisms to limit the activity and accumulation of c-MYC, so as to avoid abnormal cell behavior. c-MYC activity is normally tightly controlled by conflicting external signals including growth factors, mitogens, and β-catenin, which promote its activity, and factors such as TGF-β, which inhibit such activity (also mentioned in

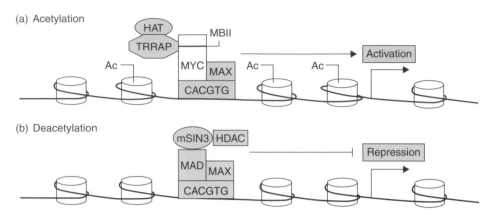

Fig. 6.7 MYC–Max and Mad–Max heterodimers regulate gene activation through chromatin remodeling. (a) MYC–MAX heterodimers binds to an E-box sequence (CACGTG) near the promoter of a *c-MYC* target gene. Coactivator TRRAP (transformation/transcription domain-associated protein), a component of a complex that contains HAT activity, is then recruited to the MYC box II (MBII) domain of c-MYC and acetylates (Ac) nucleosomal histone H4 at the E-box and adjacent regions. Nucleosomal acetylation alters chromatin structure, allowing accessibility of MYC–MAX transcriptional-activator complexes to target DNA, resulting in expression of the target gene. (b) Induction of Mad during terminal differentiation results in the Mad–Max heterodimer binding to an E-box of a *c-MYC* target gene. Corepressor SIN3 and histone deacetylases (HDACs) are then recruited to MAD, resulting in local nucleosomal histone deacetylation and repression of target-gene expression.

Chapters 4 and 5). Much of this regulation is achieved by upstream signaling pathways which influence MYC expression and activity. However, c-MYC, like another key transcription factor oncogene, c-JUN, is a highly unstable protein and rapidly removed by ubiquitin-dependent degradation. The turnover of c-MYC is at least in part dependent on phosphorylation of two highly conserved residues in MB1 that are mutated in *v-MYC* (the viral oncogenic MYC) and in various cancers; phosphorylation at Ser 62 stabilizes c-MYC, whereas phosphorylation at Thr 58 and dephosphorylation at Ser 62 are needed prior to ubiquitination of c-MYC and degradation in the proteasome. The RAS pathway has recently been identified as a key determinant of c-MYC stability. Phosphorylation of c-MYC (and c-JUN) by the enzyme GSK3β creates a high-affinity binding site for the E3 ligase FBW7, which targets c-MYC for polyubiquitination and proteasomal degradation. Conversely, the RAS–PI3K–AKT pathway is a major inhibitor of GSK3β activity and can reduce c-MYC degradation. Specifically, AKT inhibit GSK3β, during early G_1, preventing phosphorylation of Thr 58 stabilizing c-MYC. Furthermore, c-MYC can be directly phosphorylated on Ser 62, via

the RAF–MAPK cascade, which also stabilizes c-MYC.

Paradoxically, however, RAS-regulated enzymes, the protein phosphatase PPA2 and the isomerase PIN1, may also destabilize c-MYC by removing the phosphate at Ser 62. As mentioned, the final step in c-MYC degradation is ubiquitination, which targets the protein to the proteasome for degradation (Chapter 11). The human tumor suppressor FBW7/hCDC4 acts in a complex with SCF ubiquitin ligases to catalyze the ubiquitination of c-MYC. One particular isoform, the FBW7 gamma, may regulate nucleolar c-MYC accumulation and ubiquitination by binding the c-MYC box1 (MB1) domain (shown in Fig. 6.5). Recent studies have indicated how complex the regulation of c-MYC may prove to be, as not all c-MYC is necessarily subjected to the same rates of turnover. Thus, a pool of c-MYC that is metabolically stable has been identified.

The ability of c-MYC to activate transcription depends on the recruitment of several cofactor complexes including histone acetyltransferases (discussed below). In fact, acetylation of nucleosomal histones has long been recognized as a major regulator of gene transcription, but only recently has it been discovered that acetylation may also regulate subcellular localization and protein turnover. Thus, in a very recent study it has been demonstrated that degradation of c-MYC may also be regulated by acetylation by the acetytransferases mGCN5/PCAF and TIP60, which enhance protein stability. As these same enzymes, alongside others, seem important in mediating transcriptional activation by c-MYC this raises the intriguing possibility that at least in part this may involve stabilizing of protein levels.

c-MYC and cell differentiation – role of the Mad protein family

For a proliferating cell to become terminally differentiated, it is usually required to exit the cell cycle and express various genes whose protein products are associated with establishing the differentiated phenotype. For this reason, the general rule is that activation of c-MYC and subsequent cell cycle entry is incompatible with terminal differentiation. In other words, c-MYC blocks cell differentiation. Thus, in most cell types c-MYC would need to be downregulated in order for differentiation to occur. If this is the case, then one can begin to see that unregulated levels of c-MYC in a given cell might not only lead to excessive proliferation, but also keep the cell in an undifferentiated state. Of key importance to this process is another family of proteins known as the MAD/MXI1 family. The expression of various members of this family (described below) usually coincides with downregulation of c-MYC expression, and cells begin to exit the cell cycle and acquire a terminally differentiated phenotype – although there are some exceptions. In tissues that are regionally compartmentalized with respect to cell proliferation and differentiation — such as the crypts of the gastrointestinal system and stratified squamous epithelium of the skin — c-MYC protein is readily detected in immature proliferating cells, whereas MAD proteins are restricted to post-mitotic differentiating cells. Whether downregulation of c-MYC, however, is the trigger for differentiation or a consequence of this cell fate is still unclear. Intriguingly, some studies indicate that onset of differentiation does not always involve cell cycle arrest – c-MYC may play a role in advancing

cells along pathways of epidermal and hematopoietic differentiation. Whether an increase in cell growth and metabolism induced by c-MYC is important for lineage commitment awaits further investigation. With this in mind, recent studies have suggested that c-MYC activity may play an important part in maintaining stem cell pluripotency and self-renewal.

How do the MAD/MXI1 proteins antagonize c-MYC function? These proteins heterodimerize with MAX and subsequently repress transcription by recruiting a chromatin-modifying co-repressor complex to E-box sites on the same target genes as MYC/MAX, such that MYC–MAX complexes can no longer activate its target genes. MAD–MAX recruits HDACs to its specific target DNA to result in local histone deacetylation within nucleosomes, thereby decreasing the accessibility of DNA to transactivation factors (Fig. 6.7). Recent results from the laboratory of Robert Eisenmann have suggested that the situation may be more complex than previously appreciated; MYC and MAD, although possessing identical *in vitro* DNA-binding specificities, do not have an identical set of target genes *in vivo*. In particular, apoptosis is one biological outcome in which the transcriptional effects of MYC are not directly antagonized by those of MAD.

The importance of tight control over c-MYC activity by members of the MAD family is emphasized by the phenotypes of genetically altered mice in which the *MAD1* gene or the *MXI1* gene has been "knocked out". Although quite distinct, the phenotypes displayed by both knockouts display an increase in cellular proliferative capacity. Absence of MXI1 leads to hyperplasia in multiple tissues and tumor growth, whereas absence of MAD1 results in altered differentiation of granulocytes and compensating decrease in cell survival. It is possible that the highly tissue-specific defect observed in *MAD1* knockout mice is a result of functional redundancy among the various members of the MAD family. In other words, it is the expression of other members of the family in those tissues that normally would not express them, that compensates for the loss of MAD1. However, the results from *MXI1* knockout mice support the contention that MXI1 has the potential to act as a tumor suppressor in humans. Recently targeted deletion of the S-phase-specific c-MYC

antagonist MAD3 has been shown to sensitize neuronal and lymphoid cells to radiation-
induced apoptosis. Effects of disrupting other
members of the MAD family await future
investigation.

c-MYC and apoptosis – intrinsic tumor-suppressing activity

Putative cancer cells must avoid death (apoptosis) in order for tumors to arise; the net
expansion of a clone of transformed cells is
achieved by an increase in proliferative index
and by a decreased apoptotic rate. In the early
1990s, several laboratories made an intriguing
discovery: oncoproteins such as c-MYC and the
adenovirus E1A, both potent inducers of cell
proliferation, were shown to possess apoptotic
activity (see section on oncogene collaboration).
Ectopic expression of c-MYC in fibroblasts cultured in the absence (or limited supply) of survival factors (e.g. IGF-1) led to apoptosis with
the eventual loss of the entire cell population
(Fig. 6.4). Some of the research community interpreted these findings as "a conflict of growth
signals", namely that oncogenes activate apoptosis if the proliferative pathway is blocked in
some way. However, the most widely held view
of oncoprotein-induced apoptosis was that the
induction of cell cycle entry somehow "sensitized" the cell to apoptosis. In other words,
cell proliferative and apoptotic pathways are
coupled. These important findings gave birth to
the concept that oncogenic mutations that drive
uncontrolled proliferation (as is the case with c-
MYC) also possess intrinsic tumor-suppressing
activity. Such a cell population would be unable
to outgrow its environment unless apoptosis was
inhibited, for example, by excess IGF-1 signaling, overexpression of the oncogene *BCL2*, activation of RAS (see later), or loss of tumor suppressors, such as p53 (Chapter 8). As highlighted in the
section on oncogene collaboration, early *in vivo*
experiments supported this notion, but, with the
development of more sophisticated mouse models, the case may be regarded as proven. This is
graphically illustrated by the dramatic synergy
between oncoproteins such as c-MYC and mechanisms that suppress apoptosis (see *mouse models of tumor development*, below). The importance of intrinsic tumor-suppressive mechanisms

accompanying oncogenic activation cannot be
overstated – this may be a critical "failsafe"
mechanism serving to protect the organism from
an immediate threat of cancer if an oncogene
were to become deregulated.

How does the c-MYC oncoprotein induce
or "sensitize" cells to apoptosis? Most of the
key experiments used to answer this question
have come from cell culture. Although much
of the apoptotic machinery has been identified,
we are only beginning to define the mechanisms by which c-MYC engages such machinery
(described in Chapter 8). Of course, as is often
the case for cell signaling events, there is not
just one mechanism by which c-MYC engages
or activates the apoptotic machinery. The mechanism chosen most likely depends on factors
such as cell type, signals received from the cell's
environment, or whether the cell has acquired
DNA damage or not. One pathway by which
c-MYC can mediate apoptosis is through its
indirect activation of p53 via another important tumor suppressor, p19^{ARF} (Chapter 8). The
tumor suppressor p53 is a master regulator of
cell proliferation, and can trigger apoptosis, senescence, or DNA repair, in response to a variety
of cellular stresses, including DNA damage, hypoxia, and nutrient deprivation (Chapter 7). The
importance of p19^{ARF} (ARF) in c-MYC-induced
apoptosis *in vivo* was underscored in genetically altered mice in which expression of *c-MYC*
was targeted to B lymphocytes. When these mice
were cross-bred with another strain in which
p19^{ARF} was disrupted, c-MYC-induced lymphoma development was dramatically accelerated.
This outcome, similar to that seen when p53
is disrupted, shows that c-MYC strongly collaborates with loss of p53 or p19^{ARF} in murine lymphomagenesis presumably by inhibiting
c-MYC-induced apoptosis (extrapolating from
in vitro studies). In a similar fashion, deregulated expression of *BMI1* oncogene (encoding
a polycomb group protein) accelerates c-MYC-
induced lymphomas by inhibiting expression
from the *INK4a* locus, which encodes ARF and
p16^{Ink4a} (Chapter 7). We now know that p53
activates many proapoptotic proteins, such as
BAX, APAF1, NOXA, and PUMA, which are
involved in cytochrome c release from the mitochondria and/or the activation of caspases –
enzymes responsible for the destruction of the
cell (Chapter 8).

c-MYC can also promote apoptosis through effects on the expression of members of the BCL-2 family which were alluded to earlier. For example, c-MYC represses expression of BCL-2 and BCL-x_L, both of which are antiapoptotic proteins, but induces expression of proapoptotic members, such as BIM. This has the net effect of permeabilizing the mitochondria to release cytochrome c and other proapoptotic factors, which ultimately lead to apoptosis by activation of caspases or by other means (see Chapter 8).

Lastly, the survival factor, IGF-1, has been shown to inhibit c-MYC-induced apoptosis *in vitro* by blocking cytochrome *c* release from mitochondria. Survival signals mediated via the IGF-1 receptor or activated RAS can lead to activation of the AKT/PKB serine/threonine kinase (described in the RAS section below). Activated AKT then phosphorylates the proapoptotic BH3-only protein BAD resulting in its sequestration and inactivation by the cytosolic 14-3-3 proteins (see Chapter 8). Referring back to the initial experiments *in vitro*, in which cells with deregulated expression of c-MYC die by apoptosis when grown in low serum, it becomes clear how growth factor signaling is able to regulate apoptosis. Given the importance of such growth factors in determining the survival or death of a cell, it is not surprising that elevated signaling through the IGF-1 pathway occurs in many tumors. Similarly, genetic mutations that activate the PI3K pathway (see RAS section) dramatically collaborate with c-MYC during tumor progression (Fig. 6.8).

You will remember that much has been discovered about the role of the c-MYC box I and II (MBI and MBII) in mediating transcriptional activity and stability; now recent studies have begun to reveal an important role for the highly conserved c-MYC box III (MBIII) region. Situated in the central region of the protein, MBIII is important for transcriptional repression by c-MYC, and for transformation *in vitro* and lymphoma development *in vivo*. Conversely, disruption of MBIII prevents transformation by increasing the apoptotic activity of c-MYC, suggesting this region may be an important site for inherent avoidance of apoptosis for c-MYC.

Mouse models of tumor development Deregulated c-MYC expression is often associated with aggressive, poorly differentiated tumors. However, given that most human tumors are quite advanced by the time they are seen in the clinic or surgery – often possessing many genetic alterations – it is difficult to ascertain at which stage of tumor progression c-MYC became activated. This is an important point if we wish to understand what part c-MYC has to play in the initiation and evolution of a tumor, or indeed whether it would serve as a therapeutic target at later stages of progression. It is assumed from the majority of *in vitro* and *in vivo* data that the predominant role of deregulated c-MYC in tumor initiation/progression *in vivo* is through uncontrolled cell proliferation concomitant with loss of terminal differentiation (in most cases). Although this may be part of the picture, it is now known that there are other attributes afforded to c-MYC (e.g. angiogenesis – the formation of new blood vasculature in the growing tissue mass) that had remained occult in the past due to the limitations of the study systems previously available: cell culture (*in vitro*) and conventional genetically altered mouse tumor models *in vivo* (see below).

Studies employing such conventional mouse models, in which the oncogene is continuously expressed in a given cell type by means of a tissue-specific promoter, have supported the view that deregulated c-*MYC* is important for the formation of certain cancers, albeit with a long latency (see oncogene collaboration). The prolonged period of time it takes for tumors to develop implies that other mutations have occurred along the way – in genes that collaborate with c-MYC to transform these cells. Examples of such collaboration with c-MYC, have been demonstrated in mouse models where antiapoptotic proteins such as BCL-2 or BCL-x_L are overexpressed, or *ARF* or *p53* tumor-suppressor genes are lost. In these genetically altered mice, dramatically accelerated tumor development is likely to be due to the inhibition of c-MYC-induced apoptosis as described in Chapter 8, but, as will be seen below, the generation of more sophisticated mouse models has allowed us to conclusively answer this.

The importance of conventional mouse models in cancer biology is undisputed, but it is worth bearing in mind some of the limitations. The oncogene in question is continuously expressed and such expression often starts in

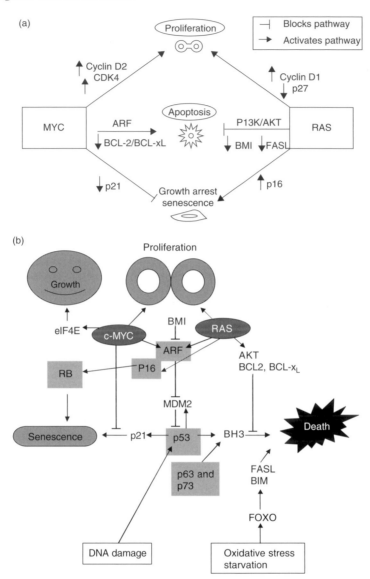

Fig. 6.8 Cooperation between RAS and MYC. (a) Activation of mitogenic proteins, such as c-MYC and RAS, triggers not only pathways that lead to cell cycling but also those that promote cell death (apoptosis) and growth arrest/senescence, respectively. Deregulated activation of c-MYC alone may preferentially lead to apoptosis rather than cell division, while activated RAS alone may lead to cellular senescence – these outcomes serve to protect the organism from cancer-inherent tumor-suppressor activity. However, when RAS and MYC are both activated, these "in-built" tumor suppressor activities are lost: RAS suppresses MYC-induced apoptosis and MYC suppresses RAS-induced growth arrest. It is such cooperation between oncogenes that promotes tumor development. (b) Activation of RAS and MYC results in potential engagement of both replication and growth but also of apoptosis and possibly senescence. If either MYC or RAS levels are excessive (as might occur during oncogenesis) or other proapoptotic signals are received then the balance may be tipped away from replication. RAS can promote senescence through either p16^{INK4a} or ARF, which activate RB or p53 pathways respectively, while MYC can inhibit growth arrest/senescence by inhibiting p21^{CIP1} and inducing TERT (telomerese reverse transcriptase). RAS may also activate p21^{CIP1} via RAF activation. Although MYC may activate the apoptotic pathway (e.g. via ARF), RAS is able to suppress apoptosis by activating the PI3K pathway and, subsequently, AKT. It can readily be appreciated how oncogenic MYC and RAS may conspire in oncogenesis. The combination of RAS and MYC acting together provides a potential means of avoiding apoptotic and senescence mechanisms activated by either acting individually. Moreover, it can also be appreciated how inactivating mutations in RB or p53 (or their pathways involving p19^{Arf}, p16^{INK4a}, p21^{CIP1} etc.) may contribute to tumorigenesis by enabling the cancer cell to avoid either senescence, or apoptosis, or both.

tissues of the developing embryo. Since the majority of human cancers arise within the adult (sporadic tumors), continuous expression during embryonic development may not recapitulate the situation of the adult – the signals derived from the tissue environment are likely to differ from the adult counterpart. Moreover, how can we determine precisely what early effects the oncogene has on that tissue? By the time the mouse is born, the oncogene is likely to have been expressed for a considerable period, during which time additional changes (e.g. genetic mutations or epigenetic changes – see also Chapter 11) may have occurred. In response to these concerns researchers have developed regulatable (or conditional) mouse models in which oncogene expression or activation is temporally controlled within the target tissue. Although the widespread expression of given oncogenes (in all cells of the target tissue rather than a single cell) does not precisely recapitulate the process likely pertaining to development of sporadic tumors, the ability to activate oncogenes at any time desired by the experimenter in the adult provides a significant advance. As mentioned earlier, it enables us to determine how a particular oncogene affects cell behavior shortly after it is expressed *in vivo*, as well as later on during tumor development (discussed below). Moreover, being able to turn off the initiating oncogene in tumors that have subsequently developed allows us to determine whether tumors will regress or not, with unprecedented opportunities to "proof of hypothesis" test new drug targets (this is discussed in detail later on in this chapter).

We will now describe a few regulatable mouse models to illustrate these points. One such model employs a switchable form of the c-MYC protein, called MYCER^{TAM} (Chapter 18), which has been used to investigate the effect of "switching on" c-MYC activation in distinct tissues of the adult: skin epidermis and pancreatic islet β cells. In this way we have been able to confirm the notion that c-MYC-induced apoptosis serves as an intrinsic tumor-suppressing mechanism *in vivo* (as described in the previous section). Furthermore, whether c-MYC remains relevant to the maintenance of established tumors in these mice can be addressed by simply "switching off" expression of the oncogene. In other words, will switching c-MYC off lead to regression of the tumor? The differences in cell behavior observed following c-MYC activation in skin epidermis and pancreatic islets was striking, and serves to highlight the importance of validating data derived from *in vitro* work with studies in the context of an intact organism. It also serves to remind us that what is learned in one cell type does not necessarily apply to all other cell types. c-MYC was activated in suprabasal keratinocytes of the skin epidermis of the transgenic mouse: cells that have exited the cell cycle and have commenced differentiation (in normal wild-type epidermis, c-MYC would be undetected as these cells are post-mitotic). Yet the effects of activating c-MYC in suprabasal keratinocytes included induction of cell cycle entry as well as impaired differentiation, and epidermal hyperplasia with areas of focal dysplasia and papillomatous lesions resembling the human premalignant skin lesions known as actinic keratosis — a precursor of squamous cell carcinoma (Fig. 6.9). A striking feature was the widespread induction of angiogenesis (growth of new blood vessels) and increased levels of vascular endothelial growth factor (VEGF), suggesting that c-MYC might act as an "angiogenic switch" (Chapter 14), or alternatively that expansion of a tissue compartment may in this case be automatically coupled to expansion of the vasculature. Importantly, c-MYC has now been shown to promote angiogenesis in multiple cell types both through its repression of the angiogenesis inhibitor thrombospondin-1 and through its induction of VEGF (see Chapter 14). A key observation was the obvious lack of c-MYC-induced apoptosis in epidermal keratinocytes *in vivo*, despite the potent induction of apoptosis by c-MYC in isolated serum-deprived keratinocytes, from the same genetically manipulated mice, cultured *in vitro*. Thus, in this tissue proliferation is the predominant outcome over apoptosis.

What then of the intrinsic tumor-suppressing activity of c-MYC discussed earlier? Although c-MYC does not induce apoptosis, the papillomatous lesions formed are completely benign (that is, they do not invade underlying dermis). Moreover, in this tissue, cells are on a "conveyor belt" moving outward and ultimately shedding from the surface of the skin. Therefore, the organism has an effective means of disposal for

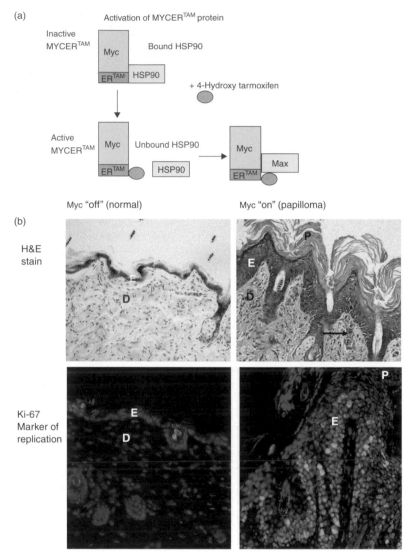

Fig. 6.9 (a) A system has been developed allowing the ectopic activation of c-MYC in various tissues *in vivo*. A transgene encoding a chimeric protein (c-MYC fused to the ligand-binding domain of a modified estrogen receptor ER^{TAM}, which now responds only to the synthetic SERM, 4-hydroxy tamoxifen- 4-OHT) is placed under a tissue-specific promoter directing expression to a predetermined tissue (Involucrin – Inv – for suprabasl keratinocytes or Insulin – pIns – for pancreatic β cells). Activation of c-MYC is achieved by administration of 4-OHT, which binds to the ER^{TAM}, thus displacing heat shock proteins that otherwise hold the protein in an inactive state. (b) Normal adult wild-type skin (or skin in transgenic mice prior to c-MYC activation) is shown in the left-hand panel. In general the epidermis (E) is only two cells thick (more in man), comprising a basal layer (containing stem cell precursors) and a suprabasal layer containing differentiating skin keratinocytes. Underlying the epidermis is the dermis. The upper panels are hematoxylin-and-eosin stained to show structural features. The lower panels show cell nuclei in blue (DAPI) and the nuclei of replicating cells in green (FITC), as they are stained with an antibody to a cell cycle marker – Ki67. The right-hand panels show the effects of activating c-MYC for 2 weeks. The epidermis is greatly expanded and now contains large numbers of replicating cells. Not only is the epidermis expanded but there are now large keratotic spires (keratinized dead or dying cells that remain nucleated – normally cells lose their nuclei at the surface and are then shed) forming papillomas (P). These papillomas become vascularized by angiogenesis (black arrow). Although not shown here, deactivating c-MYC results in complete reversal of this aberrant phenotype and restoration of normal appearing skin. (Please see our website for the color version of this figure).

such premalignant cells (the vacuum cleaner for one) and apoptosis may not be required. This makes sense — if premalignant keratinocytes do not pose a serious neoplastic risk, it may be more advantageous to maintain them rather than risk the depredations of apoptosis on a tissue whose structural integrity is so vital. Although further investigation is required, it is likely that c-MYC-induced apoptosis in intact skin is suppressed by the presence of excess survival signals, such as those arising from contacts with neighboring cells or secreted factors (Chapter 12).

We next turn to a further regulatable MYCER^{TAM} mouse model, in which the switchable form of c-MYC has been targeted to the insulin-producing β cells of pancreatic islets. In contrast to skin, the predominant effect of activating c-MYC in pancreatic β cells of adult transgenic mice is apoptosis, and not proliferation. When the "switchable" c-MYC protein is inactive, β cells exhibit the same very low incidence of either cell proliferation or apoptosis, as observed in wild-type islets of the adult. However, following c-MYC activation essentially all β cells in every islet are induced to proliferate within 24 hours. However, in complete contrast to epidermal keratinocytes *in vivo*, c-MYC activation soon triggers apoptosis of islet β cells (Fig. 6.10). The acute sensitivity to induction of apoptosis by c-MYC suggests that β cells are only modestly buffered against cell death by survival signals or intrinsic antiapoptotic mechanisms, such as BCL-2/BCL-x$_L$ expression (Chapter 8), *in vivo*. This result indeed suggested that in this tissue c-MYC acts as its own "tumor suppressor". In fact, when c-MYC-induced apoptosis was blocked (by coexpressing BCL-x$_L$), c-MYC triggered rapid and uniform carcinogenic progression. In fact, c-MYC activation now generated many of the "hallmark" features of cancer, including immediate and sustained β-cell proliferation resulting in islet expansion, loss of differentiation, angiogenesis, loss of cell–cell contacts resulting from loss of E-cadherin (see Chapter 12), and local invasion of β cells into surrounding exocrine pancreas (Fig. 6.10). The rapidity and synchrony of this neoplastic phenotype in the majority of islets suggested that there was no evident requirement for additional mutations, but rather the phenotype developed as a direct consequence of c-MYC activation in β cells. The dramatic and immediate oncogenic

progression observed in islets *in vivo* strongly supports the unorthodox notion that complex neoplastic phenomena involving the tumor cell and its interactions with normal surrounding tissues may both be induced and maintained by a simple combination of two genetic lesions – "the minimal platform". Although more mutations are likely to be necessary in the equivalent human counterparts, these results suggest that the complexity of tumor phenotype need not always be the result of an equivalent complexity of genetic alteration. This may in fact depend on the tissue type, and on the combination of oncogenes.

THE RAS SUPERFAMILY

With over 100 members in humans, the RAS superfamily is a diverse group of monomeric G proteins, participating in many normal cellular processes including proliferation, motility, and survival. RAS proteins are also widely implicated in human tumorigenesis, either through activating mutations or by overexpression, where they can contribute significantly to several aspects of the malignant phenotype, such as deregulated cell proliferation, survival, invasiveness, and the ability to induce new blood vessel formation (angiogenesis).

The RAS proteins were amongst the first proteins identified that possessed the ability to regulate cell proliferation. They were discovered as proteins encoded by retroviral oncogenes that had been hijacked from the host genome by the Kirsten and Harvey rat sarcoma viruses (Table 6.1 and Appendix 6.1). Human tumors very frequently express RAS proteins that have been activated by point mutation (Table 6.4) and around 20% of all tumors have undergone an activating mutation in one of the *RAS* genes (e.g. a single base substitution in the 12th codon often causes the replacement of a glycine with a valine). In other cases, amplification of the *RAS* gene has occurred (Table 6.2). Moreover, many tumors that lack *RAS* mutations have found other ways to activate the same pathways, for example EGFR overexpression, activating mutations in *BRAF*, amplification of *AKT* or PI3K, some of which are discussed later. Not surprisingly, therefore, there is a huge interest

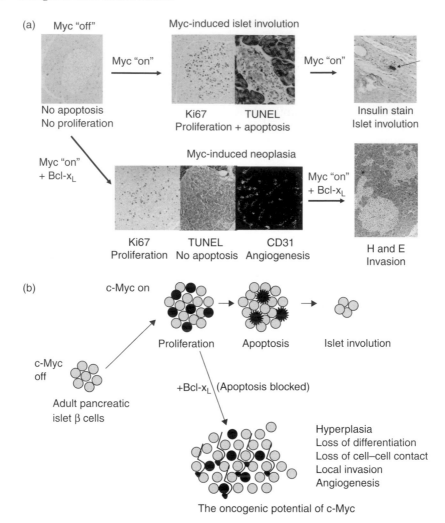

Fig. 6.10 The oncogenic potential of c-MYC is exposed in pancreatic β cells when apoptosis is blocked. (a) Activation of c-MYC in β cells of the pancreatic islets results in cell cycle entry (Ki67 positive cells – stained brown), but then also results in β-cell apoptosis (TUNEL positive cells – stained brown). In fact, the net effect is almost complete ablation of the β-cell population within 10 days, resulting in diabetes. In this tissue, unlike in skin, the net effect of deregulated c-MYC expression is apoptosis, and avoidance of neoplasia. If, however, c-MYC apoptosis of β cells is prevented by concurrent overexpression of the antiapoptotic protein BCL-xL, then the devastating oncogenic properties of unapposed c-MYC action are unmasked – culminating in relentless replication, avoidance of apoptosis, loss of differentiation, loss of cell – cell contacts, angiogenesis, and invasion – the "hallmark features" of cancer. (b) This intrinsic "tumor suppressor" function of c-MYC is illustrated schematically. (Please see our website for the color version of this figure).

in therapies that might target the RAS proteins or the signaling pathways that they control. Numerous therapeutic agents are in clinical trials at present and more are under development.

Since studies of the *RAS* oncogenes began in the early 1980s, a large superfamily of small monomeric guanine nucleotide-binding proteins, including RAC, RHO, and RAB, that bind GTP and hydrolyze it to GDP (Fig. 6.11) have been identified, and they can be divided into several families according to the degree of sequence conservation. They are thus distinguished from

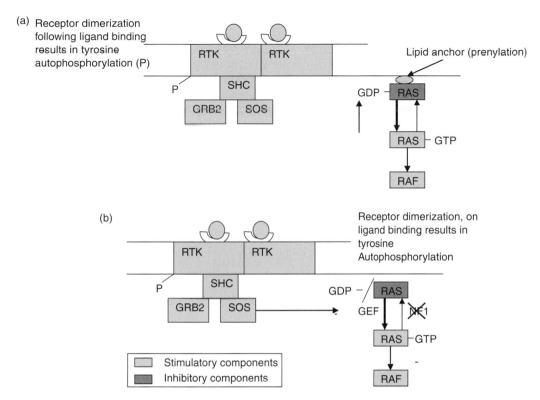

Fig. 6.11 RAS activation depends on regulation of GTP/GDP exchange. (a) In a healthy cell, RAS proteins regulate diverse cellular processes by cycling between biologically active GTP- and inactive GDP-bound conformations. In the active state, RAS proteins are bound to GTP and are thus able to engage downstream effectors that activate signaling pathways controlling several aspects of cell behavior. When bound to GDP, RAS proteins are inactive and so fail to interact with these effectors. The conversion of bound GTP to GDP, and vice versa, is catalyzed within the cell: the nucleotide exchange by guanine nucleotide exchange factors (GEFs), such as Sos, and the nucleotide hydrolysis by GTPase activating proteins (GAPs), such as NF1. It is the balance between these proteins that will ultimately determine the activation state of RAS and its downstream pathways. (b)Activation of RAS in tumors: mutations in RAS can prevent the intrinsic GTPase activity and impede exchange of GTP for GDP, thus favoring the active conformation. A similar outcome can result from loss of the NF1 tumor-suppressor protein, which normally amplifies the intrinsic GTPase activity of RAS. Although growth factor stimulation will still play some role in phenotype, these alterations can result in constitutive RAS activation.

the other main class of cellular GTP-binding proteins, namely those coupled to 7 transmembrane spanning domain receptors (GPCR), which in contrast are trimeric and link directly with receptors without intermediate adapter proteins. Different families are important for different cellular processes: the RAS family controls cell proliferation and survival; the RHO family (which includes RAC) controls the actin cytoskeleton, which is instrumental in governing cell motility; and the RAB family is involved

in regulating the traffic of intracellular transport vesicles.

The prototypical RAS proteins, H-RAS, N-RAS, and K-RAS, were originally identified in tumors as the hyperactive products of mutant *RAS* genes, which promote cancer by disrupting the normal controls on cell proliferation and differentiation. These proteins (in their non-mutated form) were subsequently found to contribute to cell cycle regulation in normal, non-transformed cells. H-RAS, N-RAS, and K-RAS share 85% identity, with divergence confined largely to the carboxyl terminus. Broadly, all three proteins have similar activities on cell cycle progression and replication though subtle differences may be observed in cellular distribution and aspects of downstream signal transduction pathways activated. Out of the three, however, K-RAS is the only one that is essential for embryonic development, as deduced from knockout studies. This is most likely due to K-RAS being expressed more ubiquitously compared to H-RAS and N-RAS.

In a healthy cell, typical members of the RAS superfamily function as regulators of diverse cellular processes by cycling between biologically active GTP- and inactive GDP-bound conformations. In the active state, RAS proteins are bound to GTP and are thus able to engage downstream effectors, such as the serine/threonine kinase RAF, to activate signaling pathways controlling several aspects of cell behavior mentioned above. However, when bound to GDP, RAS proteins are inactive and so fail to interact with these effectors (Fig. 6.11). Thus, in normal cells, the activity of RAS proteins is controlled by the ratio of bound GTP to GDP. The hydrolysis of bound GTP leads to deactivation of RAS, as it does also for the Gα subunit of receptor coupled G proteins (Chapter 5). However, the average lifetime of RAS-GTP is far longer than that of Gα-GTP; the latter is rapidly inactivated by intrinsic GTPase activity, whereas for RAS other proteins are required to speed the otherwise slow rate of cycling between the GTP- and GDP-bound forms. The conversion of RAS bound GTP to GDP, and vice versa, is catalyzed by interactions with other proteins within the cell, which are therefore key regulators of RAS signaling: activators that catalyze nucleotide exchange – guanine nucleotide exchange factors (GEFs) – and inactivators that facilitate nucleotide hydrolysis – GTPase activating proteins (GAPs). It is the balance between these proteins that will ultimately determine the activation state of RAS and its downstream pathways.

In tumor cells oncogenic *RAS* mutations result in loss or diminution in GTPase activity, with the result that RAS proteins remain constitutively in the active GTP-bound form (Fig. 6.11). In many human tumors, mutations may occur in genes encoding proteins involved in regulating RAS activity rather than in RAS

itself. One such example is loss of the tumor-suppressor gene *NF1*, which encodes a GTPase-activating protein called neurofibromin, resulting in upregulated RAS signaling. Mutations in *NF1* cause neurofibromatosis type 1 (NF1) – a condition immortalized, though maybe incorrectly, by the story of the Elephant Man. NF1 is characterized by the development of neurofibromas on peripheral nerves, which are complex, primarily benign, tumors composed of Schwann cells and other cell types. Defective NF1 also contributes to childhood chronic myelogenous leukemia.

Activation of RAS by receptor tyrosine kinases

RAS lies downstream of most known receptor tyrosine kinases (RTK). Unlike for G-protein coupled receptors, however, where G proteins are directly activated by ligand binding to the receptor, RAS activation by growth factor bound RTKs requires receruitment of adaptor proteins which link the RTK to RAS (Chapter 5). The two best known of these proteins are SOS and GRB2; the SH2 domain of GRB2 binds to activated (tyrosine phosphorylated) RTKs. The SH3 domains in GRB2 then bind to SOS (a GEF), bringing this protein into proximity with RAS, activating exchange of GDP to GTP, thereby activating RAS. It is believed that this proximity-based process is facilitated by lipid modification of RAS, which helps localize RAS to the cell membrane (see next section).

Lipid modification of RAS

The normal function of RAS proteins requires them to be post-translationally modified by the covalent attachment of a lipid isoprenyl group to their hypervariable carboxyl termini by protein farnesyl transferase (FTase) or protein geranylgeranyl transferase type I (GGTase-I) (Fig. 6.12). Interestingly, such lipid groups derive from the same intracellular sterol biosynthesis pathway that cells use to produce cholesterol. Such isoprenylation of RAS proteins appears to be an essential prerequisite for RAS function and for the transforming activity of oncogenic mutants, as it enables direction

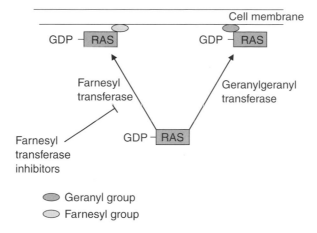

Fig. 6.12 Lipid modification of RAS. RAS activation (exchange of bound GDP for GTP) requires localization of RAS-GDP to the inner cell membrane, presumably because in this location it is more readily available for interaction with growth factor receptor associated proteins. Usually this is achieved by addition of enzymatic addition of prenyl groups, primarily farnesyls. Though if farnesyl transferase is inhibited, at least for Ki-RAS, then these can be replaced effectively by geranylgeranyl groups, which appear to serve the same or similar purpose.

of RAS proteins to the appropriate subcellular compartment, principally the inner face of the plasma membrane, and is necessary for their dynamic sorting and trafficking. In fact, the key role of isoprenylation in RAS activity is emphasized by the current interest in drugs, such as statins and farnesyl transferase inhibitors, which might impede this process, as potential anticancer agents. Aberrant localization of RAS impedes activity most likely by reducing the chance of interaction with its usual targets. Studies have shown that farnesylation of RAS is the first, obligatory step in a series of post-translational modifications leading to membrane association, which, in turn, determines the switch from an inactive to an active RAS-GTP bound form. The enzyme responsible for this farnesylation of RAS is known as protein farnesyl transferase (FTase). Theoretically, inhibiting this reaction might be able to revert the transformed phenotype, provided the rationale for the development of farnesyl transferase inhibitors (FTI) as anticancer drugs. Many FTIs have been entered into phase II studies and at least two phase III trials, but the exact mechanism of action remains uncertain. FTIs can block the farnesylation of several additional proteins, such as RHOB, prelamins A and B, centromere proteins (CENP-E, CENP-F), on the one hand, while they seem only

partially able to prevent RAS localization to the inner membrane. One potential fly in the ointment, at least for K-RAS protein, is that prenylation may also be mediated by other enzymes such as geranylgeranyl transferase (GGT), which by adding geranylgeranyl groups to RAS can also help localize it to the membrane for activation.

The view of RAS signaling being confined to the inner surface of the plasma membrane is increasingly being challenged by observations indicating that RAS proteins interact dynamically not just with specific microdomains of the plasma membrane but also with other internal cell membranes and intracellular organelles. Importantly, the location of RAS may be an important determinant of downstream signaling. For example, activation of RAS on the Golgi exhibits kinetics different from RAS activation on the plasma membrane, and compartmentalized RAS signaling seems particularly prominent in lymphocytes.

Signaling downstream of RAS

RAS is activated by a large variety of extracellular stimuli, largely through activation of RTKs. Once in the active GTP-bound form, RAS can activate a number of effector proteins, each representing distinct signaling pathways (Fig. 6.13). These effector proteins are members of the RAF family (RAF1, B-RAF, and A-RAF), phosphatidylinositol 3 kinase (PI3K) and members of the RAL family (RALGDS, RGL, and RGL/RGL2). More recently, another RAF effector, phospholipase Cε (PLCε), has been reported which activates protein kinase C (PKC) and mobilizes calcium from intracellular stores.

Without question, the best-known signaling pathway downstream from RAS is the serine-threonine kinase RAF–MEK–ERK pathway (also known as the mitogen-activated protein kinase-MAPK pathway) that regulates cell cycle progression. The PI3K–AKT pathway regulates cell survival and also stimulates RAC, a RHO family protein that is involved in regulating the actin cytoskeleton. The third well-studied RAF effector family includes the RAL proteins: RAL guanine nucleotide dissociation stimulator (RALGDS), RALGDS-like gene (RGL), and RGL2/RLF. This signaling pathway is involved in vesicle transport and cell cycle progression.

We will next discuss some of the RAS signaling pathways in the context of normal cell behavior and in tumor cells in which RAS signaling is constitutively active (oncogenic).

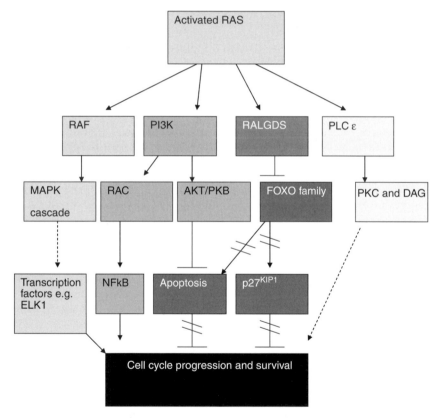

Fig. 6.13 Signaling pathways regulated by RAS. Active GTP-bound RAS will interact with several families of effector proteins, with the most important shown. RAF protein kinases initiate the MAPK cascade, which leads to ERK activation. ERK has numerous substrates including ETS family transcription factors such as ELK1 that regulate cell cycle progression. Phosphoinositide 3-kinases (PI3Ks) generate lipid messengers, such as phosphatidylinositol-3,4,5-trisphosphate, which activate the kinase AKT/PKB, involved in survival. RALGDS proteins are GEFs for RAL, an RAS-related protein. Downstream targets include forkhead transcription factors. Phospholipase C (PLC) catalyzes the hydrolysis of phosphatidylinositol-4,5-bisphosphate to diacylglycerol and inositol trisphosphate, resulting in protein kinase C (PKC) activation and calcium mobilization from intracellular stores.

RAS and cell growth and proliferation The ability of RAS family proteins to be rapidly switched "on" and "off" (a property incidentally shared by all G proteins) makes them ideally suited for their involvement in a wide range of key cellular behaviors: cell growth, differentiation, gene expression, cell survival/death, and cell movement. The prototypical RAS GTPases activate several signaling pathways that control the cell cycle (Fig. 6.13). In studies, blocking RAS activity prevents growth factor-induced G_1/S transition, whereas conversely quiescent cells can be induced to replicate by introducing RAS protein. The RAS proteins help relay signals from RTK (Chapter 5) to the nucleus by regulating a number of distinct but overlapping signaling pathways. By this means RAS can influence multiple cell processes including proliferation and growth but also survival, differentiation, senescence, or even cell death (apoptosis), which will be discussed later. Given the complexity of RAS signaling in regulation of cell proliferation, we will first introduce the key players involved in this cell process and conclude with the most intensively studied RAF–MAPK and PI3–AKT pathways.

A major role of RAS is to transduce growth factor signals required for G_1/S transition in the cell cycle. RAS activates numerous signaling pathways downstream of activated RTK growth factor signals, including the RAF–MAPK and PI3K–AKT (Fig. 6.13). Together these signals enhance cyclin D1 expression and help inactivate the retinoblastoma (RB) protein. Such actions are mediated both by transcriptional effects, including activation of transcription factors such as c-MYC (cyclin D2 is a direct c-MYC target gene) and by altering protein stability. As described in Chapter 4 several cell cycle regulatory proteins are degraded by the ubiquitin–proteasome pathway (Chapter 11). In the case of cyclin D1, phosphorylation by a key enzyme, glycogen-synthase kinase-3β (GSK3β) allows ubiquitination by the SCF (see Chapter 4) targeting cyclin D1 for degradation. Conversely, RAS promotes cyclin D1 stability by inhibiting GSK3β activity; RAS activates the PI3K–AKT pathway which inhibits GSK3β.

RAS pathways may also promote cell division by reducing levels of the CKI p27^{KIP1} (Chapter 4, 7); RAS inhibits synthesis of p27^{KIP1}, via downregulation of forkhead family transcription factors, and increases p27^{KIP1} degradation. This places p27^{KIP1} at a crucial integration point of positive and negative growth factor signaling for G_1/S transition. Put simply, a cell can decide if it will commit to replication depending on the balance of positive and negative growth signals (Chapter 5). Deregulated or oncogenic RAS is capable of both enhancing the positive signal and inhibiting some of the negative signal, thus strongly favoring cell replication.

RAS may also contribute to cell replication by enhancing cellular growth (size) by upregulating the translational machinery and protein synthesis. One of the master regulators of protein synthesis and translation control is the mammalian target of rapamycin (mTOR) protein, which controls the translational apparatus through protein phosphorylation. The mTOR–S6K pathway is regulated by signals that are transmitted by PI3K in response to mitogen stimulation and nutrient supply (see Chapter 5). Briefly, PI3K functions through AKT/PKB-mediated phosphorylation and inhibition of a suppressor complex composed of tuberous sclerosis 1 (TSC1) (also known as hamartin) and TSC2 (also known as tuberin), which are negative regulators of the mTOR pathway. Inactivating TSC gene mutations result in a predisposition to at least two cancer-related diseases: tuberous sclerosis and lymphangioleiomyomatosis.

It seems increasingly likely that other members of the RAS – and RHO – family of GTPases also influence cell cycle progression and growth. In fact, irrespective of the effects these proteins may have on cell motility etc, they share the property of increasing cellular replication and likely do so by activating similar or overlapping pathways.

RAS signaling through the RAF–MAP kinase pathway The best-known signaling pathway downstream from RAS is the serine/threonine kinase RAF–MEK–ERK pathway (also known as the mitogen-activated protein kinase MAPK pathway) which regulates cell cycle progression (Fig. 6.14). This is one of a series of kinase cascades with the generic description of MAPKKK–MAPKK–MAPK. The first mammalian effector of RAS to be characterized was the protein serine/threonine kinase RAF (a mitogen-activated protein kinase kinase kinase–MAPKKK); active GTP-bound RAS binds to, and

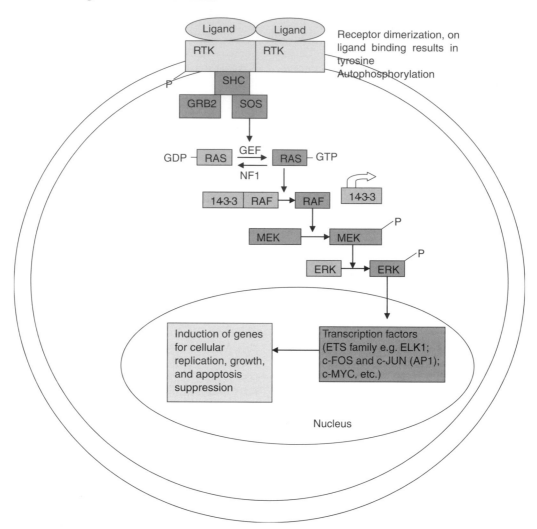

Fig. 6.14 The RAS–RAF signaling pathway. Receptor dimerization on ligand binding induces tyrosine autophosphorylation of the RTK, binding of GRB2 to the receptor via SH2 domain, and translocation of the GEF, SOS, to the cell membrane. SOS in turn promotes RAS activation by enhancing GDP–GTP exchange. A third adaptor protein SHC (also carrying an SH2 domain) may be phosphorylated by growth factor binding and can recruit the GRB2/SOS complex. Activated RAS by either of these two overlapping pathways in turn phosphorylates the serine/threonine kinase RAF, in part by a conformational effect and replacement of 14-3-3 protein. RAF phosphorylates and activates MAP kinase/ERK kinase (MEK). MEK then activates ERK, which can then translocate to the nucleus and activate transcription factors, such as ETS family, c-FOS, and c-JUN (which together form the AP1 transcription factor), c-MYC, and others.

contributes to the relocation of three closely related RAF proteins (c-RAF, B-RAF, and A-RAF) to the plasma membrane, which appears essential for their activation. RAS binding induces the transition of RAF from a closed to an open conformation, which enables RAF to bind and then phosphorylate and activate MEK1 and MEK2 (mitogen-activated protein kinase kinases–MAPKK) – dual-specificity kinases that are capable of phosphorylating and activating ERK1 and ERK2 (the mitogen-activated protein kinases–MAPK).

Substrates for ERK1/2 (extracellular signal-regulated kinases 1 and 2), include cytosolic and nuclear proteins, reflecting the fact that they can be transported into the nucleus following activation. ERK phosphorylates ETS family transcription factors such as ELK1, part of the serum response factor (SRF) that regulates the expression of FOS. In addition, ERK phosphorylates c-JUN. This leads to activation of the AP1 transcription factor, which is made up of FOS–JUN heterodimers. AP1 is a well-known and important regulator of GF signaling in the nucleus, and regulates expression of multiple key genes downstream of GF activity, including c-MYC. Together, these transcription factors influence expression of several key cell cycle regulatory proteins, including increases in cyclin D, inhibition of INK4a CKIs required for the cell to transit through G_1/S in the cell-cycle. The ERK group of MAPKs is thus essential for normal mitogen-regulated cell proliferation.

In addition to the well-described role of the RAF–MAPK pathway in transduction of the growth-promoting or mitogenic activity of numerous growth factors, it has recently also been shown to mediate antiapoptotic survival signals, distinct from those through another RAS effector – the PI3K–AKT pathway described later. In addition, this pathway may also couple cell replication/growth with induction of angiogenesis, though this may require activity of c-MYC. Some studies also suggest that ERK may promote transcriptional upregulation of angiogenic factors, and promote increased invasiveness as a result of ERK-mediated expression of matrix metalloproteinases (Chapter 12) and RAC-mediated effects on the cytoskeleton.

Thus, aberrant signaling through the RAS-MAPK pathway has the potential to promote several of the "hallmark" features characteristic of malignant transformation. However, the actual cellular responses operating *in vivo* are likely to depend upon various factors, such as cell type, cell location, and contact with other cell types. It is crucially important not to forget that a single oncogene by itself is not sufficient for cellular transformation (as discussed earlier under Oncogene collaboration) as additional oncogenic mutations are necessary – probably more needed in humans compared to mice. Later, we will indicate possible explanations as to why oncogenic *RAS* alone is unable to transform cells – due to its "intrinsic tumor-suppressive" activities (as previously also described for c-MYC). Importantly, collaboration between oncogenic *RAS* and c-MYC may overcome the tumor suppressor activity of each – promoting tumorigenesis.

Finally, aberrant activation of the RAS–MAPK pathway occurs in many human cancers through mutations that activate RAS (as discussed), or other proteins (RAF, PI3K, PTEN, AKT) which serve to regulate its downstream activity. For this reason, inhibitors of RAS, RAF, MEK, and some downstream targets have been developed and many are currently in clinical trials.

RAS signaling through PI3K–AKT pathway

In addition to the RAF–MAPK pathway described above, RAS has also been found to activate several other effector pathways, the best characterized of which are shown in Fig. 6.13. Although only briefly described here, later sections will relate some of these pathways to cell behavior in more detail.

RAS activation of the PI3K–AKT pathway is of key importance in normal cell growth and in cancer. RAS can interact directly with the catalytic subunit of type I phosphatidylinositol 3-kinases (PI3K), leading to activation of the lipid kinase. In turn, activated PI3K phosphorylates PIP2 (phosphatidylinositol-4,5-bisphosphate or PtdIns(4,5)P$_2$) to produce PIP3 (phosphatidylinositol-3,4,5-trisphosphate or PtdIns(3,4,5)P$_3$). PIP3 is a key second messenger that binds to a large number of proteins through the pleckstrin homology and other domains. In this way, PI3K controls the activity of a large number of downstream enzymes including 3-phosphoinositide-dependent protein kinase 1 (PDK1) and AKT (otherwise known as PKB – protein kinase B). PDK1 is important for the activation of a large number of protein kinases, and AKT is an important mediator of survival signals, generated by RAS, that prevent cells from dying by apoptosis (Chapter 8). Moreover, AKT is a key regulator of protein stability of multiple important cell cycle regulatory proteins, because AKT can inhibit the activity of GSK3β preventing ubiquitination of proteins such as c-MYC, c-JUN, and cyclins D and E, by E3 ligases, required for proteasomal degradation.

PI3K also activates RAC, a RHO-family protein involved not only in regulating the actin cytoskeleton important in cell motility, but also in activating transcription factors such as nuclear factor-κB (NFκB) involved in cell survival and inflammation. Regulatory subunits of NFκB are also inhibited by GSK3β so stability of this transcription factor will likely also be enhanced by PI3K–AKT signaling.

RAS, PLC$_\varepsilon$, and PKC

In addition to activation of RAS, RTK signaling can also activate the enzyme Phospholipase Cγ (PLCγ) which also plays a key role in cellular growth and proliferation. Upon growth factor binding, RTKs promote phosphorylation of PLCγ at three known tyrosine residues: Tyr771, Tyr783, and Tyr1254 and its enzymatic activity is upregulated. In turn activated PLCγ can then trigger downstream signaling (Chapter 5) by hydrolysis of PIP2 to diacylglycerol (DAG) and inositol-1,4,5-trisphosphate (IP3). Both DAG and IP3 are important second messengers in growth signaling as they may activate protein kinase C (PKC) and calcium-dependent signaling pathways. Intriguingly, another member of the PLC family, PLC$_\varepsilon$, has now been shown to be a RAS effector, linking RAS with activation of PKC.

The PKC family of serine/threonine kinases regulate multiple cell functions including growth, proliferation, differentiation, cytoskeletal organization, motility, and apoptosis. PKC signaling activates the ERK 1 and 2 MAP kinases triggering cell replication. PKC has also been implicated in mediating survival signals from various growth factors such as IGFs, and inhibition of PKC can trigger apoptosis in many cells. PKC epsilon may activate the AKT and mTOR signaling pathways also, showing the considerable overlap between signaling downstream of PLC and RAS. Aberrant PKC, for instance by drugs such as phorbol esters, has long been known to promote cancer in skin and other tissues, at least in part by driving aberrant cell cycling. Moreover, mutant PKC proteins have been found in several different human cancers.

Other RAS effectors

RAS effectors also include three exchange factors for the RAS-related RAL proteins: RAL guanine nucleotide dissociation stimulator (RAL-GDS), RALGDS-like gene (RGL) and RGL2, through which RAS activates RAL and in turn phospholipase D1, and the CDC42/RAC–GAP–RAL binding protein 1 (RALBP1). The RAL-GDS pathway – together with AKT – inhibits downstream transcription factors including the forkhead transcription factors JNK and AFX. Indeed one of the main functions of AFX is to keep cells in G$_1$ by increasing the levels of the CKI p27^{KIP1} involved in repressing S-phase entry. However, it is also involved in the induction of the proapoptotic proteins BIM and FasL (Chapter 8). Thus, by inhibiting both growth arrest and cell death, the RALGDS pathway contributes to the induction of cell proliferation and together with AKT blocks apoptosis. RAL is located in late endosomes/early lysosomes and in the plasma membrane, suggesting that it may also be involved in the control of vesicular transport systems underlying signaling events. As is also the case for the AKT pathway, the RALGDS pathway can be activated independently of RAS by multiple other signaling mechanisms.

The net action of RAS signaling pathways could account for the numerous characteristics of malignant transformation induced by mutant RAS; namely, increased proliferation, resistance to apoptosis, induction of angiogenesis, and increased invasiveness. It is likely, however, that the location and strength of the signal, and the integration with other signaling events, will determine how the cell behaves. Again, it is important to mention here that cell behavior following oncogenic RAS signaling can vary between cell types and between cells cultured *in vitro* versus their normal counterparts *in vivo*.

RAS and cellular senescence – intrinsic tumor suppression?

As discussed earlier for the *c-MYC* oncogene, cell death by apoptosis provides an important intrinsic mechanism for tumor suppression. However, it is not the only way that oncogenes can stop potentially harmful cells replicating.

Some oncogenes, exemplified by RAS, can trigger cellular senescence – a state characterized by permanent cell cycle arrest and specific changes in morphology and gene expression that distinguish the process from quiescence (see Chapter 9). Consistent with its role in tumor suppression, cellular senescence is regulated by a number of tumor-suppressor genes, the most crucial of these encoding the p53 or p21^{CIP1}, pRB or p16^{INK4a}, ARF, BMI, and PML proteins. However, unlike oncogene-induced apoptosis, the relevance of oncogene-induced senescence as a tumor suppressor mechanism was less clear because it had not been observed definitively *in vivo*. Moreover, even *in vitro* expression of oncogenic *RAS* in primary cells does not always trigger senescence – especially when expressed from its endogenous locus. Despite these concerns, studies in certain mouse models provide evidence that senescence might play a part in suppressing tumor growth. For example, chemically induced skin cancers in mice show that the initiating oncogenic mutations occur in the endogenous *H-RAS* gene of keratinocytes. However, these benign hyperplastic lesions only progress to malignant tumors when secondary mutations, in the *p53*, *p16^{INK4a}*, *p19^{ARF}*, or *p21^{CIP1}* genes (Chapter 7) occur – precisely those genes that mediate RAS-induced growth arrest in cultured keratinocytes.

How does RAS trigger cellular senescence? *In vitro* studies using human and rodent cells have shown that oncogenic RAS signals via the MAPK pathway to induce p16^{INK4a} and/or ARF (depending on cell type), which ultimately activate RB and p53, respectively (Fig. 6.8). For example, mouse embryo fibroblasts depend primarily on ARF–p53, whereas human fibroblasts also rely on p16^{INK4a}–RB functions. In turn, p53 and RB promote senescence by regulating a number of effectors, including p21^{CIP1}, PML, and various chromatin-modifying factors.

PML is a tumor suppressor first identified in a mouse model for acute promylocytic leukemia (see Chapter 11), and regulates responses of *p53* to oncogenic signals from RAS. Expression of RAS causes *p53* to accumulate and *PML* expression to increase, *PML* overexpression acetylates *p53* at lysine-382, and this makes *p53* biologically active. The outcome is senescence. RAS stimulation causes *p53* and the acetyltransferase CBP to form a trimeric *p53–PML–CBP* complex within the nuclear bodies, a site where *PML* occurs even in normal cells. Knock-out experiments have now shown that *PML* null fibroblasts lose RAS-induced *p53* acetylation, *p53*–CBP complex stability, and senescence. These data establish a strong link between *PML* and *p53* and moreover emphasize the central role of *PML* in mediating the effects of RAS in this context.

Very recent studies now strongly support the notion that diverse potentially cancer-promoting genes can trigger senescence *in vivo*, including oncogenic *RAS* and *BRAF*. The mechanism, as may have been predicted by the earlier studies discussed, involves induction of p16^{INK4a} via the ERK pathway and activation of AKT or inactivation of PTEN, which activate the ARF–p53–p21^{CIP1} pathway.

If, as now seems likely, replicative senescence operates *in vivo*, then avoidance or even escape from oncogene-induced senescence is crucial for the transformation of cells and tumor development. Perhaps this explains the collaboration between RAS and other oncogenic lesions, such as c-MYC, observed in transformation assays *in vitro* (see Oncogene collaboration). In this scenario, c-MYC may block RAS-induced senescence by inhibiting p21^{CIP1}. On the other hand, c-MYC-induced apoptosis may be suppressed by RAS acting via the survival protein AKT (Fig. 6.8). This simplified view may also explain why antiapoptotic proteins, such as BCL-2 and BCL-xL, collaborate more effectively with c-MYC compared to RAS, since they serve to block c-MYC-induced apoptosis, but probably play relatively little part in activating senescence pathways (at least in the presence of activated c-MYC). Other examples of collaboration include mouse embryo fibroblasts and skin keratinocytes, in which loss of p19^{ARF} or p53 prevents RAS from inducing cell cycle arrest and promotes transformation. In human cells, at least *in vitro*, the situation is often more complex, requiring additional mutations to overcome RAS-induced senescence, for example, loss of p16^{INK4a} (Chapter 7). Indeed, it is the various combinations of oncogenes and loss of tumor-suppressor genes that dictate whether cancer cells are more responsive or resistant to cancer therapy (see Figs. 8.17 and 8.18 in Chapter 8). Thus both senescence and apoptosis represent inherent tumor-suppressor mechanisms invoked by oncogene deregulation that

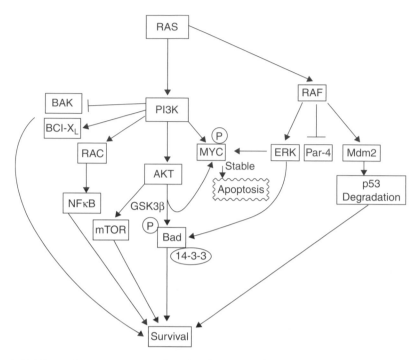

Fig. 6.15 Oncogenic RAS – pathways to survival or death. Tumor development and maintenance depends on the ability of a cell to avoid apoptosis. Although RAS is known to promote cell survival by preventing apoptosis, in some circumstances, it has been shown that RAS can also mediate proapoptotic signals. The ultimate outcome of these contradictory signals depends greatly on the cell type and context. RAS can promote cell survival through a number of signaling pathways. A key pathway that predominantly mediates survival is the PI3K signaling cascade, which activates the serine/threonine (ST) kinase AKT. In turn, AKT phosphorylates a number of substrates including Bad – a proapoptotic member of the BCL-2 family. AKT phosphorylation of Bad causes it to bind preferentially to 14-3-3 proteins in an inactive complex, thereby preventing it from sequestering and inactivating the antiapoptotic proteins BCL-2 and BCL-x_L. However, AKT may also phosphorylate the Thr 58 residue on c-MYC protein, leading to its increased stability (as discussed in the main text); c-MYC can promote apoptosis. RAS signaling through PI3K can also mediate survival by downregulating the proapoptotic BCL-2 family member Bak, and by preventing downregulation of the antiapoptotic protein BCL-x_L. In addition, PI3K can promote survival through the transcription factor NFκB. RAS signaling through RAFvERK can be either anti- or proapoptotic, depending on the circumstances. This pathway can lead to stabilization of c-MYC through phosphorylation of its Ser 62 residue (proapoptotic) or through phosphorylation of Bad as described above, which promotes survival. Lastly, although in many cases oncogenic RAS can provoke a response from p53 designed to cause cell cycle arrest or apoptosis, to complicate matters it can also suppress p53 by inducing its degradation via Mdm2 in a RAF-dependent manner. RAF may also contribute to the ability of oncogenic RAS to provide a prosurvival function by downregulating transcription of Par-4, a proapoptotic transcriptional repressor. RAF and PI3K signaling may also converge downstream of oncogenic RAS to prevent apoptosis.

may restrain tumor progression by inhibiting propagation of the mutated cell.

RAS and differentiation

The RAS pathway transduces divergent signals determining normal cell fate and importantly

may have differing effects on differentiation depending both on cell type and on the persistence or extent of RAS activity. For example, sustained activation of the RAS–MAPK pathway in some neuronal precursors can induce cell cycle arrest and may promote differentiation, whereas in primary erythroid cells differentiation is blocked and proliferation is enhanced. Differentiation is generally associated with cell cycle arrest, but does not necessarily require silencing of MAPK/ERK. In fact, suppression of cell proliferation after differentiation can be achieved by restricting nuclear entry of activated MAPK.

RAS – survival or death

Although the mechanisms by which oncogenic RAS promotes uncontrolled cellular proliferation are perhaps the best characterized and understood, it is now clear that oncogenic RAS (like c-MYC) can also deregulate processes that control cell death, otherwise known as apoptosis (Chapter 8). Large, long-lived animals like man have a high propensity for acquiring mutations; it has been estimated that point mutations resulting in activation of *RAS* occur in thousands of cells daily in the average human (Paul 1986). As the vast majority of these mutations do not result in tumor growth, it is assumed that the usual outcome of such mutations is apoptosis, differentiation, or growth arrest.

The ability of oncogenic RAS proteins to cause increased proliferation has long been considered critical to RAS transformation. However, as was discussed for c-MYC earlier, tumor development and maintenance also depends on the ability of a cell to avoid apoptosis. Recently, there has been an increasing appreciation that RAS paradoxically induces both pro- and antiapoptotic signaling. The ultimate outcome of these contradictory signals depends greatly on the cell type and context. For example, in normal cells, a high level of activated RAS has been thought to be more likely to induce a protective proapoptotic response to prevent oncogenesis in response to hyperproliferative signals. However, in cells that are already transformed (e.g. with c-MYC and RAS), the activity of oncogenic RAS is likely to mediate survival rather than death.

How does RAS mediate cell survival? The key pathways involved in RAS-mediated cell survival are outlined in Figure 6.15. One key pathway we know of is the RAS–PI3K–AKT, already discussed as an important facilitator of cell replication and growth. This pathway also phosphorylates a number of other substrates amongst which BAD – a proapoptotic member of the BCL-2 family (Chapter 8) – may be

the most important target in apoptosis prevention. AKT phosphorylation of BAD causes it to bind preferentially to 14-3-3 proteins in an inactive complex, thereby preventing it from sequestering and inactivating the antiapoptotic proteins BCL-2 and BCL-X$_L$. In addition, AKT survival signaling is potentiated by its effects on cellular energy homeostasis and its modulation of the mTOR pathway described earlier, which controls the cell's response to nutrients (metabolism).

The initiation factor of translation (eIF4E) is a downstream effector of mTOR that has recently been shown to cooperate with c-MYC in forming B-cell lymphomas in mouse models. In a recent study it was found that c-MYC could override senescence activated by eIF4E, and eIF4E could antagonize c-MYC-dependent apoptosis *in vivo*; this obviously mirrors the interactions shown between c-MYC and RAS and suggests that activation of eIF4E may be an important mediator of PI3K and AKT during tumorigenesis.

Attachment to matrix normally provides a survival signal through the activation of PI3K, which is lost in invading or metastatic cells; activated RAS can overcome this loss by restoring the PI3K signal. The process of apoptosis induced specifically by epithelial cell deprivation of matrix interaction is referred to as anoikis (Fig. 12.1). In epithelial cell types oncogenic RAS restores resistance to anoikis in part by downregulating the proapoptotic BCL-2 family member BAK in a PI3K-dependent manner and in part by preventing downregulation of the antiapoptotic protein BCL-X$_L$ in a PI3K- and RAF-independent manner. Generally, PI3K/AKT and NFκB signaling are considered to protect against apoptosis. It is perhaps not surprising then that many human tumors have mutations that lead to elevated signaling through the IGF pathway causing continuous activation of the PI3K survival pathway.

Confusingly, RAF/ERK signaling can be either anti- or proapoptotic, depending on the circumstances. Thus, the combined outcomes of RAS signaling determine whether cells live or die. RAS specifically utilizes a PI3K/AKT pathway to promote cell survival in the presence of apoptotic signals such as those induced by c-MYC (Fig. 6.8). The activity of oncogenic RAS itself can promote proapoptotic signaling that is then rescued by the antiapoptotic signals that it induces such as RAC or NFκB (Fig. 6.15). Defects or interference in the ability of RAS to activate the antiapoptotic signals can allow its proapoptotic activity to take over.

In many contexts, excessive signaling due to oncogenic RAS provokes a response from p53 designed to cause cell cycle arrest or apoptosis and thereby remove the threat of unbalanced oncogenic stimuli (described previously in RAS and cellular senescence). To complicate matters further, oncogenic RAS can suppress p53 by inducing its degradation via MDM2 in a RAF-dependent manner, and thus contribute to the resistance of RAS-transformed cells to p53-mediated apoptosis (Fig. 6.15). RAF may also contribute to the ability of oncogenic RAS to provide a pro-survival function by downregulating transcription of Par-4, a proapoptotic transcriptional repressor. Both RAS and RAF can alter the expression and activity of BCL-2 family members. For example, loss of K-RAS protein expression following targeting by an anti-K-RAS ribozyme in Capan-1 pancreatic cancer cells results in reduced levels of BCL-2 and increased apoptosis. RAF and PI3K signaling may also converge downstream of oncogenic RAS to prevent apoptosis (Fig. 6.15). One obvious convergence point is the phosphorylation of Bad. Another point of convergence for RAF and PI3K signaling downstream of oncogenic RAS is through effects on c-MYC protein degradation. Phosphorylation at Ser 62 stabilizes c-MYC, whereas subsequent phosphorylation at Thr 58 is required for its degradation. Phosphorylation on Ser 62 may be mediated via ERK, whereas Ser 62 is dephosphorylated by protein phosphatase 2A (PP2A), which is itself regulated by the PIN1 prolyl isomerase – all potentially RAS-regulated signaling molecules. By these means, RAS-regulated signaling cascades can limit the level of c-MYC accumulation and duration of action. Conversely, if these regulatory mechanisms are interfered with then this may contribute to oncogenesis.

In light of the diversity of downstream effector targets known to facilitate RAS function, it is perhaps not surprising that RAS regulation of cell survival is complex, involving the balance and interplay of multiple signaling networks. While our understanding of these events is still far from complete, and is complicated by cell type and signaling context differences, several important mechanisms have begun to emerge. It is now apparent that some of the growth-inhibitory properties of RAS are mediated via the RASSF family of RAS effector/tumor suppressors. To date, five members of this family have been identified (NORE1, RASSF1, RASSF2, RASSF3, and RASSF4). This family are involved in cell cycle arrest and in apoptosis in response to RAS. Specific effectors may regulate antiapoptotic (RAF, PI3K, and TIAM1) and apoptotic (NORE1 and RASSF1) actions of oncogenic RAS. *RASSF1A* is one of the most frequently inactivated genes described in human cancers and, as observed for other tumor-suppressor inactivation, this is often through epigenetic gene silencing through methylation of the promoter and CpG island.

RAS and cancer

Aberrant signaling through RAS pathways occurs as the result of several different classes of mutational damage in tumor cells. Mutation of the *RAS* genes is frequently observed in human cancers. Some 20% of human tumors have activating point mutations in *RAS*, most frequently in *KRAS* (about 85% of total), then *NRAS* (about 15%), then *HRAS* (less than 1%). These mutations all compromise the GTPase activity of RAS, preventing GAPs from promoting hydrolysis of GTP, thereby leading to accumulation of the active GTP-bound form of RAS (Fig. 6.11). Almost all RAS activation in tumors is accounted for by mutations in codons 12, 13, and 61. RAS can also be activated in tumors by loss of GAPs, as exemplified by the loss of neurofibromin, encoded by the *NF1* gene, mentioned earlier. Loss of one allele is inherited by individuals with type 1 neurofibromatosis, with the second "hit" presumably generated by somatic mutations in cells destined to give rise to the characteristic benign, and occasionally malignant, tumors in tissues of neural-crest origin – loss of both copies of *NF1* results in activation of RAS.

RAS signaling pathways are also commonly activated in tumors in which growth-factor receptor tyrosine kinases have been overexpressed (Chapter 5). Two notable examples are overexpression of EGFR and NEU, observed in numerous human cancers including breast, ovary and stomach. EGFR-family tyrosine kinases are also commonly activated by the autocrine production of EGF-like factors such as transforming growth factor α (TGF-α) in tumors. Also, elevated signaling through the IGF pathway occurs in many tumor types (see RAS – survival or death).

Most recently *BRAF* mutations have been found in 7% of human cancers, most notably in melanomas and colon cancer (the related *ARAF* and *CRAF* are not mutated in cancer). Recent reports have identified activating mutations in the *BRAF* gene in a large percentage (approx. 70%) of human malignant melanomas. The vast majority of these mutations represent a single nucleotide change of T-A at nucleotide 1796 resulting in a valine to glutamic acid change at residue 599 (V599E) within the activation segment of BRAF. *BRAF* and *RAS* mutations are rarely both present in the same cancers but the cancer types with *BRAF* mutations are similar to those with *RAS* mutations. This has been taken as evidence that the inappropriate regulation of the downstream ERKs (the p42/p44 MAPK) is a major contributing factor in the development of these cancers (Fig. 6.14). In fibroblasts and melanocytes *in vitro*, mutant BRAF stimulates constitutive ERK signaling leading to cell proliferation, survival, and transformed morphology and allows these cells to grow as tumors in nude mice. Conversely, if mutant BRAF is depleted in melanoma cells, ERK activity is now blocked, leading to inhibition of proliferation while inducing cell death by apoptosis (described in Chapter 8).

The PI3K pathway is frequently activated in tumors as a result of deletion of the tumor-suppressor gene *PTEN* (phosphatase and tensin homolog), which encodes a lipid phosphatase that removes the phosphate from the 3′ position of PIP3 and PIP2, so antagonizing the effects of PI3K activity. Loss of PTEN thus results in activation of the PI3K–AKT pathway. *PTEN* is deleted in almost half of human tumors, making it the second most significant tumor-suppressor gene after *p53* (see Chapter 7).

The family of RAS oncogenes promotes the initiation of tumor growth by stimulating tumor cell proliferation, but, like c-MYC, also ensures tumor progression by stimulating tumor-associated angiogenesis – growth of new vasculature (Chapter 14). Oncogenic

RAS proteins stimulate a number of effector pathways that culminate in the transcriptional activation of genes that control angiogenesis. Tumor angiogenesis is postulated to be regulated by the balance between pro- and anti-angiogenic factors. A critical step in establishing the angiogenic capability of human cells is the repression of the antiangiogenic factor thrombospondin 1 (TSP-1). This repression may be essential for tumor formation. A novel mechanism has recently been described, whereby the cooperative activity of the oncogenes *RAS* and *MYC* may lead directly to angiogenesis and tumor formation. In this case, RAS may lead to TSP-1 repression via the sequential activation of PI3K, RHO, and ROCK, leading to activation of c-MYC through phosphorylation; phosphorylation of c-MYC via this mechanism enables it to repress TSP-1 expression.

Other RAS family members

RHEB and its closest relative, RHEBL1 (RHEB-like 1, 52% identity), form a divergent branch of the RAS family. A replication-promoting role for RHEB was originally postulated following the observations that RHEB synergizes with RAF1 to transform NIH-3T3 fibroblasts, and because RHEB is upregulated in RAS-transformed cells. Ectopic expression of RHEB is sufficient for phosphorylation of both mTOR and S6K1. Analysis of cell cycle components that are influenced by TSC1/2 or mTOR indicates that potential RHEB targets might include p27^{KIP1} and cyclin E.

RAL GTPases have been implicated as important components of oncogenic transformation induced by RAS. RAL increases cyclin D1 transcription via NFκB and inhibits forkhead transcription factor FOXO4/AFX, leading to decreased p27^{KIP1} transcription (Fig. 6.13). Interestingly, despite sharing 78% identity, the human RALA and RALB proteins seem to control distinct biological functions, with RALA contributing to cell proliferation and RALB promoting cell survival.

RHO family of GTPases Some members of the RHO family of GTPases are highly expressed in human tumors. Studies suggest that RHO may play a role in cancer onset and in invasion and they are thus potential candidates for therapeutic intervention. RHO GTPases mediate a diverse range of cellular effects such as proliferation and adhesiveness (Fig. 6.16). The precise mechanisms by which RHO GTPases participate in carcinogenesis are still not fully understood. However, it is becoming more evident that the specific role of RHO overexpression in initiation, progression, and metastasis may be determined by the nature and cause of such overexpression in specific human tumors.

There are at least 20 RHO family proteins in humans, the most widely studied of which are RHOA/B, RAC1/2, and CDC42. Like all members of the small GTPases superfamily, these proteins cycle between an inactive GDP-bound and an active GTP-bound state (Fig. 6.11). RHO GTPases are involved in motility and altering the cytoskeleton (Chapter 12) and also contribute to cell cycle progression: inactivation of RHO by the *Clostridium botulinum* C3 ADP-ribosyl transferase, or microinjection with dominant-negative forms of RAC1 or CDC42, block mitogen-stimulated G_1/S transition in Swiss 3T3 fibroblasts. Conversely, microinjection of active RHOA, RAC1, or CDC42 into quiescent cells induces G_1/S transition. RHO activation allows mitogen-stimulated cells to progress through the cell cycle. However, RHO is inactive in the absence of adhesion (see SRC section) or under conditions of cell confluence, and $p21^{CIP1}$ (cell cycle inhibitor) expression remains high. Thus, RHO may function as an important "monitor" of the cellular environment and as an adhesion-dependent cell cycle checkpoint.

RHO GTPases regulate the activity of many transcription factors. However, it is only recently that newly identified RHO-regulated transcription factors and their specific effects on cell behavior have been identified.

Due to the complexity of the signaling pathways mediated by RHO proteins, we will only describe a few of these pathways (albeit simplified here) linked to the transformation of cells.

STAT transcription factors RHO GTPases (RHO and RAC1 and CDC42) can activate STAT3 (signal transducer and activator of transcription 3) transcription factor by indirectly inducing its phosphorylation. It has recently been shown that RHOA (bound to the effector protein ROCK) signals through STAT3 for transformation by inducing proliferation and anchorage-independent cell growth (Fig. 6.17). Anchorage-independent growth and the ability to avoid detachment-induced apoptosis (anoikis – Chapter 12) are hallmarks of transformed epithelial cells. RHO subfamily members contribute significantly to this process.

Active STAT3 may then promote cell proliferation by inducing the expression of c-MYC or cyclin D1. Although RAC1 and CDC42 also activate STAT3, it is not yet known whether these GTPases are able to induce transformation via STAT3.

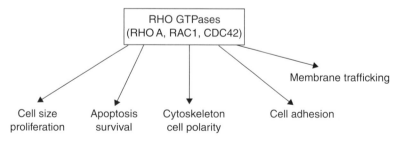

Fig. 6.16 RHO GTPases mediate a diverse range of cellular effects. The high incidence of overexpression of some members of the RHO family of GTPases in human tumors suggests that these proteins are involved in cancer onset, and that they are potential candidates for a therapeutic intervention. RHO GTPases, the most widely studied of which are RHOA/B, RAC1/2, and CDC42, mediate a diverse range of cellular effects, such as proliferation, motility, and adhesiveness, although the precise mechanisms by which RHO GTPases participate in carcinogenesis are still not fully understood. However, it is becoming more evident that the specific role of RHO overexpression in tumor initiation, progression, and metastasis may be linked to the activation of specific signaling pathways that result in transcriptional regulation.

RHOA, RAC1, and CDC42 activate another member of the STAT family of transcription factors – STAT5a (Fig. 6.17). Activation of STAT5a by RHOA by indirect phosphorylation causes cells to become more motile as well as inducing epithelial-to-mesenchymal transition (EMT), features associated with tumor cell meta static behavior (Chapter 12). STAT5a may promote such cellular effects by upregulating the expression of vimentin, which is associated with EMT, whilst downregulating E-cadherin expression with loss of adherens junctions (these mediate cell–cell adhesion). In addition to STAT5a, RHOA can induce EMT by activating another transcription factor, SRF. Whether activating both STAT5a and SRF functionally cooperate

to promote EMT downstream of RHOA awaits further investigation.

NFκB and c-MYC   RHOA, RAC1, and CDC42 proteins induce the transcriptional activity of the transcription factor, NFκB, a protein important in inflammatory responses and tumorigenesis. NFκB can promote cell proliferation, cell survival, invasion and motility of cancer cells. As it is a transcription factor, it is likely to promote such cellular effects through expression of downstream genes. For example, one target gene is the urokinase plasminogen receptor gene (*uPAR*), whose expression results in enhanced invasiveness due to degradation

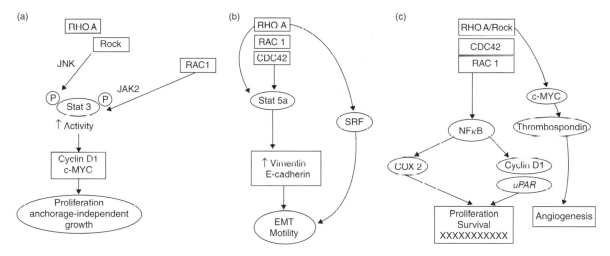

Fig. 6.17 RHO promotes tumor cell properties through activation of specific transcription factors. (a) RHO GTPases (RHOA, RAC1, and CDC42) can activate Stat3 (signal transducer and activator of transcription 3) transcription factor by indirectly inducing its phosphorylation (via JNK2). It has recently been shown that RHOA (bound to the effector protein Rock) signals through Stat3 for transformation, by inducing proliferation and anchorage-independent cell growth – hallmarks of transformed cells. Active Stat3 may promote cell proliferation by inducing the expression of c-MYC or cyclin D1 Although RAC1 and CDC42 also activate Stat3, it is not yet known whether these GTPases are able to induce transformation via Stat3. (b) RHOA, RAC1, and CDC42 proteins also activate Stat5a. RHOA activates Stat5a by indirect phosphorylation, causing cells to become more motile and inducing epithelial-to-mesenchymal transition (EMT), features associated with tumor cell behavior. Stat5a may mediate such cellular effects by upregulating the expression of vimentin that is associated with EMT, while downregulating E-cadherin expression with loss of adherens junctions (these mediate cell–cell adhesion). RHOA can also induce EMT by activating another transcription factor, SRF. (c) RHOA, RAC1, and CDC42 proteins induce activity of the transcription factor, NFκB. NFκB can promote cell proliferation, cell survival, invasion and motility of cancer cells, presumably through expression of its target genes. One target gene is the urokinase plasminogen receptor gene (*uPAR*), whose expression results in enhanced invasiveness due to degradation of the extracellular matrix. Other target genes include the cell cycle protein cyclin *D1* and the proinflammatory gene cyclooxygenase 2 (*COX-2*) that is associated with cancer. Finally, the RHOA–Rock pathway can activate c-MYC, leading to repression of the antiangiogenic factor, thrombospondin-1 (Tsp-1), which in turn allows proper angiogenesis to take place within the tumor.

of the extracellular matrix. Other target genes of NFκB include the cell cycle protein cyclin D1 (Chapter 4) and the proinflammatory gene cyclooxygenase 2 (*COX-2*). Importantly, NFκB signaling through COX-2 protein links inflammation and cancer (Fig. 6.17).

An important finding was the link between RAS (signaling via RHO GTPases) and the c-MYC transcription factor in inducing growth of tumor vasculature (angiogenesis), at least in the context of human epithelial tumors. Here, RAS signaling through the RHOA–ROCK pathway to activate c-MYC leads to repression of the anti-angiogenic factor, thrombospondin 1 (TSP-1), which in turn allows proper angiogenesis to take place within the tumor.

RHO GTPases and cancer   The link between RHO proteins and human cancer was not discovered for a long time due to the fact that no activating mutations have been found within the coding sequence of these proteins. However, overexpression of RHO proteins has been detected in many human tumors. For example, in colon and breast tumor, although mRNA levels are not elevated compared to normal tissue, RHO protein levels are much higher. In contrast, there is an increase in RHO mRNA in testicular germ tumors. Thus different human tumors display different types of RHO deregulation – although to date it does not seem to be a consequence of a genetic mutation.

SRC – THE OLDEST ONCOGENE

The SRC non-receptor tyrosine kinase is one of the oldest and most studied of all the proto-oncogenes. It is overexpressed and activated in a large number of human cancers (though mutations have not been found), although its actual role in initiation or progression of any given cancer is not fully understood. Importantly, despite a period of relative inactivity in the field, interest has resurfaced with the availability of several new compounds targeting SRC, and SRC inhibitors are entering clinical trials for the first time. So, what is SRC's role in cancer? In recent years, *in vitro* observations have led to the hypothesis that, aside from increasing cell proliferation, a

key role of SRC in cancer is to regulate cell adhesion and to promote invasion and motility.

SRC was the first oncogene discovered (see Appendix 6.1. The discovery of oncogenes), and was originally identified as the transforming agent (*v-src*) of the Rous sarcoma virus (RSV) a retrovirus that infects chickens and other animals. RSV is an acutely transforming virus that inserts its own genes into the cellular DNA, rapidly promoting the development of cancer. In fact, once infected, chickens develop large tumors within two weeks.

The SRC protein is a tyrosine kinase, which activates downstream signaling through the addition of phosphate groups to tyrosine residues on target proteins. There are at least nine different known *SRC* genes, which through different mRNA processing can encode at least 14 different proteins.

Regulation and activation of c-SRC

c-SRC (SRC) is found in most cells at a low level, but may be overexpressed or activated in certain human cancer types, including neuroblastoma, small-cell lung, colon, and breast. The fact that both high levels of SRC protein and SRC kinase activity have been observed within cancer cells indicates the potential importance of both protein levels and protein activity in various phases of tumor development.

Human SRC protein contains four SRC homology (SH domains), comprising the SH1 kinase domain, which contains the autophosphorylation site (Tyr419); the SH2 domain, which interacts with the negative-regulatory Tyr530 and binds to the platelet-derived growth factor receptor (PDGFR); the SH3 domain, which enhances interactions with the kinase domain; and the SH4 domain, which contains the myristoylation site required for membrane localization (Fig. 6.18a). Interactions between the C-terminus and the SH2 domain, and between the kinase domain and the SH3 domain, cause the SRC molecule to assume a closed configuration that prevents substrate interaction. When the C-terminal tyrosine is phosphorylated, SRC is inactive and the protein resides at a perinuclear site. However, following binding of PDGF ligand to its receptor or after integrin engagement (Chapter 12), SRC becomes active and translocates to the cell membrane

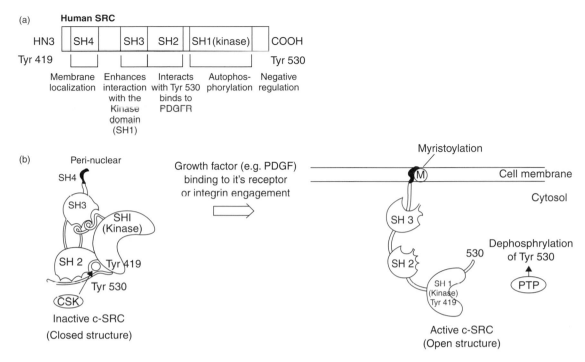

Fig. 6.18 Regulation and activation of c-SRC. (a) The functional domains of human c-SRC: human c-SRC protein contains four SRC homology (SH domains) comprising the SH1 kinase domain, which contains the autophosphorylation site (Tyr419); the SH2 domain, which interacts with the negative-regulatory Tyr530 and binds to the platelet derived growth factor receptor (PDGFR); the SH3 domain, which enhances interactions with the kinase domain; and the SH4 domain, which contains the myristoylation site required for membrane localization. (b) Activation of c-SRC: when the C-terminal tyrosine Tyr530 is phosphorylated, SRC is inactive and assumes a "closed" conformation as a result of interactions between the C-terminus and the SH2 domain, and between the kinase domain and the SH3 domain. This closed configuration prevents substrate interaction, and the inactive c-SRC protein resides at a perinuclear site. Inactivation of c-SRC by phosphorylation is performed by CSK (c-SRC tyrosine kinase). Following binding of PDGF ligand to its receptor or after integrin engagement, SRC becomes active and translocates to the cell membrane. This activity is the result of dephosphorylation of Tyr530 that leads to an "open" conformation of the protein, thus allowing substrate interaction. Protein tyrosine phosphatases (PTP) have been shown to dephosphorylate the Tyr530 of c-SRC.

(Fig. 6.18b). This activity is the result of dephosphorylation of Tyr530, which in turn leads to a change in conformation. Inactivation of SRC by phosphorylation is now known to be performed by CSK (SRC tyrosine kinase) and there is now evidence that reduced expression of CSK might have a role in SRC activation in human cancer. Conversely, protein tyrosine phosphatases (PTP) have been shown to dephosphorylate the terminal tyrosine residue of SRC. The most direct evidence for a role of SRC activation in cancer among these phosphatases is PTP1B, which is present at higher levels in breast cancer cell lines. So, like RAS, translocation and activation of SRC are inextricably linked, and the actin cytoskeleton is required for both catalytic activation and peripheral membrane targeting of this protein.

Activated SRC is translocated to the cell periphery, often to sites of cell adhesion, where myristoylation of its SH4 domain mediates attachment to the inner surface of the plasma membrane. This tethered location allows for interactions with membrane-bound RTKs and integrins associated with adhesion functions (see Chapter 12). The catalytic activity of SRC then initiates signal transduction pathways involved in cell growth, adhesion, and motility. Deregulation in cancer cells may therefore enhance tumor growth and/or stimulate migratory or invasive potential in cells that would normally be relatively non-motile. The direct binding of focal-adhesion kinase (FAK), described in Chapter 12, to the SH2 and SH3 domains of SRC also results in the open, active configuration of SRC.

Importantly, SRC can also be activated through receptor-mediated signaling: ligand-activated receptor tyrosine kinases (RTK) such as EGFR, PDGFR, ERBB2 (or HER2/NEU), FGFR. In a wide range of tumors in which SRC is overexpressed, RTKs are also overexpressed. In fact, there are various lines of evidence indicating that SRC cooperates with RTKs to promote tumorigenesis, probably by disrupting the intramolecular interactions that hold SRC in a closed configuration. For example, when EGFR and SRC are co-transfected into fibroblasts, their synergistic action results in increased proliferation, invasiveness, and tumorigenesis. In mouse models, overexpression of RTK, ERBB2 (HER2/NEU), also leads to SRC activation.

Other mechanisms of SRC regulation include ubiquitination, with subsequent degradation by the proteosome. Recent evidence indicates that this regulatory pathway is disrupted in some cancer cells, thus allowing SRC activation. Finally, although extremely rare, there are two reports of SRC activation in colon cancer and endometrial cancer resulting from naturally occurring point mutations that truncate SRC just C-terminal to the regulatory Tyr530.

The SRC phenotype

As previously mentioned, much of the experimental data on the oncogenic effects of SRC comes from infecting or transfecting fibroblasts in vitro with the highly activated *v-src* oncogene. Apart from increasing proliferation rates of normal cells, *v-src* causes fibroblasts to round up, dissaggregate, and begin to float in the culture medium. These effects result from loss of cell–cell contact (through loss of E-cadherin – see later), and decreased substrate adhesion through loss of integrin-based cytoskeletal attachments (desctribed in Chapter 12) that normally ensure that they bind to the substratum in an ordered monolayer. In addition to these cellular changes,

v-src-transformed cells become more motile and more invasive. After several weeks in culture, foci can be seen in the culture dish, where cells pile up on top of each other (as described in Oncogene collaboration). All these changes correlate with various hallmarks of a cancer cell, enabling a normal cell to become a cancer cell – increased proliferation, dissagregation from the primary tumor, invasion into the surrounding tissue, and ultimately metastasis to a distant site. Apart from the cellular effects observed in the culture dish, if *v-src*-transfected cells are injected into mice the cells grow rapidly to form visible tumors within days. Moreover, these tumor cells are capable of local invasion and metastasis to distant sites.

How activated SRC does it

Oncogenic forms of SRC can alter cell structure, in particular the actin cytoskeleton and the adhesion networks that control cell motility and invasion, and also transmit signals that regulate cell proliferation and survival (Fig. 6.19).

Adhesion, invasion, and motility, although representing independent cellular functions, are related events that require several well-orchestrated molecular interactions. In order for cells to move, whether it be normal cells or cancer cells, alterations in both cell–cell and cell–matrix adhesion is required. Invading cells, as is the case for cancer cells, require alterations in both adhesion and motility.

There are two principal subcellular structures that regulate adhesion, invasion, and motility in normal healthy cells – focal adhesions and adherens junction – both of which are regulated by SRC. These structures are described in Chapter 12, and so will not be covered in any significant detail here. Briefly, focal adhesions provide the structural and mechanical properties that are necessary for cell–matrix attachment, whilst adherens junctions enable neighboring cells to adhere to each other: cell–cell attachment (Fig. 6.20). Focal adhesions form at the sites where integrins link the actin cytoskeleton to extracellular matrix (ECM) proteins. In addition to their role in adhesion, they also participate in cell-signaling processes that regulate proliferation and gene transcription. Focal adhesions

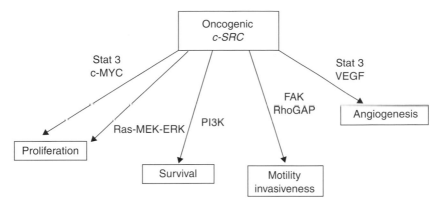

Fig. 6.19 Oncogenic SRC can affect cell behavior in various ways. Oncogenic forms of SRC can activate signaling pathways to alter cell structure, in particular the actin cytoskeleton and the adhesion networks that control cell motility and invasion. In addition, SRC can transmit signals that regulate cell proliferation and survival, as well as angiogenesis. Details of these signaling pathways are described in Fig. 6.20.

are composed of over 50 different proteins – such as talin, vinculin, α-actinin, SRC, FAK, CAS, and paxillin – that assemble into supramolecular structures (Chapter 12). These cytoskeletal proteins are recruited to focal adhesions to mediate cellular migration. They are associated with cytoskeletal stress fibers, composed of actin and myosin, which control the shape and, ultimately, the motility of the cell.

The assembly and disassembly of focal adhesions is associated with both significant cytoskeletal alterations and integrin signaling, which allow morphological changes and alterations in motility and invasiveness. So, for example, focal adhesions assemble in order to allow cells to adhere to the ECM, whereas disassembly of these structures promotes detachment from the ECM, a property that is acquired by cancer cells.

Activated SRC is now known to play a key role in focal adhesion disassembly, and it does this in various ways as shown in Figure 6.20. For example, through its binding and activation of FAK (focal adhesion kinase): activated FAK phosphorylates substrates such as p190 RhoGAP, which in turn inhibits RHOA, leading to focal adhesion disruption. SRC can also destabilize focal adhesions through its tyrosine phosphorylation of R-RAS protein: R-RAS and *v-src* form a complex that suppresses integrin activity and reduces cell–matrix adhesion.

Apart from the ability of activated SRC to promote the release of cells from the ECM, it also

acts to release cells from each other. As described in Chapter 12, in order for healthy cells to adhere to each other, they must form stable adherens junctions mediated by homotypic interactions between E-cadherin molecules on neighboring cells (Fig. 6.20). Cell motility and invasiveness depend on the loss of cell–cell adhesion mediated by E-cadherin. Activated SRC disrupts adherens junctions by inducing tyrosine phosphorylation and ubiquitination of the E-cadherin complex. As a result, cadherin molecules are internalized. Phosphorylation of FAK by SRC is also required for the disruption of E-cadherin cell–cell adhesion.

In addition to SRC's role in promoting invasion through loss of E-cadherin, SRC might affect invasion by regulating matrix metalloproteinases (MMPs) and tissue inhibitors of MMPs (TIMPs) – described in Chapter 12. MMPs are a group of enzymes that can break down extracellular matrix proteins, and are involved in physiological processes such as wound healing, as well as in promoting tumor cell invasion and angiogenesis. TIMPs are secreted proteins that play a crucial role in regulating the activity of MMPs. In cancer the regulated expression of both families of proteins can become disrupted, such that expression of MMPs is increased and/or expression of TIMPs is decreased. With respect to oncogenic SRC, activation of FAK stimulates the c-Jun N-terminal kinase (JNK) pathway, ultimately leading to the expression of MMP2 and MMP9 (Fig. 6.20).

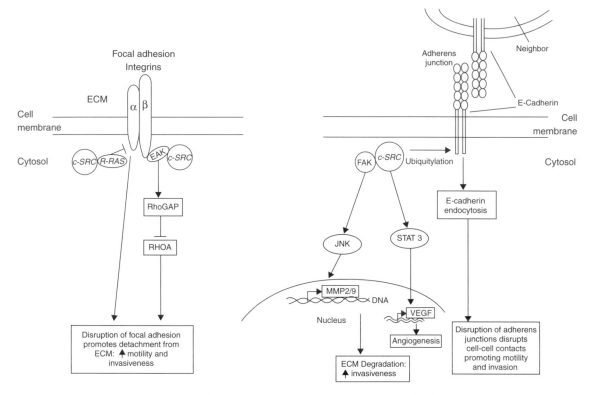

Fig. 6.20 Signaling pathways activated by oncogenic SRC to promote tumor cell behavior. The two principal sub-cellular structures that regulate adhesion, invasion, and motility in normal healthy cells – focal adhesions and adherens junction – are regulated by *c-SRC*. Focal adhesions provide the structural and mechanical properties that are necessary for cell–matrix attachment and form at the sites where integrins link the actin cytoskeleton to extracellular matrix (ECM) proteins. Adherens junctions enable neighboring cells to adhere to each other: cell–cell attachment. Activated SRC plays a key role in disassembly of focal adhesions that promotes detachment from the ECM, a property that is acquired by cancer cells. It does this through its binding and activation of FAK (focal adhesion kinase): activated FAK phosphorylates p190 RhoGap that in turn inhibits RhoA leading to focal adhesion disruption. SRC can also destabilize focal adhesions through its tyrosine phosphorylation of R-RAS protein. Activated SRC can also promote the release of cells from each other by disrupting adherens junctions by inducing tyrosine phosphorylation and ubiquitylation of the E-cadherin complex. As a result, cadherin molecules are internalized. Phosphorylation of FAK by *c-SRC* is also required for the disruption of E-cadherin cell–cell adhesion. SRC might also promote invasion by regulating MMPs, enzymes that can break down extracellular matrix proteins and thus promote tumor cell invasion and angiogenesis. SRC activation of FAK stimulates the JNK pathway, ultimately leading to the expression of MMP2 and MMP9. Lastly, SRC proteins can promote angiogenesis through the expression of VEGF, by activating the transcription factor STAT3.

Lastly, SRC proteins seem to promote angiogenesis (Chapter 14). For example, as was shown for c-MYC, *v-src* induces vascular endothelial growth factor (VEGF) expression. Although the mechanism for c-MYC-induced VEGF expression is not yet known, SRC achieves this through activating the transcription factor STAT3 (Fig. 6.20).

c-SRC and human cancer

Much redundancy is built into this complex signaling system, and hundreds of protein tyrosine kinases are now known, several of which are nearly identical to SRC. In fact, specifically blocking the action of SRC protein in normal laboratory animals has relatively little effect as other proteins fill in for the lost function, including ABL.

Although there are numerous members of the SRC-family kinases, including FYN, YES, BLK, YRK, FGR, HCK, LCK, and LYN, SRC is the one that is most implicated in cancer, particularly advanced gastrointestinal-tract cancers such as colorectal cancer. Increased SRC activity has also been demonstrated in several other gastrointestinal malignancies, including liver, pancreas, and esophagus, and also in breast and ovary. Most notably, gastrointestinal-tract cancers show a progressive increase in SRC activity with advancing tumor stage, such that metastatic lesions often have the highest levels of SRC activity. Indeed, SRC is expressed at high levels in colonic polyps and adenomas early in the course of colon cancer development, but there are also large increases in SRC-specific activity in later stages of cancer progression. This would indicate that SRC plays a role in later stages of tumorigenesis. Paradoxically though, it has been noted that the most aggressive tumors which are poorly differentiated often show lower levels of both SRC protein and SRC activity when compared to more differentiated tumors. One potential explanation is that in these aggressive poorly differentiated tumors there are high levels of receptor tyrosine kinases such as EGFR, which can synergistically activate SRC (see Chapter 5). Therefore, low levels of SRC might be compensated for by high levels of a receptor tyrosine kinase.

The actual role SRC plays in tumor progression is unclear. Much experimental data on SRC and its effects on cell behavior have derived from fibroblasts grown in culture. The fact that SRC promoted fibroblast cell division led to the notion that this oncogene contributes to tumor progression by stimulating proliferation of precancerous cells. Although this may be true for certain cell types, as is the case for fibroblasts, it is possible that SRC exerts different effects such as motility and invasion on other cell types. For example, overexpression of SRC in human colon cancer cells does not promote cell growth, but can instead enhance motility. Similarly, the cooperation of SRC and epidermal growth factor receptor (EGFR) regulates the invasiveness of colon cancer cells, but does not seem to influence proliferation.

c-SRC as a target of drug therapy SRC is now the target of new drug development for cancer, in particular because it is commonly activated in a large number of human cancers and because its mechanisms of activation are now better understood. In fact, a number of SRC inhibitors are in phase I or preclinical trials. Several compounds targeting the SRC kinase domain have been described including SKI-606 (Wyeth), SU6656 (Sugen), AP23464 (Ariad), and AP23451 (Ariad: a bone-tissue-targeted SRC kinase inhibitor developed for osteoporosis therapy). One might anticipate that the relatively benign effects of SRC "ablation" in normal cells would provide minimal toxicity. The issue of how effective such strategies will be against cancer cells remains to be seen, but some encouraging results have been obtained with colon cancer cells. However, the fact that the non-catalytic domains of SRC might also have biological effects, such as integrin assembly, raises the possibility that targeting the kinase domain might limit potential. Interestingly, recent data indicate that targeting SRC and FAK simultaneously might be effective in promoting apoptosis of colon cancer cells.

BCR ABL

The prototypic non-receptor tyrosine kinase c-ABL is implicated in various cellular processes. Its oncogenic counterpart, the BCR–ABL fusion protein, causes certain human leukemias. Myeloid leukemias are frequently associated with translocations and mutations of tyrosine kinase genes (Table 6.1). The products of these oncogenes, including BCR–ABL, TEL–PDGFR, FLT3, and c-KIT, have elevated tyrosine kinase activity and transform hematopoietic cells. Elevated activity of these tyrosine kinases is crucial for transformation, thus making the kinase domain an ideal target for therapeutic intervention.

CML is a stem cell disorder caused by a constitutively BCR–ABL tyrosine kinase. Importantly, in a regulatable transgenic mouse model of CML, it was shown that deactivating BCR–ABL caused complete reversal of the phenotype (Table 6.6) indicating that targeted therapy aimed at blocking oncogene function could lead to apoptosis or differentiation of tumor cells. Imatinib mesylate (gleevec, STI571, or

CP57148B) is a direct inhibitor of several tyrosine kinases including ABL (ABL1), ARG (ABL2), KIT, and PDGFR. This drug has had a major impact on the treatment of CML as well as other blood neoplasias and some solid tumors. However, resistance occurs in part due to amplification and/or mutations of *BCR–ABL*.

THE BCL-2 FAMILY

The BCL-2 family of proteins comprises both proapoptotic and antiapoptotic members, the balance of which determines whether or not a cell commits suicide – apoptosis. These proteins have emerged as fundamental regulators of mitochondrial-outer-membrane permeabilization (MOMP), which is necessary for cytochrome c release – a crucial event that takes place during apoptosis. Given the importance of this family in cancer – through regulating apoptosis – we have devoted a significant portion of Chapter 8 to describing these proteins.

BCL-2 (B-cell lymphoma 2) in human follicular lymphoma involves a chromosome translocation event that moves the *BCL2* gene from chromosome 18 to 14 (t14;18) linking the *BCL2* gene to an immunoglobulin locus.

A link between cell death by apoptosis (Chapter 8) and cancer emerged when BCL-2 was found to inhibit cell death rather than promote proliferation. This important work, and others, showed that cell survival is controlled separately from cell proliferation and that inhibition of apoptosis is a central step in tumor development. This unexpected discovery gave birth to the concept, now widely accepted, that impaired apoptosis is a crucial step in tumorigenesis. This concept is strongly supported by several transgenic mouse models of cancer as previously described, for example, mimicking human follicular lymphoma, in which the *BCL2* gene is linked to an immunoglobulin locus by chromosome translocation (t14;18), *BCL2* transgenes placed under the control of a B-cell promoter in transgenic mice give rise to B-lymphoid tumors. However, the stochastic onset of tumors in these mice implied that additional genetic changes were needed for lymphoma development. Further experiments, in which both c-MYC and BCL-2 were coexpressed in B cells of

transgenic mice, showed a dramatically accelerated onset of lymphomas in these mice. These results revealed a strong cooperation between c-MYC and BCL-2, whose elevated and constitutive expression enforced both cell proliferation and cell survival. This strong cooperation, or synergy, has been shown more recently to promote tumor formation in other tissues *in vivo*, such as breast. It is likely that all BCL-2 prosurvival family members are oncogenes. BCL-xL, for example, has been implicated in mouse myeloid and T-cell leukemias and insulinomas, as previously discussed.

Apoptotic defects are also important in established cancer: in both solid (e.g., prostate) and non-solid (e.g., leukemia) tumors, BCL-2 overexpression is associated with poor prognosis and often (although not always) resistance to cytotoxic therapy. Conversely, proapoptotic members of the BCL-2 family (e.g. BAX, BAK) which are likely to act as tumor suppressors, are mutated in some human gastric and colorectal cancers as well as in leukemias. It is likely that these mutations lead to loss of proapoptotic activity, thus inhibiting death of these neoplastic cells. In support of this, experimental data show that loss of BAX can enhance tumor formation and that loss of both BAX and BAK appears to synergize leading to further enhancement of tumor formation when compared to loss of either one alone.

REGRESSING TUMORS BY INACTIVATING ONCOGENES – A THERAPEUTIC TARGET?

In order to identify new therapeutic targets in oncogene-induced responses we must first answer the following key questions. How does the oncoprotein exert the distinct cellular outcomes, such as proliferation, loss of differentiation, senescence, and/or apoptosis? In the case of transcription factors such as c-MYC, are they dependent on the transcription or repression of specific distinct or overlapping sets of target genes? To what extent do cellular responses to the oncoprotein depend on cell type or, importantly, on tissue location, which will dictate the signals a given cell type receives from cellular interactions between extracellular matrix and neighboring cells? How might circulating or

Table 6.6   Use of regulatable transgenic mouse models of cancer to test tumor regression

Initial oncogenic lesion	Target tissue/cell type	Phenotype	Deactivating initiating oncogene
SV40 T antigen	Embryonic submandibular gland	(i) Atypical cells (4 weeks old) (ii) Transformed ductal cells (7 months old)	(i) Full phenotype reversal (ii) Not reversed
H-RAS (+INK4a$^{-/-}$)	Melanocytes	Melanoma	Rapid tumor reversal Larger tumors (20%) not reversed
c-MYC	Embryonic hematopoietic cells	T-cell lymphomas/ acute myeloid leukemias (5 months old)	Rapid tumor reversal 10% tumor relapse
c-MYC	Adult suprabasal epidermis (keratinocytes)	Papillomatosis (carcinoma *in situ*)	Rapid and complete tumor reversal
c-MYC	Adult suprabasal epidermis (keratinocytes)	Papillomatosis (carcinoma *in situ*)	Tumor recurrence after transient c-MYC inactivation.
BCR–ABL1	B cells	Acute B-cell leukemia	Complete tumor reversal
c-MYC	Mammary epithelium	Invasive mammary adenocarcinoma	Tumor reversal Subset not reversed (RAS activated)
c-MYC	Mammary epithelium	Invasive mammary adenocarcinoma	Many tumors fail to reverse, those that do may recur in a MYC-independent fashion
K-RAS	Lung	Adenocarcinoma	Tumor reversal
K-RAS (+p53$^{-/-}$ or INK4a$^{-/-}$ or ARF$^{-/-}$)	Lung	Adenocarcinoma	Rapid and complete tumor reversal
c-MYC (+BCL-xL)	Adult pancreatic islet β cells	Invasive islet adenocarcinoma	Rapid and complete tumor reversal
c-MYC (+BCL-xL)	Adult pancreatic islet β cells	Invasive islet adenocarcinoma	Tumor recurrence after transient c-MYC inactivation
c-MYC	Embryonic osteocytes	Malignant osteogenic sarcoma	Rapid and complete tumor reversal
c-MYC	Hepatocytes	Hepatocarcinoma	Tumor reversal, dormant tumor cells give rise to recurrent tumors, when MYC is reactivated
c-MYC	Lymphocytes	T- and B-cell lymphoma	Reversal
BCL2 (+ c-MYC)	Lymphocytes	Lymphoblastic leukemia	Apoptosis and reversal
NEU (and MMTV)	Mammary epithelium	Invasive mammary carcinoma	Essentially complete reversal even of metastatic lesions. Eventually, however, tumors spontaneously recur
WNT	Mammary epithelium	Invasive mammary carcinoma	Essentially complete reversal even of metastatic lesions. Eventually, however, tumors spontaneously recur
GLI2	Basal keratinocytes	Basal cell carcinoma	Reversal

Note: The potential for tumor regression following deactivation of the initial oncogenic lesion (c-MYC, SV40 T antigen, H-RAS, K-RAS, BCR-ABL, NEU, WNT, and GLI2) has been investigated using several regulatable transgenic mouse models of cancer. In general, the findings listed above have implications for therapy as they indicate that blocking oncogene function, even in advanced tumors, could lead to apoptosis or differentiation of tumor cells. Further to these findings, it was recently shown that transient, rather than sustained, inactivation of c-MYC is sufficient for full reversal of malignant osteogenic sarcoma in transgenic mice. If valid for other cancer models, transient inactivation of oncogenes could provide an effective cancer therapy limiting host cell toxicity. However, it has subsequently been shown in various other tissues that such transient inactivation is not sufficient for tumor reversal. In fact, in some tumors inactivation of the transgene is not followed by complete regression of all tumors, or recurrence occurs spontaneously in regressed tumors, suggesting that mutations occur which can make the tumors independent of the initiating lesion.

paracrine/autocrine signals, able to mitigate particular oncoprotein functions, be regulated? The recent genetic construction of mice described earlier, in which the expression or activation of a given gene/protein can be switched "on" or "off" *in vivo*, has provided a means to assess the roles of individual (or combinations of) genes and their protein products in tumorigenesis in the all-important context of the intact adult organism. Thus, we may now address how host and tumor cells communicate in induced tumors, as well as validate a given gene/protein as a therapeutic target by simply switching "off" expression of the oncogene and looking for phenotype reversal (such as tumor regression). Such studies should have important implications for designing future therapies targeted to cancer cells.

Restoring senescence or apoptosis that have become disrupted in cancer cells provides a promising strategy for the development of novel targeted therapeutics. A number of studies using regulatable mouse models of tumorigenesis have recently been conducted in order to assess whether a specific cancer-initiating oncogenic mutation is essential for the maintenance of an established tumor *in vivo*. In other words, whether tumor regression occurs following deactivation of the initial oncogenic lesion.

Indeed, even in some advanced tumors with metastatic lesions which contain multiple genetic and epigenetic alterations, inactivating the initiating oncogene appears to induce substantial tumor regression (Table 6.6). However, as discussed below, there are a few instances when more advanced tumors do not fully regress and, in these cases, further insight into the mechanisms of this divergence is required.

Using the "tet" system (see Chapter 18, Appendix 18.1) mice were generated that conditionally express c-*MYC* in their hematopoeitic cells. In this mouse model, c-*MYC* was expressed in this cell lineage throughout development, resulting in the occurrence of highly invasive T-cell lymphomas and acute myeloid leukemias within 5 months of age. Surprisingly, inactivation of the c-*MYC* transgene was sufficient to cause tumor regression in mice moribund with tumor burden. Remarkably, a substantial reduction in tumor mass was observed within 3 days, and after 6 weeks the majority of mice exhibited a sustained remission for as long as 30 weeks. Tumor regression following

downregulation of c-*MYC* expression coincided with tumor cells differentiating to more mature lymphocytes with the restoration of normal host hematopoiesis, as well as a proportion of tumor cells being eliminated by apoptosis. Thus, inactivation of c-*MYC* expression was sufficient to induce sustained tumor regression in the vast majority of mice, whereas 10% of mice relapsed with tumor and subsequently died. Although it is not clear as to why some tumors escape dependence upon the c-*MYC* transgene, it has been reasoned that these tumor cells ("escapers") acquire genetic lesions that in some way substitute for the requirement for c-*MYC* (Fig. 6.21).

Whether c-MYC activation is also required for solid tumor maintenance has also been addressed with regulatable mouse models of skin epidermal and pancreatic β-cell oncogenesis in which the c-MYCER^{TAM} switchable form of c-MYC is used (Chapter 18, Appendix 18.1). Benign angiogenic epidermal tumors, which develop following continuous activation of c-MYC, completely regressed over a 3-week period following deactivation of the oncoprotein. Deactivation of c-MYC coincided with keratinocytes exiting the cell cycle, resumption of normal differentiation, and complete regression of newly formed vasculature, demonstrating that continuous c-MYC activation in keratinocytes is required to maintain neo-angiogenesis. In a distinct tissue, multiple angiogenic and locally invasive β-cell tumors, which develop following continuous activation of c-MYC and the anti-apoptotic protein BCL-X$_L$ (described in c-MYC section and Fig. 6.10), were completely dependent upon active c-MYC. In this tumor model, regression appears to be mediated by rapid collapse of vasculature that triggers death of many β cells, presumably the result of hypoxia and lack of nutrients and survival factors provided by the vasculature. Remarkably, despite the continued overexpression of the antiapoptotic protein, BCL-X$_L$, in neoplastic β cells, deactivating c-MYC leads to general regression of tumors back to fully differentiated quiescent islets with restoration of normal cell–cell contacts concomitant with re-expression of E-cadherin (Plate 6.1). Interestingly, tumor regression is also accompanied by substantial infiltration of inflammatory cells that may play a role in phagocytosis and clearance of apoptotic and necrotic debris. The rapid disassembly of vasculature is consistent

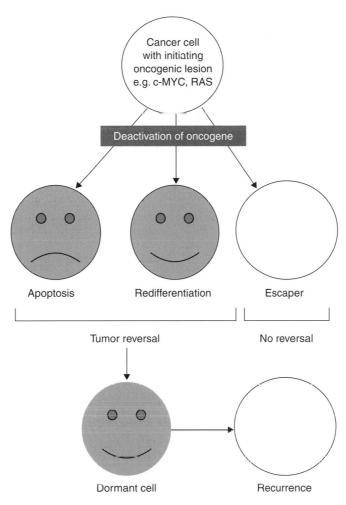

Fig. 6.21 Tumor regression following deactivation of initiating oncogenic lesion. Various transgenic mouse models of tumorigenesis have shown that subsequent deactivation of the initial oncogenic lesion leads to rapid and complete regression of most neoplastic lesions. Mechanisms of tumor regression include tumor cell apoptosis and/or differentiation and re-establishment of cell–cell contacts. Onset of vasculature collapse within tumor masses occurs rapidly and may also contribute to tumor cell apoptosis. In some cases, a subset of tumors escape reversal ("escapers") and thus are no longer maintained by the initiating oncogenic lesion. In these cases, it is likely that additional oncogenic lesions have occurred.

with a direct angiogenic role of c-MYC, although it is also possible that β-cell growth arrest induced by c-MYC deactivation allows intrinsic antineoplastic mechanisms such as the immune response to initiate regression. With regard to ectopic β cells that had invaded pancreatic ducts, vessels, and local lymph nodes by 12 weeks of c-MYC activation, extensive histological analysis of pancreatic sections failed to reveal any ectopic β cells after several weeks of c-MYC

deactivation. Moreover, other animals undergoing similar regression remained healthy after several months. Whether longer-term activation of c-MYC would lead to a small proportion of tumors escaping dependence upon c-MYC (as observed in other tumor models) requires further investigation.

Experimental studies using genetically altered mice, have shown that withdrawal of doxycycline-inducible oncogenic H-RAS(G12V) expression in transgenic mice bearing melan-omas causes apoptosis in both tumor cells and endothelial cells, followed by regres-sion of the melanomas. Similarly, withdrawal of doxycycline-inducible activated K-RAS(12D) expression from type II pneumocytes causes apoptosis and regression of both early proliferat-ive lesions and lung cancers. Similar results were obtained in animals deficient in p53 or p19^{ARF}; thus, activated K-RAS is required for tumor maintenance. Lastly, further mouse models of cancer have shown dramatic tumor reversal when the oncogene in question is switched off. For example, in a mouse model of liver cancer, invasive c-MYC-induced hepatocellular carcino-mas regress when *c-MYC* expression is turned off, as a result of apoptotis and differentiation. In another cancer model, in which both c-MYC and the antiapoptotic protein BCL-2 are expressed in lymphocytes, switching off BCL-2 leads to complete reversal of lymphoblastic leuk-emia by apoptosis. Further examples are given in Table 6.6.

As mentioned earlier, there are some instances when tumors do not fully regress (Table 6.6). Although it is not clear as to why some tumors escape dependence upon the initiating oncogene (e.g. RAS or c-MYC), it has been reasoned that these tumors acquire further genetic lesions that in some way substitute for the requirement of the oncogene. For example, in a mouse model of mammary cancer, although reversal of c-MYC-induced invasive mammary adenocarcinomas occurred, a subset of tumors failed to reverse, and were subsequently found to carry additional mutations in *RAS*. The observation that c-MYC may cause genomic instability in some model systems, *in vivo* and *in vitro*, might in the long-term contribute to c-MYC-induced neoplastic progression. Further examples of mouse models in which tumors subsequently relapse are given in Table 6.6.

In general, the above findings have implica-tions for therapy as they indicate that blocking oncogene function, even in advanced tumors, could lead to apoptosis or re-differentiation of tumor cells (Fig. 6.21). The potential for tumor regression following deactivation of initial onco-genic lesions (e.g. c-MYC, SV40 T antigen, RAS, BCR-ABL, and BCL-2) have provided valuable information and hope for future development of candidate drug molecules (Table 6.6). To this end, recent *in vivo* studies show that re-engaging apoptosis pathways that have become disrupted during tumor development can indeed have a profound therapeutic effect. For instance, the inhibition of the antiapoptotic protein, BCL-2, or the restoration of p53 function, has proven particularly lethal to some tumor types.

Regressing tumors by brief inactivation of oncogenes – a possible therapeutic strategy?

A recent study has shown that brief inactiv-ation (10 days) of c-MYC is sufficient for the sustained regression of c-MYC-induced invasive osteogenic sarcomas in transgenic mice. Sur-prisingly, subsequent reactivation of c-MYC led to extensive apoptosis rather than restora-tion of the neoplastic phenotype. One possible explanation for this outcome is that a change in epigenetic context may have taken place in this cell type, that is, presumably between the immature cell in which c-MYC was origin-ally activated and the more differentiated cell resulting from subsequent (brief) inactivation of c-MYC. Although c-MYC expression is ini-tiated in immature osteoblasts during embryo-genesis, subsequent inactivation of c-MYC in osteogenic sarcoma cells induces differentiation into mature osteocytes. Therefore, reactivation of c-MYC now takes place in a different cellular context and induces apoptosis rather than neo-plastic progression. Although the actual under-lying mechanisms are not yet known, these intriguing findings suggested the novel possibil-ity of employing transient inactivation of c-MYC as a therapeutic strategy in certain cancers, thus limiting potential toxic effects that result from prolonged therapeutic inactivation. But what about other tumor types?

Brief inactivation of c-MYC in different mouse tumor models has not resulted in sustained tumor reversal. In contrast to the osteogenic sarcoma model described above, reactivating c-MYC in islet tumors does not lead to accelerated β-cell apoptosis, but rather restores the onco-genic properties of c-MYC, rapidly re-initiating β-cell proliferation, loss of differentiation, loss of E-cadherin, local invasion, and angiogenesis. Similarly, in epidermis, reactivating c-MYC in suprabasal keratinocytes does not result in apoptosis, which remains confined to the shedding areas of parakeratosis at the skin surface, but restores the papillomatous phenotype, inducing cell proliferation and dysplasia. Another example includes a mouse model of c-MYC-induced mammary adenocarcinomas in which reactivation of c-MYC in regressed tumors led to tumor regrowth. Importantly, subsequent inactivation of c-MYC was less effective at regressing tumors, indicating that other oncogenic pathways are being activated in these cells. Lastly, in a recent mouse model of liver cancer, despite invasive c-MYC-induced hepatocellular carcinomas regressing when *c-MYC* expression is turned off, some tumor cells remain "dormant" and contribute to cancer progression if *c-MYC* expression is subsequently reinitiated.

An important point to remember is that different types of cancers are prevalent in different age groups, with the effects of oncogene activation dependent on the developmental stage of the target cell at that time. Thus, the biological consequences of activating oncogenes such as *c-MYC* are clearly influenced not only by environment but also by developmental stage. This has been elegantly shown in two recent mouse models, mammary gland and liver. c-MYC can inhibit postpartum lactation if activated within a specific 72 hour window during mid pregnancy, whereas c-MYC activation either prior to or following this 72-hour window does not. In embryonic and neonatal mice, c-MYC overexpression in the liver immediately results in hyperproliferation and neoplasia, whereas in adult mice c-MYC overexpression induces cell growth and DNA replication but without mitotic cell division, and neoplasia is considerably delayed.

Taken together, these findings suggest that a cautious approach is required in considering cancer therapies aimed at brief oncogene inactivation. First, a more comprehensive understanding of the genetic basis and environmental context of any individual tumor would be required in order to predict the likely success of such a treatment schedule. Second, at least under those circumstances where tumor cell differentiation and alteration of epigenetic context would not be predicted to reinstate apoptosis and no alternative mechanism exists for tumor cell removal, sustained inactivation of the offending oncogene would seem the desired therapeutic goal.

CONCLUSIONS AND FUTURE DIRECTIONS

Given the importance of oncogene activation in human cancers, specific targeting of oncogenic pathways provides a potentially effective therapeutic strategy. The approval in 2001 of the drug gleevec for the treatment of chronic myelogenous leukemia, by the US Food and Drug Administration, was an important milestone, because it was the first agent aimed at a specific cancer target (the BCR/ABL tyrosine kinase). Since then the HER2/Neu receptor tyrosine kinase (which is overexpressed in up to 30% of primary human breast cancers) has been successfully targeted with the neutralizing antibody trastuzumab in clinical trials, in combination with other agents, to slow disease progression. Similarly, patients with chronic myelogenous leukemia (CML) have been effectively treated with the ABL kinase inhibitor imatinib (gleevec), inducing clinical remission whilst in the CML phase. However, resistance develops in many patients, suggesting that combination of such oncogene-targeted therapies may be required.

One way in which new strategies targeting oncogenes (or other cancer-contributing mutations) can be explored is by the application of model systems. Two premises underlie the use of mouse models in oncogene research. First that mouse models can make crucial contributions to our knowledge of how oncogenes work, particularly when combined with studies of human cancers. Second that such models can help in evaluating therapeutic targets and strategies directed against specific oncogene targets, including examining timing and dosing schedules, intended either to treat and cure established cancers or to prevent them. Compelling results from rodent models and cell culture experiments suggest that several oncogenes, including c-MYC, RAS, and RAS signaling pathways such as the RAF–MAPK and PI3K, may be attractive treatment targets in cancer; with the important caveat that successful results in treating human diseases will have to be demonstrated. Intriguing paradoxes have been revealed by recent studies; thus, for example the demonstration that transient inactivation of c-MYC may be sufficient to arrest and reverse certain tumor models, but not others, raises the importance of tissue microenvironment and epigenetic context in dictating the potential therapeutic responses of any given tumor.

Even if targeting the oncogenes themselves may not be a viable strategy, various downstream targets mediating oncogenesis might. Over the last few years a large number of studies have employed gene chip microarrays, SAGE, and other high-throughput techniques to identify oncogene-regulated genes in cultured cells *in vitro*. However, all these studies, despite their undoubted utility, do not give us any direct indication of which of these many oncogene-regulated genes are actually involved in tumorigenesis *in vivo*. Yet this is of critical importance given the critical role of the tumor microenvironment as well as factors operating at the level of the organism. Taken together, these findings support the notion that a detailed understanding of the "road-map to cancer" of a given individual tumor may be the essential prerequisite to selecting optimal therapeutic strategies in the future. Thus, providing some impetus for strategies aiming to achieve one of the great hopes of post-genome-era biology, namely "individualized medicine" and "tailored therapy".

QUESTIONS

1 Oncogenes
   a. Are activated versions of proto-oncogenes.
   b. Are mutated versions of genes that usually resist cancer formation.
   c. Usually contribute to cancer formation only if both alleles are mutated.
   d. Are often involved in deregulating cell division.
   e. Are always mutated versions of cellular proto-oncogenes.
2 c-MYC
   a. Is usually mutated in human cancers.
   b. Is frequently overexpressed in human cancers.
   c. Is primarily a transcriptional activator when dimerized with Mad.
   d. May promote apoptosis as well as cell cycle transit.
   e. May cooperate with RAS in cell transformation.
3 Oncogenic *RAS*
   a. May result from increased RAS GTPase activity.
   b. May result in increased signaling through the MAPK cascade.
   c. Was first described by studies of the Rous sarcoma virus.
   d. May suppress cell death by activation of AKT.
   e. May activate replicative senescence mechanisms.

4 Regarding RAS
   a. *HRAS* is the most frequently mutated in human cancers.
   b. RAS activity in tumors is usually due to mutations in the *RAS* gene.
   c. RAS may be activated by inactivation of NF1.
   d. RAS signaling may be activated by NEU.
   e. RAS activates RAF by serine/threonine phosphorylation.
5 SRC
   a. Was the first oncogene to be described.
   b. Is a receptor tyrosine kinase.
   c. Is only involved in avian cancers.
   d. Is not known to be mutated in human cancers.
   e. May be activated by receptor tyrosine kinases.

Appendix 6.1 Historical perspectives – The Discovery of Oncogenes

The discovery of oncogenes was not only a landmark event in unraveling the molecular and genetic basis of cancer, but was a major factor in furthering knowledge concerning the regulation of normal cell proliferation. In the current era, identification of oncogene abnormalities provides a means of molecular diagnosis, and oncogenes are increasingly the targets for new types of cancer therapies.

Maybe not surprisingly, the foundations for the modern view of cancer formation were laid by chickens not humans. It was known for a hundred years or more that viruses can cause cancer, but the first such cancer-causing elements were described in viruses infecting poultry in 1909. Early studies by Peyton Rous (1911) on neoplasms in fowl demonstrated that retroviruses could induce tumors, but the relevance of these discoveries for the understanding of human cancer was disputed. Later, it was shown that the injection of these viruses was sufficient for tumor formation. The development of DNA recombinant techniques during the last three decades resulted in a dramatic acceleration of progress in this area. But the molecular basis of cancer was unearthed only with the discovery of cellular homologs to these cancer-causing elements in 1976. When the viral DNA was examined, regions bearing strong similarity to genes that already existed in the animal were identified, suggesting that viruses had at some point incorporated host genomic sequences into their genomes. By incorporating genes able to promote cellular growth and mutating them, the virus increased host cell proliferation, and consequently its own replicative potential. Some scientists predicted that, by extension, mutant cancer-causing genes must lie within all types of tumor cells, even those that lacked evidence of tumor virus infections. The availability of techniques for gene transfection led to the identification of cellular transforming genes that do not have a viral counterpart. As predicted, tumor cells were found to contain such genes, and when these mutated genes were introduced into normal fibroblasts they were sufficient to cause cancer. This gave rise to the idea that cells contain "proto-oncogenes" – genes that are concerned with normal aspects of cell behavior. The proto-oncogenes might become mutated and thus convert to "oncogenes" – genes that have the ability to convert normal cells to a cancerous state.

The experiments

Peyton Rous first described the virus responsible for transmissible growth of tumors in chickens and was the first to propose what was at the time a rather eccentric notion, namely that cancers could be caused by infectious agents (see also Chapter 3). This notion was strongly supported by work in the 1950s demonstrating that the Rous sarcoma virus (RSV)-induced tumor gave rise to infected cancer cells, and finally by experiments by Martin, who showed that temperature-sensitive mutants of chicken v-src failed to transform cells at non-permissive temperatures. The v-src gene was identified in the 1970s and was subsequently sequenced by Mike Bishop and co-workers (see below).

The idea that cancer cells bore mutant cellular genes was around already in the middle of the last century, but in the 1970s two separate strands of research helped to advance this idea into the forefront of cancer research. First, the work of Bruce Ames with cancer-causing chemicals (carcinogens) published in 1975 identified that cancers could be caused by carcinogens through induction of mutations, also suggesting that cells might normally carry genes that when mutated in some way would confer a growth advantage. A second strand followed studies by Harold Varmus and Michael Bishop, for which they were awarded the Nobel prize for medicine in 1989. In 1976,

they identified the cellular homolog c-SRC of the v-src gene, which had previously been described as the oncogene of the chicken virus referred to earlier (the Rous sarcoma virus – RSV). Earlier genetic analysis of RSV had defined the first viral oncogene (v-src) as a gene that was specifically responsible for cell transformation but was not required for virus replication, and the proposition had been put that retroviral oncogenes derive from related genes of host cells. Consistent with this proposition, normal cells of several species were subsequently found to contain retrovirus-related DNA sequences that could be detected by nucleic acid hybridization. However, it was unclear whether these sequences were related to the retroviral oncogenes or to the genes required for virus replication. It was this issue that was resolved by work from the laboratory of Varmus and Bishop. In particular, transformation-defective mutants of RSV had been isolated that had sustained deletions of approximately 1.5 kb, corresponding to most, or all, of the SRC gene. These replication-defective mutants were employed by Stehelin, then a postdoctoral researcher in the laboratory, to prepare a cDNA probe that specifically represented SRC sequences. The use of this defined probe in nucleic acid hybridization experiments allowed them to unambiguously show that normal avian cells contain SRC-related DNA sequences. It is interesting to consider the actual study in more detail. Specifically, as the viral genome is relatively compact it was possible to synthesize a radioactive cDNA probe composed of short single-stranded DNA fragments complementary to the entire genomic RNA of RSV. This probe was then hybridized to an excess of RNA isolated from a transformation-defective deletion mutant. As would be expected, fragments of cDNA complementary to the viral replication genes, which were not deleted, hybridized to the transformation-defective RSV RNA, forming RNA–DNA duplexes. In contrast, cDNA fragments complementary to SRC could not hybridize and so remained single-stranded. This single-stranded DNA was then isolated to provide a specific probe for SRC oncogene sequences. A radioactively labeled SRC cDNA was then hybridized to DNA from normal avian cells. The SRC cDNA hybridized extensively to normal chicken, quail, and duck DNA (see figure). These landmark studies identified

Appendix 6.1 (continued)

that normal cells contain DNA sequences that are closely related to the *SRC* oncogene, supporting the hypothesis that retroviral oncogenes originated from cellular genes that became incorporated into viral genomes. The discovery of the *SRC* proto-oncogene further suggested that non-virus-induced tumors might also result from mutations in related cellular genes, leading directly to the discovery of oncogenes in human tumors. By unifying studies of tumor viruses, normal cells, and non-virus-induced tumors, the results of Varmus, Bishop, and their colleagues have had an impact on virtually all aspects of cell regulation and cancer research.

Following these studies it was becoming likely that the DNA of normal cells contained proto-oncogenes that under certain circumstances could become altered so as to promote cellular growth or transformation – the somatic mutation theory of cancer. This view was given a major boost by studies from the laboratory of Bob Weinberg and colleagues, who showed that transfer of DNA from cancer cells to NIH 3T3 fibroblasts could "transform" these cells. When transfected with tumor DNA, normal cells no longer exhibited contact inhibition but would pile up in a discrete area of a Petri dish, forming a clump of cells many layers thick. Such a clump, termed a focus, contrasted to the behavior of the surrounding normal cells that stop growing once they have formed a single-cell-thick sheet termed a monolayer. The tumor DNA for these studies was prepared from mouse cells that had been exposed to methylcholanthrene (a carcinogen) and was found to induce more foci of transformed recipient cells than did control normal mouse DNA.

The similarities in behavior between the oncogenes transfected from a variety of tumor cells and those carried by retroviruses such as RSV were notable. In both cases, an agent – either a virus or a chemical carcinogen – had apparently succeeded in converting a normal cellular gene into a transforming oncogene. Later work on a human bladder cancer gene led to the discovery of an oncogene homologous to the *RAS* oncogene carried in the genome of Harvey sarcoma virus, a retrovirus of mixed rat–mouse origin whose *v-RAS*

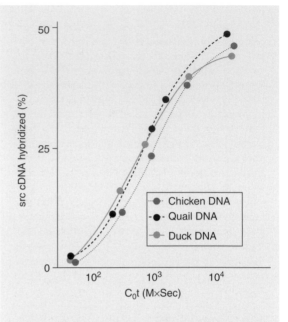

oncogene arose much like the *v-src* oncogene of Rous sarcoma virus. The resulting conclusion again was that the same normal cellular gene could be activated into a potent oncogene either by a retrovirus or by a somatic mutation of the sort inflicted by mutagenic carcinogens. However, the precise nature of the mutation was only found in 1982 when several researchers found that the difference between the normal and mutant *RAS* gene was a single base substitution in the 12th codon, which caused the replacement of a glycine by a valine. Subsequently, it has been shown that around one quarter of all human tumors, derived from a variety of organs, carry point mutations in either the 12th or 61st codon of the *RAS* gene.

These discoveries led to attempts in identifying the signal transduction pathways in which RAS operated. RAS was soon found to act like the alpha subunit of heterotrimeric G proteins; RAS switched between active GTP-bound forms to inactive GDP-bound forms. It was then found that the oncogenic form of p21RAS lacked GTPase activity, and thus mutant RAS remained trapped in an active state. Within months of the discovery of the point mutation that activated the *RAS* bladder carcinoma oncogene, other cellular genes such as *MYC* were also found in mutant form in human tumor DNA.

We now know that cancer can be provoked by a wide variety of mutant genes.

However, tumorigenesis seemed to be much more complex than would be suggested by the development of a tumor by mutations in only a single gene; first tumorigenesis seemed to involve a gestation period of many years, and second histopathological analyses of tissues strongly suggested that the process of forming a tumor involved multiple steps – implying that real human tumors carried multiple mutated genes, and that a single mutated gene, on its own, was insufficient to create a malignant cell.

The earlier results in which NIH-3T3 cells could be transformed with a single oncogene seemed to contradict this view. Nevertheless, explanations for these supposedly contradictory findings were found – they derive from an immortalized cell line and thus were likely to already carry mutations that normal cells would not. This was confirmed in a landmark study by Hartmut Land in the laboratory of Bob Weinberg. Briefly, when the cloned *RAS* oncogene was transfected into normal rat embryo fibroblasts no foci of transformants were seen. A cloned *MYC* oncogene was similarly unable to evoke transformation. However, when *RAS* and *MYC* oncogenes were co-transfected foci of transformants arose, and could give rise to tumors in nude mice – this has been termed oncogene cooperation. This oncogene collaboration indicated that cellular oncogenes did not constitute a single, analogously functioning group of genes. Instead, these two oncogenes – *RAS* and *MYC* – seemed to work in distinct, complementary ways on cell phenotype.

Only later did it become apparent that the human genes that participate in mutant form in cancer pathogenesis encompass a wider spectrum, including tumor-suppressor genes, and that those involved the maintenance of genomic integrity (see Chapters 7 and 10). In the best-studied of human cancers – colon carcinoma – mutation of a *RAS* gene represents only one of multiple distinct genetic alterations that contribute to the phenotype of the malignant tumor cells.

Other contributory approaches A further experimental approach has relied on the identification and characterization of clonal and recurrent cytogenetic abnormalities in cancer cells, especially those derived from the hematopoietic system. Numerous oncogenes have been identified by molecular cloning of chromosomal breakpoints, including translocations and inversions. Additional oncogenes have been identified through the analysis of chromosomal regions anomalously stained (homogeneously staining regions), representing gene amplification.

Recurring karyotypic abnormalities are observed in large numbers of hematologic and solid tumors and include chromosomal rearrangements, gain or loss of whole chromosomes or chromosome segments. The first such abnormality identified in a human cancer was a small chromosome in the cells of patients with chronic myelogenous leukemia – the Philadelphia chromosome. The application of chromosome banding techniques in the early 1970s enabled the precise cytogenetic characterization of many chromosomal translocations in human leukemia, lymphoma, and solid tumors. The subsequent development of molecular cloning techniques then enabled the identification of proto-oncogenes at or near chromosomal breakpoints in various neoplasms. Some of these proto-oncogenes, such as *MYC* and *ABL*, had been previously identified as retroviral oncogenes. Since then many new oncogenes have been identified by cloning of chromosomal breakpoints in cancers.

BIBLIOGRAPHY

General

Paul, J. (1986). Roles of oncogenes in carcinogenesis. In O.H. Iverson (ed.), *Theories of Carcinogenesis*. New York, Hemisphere Publishing, pp. 145–60.

Shih, C. and Weinberg, R.A. (1982). Isolation of a transforming sequence from a human bladder carcinoma cell line. *Cell*, **29**: 161–9.

Stephelin, D., Varmus, H.E., Bishop, J.M., and Vogt, P.K. (1976). DNA related to the transforming gene(s) of avian sarcoma viruses is present in normal avian DNA. *Nature*, **260**: 170–3.

c-MYC

Adams, J.M., Harris, A.W., Pinkert, C.A., et al. (2004). The *c-MYC* oncogene driven by immunoglobulin enhancers induces lymphoid malignancy in transgenic mice. *Nature*, **318**(6046): 533–8.

Amati B. (2004). MYC degradation: dancing with ubiquitin ligases. *Proceeding of the National Academy of Sciences, USA*, **101**(24): 8843–4.

Ayer, D.E., Lawrence, Q.A., and Eisenman, R.N. (1995). Mad–Max transcriptional repression is mediated by ternary complex formation with mammalian homologs of yeast repressor Sin3. *Cell*, **80**(5): 767–76.

Brenner, C., Deplus, R., Didelot, C., et al. (2005). MYC represses transcription through recruitment of DNA methyltransferase corepressor. *EMBO Journal*, **24**(2): 336–46.

Cairo, S., De Falco, F., Pizzo, M., Salomoni, P., Pandolfi, P.P., and Meroni, G. (2005). PML interacts with MYC, and MYC target gene expression is altered in PML-null fibroblasts. *Oncogene*, **24** (13): 2195–203.

Coller, H.A., Grandori, C., Tamayo, P., et al. (2000). Expression analysis with oligonucleotide microarrays reveals that MYC regulates genes involved in growth, cell-cycle, signaling, and adhesion. *Proceedings of the National Academy of Sciences, USA*, **97**(7): 3260–5.

Evan, G.I., Wyllie, A.H., Gilbert, C.S., et al. (1992). Induction of apoptosis in fibroblasts by c-MYC protein. *Cell*, **69**(1): 119–28.

Gomez-Roman, N., Grandori, C., Eisenman, R.N., and White, R.J. (2003). Direct activation of RNA polymerase III transcription by c-MYC. *Nature*, **421**(6920): 290–4.

Herbst, A., Hemann, M.T., Tworkowski, K.A., Salghetti S.E., Lowe, S.W., and Tansey, W.P. (2005). A conserved element in MYC that negatively regulates its proapoptotic activity. *EMBO Reports* **6**(2): 177–83.

Herold, S., Wanzel, M., Beuger, V., et al. (2002). Negative regulation of the mammalian UV response by MYC through association with Miz-1. *Molecular Cell*, **10**(3): 509–21.

Nilsson, J.A., Maclean, K.H., Keller, U.B., Pendeville, H., Baudino, T.A., and Cleveland, J.L. (2004). Mnt loss triggers MYC transcription targets, proliferation, apoptosis, and transformation. *Molecular and Cellular Biology*, **24**(4): 1560–9.

Patel, J.H., Loboda, A.P., Showe, M.K., Showe, L.C., and McMahon, S.B. (2004). Analysis of genomic targets reveals complex functions of c-MYC. *Nature Reviews: Cancer*, **4**(7): 562–8.

Pelengaris, S., Khan, M., and Evan, G. (2002). c-MYC: more than just a matter of life and death. *Nature Reviews: Cancer*, **2**: 764–76.

Zindy, F., Eischen, C.M., Randle, D.H., et al. (1998). MYC signaling via the ARF tumor suppressor regulates p53-dependent apoptosis and immortalization. *Genes and Development*, **12**(15): 2424–33.

Oncogene cooperation

Hahn, W.C. and Weinberg, R.A. (2002). Modelling the molecular circuitry of cancer. *Nature Reviews: Cancer*, **2**(5): 331–41.

Land, H., Parada, L.F., and Weinberg, R.A. (1983). Tumorigenic conversion of primary embryo fibroblasts requires at least two cooperating oncogenes. *Nature*, **304**(5927): 596–602.

Land, H., Chen, A.C., Morgenstern, J.P., Parada, L.F., and Weinberg, R.A. (1986). Behavior of *MYC* and *RAS* oncogenes in transformation of rat embryo fibroblasts. *Molecular and Cellular Biology*, **6**(6): 1917–25.

Strasser, A., Harris, A.W., Bath, M.L., and Cory, S. (1990). Novel primitive lymphoid tumors induced in transgenic mice by cooperation between myc and bcl-2. *Nature*, **348**(6299): 331–3.

RAS

Boguski, S. and McCormick, F. (1993). Proteins regulating RAS and its relatives. *Nature*, **366**: 643–54.

Downward, J. (2003). Targeting RAS signaling pathways in cancer therapy. *Nature Reviews: Cancer*, **3**: 11–22.

Fisher, G.H., Wellen, S.L., Klimstra, D., et al. 2001. Induction and apoptotic regression of lung adenocarcinomas by regulation of a *K-RAS* transgene in the presence and absence of tumor suppressor genes. *Genes and Development*, **15**: 3249–62.

Frame, S. and Balmain, A. (2000). Integration of positive and negative growth signals during *RAS* pathway activation *in vivo*. *Current Opinion in Genetics and Development*, **10**: 108–13.

Goldstein, J.L. and Brown, M.S. (1990). Regulation of the mevalonate pathway. *Nature*, **343**: 425–30.

Hancock, J.F. (2003). RAS proteins: different signals from different locations. *Nature Reviews: Molecular Cell Biology*, **4**: 373–84.

Johnson, L., Mercer, K., Greenbaum, D., *et al.* (2001). Somatic activation of the *K-RAS* oncogene causes early onset lung cancer in mice. *Nature*, **410**: 1111–16.

Malumbres, M. and Barbacid, M. (2003) *RAS* oncogenes: the first 30 years. *Nature Reviews: Cancer*, **3**: 7–13.

Rodriguez-Viciana, P., Warne, P.H., Khwaja, A., *et al.* (1997). Role of phosphoinositide 3-OH kinase in cell transformation and control of the actin cytoskeleton by RAS. *Cell*, **89**: 457–67.

Serrano, M., Lin, A.W., McCurrach, M.E., Beach, D., and Lowe, S.W. (1997). Oncogenic ras provokes premature senescence associated with accumulation of p53 and p16^{INK4a}. *Cell*, **88**: 593–602.

Sweet-Cordero, A., Mukherjee, A., Subramanian, A., *et al.* (2005). An oncogenic Kras expression signature identified by cross-species gene expression analysis. *Nature Genetics*, **37**(1): 48–55.

Yeh, E., Cunningham, M., Arnold, H. *et al.* (2004). A signaling pathway controlling c-MYC degradation that impacts oncogenic transformation of human cells. *Nature Cell Biology*, **6**(4): 308–18.

RAC and RHO

Coleman, M.L., Marshall, C.J., and Olson, M.F. (2004). RAS and RHO GTPases in G_1-phase cell-cycle regulation. *Nature Reviews: Molecular Cell Biology*, **5**(5): 355–66.

Sahai, E. and Marshall, C.J. (2002). RHO-GTPases and cancer. *Nature Reviews: Cancer*, **2**: 133–42.

SRC

Yeatman, T.J. (2004). A renaissance for SRC. *Nature Reviews: Cancer*, **4**(6): 470–80.

BCL-2 family

Cory, S. and Adams, J.M. (2002). The BCL2 family: regulators of the cellular life-or-death switch. *Nature Reviews: Cancer*, **2**(9): 647–56.

Egle, A., Harris, A.W., Bouillet, P., and Cory, S. (2004). Bim is a suppressor of MYC-induced mouse B-cell leukemia. *Proceedings of the National Academy of Sciences, USA*, **101**(16): 6164–9.

Cancer models

Blakely, C.M., Sintasath, L., D'Cruz, C.M., *et al.* (2005). Developmental stage determines the effects of c-MYC in the mammary epithelium. *Development*, **132**(5): 1147–60.

Pelengaris, S., Littlewood, T., Khan, M., Elia, G., and Evan, G. (1999). Reversible activation of c-MYC in skin: induction of a complex neoplastic phenotype by a single oncogenic lesion. *Molecular Cell*, **3**(5): 565–77.

Tumor reversal

Chin, L., Tam, A., Pomerantz, J., *et al.* (1999). Essential role for oncogenic RAS in tumor maintenance. *Nature*, **400**(6743): 468–72.

Felsher, D.W. (2003). Cancer revoked: oncogenes as therapeutic targets. *Nature Reviews: Cancer*, **3**: 375–80.

Jain, M., Arvanitis, C., Chu, K., *et al.* (2002). Sustained loss of a neoplastic phenotype by brief inactivation of MYC. *Science*. **297**(5578): 102–4.

Pelengaris, S., Khan, M., and Evan, G.I. (2002). Suppression of MYC-induced apoptosis in beta cells exposes multiple oncogenic properties of MYC and triggers carcinogenic progression. *Cell*, **109**(3): 321–34.

Pelengaris, S., Abouna, S., Cheung, L., Ifandi, V., Zervou, S., and Khan, M. (2004). Brief inactivation of c-MYC is not sufficient for sustained regression of c-MYC-induced tumors of pancreatic islets and skin epidermis. *BMC Biology*, **2**(1): 26.

Pao, W., Klimstra, D.S., Fisher, G.H., and Varmus, H.E. (2003). Use of avian retroviral vectors to introduce transcriptional regulators into mammalian cells for analyses of tumor maintenance. *Proceedings of the National Academy Sciences, USA*, **100**: 8764–9.

Shachaf, C.M., Kopelman, A.M., Arvanitis, C., *et al*. (2004). c-MYC inactivation uncovers pluripotent differentiation and tumor dormancy in hepatocellular cancer. *Nature*, **431**: 1112–17.

Vogelstein, B. and Kinzler, K. W. (1998). *The Genetic Basis of Human Cancer*. New York, McGraw-Hill.

7

Tumor Suppressor Genes

Martine Roussel

How much must I overcome before I triumph?   Pierre Corneille (1606–1684)

KEY POINTS

- Tumor suppressors are guardians against DNA damage induced, for example, by ultraviolet (UV) or irradiation (X-rays) or an excess of proliferative signals.
- Tumor suppressors monitor DNA damage and help repair the damage before cells divide. They can induce growth arrest at two DNA damage checkpoints: the first gap phase (G_1) and the second gap phase (G_2) of the cell cycle. This enables cells to check the integrity of the chromosomes before proceeding into DNA replication (S phase) or division (M phase). In the absence of these checkpoints, as a result of the loss of tumor-suppressor function, the damage is not repaired and cells accumulate mutations that ultimately lead to cancer.
- Genetic (gene mutation or deletion, mutations in the promoter of genes) or epigenetic alterations (promoter methylation, mutations in the promoter of genes) in tumor suppressors such as *RB*, *p53*, or the genes that regulate them, are part of the life history of all cancer cells.
- The retinoblastoma protein, pRb, and the transcription factor p53 control growth arrest, apoptosis, and differentiation, and their pathways collaborate in tumor prevention.
- pRb binds to transcription factors to regulate cell proliferation. Its phosphorylation state affects its ability to bind these factors, E2Fs being the major ones. Phosphorylation of pRb releases transcription factors that transcribe genes encoding proteins required for the initiation of DNA synthesis.
- Oncogenic and hyperproliferative stresses cause the activation of p53, which in turn restrains uncontrolled cell growth by blocking the cell cycle and inducing programmed cell death (apoptosis).
- pRb and p53 are regulated by p16^{INK4A} and p14^{ARF} (p19^{Arf} in mice).

INTRODUCTION

Tumor suppressors are the cell's guardians against DNA damage, induced for example by ultraviolet (UV) exposure from sunlight, gamma irradiation (X-rays), chemotherapeutic drugs, or an excess of inappropriate proliferative signals. They prevent incipient cells from becoming malignant by arresting their proliferation or inducing them to commit suicide (apoptotic cell death). Tumor suppressors monitor critical cellular checkpoints that govern the mitotic cycle, DNA repair, transcription, apoptosis, and differentiation. Some tumor suppressors prevent the inappropriate activation of signaling pathways involved in cell growth. The functional inactivation of tumor suppressors by mutation, deletion, or gene silencing creates an imbalance between proliferation, cell death, and differentiation programs that facilitates tumorigenesis. Some rare individuals are born with mutations in tumor-suppressor genes that predispose them to develop cancer. Tumor suppressors, to date, represent about 0.001% (30) of the total number of genes (around 30,000) that make up the entire mammalian genome. This number is likely to increase as novel technologies are used to study cancer cells. It is now clear that cells have evolved tumor suppressors as part of multiple elaborate defense systems against unscheduled cellular proliferation.

As their name indicates, tumor-suppressor proteins regulate cell proliferation by eliminating cells that are damaged (by cellular stresses such as UV, X-rays, chemotherapeutic drugs) or that abnormally proliferate in response to hyperproliferative signals. All mammalian cells express at least some tumor suppressors and thus play a critical role in cancer prevention. Knudson and others, who proposed the "two hit" hypothesis, introduced the concept of tumor suppression. Knudson theorized that the development of retinoblastoma, a rare cancer that occurs in the eyes of children, could be explained in familial cases by the acquisition from one parent (the first hit) of a mutated allele of a gene later called the retinoblastoma gene (*RB*), and the occurrence of another mutation on the second allele (second hit) during early childhood (less than 8 years of age). Nonfamilial or sporadic retinoblastoma, however, would occur as a result of mutations on both alleles in somatic cells, and in this case, the disease is not transmitted to the next generation. In each case, tumor development requires that both copies of the gene must be inactivated for cancer to occur. This concept was later confirmed by studies on the genomic DNA from patients with retinoblastoma and other types of tumors. Since the discovery of the retinoblastoma protein pRb (or pRB), the first tumor suppressor identified, many tumor suppressor proteins have been characterized and shown to conform to the "two-hit" model proposed by Knudson.

The second tumor suppressor identified was the transcription factor p53. The realization that p53 was a tumor suppressor came from early studies that showed enforced expression of wild-type p53, but not of mutant forms, induces a block in cell proliferation and transformation as well as from genetic evidence showing bi-allelic inactivation in human tumors. The results were further supported by the identification of germline p53 mutations in Li–Fraumeni syndromes. This proliferation block is achieved, in part, by the activation of the growth arrest gene encoding

$p21^{Cip1/Waf1}$, a cyclin-dependent kinase inhibitory protein, whose expression affects Rb function. p53 was later found to be activated by DNA damage. The first link between p53 and the induction of apoptotic cell death, depending on the strength of the signals that activate it, was recognized in the early 1990s.

The Rb and p53 tumor-suppressor pathways are intertwined, and mutations in genes in either pathway are likely events and even perhaps mandatory alterations in the life history of most tumor cells.

While these first two tumor suppressors continue to be the primary research interest of many cancer biologists, several other proteins have been found that share tumor-suppressing properties. Not surprisingly, many of these proteins participate in the regulation of cell proliferation and can be mapped to pathways that induce cell death (apoptosis) or growth arrest. In this chapter, the function of several tumor suppressors that participate in the Rb and p53 pathways are discussed, focusing on Rb and p53, some of their upstream or downstream regulators as well as tumor suppressors that regulate the flow of signals through critical growth stimulatory pathways. These tumor suppressors have been identified in many signaling pathways involved in multiple processes including cell proliferation in response to receptor signals, DNA repair, transcriptional regulation, and cytoskeletal structure (see Table 7.1 and discussions in Chapters 4, 5, and 10).

In addition to the common loss of both tumor-suppressor alleles, tumor suppression can manifest itself in a haploinsufficient manner, meaning that loss of function of only one allele, or expression of only half of the amount of protein, is sufficient to sensitize cells to tumor formation. In these cases, the second allele remains intact and there is no loss of heterozygosity (also called LOH). Examples of such tumor suppressors include the two cyclin-dependent kinase (cdk) inhibitory (CKI) proteins, $p27^{Kip1}$ (*CDKN1B* in human) (Kip1, **c**d**k**inase **i**nhibitor **p**rotein 1) and $p18^{Ink4c}$ (*CDKN2C* in human) (Ink4c, **in**hibitor of cd**k4**), both of which negatively regulate the activity of Rb and p53, as well as PTEN, MSH2 (mismatch repair protein 2), and NF-1 (neurofibromatosis 1) (see Table 7.1, and discussion in Chapter 4).

Table 7.1 Tumor suppressors

Gene	Nomenclature	Function	Chromosome location (human)	Tumor types
RB1	Retinoblastoma	Cell cycle regulator	13q14.1-q14.2	Retinoblastoma, sarcomas
P53/TP53		Apoptosis regulator; haploinsufficient	17p13.1	Lymphomas, sarcomas, brain and breast cancers
CDKN2A/ INK4A	Cyclin-dependent kinase inhibitory protein p16^{Ink4a}	Cell cycle regulator	9p21	Melanoma, many cancers
CDKN2A/ARF	Alternative reading frame	Cell cycle regulator	9p21	Sarcomas, lymphomas, many cancers
Tob1		Transcriptional corepressor		Liver cancers
APC	Adenomatous polyposis (familial)	Signaling	5q21-q22	Colon cancer
BRCA1	Breast cancer (familial)	DNA repair	17q21	Breast and ovarian cancers
BRCA2	Breast cancer (familial)	DNA repair	13q12.3	Breast and ovarian cancers
CDKN2C/ INK4C	Cyclin-dependent kinase inhibitory protein p18^{Ink4c}	Cell cycle regulator; haploinsufficient	1p21	Testicular cancers
CDKN1B/ KIP1	Cyclin-dependent kinase inhibitory protein p27^{Kip1}	Cell cycle regulator; haploinsufficient	17p	Breast, prostate, many cancers
MSH2	Nonpolyposis colon cancer (hereditary)	DNA mismatch repair	2p22-p21	Colon cancer
MLH1	Nonpolyposis colon cancer (hereditary)	DNA mismatch repair	3p21.3	Colon cancer
VHL	Von Hippel–Lindau syndrome	Transcription elongation regulator	3p26-p25	Renal cancers, hemangioblastoma, pheochromocytoma
PTCH	Medulloblastoma	Development and differentiation regulator	9	Gorlin syndrome; cerebellar tumor=medulloblastoma, basal cell skin cancer
PTEN	Phosphatase and tensin homolog deleted on chromosome ten	Signaling	10q23.3	Cowden disease; breast, brain, and prostate cancers; hyperkeratinosis
MEN	Multiple endocrine neoplasia type 1	Unknown	11q13	Parathyroid and pituitary adenomas, islet cell tumors
WT1	Wilms' tumors	Transcription regulator	11p13	Kidney cancer (children)
NF1	Neurofibromatosis type 1/ neurofibromin 1	Ras inactivation; haploinsufficient	17q11.2	Neurofibromas, sarcomas, gliomas
NF2	Neurofibromatosis type 2/merlin/ neurofibromin 2	Cytoskeleton	22q12.2	Schwann cell tumors, astrocytomas, meningiomas, ependymomas
TSC1	Tuberous sclerosis 1		9q34	Facial angiofibromas
DPC4	Deleted in pancreatic carcinoma 4	TGF-β/BMP signaling	18q21.1	Pancreatic carcinoma, colon cancer
ATM	Ataxia-telangiectasia mutated	Cell cycle checkpoint regulation	11q23.1	Cerebellar ataxia, cancer predisposition
DMP1	Cyclin D-binding Myb like protein 1	Positive regulator of *Arf* expression; haploinsufficient	7q21	Acute myeloid leukemia (AML) and myelodysplastic syndrome (MDS)

Tumor suppressors have been conserved throughout evolution, and several have been found in species other than mammals, including frogs, fish, and flies. The focus of this chapter, however, will be the role of the major tumor suppressors in the RB and p53 pathways in preventing human cancer, as it relates to their structure and function in mammalian cells. All the concepts and principles attributed to pRb and p53 also apply to other tumor suppressors. Indeed, about 30 tumor suppressors are known; however, the vast majority of human cancers carry *p53* and *RB* mutations or mutations in genes that regulate them.

THE "TWO HIT" HYPOTHESIS: LOSS OF HETEROZYGOSITY

In 1971, Knudson advanced his "two hit" hypothesis to explain how inherited (familial) cases of a rare cancer, called retinoblastoma, develop in the eyes of children (Fig. 7.1). Since then, this hypothesis has been confirmed in many cancers that occur as a result of the loss of a tumor suppressor. Knudson proposed that a child who inherits retinoblastoma acquires a mutated allele from one parent (the first hit) in a gene later called the retinoblastoma (*RB*) gene. Heterozygosity for the *RB* allele is not sufficient to cause tumor formation and the child is initially asymptomatic. However, during early childhood, a second mutation or deletion on the other *RB* allele (second hit) occurs, inducing the loss of function of both alleles – loss of heterozygosity (LOH) – in the retinal cells, and the development of a tumor in this tissue. Nonfamilial or "sporadic" retinoblastoma also occurs as a result of mutations on both alleles (two hits) in retinal cells, but in this case the disease is not transmitted to the next generation. Such mutations are termed somatic mutations. Regardless of the way the mutations occur, tumor development is

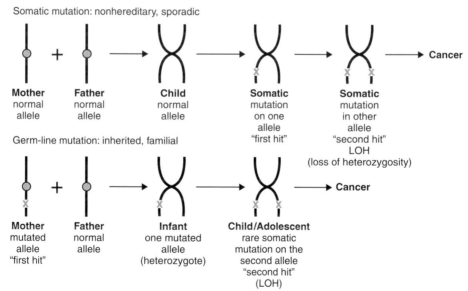

Somatic mutation: In nonhereditary and sporadic mutations, genes inherited by both parents are normal, or wild type. During early childhood or later, a mutation occurs in the allele of a gene, which represents the "first hit." This is followed by the mutation or loss of the second allele of the same gene, the second hit. Loss of function of the second allele is called loss of heterozygosity, or LOH.

Germ-line mutation: in hereditary or familial cancers, one parent, in this example the mother, transmits a mutated gene. The child is born as a heterozygote for that gene. During childhood – adolescence in retinoblastoma (*RB*) – a second mutation occurs in the second allele of the same gene inducing LOH and cancer.

Fig. 7.1 The "two hit" hypothesis.

recessive, since both copies of the gene must be inactivated for cancer to occur.

Mutation of *RB* genetically transmitted in the germ line induces bilateral and multifocal retinal tumors, whereas sporadic retinoblastoma usually occurs in one eye only as a focal tumor. This concept was validated by studies on the genomic DNA from patients with retinoblastoma as well as DNA from patients with other cancers in which *RB* or other tumor suppressors were lost.

Many human tumors contain mutations in the *RB* gene, most of which occur in somatic cells, meaning that the mutation is not transmitted from one generation to the next.

HAPLOINSUFFICIENCY IN CANCER

Haploinsufficiency is defined by the appearance of a phenotype in cells or an organism when only one of the two gene copies (also called alleles) is inactivated. For some tumor-suppressor genes, the loss of a single allele is sufficient to induce susceptibility to tumor formation. In other words, these are haploinsufficient tumor-suppressor genes. In this scenario, inactivation of one allele can be achieved by genetic (i.e. mutation or deletion) or epigenetic (i.e. transcriptional silencing by methylation, repressor complexes, or mutations in the promoter region) events.

EPIGENETIC EVENTS

Epigenetic regulation of genes can be achieved by methylation, repressor complexes, or mutations of the promoter regions or introns thereby reducing or inducing a loss of messenger RNA transcription, and subsequent reduction or loss of protein expression. Methylation occurs on the cytosine nucleotide of CpG pairs usually located in promoter regions (Box 7.1). These pairs often cluster in gene regulatory regions referred to as CpG islands (see Chapter 11). Methylation of CpG islands is associated with the formation of nuclease-resistant and compact chromatin structures, resulting in transcriptional silencing. Mutations in the promoter region can affect the efficiency with which transcription factors bind to specific DNA binding domains.

Transcriptional repression can be achieved by repressor protein complexes that bind to the promoter regions of affected genes.

Although the end result of genetic and epigenetic events is the same, that is, a given protein is not expressed, one event (epigenetic) involves heritable changes in gene function without changing the DNA sequence while the other (genetic) affects the coding regions of the gene or the promoter or noncoding regions (introns). Epigenetic events are linked to cancer development and are equally important in the process of carcinogenesis. For example, expression of the tumor suppressor $p16^{INK4A}$ or $p14^{ARF}$ is often silenced by methylation of their promoters or by active transcriptional repression by repressor protein complexes that bind to their promoters rather than deletion or mutation of the genes.

Several methods exist for detecting methylated DNA, one of which relies on a method based on the amplification of DNA fragments by the polymerase chain reaction (PCR) of high molecular weight genomic DNA treated with sodium bisulfite. Sodium bisulfite converts cytosine (C) residues to uracil (U) rather than guanine (G), but only when the C residues are not methylated. Therefore, primers can be designed that allow differentiation between sequences of non-methylated and methylated DNA by virtue of the stability of the complexes formed between the DNA and the primers. In other words, if the sequences do not match, the PCR reaction will not occur and the DNA will not be amplified. Clearly, this method requires knowledge of the sequences in the promoter region of the gene under investigation, which can be found in the genome project database or be determined by sequencing (Box 7.1).

Thus, DNA methylation and gene silencing are linked to growth and differentiation regulatory pathways that are disrupted in nearly all cancer cells.

DEFINITION OF A TUMOR SUPPRESSOR

A tumor suppressor is a protein that slows or inhibits cell proliferation and prevents damaged or abnormal cells from becoming malignant if they have been exposed to DNA-damaging agents, such as UV or X-rays, or to hyperproliferative signals induced by oncogenes

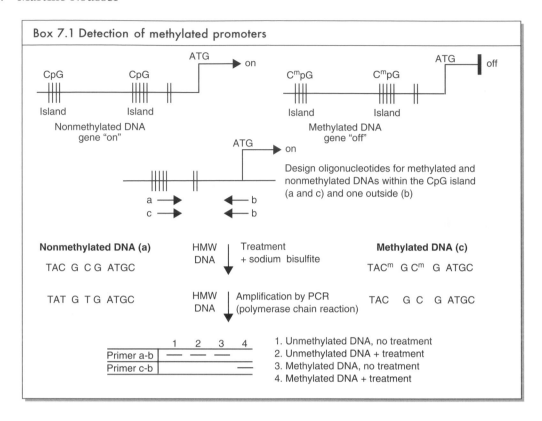

Box 7.1 Detection of methylated promoters

(see Chapter 6). Enforced expression of tumor suppressors in normal cells induces growth arrest or apoptosis, depending on the level of expression and the type of cell. Conversely, loss of tumor-suppressor function sensitizes cells to tumor formation by creating a proliferative advantage that in turn usually leads to the accumulation of genetic alterations, and ultimately aggressive cancers. Of the approximately 30,000 genes that make up the entire mammalian genome, tumor suppressors represent approximately 0.001%, although this number may increase as new technologies for studying tumor DNA become available and more tumor suppressors are identified (Table 7.1).

THE RETINOBLASTOMA PROTEIN FAMILY

Genomic locus of *RB*: mutations in human cancers

The retinoblastoma locus *RB* is located on human chromosome 13q14. It comprises 27 coding segments (exons) and spans a region of approximately 200 kilobase pairs. Retinoblastoma can occur as a result of large deletions of coding exons or of more discrete mutations (e.g. point mutations) that induce either the loss of the messenger RNA or the expression of a nonfunctional pRb protein. Most inactivating *RB* mutations lead to retinoblastoma in inherited cases, although less disruptive mutations, such as those in the promoter region of *RB* that affect its transcription, induce tumors less frequently.

pRb is important in the control of proliferation and development of not only retinal cells, but also other tissues. Children with inherited retinoblastoma due to germ-line *RB* mutations are at higher risk of developing osteosarcomas later in adolescence because they lose the remaining normal *RB* allele. Certain sporadic human tumors with *RB* mutations that have been found at high frequency include carcinomas of the bladder (33%), breast (10%), and in particular small-cell lung carcinomas (85%). Thus, the retinoblastoma protein RB plays a critical role in the control of neoplasia in a variety of tissues.

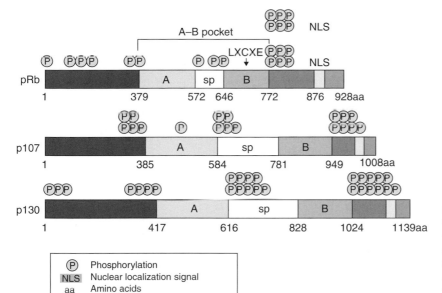

Fig. 7.2 Linear structures of the Rb family: "pocket proteins." Schematic representation of the three members of the Rb family: pRb, p107, and p130. Note the conservation of many of the domains, including the A–B pocket and the differences in length of the spacer, the N- and C-termini.

The Rb family of proteins

pRb is part of a family of three proteins that includes p107 and p130 (Fig. 7.2). A unique gene located on a different chromosome encodes each family member. While the *RB* locus resides on the long arm of human chromosome 13 (13q14), the p107 and p130 loci are located on human chromosomes 20q11.2 and 16q13, respectively. Unlike pRb, these family members appear not to function as tumor suppressors, despite their close structural similarities to each other and to pRb. Indeed, few mutations have been reported in the gene encoding p107 (a single B-cell lymphoma) or p130 (in isolated lung tumors).

The retinoblastoma gene was the first tumor suppressor to be identified and cloned. The gene encodes a nuclear phosphoprotein of approximately 110 kDa with distinct domains that are conserved between the other two family members, p107 and p130 (Fig. 7.2). The most conserved domain, termed the "A–B pocket," comprises A and B boxes linked by a spacer region. While pRb has little similarity to its family members outside the pocket domain, p107 and p130 share the highly conserved spacer domain. The spacer sequences are critical for the binding of cyclin A–cdk2 and cyclin E–cdk2 complexes. The A–B pocket is a structure rich in alpha helices to which numerous proteins are known to bind, including the E2F family of transcription factors and the cyclins. This pocket is also necessary for the binding of viral oncoproteins such as SV40 T antigen, adenovirus E1A protein, and E7 protein, all of which co-opt pRb to transform the cells in which their virus replicates. The binding of these viral proteins displaces the cellular proteins that normally interact with pRb inducing the loss of pRb function. The recently solved crystal structure of the A–B pocket in complex with the oncoprotein E7 from the human papilloma virus (HPV) has revealed that the B domain binds to an "LXCXE motif" (where L, leucine; C, cysteine; E, glutamic acid; and X, any amino acid) in its binding partners. The A box determines and stabilizes the conformation of the B domain, and the A–B interface is also highly conserved. Therefore, mutations in either domain can disrupt the structure and function of pRb. The C-terminal domain of pRb contains a nuclear localization signal (NLS) required for localization to the nucleus, and there are several phosphorylation sites in the N terminus, spacer, and C domain downstream of the B box that affect pRb function (see below).

pRb and its binding partners: role during the cell cycle and differentiation

The phosphorylation of pRb fluctuates throughout the cell cycle (Box 7.2, see also Chapter 4). When cells are out of cycle in a state of quiescence (G_0), or are in the early part of the first gap (G_1) phase of the cell cycle, pRb, p107, and p130 are underphosphorylated and in complex

---

Box 7.3   *Arf* expression is repressed by E2F-3b

The transcription factor E2F-3b is a repressor of the *Arf* gene in nonstressful conditions. It interacts with as yet unknown partners to repress transcription of *ARF*. In response to oncogenic stress, E2F-3b is displaced by E2F-1, 2, and 3a, which by recruiting histone acetylase induce the transcription of E2F-responsive genes, including *ARF*.

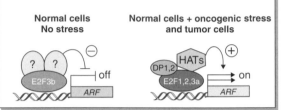

Box 7.2   Differential phosphorylation of pRb during the mammalian cell cycle

The cell cycle is composed of four main phases: mitosis (M) when cells divide; the first gap (G_1) phase during which cells are sensitive to their environment and decide to proceed through mitosis; the DNA replication (S) phase when cells duplicate their entire genome; and the second gap (G_2) phase during which cells monitor the fidelity of the duplicated genome and repair mistakes if necessary in preparation for mitosis. In addition, some cells can exit the cycle and enter a reversible state of quiescence (G_0). Early in G_1, pRb and other family members are underphosphorylated and in complex with E2F transcription factors and their partner proteins DP, repressing transcription. Upon phosphorylation by cyclin D–Cdk4–6 and cyclin E–Cdk2 complexes, pRb repression of transcription is relieved by the release of E2F–DP complexes that induce the transcription of genes required for S phase entry.

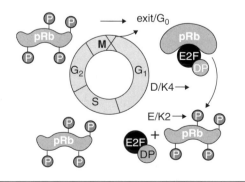

---

with other proteins, including the E2F transcription factors (E2F-1 to 5) and their subunit partners, DP1 and 2 (Fig. 7.3). While pRb interacts principally with the "activating E2Fs," E2F-1, 2, and 3a, and prevents their transcriptional activity, it interacts to some extent with E2F-4 and E2F-5 to repress gene transcription. In contrast, p107 and p130 form complexes with only E2F-4 and E2F-5, and act as repressors of transcription in G_0 and early G_1 (Fig. 7.3). Two E2F-3 proteins exist, E2F-3a, the full-length protein that acts as an activating E2F, and E2F-3b, a mutant form truncated at its N-terminus, that represses transcription of proteins, including the tumor suppressor p19[Arf] (p14[ARF] in humans) (discussed in this chapter) by interacting with as yet unknown protein partners (Box. 7.3). E2F-6, another E2F family member, is part of a repressor polycomb complex that does not interact with the Rb family of proteins. The most recently discovered E2F family member, E2F-7, like E2F-6, also acts as a transcriptional repressor and lacks an Rb-binding domain. All E2Fs form complexes with the DP1 and DP2 subunits, which confer high-affinity DNA binding on the promoter of E2F-responsive genes. As cells progress through the G_1 phase, pRb becomes progressively phosphorylated on serine and threonine residues by cyclin–cdk holoenzymes that bind to pRb via the LXCXE motif just described. Phosphorylation of pRb is initiated by the holoenzyme cyclin D–cdk4, which forms in mid-G_1 in response to mitogenic stimulation.

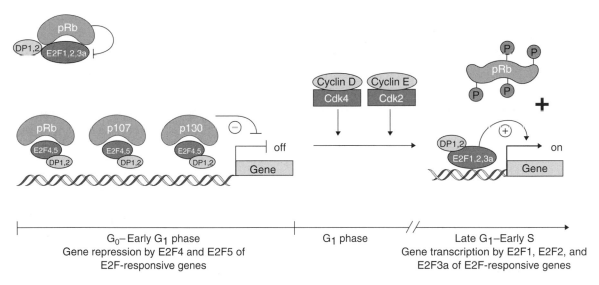

Fig. 7.3 Rb family of proteins – activation and G_1 progression. In G_0 and early G_1, pRb, p107 and p130 are underphosphorylated and in complex with E2F-4, E2F-5 and the DP1 and 2 proteins, repressing transcription. pRb also sequesters E2F-1, 2, and 3a/DP complexes without contacting DNA. In G_1, cyclin D/Cdk4–6 and cyclin E/Cdk2 holoenzymes phosphorylate pRb proteins and relieve their repressive function allowing E2Fs to induce the transcription of E2F-responsive genes required for S phase entry.

This stimulation induces the transcription of D-type cyclins, which bind cdk4 and cdk6. pRb becomes fully phosphorylated and inactivated by cyclin E cdk2 holoenzymes. Phosphorylation requires the interaction of the holoenzymes with pRb via the LXCXE motif located on the cyclins. Mutations in this motif prevent binding of the enzymatic complexes to pRb and its phosphorylation. The phosphorylation of pRb alters its conformation and growth suppressive capabilities by causing the release of bound proteins, perhaps most importantly the E2F transcription factors. Freed from pRb, the E2F-1, 2, and 3a proteins and their DP partners activate the transcription of genes required for the cell to commit and proceed to the DNA synthetic (S) phase of the cell cycle (Fig. 7.3). pRb is dephosphorylated in mitosis by the action of the phosphatase PP1alpha 2, so that the next mitotic cycle can begin.

Overexpression of wild-type but not mutated pRb blocks cells in G_1 by suppressing transcription and driving differentiation. pRb, p130, and p107 interact with many proteins, but it is their binding to members of the E2F family of transcription factors that appears to be central to their role in governing DNA replication. As many as 110 proteins have been reported to interact with pRb *in vitro* and *in vivo*, yet the ability of pRb to bind transcription factors, either repressing or stimulating their activity, appears to be the key to its ability to suppress proliferation. Unphosphorylated Rb family members and the proteins with which they interact actively repress gene expression by simultaneously recruiting histone deacetylases (HDACs), other remodeling factors, and E2F/DP1 to E2F-responsive promoters, and using the E2F/DP1 complex to position the complex onto specific promoters (Fig. 7.4). pRb recruits HDACs to the B box of the pocket domain via the LXCXE motif on HDACs and in some cases by using a binding protein that bridges HDACs to pRb, called RBP1 (RB binding protein 1). By recruiting HDACs, pRb not only forces cell cycle exit but also affects the expression of genes not involved in cell cycle regulation. p107 and p130 largely control the recruitment of HDACs to E2F-responsive promoters to repress their transcription, whereas pRb is thought to bind to other transcription factors that regulate differentiation or senescence. In contrast, once pRb is phosphorylated, gene activation proceeds by recruitment of histone acetylases to

(a)　**Repression**

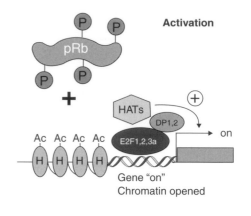

Fig. 7.4 Rb family of proteins – activation and repression of gene expression.
Gene repression: (a) pRb, p107, and p130 repress transcription (OFF) by contacting DNA via E2F-4, E2F-5, and DP complexes and recruiting histone deacetylase (HDAC). Histones are not acetylated and the chromatin is condensed. (b) Repression can be mediated by E2F-3b bound to as yet unidentified repressor proteins. (c) pRb sequesters E2F-1, 2, and 3a and their DP partners and inhibits their function without contacting DNA.
Gene activation: Phosphorylation of pRb releases E2F-1, 2 and 3a, and DP partners, which recruit histone acetylase and induce the transcription of E2F-responsive genes. Histones are acetylated; the chromatin is opened allowing gene transcription (ON).

the activating E2F/DP complexes. This in turn induces the acetylation of histones and unwinding of the DNA that facilitates access to the DNA of the transcriptional machinery (Fig. 7.4).

The Rb signaling pathway

The Rb family of proteins negatively regulates cell proliferation by repressing the transcription of the genes responsible for progression through the G$_1$ phase of the cell cycle and entry into S phase. Relief of pRb control is achieved by its phosphorylation by cyclin–cdk holoenzymes as discussed above. D-type cyclins are regulated by mitogenic signals via the

Ras/Map kinase pathway, which induces their transcription (Box 7.4, Fig. 7.5). In that sense, D-type cyclins can be considered growth factor sensors. Once expressed, D-type cyclins bind to cdk4 and cdk6, a complex that is further activated by phosphorylation by a cyclin-dependent activating kinase, CAK.

These holoenzymes are themselves regulated by inhibitory proteins, called CKIs (cyclin-dependent kinase inhibitory proteins), which comprise two families: the INK4 family and the CIP/ KIP family. Ink4 proteins bind to and specifically negatively regulate the activity of cyclin D-dependent kinases, cdk4 and cdk6. The family consists of four members, p16^{Ink4a}, p15^{Ink4b}, p18^{Ink4c}, and p19^{Ink4d}, two of which, p16^{Ink4a}

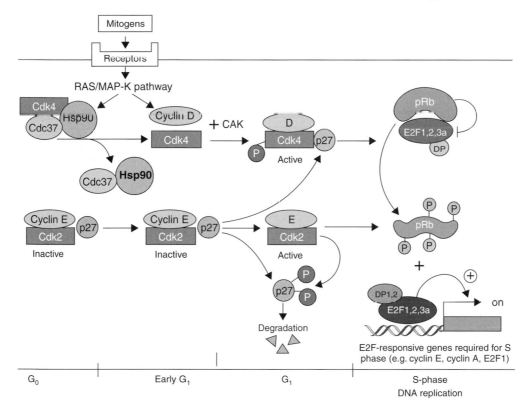

Fig. 7.5 The Rb signaling pathway. In G_0, Cdk4 is bound to the chaperone protein Hsp90, and to Cdc37 while cyclin E/Cdk2 is bound to p27^{Kip1}. Both kinases are inactive. In G_1, upon mitogen activation, receptors activate the Ras/MAP kinase pathway that regulates cyclin D transcription and the release of Cdk4 from Hsp90 and Cdc37. Cdk4 kinase activity is induced by binding to D-type cyclins, its phosphorylation by the cyclin-dependent kinase (CAK) and p27^{Kip1}, which serves as an assembly factor. The active cyclin D/Cdk4–6 holoenzyme initiates the phosphorylation of Rb. Binding of p27^{Kip1} to increased levels of cyclin D/Cdk4–6 complexes relieve the inhibition of cyclin E/Cdk2 complexes that complete the phosphorylation of Rb and phosphorylate p27^{Kip1}, which is then degraded by the proteasome machinery. Phosphorylation of Rb releases E2F1, 2 and 3a, which transactivate the expression of E2F-responsive genes, including cyclin E, cyclin A, and E2F-1 and genes required for the initiation of DNA replication (S phase).

Box 7.4 The Rb pathway

Schematic linear representation of the Rb pathway. Mitogens induce the Ras/MAP kinase pathway, which induces the transcription of D-type cyclins that form complexes with Cdk4–6. Cyclin D/Cdk4–6 phosphorylates Rb, releases E2F transcription factors that in complex with DP proteins induce entry of the cells into S phase. Cyclin D/Cdk4–6 complexes are negatively regulated by Ink4 (Inhibitors of Cdk4) and positively regulated by Cip/Kip (activating the cyclin D/Cdk4 complex) proteins.

$$\text{Mitogens} \rightarrow \text{Ras} \rightarrow \text{MAPK} \rightarrow \text{cyclin D/cdk4} \rightarrow \text{Rb} \rightarrow \text{E2F/DP1} \rightarrow \text{S phase entry}$$

Ink4 ⊥ (above cyclin D/cdk4)

Cip/Kip ↑ (below cyclin D/cdk4)

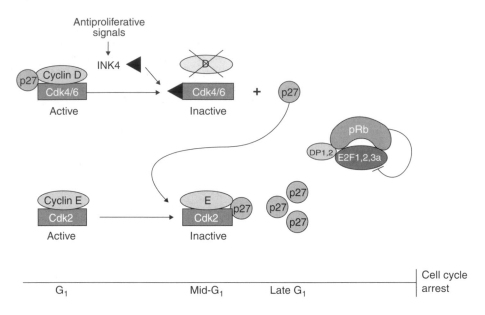

Fig. 7.6 Cell cycle exit–quiescence and senescence. In mid-G_1 and in response to antiproliferative signals, Ink4 proteins are induced. Ink4 proteins bind to Cdk4–6 and free cyclin D, which is rapidly degraded. p27^{Kip1} is reassorted to cyclin E/Cdk2 complexes to inhibit their kinase activity. Rb phosphorylation is inhibited, cells can no longer enter S phase and cells are arrested in G_1.

and p18^{Ink4c}, act as tumor suppressors. The CIP/KIP family has three members: p21$^{Cip1/Waf1}$, a p53 responsive gene (see figure above); and p27^{Kip1} and p57^{Kip2}, two tumor suppressors. All three members of this family are negative regulators of cyclin E– and cyclin A–cdk2 and of cyclin B–cdk1, while p27^{Kip1} and p21^{Cip1} also act as positive regulators of cyclin D–cdk4/6 by stimulating complex assembly (Figs 7.5, 7.6). As cyclin D–cdk4 complexes accumulate in mid-G_1, p27^{Kip1} and p21^{Cip1} are redistributed from cyclin E–cdk2 to cyclin D–cdk4/6 complexes, leading to the activation of cyclin E/cdk2, which completes the phosphorylation of pRb, as described above. At this time, p27^{Kip1} itself is phosphorylated by active cyclin E–cdk2 and targeted for proteasomal degradation, which leads to the decrease in p27^{Kip1} levels that is required for S phase entry (Fig. 7.6).

Cell cycle exit, maintenance of quiescence or differentiated states, and senescence are achieved, in part, by the expression of the CKIs. CKI expression is induced by antiproliferative signals that include cell–cell interaction (Chapter 12), or antimitotic signals, such as TGF-β (transforming growth factor) or the cytokine IL-10. Mitogens also upregulate Ink4 proteins in mid-G_1 of the normal cell cycle (Fig. 7.6; see also Chapter 5). As Ink4 proteins accumulate, they bind with high affinity to the cdk4 and cdk6 moieties, thereby displacing the D-type cyclins from the complex, leading to their rapid degradation. Similarly, p27^{Kip1} and p21^{Cip1} reassort into the active cyclin E–cdk2 complexes and inhibit their activity. This prevents pRb phosphorylation by both the cyclin D-containing and cyclin E–cdk holoenzymes and induces cell cycle exit.

In addition to its role in regulating the cell mitotic cycle, the retinoblastoma protein participates in the DNA damage response (Chapter 10), in apoptosis (Chapter 8), differentiation, and senescence (Chapter 9).

Rb and human cancers

There is now compelling evidence that several components of the Rb pathway are mutated in human cancers, leading to the suggestion that the disabling of this pathway may be an inevitable event in the formation of most or all tumor cells. Components of this pathway act either as tumor suppressors that have been inactivated (e.g. *INK4A*, *INK4B*, *KIP1*, or *RB*), or as oncogenes that have been amplified (e.g. *CDK4* or *Cyclin D1*) (see Table 7.2). While in cells containing a wild-type *RB* gene, the pRb protein represses the activating E2Fs via a sequestration mechanism in G_0 and early G_1 phase, loss of a

Table 7.2 The Rb pathway in human cancers

Cancer types	INK4A loss (%)	Cyclin D and CDK4 overexpression	RB loss (%)
Small cell lung	15	5% cyclin D1	80
Non small cell lung	58		20–30
Pancreatic	80		
Breast	31	50% cyclin D1	
Glioblastoma	60	40% cdk4	
T-ALL	75		
Mantle cell lymphomas		90% cyclin D1	

functional pRb protein unleashes the activating E2Fs, which now transcribe genes at the inappropriate time inducing either hyperproliferation or apoptosis (Box 7.5).

As discussed above, mutations in *RB* occur in several types of human cancers, not just retinoblastoma. Similarly, while loss of p16^{INK4A} functions by point mutations were initially discovered in familial melanoma, *p16^{INK4A}* mutations, deletions, or epigenetic events have been found in many other cancers. Loss of the p18^{INK4C} protein, presumably by epigenetic events affecting gene expression rather than the loss of the gene itself, is found in testicular cancers and correlates with the most aggressive forms of the disease. More recently, loss of p27^{KIP1} expression has been shown to correlate with a poor prognosis in not only colon and breast cancer, but in many other tumors, including prostate, bladder, lung, liver, ovary, stomach, and other organs.

pRb function can also be abrogated by viral proteins while leaving the gene intact. The molecular mechanisms by which these viral proteins inactivate pRb are still under intense investigation, although it is clear that pRb disruption is required for both viral replication and transformation of mammalian cells. Indeed, small DNA tumor viruses, including HPVs, have evolved several mechanisms to inhibit the function of pRb (discussed in Chapter 13). Women infected with HPV 16 and 18 develop cervical cancers several years after the infection occurs. HPV16 encodes a protein (E7) that disrupts pRb

function by mimicking the binding of cyclin D1 to pRb, thereby releasing E2Fs and possibly other proteins inhibited by pocket proteins. Indeed, pRb–E2F complexes are absent in HPV-positive tumors and in cells expressing E7 alone. E7 binds pRb via its LXCXE motif. Very much like cyclin D1, mutations in the LXCXE domain prevent E7 from binding to pRb, and block its ability to bypass pRb dependent cell cycle arrest. By releasing E2Fs, the virus, which infects nondividing cells, forces induction of S phase, and essentially hijacks its host's cell factors for its own replication. E7 also binds to p21^{Cip1}, causing the release of PCNA (**P**roliferating **C**ell **N**uclear **A**ntigen), which also induces transcription of genes required for S phase entry.

Rb family and mouse models

Mouse models in which *Rb* and its family members are deleted have provided important information regarding the functions of the Rb family in development and in adult tissues. Mice lacking two copies of *Rb* (i.e., *Rb$^{-/-}$*) in the germ line die during embryogenesis (E) between days 12.5 and 14.5 because of defects in red blood cell and skeletal differentiation, associated with apoptosis in the central nervous system (CNS) and liver (Table 7.3). More sophisticated genetic experiments were recently reported in which *Rb* was deleted conditionally or by using tetraploid aggregation. These experiments reveal an unexpected role for *Rb* in the placenta (i.e.

Box 7.5 Rb and human cancer

In normal cells expressing wild-type pRb, underphosphorylated pRb sequesters E2F-1, 2, and 3a and their DP partners. Mitogenic signals induce pRb phosphorylation, causing release of E2Fs that recruit histone acetylase (HAT) to induce the transcription of E2F-responsive genes and S phase entry. In cells in which pRb is lost or mutated, E2Fs constitutively activate transcription leading to inappropriate entry into S phase and proliferation or apoptosis.

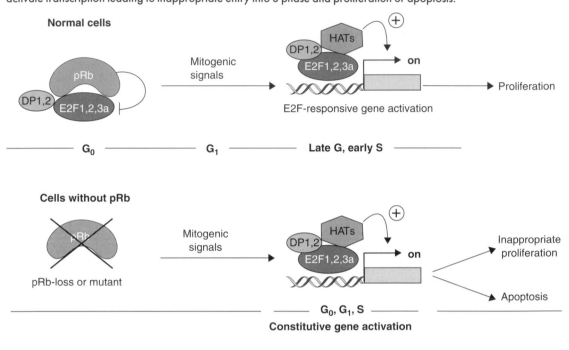

Table 7.3 Mouse models of the Rb gene family

Genotype	Phenotype
$Rb^{-/-}$	Embryonic lethality (E12.5–14.5) – CNS and liver apoptosis; erythropoiesis and skeletal defects
Rb tetraploid	Viable
$Rb^{-/-}$ (chimera)	Perinatal death; skeletal defects
$Rb^{+/-}$	Neuroendocrine tumors by 12 months; pituitary adenocarcinomas (100%) and thyroid tumors with LOH
$p107^{-/-}$	Tumor free; viable
$p130^{-/-}$	Tumor free; strain-specific survival
$p107^{-/-}, p130^{-/-}$	Perinatal death and limb defects
$Rb^{+/-}, p107^{-/-}$	Retinal hyperplasia; pituitary and thyroid tumors with LOH for Rb
$Rb^{-/-}$ (chimera), $p107^{-/-}$	Retinoblastomas; pituitary and thyroid tumors
$Rb^{+/-}, p130^{-/-}$	No retinoblastomas or retinal defects; pituitary and thyroid tumors

outside the embryonic lineages) in the control of embryonic viability and development. They also demonstrate that the loss of *Rb* in the environment, rather than in the embryo proper, can affect CNS and red blood cell development.

Mice heterozygotes for Rb function ($Rb^{+/-}$) survive embryogenesis but are highly prone to the spontaneous development of neuroendocrine tumors of the pituitary and thyroid. Examination of these tumors invariably reveals the loss of the other *Rb* allele, which is analog to the pattern of *RB* loss in human cancers. Of interest, no retinoblastoma develops in these mice, despite the similarity in etiology of both the mouse and human disease progression. In contrast, transgenic expression in the mouse retina of oncogenic proteins from DNA tumor viruses, such as SV40T antigen or E6 and E7 of the HPVs, do induce retinoblastomas. In these mouse models, the functions of both pRb and p53 are simultaneously inhibited, suggesting that both tumor suppressors are required to suppress the proliferation of retinal cells. Indeed, the loss of *Rb* induces apoptosis, whereas the loss of *p53* prevents it, thus demonstrating cooperation between these two tumor suppressors in retinal cell proliferation. Mice that lack *p107* or *p130* do not develop tumors, confirming that, despite their structural similarities to pRb, p107 and p130 are not tumor suppressors. The mouse genetic background or strain of *p107* and *p130* in which deletions have been programmed influences the phenotype associated with the loss of these genes. For example, loss of *p107* induces growth retardation and myeloid proliferation in BALB/c mice but not in C57BL/6 mice. Similarly, $p130^{-/-}$, BALB/c animals die *in utero*, whereas $p130^{-/-}$ C57BL/6 mice do not. The effects of specific backgrounds on phenotypes are an active focus of research. It refers to the identification of genes encoding modifiers of gene function that exist in some but not other strains.

Deletion of both *p107* and *p130* induces perinatal lethality associated with limb defects, which has led to the suggestion that there is a functional overlap between these two proteins in normal limb development. When mice heterozygotes for *Rb* ($Rb^{+/-}$) are bred onto a *p107*-null background (i.e., the genotype of the mice is: $Rb^{+/-}, p107^{-/-}$), they develop retinal hyperplasia. Chimeric mice containing $Rb^{-/-}, p107^{-/-}$ cells and $Rb^{+/+}, p107^{+/+}$, cells,

develop retinoblastomas and endocrine tumors with high incidence, suggesting that the development of retinoblastoma in these chimeras may be a function of the retinal cell environment in which the loss of both *Rb* and *p107* occurs.

Mouse models have thus revealed the complexity of functions associated with the Rb family of proteins, and highlighted the functional redundancy that exists during mouse development and in adult tissues. Given that pRb is expressed by all cells, it is surprising that such a restricted number of tissues are affected by its loss, until one considers the functional redundancy of this family in many tissues, where loss of one member is compensated for by another. Recent experiments in which loss of *Rb*, *p107*, and *p130* were specifically and conditionally targeted to the retina demonstrated this redundancy as 100% of mice in whom two of the three genes were lost developed retinoblastoma. Other experiments in specific cell types will further reveal how these proteins regulate proliferation, differentiation, and apoptosis, and how their disruption affects human cancer.

p53/TP53

Cloning and structure

The tumor suppressor protein p53, also known as TP53, is a transcription factor located on the short arm of human chromosome 17 (17p13.1). Its genomic locus spans 20 kbp of DNA. The p53 protein derives from a total of 11 exons that encode 393 amino acids. The protein is highly conserved throughout evolution, from *Drosophila* to human. At its N-terminus, p53 possesses a strong transactivation domain and a proline-rich region that recognizes SH3-containing proteins (Fig. 7.7). The N-terminus binds to Mdm2, which mediates its destruction by the proteasome. The mid-portion of the protein has a core DNA-binding domain that recognizes the DNA sequence: 2xPuPuPu(A/T)(T/A)GPyPyPy. Interestingly, most of the *p53* mutations in human tumors occur within the core DNA-binding domain near the protein–DNA interface and over two-thirds of the missense mutations occur in one of the three loops that bind DNA.

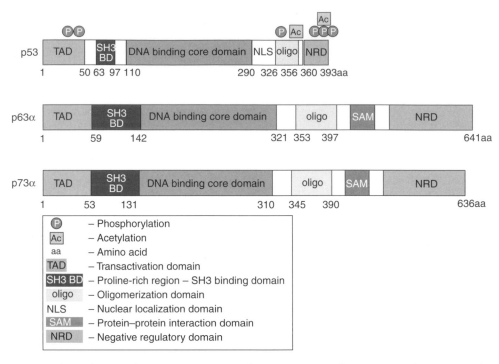

Fig. 7.7 Structure of p53 family of proteins. Schematic linear structure of p53, p63, and p73. Note that all three proteins share common domains but contain negative regulatory domains (NRD) of different lengths.

At its C-terminus, the p53 protein contains a negative regulatory region, and an oligomerization domain that mediates p53 tetramerization. The p53 protein contains several phosphorylation and acetylation sites that are involved in regulating its activity. In 1992, Vogelstein and Kinzler proposed that p53 binds as a tetramer to a p53-specific DNA-binding site to activate the transcription of p53-responsive genes. The three-dimensional structure of the tetramerization domain subsequently confirmed that the protein does indeed form a tetramer through its oligomerization domain and supported the prediction that mutations in one *p53* allele would, in a *trans*-dominant manner, abolish the ability of the tetrameric p53 to bind to its recognition sequence. Indeed, mutation of a single allele of *p53* has been shown to be sufficient to cripple p53 function.

Activation and function of p53

When overexpressed, p53 inhibits cell proliferation and transformation. Therefore, low levels of p53 are maintained in cells to allow normal proliferation. p53 is a sensor of multiple cellular stresses that include genotoxic stress (e.g., UV, X-ray, carcinogens, and cytotoxic chemotherapeutic drugs) or oncogenic stresses due to hyperproliferative signals induced by constitutive expression of oncogenes due to spontaneous mutations (Fig. 7.8).

In response to cellular stress, the p53 protein is stabilized and activated to induce a set of target genes involved in cell cycle arrest, DNA repair, or apoptosis, depending on the strength of the signal, and cell context. Phosphorylation of p53 causes its activation through three mechanisms: stabilizing the protein by disrupting its interaction with Mdm2, regulating its transactivation activity, and promoting its nuclear localization.

p53 restrains uncontrolled cell growth by blocking the cell cycle, and inducing programmed cell death (apoptosis) (Fig. 7.8 and discussed in Chapter 8). It acts as a tetramer that recognizes p53-responsive elements in the promoters of p53-responsive genes, many of which have been cataloged by gene chip microarray analysis. Interestingly, p53 positively regulates

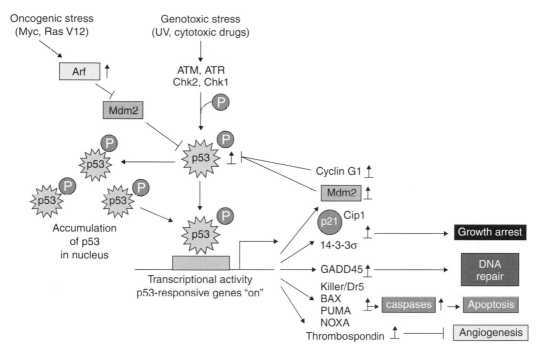

Fig. 7.8 p53 activation and function. The transcription factor p53 is activated by phosphorylation in response to hyperproliferative signals and genotoxic stress. This induces an accumulation of p53 in the nucleus and the transcription of multiple p53 target genes that induce growth arrest, apoptosis, DNA repair, or a block in angiogenesis.

the transcription of at least two of its negative regulators, Mdm2 and cyclin G. Mdm2 has a ring domain that possesses E3 ligase activity. Direct association of p53 with the N-terminal domain of Mdm2 results in the ubiquitination of p53 and its subsequent degradation by the ubiquitin-dependent proteasome pathway. Upon DNA damage, p53 is phosphorylated on serine residues in its transactivation domain, in particular on serine 15 by the Chk2 kinase. Serine 15 phosphorylation, in turn, induces a conformational change that prevents Mdm2 binding to p53 and results in the relief of the inhibitory effect of Mdm2 on p53. However, mice in which a S15A mutation was inserted in the p53 genomic locus were not tumor prone. This result in the mouse suggests that this serine 15 phosphorylation alone is not sufficient to release p53 from regulation by Mdm2. A negative feedback loop exists whereby p53 upregulates Mdm2 while Mdm2 downregulates p53. Of note, the *MDM2* gene is amplified in a significant proportion of the most common human tumors, where overexpression

of the protein interferes with p53 activity. Thus, overexpression of Mdm2 equates to loss of p53 and increased propensity to tumor formation. p53 is also regulated by Mdmx (Mdm4), although the mechanism by which this occurs is less clear. Mdmx was found to be a key regulator of p53 in mice because its loss is lethal to the embryo, but codeletion of p53 rescues the phenotype.

A similar negative feedback loop exists between cyclin G and p53. Cyclin G is thought to downregulate the activity of p53 by recruiting a PP2A phosphatase to activated Mdm2 by dephosphorylation, such that active Mdm2 may, in turn, stimulate p53 degradation. Loss of cyclin G in mice renders them susceptible to liver cancer.

Activation of p53 is regulated by upstream mediators of the DNA damage response, which include a number of kinases (ATM, ATR, DNA-PK, Chk1, and Chk2) – see also Chapter 10. ATM and ATR are members of the phosphatidyl-inositol-3-kinase (PI3K) family. These kinases are activated by DNA damage and are recruited to DNA lesions where they phosphorylate proteins including p53 (Fig. 7.8). The ATM ("ataxia-telangiectasia mutated") protein is responsible for the disorder ataxia-telangiectasia (AT) (Table 7.1, and Chapter 4). AT is a rare, autosomal recessive inherited disease, in which patients are extremely sensitive, and show abnormal responses, to ionizing radiation due to the loss of the DNA damage checkpoint induced by p53 and, as a result, suffer increased chromosomal breakage and telomere-end fusion leading to an increased rate of cancers. Normal ATM is recruited to X-ray-induced double-strand breaks and activated to phosphorylate p53 on S15, thereby preventing Mdm2 binding. In addition, ATM phosphorylates Mdm2 on S395, decreasing the ability of Mdm2 to shuttle p53 from the nucleus to the cytoplasm, thereby allowing the stabilization and nuclear accumulation of p53 protein. In individuals with AT, ATM can no longer perform these functions, which leads to the lack of functional p53.

Many of the p53-responsive genes that are induced in response to stress signals can arrest cell proliferation at the G_1/S and G_2/M transitions allowing cells to repair any defects due to DNA damage before they proceed to DNA replication or mitosis. Thus, p53 regulates these two critical cellular checkpoints. One of the major transcriptional targets of p53 is the p21^{CIP1} gene (Box. 7.6). p21^{CIP1} (also called WAF-1) is a cyclin-dependent kinase inhibitory protein whose overexpression induces G_1 arrest by preventing the cyclin E/cdk2 holoenzyme from phosphorylating pRb (see section on the retinoblastoma protein family). Other proteins with antiproliferative capabilities that are regulated by p53 include several members of the Btg family of proteins. This family of proteins is thought to suppress cell growth as part of transcriptional corepressor complexes that include histone protein deacetylase (HDAC) and protein arginine methyl transferase, which methylates histones. One member of the Btg family, named Tob1, was recently found to have tumor-suppressor activity in mice and in humans (Table 7.1). *Tob1* loss in the mouse predisposes the animal to spontaneous tumor development between 6 and 22 months of age. *Tob1*-null mice develop mostly hepatocellular adenoma and lung carcinoma, the onset of which is accelerated in

a *p53*-null background, confirming that p53 and Tob1 cooperate in murine tumor development. An assessment of 18 tumor samples from human lung cancers revealed that in 13 of the 18 samples, the *TOB1* gene was not mutated or deleted, but the expression of *TOB1* mRNA was decreased, suggesting that TOB1 is a tumor suppressor in humans. The p53 protein also triggers cell cycle arrest at the G_2/M boundary by inducing the transcription of genes that encode proteins that inhibit cyclin B-cdk1 activity and thus mitosis (e.g., 14-3-3 σ, reprimo, and b99).

Oncogenic and hyperproliferative stresses can cause the activation of p53, which, in turn, induces apoptotic cell death by activating many proapoptotic genes (Fig. 7.8). Many of these p53-dependent proapoptotic genes have been identified by differential gene expression approaches such as the serial analysis of gene expression (SAGE) and gene chip microarray analysis. The proteins encoded by these proapoptotic genes ultimately activate caspases, including caspase 9, which induces cell death (see Chapter 8). Recently, a direct transcriptional

---

Box 7.6 The p53-responsive gene p21^{CIP1}

p53 is activated by phosphorylation in response to upstream signals. Active p53 binds DNA and activates the transcription of *p21*^{CIP1} (referred to also as *p21*). p21^{CIP1} inhibits the cyclin E/Cdk2 holoenzyme by interacting with the complex, thereby inducing growth arrest.

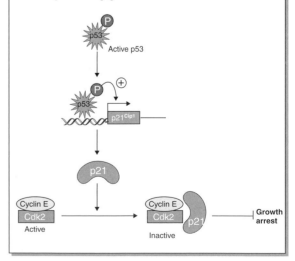

target of p53 called Bbc3 (or PUMA) was found to play a key role in p53-dependent apoptotic cell death. PUMA localizes to mitochondria, where its overexpression has been shown to effectively kill cells. Remarkably, mice that lack *Puma* in the germ line (created by deletion of the gene by homologous recombination) are deficient in all of the phenotypes associated with p53-dependent apoptosis. These phenotypes include the inability of the *Puma*-null cells to die in response to genotoxic stresses even in the presence of a functional, nonmutated p53 protein, which suggests that PUMA is the major downstream effector of p53-dependent apoptosis.

Function of p53 family members

p53 is the founding member of a family of three proteins that also includes p63 and p73. Like the Rb family described earlier, p53 is the sole member of the family with tumor suppressor activity. p63 and p73 were isolated in the last 6 years, and studies performed to date indicate that p73 and p63 are rarely mutated in human cancers. Similar to p53, p73 can induce growth arrest and apoptosis when overexpressed in some p53-null cells. p73 was recently identified as a chemosensitizer for a variety of chemotherapeutic drugs, and an important contributor to the cell's response to cytotoxic agents. Both p63 and p73 cooperate with p53 to induce apoptosis, suggesting that although these two proteins are not true tumor suppressors, they have a role in the regulation of DNA damage-induced cell death. Because *p73* and *p63* are rarely mutated or silenced in human tumors, the possibility remains that they may be inactivated by mechanisms that have yet to be elucidated.

p53 in human cancer

Sir David Lane dubbed p53 "the guardian of the genome" because its loss leads to genomic instability and increased mutagenesis. Unlike normal cells, tumor cells often have high levels of p53 that is almost always mutated. Of note, p53 was first identified in tumor cells where it was expressed at high levels but was unable to arrest cell proliferation. Only later did scientists realize that the gene everyone had isolated was in fact mutated *p53*, at which point p53 was recognized as a bona fide tumor suppressor. Since its discovery in 1979, and its definition as a tumor suppressor in 1989, more than 19,000 *p53* mutations have been identified in approximately 150 different tumor types. Indeed, p53 inactivation is found in more than half of all human cancers.

It is believed that the most important function of p53 in protecting us from cancers is its ability to induce cell death. Loss of p53 function abolishes growth arrest or apoptosis, which prevents cells from properly responding to stress or damage, leading to genomic instability and the accumulation of additional genetic abnormalities. Inactivation of p53 can occur via numerous mechanisms: (i) missense mutations acting in a dominant-negative manner to reduce the function of the p53 tetramers; (ii) deletion of one or both p53 alleles; (iii) binding of p53 to viral proteins, such as E6 from HPV or SV40 T antigen; and (iv) degradation by Mdm2, the negative regulator of p53.

The updated p53 mutation database contains 3200 variants for approximately 18,000 tumors listed, an impressive number that highlights the importance of this gene in cancer (http://www.iarc.fr/p53). Most of these p53 mutations have been mapped to the core DNA-binding domain, although several mutations outside the core are also relevant to specific types of cancers (Fig. 7.9).

Mutations in p53 can be inherited via the germ line, as is the case with Li–Fraumeni syndrome, where families are predisposed to diverse types of cancer, including breast cancers, bone, soft tissue, brain, adrenal, and colorectal carcinoma, or less frequently melanoma. Interestingly, in families affected by this syndrome, the *p53* mutations are transmitted from one generation to the next, and the cancers generally occur at an increasingly earlier age.

A germ-line mutation in the oligomerization domain of p53, where the arginine (R) residue at position 337 is mutated to histidine (H), was found initially in young Brazilian patients, although a few cases have been reported in other countries. These patients develop adrenocortical carcinoma early in life despite no family history or increased predisposition for cancer. In individuals with this *p53* mutation, the mutant protein forms tetramers that are destabilized in basic conditions. It is believed that elevated

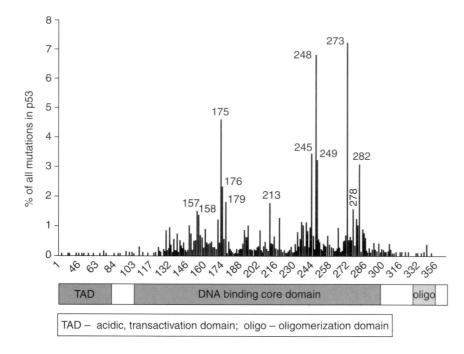

Fig. 7.9 *p53* mutations in human cancer. Mutations in the p53 gene have been found throughout the entire coding sequence but with increased frequency in the DNA-binding domain. Mutations affect p53 transcriptional activity. Reproduced from the IARC database (www.iarc.fr/p53), R9 July 2004. Olivier M, Eeles R, Hollstein M, Khan MA, Harris CC, Hainaut P. The IARC TP53 Database: new online mutation analysis and recommendations to users. *Hum Mutat* 2002 Jun: 19(6): 607–14.

pH is a characteristic of adrenal cells, although a causal relationship between the pH of the adrenal cells and the adrenocortical carcinomas developed by individuals with this *p53* mutation has not yet been confirmed.

Nonhereditary or somatic *p53* mutations or *p53* loss have been described in many different types of cancers, including tissue-specific cancers, such as liver (hepatocellular carcinomas), bone (osteogenic sarcomas), muscle (rhabdomyosarcoma), lung, colon, bladder, cervix and anus, pancreas, esophagus, brain, and skin (squamous cell carcinoma) (NCBI OMIM website http//www3.ncbi.nlm.nih.gov:80/htbin-post).

Although p53 function is most often lost as a result of mutations or deletions, p53 can be inactivated while retaining a normal sequence. This is achieved by its rapid degradation by the negative regulator MDM2, which is often overexpressed in human tumors, or by viral proteins expressed by viruses that are etiologically linked to specific cancers. For example, functional loss of p53 in cervical cancers is rarely associated with *p53* mutations, but is often linked to HPV, a high risk human papilloma virus that encodes the proteins E6 and E7 that neutralize the function of p53 and pRb, respectively. E6 associates with p53 and with E6AP [a ubiquitin E3 protein ligase (E3A)], which catalyzes ubiquitination and degradation of p53. Other tumors lose p53 function via the overexpression of high levels of MDM2 protein that, as described earlier, binds to p53 catalyzing its degradation by the proteasome-dependent degradation machinery.

p53 family and mouse models

Each p53 family member has been deleted by homologous recombination in the mouse, providing insights into its role in tumorigenesis and development (Table 7.4). Mice lacking one or two copies of *p53* are highly tumor prone and develop a range of cancers, including lymphomas and sarcomas, as well as lung and brain tumors. This spectrum of tumors resembles the one seen in patients with Li–Fraumeni syndrome, characterized by the inheritance of a *p53* mutant allele and LOH, or the loss of the second allele in the tumor cells. These *p53*-null mice have proven useful for the testing of potential new therapies, since human tumors often develop resistance to therapy as a result of the loss of p53 function. These *p53* mutant mice can also be used to screen potential carcinogenic compounds. Since

Table 7.4 Mouse models of the p53 gene family

Genotype	Phenotype
$p53^{-/-}$	Highly tumor prone; mostly lymphomas and sarcomas with other tumors with less penetrance
$p53^{+/-}$	Tumor prone with LOH
p53 R172H	Higher penetrance of carcinomas and metastases
"Super p53"	No tumors
$p63^{-/-}$	No tumors. Limb defects, craniofacial and epithelial development defects; no skin
$p73^{-/-}$	No spontaneous tumors; hydrocephaly and chronic infections

they are "sensitized" to cancer, the test is more sensitive and fewer mice are needed. As expected, tumors from these animals are resistant to treatment with gamma-irradiation or adriamycin (a chemotherapeutic drug), since in the absence of functional p53, the cells can no longer trigger their stress-induced and p53-dependent apoptotic program. Since their generation, *p53*-null mice have been used extensively to define collaboration between the p53 apoptotic pathway and other tumor suppressors or oncogenes. In combination with other mice that have also been genetically engineered to lose other tumor suppressors or to express oncogenes, the *p53*-null mice develop a wide range of tumors at an accelerated rate.

More sophisticated models of cancers have now been generated in which p53 missense mutations found in human cancers have been reprogrammed in the mouse genomic locus to mimic more closely what happens in human tumors. In addition, p53 as well as other tumor suppressors are being deleted in a temporal and tissue-specific manner by using the Cre–Lox system. For example, a single amino acid substitution in the mouse *p53* allele at residue R 172 to H (p53R172H), a mutation equivalent to the same substitution at residue 175 in human tumors in combination with loss of a G nucleotide at the splice junction that reduces the level of the

mutated protein to wild-type levels, induces an increase in the number of carcinomas and in metastases not found in the $p53^{+/-}$ mice (Table 7.4). This suggests that these point mutations induce a gain rather than a loss of p53 function since the tumor spectrum is worse when the protein is mutated compared to when it is lost. These more refined mouse models may better reflect events in human cancers. Finally, the group of Serrano demonstrated the concept that functional p53 protects from cancer. This group found that the addition of one copy of *p53* to mice with the normal gene protected the animals from developing tumors. These investigators inserted in the mouse genome a large fragment of genomic DNA containing the *p53* gene and its upstream and downstream sequences required for proper regulation of gene expression. The concept is that in these "super-p53" animals, the probability that both genes become mutated is infinitely low, and therefore the mice are completely protected from the development of cancer induced by chemotherapy or other stresses.

INK4A/ARF

The INK4A/ARF locus

The *INK4A/ARF* locus, identified on the short arm of human chromosome 9 (9p21), is frequently mutated or deleted in human tumors. ARF negatively regulates p53 by sequestering HDM2 (Mdm2 in the mouse), thus preventing HDM2 from degrading p53. Cloning of this locus both in mice and in humans, revealed that the same exon encodes two proteins by using an alternative reading frame for each protein (Fig. 7.10). p16INK4A derives from a transcript that comprises three exons, exon 1α, exon 2, and exon 3. p19Arf in the mouse (p14ARF in humans) derives from a transcript that begins approximately 13 kbp upstream of exon 1α, at the alternative exon 1β, and splices into the same acceptor site on exon 2 used by p16INK4A but in an alternative reading frame (hence its name, ARF). This reading frame contains a stop codon 105 amino acids downstream in exon 2 that terminates the protein. Another INK4 member, *INK4B*, is also genetically linked and located 5' to exon 1β. This unusual organization is conserved in humans and mice. A puzzling question remains

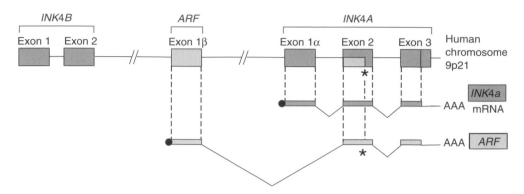

Fig. 7.10 The *INK4A/ARF* genomic locus. Schematic representation of genomic organization of the genes that are encoded by the locus. p16^{INK4a} is encoded by three exons, exon 1α, exon 2, and exon 3. Exon 2 also encodes another protein, but in an alternative reading frame, i.e., p19^{Arf} (p14^{ARF} in humans). *ARF* transcription is initiated from exon 1β. Another protein, p15^{INK4b}, is encoded by two exons that are genetically linked to the *INK4a/ARF* locus.

as to the origin of ARF. Indeed, to date, ARF has been found in all mammals, opossum, and birds (chicken), but it has not been identified in amphibians (*Xenopus*) or fish (*Fugu*), where only *INK4A* and *INK4B* homologs exist at the locus. Further, ARF has not been found in lower organisms, such as the worm *Caenorhabditis elegans*. In birds, *Arf*, rather than *Ink4a*, is present at the locus suggesting that in evolutionary terms *Ink4a* may have arisen more recently than *Arf*, from the duplication of an ancestral *Ink4 a, b* gene. Because only the first 37 amino acids of ARF are functionally relevant to its growth suppressive function, it is possible that ARF exists in lower organisms, but that its sequence is not sufficiently conserved to allow its identification from the *Drosophila* or *C. elegans* genome databases. p19^{Arf} was recently found to inhibit ribosomal processing. This suggests that the function of the primordial *ARF* gene may be to negatively regulate growth rather than to induce growth arrest. This novel function of ARF is intriguing as it places *ARF* in opposition to the oncogene *MYC*, which drives the expression of ribosomal proteins (discussed in Chapter 6).

Ink4a and Arf functions

The p16^{Ink4a} and p19^{Arf} proteins suppress tumor development by regulating the two best-known tumor suppressors in cancer, pRb and p53 (Fig. 7.11 and elsewhere in this chapter).

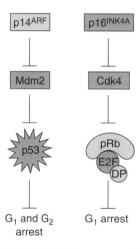

Fig. 7.11 The *INK4a/ARF* locus encodes two tumor suppressors. The same exon 2 encodes two tumor suppressor proteins, p14^{ARF} (p19^{Arf} in the mouse) and p16^{INK4a}. Whereas the tumor suppressor p14^{ARF} activates p53 by binding to and sequestering its negative regulator Mdm2, the tumor suppressor p16^{INK4a} binds to and inhibits cyclin D/Cdk4 activity and pRb phosphorylation. Expression of p14^{ARF} induces G$_1$ and G$_2$ arrest while enforced expression of p16^{INK4a} induces G$_1$ arrest. Thus, remarkably, the *INK4a/ARF* locus encodes two tumor suppressors that regulate two of the most important tumor suppressors in human cancers.

Expression of p16^{Ink4a} induces G$_1$ arrest by binding to and inhibiting the cyclin D-dependent kinases, cdk4 and cdk6, thus preventing the phosphorylation of pRb and attenuating its

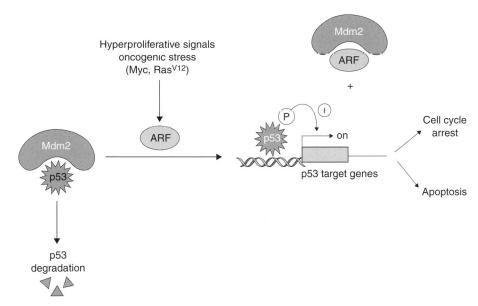

Fig. 7.12 ARF activates p53 by sequestration of Mdm2. Mdm2 is a negative regulator of p53 that induces its degradation. In response to hyper-proliferative signals due to oncogene activation, including overexpression of Myc or expression of a constitutively active RasV12 protein, ARF is induced, it binds to Mdm2, which results in the accumulation and activation of p53 leading to cell cycle arrest or apoptosis.

function (see section on pRb above). The p19^{Arf} protein does not affect Rb but does interact with Mdm2. ARF sequesters Mdm2, the negative regulator of p53, thereby preventing p53 degradation, which leads to an increase in p53 protein levels in the nucleus and to transcriptional activity (Fig. 7.14). This in turn induces cell cycle arrest or apoptosis (see section on p53 and human cancer).

ARF function

ARF is a nucleolar protein that limits cell cycle progression in response to an excess of proliferative signals induced by oncogenes, such as *MYC, E2F-1, RasV12,* and *V-ABL* (Fig. 7.13). The molecular mechanisms by which oncogenic signals induce *Arf* expression are not completely understood. To date, for example, there is no evidence that Myc transactivates *Arf* by directly binding to its promoter region, although Myc rapidly induces p19^{Arf} protein. E2F-1 was reported to bind directly to the *Arf* promoter and to directly regulate its mRNA expression. Recent genetic experiments with the Eμ-Myc mouse model for Burkitt lymphomas have demonstrated that E2F-1 is not required for Myc-induced Arf-dependent apoptosis but rather for Myc-induced proliferation by regulating the

levels of p27^{Kip1}. In contrast, the transcription of p19^{Arf} is directly upregulated by DMP1, a transcription factor that directly binds to the *Arf* promoter. DMP1 is a tumor suppressor in acute myeloid leukemia and myelodysplastic syndrome in man (Table 7.1). Mice that lack *Dmp1* develop tumors spontaneously due to their low cellular levels of p19^{Arf} protein. Once induced, p19^{Arf} activates p53 by sequestering Mdm2 and causing its relocalization to the nucleolus or the nucleoplasm, depending on the cell context. Free from Mdm2, p53 is stabilized and transcriptionally active (Fig. 7.12).

As with *Ink4a*, expression of Arf is induced in response to stress signals that include oncogenic stress (as described above) and "culture shock," which occurs when mouse embryo fibroblasts are dissociated and established in culture. Of interest, *Arf* mRNA and protein are undetectable in embryos, but can be detected by PCR amplification in adult tissues, suggesting that ARF expression is actively repressed in most normally proliferating cells. The active repression of ARF is thought to be mediated by several proteins, which include Tbx2 and Tbx3 (both transcriptional repressors), Twist and Dermo (two proteins that were identified in a screen for proteins that bypass Myc-induced apoptosis), Jun D (a transcription factor), Bmi1, a member of the polycomb transcriptional repressors, and

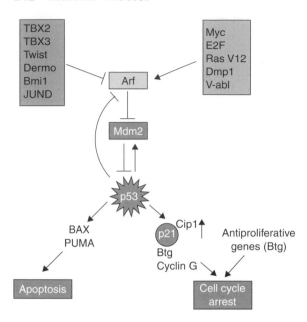

Fig. 7.13 The Arf/Mdm2/p53 signaling pathway. Transcription of the tumor suppressor p19[Arf] is induced by Myc, E2F and others (in green box). *Arf* expression is actively repressed by repressor protein complexes, including Bmi1 (in red box). ARF activates p53 by sequestration and inhibition of Mdm2 to induce cell cycle arrest or apoptosis. p19[Arf] also induces growth arrest in a p53-independent manner by inducing many antiproliferative genes (including the Btg proteins), which in concert induce cell cycle arrest.

E2F-3b (see sections on pRb above and Box 7.3) (Fig. 7.13). Genetic studies have demonstrated that the defects induced by the loss of *Bmi1* can be rescued in *Ink4a/Arf*-null mice, which suggests that *Arf* and *Ink4a* are downstream targets of Bmi1-dependent repression, and that *Arf* repression by Bmi1 is important for mouse development.

Enforced *Arf* expression induces acute G_1/S and G_2/M cell cycle arrest in a p53-dependent manner. However, *Arf* can also induce growth arrest in the absence of p53 or Mdm2, albeit with slower kinetics (Fig. 7.13). Mice lacking *Arf*, *Mdm2*, and *p53* develop more tumors per animal and more aggressive tumors than those lacking any one of these genes alone. These genetic experiments have underscored the nonlinearity of this pathway and suggest that each member of the pathway interacts with other proteins to protect cells against cancer. When the effect of enforced *Arf* expression in mouse embryo fibroblast cells lacking *p53* and *Mdm2* was compared to that in cells expressing wild-type levels of p53 and Mdm2, a family of genes induced by ARF in the absence of p53 was identified. This family is called the Btg family, and one of its members, Tob1, is a bona fide tumor suppressor, since its loss in mice has been shown to induce spontaneous tumors (see section on p53 target genes in this chapter). Thus, ARF has

p53- and Mdm2-independent functions, some of which are beginning to emerge. Mice lacking *Arf* are blind due to the hyperproliferation of the hyalovasculature that irrigates the lens during the first 10 days following birth. These mice revealed a p53-independent function for Arf in the regression of the vasculature in the eye. This phenotype in mice mimics a human condition called PHPV (**P**ersistent **H**yperplastic **P**rimary **V**itreous) found in children.

INK4A/ARF in human cancers

The *INK4A/ARF* locus is often targeted by mutation, deletion, or epigenetic gene silencing in human cancers with frequencies approaching those seen for *p53*. While mutations that do not affect *ARF* have been reported in the *INK4A* gene, specifically in familial melanoma, no point mutations have been reported for *ARF* in human cancers. Instead *ARF* is either deleted with *INK4A* and *INK4B* with the loss of the entire locus, or its expression is silenced by methylation of, or repressor complexes on, its promoter. Interestingly, cases in which *ARF* expression is silenced while *INK4A* remains intact and vice versa have been reported in certain cancers. In several types of human cancers, repressors of *ARF* expression are found overexpressed. For example, TWIST is overexpressed in rhabdomyosarcoma, a type of muscle tumor, and TBX2 is overexpressed in breast cancers. In these scenarios, the *ARF* locus remains intact, but p14^{ARF} expression is suppressed to a level analogs to that achieved by gene silencing or deletion.

Mouse models of *Ink4a* and *Arf* loss

The tumor suppressor properties of p16^{Ink4a} and p19^{Arf} have been assessed in knockout mice, in which each gene was deleted independently or together (Table 7.5). Mice lacking *Ink4a* and *Arf* (like *Arf*-null animals) are highly susceptible to tumors and die of cancer by 14 months of age. Mice lacking *Ink4a* alone, by deletion of exon 1α (Fig. 7.10), however, develop tumors spontaneously in only 25% of cases, and develop lung

and skin cancers after exposure to carcinogens. In contrast, mice carrying an *Ink4a* mutant that no longer can bind its catalytic partner, cdk4, do not develop tumors spontaneously in the first year of life. In this regard, the latter knockout mouse does not mimic tumor development in humans, because in humans mutations in *INK4A* (*CDKN2A*) that abolish p16^{INK4A} binding to CDK4 and CDK6 are a characteristic of inherited familial melanoma. Melanoma is a skin cancer caused by prolonged and repeated exposure to UV and thus mutation of *Ink4a* was not sufficient in itself to cause skin cancers in mice. Indeed, melanomas occur in rodents lacking *Ink4a* function only when tumors are induced experimentally in *Ink4a*-null mice by either UV exposure, breeding with RasV12 transgenic mice, or when *Ink4*-null mutant mice retain a single copy of the *Arf* gene (Table 7.5).

THE p53 AND Rb PATHWAYS IN CANCER

As described elsewhere in this chapter, *p53* and *RB*, as well as their upstream regulators and downstream targets, are often targeted in cancers, which suggests that genetic anomalies in these pathways may be a part of the life history of most if not all cancer cells. In mice and in many human tumors, *p53* mutations or *Ink4a/Arf* deletions are often mutually exclusive, suggesting that when the function of one tumor suppressor is disrupted, there is no selective pressure to inhibit the other. The p53 and Rb pathways were functionally linked by the discovery of the tumor suppressor *Arf* (Figs 7.10 and 7.14). Following this discovery, it became clear that enforced *Arf* expression induces acute growth arrest in both the G$_1$ and G$_2$ phases of the cell cycle, but only in cells containing functional p53. In the presence of p53, enforced *Arf* expression induces growth arrest within 24 hours of its expression. Later experiments showed that Arf could induce growth arrest in the absence of p53 and Mdm2, but with slower kinetics, taking 3 days instead of one. Unlike p53, Arf is not induced by DNA damage, although it can act as a modifier of ATM function, since loss of *Arf* has been shown to rescue *ATM*-null induced senescence in mouse embryo fibroblasts, but not CNS

Table 7.5 Mouse models of *Ink4a* and *Arf* genes

Genotype	Phenotype
Ink4a$^{-/-}$	Spontaneous tumors in the first year of life with 25% penetrance. Lung and skin tumors after carcinogen treatment. Low incidence of spontaneous melanoma.
Ink4a mut/− (knockin of an *Ink4a* mutant)	No spontaneous tumor in the first year of life.
Arf$^{-/-}$	Spontaneous tumors by 38 weeks (lymphomas and carcinomas); accelerated onset with carcinogen (X-rays) treatment
Ink4a$^{-/-}$, *Arf*$^{-/-}$	Spontaneous tumors by 34 weeks (lymphoma and sarcomas); accelerated onset with carcinogen treatment
Ink4a-null mutant, *Arf*$^{+/-}$	Melanomas and other tumors

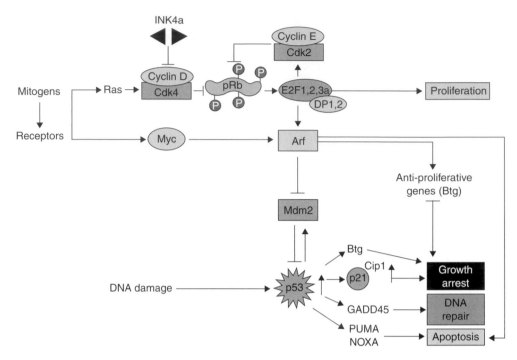

Fig. 7.14 Coupling of pRb and p53 signaling pathways. ARF expression is induced in response to hyperproliferative signals. Ras induces cyclin D transcription. Cyclin D forms active complexes with Cdk4–6, which phosphorylate pRb and release E2Fs. E2Fs and Myc induce ARF expression. ARF activates p53 by sequestering Mdm2. p53 in turn activates the transcription of p53-responsive genes responsible for growth arrest, apoptosis (cell death), or DNA repair. Adapted from Sherr (1998). Reproduced by permission of Cold Spring Harbor Laboratory Press.

apoptosis sensitivity. Deletion of both *p53* and *Rb* in just the cerebellum of mice was achieved by breeding mice whose *Rb* and *p53* genes were flanked by Lox sites, which are recognized by Cre recombinase, with mice expressing a transgene in which Cre recombinase was expressed from a cerebellar-specific promoter (GFAP-Cre mouse). These mice develop medulloblastoma, one of the most frequently seen brain tumors in children, demonstrating that in this tissue, *p53* and *Rb* function are required for tumor surveillance.

SENESCENCE AND IMMORTALIZATION: ROLE OF Rb AND p53

Senescence represents an arrest of proliferation from which cells rarely escape (see Chapter 9). It is, in itself, a potent antitumor

mechanism. Senescent cells are usually large and flat, upregulate a senescence-associated activity, β-galactosidase, and although metabolically active, do not divide even in the presence of serum and mitotic signals. These features distinguish senescence from quiescence, from which cells can re-enter the mitotic cycle upon mitogenic stimulation. Although "replicative" senescence is triggered by telomere attrition, this state can be induced by activated oncogenes, DNA damage, oxidative stress, and suboptimal culture conditions. Cells that reach senescence upregulate many cell cycle inhibitory proteins, including p16^{Ink4a} and p19^{Arf} (p14^{ARF} in humans), which, as we have discussed in this chapter, regulate Rb and p53, respectively. p16^{Ink4a} and p14^{ARF} accumulate in senescent cells and can induce senescence when overexpressed. Conversely, loss of p19^{Arf}, p53, or the three Rb family members (pRb, p107, and p130) induces the immortalization of mouse embryo fibroblasts. Interestingly, loss of *Rb* in the germ line does not induce immortalization because of compensating high levels of p107 that substitute for pRb, thus leading to cell senescence. Recently, it was demonstrated that acute loss of *Rb* in cells from conditional *Rb* knockout mice relieves them from the senescence imposed by unphosphorylated pRb, stimulates S-phase entry, and eventually leads to immortalization, presumably by acquisition of mutations in the p53 pathway. These results suggest that both pRb and p53 play critical roles in the regulation of the senescence process. pRb, whose activity is induced by high levels of p16^{Ink4a}, is thought to be critical to promote senescence or permanent growth arrest by altering the chromatin state of growth regulatory genes. Much experimental data now implicate the Rb and p53 pathways in cellular senescence, confirming their critical roles in preventing tumor development.

TUMOR SUPPRESSORS AND THE CONTROL OF CELL PROLIFERATION

The phosphatase PTEN

The next most frequently mutated gene in many types of cancers is that encoding the tumor-suppressor protein PTEN, a protein phosphatase that negatively regulates the activity of the phosphoinositol-3-kinase (PI3K) pathway (Fig. 7.15, discussed in Chapter 5). PTEN, like other tumor suppressors involved in ligand-dependent signaling, negatively regulates cell proliferation via a pathway that is a major regulator of MTOR signaling through translation regulation. Many growth factors, by binding to their cognate receptors, induce signals that affect the PI3K pathway, critical for the activation of the serine/threonine kinase AKT. AKT, in turn, phosphorylates and modulates the activity of a number of proteins important in cell cycle

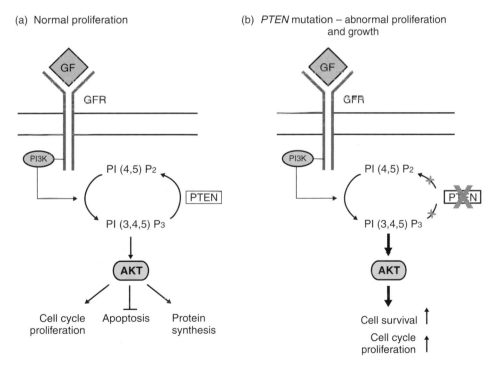

Fig. 7.15 The PI3K–AKT survival pathway – PTEN phosphatase. Growth factors by binding to their respective receptors (GFR) activate the phosphatidylinositol-3 kinase (PI3K), which converts phosphatidyl inositol 4,5 biphosphate (PIP$_2$) to phosphatidylinositol 3,4,5 triphosphate (PIP$_3$). PIP$_3$ activates AKT, which blocks apoptosis and induces protein synthesis and cell proliferation. PTEN is a phosphatase that antagonizes PI3K activity. In many human tumors, PTEN mutations lead to increased cell survival and cell proliferation.

control and cell survival. Consistent with being a tumor suppressor, enforced expression of *PTEN* in tumor cell lines induces growth arrest and/or apoptosis and in some cases prevents motility. Loss of *PTEN* function is common in several cancer types. Somatic inactivating mutations in *PTEN* are frequently found in glioblastoma, endometrial carcinoma, and prostate adenocarcinoma, but less frequently in other tumors such as melanoma, renal cell carcinoma, and head and neck squamous cell carcinoma. In addition, *PTEN* mutations are found in sporadic cancers of the breast, thyroid, lung, stomach, and blood. Inherited *PTEN* germ line mutations characterize three familial cancer syndromes: Cowden syndrome, Lhermitte–Duclos disease, and Bannayan–Zonana syndrome. In mice, deletion of *PTEN* in the germ-line induces early embryonic lethality, but conditional deletion of *PTEN*, for example in the cerebellum, leads to a phenotype that mimics Lhermitte–Duclos disease. Loss of *PTEN* is often associated with loss of the *INK4A* and *ARF* genes, suggesting that these key tumor suppressors functionally collaborate in cancer prevention.

Patched and Wnt signaling pathways

Patched (PTC), the receptor for sonic hedgehog (Shh), and Wnt, the ligand for Frizzled, were discovered in *Drosophila* and found to regulate patterning during development (Figs 7.16, 7.17; also discussed in Chapters 5

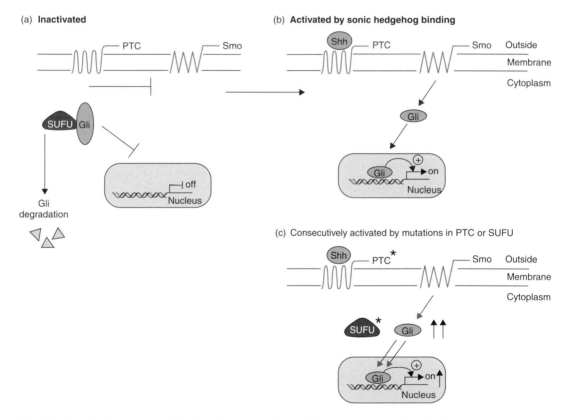

Fig. 7.16 The Patched pathway. (a) In the absence of ligand (Sonic hedgehog), Patched, a seven-transmembrane protein, inhibits Smoothened activity and the transcription factor GLI is inhibited by SUFU. (b) Upon SHH binding, Patched is activated, thereby relieving its suppression on Smoothened. Activated Smoothened induces the translocation of GLI to the nucleus and transcription of GLI-responsive genes. (c) Mutations in Patched or in SUFU, a negative regulator of GLI, activate the pathway constitutively in the absence of SHH.

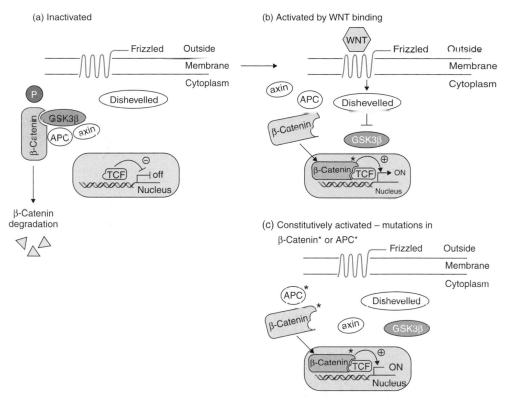

Fig. 7.17 The WNT pathway. (a) In the absence of its ligand, the seven-transmembrane receptor Frizzled is inactive. The kinase GSK-3β phosphorylates beta-catenin in complex with APC. β-Catenin is rapidly degraded, (b) When activated by binding of its ligand, Wnt, activated Frizzled activates Dishevelled, which inhibits GSK-3 β. This results in the dissociation of the APC complex, accumulation of beta-catenin and its translocation to the nucleus, where in complex with the transcription factor TCF it induces transcription of TCF-responsive genes. (c) Mutation of APC or β-catenin induces the constitutive translocation of β-catenin in the nucleus and transcription.

and 12). Several genes that regulate both pathways are disrupted in cancers. PTC is an unusual receptor because in the absence of its ligand, SHH, it suppresses the activity of another transmembrane protein, Smoothened, which itself represses the function and the translocation of a transcription factor Gli1 (Fig. 7.16b). Gli1 is the major transcription factor responsible for the expression of genes required for SHH-dependent proliferation, including Myc, and cyclins D1 and E. Gli itself is negatively regulated by a protein, SUFU (Suppressor of Fused), that is also a tumor suppressor in cerebellar granule neural precursors (Fig. 7.16b). Cells expressing the constitutively activated Smoothened (SMO) protein due to the inactivation of the *ptc* gene or carrying mutations in SUFU, no longer require SHH to induce Gli-mediated transcription and, as a consequence, receive unregulated signals for growth, which leads to cancer development (Fig. 7.16c). Germ-line mutation of *PTC1* is responsible for a familial cancer, Gorlin syndrome, which predisposes patients to a childhood cerebellar tumor (medulloblastoma) and a high incidence of basal cell carcinoma, a skin cancer (Fig. 7.16c). *PTC* mutations are also found in rhabdomyosarcoma, a muscle tumor. Mice lacking one copy of *Ptc* develop medulloblastoma, that loose the wild allele demonstrating that *Ptc* is a bona-fide tumor suppressor. Interestingly, SUFU mutations are also associated with medulloblastoma, and like *Ptc* mutations cause the constitutive activation of SHH signaling.

The Wnt family is composed of conserved secreted proteins that play a crucial role in patterning during development by regulating cell–cell contacts. WNT binds to its receptor Frizzled, and activates the Dishevelled protein, which regulates the transcriptional activator β-catenin through a complex containing the adenomatous polyposis coli (APC) tumor-suppressor protein, the kinase GSK-3β, and axin (Fig. 7.17a,b). This complex prevents β-catenin from entering the nucleus. Mutations in *APC*, *AXIN* and the β-catenin gene are responsible for familial adenomatous polyposis (FAP) and colorectal cancers, as well as sporadic medulloblastoma (Fig. 7.17c).

The GTPase NF-1

NF-1 is a GTPase that negatively regulates Ras signaling. It is found mutated in patients with neurofibromatosis type I, hence its name, but also gliomas. Ras is a critical mediator of proliferative signals in response to receptor activation, which explains why its unrestricted action induces hyperproliferative signals (see Chapter 6, section on Oncogenic *Ras*). Ras regulates cell proliferation and cell survival and links the cell cycle and the p53 pathway. Also one should note that PTEN negatively regulates the survival pathway activated by Ras (see Fig. 7.18).

TUMOR SUPPRESSORS AND CONTROL OF THE DNA DAMAGE RESPONSE AND GENOMIC STABILITY

Besides controlling proliferation and repairing DNA damage, cells must also maintain the integrity of their genome to avoid the accumulation of undesirable mutations, gene duplication, or chromosomal rearrangements (see also Chapter 10). In addition to the roles of p53 and the AT kinase ATM in these processes, other tumor-suppressor proteins also respond to DNA damage. They include the BRCA-1 and BRCA-2 proteins, which are recruited to DNA during homologous recombination and are found mutated in familial breast and ovarian cancers, and Chk2, a protein kinase activated by ATM that controls the G₁ checkpoint in the cells of patients with Li–Fraumeni syndrome, in which germ-line p53 is mutated (also discussed in

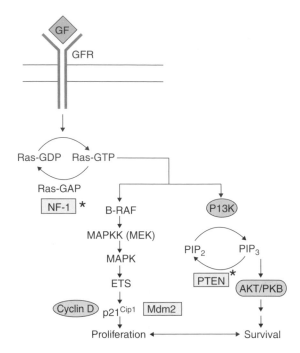

Fig. 7.18 The RAS signaling pathway. Ras is activated by growth factor receptor signaling. Activated RasGTP is inactivated to RasGDP by a GAP protein such as NF-1, a tumor suppressor protein mutated in neurofibromatosis. Activated RasGTP induces two pathways; the RAF/MAP kinase and the PI3K pathways. The RAF/MAP kinase pathway induces proliferation by activating the transcription of D-type cyclins and Mdm2. The PI3K pathway activates AKT to enforce cell survival. The phosphatase PTEN opposes the activity of PI3K.

Chapter 10). *MSH2* and *MLH1*, involved in DNA mismatch repair, are mutated in hereditary non-polyposis colorectal cancer, also called Lynch syndrome. Mutations in these two genes also induce other types of tumors, including endometrial, gastric, ovarian, and bladder cancers. The protein NSB1 and those in Fanconi anemia (FA) are involved in DNA repair and the control of DNA replication (S phase). NBSI is inactivated in Nijmegen breakage syndrome, and also inactivated in lymphoreticular malignancies.

QUESTIONS REMAINING

1 To know how many tumor suppressors exist, which ones regulate with specific cell type, and how loss

of one influences the loss of another in cancer progression.

2 The exact order with which genetic and epigenetic events influence cell proliferation and cell death by affecting tumor suppressors. Knowing exactly what happens to a cell from the time it receives hyper-proliferative signals to the time it loses a tumor suppressor would help identify the best targets for therapeutic intervention.

CONCLUSIONS AND FUTURE DIRECTIONS

Since the identification of the first tumor suppressor in the early 1980s, tumor suppressors have been recognized as critical contributors to cancer prevention. They participate in signaling pathways that regulate cell proliferation, DNA repair, programmed cell death (apoptosis), and the cell's architecture (cytoskeleton). They often collaborate to repress tumor formation and are found genetically altered in most cancers. Although many tumor suppressors have been characterized in the last 10 years, they are significantly outnumbered by the proto-oncogenes, which positively regulate cell growth. However, more tumor suppressors will likely emerge now that the complete genome sequence of humans, and other organisms, is available. Genetic experiments in the mouse have and will continue to provide confirmation of tumor-suppressor functions and insight into many types of human cancers. Indeed, mouse models for cancers are now used in preclinical trials to evaluate these inhibitory molecules as potential therapeutic agents. The advent of RNA interference, a new technology that allows specific, isolated gene suppression *in vivo*, will no doubt lead to the identification of new tumor suppressors that may serve as novel targets for therapeutic interventions in cancers and improve our current options for the prevention and treatment of these diseases.

QUESTIONS

1 A tumor suppressor is defined by:
  a. Loss of function by mutations.
  b. Loss of function by promoter methylation.
  c. Loss of function by deletion of the gene.
  d. Loss of function by genetic alterations.
  e. Loss of function by epigenetic alterations.

2 The role of tumor suppressors is to:
  a. Increase cell division.
  b. Restrain cell proliferation.
3 Tumor suppressors are induced by:
  a. Gamma-irradiation.
  b. DNA damage.
  c. Chemotherapeutic drugs.
  d. Ultraviolet (UV) damage.
  e. Excess of inappropriate proliferative signals.
4 Retinoblastoma in humans requires the loss of function of:
  a. Only one allele.
  b. Both alleles.
5 Epigenetic gene regulation is defined by:
  a. Silencing of gene transcription by promoter methylation.
  b. Gene mutation.
  c. Gene deletion.

ACKNOWLEDGMENTS

I would like to apologize to all my colleagues whose critical scientific contributions could not be cited due to space restriction. My thanks to the members of my laboratory who have encouraged me and provided keen insights and help during the writing of this chapter. I would specifically like to thank Drs Suzanne Baker, David Lasparga, and Gerard Zambetti, who offered many helpful comments and key criticisms. My thanks also go to Dr Susan Watson, who provided invaluable scientific editing. It would be remiss if I did not thank my collaborator and husband Dr Charles J. Sherr and my son Jonathan R. Sherr, who encouraged and supported me throughout. This work was possible thanks to funding from NCI grants CA71907 and CA96832, a Cancer Center support grant CA21765, the Children's Brain Tumor Foundation, and the American Lebanese-Syrian Associated Charities (ALSAC) of St Jude Children's Research Hospital.

BIBLIOGRAPHY

Two hits hypothesis/loss of heterozygosity

Knudson, A.G., Jr. (1971). Mutation and cancer: statistical study of retinoblastoma. *Proceedings of the National Academy of Sciences of USA*, **68**: 820–3.

Epigenetic events

Jones, P.A. (2003). Epigenetics in carcinogenesis and cancer prevention. *Annals of the New York Academy of Sciences*, **983**: 213–19.

The retinoblastoma protein

Classon, M. and Harlow, E. (2002). The retinoblastoma tumor suppressor in development and cancer. *Nature Review Cancer*, **2**: 910–17.

Trimarchi, J.M. and Lees, J.A. (2002). Sibling rivalry in the E2F family. *Nature Reviews: Molecular Cell Biology*, **3**: 11–20.

The p53 tumor suppressor

Lane, D.P. (1992). p53, guardian of the genome. *Nature*, **358**: 15–16.

Levine, A.J., Momand, J., and Finlay, C.A. (1991). The p53 tumor suppressor gene. *Nature*, **351**: 453–6.

Parant, J.M. and Lozano, G. (2003). Disrupting TP53 in mouse models of human cancers. *Human Mutation*, **21**: 321–6.

INK4/ARF

Hanahan, D. and Weinberg, R.A. (2000). The hallmarks of cancer. *Cell*, **100**: 57–70.

Lowe, S.W. and Sherr, C.J. (2003). Tumor suppression by *Ink4a-Arf*: progress and puzzles. *Current Opinions in Genetics and Development*, **13**: 77–83.

Ruas, M. and Peters, G. (1998). The p16^{INK4a}/CDKN2A tumor suppressor and its relatives. *BBA Reviews in Cancer*, **1378**: F115–77.

PTEN

Steelman, L.S., Bertrand, F.E., and McCubrey, J.A. (2004). The complexity of PTEN: mutation, marker and potential target for therapeutic intervention. *Expert Opinion on Therapeutic Targets*, **8**: 537–50.

Senescence

Campisi, J. (2001). Cellular senescence as a tumor-suppressor mechanism. *Trends in Cell Biology*, **11**: S27–31.

Sherr, C.J. and DePinho, R.A. (2001). Cellular senescence: mitotic clock or culture shock? *Cell*, **102**: 407–10.

PTEN and PI3K pathway

Baker, S.J. and McKinnon, P.J. (2004). Tumor-suppressor function in the nervous system. *Nature Reviews Cancer*, **4**: 184–196.

Parsons, R. (2004). Human cancer, PTEN and the PI-3 kinase pathway. *Seminars in Cell and Development Biology*, **15**: 171–6.

Sansal, I. and W.R. Sellers. (2004). The biology and clinical relevance of the PTEN tumor suppressor pathway. *Journal of Clinical Oncology*, **22**: 2954–63.

8

Apoptosis

Stella Pelengaris and Mike Khan

Death is a very dull, dreary affair, and my advice to you is to have nothing whatever to do with it.
William Somerset Maugham

I don't want to achieve immortality through my work... I want to achieve it through not dying.
Woody Allen

KEY POINTS

- Apoptosis is a specialized form of cell death that, in contrast to necrosis, does not generally provoke inflammation, unless something goes wrong with the clearing up.
- Apoptosis is a major barrier to oncogenesis, and suppression of apoptosis is a prerequisite for cancer formation.
- "Extrinsic" and "intrinsic" pathways for apoptosis have been described involving distinct upstream apical caspases: the extrinsic pathway involves activation of death receptors (e.g. FAS), while the intrinsic pathway involves the release of cytochrome c from the mitochondria.
- Once apoptotic signals are received by the cell, the final stages that lead to dismantling of the cell are executed by a subfamily of proteases known as caspases. Caspases are cysteine proteases that cleave hundreds of cellular proteins and ultimately lead to a series of morphological changes characteristic of apoptotic cell death.
- Inhibitors of apoptosis proteins (IAPs) are a family of proteins that are able to inhibit apoptosis by directly binding and inhibiting specific caspases. One of the mechanisms by which tumor cells are believed to acquire resistance to apoptosis is by overexpression of IAPs.
- The BCL-2 family of proteins comprises both proapoptotic and antiapoptotic members, the balance of which determines whether or not a cell commits suicide – apoptosis. These proteins have emerged as fundamental regulators of mitochondrial outer membrane permeabilization (MOMP), which is necessary for cytochrome c release. Importantly, upregulation of antiapoptotic BCL-2 family proteins is common to many human tumors.
- The tumor suppressor p53 can prevent cells from becoming malignant by inducing growth arrest or inducing apoptosis. There are several proapoptotic transcriptional targets of p53, such as BAX, NOXA, and PUMA, that promote cytochrome c release from the mitochondria.
- Oncogenic proteins, such as c-MYC, that possess mitogenic action also induce apoptosis unless a survival signal is also received – this mechanism may operate as a "failsafe" to prevent cancer formation if an oncogene becomes deregulated.
- Anticancer therapy can induce cellular senescence, and/or cell death by apoptosis or by nonapoptotic mechanisms, such as necrosis, autophagy, and mitotic catastrophe. However, since defects in apoptosis cause resistance to such therapy, restoring or activating apoptosis in tumors is an active area of cancer research.
- Cells in the final throes of apoptosis, display "eat me" signals, like phosphatidylserine, that are recognized by phagocytes such as macrophages that then dispose of the corpse, without provoking an inflammatory response.

INTRODUCTION

Cell death is inevitable and in the vast majority of cases under normal physiological conditions occurs by apoptosis. A staggering 50 billion or more cells die each day in the human adult; just to balance the books each one of us replaces around 70 kg of cells every year. Apoptosis, a characteristic form of cell death, is a fundamental process that occurs during development, maintenance of tissue homeostasis, and the elimination of damaged cells. However, deregulation of apoptosis may lead to diseases such as neurodegenerative disorders, diabetes, or cancer. It is now widely accepted that putative cancer cells must avoid apoptosis in order for tumors to arise. This discovery has led to a worldwide research effort into elucidating signal transduction pathways that mediate apoptosis as well as the mechanisms that have enabled cancer cells to inhibit apoptosis (e.g. through the loss of p53 tumor suppressor or overexpression of BCL-2 protein). The ultimate goal is to develop candidate drug molecules to target cancer such that their sensitivity to apoptosis is restored.

Although studies on cell death are thought to have been performed centuries ago by Aristotle, and later by Galen, who described the regression of larval and fetal structures during development (and probably first used the term "necrosis") it was not until after the formulation of the "cell theory" by Jacob Schleiden and Theodore Schwann in 1838, that the nineteenth-century pathologists, in particular those of the "German school," started to take interest in the process of cell death as a physiological phenomenon. Initially, it was appreciated that cells can die (in fact this was implicit from the time it was first realized that cells existed and were alive), but such death was assumed to be a passive response with cells as the victims of circumstances, such as poisons and trauma, including death of the organism, largely beyond their control. This view began to change in the latter half of the nineteenth century though it would be over a century later that cell death was first recognized by modern biology as a normal feature of multicellular organisms, and moreover that this might be a process involving the active participation of the cell. In 1842, Carl Vogt suggested that cell death could be an important part of normal development based on observations in amphibian metamorphosis. This was followed by Rudolf Virchow, widely regarded as the father of modern pathology, who in 1858 described what he called "degeneration, necrosis, and mortification." It was Flemming who in 1885 proposed that cells might actually die spontaneously – during studies of regressing ovarian follicles he observed nuclei that appeared to be breaking apart, a process that he called "chromatolysis." Intriguingly, Flemming's sketches, appearing nearly a century before the concept was introduced, are probably the earliest clear example of cells undergoing what we now call apoptosis. In 1914, Ludvig Gräper suggested that mitosis would need to be balanced by processes such as "Flemming's chromatolysis" that could keep proliferation in check. By the 1950s embryologists, such as Glucksmann, had clearly described physiological cell death, and DeDuve first proposed the concept of cell suicide. The term programmed cell death was introduced in 1964, to reflect the view that cell death during development was not an accidental occurrence but rather was part of a locally and temporally orchestrated plan.

Probably the key event in recent times was when in 1971, John F. Kerr, Andrew Wyllie, and Sir Alistair Currie introduced the term "apoptosis" (Greek: falling of leaves) to describe this phenomenon of cell suicide and thereafter founded the field of modern cell death research – although it was not until almost 20 years later that the idea that cells carried within them an intrinsic "suicide" program became generally accepted (see Box 8.1). The physiological cell suicide concept was proved molecularly in the 1990s by Horvitz (Nobel Prize winner – 2002) and colleagues, who showed in the worm *Caenorhabditis elegans* an intrinsic signaling pathway controlling the cell death of a group of specific neuronal cells during development (Fig. 8.1, Key Experiment).

Apoptosis is an ordered cell death process in which the entire cell is dismantled within the context of membrane-enclosed vesicles thereby preventing the release of intracellular components from the dying cell that would otherwise provoke an immune response. Apoptotic cells and their membrane-bound apoptotic fragments are rapidly phagocytosed by macrophages or neighboring cells before leakage of any cell contents (Plate 8.1). This contrasts with cells that

Box 8.1 Apoptosis – the birth of a new concept in biology

Developmental biologists have long been familiar with cell death in carving a vertebrate's digits and in insect metamorphosis. But today's cell-death community credits a paper by University of Edinburgh researcher Andrew Wyllie and his colleagues as the seminal work in the field (Kerr *et al.* 1972). They coined the term "apoptosis," writing that it plays "a complementary but opposite role to mitosis in the regulation of animal cell populations."

The paper created little excitement initially – it remained dormant for 10–15 years. Then it was gradually rediscovered and gained recognition as a generally important mechanism. What catapulted apoptosis into "hot topic" status was its meticulous demonstration in *C. elegans*, the tiny, transparent nematode worm, whose cell-death program removes precisely 131 of 1090 cells to form the adult (Sulston and Horvitz, 1977). This allowed Horvitz and colleagues to look at the process in the worm – the fact that you have a certain number of cells and can trace their developmental fates under a microscope, and ultimately isolate genes – and made the field flourish (see Fig. 8.1). The three critical genes regulating this cell fate included *CED-3* and *CED-4*, which are required for the execution of the death program such that inactivation of either gene by mutation prevents all 131 deaths. The third gene, *CED-9*, acts to suppress *CED-3/CED-4*-dependent apoptosis – inactivation of *CED-9* leads to massive ectopic cell death.

Meanwhile, little was known about cell death in other animals. When David Hockenbery, Stanley Korsmeyer, and colleagues discovered that the proto-oncogene *BCL2* (the mammalian homolog of the worm's *ced-9* gene) blocks programmed cell death, this and other work on *BCL2* refocused attention on apoptosis, contributing to the second wave of interest in the early 1990s.

essential role in animal development and tissue homeostasis in the adult. During development, apoptosis occurs during the sculpting of somatic structures – for example, the loss of interdigital webs during the formation of the digits, and the hollowing out of solid structures to create lumina such as the gut. In the developing nervous system, only half of the neurons formed receive sufficient survival signals from their target cells; the remaining neurons die by apoptosis. Although this appears an inefficient and wasteful process, in fact this developmental mechanism is believed to be vital for ensuring the correct innervation of target cells by appropriate neurons. Similarly, during development of the immune system, up to 95% of all cells die by apoptosis because they make unproductive or autoreactive antigen receptors. Apoptosis in development is often referred to as "programmed cell death." An example of when the regulation of apoptosis plays a key role in the adult is following childbirth – in preparation for lactation there is growth of breast tissue (proliferation > apoptosis). However, following lactation, dramatic regression of tissue occurs by apoptosis.

Apoptosis continues throughout the life of an animal in order to maintain tissue homeostasis, that is, a balance between cell proliferation and cell death. In fact, in an adult human, billions of cells die every hour! Maintaining tissue homeostasis is particularly important in those tissues that have a high cell turnover, such as epithelial (e.g. skin epidermis and gut) and hematopoietic tissues. It is therefore not surprising that perturbing the balance between cell proliferation and cell death can have a profound effect on the host. For example, excessive apoptosis is associated with degenerative diseases such as Alzheimer's disease, spinal muscular atrophy, Huntington's disease, and Parkinson's disease. Conversely, suppression of apoptosis is essential during the development of tumors – although highlighted at various points throughout the chapter, this deleterious effect is discussed in Chapter 6 in relation to oncogenes, the so-called cancer-causing genes. Here, we will describe the process of apoptosis, the signal transduction pathways controlling this form of cell death, and highlight ways in which normal apoptosis regulation becomes derailed in cancer. We will conclude by discussing the potential of targeting cell death pathways for developing cancer therapies.

die by necrosis, a process in which the cell and its organelles swell and rupture, releasing cellular contents into the surrounding tissue and therefore usually inciting an inflammatory response. Cells that die in this way often do so in response to profound physical, chemical, or genotoxic insult. In its physiological setting, the elimination of cells by apoptosis plays an

KEY EXPERIMENT Studies from H. Robert Horvitz, Nobel Prize Winner (2002) and colleagues

Cell lineage studies in the developing worm (John Sulston and Robert Horvitz).

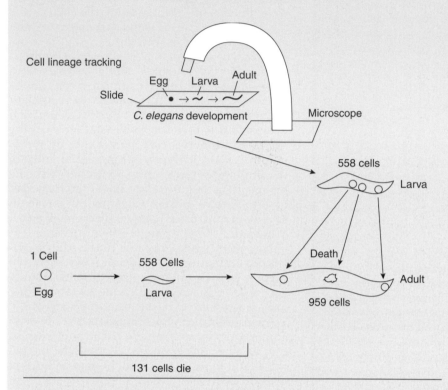

Fig. 8.1a The rediscovery of apoptosis as a generally important mechanism using *C. elegans*, the tiny, transparent nematode worm.

Newly hatched C. *elegans* larva were placed on a glass microscope slide dabbed with a sample of the bacterium *Escherichia coli* (nematode food). Then, using Nomarski differential interference contrast optics, individual cells within the living animal could be closely observed and followed as they migrated, divided, and died. In this way, the fate of every single cell from the larval stage to the adult worm could be determined and a cell lineage map generated.

Subsequently, the pattern of cell divisions between the single-celled fertilized egg and the newly hatched larva were tracked. This was far more difficult than tracking cell fates in larvae, in part because the process of embryonic morphogenesis involves a major cellular rearrangement of a ball of cells to generate a worm-shaped larva. Imagine watching a bowl with hundreds of grapes, trying to keep your eye on each grape as it and many others move! Nevertheless, all 558 nuclei were followed.

Together, these studies defined the first, and to date only, completely known cell lineage of an animal.

One interesting aspect of the cell lineage studies was that in addition to the 959 cells generated during worm development and found in the adult, another 131 cells generated during development were not present in the adult. These cells were absent because they had undergone programmed cell death.

In fact, 105 of the 131 cell deaths in C. *elegans* are in the nervous system.

Identifying genes involved in cell death

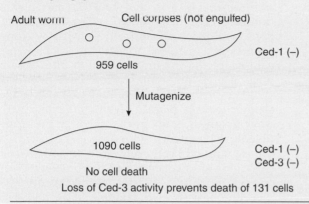

Fig.8.1b
Identifying cell
death genes.

The next step was to identify genes responsible for causing or preventing programmed cell death during worm development. Researchers mutagenized a *CED-1* (**cell d**eath abnormal) mutant worm in which the process of engulfment, or phagocytosis, which normally removes dying cells from the body of the animal, is defective. In *CED-1* mutants, although programmed cell death still occurs, dead or dying cells are not engulfed and so cell corpses persist and can be easily visualized in living individuals using Nomarski optics. By mutagenizing *CED-1* animals, it was hoped that mutants possessing abnormalities in the pattern of programmed cell deaths (as seen with Nomarski optics) would be generated. In this way, mutants could be identified in which the process of programmed cell death had not been initiated or in which the pattern of programmed cell deaths was altered. Indeed, a mutant was found in which no cell corpses could be seen. The gene defined by this mutant was called *CED-3*. It was shown that if *CED-3* activity is reduced or eliminated by mutation, essentially all 131 cells that normally die instead survive — hence the absence of cell corpses. *CED-3* protein, it was later discovered, was in fact a caspase.

A second mutant was then discovered (using different mutant animals) that prevented cell death. These worms proved to be defective in a new gene with properties essentially identical to those of *CED-3*. This killer gene was named *CED-4*. Some years later, in 1997, a protein similar to CED-4 was identified, called APAF-1 (**a**poptotic **p**rotease **a**ctivating **f**actor), with a domain with significant similarity to CED-4. APAF-1 is a proapoptotic human protein similar in both sequence and function to the *C. elegans* programmed cell death killer *CED-4*.

In a similar manner, the *CED-9* gene was discovered, and was later shown to be homologous to the human proto-oncogene, *BCL*-2 whose protein product serves to protect cells against cell death.

Genetic pathway for programmed cell death

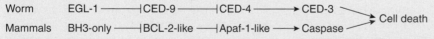

Fig.8.1c Genetic
pathway for
programmed cell death.

From these studies, the core molecular genetic pathway for programmed cell death in *C. elegans* has been identified; it shows great similarity to the mammalian pathway: EGL-1 (**egg**-laying abnormal) is similar to mammalian "BH3-only" proapoptotic proteins; CED-9 is similar to mammalian antiapoptotic BCL-2 protein; CED-4 is similar to APAF-1 (important for the activation of procaspase-9 on the apoptosome); CED-3 is similar to caspases.

APOPTOSIS AS A BARRIER TO CANCER FORMATION

As described in earlier chapters, cancers arise as a result of accumulation of multiple genetic and epigenetic lesions that allow cells to proliferate uncontrollably irrespective of exogenous mitogens, resist apoptosis, recruit a blood vasculature, invade surrounding tissues, and eventually metastasize.

Apoptosis is widely accepted as a tumor-suppressive mechanism: putative cancer cells must avoid cell death for tumors to arise. As described earlier, apoptosis occurs in normal adult tissues to eliminate cells that are damaged (e.g. DNA damage, hypoxia, nutrient limitation) or displaced (cells that have moved out of their normal environment – "anoikis," see Chapter 12) in order to ensure proper tissue homeostasis. However, we now know that cells acquiring a growth-promoting mutation – for example, overexpression of the oncogenes *c-MYC*, *E1A*, or *E2F* – are "sensitized" to apoptosis (Chapters 6 and 10, Figs 6.15 and 10.10). The importance of this cannot be overestimated, as such cells can be eliminated from the body and thus avoid becoming cancerous cells. The notion that a growth-deregulating oncogenic mutation *in vivo* could also possess an "in-built" tumor-suppressor function to hinder expansion of potentially malignant cells, is indeed an important one in the study of cancer biology. Apart from elucidating mechanisms that have enabled cancer cells to escape apoptosis, it also provides an avenue for future development of candidate drug molecules to target cancer (see the section on "Exploiting cell death and senescence in cancer control").

An emerging concept is that highly conserved pathways regulating cellular growth/replication and apoptosis may be coupled at numerous levels, even under normal physiological condition and that this plays a key role in the regulation of appropriate cell numbers and organ size during development and in maintaining these in the adult. In addition to the likely role of c-MYC, E2F, and others in this process, recent studies in the fruitfly *Drosophila* have revealed another important signaling pathway involving the large tumor suppressor LATS/WARTS protein kinase, which alongside two other proteins, HIPPO and SALVADOR, is also a regulator of both cell cycle exit and apoptosis. At least two mammalian homologs of LATS/WARTS are known, LATS1 and LATS2, and disruption of LATS1 or proteins, such as the HIPPO homolog MST2 or the MATS1 (Mob as tumor suppressor), required to activate LATS1, prevent apoptosis and increase tumor formation in mice. Moreover, some of these have now been shown to be disrupted in human cancers. As this chapter focuses mainly on the molecular pathways involved in apoptosis, we urge the reader to look at Chapters 4–7 for a broader overview of the regulation of cell cycle, growth, and the contributions of oncogenes and tumor suppressors, and cell death (or loss of cells) and cancer.

As the best-studied example of an oncogene with intrinsic tumor-suppressor activity, the various signaling pathways that c-MYC may activate to promote apoptosis are described later.

APOPTOSIS VERSUS NECROSIS

Apoptosis typically affects single cells that are aged, dysfunctional, or damaged by external stimuli. Unlike necrosis, it is an active, energy-requiring process leading to a characteristic series of morphological changes that accompany the degradation of the cell. Key differences between the two forms of cell death are described below.

During apoptosis, gross morphological changes – as shown in Plate 8.1 – can be seen under a microscope. Early features of apoptosis occur within minutes of the apoptotic trigger and during this phase, mitochondria, lysosomes, and cellular membranes remain fully intact. These features include: *chromatin condensation, DNA fragmentation (multiples of 180 bp), cell shrinkage, and dilatation of the endoplasmic reticulum.*

Later features of apoptosis are completed within hours, depending on cell type and tissue: *budding of cell membrane leading to the packaging of cellular components into vesicles – known as "apoptotic bodies," which are phagocytosed by macrophages and neighboring cells thus avoiding an inflammatory response* (Plate 8.1).

In contrast, necrosis affects groups of cells or whole tissue after extended damage induced by external stimuli, such as trauma, ischemia, and

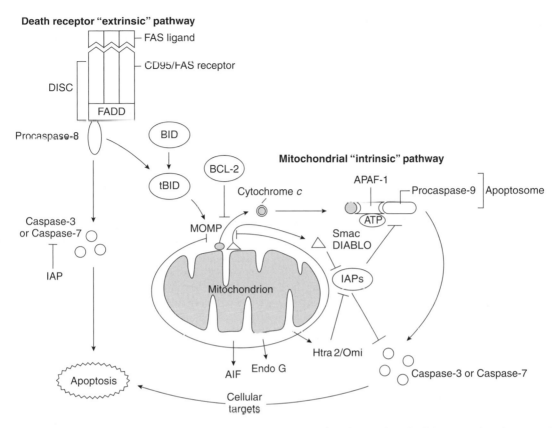

Fig. 8.2 The two pathways of apoptosis: death receptor (extrinsic) and mitochondrial (intrinsic) pathways. Here, the death receptor pathway is triggered following binding of FAS ligand (FASL) to the FAS receptor (FAS). FASL binding induces clustering of FAS, which in turn recruits the adaptor protein FADD (FAS-associated death domain) to form a complex called the death-inducing signaling complex (DISC). DISC then recruits and activates the initiator procaspase-8. Subsequently, activated procaspase-8 triggers a caspase cascade, activating downstream executioner caspases, such as caspase-3 and caspase-7, which ultimately kill the cell. However, caspase-8 can also activate the proapoptotic protein, BID (by cleavage to become truncated tBid), which promotes release of cytochrome *c* from the mitochondria – the "intrinsic" pathway.

The mitochondrial pathway can be triggered by a variety of cellular stresses (e.g. DNA damage, hypoxia, depleted survival factors, or deregulated oncogenes). Once released into the cytosol, cytochrome *c* associates with APAF-1 to create the apoptosome, a complex that activates procaspase-9. In the presence of cytochrome *c* and the nucleotide dATP/ATP, procaspase-9 is autocatalytically activated and can now go on to activate downstream executioner caspases, such as caspase-3 and caspase-7. Other proteins that are released from the mitochondria include Smac/DIABLO (and probably Htra2/Omi), which bind to, and inhibit, IAPs thus preventing caspase-9 and caspase-3 inhibition.

high-dose irradiation. This energy-independent form of cell death occurs over 12–24 hours, with loss of function of mitochondria and endoplasmic reticulum resulting in a dramatic breakdown of energy supply leading to cellular, nuclear, and organellar swelling, and ultimately rupture of the plasma membrane. This final event allows the release of lysosomal enzymes that attack neighboring cells and surrounding tissue thus triggering an inflammatory response. In contrast to apoptosis, there is nonspecific degradation of DNA.

THE PATHWAYS TO APOPTOSIS

Generally, there are two pathways that trigger apoptosis – the extrinsic and intrinsic apoptotic pathways. The extrinsic pathway, otherwise known as the "death receptor pathway," is activated by the engagement of death receptors on the cell surface. In contrast, the intrinsic pathway involves the release of cytochrome *c* (and other proteins) from the mitochondria. Whichever pathway is taken (Fig. 8.2), both lead to the activation of various caspases – enzymes responsible for the demise of the cell. These crucial enzymes are discussed later in their own section. It is important to emphasize that in vertebrates the majority of apoptosis proceeds through the mitochondrial pathway.

Death receptor pathway (extrinsic)

This pathway is initiated by extracellular hormones or agonists that belong to the tumor necrosis factor (TNF) superfamily, including TNFα, FAS/CD95 ligand, and APO2 ligand/TRAIL. These agonists recognize and activate their corresponding receptors (TNF/NGF receptor family), such as TNFR1, FAS/CD95, and APO2. For example, as shown in Fig. 8.2, binding of FAS ligand (FASL) to the FAS receptor (FAS) induces clustering of FAS, which in turn recruits the adaptor protein, FADD (FAS-associated death domain) to form a complex called the death-inducing signaling complex (DISC). DISC then recruits and activates the initiator procaspase-8, probably by bringing the procaspases close enough in proximity – "close proximity" model – so that they can cleave against each other. Subsequently, these activated initiator caspases trigger a caspase cascade activating downstream executioner caspases, such as caspase-3 and caspase-7, that ultimately kill the cell. When large amounts of caspase-8 are formed at the DISC, then apoptosis proceeds via direct cleavage of procaspase-3 independently of the mitochondria (see mitochondrial pathway). However, caspase-8 can also activate the proapoptotic protein, BID, which promotes release of cytochrome *c* from the mitochondria – the "intrinsic" pathway described below (Fig. 8.2). This "crosstalk" between the death receptor and mitochondrial pathways can occur in cells when DISC formation and active caspase-8 are insufficient to activate procaspase-3 independently of the mitochondria.

Apoptotic cell death mediated via death receptors is critical for normal immune system function, for example, mutations on *FAS* and *FASL* in humans can lead to a complicated immune disorder known as the autoimmune lymphoproliferative syndrome (ALPS), a similar phenotype to that seen in mice with *FAS* and *FASL* mutations.

Importantly, in some cell types, apoptotic response to anticancer therapy implicates the death receptor FAS (see previous section). Various anticancer drugs can activate the death receptor pathway by enhancing the expression of FAS and FASL. Interaction of FASL with FAS at the cell surface defines an autocrine/paracrine pathway similar to that observed in activation-induced cell death in T lymphocytes. However, it is possible that FASL is not crucial for drug-induced apoptosis since apoptosis is not suppressed by antagonist antibodies or molecules that prevent FASL binding to FAS. It is likely that certain anticancer drugs exert their apoptotic effects by inducing clustering of FAS receptor at the cell surface of tumor cells in the absence of FASL.

The mitochondrial pathway (intrinsic)

It is now known that the mitochondria play a central role in most cases of apoptotic cell death in vertebrate cells. The key and defining event in the mitochondrial pathway is MOMP (mitochondrial outer membrane permeabilization), as a result of which various proteins normally confined to the mitochondrion are released and activate the subsequent dismantling of the cell (Fig. 8.2). The first described of these proteins appearing in the cytoplasm in apoptotic cells was cytochrome *c*, an observation which was rather puzzling at first, not least because cytochrome *c* is an essential protein in energy production within cells and is usually located inside the mitochondria (in the space between the inner and outer mitochondrial membranes). However, the role of cytochrome *c* in activating apoptosis was confirmed by two findings: the first was the identification of its downstream binding partner, APAF-1 (apoptotic protease-activating factor 1) present in the cytosol, and the second was the demonstration that the antiapoptotic protein, BCL-2 (described later), inhibits cell death by preventing release of cytochrome *c* from the mitochondria.

In a recent review article by Doug Green in the journal *Science* a very apt analogy was used to describe the mitochondrial pathway – "we have an upstairs/downstairs situation where at first pass most of the aristocratic decisions are made before MOMP (upstairs) and the workmanlike consequences occur thereafter (downstairs)". A key question is what happens to promote MOMP and thence the release of cytochrome *c* and other proteins from mitochondria? Without a doubt, the BCL-2 family of proteins are key regulators of MOMP and will be discussed in more detail later under their own heading, but it is worth continuing the upstairs/downstairs analogy at this point. As we will see, BAX and BAK are the butlers of this piece; they operate above and below stairs and most importantly we know by literary convention that they did it. We also know something of the motive; the normally affable BAX and BAK are driven to "commit MOMP" by imbalances between proapoptotic and antiapoptotic BCL-2 family members upstairs. Stress of various kinds can trigger activity of proapoptotic (BH3-only) proteins in part by removing the calming embrace of antiapoptotic BCL-2 and BCL-X$_L$. The nature of the upstairs activating signals, which lead to changes in expression and/or activation of various BCL-2 family members, will vary depending on cell type and what stresses the cell is under, for example, hypoxia, depleted survival factors, DNA damage, or if the cell has acquired an oncogenic mutation (e.g. in c-MYC leading to deregulated expression).

Once MOMP occurs, cytochrome *c* can associate with APAF-1 in the cytoplasm to create the apoptosome, a complex that activates procaspase-9. In the presence of cytochrome *c* and the nucleotide dATP/ATP, procaspase-9 forms oligomers and becomes autocatalytically activated; active caspase-9 can now go on to activate downstream executioner caspases, such as caspase-3 and caspase-7 (Fig. 8.2), by cleavage at specific sites. Since formation of the apoptosome plays a crucial role in mediating apoptosis, it deserves a more thorough explanation – see below. The regulation of apoptosis downstairs is complex and involves numerous activation steps and also protein stability. Several different inhibitors of caspases, the inhibitor of apoptosis proteins (see "The IAP family" below),

have been described that inhibit apoptosis by targeting caspases for proteasome degradation. Conversely, various proteins released from the mitochondrial intermembrane space, alongside cytochrome *c* (Fig. 8.2), allow MOMP to circumvent the action of the apoptosis-inhibiting IAPs. Thus, MOMP releases two inhibitors of IAPs, DIABLO/Smac (second mitochondria derived activator of caspase) and Omi/HtrA2. In fact, other important proteins are released by MOMP, including two that can contribute to apoptosis even in the absence of caspase activation, AIF (apoptosis-induction factor), involved in chromatin condensation and large-scale DNA degradation, and endonuclease G (EndoG), that might aid the caspase-activated DNase (CAD) in nucleosomal DNA fragmentation. In some circumstances MOMP may even result in release of a small proportion of procaspase molecules.

THE APOPTOSOME – "WHEEL OF DEATH"

The three-dimensional structure of the apoptosome has now been solved using cryoelectron microscopy technology. The structure has given insight into how the apoptosome assembles, how it might activate procaspase-9, and why activation of this procaspase is distinct from the conventional caspase activation mechanism (i.e. proteolytic cleavage of effector caspase), which is discussed in more detail in the section entitled "Caspases – the initiators and executioners of apoptosis."

Once cytochrome *c* is released from mitochondria into the cytosol, it associates with APAF-1 to create the apoptosome, a complex that activates procaspase-9. In mammalian cells, APAF-1 is cytosolic and, importantly, its activity is restrained by its long carboxy-terminal extension containing 13 repeats of the WD40 motif (Fig. 8.3). The WD40 motif is a conserved protein domain approximately 40 residues long that has a characteristic tryptophan-aspartate motif. In an individual APAF-1 molecule, two groups of WD40 repeats in the carboxy-terminal region are thought to keep the protein inactive until cytochrome *c* engages the repeats (Fig. 8.4). After association with cytochrome *c*, APAF-1 switches from a rigid conformation ("closed") to a more flexible one ("open") such that the

Fig. 8.3 Functional domains of the APAF-1 protein. APAF-1 contains an N-terminal CARD by which it interacts with procaspase-9; a nucleotide binding domain (NBD); and a long carboxy-terminal extension containing 13 repeats of the WD40 motif. The WD40 motif is a conserved protein domain, approximately 40 residues long, that has a characteristic tryptophan-aspartate motif, and is thought to negatively regulate APAF-1.

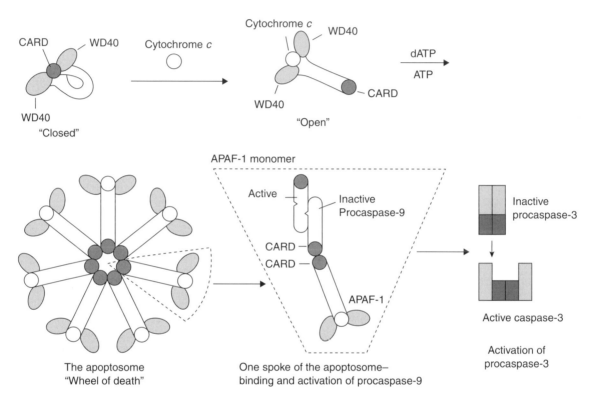

Fig. 8.4 Formation of the apoptosome. In an individual APAF-1 molecule, two groups of WD40 repeats in the carboxy-terminal region are thought to keep the protein inactive until cytochrome *c* engages the repeats. Association with cytochrome *c* causes APAF-1 to convert from a "closed" conformation to a more "open" one, thus allowing the nucleotide dATP/ATP binding activity to be greatly facilitated. This binding triggers formation of the active seven-span symmetrical "wheel of death" – the apoptosome – via interaction among the N-terminal CARD of the individual APAF-1 molecules. The apoptosome subsequently recruits procaspase-9 into its central hub through CARD–CARD domain interaction – between procaspase-9 and APAF-1 molecules. An inactive procaspase-9 monomer on one spoke of the apoptosome is thought to recruit another monomer to create a dimer with a single active site. This active caspase activates downstream executioner caspases, such as caspase-3 and caspase-7.

nucleotide dATP/ATP binding activity is greatly facilitated. This binding triggers formation of the active seven-span symmetrical wheel-like structure ("wheel of death") – the apoptosome – via interaction among the *N*-terminal caspase recruitment domains (CARD) of the individual

APAF-1 molecules (Fig. 8.4). The apoptosome subsequently recruits procaspase-9 into its central hub through CARD–CARD domain interaction – between procaspase-9 and APAF-1 molecules – which brings about a conformational change of procaspase-9. This enzyme is now in its active form and can go on to activate downstream executioner caspases, such as caspase-3 and caspase-7, that will eventually lead to cell death.

CASPASES – THE INITIATORS AND EXECUTIONERS OF APOPTOSIS

Once apoptotic signals are received by the cell, the final stages that lead to dismantling of the cell are executed by a subfamily of proteases known as caspases. Caspases are highly selective cysteine proteases that have a preference for cleaving proteins after aspartate residues. This specificity ensures that apoptosis is primarily a set of limited proteolytic cleavages rather than a generalized degradative process, resulting in caspase-dependent cleavage of hundreds of cellular proteins and ultimately leading to a series of morphological changes characteristic of apoptotic cell death.

Caspases can be grouped into two categories: "initiator" caspases and "effector" (or executioner) caspases. As mentioned earlier (Fig. 8.2), initiator caspases (e.g. caspase-2, -8, and -9) are activated as they bind to their appropriate adaptor molecules (APAF-1, FADD; see below) after which they cleave and thereby activate downstream ("effector") procaspases and various proteins. The downstream "effector" (or executioner) caspases (e.g. caspase-3, -6, and -7) go on to degrade many cellular substrates (Box 8.2) that result in the characteristic features of apoptotic cell death.

Caspases are synthesized and stored as inactive precursors, procaspases, in order to protect the cell from ultimate proteolytic events. Each procaspase consists of a prodomain at its amino (NH$_2$)-terminus, a large subunit, and a small subunit (Fig. 8.5). The length of the prodomain varies among caspases (Table 8.1) – long prodomains are characteristic of initiator caspases and contain a death effector domain (DED – procaspase-8, -10) or caspase recruitment domain (CARD – procaspase-9). These domains mediate the homophilic interactions (i.e. CARD–CARD and DED–DED) between procaspases and their adaptor proteins (e.g. APAF-1 with procaspase-9; FADD with procaspase-8). An example of this can be seen in Fig. 8.4 during formation of the apoptosome.

Box 8.2 Cellular substrates of effector caspases (in particular caspase-3)

Effector caspases: (i) cleave structural components of the cytoskeleton and the nuclear membrane such as actin, cytokeratins, and lamins; (ii) cause phosphatidylserine to be exposed on the outside of the cellular membrane thus promoting phagocytosis by macrophages and neighboring cells; (iii) counteract the apoptosis-inhibiting effect of BCL-2 and BCL-x$_L$ proteins; (iv) inhibit genes that regulate the repair of DNA lesions during cell cycle such as *MDM2* and *RB*; (v) inactivate enzymes responsible for stability, integrity, and repair of DNA, such as poly-(adp-ribose)polymerase (PARP), DNA-PK, and DNA replication factor 140; and (vi) cleave ICAD (inhibitor of the caspase-activated DNase) and inactivate CAD, which causes the characteristic oligonucleosomal fragmentation of the chromatin.

It is now recognized that activation of initiator caspases occurs following oligomerization of adaptor proteins (APAF-1, FADD), which in turn leads to recruitment and autoactivation of procaspases (procaspase-8 and -9, respectively). This mode of activation differs from that required for executioner procaspases, which typically require proteolytic cleavage. Executioner procaspases (procaspases-3, -6, -7) have short prodomains that seem to inhibit caspase activation. These caspases exist constitutively as homodimers – both before and after activation cleavage. Executioner procaspases are activated by initiator caspases, for example, procaspase-3 and -7 can be activated by caspases-6, -8, -9, and -10. During activation, the executioner procaspase is cleaved at specific Asp residues to yield a short inhibitory prodomain, and a large and a small subunit, termed homodimers p20 and p10, respectively (Fig. 8.5). Subsequently, p20 and p10 homodimers (refers to subunit size, in kDa) interact in a heterodimer – the association of two heterodimers forms the proteolytic tetramer with the two adjacent p10 subunits surrounded by two large subunits. The catalytic activity of an executioner caspase is increased by several orders after such cleavage.

Unlike executioner procaspases, our understanding of how initiator procaspases are

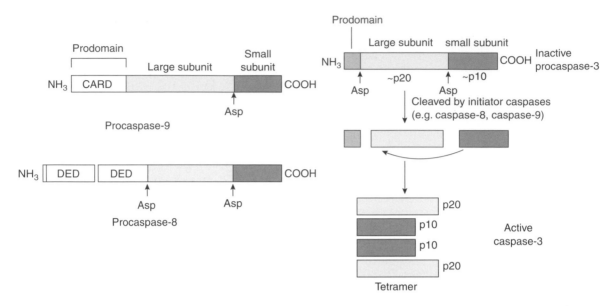

Fig. 8.5 Initiator and executioner caspases. Caspases are highly selective **c**ysteine proteases that have a preference for cleaving proteins after aspartate residues (arrows). Initiator procaspases, such as procaspase-8 and procaspase-9, possess long prodomains that contain a protein interaction domain, such as the death-effector domain (DED) and CARD, respectively. These domains mediate the homophilic interactions (i.e. CARD–CARD and DED–DED) between procaspases and their adaptor proteins (e.g. APAF-1 with procaspase-9; FADD with procaspase-8) and are indispensible to the activation of initiator caspases. Executioner procaspases, such as procaspase-3, have short prodomains that seem to inhibit caspase activation. Procaspase-3 is activated by initiator caspases: it is cleaved at specific Asp residues to yield a short inhibitory prodomain and a large and a small subunit, termed homodimers p20 and p10, respectively. Subsequently, p20 and p10 homodimers interact in a heterodimer – the association of two heterodimers forms the proteolytic tetramer with the two adjacent p10 subunits surrounded by two large subunits.

Table 8.1 Procaspases subgrouped on the basis of domain composition

Long prodomain (DED, CARD) – apoptotic initiator type:	Caspases-2, 8, 9, 10
Short prodomain – apoptotic effector (executioner) type:	Caspases-3, 6, 7
Long prodomain (CARD) – cytokine activator type:	Caspases-1, 4, 5

Note: The human genome encodes 11 caspases, and these can be divided into subgroups depending on inherent substrate specificity, domain composition, or presumed role *in vivo*.

activated is far from complete. With regard to the activation of procaspase-9 by the apoptosome (described previously), two models have been proposed: the "induced proximity" model and the "proximity-induced dimerization" model. The first model states that the initiator procaspases autoprocess themselves when they are brought into close proximity of each other. However, the mechanism underlying such caspase activation is not known. The second model states that procaspase-8 and procaspase-9 are activated on dimerization, which is facilitated by the apoptosome and DISC complexes, respectively. Procaspase-9 can be activated without processing, apparently because of the flexibility allowed by the unusually long linker between its large and small subunits. An inactive procaspase-9 monomer on one spoke of the apoptosome is presumed to recruit another monomer (homodimerization) in an antiparallel fashion to create the asymmetric dimer having a single active site (Fig. 8.4). This is a unique feature as only one of the two constituent polypeptides becomes active. Importantly, most of the active caspase-9 remains complexed to APAF-1 and the released caspase-9 has minimal activity thus indicating that APAF-1 is not merely a scaffold but, within the apoptosome, acts as a holoenzyme. In fact, after associating with the apoptosome, there is a dramatic increase in the catalytic activity of caspase-9 (up to 2000-fold) of both processed (cleaved) and unprocessed caspase-9. Thus, for caspase-9 at least, activation has little to do with cleavage; rather, it occurs through apoptosome-mediated enhancement of the catalytic activity of caspase-9.

THE IAP FAMILY

Caspases kill cells by irreversibly cleaving a host of cellular components. It is important then for healthy cells to prevent aberrant activation of these enzymes and thus unwanted suicide. To achieve this, cells have evolved a system of checks and balances, at the heart of which is the IAP (inhibitors of apoptosis proteins) family of proteins.

IAPs are a family of structurally related proteins that were initially identified in the genome of baculovirus on the basis of their ability to suppress apoptosis in infected host cells. These proteins were subsequently found in mammals and fruitflies, but not in nematodes. To date, eight human IAPs have been identified: c-IAP1, c-IAP2, NAIP (neuronal apoptosis inhibitory protein), Survivin, XIAP (crosslinked IAP), Bruce/Apollon, ILP-2 (IAP-like protein-2), and Livin/ML-IAP (melanoma IAP). These proteins are able to inhibit apoptosis induced by a variety of stimuli, a process mediated mainly by direct binding and inhibition of specific caspases (Figs 8.2 and 8.9). Interestingly, although caspase-9 binds to several IAPs, it is primarily inhibited by XIAP.

The hallmark of IAPs is the baculoviral IAP repeat (BIR) domain, an 80-amino-acid zinc-binding domain. XIAP, c-IAP1, and c-IAP2 contain three BIR domains, each domain having a distinct function. For example, in XIAP, the third BIR domain (BIR3) potently inhibits the activity of processed caspase-9, whereas the linker region between BIR1 and BIR2 specifically targets caspase-3 and caspase-7 (Fig. 8.6). On the other hand, Survivin contains a single BIR domain and does not inhibit caspase activity *in vitro*.

Inhibition of IAPs

As mentioned earlier, during apoptosis one of the proteins released from the mitochondria is Smac/DIABLO. This protein binds to IAPs and thus relieves IAP-mediated caspase inhibition (Figs 8.2 and 8.9). Another mitochondrial protein, Htra2/Omi, released during apoptosis, has been shown to antagonize XIAP-mediated inhibition of caspase-9 at high concentrations. However, whether this occurs under physiological settings is unclear.

One of the mechanisms by which tumor cells are believed to acquire resistance to apoptosis is by overexpression of IAPs. Of all IAPs, Survivin involvement in tumors has been most extensively studied. Fetal and embryonic tissues show high expression of Survivin, while it is undetectable in normal, fully differentiated tissues. This is in marked contrast to high levels of expression observed in a wide range of malignancies. High levels of Survivin that correlated with clinical status have been reported in colorectal cancer, esophageal cancer, and soft tissue sarcoma.

Survivin expression has also been associated with poor prognosis in several CNS tumors including glioma. Although the prognostic relevance is still unclear, high levels of Survivin were also detected in carcinoma of

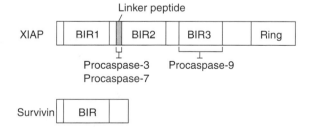

Fig. 8.6 Proteins of the IAP family. XIAP, an IAP family member, contains three baculoviral IAP repeat (BIR) domains. The third BIR domain (BIR3) potently inhibits the activity of processed caspase-9, whereas the linker region between BIR1 and BIR2 specifically targets caspase-3 and caspase-7. On the other hand, Survivin contains a single BIR domain and does not inhibit caspase activity *in vitro*.

the stomach, pancreas, liver, uterus, and in pheochromocytoma.

KEY PLAYERS IN REGULATING CYTOCHROME c RELEASE – BCL-2 FAMILY, fp53

BCL-2 family

The BCL-2 family of proteins comprises both proapoptotic and antiapoptotic members, the balance of which determines whether or not a cell commits suicide – apoptosis. These proteins have emerged as fundamental regulators of MOMP, described earlier, which is necessary for cytochrome c release from the mitochondria. Importantly, upregulation of antiapoptotic BCL-2 family proteins is common to many human tumors (see also Chapter 6), although loss of function of proapoptotic proteins such as BAX and BAK can also occur (see below). Dysregulation of these proteins can confer a survival advantage, preventing cell death in response to a variety of apoptotic stimuli that usually

function to prevent tumorigenesis (such as DNA damage, hypoxia, detachment from extracellular matrix). As discussed in the last section of this chapter, p53-inactivating mutations, as well as blockade of apoptosis by BCL-2 family proteins, reduces sensitivity to antineoplastic agents that kill cells via apoptosis, thereby limiting their clinical efficacy.

In mammals, the BCL-2 family consist of at least 20 members, all of which share at least one conserved BCL-2 homology (BH) domain (Fig. 8.7).

Prosurvival (antiapoptotic) members: BCL-2, BCL-x_L, A1, BCL-w, MCL-1　These proteins have four BH domains (BH1 to BH4). A hydrophobic groove is formed by residues from BH1, BH2, and BH3, which can bind the BH3 domain of an interacting BH3-only (proapoptotic) relative. In this way, BH3-only proteins (with the possible exception of BID) are thought to promote apoptosis by binding to and neutralizing their antiapoptotic relatives. Importantly, BH3-only proteins cannot kill in the absence of BAX and BAK, strongly indicating that they must act upstream in the

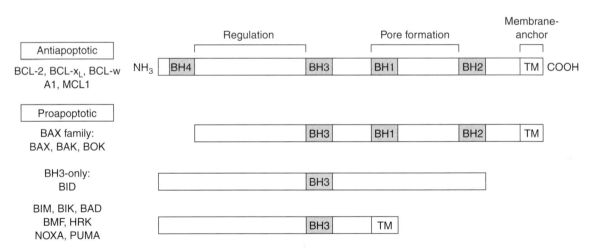

Fig. 8.7 The BCL-2 family. The BCL-2 family of proteins can be subdivided into three categories according to their structure and function: the antiapoptotic members promote cell survival, whereas members of the BAX family and 'BH3-only' categories are proapoptotic and thus promote apoptosis. The BCL-2 family members possess at least one of four conserved motifs known as BCL-2 homology domains (BH1–BH4). Most members have a carboxy-terminal hydrophobic domain (TM) that aids association with intracellular membranes. Pore formation is enabled by residues of BH1, BH2, and BH3. During apoptosis, the proapoptotic members are activated, and presumably undergo a conformational change leading to exposure of the BH3 domain – an interaction domain that is necessary for their killing action. In this way, BH3-only proteins promote apoptosis by directly binding to their antiapoptotic relatives.

same pathway. Pro-survival proteins may be divided into two subclasses (Mcl-1 and A1, and another comprising BCL-2, BCL-x$_L$, BCL-w). Interestingly BAK activation may be restrained by at least two of these, one member from each subclass, namely BCL-x$_L$ and MCL-1.

Proapoptotic members These consist of two groups:

1. *BAX family*: BAX, BAK, BOK. These proteins have three BH domains (BH1 to BH3) that are similar in sequence to those in BCL-2, and are thus sometimes referred to as BH123 or multidomain proteins. In contrast to BH3-only members (see below), which act as direct antagonists of the antiapoptotic members, the BAX family proteins act further downstream probably in mitochondrial disruption. BH123 proteins are key effectors of MOMP, as cells from mice lacking the two major forms, BAX and BAK, are resistant to multiple apoptotic stimuli and fail to undergo MOMP.

2. *BH3-only family*: BID, BIM, BIK, BAD, BMF, HRK, NOXA, PUMA. As the name implies, these proteins have only the short BH3 motif — an interaction domain that is both necessary and sufficient for their killing action. These proteins are direct antagonists of the antiapoptotic members, and may also serve at least in part to activate BAX and BAK, though this latter remains contentious. Broadly, BIM, PUMA, and truncated BID bind to all pro-survival BCL-2 family proteins; NOXA binds only to MCL-1 and A1; BAD and BMF bind only to BCL-2, BCL-x$_L$ and BCL-w.

Many members of the BCL-2 family (except A1 and many of the BH3-only proteins) have a carboxy-terminal hydrophobic domain that aids association with intracellular membranes.

How do the BCL-2 family regulate apoptosis?

Antiapoptotic members, BCL-2, and its closest homologs BCL-x$_L$ and BCL-w, are potent inhibitors of apoptosis in response to many stress signals (the less well-studied A1 and MCL-1 seem to have weaker antiapoptotic activity). The widespread view is that these members block apoptosis by guarding mitochondrial integrity and thus prevent the release of cytochrome *c*, as well as other proapoptotic

molecules (Smac/DIABLO and Omi/Htra2) – see next section. Their hydrophobic carboxy-terminal domain (Fig. 8.7) aids association with the cytoplasmic face of three intracellular membranes: the outer mitochondrial membrane, the endoplasmic reticulum (ER), and the nuclear envelope. BCL-2 is an integral membrane protein in all cells, whereas BCL-x$_L$ and BCL-w only become tightly associated with the membrane after a cytotoxic signal, possibly through conformational change. These antiapoptotic (or prosurvival) proteins prevent the release of cytochrome *c* from the mitochondria and hence prevent formation of the apoptosome. In addition, the release of the protein Smac/DIABLO from the mitochondria is prevented, which would otherwise have acted to inhibit IAPs – proteins that prevent caspase-9 and caspase-3 activation (discussed above) – see Figs 8.2 and 8.9.

Proapoptotic proteins, BAX and BAK, are thought to function mainly at the mitochondrion and are widely believed to induce the permeabilization of the outer mitochondrial membrane (MOMP) allowing efflux of cytochrome *c* (Figs 8.2 and 8.9). The mechanism for MOMP is, however, controversial – see below. In healthy cells, BAX is a cytosolic monomer but changes conformation during apoptosis, presumably flipping out its carboxy-terminus and then forming homo-oligomers within the outer mitochondrial membranes. BAK, however, is an oligomeric integral mitochondrial membrane protein but also changes conformation during apoptosis and might form larger aggregates. It is unclear, though, how the oligomers form.

In healthy cells, individual BH3-only proteins are held in check (see Fig. 8.8), whereas NOXA, PUMA, and HRK are controlled primarily at the transcriptional level (see "Tumor-suppressor p53" below). BH3-only protein, BID, promotes death by activating BAX and BAK – the cleaved BID protein (e.g. by caspase-8 or granzyme-B) migrates to the mitochondria following lipid modification, and if BAX or BAK is present, BID rapidly triggers cytochrome *c* release and apoptosis (Fig. 8.9). It is possible that BID acts by inducing oligomerization of BAX and BAK; however, experiments have shown that the oligomers do not contain BID.

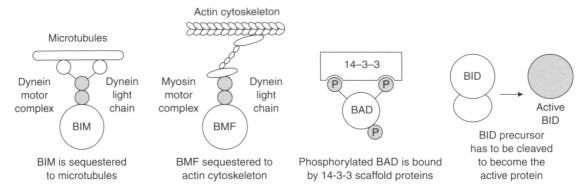

Fig. 8.8 BH3-only proteins kept inactive in healthy cells. BIM and BMF are sequestered by binding to dynein light chains that are associated with the microtubules and actin cytoskeleton. BAD is kept inactive when phosphorylated by kinases, such as AKT and protein kinase A, through being bound by 14-3-3 scaffold proteins (proteins that provide a platform for the assembly of other proteins). BID is inactive until proteolytically cleaved, for example, by caspase 8 or granzyme B.

MITOCHONDRIAL OUTER MEMBRANE PERMEABILIZATION (MOMP)

Proapoptotic BCL-2 members, BAX and BAK, are believed to contribute to the permeabilization of the outer mitochondrial membrane (MOMP) thus allowing the release of cytochrome *c* and other proteins involved in apoptosis (described earlier). Although the mechanism is unclear, one model is that BAX and BAK form channels in the mitochondrial membrane (Fig. 8.9). In support of this model, BAX oligomers can form pores in liposomes that allow passage of cytochrome *c*. Alternatively, BAX might interact with components of the existing permeability transition pore, for example, the voltage-dependent anion channel, to create a larger channel – no such complexes, however, have been found as yet.

How do antiapoptotic BCL-2 proteins antagonize the BAX-like proteins in order to prevent apoptosis? The answer is still uncertain – although it has been shown experimentally that high concentrations of antiapoptotic BCL-2 proteins prevent BAX oligomerization and channel formation, it is not known whether this occurs in a physiological setting. It is also possible that BCL-2 and BCL-x$_L$ bind to APAF-1 to prevent it from activating caspase-9.

Lastly, MOMP may also be activated by a different mechanism, not involving BAX/BAK, but rather the formation of a permeability transition pore complex (PTCP) consisting of two putative pore-forming proteins – VDAC (voltage-dependent anion channel) in the outer membrane and the ANT (adenine nucleotide translocator) in the inner membrane. However, the role of this in physiological apoptosis induction in vertebrates is controversial.

Finally, given that replication and apoptosis seem to be inextricably linked it is tempting to speculate that this might in some way involve the (division) fission of mitochondria, which is an obligatory feature of cell division. Although, this will require some understanding of how under some circumstances mitochondrial dynamics and fission are associated with MOMP and under others only with normal cell replication. Interestingly, MOMP and fission are in some ways connected. Thus, BAX interacts with proteins involved in fission such as endophilin B1, but this is a very new and not well understood area. Moreover, many cases of MOMP appear not to involve fission of mitochondria, or fission occurs as a later downstream consequence of apoptosis rather than as a contributory cause. This will likely be an interesting area of research in the next few years.

New concepts

It has widely been believed that all stress-induced apoptosis (regulated by the BCL-2 family) proceeds through mitochondrial disruption and activation of procaspase-9 via

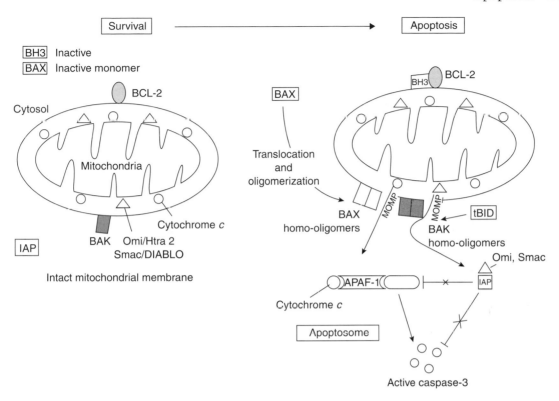

Fig. 8.9 Regulation of MOMP by the BCL-2 family. In healthy cells, mitochondrial membranes remain intact. Antiapoptotic members BCL-2 and its closest homologs, BCL-x$_L$ and BCL-w, are potent inhibitors of apoptosis in response to many stress signals. They protect mitochondrial integrity and thus prevent the release of cytochrome *c*, as well as other proapoptotic molecules (DIABLO/Smac and Omi/Htra2). BCL-2 is an integral membrane protein in all cells, whereas BCL-x$_L$ and BCL-w only become tightly associated with the membrane after a cytotoxic signal, possibly through conformational change.

During apoptosis, proapoptotic "BH3-only" proteins neutralize the activity of BCL-2. Then, proapoptotic members, BAX and BAK, form homo-oligomers within the mitochondrial membrane and induce MOMP allowing efflux of cytochrome *c* and the formation of the apoptosome. BAX is normally a cytosolic monomer but translocates to the mitochondrial membrane during apoptosis and changes conformation forming homo-oligomers within the outer mitochondrial membranes. BAK, however, is an oligomeric integral mitochondrial membrane protein but also changes conformation during apoptosis and might form larger aggregates. The release of DIABLO/Smac and Omi/Htra2 from the mitochondria permits them to bind to and inhibit IAPs – proteins that prevent caspase-9 and caspase-3 activation.

APAF-1. Recent data, however, argue that the apoptosome and cytochrome *c* release is not always necessary. For example, the absence of APAF-1 or caspase-9 in the hematopoietic system caused no effect on cell numbers but these cells underwent apoptosis in response to diverse stimuli with activation of various initiator caspases.

It was surprising that the predominant phenotype shared by the knockout of *Apaf-1*, *caspase-9*, and *caspase-3* genes in mice was the severe developmental defect in the central nervous system that results in the protrusion of brain tissue from the forehead and perinatal lethality. These findings indicate an essential role of the apoptosome in brain development but surprisingly not in other tissues. Given that apoptosis is known to occur in many tissues during development, such as the immune system, this suggests that other tissue-specific pathways are involved in the development of these organs, such as the "death receptor" mediated pathway. Given that the conventional knockout mice mentioned above resulted in lethality following birth, it is still possible that cytochrome-*c*- mediated apoptosis is involved in the homeostasis of many adult tissues – we await the generation of more tissue- and time-specific genetically modified mouse models.

Endoplasmic reticulum(ER) stress

The ER serves many specialized functions in the cell including calcium storage and gated release, biosynthesis of membrane and secretory proteins, and production of lipids and sterols. When misfolded or unfolded proteins accumulate in the ER (ER stress), an intracellular signaling pathway termed the unfolded protein response (UPR) is activated, with the presumed purpose of alleviating the stress. If the stress-inducer is not corrected then the UPR activates apoptosis. ER stress may result from defects in chaperone-mediated folding and numerous diseases may in part involve such defects, including diabetes. Moreover, cancer cells exposed to hypoxia or treated with some chemotherapeutic agents also trigger the UPR. The proteins involved in triggering the ER stress-responsive cell death pathway are not well understood. UPR gene expression is induced following exposure to ER stress agents in the presence or absence of caspase-9 and -12 and even in cells that overexpress BCL-x_L. However, ER stress-induced apoptosis is caspase-dependent and is prevented at least in some cells by overexpression of BCL-x_L or a dominant negative caspase-9. There are clearly other parallels between ER stress-induced apoptosis and that induced by other triggering stimuli, including an inhibitory effect of the PI3K-AKT signaling pathway. BIM is translocated to the ER in response to ER stress and plays a key role in activation of caspase-12 and initiation of the ER stress-specific caspase cascade. The proapoptotic BH3-only protein BIK, discussed earlier, targets the membrane of the ER but appears not to play a role in UPR. However, BIK is induced by genotoxic stress and overexpression of E1A or p53 and may link the ER with induction of MOMP. BAX and BAK also operate at both mitochondria and ER to regulate MOMP and the intrinsic apoptotic pathway. They play a key role in activation of the intrinsic pathway of apoptosis. In a recent study from the laboratory of Stanley Korsmayer employing BAX/Bak double knockout cells,

no activation of CARD-containing initiator caspases was detected following either induction of ER stress or DNA damage. Nor were effector caspases activated, effectively ruling out a major role of some alternative activating pathway.

In response to ER stress, cells activate a translational control program known as the integrated stress response (ISR). Stress-induced genes are activated through assembly of transcription factors on ER stress response elements (ERSEs) in target gene promoters. Transcriptional regulators binding this region include TFII-I and ATF6 and the Sp proteins Sp1, Sp3, and Sp4. Various ISR targets are known and include markers of ER stress such as C/EBP homologous protein (CHOP) and ATF4 and their target TRB3. The ISR adapts cells to ER stress and may prevent apoptosis. Conversely, inactivating ISR proteins, such as PERK and eIF2α, increases apoptosis. ISR targets ATF4 and CHOP have been shown in hypoxic areas of human tumors, suggesting that ISR is involved in tumor cell adaptation to hypoxia, and raising the possibility of targeting this therapeutically in cancer.

Stress-inducible heat shock proteins

As will be referred to in subsequent chapters, diverse stresses such as DNA damage and nutrient starvation can trigger apoptosis, but must also activate survival pathways at least under conditions where the cell can cope with or repair the damage. Increasing attention is turning to the role of the heat shock proteins (HSPs) as regulators of apoptosis following damaging stimuli, and thereby key contributors to the recovery of cells from these insults. One way in which HSP can protect the cell is as chaperones for for other proteins; HSP may be able to repair misfolded proteins, which may contribute to avoidance of cell death that might otherwise result. The HSP27, 70, and 90 subfamilies are strongly implicated in avoidance of apoptosis following diverse triggers including DNA damage and ER stress. HSP can prevent release of cytochrome c and activation of caspases. HSP27 and HSP70 can inhibit activation of BID by caspase-8 and thereby the extrinsic pathway of apoptosis, and both HSP70 and HSP90 can prevent assembly/activity of the apoptosome thereby inhibiting the intrinsic pathway. HSP may also prevent MOMP by interacting with BAX. Not surprisingly, HSP levels are increased in many cancers and are associated with resistance to chemotherapy. It is tempting to speculate that this may relate directly to the ability of such stress-inducible HSPs to block cancer cell apoptosis and facilitate their survival in the face of DNA damage and hypoxia. GeldanaMYCin is a therapeutic agent targeted to HSPs, and by binding to HSP90 prevents interactions with target proteins. Some proteins regulated by binding to HSP90 include AKT, RAS, and p53, suggesting that this may prove a potentially useful theraputic strategy in cancer.

TUMOR-SUPPRESSOR p53

The *p53* tumor-suppressor gene is one of the most frequent targets for mutation in human tumors – the majority of human tumors have either mutated p53 itself, or have incurred mutations that disable function of the p53 pathway. Numerous mouse models have shown a critical role for p53 (and regulators of p53) in tumor suppression, such that inactivating p53 greatly facilitates tumor progression (see Chapter 7). In addition, p53 also has a profound impact on sensitivity to anticancer cytotoxic agents, such that disabling p53 function during tumorigenesis also reduces the sensitivity of these tumors to such antineoplastic agents, thereby limiting their clinical efficacy (see last section).

As described in Chapter 7, p53 guards against DNA damage, stresses, or a surfeit of proliferative signals, preventing cells from becoming malignant by inducing growth arrest or inducing apoptosis. How does p53 promote apoptosis? The precise mechanism of p53-mediated death is likely to differ depending on cell type and on the apoptotic trigger. There are several proapoptotic transcriptional targets of p53, such as BAX and BH-3 proteins, NOXA, and PUMA, which promote cytochrome *c* release from the mitochondria (Fig. 8.10) as described above under "BCL-2 family." During apoptosis, BAX (normally an inactive monomer) can be induced to oligomerize and migrate from the cytoplasm to the mitochondria

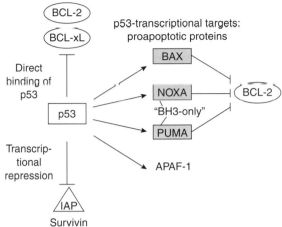

Fig. 8.10 p53 and apoptosis. The tumor-suppressor protein, p53, can promote apoptosis in a number of ways. There are several proapoptotic transcriptional targets of p53, such as BAX and BH-3 proteins, NOXA, and PUMA, that promote cytochrome *c* release from the mitochondria. p53 can also downregulate the transcription of genes that inhibit apoptosis, including the IAP-family member, Survivin. p53 protein has also been shown to antagonize the antiapoptotic proteins BCL-2 and BCL-x$_L$, at the mitochondrial outer membrane, by directly binding to them.

by various BH3-only proteins (e.g. BID). Once inserted into the outer mitochondrial membrane, BAX induces cytochrome *c* release by the creation or alteration of membrane pores (Fig. 8.9).

Aside from transcriptional activation, p53 can downregulate the transcription of genes that inhibit apoptosis, including the IAP family member Survivin (see IAP section). p53 protein has also been shown to antagonize the antiapoptotic proteins BCL-2 and BCL-x$_L$ at the mitochondrial outer membrane, by directly binding to them.

Interestingly, APAF-1 is a transcriptional target of both p53 and the oncogene *E2F*. Given that loss of p53 and caspase-9 promoted fibroblast transformation, it may be that APAF-1 and caspase-9 mediate the apoptotic actions of wild-type p53, at least in certain cell types. Importantly, the APAF-1 gene is lost or silenced in some melanomas indicating that loss of apoptosome function can be an important step in tumor development.

ONCOGENIC STRESS: c-MYC–INDUCED APOPTOSIS

Deregulated expression of *c-MYC* is present in most, if not all, human cancers and is associated with a poor prognosis. The *c-MYC* proto-oncogene is essential for both cellular growth and proliferation, but paradoxically may also promote cell death (Chapter 6). The study of this "dual potential" of c-MYC over the last two decades has provided a paradigm for exploring the role of other proteins with mitogenic activity, such as E2F and RAS, many of which have now also been shown to have such intrinsic "tumor-suppressor" properties. In fact, it may be a general feature of proteins that promote the cell cycle that the oncogenic potential of deregulated expression is restrained by concurrent activation of processes, such as apoptosis or senescence, which effectively prevent propagation of the "damaged" cell. By implication, these in-built "failsafe" mechanisms must be overcome during tumorigenesis. However, once prevented, for instance by inactivation of the p53 or Rb pathways (Chapter 7) or upregulation of antiapoptotic proteins, the potentially devastating oncogenic potential of proteins such as c-MYC is unmasked (Chapter 6).

Some 15 years ago, several laboratories made an intriguing discovery: oncoproteins such as c-MYC and the adenovirus E1A – both potent inducers of cell proliferation – were shown to possess apoptotic activity. The most widely held view of oncoprotein-induced apoptosis is that the induction of cell cycle entry sensitizes the cell to apoptosis: cell proliferative and apoptotic pathways are coupled. However, the apoptotic pathway is suppressed so long as appropriate survival factors deliver antiapoptotic signals. Given this, the predominant outcome of these contradictory processes will depend on the availability of survival factors.

Since these early experiments, other promoters of cell proliferation (e.g. E2F) have been found to possess proapoptotic activity. The notion that cells acquiring growth-deregulating mutations *in vivo* possess an "in-built" tumor-suppressor function, which hinders expansion of potentially malignant cells, is a fascinating one. Such a cell population would be unable to outgrow its environment unless apoptosis was inhibited. Indirect evidence supports this

idea, as shown by the dramatic synergy between oncoproteins such as c-MYC and mechanisms that suppress apoptosis, for example, overexpression of antiapoptotic proteins such as BCL-2 or BCL-x_L, or loss of ARF or p53 tumor suppressors. Interestingly, stimulation of apoptosis by c-MYC may not invariably be linked directly to cell cycling but might also be an indirect response to DNA damage. Thus, recent data *in vitro* link c-MYC to the accumulation of reactive oxygen species (ROS), which can damage DNA and generate double-strand breaks (DSBs) – see also Chapter 10; the mechanisms are not fully understood, but may include inhibition of NF-κB. Interestingly, E2F1 may also be able to promote DSBs and apoptosis but independently from c-MYC and without ROS. The consequences of DNA damage – either apoptosis or growth arrest – may be critically dependent on cell type.

As already mentioned, at least two separate pathways are involved in the induction of apoptosis – an "intrinsic" pathway regulated by various factors such as DNA damage, stress, and imbalances between growth-promoting and survival-promoting factors that act through mitochondrial permeability; and an "extrinsic" pathway, utilized, for example, by cytotoxic T cells, involving ligation of cell surface death receptors (e.g. FAS and TNFR) and activation of the DISC. This distinction may, however, become "blurred" at least in some cells, where activation of caspase-8 in the extrinsic pathway may also co-opt the intrinsic pathway through activation of the BH3-only protein BID. Both pathways have been implicated in mediating the action of c-MYC on apoptosis in various studies and are in turn subject to various regulatory influences (summarized in Plate 8.2). However, with the exception of the p19^{ARF} pathway, the relative contributions of these various potential mechanisms for any given tissue, cell type, or critically within an individual developing tumor *in vivo* are generally not well known.

How does the c-MYC oncoprotein induce or "sensitize" cells to apoptosis? Most of the key experiments used to answer this question have come from cell culture. Although much of the apoptotic machinery has been identified, we are only beginning to define the mechanisms by which this is engaged by c-MYC. As is often the case for cell signaling events, there is not just one mechanism by which c-MYC may trigger apoptosis. The mechanism chosen, or even whether apoptosis will occur at all, most likely depends on factors, such as cell type and developmental stage, signals received from the cell's environment, or whether the cell has acquired DNA damage or not.

In different experimental models, expression of c-MYC may make cells vulnerable to a wide range of proapoptotic stimuli – such as hypoxia, DNA damage, and depleted survival factors – as well as enhancing sensitivity to signaling through the extrinsic pathway triggered by CD95, TNF, or TRAIL death receptors. In fact, it has recently been shown that c-MYC can lead to downregulation of the inhibitor of caspase activation, FLICE inhibitory protein (FLIP). FLIP is an inhibitor of the extrinsic pathway, which normally competes with caspase-8 for binding to the DISC – releasing FLIP inhibition may at least in part explain the ability of c-MYC to sensitize cells to death receptor stimuli.

First insight into c-MYC's role in apoptosis came from studies showing that c-MYC could induce the release of cytochrome *c* from the mitochondria during apoptosis, and that ectopic addition of cytochrome *c* sensitized cells to undergoing apoptosis. As already discussed, once released into the cytoplasm, cytochrome *c* associates with another protein called apoptotic protease-activating factor 1 (APAF-1) to create the apoptosome, which acts as a scaffold for activating procaspase-9. There are various ways in which c-MYC may provoke MOMP and the release of cytochrome *c*. One of the best-known methods involves the expression of members of the BCL-2 family, discussed earlier. For example, c-MYC represses expression of the antiapoptotic proteins BCL-2 and BCL-x$_L$, and may induce expression/activation of proapoptotic members, such as BIM and BAX. This has the net effect of permeabilizing the mitochondria to release cytochrome *c* and other proapoptotic factors, which ultimately leads to activation of caspases. Moreover, c-MYC may actually increase expression of potentially dangerous mitochondrial genes, including cytochrome *c*, via the transcription factor nuclear respiratory factor-1 (NRF-1), a potential c-MYC target gene.

How might interfering with BCL-2 family members provoke MOMP? During apoptosis, BAX (normally an inactive monomer) can be induced to oligomerize and migrate from the cytoplasm to the mitochondria by various BH3-only proteins (those BCL-2 family members containing only the single BCL-2 homology-3 domain), such as BID. BAX then inserts into the outer mitochondrial membrane to create or alter the behavior of membrane pores, that is, MOMP, through which cytochrome *c* can escape into the cytoplasm. Conversely, antiapoptotic proteins such as BCL-2 and BCL-x$_L$ reside in the outer mitochondrial membrane and suppress apoptosis by blocking MOMP, possibly through sequestration of activated BAX and also by preventing activation of the apoptosome. In this scenario, the balance of anti- and proapoptotic molecules present within a c-MYC-activated cell would determine whether it lives or dies.

With this in mind, survival factors such as IGF-1, have been shown to inhibit c-MYC-induced apoptosis *in vitro* by blocking cytochrome *c* release from the mitochondria. In this case, survival signals, for instance those mediated via the IGF-1 receptor or activated RAS, can activate the PI3 kinase pathway and the AKT/PKB serine/threonine kinase – see Chapters 5 and 6. Activated AKT can phosphorylate the proapoptotic BH3-only protein BAD resulting in its sequestration and inactivation by the cytosolic 14-3-3 proteins. Referring back to the initial experiments *in vitro*, in which cells with deregulated expression of c-MYC die by apoptosis when grown in low serum, it becomes clear how growth factor signaling is able to regulate apoptosis. Given the importance of such growth factors in determining the survival or death of a cell, it is not surprising that elevated signaling through the IGF-1 pathway occurs in many tumors. Similarly, genetic mutations that activate the PI3K pathway dramatically collaborate with c-MYC during tumor progression.

An important mechanism that links c-MYC and apoptosis is through the p19^{ARF} (ARF) tumor-suppressor protein, which acts in a checkpoint that guards against unscheduled cellular proliferation in response to oncogenic signaling (Chapter 7; see also in Chapter 10 – "Checkpoints, apoptosis, and oncogenic stress"). ARF may contribute to c-MYC apoptosis indirectly via activation of p53, or may engage apoptosis directly and independently of p53. The importance of ARF in c-MYC-induced apoptosis *in vivo* was underscored in genetically altered mice in which expression of *c-MYC* was targeted to B lymphocytes. When these mice were crossbred with another strain in which ARF was disrupted, MYC-induced lymphoma development was dramatically accelerated. This outcome, similar to that seen when p53 is disrupted, shows that c-MYC strongly collaborates with loss of p53 or ARF in murine lymphomagenesis presumably by inhibiting c-MYC-induced apoptosis (extrapolating from *in vitro* experiments). In a similar fashion, deregulated expression of the *Bmi1* oncogene (which encodes a polycomb group protein), accelerates c-MYC-induced lymphomas by inhibiting expression from the *Ink4a* locus, which encodes both ARF and another important tumor-suppressor $p16^{INK4A}$ (Chapter 7). One key problem has been to identify the means by which oncogenic c-MYC but not normal c-MYC could trigger ARF (how do cells distinguish between normal cell cycles and aberrant 'cancer' cell cycles. A series of studies over the last few years have begun to unravel the means by which oncogenes such as c-MYC trigger MOMP.

Although much remains to be learned we can now attempt a synthesis of how deregulated expression of c-MYC triggers MOMP. An apical event in this seems to be activation of a DNA damage response, particularly involving the PI3-like kinases, ATM and ATR (Chapter 10). It also seems likley that in many cases this response might be triggered because c-MYC promotes DNA damage such as DNA double-strand breaks (DSBs), in part by excess production of ROS. However, it is also plausible that ATM is activated by a more ill-defined induction of "cellular stress," not necessarily requiring active DNA damage. ATM/ATR in turn activate various checkpoint pathways that culminate in apoptosis (with other oncogenes growth arrest may be a more prominent response). This is discussed in further detail in Chapter 10, but in summary – phosphorylated ATM activates various downstream targets including another kinase CHK2. Together these events phosphorylate and activate p53 and p53-targets such as PUMA and NOXA, which contribute to MOMP as previously outlined. ATM can also activate another BH3-only protein, BID, which can mediate both growth arrest in its full form and apoptosis in its truncated form. As c-MYC can reduce expression of pro-survival proteins, BCL-2 and BCL-x_L, this together with the inhibitory action of the various BH3-only proteins allows activation of BAK and/or BAX and thereby MOMP. Although, it is easy to see how ARF activation would contribute to apoptosis by activating p53, what still remains unclear is how ARF is activated by deregulated c-MYC, though it is tempting to speculate that this might in some way depend on ATM or members of the MRN complex, such as NBS1 protein known to be upregulated by c-MYC, but this is conjectural.

We now know that p53 activates many proapoptotic proteins, such as BAX, Noxa, and PUMA, which are involved in cytochrome *c* release from the mitochondria, and APAF-1, involved in the activation of caspases (enzymes responsible for the destruction of the cell).

It seems likely that specific regions of c-MYC are responsible for its proapoptotic activity, at least in some systems. Thus, factors such as Tiam1, which bind to the c-MYC box II (MBII) region of c-MYC (see Fig. 6.5), can inhibit the proapoptotic activity of MYC; conversely in other studies disruption of the MBIII region, important for transcriptional repression by c-MYC, increases the apoptotic activity of c-MYC, and may prevent transformation.

As mentioned earlier, c-MYC apoptosis may not always be a positive action as it undoubtedly is in restraining cancer. An example is the pancreas, which can adapt islet β-cell mass to meet changes in demand for insulin, as described in late pregnancy and obesity – pancreas plasticity. Individuals that fail to adapt become diabetic with time. During the progression to diabetes both functional defects and β-cell apoptosis contribute to defective insulin secretion, generally described as β-cell failure. Several potential contributors to this have been identified, including c-MYC. The notion that c-MYC may be involved in β-cell failure is supported by various studies. First, replicating β-cells seem to be most sensitive to apoptosis, which indirectly implicates proteins such as c-MYC known to be involved in β-cell replication. Moreover, c-MYC is upregulated in β-cells exposed to rising blood glucose levels. Recently, studies with various transgenic mouse models have demonstrated that the immediate consequences of c-MYC activation in β-cells *in vivo* include replication, but also β-cell apoptosis and loss of β-cell differentiation. In fact, even when apoptosis is prevented by simultaneous overexpression of the antiapoptotic protein BCL-x$_L$, c-MYC activation still results in loss of β-cell differentiation and hyperglycaemia, which is only reversed after considerable expansion in β-cell mass has taken place. Thus, a possible model is that β-cells activate c-MYC in response to a need for adaptive growth or later in the process because of rising blood glucose. Elevated c-MYC may then contribute to loss of function and apoptosis of β-cells. Various other processes have been described as underlying glucose toxicity and include activation of the extrinsic pathway of apoptosis by cytokines, such as IL-1β or FAS/FASL interactions and ROS. Intriguingly, c-MYC has been shown to trigger apoptosis through all of these pathways in other systems.

It is clear that the favored mechanism for c-MYC-induced apoptosis may be dictated by cell type as well as by tissue location and moreover modified by the presence or absence of additional mutations in other pro- and antiapoptotic genes. In order to further illuminate the complex interactions surrounding c-MYC and apoptosis, a role so sensitive to tissue location *in vivo*, we eagerly await the outcome of further studies in which apoptotic pathways are selectively manipulated in mouse mutants or by employing new pharmacological tools and phenotype determined in the all important context of the intact organism.

EXPLOITING CELL DEATH (AND SENESCENCE) IN CANCER CONTROL

As previously pointed out in this chapter, the resistance of tumor cells to apoptosis is a crucial aspect of tumor development. Unfortunately, it is this same aspect that can cause some cancer cells to become resistant to anticancer therapies, which will ultimately lead to the death of the individual. In other words, some of the mutations that allowed tumor progression can also confer chemoresistance, as we will see below. It is also important to mention here that apart from apoptosis, there

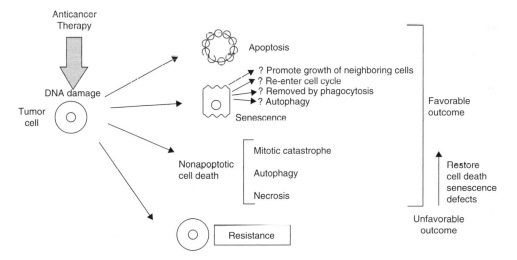

Fig. 8.11 Potential responses to anticancer therapy. Anticancer therapy causes DNA damage to tumor cells. Depending on cell type and the genetic mutations acquired, a given tumor cell may respond to such damage in one of the following ways: die by apoptosis, mitotic catastrophe, autophagy, or necrosis (nonapoptotic cell death). Elimination is of course the most favorable outcome. Alternatively, the cell may enter a state of "permanent" cell cycle arrest (cellular senescence). However, it remains debatable as to whether the senescent cell promotes growth of neighboring tumor cells by secreting growth factors, or whether it will eventually re-enter the cell cycle. The most unfavorable outcome is obviously resistance to therapy as the cell is unable to undergo apoptosis or cellular senescence.

are indeed alternative outcomes in response to drug-induced DNA damage: cells may still be induced to die by nonapoptotic mechanisms, such as necrosis, autophagy, and mitotic catastrophe, or enter a permanent state of cell cycle arrest – "cellular senescence." All these potential cellular responses to anticancer therapy are outlined in Fig. 8.11 and will be discussed next.

Many tumor cells are killed in response to cytotoxic therapies, such as chemotherapy, gamma irradiation, or immunotherapy, predominantly mediated by triggering apoptosis (Chapter 16). Such anticancer agents do not simply destroy cells directly, but rather trigger the cell's own apoptotic response programs. Although the underlying mechanisms for initiation of this apoptotic response are not always known and may differ depending on the

cytotoxic agent, damage to DNA or to other critical molecules is considered to be a common initial event, which is then propagated by the cellular stress response (mitochondrial pathway – described earlier). Therefore, if a cell acquires defects in stress response programs during tumor formation, it may now be more resistant to drug-induced apoptosis. For example, preclinical data show that tumor cells that have acquired mutations in *p53* are also chemoresistant. Furthermore, *p53* defects have been shown to mediate multidrug resistance to numerous anticancer drugs. Similar observations have been made in clinical investigations. Figure 8.12 outlines potential pathways to such resistance.

T cells or natural killer (NK) cells may contribute to tumor cell killing by releasing cytotoxic compounds such as granzyme B or by using

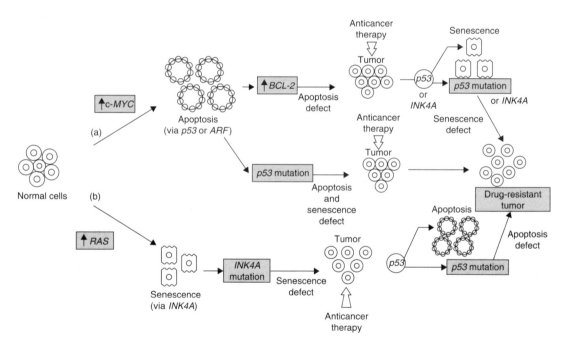

Fig. 8.12 Pathways to resistance. Whether a tumor cell responds favorably (cell death or cellular senescence) or unfavorably (resistant) to anticancer therapy will depend on the genetic mutations acquired during tumor development. For example: (a) in the early stages of tumorigenesis, if the initiating activating oncogene is *c-MYC* (where apoptosis is the "in-built" fail-safe mechanism), an apoptosis defect such as *BCL-2* overexpression will now prevent cells from dying by apoptosis. Despite this, such tumor cells may respond favorably by entering a state of cellular senescence. Further mutations, however, may also prevent cellular senescence, for example, inactivating *p53* or *INK4A* mutations, leading to resistance. Another example (b) shows tumor cells that have acquired an initiating oncogenic *RAS*, where cellular senescence is the fail-safe mechanism. Inactivating mutations in *INK4A* tumor-suppressor gene prevent senescence in response to anticancer therapy. However, cells can still undergo apoptosis due to the presence of intact p53. Not surprisingly, the loss of p53 prevents tumor cells dying by apoptosis and the tumor is now resistant to anticancer therapy.

"death ligands" such as FASL, which bind to FAS (CD95) receptors on tumor cells and subsequently activate downstream apoptotic pathways (Chapter 13).

Since defects in apoptosis are likely to be universal lesions in tumor progression, restoring or activating apoptosis in tumors is an active area of cancer research. Preclinical trials have validated the antitumor efficacy of this approach, and clinical trials are currently ongoing for various apoptosis-activating strategies. However, as seen above, apoptosis involves many regulators and effectors that vary from tumor to tumor. This diversity can hamper determination of effective targets for individual tumors, and can contribute to drug resistance during treatment. The way forward for effective cancer treatment strategies is therefore likely to require combinatorial approaches that, together with individual tumor typing and matching with type-specific treatments, will allow effective tumor-specific therapeutics, or personalized medicine.

It was recently discovered that, in addition to inducing apoptosis, anticancer therapies can also induce cellular senescence – a permanent state of growth arrest (this cellular state is described more fully in Chapter 9). What is the fate of drug-induced senescent tumor cells? Do they eventually re-enter the cell cycle or die? The answer to this remains unclear at present. Although senescent tumor cells are locked into a nondividing state, it is conceivable that such cells could still promote growth of other tumor cells in their vicinity, as they remain metabolically active and can secrete growth-regulating cytokines into the environment. It is also possible that senescent cells will ultimately be cleared by phagocytosis (although unlike apoptotic cells, no "eat-me" signals have been identified as yet), or indeed be able to eliminate themselves by autophagy (described below).

Drug-induced cellular senescence has been elegantly shown *in vivo* using a mouse lymphoma model (Eμ-MYC) in which the oncogene c-*MYC* is constitutively expressed in the B-cell lineage. This mouse model has allowed the study of treatment responses at naturally occurring tumor sites. For example, if the antiapoptotic gene, *BCL-2*, is introduced into the B-cell lineage of Eμ-MYC mice, animals fail to undergo remission following therapy – that is, a reduction in tumor mass below detectable levels – due to

resistance to undergo apoptosis. Instead, they preserve a constant tumor burden as a consequence of tumor cells entering senescence. Importantly, those tumors that started to grow again (progressive disease) were frequently found to have acquired defects in either *p53* or *INK4A* genes, indicating that these two gene products control drug-induced senescence *in vivo*. Thus, tumors that have acquired defects in both the apoptosis and senescence programs, such as overexpression of *BCL-2* plus loss of *p53*, are chemoresistant and so severely impair the long-term survival of the host. Importantly, sequential disruption of apoptosis and senescence-controlling genes during tumor formation and subsequent therapy has also been reported in human cancers. For example, *INK4A* defects correlate with relapses in non-Hodgkin's lymphomas, while patients diagnosed with follicular lymphomas that overexpress *BCL-2*, due to the characteristic t(14;18) translocation, typically achieve long-term disease control following chemotherapy. Poor prognosis is associated with acquisition of *p53* or *INK4A* mutations that were not detectable at diagnosis. This pathway to resistance is outlined in Fig. 8.12(a).

Despite the increasing evidence that drug-induced cellular senescence can serve as a cancer therapy, additional mouse models for tumors other than lymphomas/leukemias are needed, as well as a thorough investigation of human tumor specimens. Notably, cellular senescence has been seen in tissue specimens taken from human breast tumors following chemotherapy.

So, on a final note, for cellular senescence to be an effective anticancer response, its therapeutic potential strongly relies on the irreversibility of this process. Unlike apoptosis, which eliminates a potentially harmful cell, cellular senescence only represents an altered state of a tumor cell, which if not eliminated (e.g. by phagocytosis or autophagy) may at some point revert to a dividing cell and cause tumor relapse. Therefore, at present the potential use of cellular senescence in treating cancers is not clearly known, and is likely to depend on the various genetic lesions within the tumor cells.

Nonapoptotic cell death

Apart from apoptosis and senescence, there are indeed alternative outcomes in response to drug-induced DNA damage: cells may still be induced to die by nonapoptotic mechanisms, such as necrosis, autophagy, and mitotic catastrophe (Fig. 8.11). In normal cells, unwanted proteins are degraded by two distinct mechanisms: (i) the familiar ubiquitin-mediated proteolysis that takes place in the proteasomes; and (ii) autophagy, a mechanism by which long-lived proteins and organelle components are directed to, and degraded within, lysosomes. Cells that undergo excessive autophagy are induced to die in a nonapoptotic manner. However, the precise function of cell death by autophagy in mammals is not yet fully understood – there is evidence that lysosomal degradation of organelles is required for cellular remodeling due to differentiation, stress, or damage following exposure to cytotoxins. Interestingly, there is some evidence linking defects in the autophagic pathway to cancer. For example, the *ras* and *MYC* oncogenes might induce tumorigenesis partly by inhibiting autophagy, while the antiapoptotic signaling pathway (PI3K-AKT, described in Chapter 6), commonly activated in human cancer cells, might also suppress autophagy.

Studies have shown that tumor cells treated with anticancer drugs also undergo death by autophagy. For example, breast carcinoma cells treated with tamoxifen (the estrogen-receptor antagonist) accumulate autophagic vacuoles shortly before dying.

Tumor cells may also undergo mitotic catastrophe, a type of cell death that is caused by aberrant mitosis. Mitotic catastrophe is associated with the formation of multinucleate giant cells that contain uncondensed chromosomes. Cells that have acquired a defective G_2 checkpoint of the cell cycle (this checkpoint is responsible for blocking mitosis when a cell has sustained DNA damage – described in Chapters 4 and 10) can enter mitosis prematurely, before DNA replication is complete or DNA damage has been repaired. This aberrant mitosis causes the cell to undergo death by mitotic catastrophe. A key molecule that regulates mitotic progression is Survivin. Although a member of the IAP family, the primary function of Survivin is not in the control of apoptosis, but rather of mitotic progression. At present, there are few anticancer therapies aimed to induce tumor cell death by mitotic catastrophe. However, targeting this pathway might represent a novel approach to eliminating tumors, perhaps through targeting molecules such as Survivin.

Although the molecular pathways that underlie nonapoptotic forms of cell death are ill-defined at present, it might be possible to develop new cancer therapies that will induce tumor cell death by necrosis, autophagy, or mitotic catastrophe. Importantly, one key observation made is the fraction of tumor cells that undergo nonapoptotic death increases if apoptosis is inhibited in response to anticancer agents or radiation. With this in mind, a strategy that deliberately induces nonapoptotic cell death could serve as a useful adjunct to standard cancer therapies and might make a significant contribution to a favorable treatment outcome.

It is likely that apoptotic and nonapoptotic programs of cell death occur to varying extents in different tumor cell types, and on the mode of cancer treatment. If this is the case, then the combined action of both cell death programs will probably be required to achieve a reduction in tumor size and minimize the chance of relapse.

QUESTIONS REMAINING

1 Appropriate cell number and organ size in a multicellular organism are determined by coordinated cellular processes such as growth, proliferation, senescence, and death. Given that all of these processes are in themselves subject to regulation by multiple external factors, including nutrient availability, growth factors, and cell–cell and cell–matrix interactions, as well as intrinsic signaling pathways, one of the key challenges is to more fully understand how tissue growth is constrained during development and maintained in the adult.

2 In the context of the balancing act between growth/replication and cell death, we need to progress in our understanding of how key proteins such as c-MYC, E2F, E1A, LATS, and others are able to participate in signaling pathways regulating both opposing cellular outcomes.

3 As disturbing the balance of these processes at any of a myriad of potential target sites can result in cancer on the one hand or degenerative diseases on the other, we also need to further our understanding of the underlying apoptotic machinery and how individual components may be successfully manipulated in the treatment of cancer and other diseases.

CONCLUSIONS AND FUTURE DIRECTIONS

Cell death is now known to be at least as important as growth and replication in achieving appropriate cell numbers and tissue size during development and in maintaining these during adult life. It seems remarkable now that modern biology has only been focused on this process for the last 15 years yet, despite this, progress has been nothing short of spectacular, as evidenced by the astounding number of key publications in this area. Much is now known about some of the key intracellular proteins that can regulate the delicate balance between life and death of the cell, and some of the environmental influences, which in turn regulate these. Not surprisingly, many of these regulatory processes are the subject of mutations or epigenetic changes in cancer and are increasingly being targeted in cancer treatments. Thus, we now have the opportunity to use various high-throughput techniques (Chapter 18) on human cancer samples to detect abnormalities in many of the key signaling pathways in apoptosis and even begin to predict how these will influence disease progression and therapeutic response. However, although it may prove relatively straightforward in more extreme cases, such as overexpression of antiapoptotic BCL-x_L or BCL-2, or proapoptotic BAX, to conclude that the cancer cell will either be resistant to or sensitive to cell death respectively, in other cases more subtle alterations in the balance of multiple factors may achieve the same cellular outcome, but be much harder to detect or predict. Moreover, the abnormalities in any given cancer cell will also have to be placed in the environmental context of that individual tumor, as this will determine the balance of survival cues acting on the cell and also strongly influence the outcome.

QUESTIONS

1 Apoptosis:
  a. Usually provokes an inflammatory response.
  b. Involves nuclear condensation.
  c. Involves membrane blebbing.
  d. Is usually indicative of disease.
  e. Is suppressed in cancer.

2 Apoptosis:
  a. Is suppressed by inactivation of the p53 pathway.
  b. Is suppressed by inactivation of telomerase.
  c. Is executed by activation of caspase enzymes.
  d. Is inhibited by cytoplasmic cytochrome *c*.
  e. Is the most ubiquitous form of cell death in most organisms.

3 Apoptosis may be triggered by:
  a. FAS–FASL binding.
  b. Mitochondrial insertion of BCL-x_L.
  c. Overexpression of c-*MYC*.
  d. Loss of cell–cell contacts.
  e. Active AKT.

4 c-MYC may promote apoptosis in some cell types by:
  a. Increased MOMP.
  b. Induction of FAS–FASL.
  c. Oxidative stress.
  d. Upregulation of WRN.
  e. Upregulation of BCL-x_L.

5 Cancer cells may avoid apoptosis by:
  a. Inactivating checkpoint proteins.
  b. Producing survival factors.
  c. Activating the PI3K pathway.
  d. Upregulating BAX.
  e. Inactivating survivin.

BIBLIOGRAPHY

Historical and general

Ellis, H.M. and Horvitz, H.R. (1986). Genetic control of programmed cell death in the nematode *C. elegans*. *Cell*, **44**: 817–29.

Kerr, J.F.R., Wyllie, A.H., and Currie, A.R. (1972). Apoptosis: a basic biological phenomenon with wide-ranging implication in tissue kinetics. *British Journal of Cancer*, **26**: 239–57.

Kroemer, G. and Martin, S.J. (2005). Caspase-independent cell death. *Nat Med.* **11**(7): 725–30.

Levi-Montalcini, R. (1987). The nerve growth factor 35 years later. *Science*, **237**: 1154–62.

Raff, M. (1998). Cell suicide for beginners. *Nature*, **396**(6707): 119–22.

Sulston, J. and Horvitz, H.R. (1977). Postembryonic cell lineages of the nematode, *Caenorhabditis elegans*. *Developmental Biology*, **56**: 110–56.

Apoptosis and cancer

Brown, J.M. and Attardi, L.D. (2005). The role of apoptosis in cancer development and treatment response. *Nature Reviews: Cancer*, 5(3): 231–7.

Christofori, G., Naik, P., and Hanahan, D. (1994). A second signal supplied by insulin-like growth factor II in oncogene-induced tumorigenesis. *Nature*, 369(6479): 414–8.

Lowe, S.W., Cepero, E., and Evan, G. (2004). Intrinsic tumor suppression. *Nature*, 432(7015): 307–15.

Zhivotovsky, B. and Kroemer, G. (2004). Apoptosis and genomic instability. *Nature Reviews Molecular Cell Biology*, 5(9): 752–62.

Linking cell growth and apoptosis, c-MYC

Evan, G.I., Wyllie, A.H., Gilbert, C.S. *et al.* (1992). Induction of apoptosis in fibroblasts by c-MYC protein. *Cell*, 69(1): 119–28.

Lai, Z.C., Wei, X., Shimizu, T., *et al.* (2005). Control of cell proliferation and apoptosis by mob as tumor suppressor, Mats. *Cell*, 120(5): 675–85.

Pelengaris, S., Khan, M., and Evan, G. (2002a). c-MYC: more than just a matter of life and death. *Nature Reviews Cancer*, 2: 764–76.

Pelengaris, S., Khan, M., and Evan, G.I. (2002b). Suppression of MYC-induced apoptosis in beta cells exposes multiple oncogenic properties of MYC and triggers carcinogenic progression. *Cell*, 109(3): 321–34.

Zindy, F., Eischen, C.M., Randle, D.H. *et al.* (1998). MYC signaling via the ARF tumor suppressor regulates p53-dependent apoptosis and immortalization. *Genes Development*, 12(15): 2424–33.

The apoptotic machinery

Chipuk, J.E., Bouchier-Hayes, L., Kuwana, T., Newmeyer, D.D., Green, D.R. (2005). PUMA couples the nuclear and cytoplasmic proapoptotic function of p53. *Science*, 309(5741): 1732–5.

Cory, S. and Adams, J.M. (2002). The BCL2 family: regulators of the cellular life-or-death switch. *Nature Reviews: Cancer*, 2: 647–656.

Garrido, C. and Kroemer, G. (2004). Life's smile, death's grin: vital functions of apoptosis-executing proteins. *Current Opinion in Cell Biology*, 16(6): 639–46.

Green, D.R. and Kroemer, G. (2004). The pathophysiology of mitochondrial cell death. *Science*, 305(5684): 626–9.

Hengartner, M.O. and Horvitz, H.R. (1994). *C. elegans* cell survival gene *CED-9* encodes a functional homolog of the mammalian proto-oncogene *bcl-2*. *Cell*, 76: 665–76.

Hockenbery, D., Nunez, G., Milliman, C., Schreiber, R.D., and Korsmeyer, S.J. (1990). BCL-2 is an inner mitochondrial membrane protein that blocks programmed cell death. *Nature*, 348(6299): 334–6.

Lowe, S.W., Schmitt, E.M., Smith, S.W., Osborne, B.A., and Jacks, T. (1993). p53 is required for radiation-induced apoptosis in mouse thymocytes. *Nature*, 362: 847–9.

Martin, S.J. and Green, D.R. (1995). Protease activation during apoptosis: death by a thousand cuts? *Cell*, 82: 349–52.

Nachmias, B., Ashhab, Y., and Ben-Yehudam, D. (2004). The inhibitor of apoptosis protein family (IAPs): an emerging therapeutic target in cancer. *Seminars in Cancer Biology*, 14(4): 231–43.

Salvesen, G.S. and Dixit, V.M. (1999). Caspase activation: the induced-proximity model. *Proceedings of the National Academy of Sciences USA*, 96(20): 10964–7.

Spierings, D., McStay, G., Saleh, M., Bender, C., Chipuk, J., Maurer, U., Green, D.R. (2005). Connected to death: the (unexpurgated) mitochondrial pathway of apoptosis. *Science*, 310(5745): 66–7.

Vaux, D.L., Weissman, I.L., and Kim, S.K. (1992). Prevention of programmed cell death in *Caenorhabditis elegans* by human *bcl-2*. *Science*, 258: 1955–7.

Zou, H., Henzel, W.J., Liu, X., Lutschg, A., and Wang, X. (1997). APAF-1, a human protein homologous to *C. elegans* CED-4, participates in cytochrome *c*-dependent activation of caspase-3. *Cell*, 90(3): 405–13.

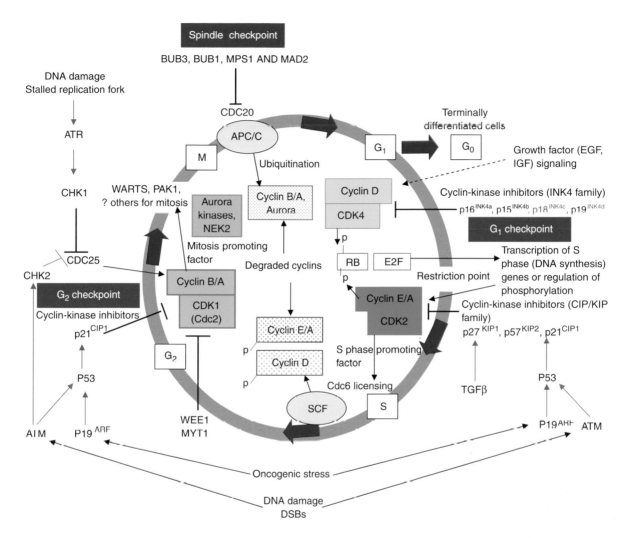

The cell cycle consists of 4 phases, G_1, S, G_2 and M. Entry and progression through each stage is regulated by positive and negative factors. The key step in cell cycle entry is the G_1/S transition (passage through the restriction point), because after this the cell cycle will proceed even if growth factor stimuli are no longer present unless errors in DNA replication, chromosome segregation, or other factors trigger various checkpoints that arrest the cycle. Various checkpoints are shown in red text and red stop arrows. TGFβ can inhibit G_1/S transition via p27^{KIP1}, whereas stimulatory growth factors such as IGF1 activate signaling pathways such as the RAS -MAPK-ERK, that culminate in expression of c-MYC. C-MYC in turn enhances expression of cyclin D and CDK4 and promotes degradation of CKI, such as p27^{KIP1}. Together with Cyclin E-CDK2, the net effect is phosphorylation of RB and release of E2F for transcription of key genes needed for S phase. Key CKI proteins active at the G_1/S transition are the INK4 and CIP/KIP families which inhibit Cyclin D-CDK4/6 and Cyclin E-CDK2 holoenzymes respectively. A recent study in mice suggests that in some cases CDK1 (Cdc2) may replace CDK2 as a cyclin D partner. During S phase progression the SCF ubiquitin ligase complex targets D and E type Cyclins for proteasome degradation. Cyclin E-CDK2 may also help the licensing of prereplicative complexes, for DNA synthesis by stabilizing Cdc6. The cyclin B-CDK1 complex is inactive through G_2, due to phosphorylation by WEE1 and MYT1; at the end of G_2 CDK1 is dephosphorylated and activated by CDC25. The G_2 checkpoint is activated by DNA damage such as DNA double strand breaks which trigger various kinases and activate p53 and p21^{CIP1}. If DNA damage is successfully repaired the cell cycle maybe re-entered; if not then the cell undergoes apoptosis or growth arrest. The APC/C is a multi-subunit ubiquitin ligase that is analogous to SCF earlier in the cycle targets B-type cyclins for proteasome degradation during passage through anaphase and telophase, and also limits accumulation of B-type cyclins during G_1. Inactivation is required for timely S-phase entry. The APC/C also targets cohesins for degradation allowing separation of sister chromatids in a CDK-dependent manner at the metaphase-to-anaphase transition. The APC/C is itself regulated by the spindle or kinetochore checkpoint, which has an important role in maintaining genomic stability by preventing sister chromatid separation until all chromosomes are correctly aligned on the mitotic spindle. The spindle checkpoint regulates the APC/C by inactivating CDC20, an important co-activator of the APC/C.

Plate 4.1 The cell cycle – showing interactions between Cyclin-CDKs and CKI.

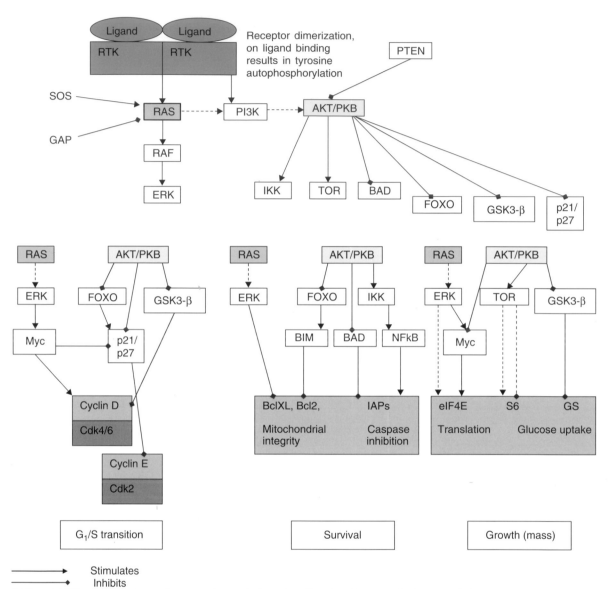

	Stimulates
	Inhibits

RAS and AKT co-ordinately regulate growth, survival, and proliferation. Key sites for integrated signaling are occupied by the kinases ERK, TOR, and GSK3-β, the CDK inhibitors p21^{CIP1} and p27^{KIP1} and the transcription factors c-MYC, FOXO, and NFkB. Oncogenic RAS may also activate ARF and p16^{INK4a} (not shown), suggesting that in some way oncogenic cell cycles trigger additional responses which may seek to minimize the effect of such mutations.

After Joan Massague, 2004, Nature 432, 298-306

Plate 5.1 Integrated networks involving RAS and PI3K act downstream of receptor tyrosine kinases to regulate replication, survival, and growth.

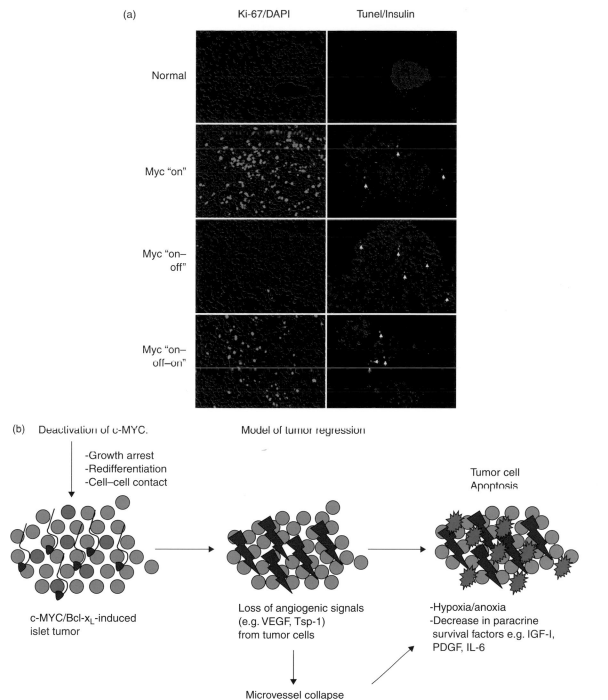

Plate 6.1 Tumor reversal. (a) Deactivation of c-MYC in pancreatic islet tumors results in complete reversal in the overwhelming majority of tumors by 4 weeks after deactivation. Here we show the effects of transient inactivation of c-MYC. After a period of 5 days of inactivation, β-cells have all stopped replicating (absence of Ki-67 staining – green) and have acquired a more differentiated phenotype (enhanced expression of insulin – red). Moreover, some β-cells are undergoing apoptosis during the period of c-MYC inactivation (TUNEL positive – green). Those cells undergoing apoptosis while c-MYC is on are not β-cells as they do not express β-cell markers. (b) The processes of tumor reversal are shown schematically.

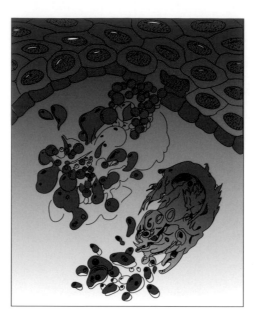

Plate 8.1 Cell dying by apoptosis. Once the apoptotic program has been initiated within a cell, several morphological characteristics can be seen using high-power microscopy: (1) the cell loses contact with its neighbors; (2) then blebs (b) appear on the cell surface, and chromatin (c) condenses at the edges of the nucleus; (3) the nucleus, and then the cell itself, break up; and finally, (4) cell fragments are rapidly ingested by neighboring cells or macrophages in the vicinity.

Plate 8.2 (*Continued*) FADD then recruits procaspase 8 resulting in autoactivation of the procaspase, which cleaves and activates executioner caspases. Activation of caspase-8 may be negatively regulated by FLIP, which competes for binding to the DISC. Caspase-8 may also activate the proapoptotic protein, BID, which may promote MOMP, thereby linking the extrinsic and intrinsic pathways of apoptosis. Recently, c-MYC has been reported to mediate apoptosis in some cells by a mechanism involving generation of reactive oxygen species (ROS) and suppression of the survival promoting activities of NFkB (d). In some cell types, c-MYC can induce expression of the BH3-only proapoptotic protein Bim and suppress expression of the antiapoptotic proteins BCL-2 and BCL-xL (e). c-MYC may increase expression of the p53 family member p73, which might act to direct p53 itself towards proapoptotic genes. (f) One question that remains unanswered is how activation of oncogenes such as c-MYC results in expression of ARF, which is not usually activated during normal cell replication. One possibility is through inhibition of proteins like Bmi-l (g), through this speculative. Lastly, the inhibitor of caspase activation, FLICE inhibitory protein (FLIP) is transcriptionally repressed by c-MYC, providing a mechanism whereby c-MYC could sensitize cells to a wide range of death receptor stimuli (h). Survival signals that serve to block c-MYC-induced apoptosis (i) include signaling via the IGFI receptor and P13 kinase (P13K) or activated RAS, which can lead to activation of AKT serine/threonine kinase and subsequent phosphorylation of proapoptotic protein BAD. Phosphorylated BAD is sequestrated and inactivated by cyctosolic 14-3-3 proteins. This pathway may also block apoptosis in other ways. AKT also activates mTOR and its downstream target the elongation factor eIF-4E. In at least one model, expression of eIF-4E could block c-MYC-induced apoptosis and cooperate with c-MYC in tumorigenesis. Antiapoptotic proteins, such as BCL-2 and BCL-X$_L$, reside in the outer mitochondrial membrane and block cytochrome *c* release possibly through sequestration of BAX. The PTEN tumor suppressor acts as a negatively regulator of AKT activation.

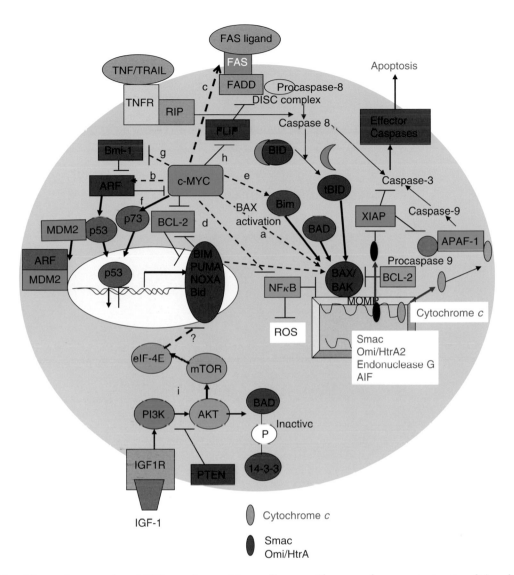

Plate 8.2 c-Myc and apoptosis. c-MYC protein sensitizes cells to a wide range of proapoptotic stimuli (e.g. hypoxia, DNA damage, depleted survival factors, signaling through CD95, TNF, and TRAIL receptors). Moreover, as c-MYC can also transcriptionally activate other "dual function" mitogenic proteins, such as some E2F family members, the potential repertoire of mediators of MYC's role in apoptosis may be even more complex than suggested in this schematic. During apoptosis c-MYC induces release of cytochrome *c* from the mitochondria into the cytosol possibly through activation of proapoptotic molecule, BAX(a). Activated BAX within the mitochondrial membrane leads to creation or alteration of membrane pores, resulting in mitochondrial outer membrane permeabilization (MOMP). Once released into the cytosol, cytochrome *c* associates with APAF-1 protein and procaspase 9 to form the apoptosome ("wheel of death"). In the presence of ATP, caspase 9 is activated leading to activation or downstream effector caspases including caspase 3, which ultimately leads to degradation of cell components and demise of the cell. Also released by MOMP, Smac and Omi can inhibit the action of IAPs, such as XIAP, that otherwise normally prevent activation of the apotosome and also of effector caspases. Other pathways involving c-MYC-induced cytochrome *c* release and apoptosis include indirect activation of p53 tumor suppressor via p19ARF leading to transcription of BAX. ARF can also inhibit c-MYC by blocking transcriptional activation of genes involved in growth while not affecting c-MYC mediated gene repression, resulting in increased apoptosis. This feedback mechanism operates independently of p53 (b). In some tumor models, deregulated expression of the inhibitor of ARF, Bmi-1 can cooperate with MYC in tumorigenesis. Ligation of death receptor CD95/FAS triggers association of the intracellular adaptor protein FADD with the CD95 receptor and forms the DISC (c).

Normal mouse chromosomes

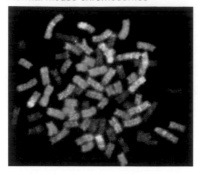

Chromosomal polyploidy

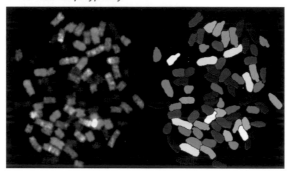

Plate 10.1 Polyploidy in mouse cells as shown by spectral karyotype (SKY) – images by courtesy of Maria Blasco. SKY provides visualization of all of an organism's chromosomes together, each labelled with a different color, and is readily employed to determine abnormalities in chromosome number. The left hand images show the colors of the different chromosome pairs as seen under microsocopy. The right hand panel for the polyploid cells has been artificially colored to show more clearly the abnormal chromosome numbers – several chromosomes are present in excess numbers (polyploidy).

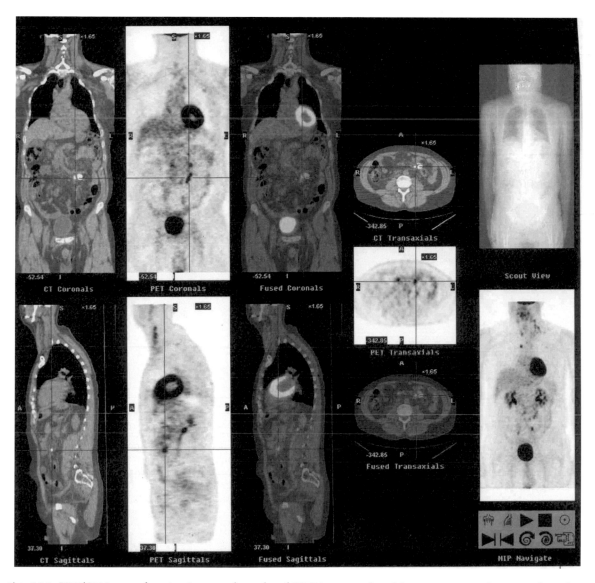

Plate 15.1 PET/CT image showing increased uptake of FDG in areas of nodal metastases in the cervical, coeliac, and mesenteric nodes. Increased uptake is correlated with the CT abnormality.

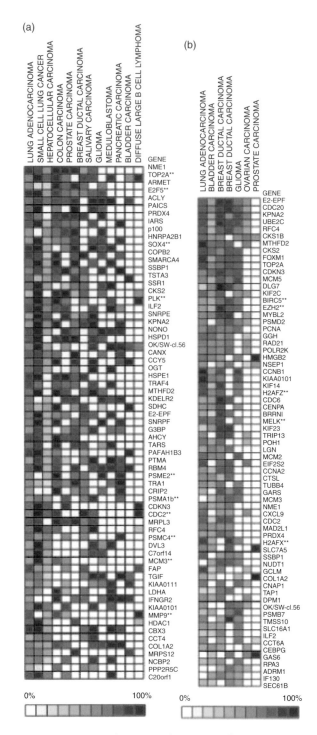

Plate 18.1 (a) Genes that play an important role in neoplastic transformation in various cancers. White boxes indicate not present or no significant changes. Red boxes indicate significant overexpression in cancer versus normal tissue. Shades of red indicates the percentage of cancer samples that had an expression value greater than the 90th percentile of normal samples. More information on individual genes can be found through the gene ontology website at www.geneontology.org (b) Sixty-nine genes that are overexpressed in undifferentiated cancer relative to well-differentiated cancers. EZH2, H2AFX, and H2AFZ are particularly interesting genes as they play an active role in transcriptional regulation. Reproduced from Rhodes et al. 2004, *PNAS*, 101, no. 25, pp. 9312 (a) and 9313 (b).

9

Telomeres and Senescence

Maria Blasco and Mike Khan

Senescence begins and middle age ends the day your descendants outnumber your friends.

Ogden Nash

KEY POINTS

- Cultured cells can divide only so many times before they growth arrest and stop dividing – under usual circumstances a largely stable and irreversible state termed "replicative senescence." Senescence may also follow enforced expression of oncogenes and activation of DNA damage responses.

- Senescence involves the attrition or impaired function of important structures at the ends of chromosomes (telomeres), the activity of DNA damage responses, and activation of the ARF–p53–p21^{CIP} and RB–p16^{INK4a} pathways, which collaborate to stop cellular proliferation. DNA damage responses may be triggered by the activity of DNA-damaging agents or replication errors, but can also result from the cumulative effect of repeated cycles of replication on telomere length and function.

- Telomerase activity is necessary to maintain telomere integrity, which in turn prevents chromosome ends from being detected and processed as damaged DNA.

- Increasing generations of a telomerase-deficient mouse model result in progressive telomere loss, which eventually leads to premature aging phenotypes in these mice.

- In contrast, more than 90% of all types of human tumors reactivate telomerase.

- Cancer cells need telomerase to maintain functional telomeres and to divide indefinitely and this has opened the possibility of novel therapeutic approaches based on using telomerase as a target.

- Importantly, senescence may also result from mechanisms independent of telomerase activity and telomere attrition, particularly in response to environmental factors and DNA damage. This is characterized by activation of the RB and p53 pathways including the p16^{INK4a}, p21^{CIP1}, p27^{KIP1} cyclin dependent kinase inhibitors (CKI) and disruption of lysosomal function through enhanced activity of the senescence-associated beta galactosidase (SA-β-Gal).

- Oncogene-induced senescence has been shown to restrain transformation in human cancer cells and cancer progression in animal models *in vivo*, and together with apoptosis is a key innate restraint on the potential of oncogene deregulation to give rise to cancer.

- Signaling through the p53 and RB tumor suppressor pathways is also required to maintain stable growth arrest in senescent cells *in vitro*. In fact, at least in cultured cells, senescence may be reversible by inactivating these pathways though this has not as yet been shown to occur *in vivo*. However, the possibility should be considered that senescent cells could still contribute to cancer at a future date by re-entering the cell cycle.

INTRODUCTION

Most human cells in culture can only reproduce a limited number of times before they lose the ability to divide and become what is called "senescent." This process has been associated with various aspects of longevity determination and aging, and it may also play an important role in preventing cancer.

Seminal work by Hermann Muller and Barbara McClintock identified that the ends of chromosomes are capped by structures, the telomeres, which serve to prevent chromosome fusions. As the mechanics of DNA replication were described it was identified that DNA polymerase, the enzyme responsible for DNA replication, could not fully synthesize the 3′ end of linear DNA. This was described by James Watson as the **end-replication problem**. At about the same time in Russia, Alexey Olovnikov proposed that the end-replication problem would result in telomere shortening with each round of replication and that this mechanism could account for replicative senescence, which was compatible with studies by Leonard Hayflick and colleagues showing that control of replicative senescence was situated in the nucleus. Olovnikov's model is supported by numerous studies, which show that telomere shortening is a major contributor to replicative senescence and that telomere length may act as a "mitotic clock" to determine the number of cell divisions allowed for a given cell (see Box 9.1).

Telomerase is a reverse-transcriptase enzyme that elongates the telomeres, thus counteracting the normal telomere attrition occurring every time DNA is replicated. Telomerase has two components: an RNA component and a catalytic subunit and is strongly implicated in the avoidance of senescence by "immortal" tumor cell lines. Studies have also confirmed that expression of the catalytic subunit of human telomerase (hTERT) in various cell types avoids replicative senescence by maintaining telomere length. Telomerase is not the only mechanism capable of elongating the telomeres, although the mechanisms underlying this "alternative lengthening of telomeres" (ALT) remain unknown. Whatever the mechanisms, stabilizing telomeres is a prerequisite for immortality and likely for tumor development, normally by telomerase. In contrast, telomerase

---

Box 9.1 Hayflick limit

Hayflick and colleagues defined the stages of cell culture in three phases. Phase I is the primary culture, when cells from the explant multiply to cover the surface of the culture flask. Phase II represents the time when cells divide in culture; once cells cover the culture surface, they stop multiplying, with growth continuing only if cells are subcultivated. Subcultivation requires removing culture medium and detaching cells from the culture substrate with a digestive enzyme called trypsin, which dissolves the substances keeping cells together. Cells can then be replated at lower density and further culture medium added. Cells reattach to the new culture substrate and start dividing again until a new subcultivation is required. These procedures are termed "passages." Lastly, Phase III begins when cells start dividing slower. Eventually they stop dividing and die. Hayflick and Moorhead noticed that cultures stopped dividing after an average of 50 cumulative population doublings. This phenomenon is known as Hayflick's limit, Phase III phenomenon, or replicative senescence.

---

inhibition can induce senescence in cancer cells, but if cells with shortened telomeres manage to divide, this may also under these circumstances be a strong supporter of cancer progression due to the increased rate of subsequent chromosomal aberrations (Fig. 9.1). Most, if not all, human somatic tissues have no detectable telomerase activity; in the bone marrow, hematopoietic cells express telomerase. Telomerase activity is higher in primitive progenitor cells and then downregulated during proliferation and differentiation. Telomerase activity has been detected in several normal human somatic proliferating cells: for instance, skin and colorectal tissues. Expression of hTERT appears to be regulated by the proto-oncoprotein c-MYC, thus providing a potential link between cell cycle progression and telomere maintenance.

Although telomere length regulates replicative senescence it is not the sole regulator of cell senescence. Telomeres are not linear, but form duplex loops, called t-loops (Fig. 9.2), the formation of which is dependent on telomeric repeat-binding factors TRF1 and TRF2, and may be crucial for stabilizing telomere caps. Capping

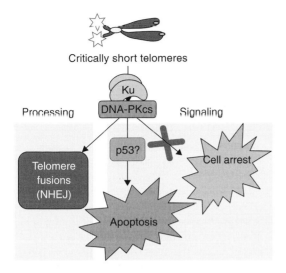

Fig. 9.1 The nonhomologous end joining (NHEJ) complex DNA-PK has been involved both in mediating telomere fusions and apoptosis triggered by critically short telomeres. In particular, these consequences of telomere dysfunction are rescued in DNA-PKcs and Ku86-deficient backgrounds. However, cell cycle arrest due to short telomeres is not rescued by the absence of DNA-PKcs and Ku86 proteins. Blasco *et al.* have suggested that p53 may be downstream of DNA-PK in signaling apoptosis due to short telomeres.

may protect the telomeres from being recognized as DNA damage and may be as important as telomere length in avoiding telomere dysfunction and preventing apoptosis and senescence.

Senescence can occur in fibroblasts in the absence of cell division and short telomeres. Although one hypothesis is that telomere dysfunction occurs in confluent cells despite lack of telomere shortening, it is also apparent that viral genes such as *SV40T-antigen*, adenoviral *E1A* and *E1B*, or the human papillomavirus *E6* and *E7* genes can also immortalize cells. The E1B and E6 proteins bind and inactivate the tumor-suppressor protein p53 while E1A and E7 bind and inactivate the retinoblastoma protein, RB (or pRB). Immortalization depends on inactivation of both p53 and RB, or their pathways.

At present, it is accepted that the p53 and RB pathways are largely responsible for inducing senescence. Immortalization with viral protein oncogenes results in an extended life span after which cells enter a stage called crisis, where cells proliferate but the rate of

apoptotic cells gradually increases and thus cell numbers eventually diminish. Since both p53 and RB/p16^{INK4a} pathways are inactive and chromosomal instability and fusions are abundant, crisis is thought to emerge due to extremely short telomeres. Occasionally, immortal cells emerge from crisis with stabilized telomeres, normally involving telomerase activation. In a sense, crisis can be seen as the ultimate consequence of telomere dysfunction. Whatever changes occur during telomere dysfunction, the mechanisms triggering growth arrest appear to involve the recognition of dysfunctional telomeres as DNA damage and that the p53 and RB/p16^{INK4a} pathways collaborate to stop cellular proliferation.

In a recent publication, an intriguing link between telomere crisis and progression of a form of breast cancer has been identified. It was found using an *in vitro* model that ductal hyperplasia may be associated with a critical shortening of telomeres in rapidly dividing cells resulting in crisis. Importantly, although the majority of such cells undergo apoptosis or presumably senescence, thus avoiding further risk of progression to cancer, a small number of cells may escape these usual consequences of chromosome instability. Instead rare cells may reactivate telomerase, thus preventing further telomere attrition and allowing such cells to progress to ductal carcinoma *in situ* and presumably thence to invasive cancer. Not surprisingly, the rate of genetic aberrations and chromosomal rearrangements was highest during the crisis period. It is speculated that such a transition through telomere crisis may be a crucial event in the development of most breast carcinomas.

SENESCENCE

In senescence, terminal differentiation is induced by a variety of stimuli including alterations of telomere length and structure, some forms of DNA damage (e.g. oxidative stress), and activation of certain oncogenes.

Senescence differs from other physiologic forms of cell cycle arrest such as quiescence in that it is largely irreversible, barring the inactivation of p53 and/or RB and it is associated with

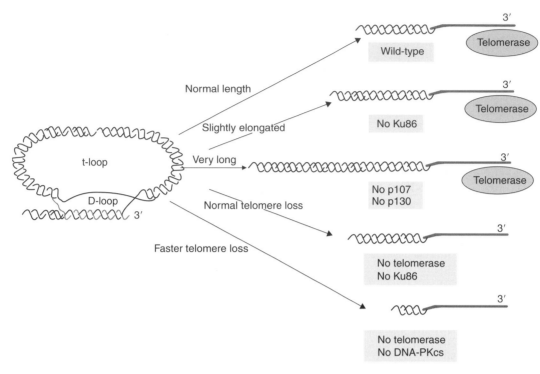

Fig. 9.2 Different activities modulate telomere length, such as the telomere-binding proteins TRF1 and TRF2. More recently, activities involved in DNA repair (Ku86 and DNA-PKcs), as well as in cell cycle regulation (p107 and p130) have been also shown to have a direct impact in regulating telomere length. Ku86 is a negative regulator of telomerase, while DNA-PKcs cooperates with telomerase in maintaining telomere length. Simultaneous absence of p107 and p130 results in a fast elongation of telomeres in the absence of changes in telomerase activity.

distinctive phenotypic alterations such as cellular flattening, expression of SA–β–Gal activity and the establishment of an unusual form of heterochromatin, known as senescence-associated heterochromatic foci (SAHF). As discussed, senescence requires activation of the p16^{INK4a}-RB and/or p53 pathways. The CKI, p16^{INK4a}, inactive in many cancers, inhibits the activities of CDK4 and CDK6, thus preventing cell cycle progression (Chapters 4 and 7). Exogenous expression of p16^{INK4a} induces senescence in immortal cells lacking functional p53 and conversely cells may be immortalized by disruption of both p16^{INK4a} and p53. RB is needed for growth suppression mediated by p16^{INK4a} suggesting that it acts upstream of RB in regulating senescence. Telomere dysfunction can activate p16^{INK4a}. A second CKI, p21^{CIP1}, blocks the cell cycle by inhibiting CDK2, CDK4, and CDK6, thereby preventing RB phosphorylation, and can induce senescence independently of p16^{INK4a}. p21^{CIP1} is induced by p53 and is an important mediator of replicative senescence induced by p53 (Fig. 9.3). The p53 pathway is an important mediator of senescence induced by dysfunctional telomeres likely as a result of triggering of DNA damage responses (Chapter 10).

Activity of p53 is regulated in part by MDM2-mediated degradation. Stabilization of p53 occurs in response to DNA damage and also to inappropriate activation of oncogenes such as *c-MYC* via induction of ARF (p19^{Arf} in the mouse), which can bind MDM2, thereby inhibiting the destruction of p53. Either p53–p21^{CIP1} or p16^{INK4a} can promote RB hypophosphorylation and initiate senescence. Some senescence-inducing stimuli (e.g. activation of the *RAS* oncogene) appear to induce both pathways, while others (e.g. DNA damage) appear to preferentially activate one or the other.

Recent studies confirm that diverse potentially cancer-promoting genes can trigger

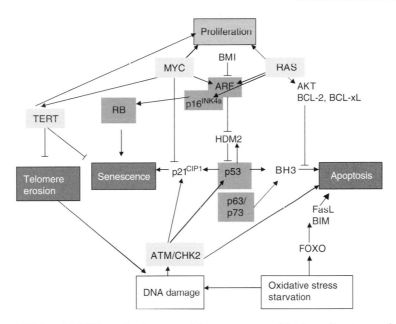

Fig. 9.3 Activation of RAS and MYC results in potential engagement of both replication and growth but also of apoptosis and possibly senescence. If either MYC or RAS levels are excessive (as might occur during onogenesis) or other proapoptotic signals are received then the balance may be tipped away from replication. RAS can promote senescence through either p16^{INK4a} or ARF, which activate RB or p53 pathways respectively. Senescence may also result from telomere erosion or dysfunction which may activate DNA damage responses that in part are the same as those also activated by DNA DSBs. Functional telomeres must normally prevent DNA ends being recognized and processed as DSBs. MYC can inhibit growth arrest/senescence by inhibiting p21^{CIP1} and inducing TERT (telomeres reverse transcriptase), which prevents erosion of telomeres and may activate replication by other means also. Although MYC may activate the apoptotic pathway (e.g. via ARF), RAS is able to suppress apoptosis by activating the PI3K pathway and, subsequently, AKT. It can readily be appreciated how oncogenic MYC and RAS may conspire in oncogenesis. The combination of RAS and MYC acting together provides a potential means of avoiding apoptotic and senescence mechanisms activated by either acting individually. Moreover, it can also be appreciated how inactivating mutations in RB or p53 (or their pathways involving p19^{Arf}, p16^{INK4a}, p21^{CIP1}, etc) may contribute to tumorigenesis by enabling the cancer cell to avoid either senescence or apoptosis or both.

senescence *in vivo*, including oncogenic Ras and BRAF, which appear to induce p16^{INK4a} via the ERK pathway and activation of AKT or inactivation of PTEN, which activate the ARF–p53–p21^{CIP1} pathway. The actual mechanisms involved will likely vary between different tumors but likely share a common final denominator, namely, prevention of RB phosphorylation and therefore inhibition of G_0/G_1 to S transition. Intriguingly, RB may also promote senescence by interacting with various factors that can result in histone methylation and formation of transcriptionally inactive heterochromatin. This may stably silence S phase genes in senescent cells. This raises some questions over the use of drugs that can reverse such epigenetic changes (Chapter 11). Recently it has been

shown in a mouse model of lymphoma that inactivation of a key histone methyltransferase often associated with RB (Suv39h1) prevented senescence and fostered cancer progression. Many drugs targeting epigenetic modification (inhibitors of histone deacetylases and DNA methyltransferases) with the aim of reversing epigenetic silencing of tumor suppressors (Chapter 11) could conceivably also reverse or prevent senescence.

Cellular senescence, immortality, and cancer

In 1961, Leonard Hayflick and Paul Moorhead made the unexpected discovery that human cells derived from embryonic tissues can only divide a finite number of times in culture – see Box 9.1. Hayflick and Moorhead worked with a connective tissue cell, the fibroblast, but similar findings have been observed for other cell types including skin keratinocytes, endothelial cells, and lymphocytes. Importantly, senescence also occurs in cells derived from embryonic tissues and in cells taken from mice and other animals. Early results suggested a relation between the number of population doublings cells undergo in culture and the longevity of the species from which the cells were derived. For example, cells from the Galapagos tortoise, which can live over a century, divide about 110 times, while mouse cells divide roughly 15 times. If cells are taken from individuals with premature aging syndromes, such as Werner's syndrome (the gene responsible for Werner's syndrome is a member of the RecQ subfamily of DNA helicases), they exhibit far fewer doublings than normal cells. Some cells never reach replicative senescence and are said to be "immortal." These cells include embryonic germ cells and most cell lines derived from tumors. Replicative senescence in human fibroblasts is characterized by growth arrest, that is, cells stop dividing, they are larger and morphologically heterogeneous and are more sensitive to cell–cell contact inhibition. Senescent cells growth arrest at the G_1/S transition of the cell cycle, which is in general irreversible (inactivation of p53 and RB pathways notwithstanding) and cells remain refractory to growth factors even though senescent cells can remain metabolically active for long periods of time.

Replicative senescence may limit the growth potential, but not necessarily the survival, of a dividing cell as a consequence of telomere attrition. Senescent cells exhibit a particular phenotype when arrested in the G_1 phase of the cell cycle. They appear flattened and enlarged with increased cytoplasmic granularity, and enhanced activity of senescence-associated lysosomal hydrolase β-galactosidase when assessed at an acidic pH. In 1995, Judith Campisi and her team discovered that the enzyme β-galactosidase is abnormal in senescent cells; SA-β-Gal activity becomes active at a higher pH (pH 6 rather than pH 4). Early reports showed that lysosomes increase in number and size in senescent cells, and recently it has been shown that the increased autophagy with aging in culture

may be associated with an increase of lysosomal mass. While refractory to mitogenic stimuli, senescent cells remain viable and metabolically active and possess a typical transcriptional profile that distinguishes them from quiescent cells. At the protein level, numerous regulators of cell cycle progression, checkpoint control, and cellular integrity such as p53 or p16^{INK4a} have been found to be induced in response to various prosenescent stimuli. Although the molecular mechanisms underlying the senescent phenotype remain largely unknown, there is increasing evidence that formation of heterochromatin in the vicinity of promoters that control gene expression related to cell cycle progression might be implicated in the maintenance of an irreversible growth arrest. Normal human cells are diploid, yet with each passage in culture, the proportion of polyploid cells increases. Mutations in mitochondrial DNA also increase with age. Senescent cells are less able to express heat shock proteins, which may also increase their vulnerability to death triggers. The expression of multiple genes is altered during cellular aging. Normally, cell culture conditions include 20% oxygen and these were the conditions initially used by Hayflick and Moorhead. However, if human fibroblasts are cultured at reduced oxygen (3%) closer to physiological conditions, they achieve a further 20 doublings. In contrast, different types of human cells cultured above 20% oxygen survive fewer doublings. Importantly, the same effect is not observed in tumor cell lines.

Cell senescence can be found without telomere shortening *in vivo*. One view is that replicative senescence is not a major factor *in vivo* and that cellular senescence instead depends on stress factors that trigger the RB and p53 tumor-suppressor pathways. From a simple mathematical perspective, the existence of replicative senescence under normal physiological conditions *in vivo* has been questioned. The reasoning is along the following lines: assuming human fibroblasts endure 50 population doublings, 2^{50} would seem more than enough cells to last several lifetimes under normal conditions. However, in cancer the issue has recently been somewhat clarified. First deregulated expression of oncogenes seems to trigger cell cycles that in some way are detected by the cell as aberrant (cancer cell cycles). These cancer cell cycles trigger DNA damage responses, likely independently of telomere attrition, which culminate in growth arrest or apoptosis. These responses variously involve activation of ARF–p53 or p16^{INK4A}–RB. Whether detection of DNA damage or activation of DNA damage responses is essential for this remains to be shown.

Stem cells participate in tissue homeostasis by replacing differentiated cells that have undergone cell death or have been lost (e.g. shed from the skin surface). Human stem cells express the enzyme telomerase suggesting that these actively dividing cells could avoid telomere attrition by replacing telomeric repeats. However, somatic stem cells can show senescence and telomere-independent mechanisms of cellular senescence are increasingly being described.

The purpose of cellular senescence is unclear but it may act as an anticancer barrier. Loss-of-function mutations in *WRN* lead to genomic instability, an elevated cancer risk, and premature cellular senescence. Interestingly, WRN is normally upregulated by c-MYC, suggesting that WRN may act downstream of c-MYC to prevent cellular senescence during tumorigenesis (Chapter 6). Epidemiological studies have indicated for a long time now that cancer rates accelerate dramatically after the age of 50 years, and observation usually attributed an accumulation of harmful genetic mutations. However, Judith Campisi (2005) and others have suggested that the accumulation of cells with a senescent phenotype (persisting in certain tissues after undergoing changes in morphology, behavior, and function) might also contribute to cancer. Certainly, even in animal models it has been shown that telomere attrition may not invariably be an anticancer mechanism. Thus, if cells are able to avoid replicative senescence (e.g. by losing p53) despite telomere shortening then they are at increased risk of undergoing major chromosomal aberrations during cell division, which may actually support cancer progression. Moreover, senescent cells may secrete many different molecules, some of which have been shown by Campisi and colleagues to have a "field effect" that could encourage malignant behavior in nearby cells in a paracrine fashion. This has resulted in speculation that cellular senescence evolved as a cancer suppression mechanism at a time when the life expectancy for humans was far shorter than it is today.

Given that most people now can expect a much greater life expectancy (at least in developed countries), senescence may be an example of "antagonistic pleiotropy." Namely, a trait selected to optimize fitness early in life turns out to have unselected deleterious effects later on. In fact, such mechanisms, including fetal programming, have been proposed as contributors to many of the more prevalent diseases of modern life including not only cancer but also obesity, diabetes, and atherosclerotic diseases. One version of this is the so-called Barker hypothesis, whereby exposure of a fetus to malnutrition during ontogeny may in some way foster the birth of a child "programmed" to successfully cope with malnutrition through adult life. The subsequent exposure of such an "adapted" individual to an unexpected energy-rich diet of burgers, pizzas, and sugar-containing drinks, for which that individual is now maladapted, may then result in disease, albeit usually at an older age.

Senescent cells may promote cancer behaviors in other cells – the dangers of old stroma

As discussed, cellular senescence is an effective barrier to neoplastic transformation as it removes risky cells from the cell division cycle. Intriguingly, however, senescent cells are now known to secrete various factors some of which may actually stimulate the proliferation of non-senescent cells, potentially including premalignant cells – this is a further example of antagonistic pleiotropy. In a recent study from the laboratory of Judith Campisi, premalignant mammary epithelial cells exposed to senescent human fibroblasts in mice irreversibly lose differentiation and undergo malignant transformation. One of the secreted factors from senescent cells that may contribute to this behavior is the matrix-degrading enzyme, MMP-3. Thus, senescent cells in stromal tissues may contribute to age-related tumorigenesis by secreting factors that can alter the differentiation state of other cell types, such as epithelial cells in the evolving tumor. Intriguingly, MMP3 produced by stromal cells has recently been shown to influence various processes in a cancer (or would be cancer) cell. In addition to fostering invasion and metastases at later stages in tumor development, MMPs may exert an early influence on cell transformation and cancer development. MMP3 can activate a member of the RAC family (RAC1b) promoting formation of reactive oxygen species (ROS) with resultant DNA damage and genomic instability. Moreover, MMP3n also induces the epithelial to mesenchymal transition (EMT), a major factor in cancer formation and spread, possibly by cleaving E-cadherin and triggering signaling via the WNT-β-catenin pathway.

Premature senescence, reversibility, and cancer

Premature senescence recapitulates cellular and molecular features of replicative senescence. Extrinsic factors such as anticancer agents, γ-irradiation, or UV light have been shown to induce premature senescence as a DNA damage-mediated cellular stress response. DNA lesions are sensed and transduced via protein complexes involved in DNA maintenance and repair including the ataxia telangiectasia mutated kinase (ATM) amongst others (Chapter 10). Such kinases directly or indirectly phosphorylate the p53 protein, at certain residues, which can then participate in apoptosis or in the induction of DNA damage-mediated senescence. p16^{INK4a} has been implicated in both response to DNA damage and control of stress-induced senescence, although how p16^{INK4a} induces permanent G$_1$ arrest is not known. In fact, p16^{INK4a} is essential for the maintenance of cellular senescence. The p21^{CIP1} protein acts as a molecular switch that triggers telomere-initiated senescence. Interestingly, the mechanisms by which dysfunctional telomeres lead to p21^{CIP1} activation are similar, but not identical, to cellular responses to DNA damage. The p16^{INK4a} protein acts independently from telomeres.

Damage to DNA, the prime target of anticancer therapy, triggers programmed cellular responses. In addition to apoptosis, therapy-mediated premature senescence has been identified as another drug-responsive program that impacts the outcome of cancer therapy. The best-known signaling pathway for drug-induced senescence is triggered in response to DNA damage. Chemotherapeutic agents such as the topoisomerase inhibitors doxorubicin and cisplatin induce DNA damage and particularly DNA double-strand breaks (DSBs). DNA damage is followed by checkpoint-mediated cell cycle arrest culminating in either repair or if irreparable in apoptosis or senescence (Chapters 4 and 10). As discussed in Chapter 10, the main sensing mechanism for DNA damage involves ATM and ATR, members of the PI3K-related kinases, and their downstream targets including H2AX, NBS1, and BRCA1, which in turn lead to activation of Chk1 and Chk2, the histone H2AX, NBSI and BRCAI. A key event in drug-induced cell cycle arrest is the stabilization of p53 and enhanced transcription of p21^{CIP1}. Chemotherapeutic drugs may trigger both senescence and apoptosis, but the former may be achieved at lower administered doses. Cancer cells must escape both senescence and apoptosis in order to become drug resistant.

The possibility that senescence may be one aspect of the inherent tumor-suppressor activity of some oncogenes such as RAS was discussed in Chapter 6. Cellular senescence is an attractive therapeutic goal, but the value of this strongly relies on senescence being irreversible. Unlike apoptosis, senescence still leaves a viable cell, with reversion of a senescent cell into a dividing cell posing the threat of a tumor relapse, unless senescent cells are in some way cleared by other processes, such as shedding from the skin surface or phagocytosis. In an *in vivo* model of drug-induced senescence in mouse lymphomas, repeated anticancer therapy eventually selected against senescence-controling genes such as *p16^{INK4a}* or *p53*, resulted in tumor relapse and progression to a more aggressive tumor.

Several studies suggest that senescence can be reversible, if key proteins involved in maintenance of senescence are lost. Thus, replicative senescence in human lung fibroblasts was reversed by dual functional inactivation of p53 and RB, by the expression of simian virus 40 large T antigen (SV40T), or by a combination of p53 inactivation and knockdown of p16^{INK4a} expression using siRNA. Inactivation of p53 in senescent mouse embryonic fibroblasts enabled cells to resume proliferation. Overall, this supports the notion that p53 and p16^{INK4a}–RB variously cooperate not only in the induction but also in the maintenance of premature senescence *in vivo*. However, we still require experiments demonstrating successful reversal of cellular senescence *in vivo*. While the acquisition of spontaneous mutations that disable *p53* or *RB* in a resting cell without DNA replication seems rather unlikely, epigenetic changes, for example, promoter methylation to silence p16^INK4a expression, might occur in senescent cells. Moreover, in any experiments conducted to address this important question, the adverse and potentially protumorigenic effects that such senescent cells might have on their nonsenescent neighbors would also need to be examined.

STEM CELLS AND CANCER

Mammalian aging occurs in part because of a decline in the restorative capacity of tissue stem cells. These self-renewing cells are rendered malignant by a small number of oncogenic mutations, and overlapping tumor-suppressor mechanisms (e.g. p16^INK4a–RB, ARF–p53, and the telomere) have evolved to ward against this possibility. These beneficial antitumor pathways, however, appear also to limit the stem cell life span, thereby contributing to aging.

Tissue homeostasis in the adult requires that cells that have died or have been shed (blood cells, skin cells, liver cells, etc.), are replaced by newly generated cells. Even in organs once thought to be postmitotic such as the pancreatic islet and the brain renewal has been demonstrated to occur and intriguingly in the former may usually actually take place from the existing mature β cells rather than from a stem cell precursor.

Adult mammals exhibit extensive proliferation and renewal even in the absence of disease. Thus, the intestinal lining replaces itself entirely on a weekly basis, and the bone marrow produces trillions of new blood cells daily. This obviously creates a potential risk of cancer, given the likely rate of somatic mutation in the stem cells giving rise to these new cells in some tissues. It may be unsurprising, therefore, that some 1% of neonatal cord blood collections contain significant numbers of myeloid clones harboring oncogenic fusions such as the AML–ETO fusion associated with acute leukemia, and

up to one in three adults possess detectable IgH-BCL2 translocations, which are commonly associated with follicular lymphoma. Given that cancers are far less frequent, this gives testament to the efficiency of cancer restraining mechanisms such as the tumor suppressors. Various overlapping tumor-suppressor barriers appear to be most important and include the p16^INK4a–RB pathway, the ARF–p53 pathway, and telomeres. A common endpoint for these major tumor suppressors is senescence or apoptosis.

TELOMERES

Telomerase activity is necessary to maintain the integrity of telomeres, which in turn prevent chromosome ends from being detected and processed as damaged DNA. Increasing generations of primary cells result in telomere attrition and the reproducible loss of proliferative potential, something that has been termed "replicative senescence." Similarly, increasing generations of a telomerase-deficient mouse model result in progressive telomere loss, which eventually leads to premature aging phenotypes in these mice. In contrast, more than 90% of all types of human tumors reactivate telomerase. Cancer cells need telomerase to maintain functional telomeres and to divide indefinitely and this has opened the possibility of novel therapeutic approaches based on using telomerase as a target.

Telomeres and replicative senescence

Telomeres cap the ends of chromosomes by forming a higher-order chromatin structure that protects the 3' end from degradation and DNA repair activities (see Figs 9.4 and 9.5 for FISH visualizations of telomeres, mouse chromosomes are shown in Fig. 9.6). Mammalian telomeres are composed of TTAGGG repeats bound to specialized proteins (see Fig. 9.7 for the current model for telomere structure and telomere-binding proteins). Loss of telomere capping, either due to TTAGGG exhaustion or disruption of telomere structure, results in end-to-end chromosomal fusions and loss of cell viability.

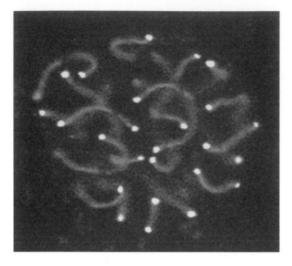

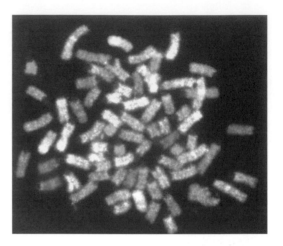

Fig. 9.6 Mouse chromosomes. Spectral karyotyping (SKY) to visualize mouse chromosomes. (Please see the website for the color version of this figure.)

Fig. 9.4 Meiotic telomeres. Fluorescence *in situ* hybridization (FISH) showing a meiocyte nucleus where chromosomes are stained in red with a synaptonemal complex protein, and telomeres are stained in yellow with a PNA-telomeric probe. (Please see the website for the color version of this figure.)

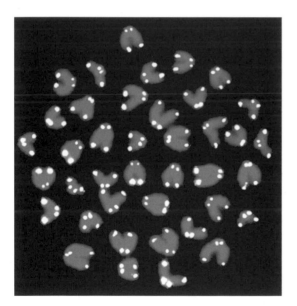

Fig. 9.5 Metaphase chromosomes. Fluorescence *in situ* hybridization (FISH) of a mouse metaphase nucleus showing the chromosomes in blue (DAPI staining), and the telomeres in yellow (staining with a PNA-telomeric probe). (Please see the website for the color version of this figure.)

As cells proliferate, TTAGGG repeats are lost from telomeres unless they have activated telomerase, a reverse transcriptase that adds TTAGGG repeats onto preexistent telomeres. Most normal somatic cells do not have sufficient telomerase activity and suffer telomere attrition. Global telomere shortening eventually results in individual chromosomes with critically short telomeres that have lost their functionality. The most dramatic consequences of telomere dysfunction are the appearance of chromosomal fusions and the loss of cell viability, which occur with increasing passages of cultured primary cells and are thought to trigger what is known as "replicative senescence." Loss of viability due to critically short telomeres has been demonstrated in telomerase-deficient mice that show degenerative pathologies in various organ systems. These pathologies are coincidental with decreased proliferative index and/or increased apoptosis of the affected cell types. Telomerase reintroduction results in elongation of the population of short telomeres and prevents telomere fusions, as well as pathologies, in the context of these mice. Similarly, telomerase reconstitution in most human normal primary cells is sufficient to immortalize them.

Telomeres and cancer

More than 90% of all human tumors reactivate telomerase, suggesting that telomerase is

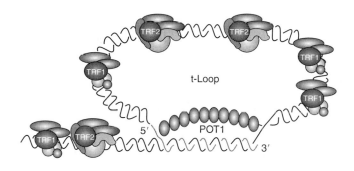

Telomere length control

Telomere capping

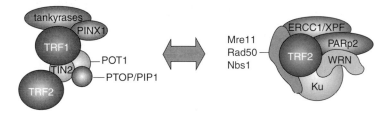

Fig. 9.7 Telomere-interacting proteins. A number of proteins are bound to telomeres and have been shown to be involved in regulating telomere length and telomere capping.

necessary to sustain viability of cancer cells. While normal cells lack telomerase activity, tumor cells depend on telomerase-mediated telomere maintenance to keep proliferating. This fact has made telomerase one of the most universal tumor markers, as well as an attractive target for therapeutic strategies. However, as with so many cancer-related processes, lack of telomerase activity or failure of telomere maintenance may also contribute to cancer progression. Thus, short telomeres in the absence of telomerase activity can trigger chromosomal instability, which could also favor tumor growth if permissive mutations are selected and if telomerase or other telomere-maintenance mechanisms are reactivated.

The consequences of critical telomere loss

Cancer cells generally have short telomeres and high levels of telomerase activity. Telomerase is able to rescue short telomeres, thus preventing activation of the DNA damage responses and allowing viability of cancer cells, which otherwise would arrest division or undergo apoptosis due to telomere damage and catastrophic chromosomal rearrangements. In agreement with this, mice deficient for telomerase activity and with short telomeres are resistant to developing

carcinogen-induced skin tumors, coincidental with p53 upregulation. If p53 is defective, however, the DNA damage response triggered by short telomeres is abrogated and telomerase-deficient mice show a higher frequency of chromosomal aberrations and a higher incidence of epithelial tumors. Therefore, simultaneous absence of telomerase and p53 may favor survival of cancer cells harboring chromosomal aberrations. In contrast, if either RB/p16^{INK4a} or APC tumor-suppressor pathways are abrogated, short telomeres have a negative impact on tumorigenesis, suggesting that in the mouse p53 is the main tumor suppressor that mediates cell arrest or apoptosis due to telomere dysfunction. Similarly, p53 but not RB/p16^{INK4a} is important in signaling telomere dysfunction produced by TRF2 mutation in mouse cells. In human cells, however, both pathways are important. Both ATM and the nonhomologous end-joining (NHEJ) have been proposed to signal dysfunctional telomeres. Abrogation of the NHEJ activities Ku86 or DNA-PKcs, rescues both end-to-end fusions and apoptosis, but not cell cycle arrest, triggered by critically short telomeres. These findings suggest that NHEJ could be involved in signaling apoptosis due to critical telomere shortening. Although it has been described that DNA-PKcs does not seem to phosphorylate p53 *in vivo* it is possible, however, that DNA–PKcs may signal through p53 in the particular case of telomere dysfunction, since DNA–PKcs and Ku86 are located at mammalian telomeres. All together these findings suggest that a dysfunctional telomere is probably detected as a DSB and signaled as such, and predict that mutations that abrogate NHEJ may affect the outcome of telomere dysfunction (Fig. 9.1). Recent data suggest that short and/or dysfunctional telomeres are indeed recognized as DSBs, as suggested by increased γ-H2AX at telomeres.

Telomere dysfunction in the absence of critical telomere loss

Mutation of telomere binding proteins can disrupt telomeric capping in the absence of significant telomere shortening. A dominant negative mutant of TRF2, which impairs the binding of TRF2 to TTAGGG repeats, results in telomere fusions with long telomeres at the fusion point and in the absence of a significant loss of telomeric sequences. Similarly, mice deficient for either Ku86 or DNA-PKcs, both of which are essential components of the NHEJ machinery for DSB repair, also result in end-to-end telomere fusions in the absence of telomere shortening. These findings suggest a role for NHEJ activities in telomere capping, similar to that proposed for TRF2. Indeed, TRF2 and DNA-PKcs mutation share the outcome that the resulting end-to-end fusions involve telomeres produced via leading-strand synthesis, suggesting that these two activities could be required for the postreplicative processing of the G-rich strand and for the formation of a proper telomere cap.

The role of telomere binding proteins in maintaining telomere function predicts that they may be also important in regulating cellular senescence, as well as cancer and aging. Recently, a mouse model for TRF2 overexpression (K5-TRF2 mice) has demonstrated an important role for this protein in organismal cancer and aging (Muñoz *et al.* 2005). In the case of primary mouse embryonic fibroblasts (MEFs), either telomere shortening to a critical length in the absence of telomerase (late generation Terc$^{-/-}$ cells), or long and dysfunctional

telomeres due abrogation of Ku86 (Ku86$^{-/-}$ cells), lead to similar frequencies of telomeric fusions. In both cases, dysfunctional telomeres trigger premature senescence-like arrest and result in decreased immortalization frequencies of MEF. In addition, both Terc- and Ku86-deficient mice show premature aging phenotypes and are tumor resistant when in a p53 wild-type background, suggesting that dysfunctional telomeres have an impact in both aging and cancer in the mouse. In turn, a dominant negative mutation of TRF2, which disrupts telomere capping, can influence the average telomere length at which senescence is triggered in human primary cells. Therefore, both telomere length and telomere state influence telomere function, which in turn impacts on senescence and immortalization both in human and mouse cells.

Multiple factors regulate telomere length and function

The telomere-binding proteins TRF1 and TRF2 can also influence telomere length. Similarly, simultaneous deletion of telomerase and Ku86 in doubly deficient Terc/Ku86 mice demonstrated that Ku86 acts as a negative regulator of telomerase-mediated telomere elongation (Fig. 9.2). The study of double Terc$^{-/-}$/DNA-PKcs$^{-/-}$ mice, in contrast, showed that DNA-PKcs is required for telomere length maintenance. In particular, absence of DNA-PKcs leads to a faster rate of telomere loss and to an earlier appearance of phenotypes in telomerase-deficient mice (Fig. 9.2). These findings suggest that telomere-binding proteins simultaneously regulate telomere capping and telomere length.

A connection between the members of the retinoblastoma family, RB, p107, and p130, and the mechanisms that regulate telomere length has been recently made (Fig. 9.2). In particular, MEFs doubly deficient in p107 and p130 (DKO), or triply deficient in p107, p130, and RB (TKO), have dramatically elongated telomeres compared to those of wild-type or RB-deficient cells in the absence of changes in telomerase activity. These findings reveal a connection between the RB family and telomere length control in mammalian cells, which could be at the basis of the lifespan extension exerted by a number of viral oncoproteins that inactivate the

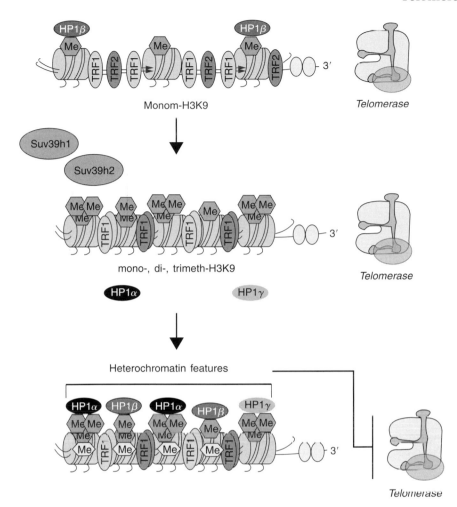

Monom-H3K9

mono-, di-, trimeth-H3K9

Heterochromatin features

Telomerase

Telomerase

Telomerase

Fig. 9.8 Telomere heterochromatin. Model for the assembly of telomeric heterochromatin. The Suv39h histone methyltransferases trimethylate lysine 9 at histone 3 (H3K9), which in turn creates a binding site for the heterochromatin protein 1 (HP1). These modifications confer heterochromatic features to telomeric chromatin, which in turn may impede the access of telomerase or other telomere elongating activities to the telomere, thus regulating telomere length.

RB family. It is also possible that the RB family proteins p107 and p130 could have a direct role in regulating telomere structure as suggested by role of these proteins in directing H4K20 histone methylation to these regions (Gonzalo *et al*. 2005).

More recently, it has been described that activities that modify chromatin architecture, can also influence telomere length. In particular, mammalian telomeres contain histone modifications characteristic of constitutive repressive chromatin domains, such as those of pericentric heterochromatin, and in the absence of these modifications telomeres suffer abnormal elongation (see Fig. 9.8 for a model on the regulation of telomere length by a heterochromatic higher-order structure).

Roles of telomerase beyond net telomere elongation

Telomerase activity is upregulated during mouse tumorigenesis, despite mice having very long telomeres. A selection for tumor cells that are telomerase-positive could be indicative of a novel role for telomerase in promoting survival independently of telomere length. The study of mice that either lack or have constitutive telomerase activity has provided evidence for this. In particular, telomerase-deficient mice with long telomeres are more tumor resistant than wild-type mice. Further evidence for a role of telomerase in promoting tumor growth independently of telomere length, was first obtained from studying mice that express constitutive levels of the catalytic component of mouse telomerase, Tert, in basal keratinocytes. The epidermis of these mice is more responsive to mitogenic stimuli and shows a higher incidence of carcinogen-induced tumors than wild-type controls. In addition, these mice are more susceptible to develop tumors as they age, and this is further aggravated when in a p53$^{+/-}$ genetic background. Interestingly, many human cancers also show high telomerase activity and p53 mutations, suggesting cooperation between high telomerase expression and p53 mutation both in humans and mice. Mice with transgenic telomerase expression under a β-actin promoter also have an increased incidence of spontaneous mammary epithelial tumors as they age. Additional data obtained in epithelial cells and cells of neural origin, in which telomerase is reactivated by forced Tert expression, also indicate that telomerase activity has an active role in promoting growth and survival.

The mechanisms by which having a constitutive telomerase activity may be promoting cell growth and survival are still unclear, although they seem to require a catalytically active enzyme. Yeast lacking telomerase show genome-wide changes of gene expression, supporting the view that telomerase may have other roles besides that of maintaining telomere length. A more recent genome-wide analysis showed that telomerase modulates the expression of growth-controling genes and enhances cell proliferation. Telomerase could also have a role in suppressing or processing DNA damage in the genome, thus favoring cell survival and proliferation. In this regard, telomerase is modified by activities involved in DNA damage signaling, such as AKT, PI3K, c-Abl, and p53, among others.

In summary, telomerase activation could favor tumorigenesis at least by two different mechanisms: by signaling proliferation and promoting growth independently of telomere length, and by rescuing tumor cells with critically short telomeres.

Short telomeres and DNA damage repair

Telomeres are important biological determinants of sensitivity to DNA-damaging agents. In particular, late generation Terc$^{-/-}$ mice show an enhanced mortality when irradiated with γ-irradiation. Intriguingly, the main DSB-DNA repair pathways are not significantly altered in these mice. More recently, it has been demonstrated that shortened telomeres join with DNA breaks and interfere with their correct repair in the context of the telomerase knockout mouse (Fig. 9.1). A correlation between telomere length and radiosensitivity has also been found for cultured cells. In addition, telomere shortening has been shown to alter the chemotherapeutic profile of transformed cells and p53 is required for this. Telomeric dysfunction even if telomeres are sufficiently long, also results in increased radiosensitivity. These findings have important implications for the therapy of cancer, as tumors treated with telomerase inhibitors could lead to telomere loss, thus increasing the sensitivity of these tumors to radiotherapy or to other genotoxic agents.

Telomerase inhibition in cancer therapy

The fact that cancer cells depend on telomerase-mediated rescue of short telomeres, opened the possibility of using telomerase inhibitors to selectively halt tumor growth without interfering with the viability of normal cells. In turn, a telomerase inhibitor could also impair survival/growth of tumor cells with long telomeres by interfering with its role in promoting tumorigenesis independently of telomere length. A telomerase inhibitor may also reduce the angiogenic potential of the tumor.

Telomerase inhibitors include antisense oligonucleotides against the telomerase RNA component, dominant negative mutants of the Tert subunit, molecules directed against G-quadruplex formation, as well as small pharmaceutical compounds highly specific for telomerase. Although the efficiency of these inhibitors has yet to be tested in clinical trials of human cancer, there is evidence from both cell culture models and mouse models to suggest that they would be efficient in stopping tumor growth. In addition, as just discussed above, these inhibitors may be combined with genotoxic agents to potentiate their effects. The efficiency of a telomerase-directed drug may depend on the status of p53 in the tumor. However, when telomerase was inhibited in various human cancer cell lines, the status of p53 was irrelevant for the final outcome, suggesting that p53-independent pathways signal telomere damage.

In mouse tumors, inhibition of telomerase using a dominant negative mutant of Tert leads to a fast selection of tumor cells that upregulate the endogenous *Tert* gene, thus compensating for the presence of the inhibitor and allowing growth. Whether this also happens in human tumor cells has yet to be determined. A telomerase inhibitor could also result in the selection of tumor cells that maintain telomeres in the absence of telomerase, a phenomenon termed ALT. Although only 5% of human tumors seem to be sustained by ALT, inhibition of telomerase could generate a selection pressure to activate ALT within the tumor, which in turn would no longer be responsive to the inhibitor.

Finally, the efficiency of a telomerase inhibitor will depend on its toxicity on cells that express telomerase, such as germ cells and stem cells. Since these cells have longer telomeres than tumor cells, it has been reasoned that they will not be affected by temporary telomerase inhibition, although there is no direct evaluation of this. Data from the telomerase knockout model indicate that these cell types are not affected in the absence of telomerase until their telomeres shorten below a critical length after several mouse generations. Similarly, patients who suffer dyskeratosis congenita, a human disease that is characterized by a faster rate of telomere shortening due to mutations in either telomerase or in Dyskeryn (required for an active telomerase complex), develop normally during the first years of life before the disease is manifested, suggesting that temporary telomerase inhibition during a chemotherapy treatment will not have deleterious effects for the organism.

QUESTIONS REMAINING

1 To define more accurately for which cancers targeting telomerase may be an effective cancer treatment.
2 To further understand the role of various telomere interacting proteins in telomere maintenance.
3 To further understand alternative mechanisms (other than telomerase) for telomere maintenance.

4 To further define the mechanisms underlying replicative senescence.
5 To explore the potential "field" effect of senescent cells on other cell types during tumorigenesis.
6 To explore the potential for inducing irreversible growth arrest in cancer cells as an effective therapeutic option.

CONCLUSIONS AND FUTURE DIRECTIONS

Telomerase and telomere biology are intense fields of research in oncology. Much has been learned in the last 10 years about their roles in cell immortalization and tumorigenesis. On the one hand, work done both with cultured human cells and with a telomerase-deficient mouse supports telomerase as an attractive target for development of new compounds with antitumorigenic activity. In particular, the fact that short telomeres in the absence of telomerase trigger a rapid loss of cell viability, as well as aggravate the cytotoxic effects of DNA-damaging agents, suggests that telomerase inhibitors would be effective when combined with genotoxic agents in halting cancer growth. On the other hand, new roles of telomerase in tumorigenesis, which are independent of its role in net telomere elongation, further support the efficiency of telomerase based therapeutic approaches in cancer. Therefore, we should expect that strategies that effectively inhibit telomerase are likely to be tested in cancer clinical trials in the near future. However, much is still unknown about the functional interactions between the different activities at the telomere, as well as about the pathways by which telomerase activity is regulated and by which it regulates other processes.

QUESTIONS

1 Replicative senescence may be induced by:
   a. Telomere attrition.
   b. Inactivation of RB signaling.
   c. Oncogenic *RAS*.
   d. DNA damage.
   e. Telomerase activity.

2 Telomeres:
   a. Protect the ends of chromosomes from DNA repair processes.
   b. Always shorten with every cell division in all cells.
   c. Are maintained entirely by action of telomerase.
   d. Can be maintained in the absence of telomerase.
   e. Are linear structures at the ends of chromosomes.
3 Telomerase:
   a. Is a reverse transcriptase.
   b. Comprises a catalytic and an RNA component.
   c. In the adult is active only in stem cells and cancer cells.
   d. Activity promotes senescence.
   e. May be activated by c-MYC.
4 Cellular senescence:
   a. Is always irreversible.
   b. Might be reversed by mutations in the RB pathway alone.
   c. Might be prevented by c-MYC.
   d. Can occur in the absence of telomere attrition.
   e. Only occurs after 50 or more cell divisions.
5 Telomeres:
   a. Of adequate length are always effective at preventing genetic instability.
   b. Require several interacting proteins in order to function effectively.
   c. When critically short may cause premature aging.
   d. When critically short always act to restrain cancer progression.
   e. Are much longer in mouse compared to human cells.

BIBLIOGRAPHY

Senescence

Beausejour, C.M., Krtolica, A., Galimi, F. *et al.* (2003). Reversal of human cellular senescence: roles of the p53 and p16 pathways. *EMBO Journal*, **22**: 4212–22.

Campisi, J. (2005). Suppressing cancer: the importance of being senescent. *Science*, **309**(5736): 886–7.

Campisi, J. (2005). Senescent cells, tumor suppression, and organismal aging: good citizens, bad neighbors. *Cell*, **120**(4): 513–22.

Hayflick, L. and Moorhead, P.S. (1961). The serial cultivation of human diploid cell strains. *Experimental Cell Research*, **25**: 585–621.

Sage, J., Miller, A.L., Perez-Mancera, P.A., Wysocki, J.M., and Jacks, T. (2003). Acute mutation of

retinoblastoma gene function is sufficient for cell cycle re-entry. *Nature* **424**: 223–8.

Serrano, M., Lin, A.W., McCurrach, M.E., Beach, D., and Lowe, S.W. (1997). Oncogenic ras provokes premature cell senescence associated with accumulation of p53 and p16INK4a. *Cell*, **88**: 593–602.

Telomeres

Blasco, M.A. (2002). Telomerase beyond telomeres. *Nature Reviews: Cancer*, **2**: 627–33.

Blasco, M.A., Lee, H.-W., Hande, P. *et al.* (1997). Telomere shortening and tumor formation by mouse cells lacking telomerase RNA. *Cell*, **91**: 25–34.

Chin, K., de Solorzaw, C.O., Knowles, D. *et al.* (2004). *In situ* analyses of genome instability in breast cancer. *Nature Genetics* **36**(9): 984–8.

Chin, L., Artandi, S. E., Shen, Q. *et al.* (1999). p53 deficiency rescues the adverse effects of telomere loss and cooperates with telomere dysfunction to accelerate carcinogenesis. *Cell*, **97**: 527–38.

d'Adda di Fagagna, F., Reaper, P.M., Clay-Farrace, L. *et al.* (2003). DNA damage checkpoint response in telomere-initiated senescence. *Nature*, **426**: 194–8.

García-Cao, M., Gonzalo, S., Dean, D., and Blasco, M.A. (2002). Role of the RB family members in controlling telomere length. *Nature Genetics*, **32**: 415–19.

González-Suárez, E., Samper, E., Flores, J.M., and Blasco, M.A. (2000). Telomerase-deficient mice with short telomeres are resistant to skin tumorigenesis. *Nature Genetics*, **26**: 114–17.

Goytisolo, F.A. and Blasco, M.A. (2002). Many ways to telomere dysfunction: *in vivo* studies using mouse models. *Oncogene*, **21**: 584–91.

Hemann, M.T., Strong, M.A., Hao, L.Y., and Greider, C.W. (2001). The shortest telomere, not average telomere length, is critical for cell viability and chromosome stability. *Cell*, **107**: 67–77.

Henson, J.D., Neumann, A.A., Yeager, T.R., and Reddel, R.R. (2002). Alternative lengthening of telomeres in mammalian cells. *Oncogene*, **21**: 598–610.

Karlseder J., Kachatrian, L., Takai, H. *et al.* (2003). Targeted deletion reveals an essential function for the telomere length regulator Trf1. *Molecular Cell Biology*, **23**(18): 6533–41.

Lee, H.W., Blasco, M.A., Gottlieb, G.J., Horner, J.W., Greider, C.W., and DePinho, R.A. (1998). Essential role of mouse telomerase in highly proliferative organs. *Nature*, **392**: 569–74.

Smith, L.L., Coller, H.A., and Roberts, J.M. (2003). Telomerase modulates expression of growth-controlling genes and enhances cell proliferation. *Nature Cell Biology*, **5**: 474–9.

Takai, H., Smogorzewska, A., and de Lange, T. (2003). DNA damage foci at dysfunctional telomeres. *Current Biology*, **13**: 1549–56.

Oncogene-induced senescense

Braig, M., Lee, S., Loddenkemper, C. *et al.* (2005). Oncogene-induced senescence as an initial barrier in lymphoma development. *Nature*, **436**(7051): 660–5.

Chen, Z., Trotman, L.C., Shaffer D. *et al.* (2005). Crucial role of p53-dependent cellular senescence in suppression of Pten-deficient tumorigenesis. *Nature*, **436**(7051): 725–30.

Michaloglou, C., Vredeveld, L.C., Soengas, M.S. *et al.* (2005). BRAFE600-associated senescence-like cell cycle arrest of human naevi. *Nature*, **436**(7051): 720–4.

10
Genetic Instability, Chromosomes, and Repair

Stella Pelengaris and Mike Khan

Without mutation, there can be no evolution. We may suffer for its faults, but if DNA were perfectly stable, we wouldn't even be here
Phil Hanawalt, scientist

We share half our genes with bananas, something which is more apparent to me in some of my colleagues as opposed to others!
David Horobin

KEY POINTS

- Numerous "caretaker genes" have evolved to tackle the surveillance and maintenance of DNA integrity, the loss of which gives rise to genomic instability and defects in mismatch repair – a major cause or effect of carcinogenesis.
- These "caretaker genes" are major barriers to the initiation and progression of cancer and represent the third major class of genes subject to "epimutations" in cancer, alongside the oncogenes and tumor suppressor genes.
- Cellular DNA is continually being damaged by exogenous as well as endogenous factors including: ultraviolet (UV) light, ionizing radiation and other genotoxic agents and factors relating to oxygen metabolism and the generation of reactive oxygen species (ROS) within cells.
- DNA damage is either *cytotoxic* (interstrand crosslinks and double-strand breaks) interfering with transcription, replication or chromosome segregation, or *mutagenic*, which if not removed before replication could cause miscoding resulting in mutations and subsequent carcinogenesis.
- DNA-damage signaling mechanisms seem to have evolved primarily to respond to cytotoxic lesions (and many of these are overlapping with those involved in telomere maintenance – Chapter 9), whereas DNA-repair mechanisms have evolved to deal with both classes of DNA lesions.
- DNA damage sensing likely takes place primarily during replication or transcription as the DNA strands may be more readily accessible to key "sensing" proteins at these times.
- DNA-damage signaling proteins, activated following detection of DNA damage, keep damaged cells arrested at specific points of the cell cycle in order for cells to attempt DNA repair before entering S phase or mitosis. However, if effective DNA repair is impossible or damage extensive these signaling processes can also trigger irreversible growth arrest or apoptosis of the cell (see Chapters 8 and 9). The p53–p21^{Cip1} tumor suppressor pathway plays a central role in this.
- In response to diverse stresses, exemplified by DNA damage, the well-known tumor-suppressor, p53, prevents propagation of a potential cancer cell by inducing genes which arrest the cell cycle or promote apoptosis. Interestingly, p53 may also regulate genes required for sucessful DNA repair and moreover can reduce levels of reactive oxygen species (ROS) thus protecting the genome from damage. This duality of p53 roles, as both protector and executioner, may in part be explained by the differences in genes regulated by p53 at low

---

and persistent high levels respectively. At low levels and early after activation, p53 mediates growth arrest, survival, and DNA repair, whereas high or persistent p53 activation mediates increased levels of p21^{Cip1}, ROS and apoptosis.

- Failure of DNA damage responses, due to epigenetic or mutational changes in "caretaker genes" or tumor suppressors, can result in inappropriate survival and replication of cells with damaged DNA and is amongst the cardinal features of cancer cells. The result will be genomic instability, which is generally classed as chromosomal instability (CIN), recognized by gross chromosomal abnormalities, or microsatellite instability (MIN) associated with a "mutator" phenotype.

- A question that has long vexed cancer biologists is do mutations arise by chance (stochastically) and are then selected for, or does cancer development depend on processes accelerating the rate of acquisition of mutations? The answer for different cancers is likely a combination of both at different stages of progression.

- Aberrant DNA damage responses arising through epigenetic or genetic changes may be present in premalignant colon cancer lesions, suggesting that in this example of a multistage cancer pathway a propensity to "epimutations" may come first.

INTRODUCTION

With around 10^{14} cells, each with 2 m of DNA, the adult human contains a total length of DNA that if stitched end to end would be around 200,000,000,000 kilometres long and comprise a stretch of 10^{24} bases, all subject to daily threat of damage and mutation. The herculean task of detecting and repairing this DNA damage falls to each individual cell, which is solely responsible for its own 2 meters and around 6×10^9 base pairs, divided into separate chromosomes. Given that to all intents and purposes all diploid cells contain the same DNA as each other it might seem wasteful that this task couldn't in some way be coordinated between multiple cells. Well in one circuitous way it is. DNA damage is most dangerous during cell replication (also because DNA is unraveled this is usually when it is detected) as the damage could be propagated, whereas even fatal damage to the DNA of most cells that cannot replicate is largely irrelevant to the organism as you have a myriad more and they can usually be replaced. However, propagation of DNA damage can give rise to a new clone of cells that if expanded could compromise the survival of the whole organism. Coordination arises indirectly by means of one of the most startling examples of selfless cellular behavior observed in the context of a multicellular organism. Namely, that a cell which determines that it has too much DNA damage to repair effectively either commits suicide or permanently declines to reproduce itself in order to protect other cells from the potential

danger. In either case producing a new cell of the required type now falls to another cell without DNA damage and for the organism a perfect outcome – usually. Not surprisingly, defects in any of these processes – detection, repair, apoptosis or growth arrest – can result in cells with DNA damage reproducing themselves – the platform for cancer.

Under the influence of extrinsic and intrinsic insults, mammalian cells are continuously acquiring DNA lesions. Extrinsic factors include exposure to UV light, anticancer drugs and ionizing radiation, and intrinsic factors include reactive oxygen species (ROS), replication errors or stalled replication forks. Fortunately, DNA repair mechanisms are present to ensure that the genome remains intact or if not is not propagated; the damaged cell withdraws from the cell cycle or commits suicide. Normal cells can prevent the occurrence of mutations at the nucleotide sequence level and the chromosome level. These mechanisms include enzymes that repair damaged DNA, and signal transduction pathways (checkpoints) that induce cell cycle arrest or apoptosis when individual stages in the cell cycle are not appropriately completed. DNA damage triggers complex signaling cascades, which may be divided into: (a) DNA damage sensors, (b) checkpoint transducers, and (c) checkpoint effectors. The checkpoints that are activated during the cell cycle when mutations are detected have already been discussed in depth in Chapter 4. Through the operation of these checkpoints, cells with damaged

DNA are blocked from cell cycle progression and DNA replication until the damage is repaired. In addition, cells with chromosomes not properly attached to the mitotic spindles are prevented from undergoing mitosis. In contrast to normal cells, most tumor cells acquire genomic instability resulting in multiple mutations – some of which may promote tumor development. The gene products that control genomic stability in normal cells have been defined to a certain extent, some of which will be discussed later. Not surprisingly perhaps, defects in those genes that control genomic stability can contribute to tumor development.

A complete set of chromosomes in a human somatic cell contains around 6×10^9 base pairs of DNA. Although some of these comprise sequences representing genes and their regulatory elements, large stretches of DNA appear not to encode anything and exhibit extensive mutation (though this view is partly being challenged by the appreciation that regions previously believed to encode nothing, might encode small RNA molecules that contribute to regulation of gene expression – see Chapter 11). DNA in every cell of the human body is spontaneously damaged more than 10,000 times every day. Such damage may result from potentially avoidable exposure to genotoxic environmental factors such as UV light and ionizing radiation, but also through endogenous factors such as accumulating reactive oxygen species (ROS) derived from oxidative metabolism. In dividing cells, errors may also be acquired during replication and mitosis. Specific sequences can also be particularly prone to rearrangement. For example, some repetitive DNA tracts such as microsatellites and minisatellites are naturally unstable and, in humans, alterations in these sequences are often associated with various genetic diseases and cancer. Different types of DNA damage are found including: conversion of bases by loss of an amino group (thus cytosine can be converted to uridine); mismatches during DNA replication (such as incorporation of uridine instead of thymine); breaks in the DNA backbone (either single or double strand breaks), and crosslinking between bases on the same or opposite strands (may also be induced by chemotherapy during cancer treatment).

Clearly, to some extent extrinsic DNA damaging agents may be avoided or exposure minimized (for example, avoiding sunburn!). Importantly, it may also be possible to minimize the damage caused by some intrinsic factors; thus, in part the development of ROS may be reduced by healthy lifestyle choices, including diets low in saturated fat and cholesterol and high in antioxidants – dietary modifications even more convincingly linked to reducing risk of coronary heart disease. However, given that oxidative metabolism is a "non-negotiable" and pivotal aspect of the life of a eukaryotic cell, any processes favoring release of superoxides from mitochondria may culminate in DNA damage, no matter how healthy the diet. During tumorigenesis processes both intrinsic and extrinsic to the cancer cell can provoke oxidative stress. Thus, deregulated expression of oncogenes such as c-*MYC* can promote mitochondrial biogenesis and formation of ROS, which may be a contributory factor in the well-recognized propensity of oncogenic MYC to promote genomic instability. Recent studies suggest that neighboring stromal cells might indirectly provoke DNA damage in adjacent nascent cancer cells; MMPs (matrix metalloproteinases) released by stromal cells may induce release of ROS from mitochondria in cancer cells.

Given that DNA damage is unavoidable and inevitable, cells have evolved a highly sophisticated machinery to replicate and repair DNA accurately and efficiently or if this fails then "damage limitation" mechanisms operate to trigger apoptosis, growth arrest or replicative senescence in the damaged cell (Fig. 10.1). In other words if the DNA cannot be fixed then at least the potential cancer risk can be removed by the suicide of the damaged cell or at least by preventing propagation of the damage.

Given that the DNA in a diploid human cell is very long (about 2 meters in length), very convoluted and very compacted (packaged into a compact chromatin structure by histones), monitoring and repairing damaged DNA is not a trivial undertaking and involves a large and ever expanding group of proteins encoded by what have been dubbed "caretaker" genes. Importantly, these caretaker genes may themselves be aberrantly expressed by mutation or epigenetic factors. The resultant disruption of mechanisms that regulate cell-cycle checkpoints, DNA repair, and apoptosis result in genomic instability – a major contributor to tumorigenesis.

Cellular choices following DNA damage

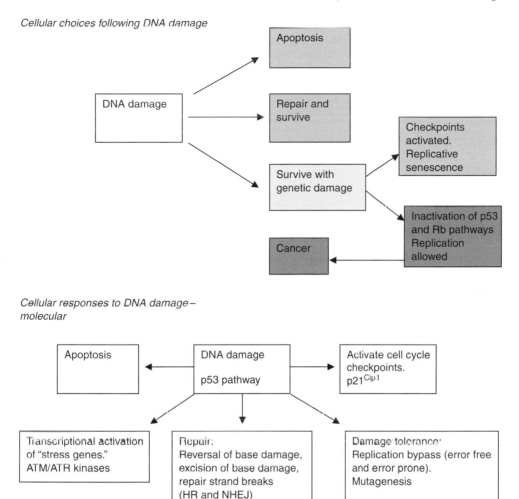

Cellular responses to DNA damage – molecular

Fig. 10.1 Cellular responses to DNA damage. Following DNA damage the cell may undergo one of three fates, repair and survive, fail to repair and apoptosis, or replicative senescence. Rarely the cell with damaged DNA may still be able to replicate and thus pose a threat of cancer. The molecular processes activated by DNA damage are shown in the second figure.

Moreover, DNA damage responses take place within the context of the cell cycle and therefore impact on and integrate with various checkpoints, discussed in Chapters 4, 7, and 8. Not surprisingly, therefore, many tumor suppressor pathways, frequently subject to 'epimutations' in cancer, such as the ARF–p53–p21^{Cip1} and p16^{INK4a}–RB are also involved in DNA damage responses, alongside the caretaker gene products more specifically dedicated to this task.

Although there is considerable redundancy, broadly two major pathways are activated in response to different types of DNA damage. The ATR–CHK1 pathway is triggered by single stranded RNA and stalling of DNA replication forks sensed by among others the ATRIP protein and the RAD9/RAD1/HUS1 cell-cycle checkpoint complex (9-1-1 complex). 9-1-1 forms a clamp-like complex that is recruited to the lesion after DNA damage and interacts with and activates proteins involved in base excision repair (BER) as well as with various checkpoint proteins. The ATM–CHK2 (and parallel DNA-PKcs) pathway is triggered primarily by DNA double-strand breaks, which are sensed by as yet unconfirmed means but likely include the MRE11–RAD50–NBS1 complex (MRN complex) and KU80 protein. An area of great interest here is how the ends of chromosomes avoid being recognized as a DSB, particularly in light of the presence of proteins such as KU at the telomere. MRN may also play a role in activation of some ATR responses, somewhat complicating the picture.

Numerous different repair systems are of particular importance in avoidance of genomic instability and cancer. The two repair systems most frequently discussed with respect to cancer are the DNA mismatch repair system, which corrects DNA sequence errors generated during DNA replication, during recombination, and by mutagenic agents, such as the drug cis-platinum; and the repair of DNA double-strand breaks, induced by ionizing radiation and drugs such as etoposide. Postsynthesis mismatch repair improves the fidelity of DNA replication by several orders of magnitude. DSBs are amongst the most dangerous of all DNA damage and much space will be devoted to these in the chapter. Various proteins involved in mismatch and DSB repair also act to couple DNA repair to cell cycle checkpoint regulation and apoptosis. It is worth noting here that promoting extreme genetic damage/instability with drugs/radiation may also paradoxically be a means of treating cancer – effective as long as growth arrest and/or apoptosis of cancer cells are appropriately triggered as a result. Both the ATM–CHK2 and ATR–CHK1 checkpoint pathways are activated after treatment with DNA-damaging agents; the relative contributions depend on both the agents used and the nature of the damage ensuing; the result for the cell is either cell cycle arrest (in G_1, S, or G_2/M) or death. These outcomes are all driven by various targets of the ATM/ATR kinases and include the tumor suppressor p53 (and in turn key p53 targets in particular $p21^{CIP1}$ for inducing growth arrest in G_1, and BAX, PUMA, and NOXA, which promote apoptosis) and the proapoptotic BCL-2 family member, BID, which may mediate S phase arrest (alongside various other ATM/ATR targets) as well as contributing to cell death and inhibition of CDC25, which causes arrest at G_2/M. Coordinated action between proteins involved in cell cycle arrest, apoptosis, and DNA repair is evident at many levels, suggesting that a

delicate balance between factors dictates if a cell with damaged DNA survives and repairs this or undergoes apoptosis, presumably with the survival option reviewed at various points in the process.

Cancers are generally accepted as clonal in origin, with the cancer cell arising through the stepwise accumulation of gene alterations in a process akin to evolution and natural selection. It has been suggested, at least for some cancers, that the iterative process of mutations and selections driving tumorigenesis may be driven by the propensity of such nascent cancer cells to mutate. The acquisition of genomic instability may well be a crucial step in the progression of human cancer as mutations in other genes might take place at a greatly accelerated rate. In fact, it is worth noting that although cancers may be clonal their chromosomes and genes may not be – in other words genes and chromosomes may be quite heterogeneous, largely due to their inherent genomic instability.

Importantly, debate still continues over the role of genomic instability in initiating cancers. *What is generally uncontested is that the majority of human cancers when studied exhibit genomic instability – but when in the life of the cancer has this occurred – cause or effect of cancer?* Possibly the best studied major cancers with regards to the role of genomic instability in pathoetiology are colorectal and breast (see below). A recent study using an *in vitro* model of breast cancer has suggested that genomic instability (in this case related to telomere shortenening – Chapter 9) might take place at the transition from benign to malignant cancer, but whether this is a generalizable finding or not remains to be clarified. It is, however, accepted that genomic instability plays a major role in the **progression** of most cancers at some stage in their life.

In addition to acquired problems mentioned above, inherited factors may contribute to DNA damage and cancer. Thus, mutations or genetic variation at several alleles may either increase susceptibility to DNA damaging agents such as UV light or compromise the effectiveness of DNA repair. The study of syndromes with inherited defects in DNA repair has greatly contributed to our understanding of how this is normally effected in mammalian cells and some of these will be discussed below.

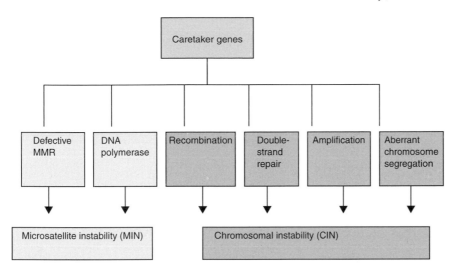

Fig. 10.2 Potential causes of the "mutator" phenotype. Mutations in multiple pathways can result in the mutator phenotype in cancer cells. From Loeb, L.A., Loeb, K.R., and Anderson, J.P. (2003). Multiple mutations and cancer. *Proceedings of the National Academy of Sciences USA*, **100**(3): 776-81. Epub: January 27, 2003.

TYPES OF GENOMIC INSTABILITY

An increased tendency to spontaneously acquire and propagate gross chromosomal abnormalities or smaller scale mutations in DNA is described as genomic instability and occurs in a majority of cancers. Mechanisms of genomic instability are linked to a wide range of key cellular processes including cell cycle regulation, DNA damage and repair, aging, and telomere function. Genomic instability is not simply the presence of a number of defined mutations – this occurs in all cancers – but, rather is the result of defects in DNA damage responses that greatly increase the rate at which chromosomes and DNA are damaged and that allow cells with such damage to survive and to replicate.

Genomic instability comprises several processes, of which chromosomal instability (CIN) recognized by gross chromosomal abnormalities (deletion and duplication of chromosomes or chromosome parts and aneuploidy), and microsatellite instability (MIN) (alterations in the length of short repetitive sequences – microsatellites) associated with a 'mutator' phenotype (described below) have received the most attention (Fig. 10.2).

Aneuploidy

Aberrant chromosome numbers (aneuploidy) are present in cell populations that undergo multiple cell divisions including yeast strains, cell lines, and tumor cells, and may result from gene deregulation and genome instability. It has been appreciated for almost a century that advanced human cancers invariably contain cells with abnormal numbers of chromosomes, but the role of this in cancer pathogenesis and the likely causes are only recently being elucidated and in some cases remain controversial. CIN manifests as gross chromosomal abnormalities such as deletion and duplication of chromosomes or chromosome parts, chromosomal rearrangements and mitotic recombinations. The loss or gain of whole chromosomes results primarily from defects of segregation during mitosis, including chromosomal nondisjunction and failure of the mitotic (spindle) checkpoint pathways (Chapters 3 and 4). Normally, unattached kinetochores or those with inappropriate tension during mitosis activate this checkpoint blocking mitosis. Mutations in key genes involved in this checkpoint, such as BUB1, may disrupt the mitotic checkpoint and promote CIN. Chromosome numbers can readily be visualized in skilled hands, by the application of a technique known as spectral karyotyping (SKY). An example of normal human chromosome numbers is shown in Fig. 10.3, conversely, polyploidy (an increase in chromosome number) in mouse cells is shown in Plate 10.1. Other types of genomic instability are characterized by an increased rate of small-scale mutations, such as MIN. Amplification of individual loci on chromosomes is another form of genetic instability that causes overexpression of proteins, as often occurs for the oncogene c-*MYC* (Chapter 6).

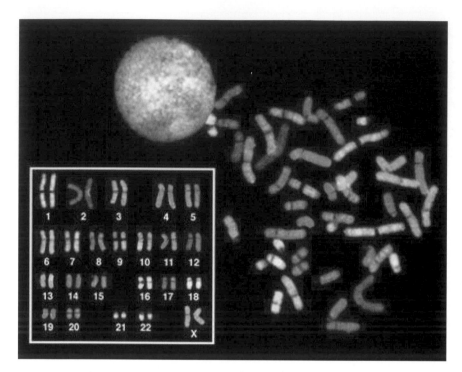

Fig. 10.3 Normal human SKY profile. Images courtesy of the NIH Human Genome Resource.

As you will no doubt have become accustomed to reading by now, there has been a long-standing debate about genetic instability; namely, cause or effect of cancer? The CIN hypothesis contends that aneuploidy is the catalyst for transformation, whereas the gene mutation hypothesis asserts that cancer is driven by mutations to proto-oncogenes and tumor-suppressor genes, with aneuploidy a side effect of tumorigenesis. The role of point mutations in human cancer is well established and MIN has been widely accepted as causal in tumorigenesis. However, the contribution of massive genomic changes resulting from CIN and aneuploidy is less certain. Aneuploidy is required for sporadic carcinogenesis in mice and may even collaborate with specific gene mutations during tumorigenesis. It has been argued that CIN contributes to cancer initiation because chromosome loss can unmask a mutated tumor suppressor gene. However, at the same time, CIN is costly for the cell because it destroys the genome and therefore compromises clonal expansion. Interestingly, either CIN or MIN alone may be a sufficient driver of tumorigenesis as, in general, individual tumors manifest one or other not both.

Mutator phenotype

The concept of a so-called mutator phenotype of cancers developed from the observation that mutations are rare in normal cells but common in cancer cells (Fig. 10.2). The mutator phenotype model proposes that the mutation rate in the early stages of tumorigenesis must be greater than the normal spontaneous mutation rate of human somatic cells (1.5×10^{-10} per base pair per cell generation), because this mutation rate is deemed insufficient to produce the multiple mutations identified in many cancers. The mutator phenotype implies that genetic instability provides the impetus for accumulation of multiple mutations by the cancer cell. However, only a small number of such mutations may actually confer a growth advantage (and will be selected for), whereas the rest (possibly the majority) have no bearing on the cancer and are merely "baggage" that is not selected for. This differs from the traditional view of cancer progression through clonal evolution driven by mutation and natural selection in so far as the traditional view implies that most mutations will confer some growth advantage and are selected for individually (or in small numbers) – there is little "baggage" in this model. Whatever the answer, most cancers probably develop through a combination of stochastic mutations (and epigenetic factors) as well as those arising through genomic instability (and maybe the epigenetic equivalent – methylator phenotype, discussed in Chapter 11), which are then subject to natural selection and clonal expansion. However, as discussed below, particularly in colorectal and breast cancer there is increasing evidence that genetic instability may occur at the transition from premalignant to malignant tumors.

All this interesting debate not withstanding, the most important issue (and probably one of the most important questions in the whole of cancer biology) is whether the plethora of mutations observed in tumors are all mission-critical or rate-limiting for tumor growth or whether many of the mutations are irrelevant both to the evolution and maintenance of the tumor. In other words are most mutations in established cancers associated with tumorigenesis in a noncausal way? This has a direct bearing on the design of cancer therapies – as new therapeutics should be designed to target cancer-relevant genes/proteins, which must first be distinguished from the rest.

The notion of a "mutator" phenotype is based on seminal studies with one form of a progressive multistage human cancer, a familial colon cancer known as hereditary nonpolyposis colon cancer (HNPCC). In HNPCC, mutation in one allele of the genes involved in DNA mismatch repair is inherited. Loss of the second wild-type allele occurs frequently promoting the formation of colorectal carcinoma. Even sporadic colon cancers often exhibit MIN and this is an almost ubiquitous finding in HNPCC. First described in the bacterium *E. coli*, the mismatch repair genes *mutS* or *mutL* were implicated in MIN, and subsequently mutations in two homologous human genes in the DNA mismatch repair pathway were found in HNPCC by linkage studies (Chapter 18), *MSH2*, encoded on chromosome 2, or *MLH1*, encoded on chromosome 3. These genes may also be inactivated or suppressed by epigenetic factors, though interestingly in human cancers this seems to involve the *MLH1* gene (Chapter 11). MIN occurs in several other nonhereditary (sporadic) cancers

including pancreas, colon, and ovary. It is likely that many other genes, not necessarily involved in mismatch repair, may also contribute to MIN. The human genome comprises vast numbers of microsatellites, but instability also manifests in coding regions of various growth-regulating genes, including apoptosis-regulators such as BAX. Thus, MIN may be a "readily measurable" marker of the existence of mutagenic mechanisms, but it must be noted that most sporadic tumors do not display high levels of MIN.

TELOMERE ATTRITION AND GENOMIC INSTABILITY

Increasing evidence indicates that dysfunctional telomeres likely play a causal role in the process of malignant transformation, in at least a fraction of human cancers, by initiating chromosomal instability. Critical telomeric shortening can lead to telomere "uncapping" and may occur at the earliest recognizable stages of malignant transformation in epithelial tissues. The widespread activation of the telomere synthesizing enzyme telomerase in human cancers not only confers unlimited replicative potential but also prevents intolerable levels of chromosomal instability. See Chapter 9 for a detailed discussion of this area. There is a considerable overlap between DNA damage responses, particularly to DSBs, and proteins involved in telomere maintenance – see below.

THE DNA DAMAGE RESPONSE

As outlined in Chapter 4, a cell responds to DNA damage by activating a DNA-damage response comprising two key overlapping networks of proteins involving two key PI3K related proteins – the ataxia-telangiectasia mutated (ATM) and ATM- and Rad3-related (ATR) kinases. Two major signaling pathways involving these proteins have been described, the ATM–CHK2–p53 and the ATR–CHK1, which play key roles in checkpoint activation and maintenance of genomic stability following DNA damage. The ATM–CHK2–p53 pathway responds to more substantive damage such as DNA double-strand

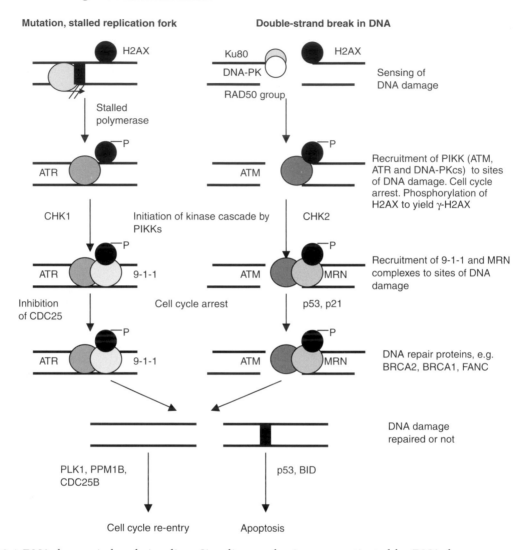

Fig. 10.4 DNA-damage-induced signaling. Signaling mechanisms are activated by DNA-damage sensors, ultimately leading to a variety of potential cellular responses discussed earlier.

breaks (DSBs), whereas the ATR–CHK1 pathway is activated by less extensive DNA damage, likely including mutations and lesions causing stalling of polymerases. However, there is likely considerable overlap and redundancy. A simplified schematic of DNA damage signaling and repair is shown in Fig. 10.4. The effect of such DNA damage response signaling on the cell is to activate cell-cycle checkpoints leading to arrest in G$_1$ or G$_2$ in order for DNA to be repaired before the cell cycle can continue. In certain cases, such as when DNA damage is extensive

and/or irreparable, the outcome will be cell death (apoptosis) or irreversible growth arrest (senescence). The development of the concept of cell cycle checkpoints is attributed to the pioneering work of Lee Hartwell and others starting in the 1960s and has been discussed in Chapter 4. However, given the key role played in avoidance of CIN, the mitotic (spindle) checkpoint will be discussed later in this chapter also. The DNA damage checkpoints coordinate a block in cell proliferation with the DNA repair process. However, once repair is effected (if possible) then the checkpoint is "released" and cell division can recommence; Polo-like kinase-1 and Cdc25B are essential components in this reactivation of cell division.

This is undoubtedly a complex area and important recent data show that DNA damage responses may be invoked in very early human tumors (even before genomic instability and malignant conversion). During the evolution of cancer, the incipient tumor cell experiences 'oncogenic stress' that triggers a 'failsafe' response to eliminate these potentially dangerous cells. In this case, it is possible that DNA damage responses are in some way activated by deregulated expression of a single oncogene (e.g. E2F, cyclin E, or c-MYC) possibly because resultant aberrant cell cycles may be more likely to involve DNA damage – see section on oncogenic stress below.

DNA double-strand breaks (DSBs) are by far the most dangerous type of DNA lesion because if they are inefficiently or inaccurately repaired they can result in genomic instability and cancer. Eukaryotic cells have two conserved mechanisms to detect and repair DSBs; homologous recombination (HR) repairs the break using genetic information recovered from the undamaged sister chromatid or chromosomal homolog, whereas nonhomologous end joining (NHEJ) involves the direct ligation of DNA ends.

SENSING DNA DAMAGE

A major focus of recent attention has been the rather tricky issue of how the cell senses DNA damage. As mentioned already the DNA is normally tightly packed into chromatin by histones – so how is the damaged DNA accessible by the damage-sensing machinery? An expanding body of evidence suggests that sensing may actually take place during the two occasions that DNA is normally stripped of its histone coat and the two strands are separated, namely, during either DNA replication or transcription. It might be argued that as DNA replication occurs only during the S phase of the cell cycle, whereas gene transcription is continuous (except during mitosis when chromatin is condensed and gene expression is silenced) the latter would be a more suitable occasion to perform such

damage sensing. As discussed in Chapter 3, eukaryotic gene transcription involves three different RNA polymerases: RNA polymerase I transcribes ribosomal DNA into ribosomal RNA for ribosome formation; RNA polymerase II transcribes genes into mRNA; and RNA polymerase III synthesizes transfer RNA and small nuclear RNA. Of these, RNA polymerase II reads a much greater proportion of the genome than the others and may, moreover, recruit DNA-repair proteins to undertake a form of nucleotide excision repair (NER) termed transcription-coupled repair (TCR). This is essential as if RNA polII progress remains stalled because DNA repair is not effected then the cell may trigger other DNA damage sensors and activate the p53 pathway already described in Chapter 7 and even undergo apoptosis.

Potential candidates for the actual DNA damage sensor are the so-called 9-1-1 complex (comprising RAD1–RAD9–Hus1) and the RAD17 protein. However, it is still unclear how stalling of DNA or potentially RNA polymerases may activate the DNA damage responses. Some proteins involved in transducing the DNA damage signal are described in detail in the next section.

Signaling DNA damage

Not surprisingly, mutations that compromise the DNA damage response (e.g. in the ATR–CHK1 or ATM–CHK2–p53 pathway) might allow cell proliferation, survival, increased genomic instability, and thus tumor progression. Examples of such mutations occurring in human cancers are described later. The signaling pathways involved in DNA damage responses are complex and most certainly still incompletely described, but the major proteins involved are detailed in Fig. 10.5. Although it is known that various targets of ATM- and ATR-mediated phosphorylation participate in transmitting the DNA damage signal to CHK1 and CHK2 proteins, much is still not known. There is redundancy in the system, and both CHK1 and CHK 2 can inactivate CDC25 thus leading to cell cycle arrest. Moreover, both CHK 2 and ATM may act on the same substrates providing a potential salvage mechanism if either were mutated. Phosphorylation of the histone variant H2AX by

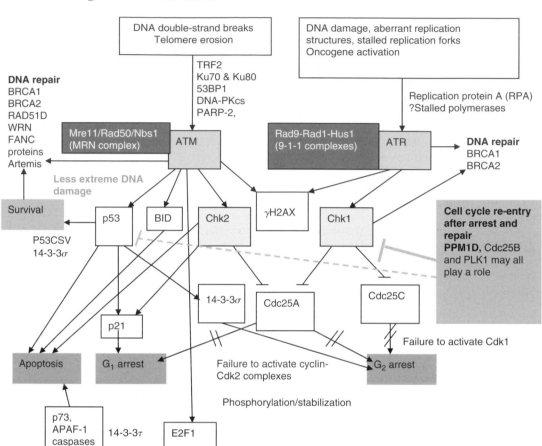

Fig. 10.5 Signaling involved downstream of different types of DNA damage. To combat recurrent threats of genomic instability, numerous distinct enzyme systems sense DNA damage and coordinate its repair. Part of this coordination involves the activation of signal transduction cascades that target repair proteins, trigger DNA-damage-dependent cell cycle checkpoints and profoundly affect chromatin neighboring a DSB. Ultimately DNA damage results in cell cycle arrest and attempted repair. If repair is successful, the cell will survive and may reenter the cell cycle. There is obvious overlap between telomere maintenance and DNA damage response. Proteins directly involved in telomere maintenance and DNA damage response include Ku, DNA-PKcs, RAD51D, PARP-2, WRN and MRE11/ RAD50/NBS1 complex. The kinases ATM and ATR are key mediators of the DNA-damage response and activate downstream effectors such as γH2AX, CHK2, and CHK1. Ultimately the activity of these kinases results in activation of p53-p21, inhibition of Cdc25 and activation of the proapoptotic BH3-only protien BID leading to growth arrest or apoptosis. 14-3-3σ is a p53 target which may also help maintain G2 arrest. Other 14-3-3 family members may seuester CDC25. Recent studies suggest that P53CSV and/or 14-3-3σ may prevent apoptosis if DNA damage is repairable and activation of another protein the p53-regulated serine/threonine phosphatase, PPM1D, may allow cell cycle reentry if DNA damage is repaired. Cdc25B and PLK1 may also help restart the cell cycle.

CHK proteins yields γH2AX, which is essential for retention (if not recruitment) of the MRE11–RAD50–NSB1 (MRN) complex, which rapidly accumulates at sites of DNA damage. The MRN complex plays a key role in genome maintenance in humans and can tether the broken DNA ends in a DNA strand break and can recruit various repair proteins in order to coordinate the proper processing of the DNA ends. BRCA1, MDC1, and 53BPI (p53 binding protein 1) are also recruited following H2AX phosphorylation. Chromatin modifications are strongly associated with the repair of DSBs and γH2AX helps prevent aberrant repair of damaged DNA. Mice deficient for both H2AX and p53 rapidly develop tumors. Moreover, even H2AX haploinsufficiency causes genomic instability in normal cells and, on a p53-deficient background, leads to the early onset of lymphoma. H2AX also maps to a cytogenetic region frequently altered in human cancers. Importantly, other histone modifications such as acetylation of histone H4 also have an important role in the response to DSBs, suggesting that DNA-damage-induced histone modifications are not confined to phosphorylation of H2AX.

ATM is itself activated by phosphorylation, and although much is known about downstream targets, such as CHK2, the arbiters of ATM phosphorylation are not well known, but might include p53BP1. The tumor suppressor protein p53 is a key target of the ATM–CHK2 pathway. Activated p53 induces cell cycle arrest, which is described in detail in Chapter 7. CHK 2 may also activate $p21^{CIP1}$ independently of p53, contributing to cell cycle arrest in G_1. Given that the p53 pathway functions in the checkpoint response to DSBs, it is not surprising that there is a selective pressure for premalignant tumor cells to mutate and thus inactivate the p53 gene – as is the case for most human cancers.

The signaling pathways involved in DNA damage responses are complex and most certainly still incompletely described, but the major proteins involved are detailed in Fig. 10.5.

14-3-3 proteins bind to and influence the activities of a diverse group of molecules involved in signal transduction, cell cycle regulation, and apoptosis, including p53, RAF, PKC, and BAD. Interactions between 14-3-3 and target proteins are strongly influenced by the phosphorylation state of 14-3-3 and the target protein. How the family of 14-3-3 proteins can interact with many different proteins and influence critical cellular pathways is a topic of intense interest. The association of 14-3-3 with such a wide variety of proteins is partly explained by the discovery of two 14-3-3 consensus-binding motifs, RSXpSXP and RXY/FXpSXP (pS = phosphoserine) within the 14-3-3-binding partners. In mammalian cells, seven isoforms have been identified (14-3-3 β, γ, ε, η, σ, τ/θ, and ζ). These isoforms are all highly conserved and have distinct functions. Several isoforms of 14-3-3 are involved in the G_2/M cell cycle checkpoint. 14-3-3σ sequesters the cyclin B1–CDK complexes in the cytoplasm after DNA damage. This prevents initiation of mitosis, whereas the β and ε isoforms bind Cdc25C. 14-3-3σ is a direct transcriptional target of p53 and may be a key factor in G_2/M arrest following DNA damage. The expression of the 14-3-3σ gene also prevents apoptosis and by arresting cells in G_2 allows the repair of damaged DNA. The 14-3-3σ gene is important in cancer and is silenced in a large number of invasive breast cancers. After 14-3-3σ was cloned, the expression level was determined to be significantly reduced both in v-HaRas-transformed mammary epithelial cells and mammary carcinoma cells. Together these observations support the idea that loss of 14-3-3σ expression contributes to malignant transformation and that 14-3-3 may operate as a tumor suppressor in humans.

REPAIRING DNA DAMAGE

In humans, at least 150 genes encode proteins known to be involved in DNA repair. After DNA damage, at the molecular level, the protein kinases ATR and ATM, their respective substrates CHK1 and CHK2, as well as molecules required for recruitment of ATR and ATM to sites of DNA damage, such as histone γH2AX and BRCA1 (see Fig. 10.5), are all critical in activating checkpoints (and thereby key proteins involved in these checkpoints as described in Chapters 4 and 7). Various other proteins are involved in ensuring that chromosomes have been accurately partitioned during mitosis, including those operating in the spindle checkpoint described later. Mutations in the 'caretaker genes' encoding many of these proteins involved in DNA repair and in chromosome segregation and apoptosis have all been implicated in genomic instability. In fact, the caretaker genes represent the third major class of genes mutated in cancer, alongside the oncogenes and tumor suppressor genes.

Various intracellular mechanisms are involved in repairing subtle mistakes made during normal DNA replication or following exposure to DNA-damaging stimuli. They include direct chemical reversal of damage (not requiring breaking of the DNA backbone), which can reverse damage due to alkylation, and includes the product of the human *MGMT* gene. But probably the most important mechanisms are the three forms of excision repair: nucleotide excision repair (NER), base excision repair (BER), and mismatch repair (MMR), all of which may contribute to avoiding the 'mutator' phenotype. As an illustration of the likely importance of these mechanisms it has been estimated that on average somatic cells in humans may need to remove around 10–20,000 damaged bases every day.

The human genome, comprising three billion base pairs coding for 30,000–40,000 genes, is under constant assault by endogenous reactive metabolites, therapeutic drugs, and a plethora of environmental mutagens that impact its integrity. Thus it is obvious that the stability of the genome must be under continuous surveillance. This is accomplished by DNA repair mechanisms, which have evolved to remove or to tolerate precytotoxic and premutagenic DNA lesions in an error-free, or in some cases, error-prone way. Defects in DNA repair give rise to hypersensitivity to DNA-damaging agents, accumulation of mutations in the genome, and finally to the development of cancer and various metabolic disorders. The importance of DNA repair is illustrated by DNA repair deficiency and genomic instability syndromes (described later), which are characterized by increased cancer incidence and multiple metabolic alterations.

Various processes contribute to DNA repair and are summarized below.

Base excision repair (BER)

This requires removal of the damaged base by one of several DNA glycosylases responsible for identifying and removing specific base damage, followed by removal of the deoxyribose phosphate in the backbone, producing a gap. This gap can then be filled by replacement with the correct nucleotide by DNA polymerase β and the strand break ligated by DNA ligases (see Fig. 10.6).

Nucleotide excision repair (NER)

NER differs from BER in that different enzymes are employed, which invariably remove not only the damaged base but also a number of adjacent bases even if these are undamaged. NER operates most effectively in cells during DNA transcription, particularly in the sense strand. In this mechanism following recognition of DNA damage, the DNA is unwound by transcription factor IIH (TFIIH) and cuts are made on both the 3' side and the 5' side of the damaged area enabling the section of DNA containing the altered base to be removed. This removal is followed by DNA synthesis, employing the intact opposite strand as a template, a process requiring the DNA polymerases delta and epsilon. This is again followed by the action of a DNA ligase. Defects in this process are characteristic of xeroderma pigmentosum (XP), which can be caused by mutations in any one of several genes involved in NER. These include *XPA*, which encodes a protein that binds the damaged site and helps assemble the other proteins needed for NER, *XPB* and *XPD*, which are part of TFIIH, and *XPF* and *XPG*, which cut the DNA backbone on the 5' or 3' side of damage respectively (see Fig. 10.7).

Mismatch repair (MMR)

Mismatch repair deals with correcting mismatches of the normal bases; that is, failures to maintain normal Watson–Crick base pairing (A-T and C-G). In addition to a specialized dedicated machinery MMR can also utilize enzymes involved in BER and NER. The mismatch is recognized by numerous proteins, including the *MSH2* gene product, and is followed by mismatch excision using the *MLH1* gene product; mutations in both genes are known to predispose to colon cancer. The excision is again repaired by DNA polymerase delta and epsilon (see Fig. 10.8).

Repairing single-strand breaks (SSBs)

Breaks in a single strand of the DNA molecule are repaired using the same enzyme systems that are used in base-excision repair (BER).

Double-strand breaks (DSBs)

DSBs can arise naturally through the action of oxygen radicals generated by normal metabolic

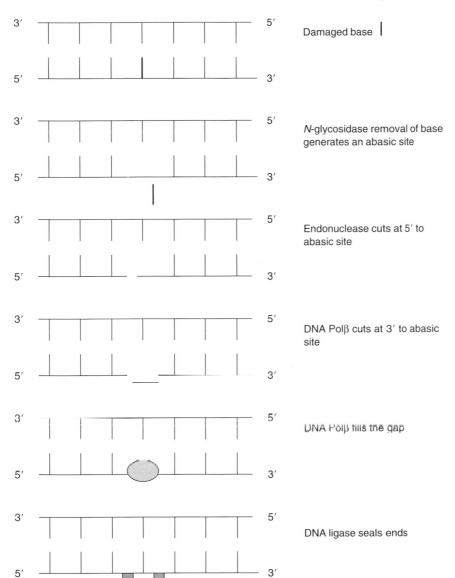

Damaged base

N-glycosidase removal of base generates an abasic site

Endonuclease cuts at 5' to abasic site

DNA Polβ cuts at 3' to abasic site

DNA Polβ fills the gap

DNA ligase seals ends

Fig. 10.6 Base excision repair.

processes but can also be caused by exogenous agents such as gamma irradiation and some chemotherapeutic drugs. DSBs represent dangerous chromosomal lesions that can lead to mutation, neoplastic transformation, or cell death. Mammalian cells possess potent and efficient mechanisms to repair DSBs, and thus complete normal development as well as mitigate oncogenic potential and prevent cell death. Failure of DSB repair can lead to CIN and is often associated with tumor formation or progression.

Two primary mechanisms are employed to repair a complete break in a DNA molecule (Fig. 10.9a,b). Direct joining of the broken ends requires proteins such as DNA-dependent protein kinase (DNA-PK), Ku70, and Ku80, which recognize and bind to the exposed DNA ends in order to bring them into apposition for joining. This does not necessarily require the presence of complementary nucleotides and is therefore also referred to as nonhomologous end-joining (NHEJ) (Fig. 10.9a). DNA-dependent protein

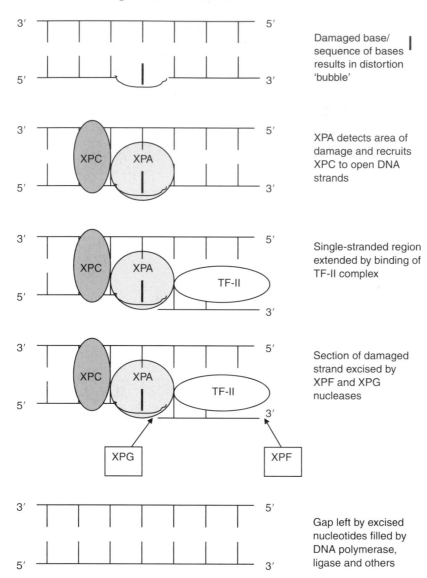

Damaged base/sequence of bases results in distortion 'bubble'

XPA detects area of damage and recruits XPC to open DNA strands

Single-stranded region extended by binding of TF-II complex

Section of damaged strand excised by XPF and XPG nucleases

Gap left by excised nucleotides filled by DNA polymerase, ligase and others

Fig. 10.7 Nucleotide excision repair.

kinase (DNA-PK) is a multisubunit protein made up of DNA-PKcs, a very large catalytic subunit (cs) of 465 kDa, and Ku, a DNA-binding heterodimer that comprises the Ku70 and Ku80 subunits. DNA-PK is activated by DNA ends and plays a major role in DSB repair. As discussed in the previous chapter many of the proteins involved in DSB repair are also involved in telomere maintenance, including Ku and the protein PARP (see Chapter 9). Errors in direct joining may cause or contribute to translocations in cancers such as

Burkitt's lymphoma and CML (Chapter 6). DNA-PKcs, like ATM and ATR, has important kinase activity involved in activating downstream signaling after DNA damage.

The other means of repairing DSBs is by homologous recombination (Fig. 10.9b), whereby the broken ends are repaired using the information on the intact sister chromatid (in G_2 after chromosome duplication), or on the homologous chromosome (in G_1 before chromosome duplication). Two of the proteins used in homologous

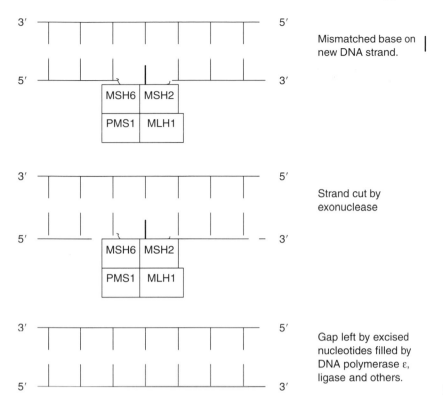

Mismatched base on new DNA strand.

Strand cut by exonuclease

Gap left by excised nucleotides filled by DNA polymerase ε, ligase and others.

Fig. 10.8 Mismatch repair.

recombination are encoded by the genes *BRCA1* and *BRCA2*, mutations in which contribute to breast and ovarian cancers (described later). In human cells a protein complex comprising MRE11, RAD50, and NBS1 (MRN) has exonuclease and endonuclease activities that may be vital for processing the DNA ends at the site of a DSB prior to religation. NBS1 is defective in a human chromosomal instability disorder associated with an increased risk of cancer, called Nijmegen breakage syndrome. The MRE11 repair protein is involved in both HR and NHEJ, whereas the function of Ku70-80 is unique to NHEJ. Human cohesin subunits physically interact with RAD50, suggesting links with mitotic checkpoint control.

The *BRCA1* gene was first cloned about 10 years ago, and since then it has become clear that it and *BRCA2* are the major causes of hereditary predisposition to breast or ovarian cancers. Since that time, BRCA1 has been linked to several key nuclear functions connected with the prevention of genomic instability. In

particular, BRCA1 functions in concert with Rad51, BRCA2, and other genes to control double-strand break repair (DSBR) and homologous recombination.

Protecting telomeres from repair processes

Chromosome end fusions (covalent connections between the C-strand of one telomere and the G-strand of another) occur when telomeres have become too short and when TRF2 is inhibited (Chapter 9). They predominantly involve ends of two different chromosomes; the resulting dicentric chromosomes can become attached to both spindle poles and lead to a problem for chromosome segregation in anaphase. The fusion of damaged telomeres requires the same factors as normal NHEJ, but with modifications due to the longer 3′ overhang. One way in which chromosome ends are usually protected from NHEJ is by formation of t-loops, as without an accessible end, the NHEJ proteins will not be able to

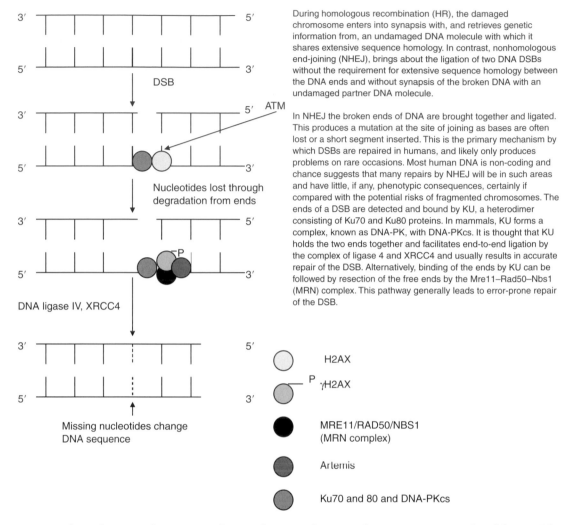

During homologous recombination (HR), the damaged chromosome enters into synapsis with, and retrieves genetic information from, an undamaged DNA molecule with which it shares extensive sequence homology. In contrast, nonhomologous end-joining (NHEJ), brings about the ligation of two DNA DSBs without the requirement for extensive sequence homology between the DNA ends and without synapsis of the broken DNA with an undamaged partner DNA molecule.

In NHEJ the broken ends of DNA are brought together and ligated. This produces a mutation at the site of joining as bases are often lost or a short segment inserted. This is the primary mechanism by which DSBs are repaired in humans, and likely only produces problems on rare occasions. Most human DNA is non-coding and chance suggests that many repairs by NHEJ will be in such areas and have little, if any, phenotypic consequences, certainly if compared with the potential risks of fragmented chromosomes. The ends of a DSB are detected and bound by KU, a heterodimer consisting of Ku70 and Ku80 proteins. In mammals, KU forms a complex, known as DNA-PK, with DNA-PKcs. It is thought that KU holds the two ends together and facilitates end-to-end ligation by the complex of ligase 4 and XRCC4 and usually results in accurate repair of the DSB. Alternatively, binding of the ends by KU can be followed by resection of the free ends by the Mre11–Rad50–Nbs1 (MRN) complex. This pathway generally leads to error-prone repair of the DSB.

Fig. 10.9a Nonhomologous end-joining results in a change in the original DNA sequence – either deletions (shown here) or insertions.

form the synaptic complex needed for processing and ligation of the ends. Telomeres also need to be protected from inappropriate homologous recombination (HR). There are three types of HR that have detrimental outcomes at chromosome ends, telomere sister chromatid exchange (T-SCE), t-loop HR, and recombination between a telomere and interstitial telomeric DNA. One puzzling feature of telomere-interacting proteins has been emphasized in a recommended review by Titia DeLange. This relates to the role of telomere-associated DNA repair factors, DNA-PKcs and the Ku70/80 heterodimer, which promote NHEJ but are also associated with

telomeres – in this case how is the inappropriate triggering of DSB repair avoided? Moreover, how DNA-PKcs and Ku contribute to telomere function is not known; but what seems clear is that they would need careful control in order to avoid NHEJ of chromosome ends.

CHECKPOINTS (see also Chapter 4)

Aneuploidy may not inevitably arise by selection during tumorigenesis but may instead arise as an inadvertent consequence of defects that

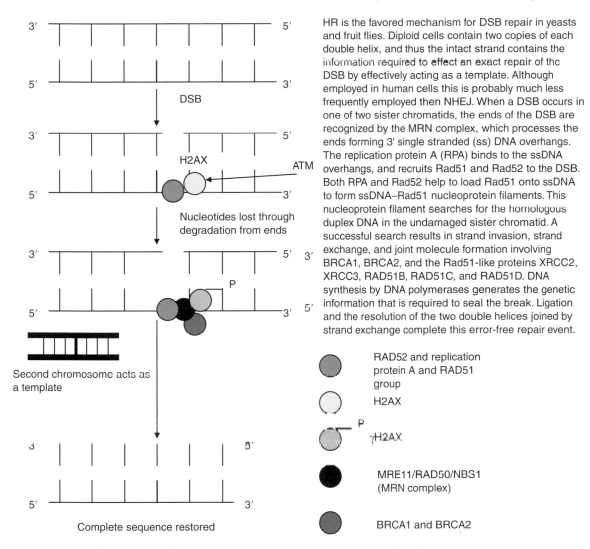

HR is the favored mechanism for DSB repair in yeasts and fruit flies. Diploid cells contain two copies of each double helix, and thus the intact strand contains the information required to effect an exact repair of the DSB by effectively acting as a template. Although employed in human cells this is probably much less frequently employed then NHEJ. When a DSB occurs in one of two sister chromatids, the ends of the DSB are recognized by the MRN complex, which processes the ends forming 3' single stranded (ss) DNA overhangs. The replication protein A (RPA) binds to the ssDNA overhangs, and recruits Rad51 and Rad52 to the DSB. Both RPA and Rad52 help to load Rad51 onto ssDNA to form ssDNA–Rad51 nucleoprotein filaments. This nucleoprotein filament searches for the homologous duplex DNA in the undamaged sister chromatid. A successful search results in strand invasion, strand exchange, and joint molecule formation involving BRCA1, BRCA2, and the Rad51-like proteins XRCC2, XRCC3, RAD51B, RAD51C, and RAD51D. DNA synthesis by DNA polymerases generates the genetic information that is required to seal the break. Ligation and the resolution of the two double helices joined by strand exchange complete this error-free repair event.

RAD52 and replication protein A and RAD51 group

H2AX

P γH2AX

MRE11/RAD50/NBS1 (MRN complex)

BRCA1 and BRCA2

Fig. 10.9b Homologous recombination results in more accurate repair of DSBs but requires a more complex process.

uncouple mitotic control from normal cell cycle progression. Data obtained from a variety of model organisms have revealed that disruption of the cell cycle controls required for homeostasis results in the acquisition of genomic instability. Defects at the G_1–S transition have been discussed in previous chapters so in this chapter we will concentrate on the role of aberrant checkpoint signaling during mitotic progression in cancer. Cells contain numerous pathways designed to protect them from the genomic instability or toxicity that can result when their DNA is damaged. The

p53 tumor suppressor is particularly important for regulating passage through G_1 phase of the cell cycle (Chapter 7), while other checkpoint regulators are important for arrest in S and G_2 phase. Tumor cells often exhibit defects in these checkpoint proteins, many of which have already been discussed earlier. This is illustrated by the predisposition to cancer seen in various inherited syndromes where checkpoint or caretaker genes are mutated: *p53* and *CHEK2* mutations in Li Fraumeni syndrome, inactivating *ATM* mutations in ataxia telangiectasia, defective DNA damage repair with breast cancer susceptibility genes 1 and 2 (*BRCA1/2*), and *NBS1* mutations in familial breast cancer and Nijmegen breakage syndrome, respectively. Other checkpoint genes, such as *53BP1*, meiotic recombination 11 (MRE11), and *H2AX*, have been identified in rodent cancer models.

Although some key gaps remain in our understanding, DNA damage is rapidly followed by binding of protein complexes to areas of strand breaks in DNA. These complexes include the MRN and DNA-PK discussed in the DNA repair section. The protein kinases ATM and ATR (and DNA-PKcs), as well as their downstream substrates CHK 1 and CHK 2, are important for checkpoint activation in response to DNA damage. Histone H2AX, ATRIP, as well as the BRCT-motif-containing molecules 53BP1, MDC1, and BRCA1, function as molecular adaptors or mediators in the recruitment of ATM or ATR and their targets to sites of DNA damage (Fig. 10.5). The increased chromosomal instability and tumor susceptibility apparent in mutant mice deficient in both p53 and either histone H2AX or proteins that contribute to NHEJ indicate that DNA damage checkpoints play a pivotal role in tumor suppression. During tumorigenesis, tumor cells frequently lose checkpoint controls and this facilitates the development of the tumor. However, these defects also represent an Achilles heel that can be targeted to improve current therapeutic strategies.

Mitotic entry depends on activity of the cyclin B–CDK1 complex. As described in Chapter 4, during G_2, CDK1 is kept inactive via inhibitory phosphorylation by WEE1 and MYT1 kinases; this phosphorylation is removed at the onset of mitosis by the CDC25 phosphatases, in turn themselves activated by Polo-like kinase-1 (PLK1). Cyclin B–CDK1 is the main target of the G_2 DNA damage checkpoint; following DNA damage, CDC25A, CDC25B, and CDC25C are inhibited by phosphorylation and/or ubiquitin-mediated degradation, PLK1 is inhibited and cyclin B–CDK1 complexes are sequestered in the cytoplasm.

The mitotic checkpoint (also referred to as the spindle assembly checkpoint) normally ensures accurate progression through mitosis in order that each new cell receives one copy of every chromosome. If even a single chromosome is improperly attached to the mitotic spindle (sensed by unattached kinetochores) then the mitotic checkpoint will arrest the cell, inhibiting activity of the APC/C ubiquitin E3 ligase (and its CDC20 cofactor – sometimes referred to as APC/C^{CDC20}) discussed in Chapter 4. Major proteins targeted for proteasomal degradation by the APC/C^{CDC20} include cyclin B and securin, which are both key regulators of mitosis. Unattached kinetochores generate a durable and powerful inhibitory signal (a single unattached kinetochore can delay anaphase by 3 hours). This signal is believed to comprise a series of proteins first identified in yeast but all with homologs in mammalian cells. These proteins include the BUB (BUB3 and BUB1) and MAD (MAD1 and MAD2) family and MPS1. Additional proteins involved in the mammalian checkpoint include the kinases BUBR1 and MAPK (see Chapter 5) and CENPE. Not surprisingly, this is complex and still not fully understood, and the identity of the 'inhibitory signal' is still not known. However, MAD2, BUBR1, BUB3, and MPS1 are readily released by unattached kinetochores. MAD2 and BUBR1 can bind to CDC20 and it is possible, but unproven, that they could thereby prevent it from activating APC/C – thus, cyclin B and securin will remain undegraded. Time will tell whether these proteins or some variant conformation of them are the inhibitory signal for mitosis. The failure of cell cycle regulatory checkpoints is a common event in human cancer. Human tumor cells have been shown to contain mutations of BUB1, BUBR1, MAD1, and MAD2 and also members of the so-called ZW10-ROD-zwilch complex (involved in recruiting MAD1-MAD2 to unattached kinetochores). Recently, haploinsufficiency of MAD2 has been associated with tumorigenesis. The actual effects of these mutations on the checkpoint is still untested but the suggestion is that even a weakened (not necessarily absent) checkpoint response may be enough to facilitate tumorigenesis. In keeping with this view, various mouse models with inactivating mutations in single components of the mitotic checkpoint have a modest increase in risk of some cancers but often with a long latency.

It is worth remembering that often the primary defect giving rise to CIN may not be in the checkpoint proteins. Although defects in mitotic checkpoint proteins are very commonly seen in cancers, MAD2 haploinsufficiency notwithstanding, this is not usually due to mutations in the encoding genes, but rather predominantly a secondary effect on protein levels due to aberrant activation of oncogenes or tumor suppressors. Loss of the tumor suppressors APC, p53, BRCA1 and 2, or deregulated expression of oncogenic proteins such as c-MYC, E6, and E7 may all result in microtubule instability or centrosome amplification (both causes of CIN). Deregulation of the cell cycle may be a common feature in these cases, but whatever the underlying mechanism, overexpression of c-MYC or loss of RB may provoke CIN. In the latter case this is associated with activation of E2F and paradoxically also of MAD2.

In order for such DNA damage to propagate, cells must overcome the checkpoints/tumor suppressor pathways discussed previously; the most obvious explanation is that these aforementioned components are in some way inactivated. This is undoubtedly the explanation in many cases and in fact the identification of new lesions that can cooperate with c-MYC and allow propagation may help identify new genes/proteins involved in checkpoint control. The Werner syndrome protein WRN may allow c-MYC activated cells with genomic instability to propagate. Genomic instability may be caused by specific genetic alterations even before malignant conversion. Mutations in hCDC4 (also known as Fbw7 or Archipelago) have recently been identified in human colorectal cancers and their precursor lesions, and targeted inactivation of hCDC4 in karyotypically stable colorectal cancer cells results in micronuclei and chromosomal instability, largely through cyclin E-dependent metaphase defects and subsequent transmission of chromosomes. The checkpoint may also be overridden by mutations in Aurora-A (Chapter 4), which can prevent MAD2 from effectively preventing APC/C activation.

Most solid cancers are aneuploid (have lost or gained chromosomes or bits of them) indicating how common defects in this checkpoint must be in cancer. Yet, many traditional chemotherapeutic agents act by deliberately disrupting the mitotic spindle and preventing segregation of chromosomes. The apparent conflict is resolved if one imagines that in the context of therapeutic disruption, one is relying on the cell doing the decent thing when its mitotic checkpoint is triggered and thence undergoing apoptosis or growth arrest. However, aneuploidy may be tolerated by some cancer cells that have acquired the means to evade these barriers. It is tempting to state that such cancers may be more resistant to drugs that impair chromosome segregation – thus it is more likely that a proportion of cells, those where damage is not so severe as to directly compromise viability, will survive chemotherapy and the cancer will continue or recur.

As stated, most solid tumors are aneuploid, but most also contain mutations in oncogenes and tumor suppressors; thus, we must again revisit our constant companion – the chicken-egg conundrum. Namely, in this case is aneuploidy a cause or consequence of cancer? Whatever the answer, there are many possible explanations for the near-ubiquitous presence of aneuploidy. CIN would certainly increase the chances of acquiring other mutations. Thus, directly, loss of a chromosome containing a haploinsufficient tumor suppressor might be enough to promote tumor progression. Alternatively, for instance, the chance of loss of heterozygosity (LOH), discussed in Chapter 3, whereby a cell with a single mutant copy of a tumor suppressor gene loses the remaining normal copy, would be greatly increased by deletion of that chromosome or duplication of the one with the mutant gene. This is exemplified by the frequent loss of PTEN activity by LOH in cancers thus taking the brakes off PI3K-AKT signaling.

Checkpoints, apoptosis, and oncogenic stress

Failure to repair DNA lesions properly after the induction of cell proliferation arrest can lead to mutations or large-scale genomic instability; such changes may have tumorigenic potential, if the cell cycle is reactivated inadvertently (possibly by mutations in PLK1 or CDC25B, or by other causes of premature checkpoint abrogation). Elevated levels of PLK1 correlate with metastatic potential and poor prognosis. Thus, in many ways it may be preferable for cells with damaged DNA to be eliminated via apoptosis. As described in Chapter 8, loss of this apoptotic response is actually one of the hallmarks of cancer. Defects in DNA-damage-induced apoptosis contribute to tumorigenesis and to the resistance of cancer cells to a variety of therapeutic agents. The intranuclear mechanisms that signal apoptosis after DNA damage overlap with those that initiate cell cycle arrest and DNA repair, and the early events in these pathways are highly conserved. In addition, multiple independent routes have recently been traced by which nuclear DNA damage can be signaled to the mitochondria, tipping the balance in favor of cell death rather than repair and survival (see Chapter 8). A conflict in cell cycle progression or DNA damage can lead to mitotic catastrophe when the DNA structure checkpoints are inactivated, for instance, when the checkpoint kinase CHK2 is inhibited. In these circumstances, cells undergo p53-independent apoptosis during metaphase. Suppression of apoptosis leads to the generation of aneuploid cells. DNA damage that fails to arrest the cell cycle together with inhibition of apoptosis can favor the occurrence of cytogenetic abnormalities and tumorigenesis.

Inappropriate deregulated expression of oncoproteins driving cell cycle progression can generate a signal that is detected by the cell and in some way is distinguished from a "normal" cell cycle involving more usual levels/persistence of the same protein (proto-oncoprotein counterpart). This has been referred to as oncogenic stress and could act as an anticancer barrier since the early tumor cell would arrest or die by apoptosis. A series of recent studies have thrown some light on how this key antitumor mechanism operates and have identified important links between DNA damage responses and activation of tumor suppressors (summarized in Fig. 10.10). Thus, it now appears that the tumor suppressor ARF–p53–p21 (see Chapters 6, 7), BIM (Chapter 8), and ATM/ATR kinase pathways may all cooperate in various contexts in promoting apoptosis in response to elevated expression of oncogenes such as c-*MYC*. Recent studies of a variety of human tumors demonstrated that the early precursor lesions, even prior to genomic instability, express phosphorylated kinases ATM and CHK2, and phosphorylated histone H2AX and p53 – all markers of an activated DNA damage response. Similar checkpoint responses could be evoked in cultured cells by deregulated expression of different oncogenes.

This is particularly intriguing as it suggests yet another potential tumor suppressive mechanism triggered by inappropriate activity of an oncogene; apoptosis and senescence have already been described in this context in previous chapters. Moreover, there may be connections. DNA damage responses to oncogene activation could act as an anti-cancer barrier in the same way as activation of ARF–p53 or BIM (another parallel apoptotic pathway activated by oncogenic MYC), namely by in turn promoting either growth arrest or apoptosis of the cancer cell. ARF and BIM are activated by deregulated expression of oncogenes such as c-*MYC*, and there is likely overlap with DNA damage responses in this activity also. Interestingly, in a recent study, the situation has become even more tortuous: E2F family members all induce S phase genes and cell cycle progression but only E2F1 has so far been shown to have proapoptotic activity also. In a recent study a putative mechanism has been unveiled that shows major parallels with c-MYC. Inactivation of RB and resultant deregulation of E2F1 in cultured cells led to accumulation of DSBs, but not ROS. Moreover, E2F1 did not contribute to c-MYC-associated DSBs suggesting that MYC and E2F1 (a c-MYC target) have independent effects on DNA damage. It is tempting to speculate that E2F1 induces apoptosis by activating DNA damage responses and oncogenic stress, but independently of upstream proteins such as c-MYC (important as a cell with mutated RB and deregulated E2F1 will not necessarily have elevated c-MYC from the outset).

As mentioned, ARF is a crucial mediator of oncogenic stress signals. In addition to activation of the p53 pathway, ARF can also suppress cell growth in a p53/MDM2-independent manner, illustrating the central role of ARF. Recent studies suggest that both the p53 dependent and independent tumor suppressor actions of ARF involve inhibiting activity of the ARF-BP1 ubiquitin ligase, which is bound by and inhibited by ARF. Thus, inactivating ARF-BP1 suppresses the growth of p53-null cells and in p53 wild-type cells, ARF-BP1 directly binds and ubiquitinates p53 (analogs to MDM2), and inactivation of endogenous ARF-BP1 is crucial for ARF-mediated p53 stabilization. Numerous apoptosis-related genes are regulated by p53, including those encoding death receptors such

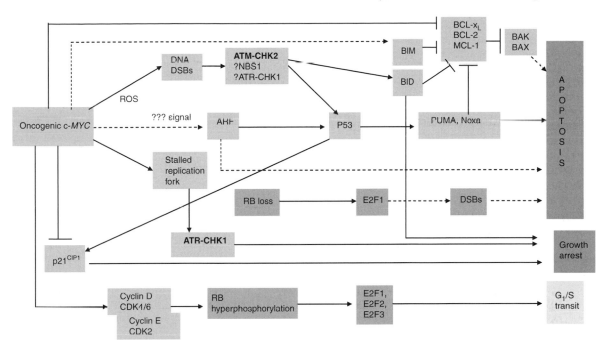

Fig. 10.10 Oncogenic stress, induced by either oncogenic c-*MYC* or loss of RB. During tumorigenesis, the developing cancer cells experience "oncogenic stress", which evokes a counter-response to eliminate them. Although, some of the key signaling proteins involved in oncogenic stress have been known for some time, notably ARF, the nature of all the key signals distinguishing between regular cell cycles and cancer cycles are still not fully characterized. This figure summarizes what is currently known. Deregulated expression of c-*MYC* can activate various BH3-only proteins such as BIM and BID, which by sequestering antiapoptotic BCL-2 proteins cooperates with the ARF–p53–PUMA/Noxa/BAX pathway in inducing apoptosis. At least in part the activation of p53 and BID may be via DNA damage responses activated following c-*MYC* activation, suggesting that this may play a key role in activating these "failsafe" mechanisms not otherwise activated by c-*MYC* during normal cell cycles. BID may also contribute to arrest in S phase. Oncogenic c-*MYC* can result in excess production of ROS and DNA DSBs, which may be key triggers for MYC-induced apoptosis in this context. Downregulation of p21^{CIP1} by c-MYC may also push cells from growth arrest towards apoptosis. DNA damage results in transcriptional induction of p53 target genes, including p21^{CIP1} as well as the proapoptotic Bcl-2 family member p53 upregulated modulator of apoptosis (PUMA), Noxa, and BAX. p21^{CIP1} and PUMA mediate cell cycle arrest and apoptosis, respectively. Importantly, p21^{CIP1} may trigger cell cycle arrest at the expense of apoptosis, therefore suppression of p21^{CIP1} by c-MYC may result in the predominance of apoptosis over cell cycle arrest as an inherent tumor suppressor activity. It is possible that other CKIs such as p16^{INK4a}, which can also growth arrest cells, might also prevent apoptosis by reducing MOMP. Loss of RB, which also deregulates cell cycle control, can also culminate in DSBs and apoptosis, probably via aberrant activation of E2F1. Dashed lines indicate the presence of intermediate proteins (known or unknown).

as *FAS* and those of proapoptotic Bcl-2 proteins such as *BAX*, *BID*, *NOXA*, and *PUMA*. However, p53 also accumulates in the cytoplasm, where it directly activates BAX to promote MOMP (Chapter 8). MOMP causes release of cytochrome *c* from mitochondria triggering activation of caspase 9 and apoptosis.

In a recent study from Doug Green's group, DNA damage-induced apoptosis was found to include both nuclear and cytoplasmic functions of p53. After p53 stabilization, p53 accumulated in the nucleus and promoted expression of proapoptotic genes, such as *BAX* and *PUMA*, and accumulated in the cytoplasm where it bound to Bcl-xL. Intriguingly, following nuclear activity of p53, PUMA displaced p53 from Bcl-xL enabling direct activation of BAX and MOMP. This study provides an explanation as to why loss of Bcl-xL might sensitize to DNA damage, whereas *Puma* deficiency promotes resistance to numerous p53-dependent apoptotic stimuli.

This is likely to prove a fertile research area in the next few years as we learn more about how oncogenes such as c-*MYC* actually trigger apoptosis as well as cell cycle progression.

MICROSATELLITES AND MINISATELLITES

Microsatellites are short sequences of 1–5 bp repeated in tandem throughout the genome, and because of their polymorphic nature, they have been widely used as genetic markers. Mutations can occur within the sequence of these microsatellites as a result of the slippage of DNA polymerase during DNA synthesis, resulting in small expansions or deletions within the repeat sequence manifesting as loss or gain of simple repetitive units. This is a common phenomenon, and in healthy cells this is repaired by the cell DNA mismatch repair (MMR) system. In cells defective in MMR (e.g., as a result of mutation or hypermethylation of one or more of the MMR genes), these expansions/deletions are not repaired. Microsatellite analysis of DNA from such MMR-deficient cells may show the appearance of a new microsatellite PCR product on an electrophoretic gel, either higher or lower than the normal alleles, and the concomitant loss of intensity of one of the normal alleles. This is defined as allelic shift, and if this occurs at a substantial proportion of microsatellite loci, it is representative of microsatellite instability (MSI), often taken as diagnostic of loss of DNA MMR.

The human genome contains a unique class of domains, referred to as AT islands, which consist typically of 200–1000 bp long tracts of up to 100% A/T DNA. AT islands are inherently unstable (expandable) minisatellites that are found in various known loci of genomic instability, such as AT-rich fragile sites. The cellular function of AT islands may differ in cancer and normal cells, but appears to include DNA replication. AT islands are preferentially targeted by the potent DNA-alkylating antitumor drug, bizelesin, and may be a major factor in cytotoxicity of these agents.

Promyelocytic leukemia nuclear bodies

Promyelocytic leukemia nuclear bodies (PML NBs) are generally present in all mammalian cells, and mice lacking PML NBs exhibit chromosome instability and are sensitive to carcinogens. PML NBs are implicated in the regulation of transcription, apoptosis, tumor suppression, and the antiviral response and may be sites for formation of multisubunit complexes involved in post-translational modification of regulatory factors, such as p53, and potentially also in DNA repair. Several repair factors transit through PML NBs, which may act as general sensors of cellular stress, and which rapidly disassemble following DNA damage into large supramolecular complexes, dispersing associated repair factors to sites of damage. The PML tumor-suppressor protein potentiates p53 function by regulating post-translational modifications, such as CBP-dependent acetylation and Chk2-dependent phosphorylation in the PML NB. PML enhances p53 stability by sequestering the p53 ubiquitin-ligase MDM2 to the nucleolus. Following DNA damage, PML and MDM2 accumulate in the nucleolus in an ARF-independent manner. Nucleolar localization of PML is dependent on ATR activation and phosphorylation of PML by ATR.

SUMO AND DNA REPAIR

All eukaryotic cells contain multi-protein complexes, such as cohesin and condensin, that include members of the structural maintenance of chromosomes (SMC) proteins, together with several non-SMC subunits. The third complex, currently known as the Smc5/6 complex, is poorly understood, though genetic analysis points to an essential role in genome integrity, and this complex has been shown to have SUMO ligase activity in fission and budding yeasts. Small ubiquitin-like modifier (SUMO) is a small protein distantly related to ubiquitin and is also discussed in Chapter 11. Like ubiquitin, SUMO is covalently conjugated to lysines on target proteins using an E1 SUMO activator and an E2 SUMO conjugating enzyme, generally referred to as Ubc9. E3 SUMO ligases have also been described, but there are far fewer such enzymes than the large family of E3 ubiquitin ligases. Modification by SUMO does not lead to protein degradation but may in fact compete with ubiquitin. The main role of sumoylation appears to be in signaling pathways, transcriptional activities, and the regulation of subcellular localization. Mammalian cells have three distinct SUMO genes/polypeptides, providing further complexity. Many sumoylated proteins have been identified by either directed or proteomic approaches, and include DNA repair proteins such as the HR proteins Rad51 and Rad52. Human homologs of both interact with SUMO in yeast two-hybrid assays. Cohesin and condensin are both implicated in DNA repair and contain SMC proteins

CHAPERONES AND GENOMIC INSTABILITY

Molecular chaperones, otherwise known as heat shock proteins, are key regulators of protein homeostasis within cells. Specifically, chaperones direct folding, localization, and proteolytic turnover of many key regulators of cell growth, differentiation, and survival. Moreover, they may contribute to inactivation of various nuclear receptors prior to ligand binding. Not surprisingly, therefore, chaperones might also be involved in facilitating or hindering malignant transformation at the molecular level. Thus, chaperones can serve as biochemical buffers at the phenotypic level for the genetic instability that is characteristic of many human cancers. Chaperone proteins could even enable progressing tumor cells to survive mutations that would otherwise be lethal. The recent discovery and study of several natural product antitumor agents that selectively inhibit the function of Hsp90 has confirmed the utility of these agents as relatively nontoxic inhibitors of multiple aberrant signaling molecules.

Viral oncogenes and instability

High-risk human papillomavirus (HPV)-associated carcinogenesis of the uterine cervix is a particularly useful model for basic mechanisms of genomic instability in cancer. Cervical

carcinogenesis is associated with the expression of two high-risk HPV-encoded oncoproteins, E6 and E7. Aneuploidy, the most frequent form of genomic instability in human carcinomas, develops as early as in nonmalignant cervical precursor lesions. In addition, cervical neoplasia is frequently associated with abnormal multipolar mitotic figures, suggesting disturbances of the cell-division process as a mechanism for chromosome segregation defects. Spindle poles are formed by centrosomes, and the high-risk HPV E6 and E7 oncoproteins can each induce abnormal centrosome numbers. These two HPV oncoproteins, however, induce centrosome abnormalities through fundamentally different mechanisms and, presumably, with different functional consequences. High-risk HPV E7, which targets the RB tumor suppressor pathway, can provoke abnormal centrosome duplication in phenotypically normal cells. On the contrary, cells expressing the HPV E6 oncoprotein, which inactivates p53, accumulate abnormal numbers of centrosomes in parallel with multinucleation and nuclear atypia. These two pathways are not mutually exclusive, since co-expression of HPV E6 and E7 has synergistic effects on centrosome abnormalities and chromosomal instability. Taken together, these findings support the general model in which chromosomal instability arises as a direct consequence of oncogenic insults and can develop at early stages of tumor progression.

RADIATION-INDUCED GENOMIC INSTABILITY

Radiation-induced genomic instability encompasses a range of measurable end points such as chromosome destabilization, sister chromatid exchanges, gene mutation and amplification, late cell death, and aneuploidy, all of which may be causative factors in the development of clinical disease, including carcinoma. Clinical implications of genomic instability can be broadly grouped into two main areas: as a marker for increased cancer risk/early detection, and as a consequence of radiation therapy (IR) that may be causative of, or a strong marker for, the induction of a therapy-induced second malignancy.

CANCER SUSCEPTIBILITY SYNDROMES INVOLVING GENETIC INSTABILITY

Ataxia-telangiectasia

Ataxia-telangiectasia is a syndrome of cancer susceptibility, immune dysfunction, and neurodegeneration that is caused by mutations in the A-T-mutated (*ATM*) gene. ATM has been implicated as a critical regulator of cellular responses to DNA damage, including the activation of cell cycle checkpoints and induction of apoptosis. Although defective cell cycle-checkpoint regulation and associated genomic instability presumably contribute to cancer susceptibility in A-T, the mechanism of neurodegeneration in A-T is not well understood. In addition, although ATM is required for the induction of the p53 transcriptional program in response to DNA damage, the identities of the relevant transcription factors that mediate ATM-dependent changes in gene expression remain largely unknown.

Nijmegen breakage syndrome

Nijmegen breakage syndrome (NBS) is an autosomal recessive hereditary chromosomal instability syndrome associated with a predisposition to tumor formation. Affected individuals also exhibit microcephaly, a "bird-like" facial appearance, growth retardation, immunodeficiency, and radiosensitivity. The gene defective in NBS has been cloned, and the gene product, NBS1 (nibrin), is a member of the MRN DSB repair complex (Mre11, Rad50, and NBS1). There are obvious similarities between NBS and ATM-mutated fibroblasts, with respect to radiation sensitivity, and this is explained by recent results suggesting that ATM kinase can phosphorylate NBS1. Although NBS1 is a putative tumor suppressor, it is also expressed in highly proliferating tissues developmentally and is located at sites of DNA synthesis through interaction with E2F. Moreover, the MRN complex can prevent DSB accumulation during chromosomal DNA synthesis. *NBS1* has been shown by some studies to be a c-MYC target gene and constitutive expression of NBS1 in cell lines enhances transformation, at least in part by activating the PI3K-AKT pathway. Thus, NBS1 can act as an oncogene by activating PI3K-AKT and, if depleted, as a tumor suppressor by disrupting function of the MRN complex.

Bloom's syndrome

RecQ helicases are conserved from bacteria to humans and are required for the maintenance of genome stability in all organisms. RecQ helicases are caretakers of the genome and influence stability through their participation in DNA replication, recombination, and repair pathways. Mutations in three of the human RecQ family members give rise to genetic disorders characterized by genomic instability and a predisposition to cancer. Bloom's syndrome (BS) is associated with cancer predisposition and genomic instability. A defining feature of Bloom's syndrome is an elevated frequency of sister chromatid exchanges. These arise from crossing over of chromatid arms during homologous recombination, a ubiquitous process that exists to repair DNA double-stranded breaks and damaged replication forks. Whereas crossing over is required in meiosis, in mitotic cells it can be associated with detrimental loss of heterozygosity. BLM forms an evolutionarily conserved complex with human topoisomerase IIIα (hTOPO IIIα), which can break and rejoin DNA to alter its topology. Inactivation of homologs of either protein leads to hyper-recombination in unicellular organisms. The BS gene product, BLM, associates with the stress-activated ATR kinase discussed already, which functions in checkpoint signaling during S-phase arrest.

Xeroderma pigmentosum

Xeroderma pigmentosum (XP) is an autosomal recessive disease characterized by sun sensitivity, early onset of freckling, and subsequent neoplastic changes on sun-exposed skin. Skin abnormalities result from an inability to repair UV-damaged DNA because of defects in the nucleotide excision repair (NER) machinery. Xeroderma pigmentosum is genetically heterogeneous and is classified into seven complementation groups (XPA–XPG) that correspond to genetic alterations in one of seven genes involved in NER. The variant type of XP (XPV), first described in 1970 by Ernst G. Jung as "pigmented xerodermoid", is caused by defects in the post-replication repair machinery while NER is not impaired. The XPV protein, polymerase (pol)η, represents a novel member of the Y family of bypass DNA polymerases that facilitate DNA translesion synthesis. The major function of (pol)η is to allow DNA translesion synthesis of UV-induced TT-dimers in an error-free manner; it also possesses the capability to bypass other DNA lesions in an error-prone manner. Xeroderma pigmentosum V is caused by molecular alterations in the POLH gene, located on chromosome 6p21.1-p12. Affected individuals are homozygous or compound heterozygous for a spectrum of genetic lesions, including nonsense mutations, deletions, or insertions, confirming the autosomal recessive nature of the condition.

Fanconi anemia

This is a chromosome instability disorder, in which cells have a predisposition to chromosomal breaks, characterized by congenital abnormalities, retarded growth, early predisposition to cancer, and bone marrow failure. The pathogenesis is believed to be due to defective repair of strand breaks, but the genetic cause is complicated with defects in multiple distinct genes causing the condition in different families – *FANCA, FANCB, FANCC, BRCA2* (also called *FANCD1), FANCD2, FANCE, FANCF, FANCG, FANCI, FANCJ* and *FANCL*. *BRCA2* was the first such gene with a clearly defined role in DNA repair (homologous recombination). More recently, *FANCJ*, which encodes a DNA-unwinding enzyme (helicase) called BRIP1, mutated in some cases of breast cancer, has been shown to interact with the BRCA1 tumor suppressor. BRIP1, like BRCA1, is involved in binding and unwinding forked duplexes in DSB repair. Ubiquitination of FANCD2 may be an important shared action of the Fanconi anemia core complex proteins (FANCA, FANCB, FANCC, FANCE, FANCF, FANCG, and FANCL). Although speculative, FANCD2 may then be able to stabilize structurally blocked and broken replication forks, thereby contributing to both translesion synthesis and homologous recombination. Crosslinks can result in stalling of DNA replication forks, and

must be unhooked by incision on both sides of an adducted base in order to break one parental strand at the replication fork. BRIP1 helicase activity might help expose the incision substrate by unwinding parental strands adjacent to the crosslink. The broken fork is subsequently restored by homologous recombination involving BRCA2 and potentially other proteins, such as BRIP1, that enable Rad51 to carry out strand transfer.

Werner's syndrome

This is an autosomal recessive disease characterized by premature aging, elevated genomic instability and increased cancer incidence, resulting from inactivation of the *WRN* gene. Telomere attrition is implicated in the pathogenesis of Werner syndrome. c-MYC overexpression directly elevates transcription of the *WRN* gene, whose presence is required to avoid senescence during c-MYC proliferative stimuli. Recent work suggests that WRN may counteract the effects of genomic instability promoted by deregulated c-MYC activation and thereby enable propagation of mutant cells, which in the absence of WRN would be ablated.

GENOMIC INSTABILITY AND COLON CANCER

Colorectal cancer results from the progressive accumulation of genetic and epigenetic alterations that lead to the transformation of normal colonic epithelial cells to colon adenocarcinoma cells (see Fig. 3.4). The loss of genomic stability appears to be a key early pathogenetic event, which may foster the subsequent occurrence of alterations in tumor suppressor genes and oncogenes; in particular, activation of Wnt target genes constitutes the primary transforming event in colorectal cancer. At least three forms of genomic instability have been identified in colon cancer: (1) microsatellite instability, (2) chromosome instability (i.e. aneuploidy), and (3) chromosomal translocations. Microsatellite instability occurs in approximately 15% of colon cancers and results from inactivation of the MMR system by gene mutations or hypermethylation of the *MLH1* promoter. MSI promotes tumorigenesis through generating mutations in target genes that possess coding microsatellite repeats, such as *TGFBR2* and *BAX*. CIN is found in the majority of colon cancers and may result primarily from defects in DNA replication checkpoints and mitotic-spindle checkpoints. Mutations of the mitotic checkpoint regulators BUB1 and BUBR1 and amplification of STK15 have been identified in a subset of CIN colon cancers.

KEY EXPERIMENT

While Watson and Crick had suggested that DNA replication would be semiconservative, proof was provided by the experiments of M.S. Meselson and F.W. Stahl.

The bacterium *E. coli* was grown in culture medium containing (NH_4^+) as the source of nitrogen for DNA (as well as protein) synthesis. In these elegant studies, two different ammonium ions containing isotopes of either nitrogen, ^{14}N, the common form, or ^{15}N were employed. Not surprisingly, after growing *E. coli* for several generations in a medium containing $^{15}NH_4^+$, they found that the DNA of the cells was heavier than normal because it contained ^{15}N. The difference in density could readily be measured by isolating the DNA and precipitating it in an ultracentrifuge – the denser DNA accumulates at a lower level in the centrifuge tube. If, however, *E. coli* was then switched from growing in $^{15}NH_4^+$ to a medium containing $^{14}NH_4^+$ and was allowed to undergo a single cell division, the DNA was now found to be intermediate in density between that of the previous generation and normal – in other words, the newly synthesized DNA contained equal amounts of both ^{14}N and ^{15}N. Now, if *E. coli* was allowed to divide once more in normal $^{14}NH_4^+$ containing medium, two distinct densities of DNA were formed, half the DNA was normal and half was intermediate. This confirmed that DNA molecules are not degraded and then made anew from free nucleotides between cell divisions, but, instead, each original strand remains intact with a new complementary strand added by new synthesis. This is termed semiconservative replication as each new DNA molecule contains both an original strand and a new one.

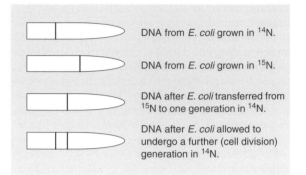

DNA from *E. coli* grown in ^{14}N.

DNA from *E. coli* grown in ^{15}N.

DNA after *E. coli* transferred from ^{15}N to one generation in ^{14}N.

DNA after *E. coli* allowed to undergo a further (cell division) generation in ^{14}N.

BREAST CANCER AND THE *BRCA* GENES

Knudson's classic "two-hit" model describing the inactivation of a tumor suppressor gene (Chapter 7) appears to fit the behavior of both the *BRCA1* and *BRCA2* genes. In familial cancers offspring inherit a germline mutation, thus this "first hit" is present in all cells of the body. A somatic mutation represents the "second hit" on a given cell, resulting in the loss or inactivation of the wild-type allele, thus rendering both copies of the gene inactive. In sporadic cancers, inactivation of a tumor suppressor gene is accomplished by two somatic mutations that eliminate or inactivate the alleles on both chromosomes. Knudson's model accurately accounts for the early onset of familial cancers caused by a preexisting germline mutation, while the accumulation of two somatic mutations in a single cell may take several decades to produce a sporadic cancer.

BRCA1 is involved in double-stranded DNA repair and normal DNA recombination. Splice variants of BRCA1 mRNA have been identified that exist normally in nonmalignant breast cells. These alternatively spliced mRNAs code for truncated proteins; however, it is yet undetermined whether these truncated proteins exhibit tumor suppressor activity. The *BRCA2* gene is similar to the *BRCA1* gene and also plays a role in DNA repair. Both genes have more than 20 exons, with a very large exon 11 and encode highly charged proteins with a putative granin domain.

It is estimated that mutations in *BRCA1* alone account for 50% of all familial early-onset female breast cancers, while mutations in *BRCA2* may be responsible for up to 35% of the remaining hereditary breast cancers. Loss of heterozygosity at the *BRCA2* locus has been observed in 30% to 40% of sporadic breast and ovarian cancers; however, very few somatic mutations have been found in the remaining allele. This suggests that either *BRCA2* is an infrequent target for somatic inactivation or that intron sequences or genomic deletions may be the target of somatic mutation.

BRCA1 carriers have a 60% to 85% lifetime risk of developing breast cancer and a 20% to 50% risk of developing ovarian cancer, in contrast to the 2% to 3% risk in the general population. *BRCA2* carriers have similar risks, although the average age of onset for ovarian cancer appears to be similar to the age of sporadic cases. In addition, male *BRCA2* carriers are at increased risk for male breast cancer, with an approximately 6% to 8% lifetime risk, as compared with the general male population risk of 0.1%.

Women affected with breast cancer who are *BRCA* carriers also have increased risk of another primary breast cancer in the remaining or contralateral breast of 30% to 50% by age 70. Thus carriers who are also survivors of breast or other cancers warrant as much counseling for adherence to surveillance as nonaffected carriers.

DRUGS AND DNA DAMAGE

Most agents used to treat cancer (see Chapter 16) clinically target genomic DNA. The selectivity of these anticancer drugs for cancers is likely due to the presence of tumor-specific defects in cell cycle checkpoints, compromising DNA repair or enhancing apoptotic responses in the tumor.

Drugs are now being developed to target components of these checkpoints. Activation of the ATM–Chk2 pathway is important in this context as it responds to more serious DNA damage such as DSBs. Chk2 can activate both apoptosis (via p53, E2F1, and PML) and cell cycle checkpoints (via Cdc25A and Cdc25C, p53, and BRCA1). Thus, Chk2 inhibitors might be used to enhance the tumor selectivity of DNA-targeted agents in p53-deficient tumors, and for

the treatment of tumors whose growth depends on enhanced Chk2 activity.

QUESTIONS REMAINING

1 Is the presence of MIN or CIN a general feature of cells destined to become cancer cells and at what stage in carcinogenesis do these features of genetic instability arise?
2 Is the mutator phenotype a generally applicable model to cancers other than colon?
3 To further define the genetic and epigenetic changes in various genes that may contribute to defective DNA repair and chromosomal segregation.
4 How might knowledge about defective mismatch repair and other processes be translated into more effective anticancer therapies. For example, might further increasing genetic instability with agents such as taxols preferentially lead to death of cancer cells.

CONCLUSIONS AND FUTURE DIRECTIONS

Genomic stability is essential for normal cellular function, and highly conserved pathways have evolved to repair DNA damage and prevent genomic instability or failing that to trigger senescence or death of the cell. Should these mechanisms fail then this may promote genomic instability – a major driver of cancer.

Genomic instability is classed as microsatellite instability (MIN), associated with mutator phenotype, or chromosome instability (CIN), manifesting as gross chromosomal abnormalities. Three DNA damage repair processes are implicated in development of a mutator phenotype, NER, BER, and MMR.

The proteins ATR and RAD1, bind to chromatin and are needed to phosphorylate several other proteins that are critical to the DNA damage response, including BRCA1, which when mutated leads to hereditary breast cancer. Defects in DNA MMR and CIN pathways are involved in several hereditary cancer predisposition syndromes including HNPCC, Bloom syndrome, and ataxia-telangiectasia. Epigenetic factors as well as genetic factors may compromise the operation of cell cycle checkpoints and sensing of DNA damage. Telomere attrition can also produce genetic instability as discussed in the previous chapter.

At least three forms of genomic instability have been identified in colon cancer: MIN, CIN, and chromosomal translocations. MIN occurs in approximately 15% of colon cancers due to defective MMR (mutations or hypermethylation of the *MLH1* promoter) and generates mutations in genes with coding microsatellite repeats, such as *TGFBR2* and *BAX*. CIN is found in the majority of colon cancers as result of deregulation of the DNA replication checkpoints and mitotic-spindle checkpoints.

Much still needs to be learned about mechanisms that induce and influence genomic instability in particular to understand more about how DNA damage is sensed and thence how mutations of the various mitotic checkpoint regulators (including BUB1, BUBR1, and others) may culminate in instability and cancer. It is hoped that, eventually, studies will lead to a deeper understanding of the pathogenesis of genetic instability and potentially new therapies exploiting this characteristic of cancer cells.

QUESTIONS

1 DNA damage:
   a. Results exclusively from extrinsic factors.
   b. Mostly results in apoptosis.
   c. May be detected by RNA polymerase.
   d. May be repaired.
   e. May be increased by certain gene mutations.
2 DNA repair:
   a. Is usually unimpaired in cancer cells.
   b. If ineffective invariably results in apoptosis or senescence.
   c. Usually takes place while the cell cycle is arrested.
   d. Does not occur within microsatellites.
   e. Cannot correct double-strand breaks.
3 Chromosomal instability (CIN):
   a. Always implies abnormal numbers of chromosomes.
   b. May result from double-strand breaks.
   c. May result from premature mitosis.
   d. Can be prevented by a functioning spindle checkpoint.

   e. Is always a consequence rather than a cause of cancer.
4 Microsatellite instability (MIN):
   a. Is associated with a "mutator phenotype."
   b. Is characterized by aneuploidy.
   c. Is seen only in inherited forms of colon cancer.
   d. May arise through epigenetic silencing of mismatch repair genes.
   e. Is found in most sporadic tumors.
5 Genetic instability:
   a. Is present in most established human cancers.
   b. Always occurs once a cancer has developed.
   c. May result from mutations in single genes.
   d. May result from telomere attrition.
   e. Usually manifests as both CIN and MIN in the same tumors.

BIBLIOGRAPHY

Albertson, D.G., Collins, C., McCormick, F. and Gray, J.W. (2003). Chromosome aberrations in solid tumors. *Nat Genet.*, **34**(4): 369–76.

Bassing, C.H., Suh, H., Ferguson, D.O., Chua, K.F., Manis, J., Eckersdorff, M., Gleason, M., Bronson, R., Lee, C. and Alt, F.W. (2003). Histone H2AX: a dosage-dependent suppressor of oncogenic translocations and tumors. *Cell*, **114**(3): 359–70.

Boyd, S.D., Tsai, K.Y. and Jacks, T. (2000) An intact HDM2 RING-finger domain is required for nuclear exclusion of p53. *Nature Cell Biol.*, **2**: 563–8.

Castedo, M., Perfettini, J.L., Roumier, T., Valent, A., Raslova, H., Yakushijin, K., Horne, D., Feunteun, J., Lenoir, G., Medema, R., Vainchenker, W. and Kroemer, G. (2004). Mitotic catastrophe constitutes a special case of apoptosis whose suppression entails aneuploidy. *Oncogene*, **23**(25): 4362–70.

Chen, D., Kon, N., Li, M., Zhang, W., Qin, J. and Gu, W. (2005). ARF-BP1/Mule is a critical mediator of the ARF tumor suppressor. *Cell*, **121**(7): 1071–83.

Chipuk, J.E., Bouchier-Hayes, L., Kuwana, T., Newmeyer, D.D. and Green, D.R. (2005). PUMA couples the nuclear and cytoplasmic proapoptotic function of p53. *Science*, **309**(5741): 1732–5.

d'Adda di Fagagna, F., Reaper, P.M., Clay-Farrace, L., Fiegler, H., Carr, P., von Zglinicki, T., Saretzki, G., Carter, N.P. and Jackson, S.P. (2003). A DNA damage checkpoint-mediated response in telomere-initiated cellular senescence. *Nature*, **426**(6963): 194–8.

Goldberg, M., Stucki, M., Falck, J., D'Amours, D., Rahman, D., Pappin, D., Bartek, J. and Jackson, S.P. (2003). MDC1 is required for the intra-S-phase DNA damage checkpoint. *Nature*, **421**(6926): 952–6.

Grandori, C., Wu, K.J., Fernandez, P., Ngouenet, C., Grim, J., Clurman, B.E., Moser, M.J., Oshima, J., Russell, D.W., Swisshelm, K., Frank, S., Amati, B., Dalla-Favera, R. and Monnat, R.J. Jr. (2003). Werner syndrome protein limits MYC-induced cellular senescence. *Genes Dev.*, **17**(13): 1569–74.

Green, D.R. (2005). Apoptotic pathways: ten minutes to dead. *Cell*, **121**(5): 671–4.

Hasty, P., Campisi, J., Hoeijmakers, J., van Steeg, H. and Vijg, J. (2003). Aging and genome maintenance: lessons from the mouse? *Science*, **299**: 1355–9.

Hernando, E., Nahle, Z., Juan, G., Diaz-Rodriguez, E., Alaminos, M., Hemann, M., Michel, L., Mittal, V., Gerald, W., Benezra, R., Lowe, S.W. and Cordon-Cardo, C. (2004). Rb inactivation promotes genomic instability by uncoupling cell cycle progression from mitotic control. *Nature*, **430**(7001): 797–802.

Jackson, A.L. and Loeb, L.A. (1998). The mutation rate and cancer. *Genetics*, **148**: 1483–90.

Lindahl, T. (1993). Instability and decay of the primary structure of DNA. *Nature*, **362**: 709–15.

Ljungman, M., Zhang, F.F., Chen, F., Rainbow, A.J. and McKay, B.C. (1999). Inhibition of RNA polymerase II as a trigger for the p53 response. *Oncogene*, **18**: 583–92.

McGlynn, P. and Lloyd, R.G. (2002). Recombinational repair and restart of damaged replication forks. *Nature Rev. Mol. Cell Biol.*, **3**: 859–70.

Mellon, I., Spivak, G. and Hanawalt, P.C. (1987). Selective removal of transcription-blocking DNA damage from the transcribed strand of the mammalian *DHFR* gene. *Cell*, **51**: 241–9.

Mitchell, J.R., Hoeijmakers, J.H. and Niedernhofer, L.J. (2003). Divide and conquer: nucleotide excision repair battles cancer and ageing. *Curr. Opin. Cell Biol.*, **15**: 232–40.

Rajagopalan, H., Lengauer, C. (2004). Aneuploidy and cancer. *Nature*, **432**(7015): 338–41.

Rouse, J. and Jackson, S.P. (2002). Interfaces between the detection, signaling, and repair of DNA damage, *Science*, **297**: 547–51.

Savitsky, K. *et al.*, (1995). A single ataxia telangiectasia gene with a product similar to PI-3 kinase. *Science*, **268**: 1749–53.

Scully, R. and Livingston, D.M. (2000). In search of the tumor-suppressor functions of BRCA1 and BRCA2. *Nature*, **408**: 429–32.

11

Cancer is More Than a Genetic Condition

Stella Pelengaris and Mike Khan

"So divinely is the world organized that every one of us, in our place and time, is in balance with everything else." Johann Wolfgang von Goethe, German dramatist, novelist, poet, and scientist (1749–1832).

KEY POINTS

- It has long been known that gene expression is regulated by activators and repressors of transcription operating on noncoding regulatory elements within a given gene/s.
- More recently, several further levels of control of gene expression have been described. Importantly, in cancer not all dysregulation of gene expression results from mutations in coding DNA. Thus, gene expression can also be regulated by:
  a. Chromatin modifications that are transmitted from one somatic cell to all its descendants. Such control is referred to as "epigenetic," as the DNA sequence is not altered.
  b. Post-transcriptional modification of mRNA.
  c. Stability and processing of mRNA.
  d. Regulation of genes and mRNAs by small RNA molecules encoded by areas of the genome previously believed to be noncoding "nonsense."
- Patterns of DNA methylation and chromatin structure are often markedly disturbed in cancer cells potentially resulting in inappropriate silencing of tumor suppressor genes (hypermethylation) or conversely activation of oncogenes (hypomethylation).
- Some cancers may be epigenetically unstable. Such cancers exhibit hypermethylation of usually unmethylated regions such as the CpG islands, found in up to half of all human gene promoters (CpG island methylator phenotype – CIMP), resulting in aberrant silencing of hundreds of genes. CIMP is analogs to the mutator phenotype resulting from defects in DNA repair described in Chapter 10.
- The search for gene mutations and environmental agents that might provoke aberrant methylation is likely to be an important area in the next few years.
- One class of small RNAs, the miRNAs, have recently been shown to be aberrantly expressed in a number of human cancers and may potentially be used for profiling. Moreover, one miRNA at least acts as an oncogene, cooperating with c-MYC in tumorigenesis in a lymphoma model *in vivo*.
- Functional consequences of changes in gene expression are ultimately determined by changes in expression of the proteins they encode. Importantly, however, both levels and activity of a protein are determined by numerous factors operating after translation. Thus, protein activity, location, and stability are regulated by various post-translational modifications including phosphorylation, ubiquitination, and sumoylation, all of which may become aberrant in cancer cells.

- The ubiquitin–proteasome pathway is involved in degradation of numerous cancer-relevant intracellular proteins, including regulators of apoptosis and cell cycling. Mutations in ligases and other genes may compromise protein stability/degradation and contribute to tumorigenesis. Inhibitors of the proteasomal degradation of proteins are being examined as therapeutic agents against cancer.
- Aside from degradation the biological activity of proteins may also be influenced by localization and partitioning within the cell.

INTRODUCTION

With the completion of genome-sequencing projects, one of the major challenges in modern biology is to understand what all the tens of thousands of identified genes actually do and moreover how their activity is regulated. To face this considerable challenge will require a much greater understanding of all the many processes that contribute to the regulation of gene expression and the ultimate formation of functional protein products, including determining how epigenetic controls are imposed on genes.

Without laboring the point, as this has been covered in depth already, tumor cells are characterized by aberrant responses to cellular signals that normally regulate cell replication, differentiation, cell adhesion/motility, and apoptosis. As already discussed in the preceding chapters, these cellular processes are regulated by the production of proteins, through the transcription and translation of genes, the regulation of which are subject to potentially cancer-causing mutations encoded in the DNA of the cancer cells or other cells in their microenvironment. But, changes in gene/protein expression do not only arise as a result of gene mutations. In fact, protein levels can also be regulated by numerous other relatively more recently described factors including changes in the methylation and structure of the DNA and associated histone proteins, by small regulatory molecules of RNA and the often overlooked control of the stability, transport and partitioning of the protein.

In many ways what is now referred to as epigenetics is not new science – it has long been recognized that traditional genetics theory, which implied a one-to-one relationship between genotype and phenotype, could not readily explain processes such as cell differentiation; here multiple often very different cell phenotypes are produced yet all bearing ostensibly the identical genome. Thus, it was hypothesized that each undifferentiated cell underwent a crisis that determined its fate, which was not inherent in its genes, and was therefore, from the Greek, *epigenetic* – "in addition to" – the genetic information encoded in the DNA (see Box 11.1 for a historical

---

Box 11.1   Historical overview of epigenetics

It was proposed some time ago that genetics, which seemed to imply a one-to-one correspondence between genotype and phenotype, could not readily explain processes such as cell differentiation (in other words how could such dramatically different cells arise given they all had the same genes). A theory was proposed that each undifferentiated cell experienced a "crisis" that determined its fate and that this event was not inherent in its genes, and was therefore (borrowing from the Greek) epigenetic. The biologist C.H. Waddington is sometimes credited with coining the term "epigenetics" in 1942, when he defined it as "the branch of biology which studies the causal interactions between genes and their products which bring the phenotype into being." However, the term goes back to at least 1896.

The notion that characteristics that were acquired during an organism's lifetime could be passed on to the offspring is called Lamarckian, after Jean-Baptiste Lamarck. This view was, until very recently, thought to be totally at odds with modern genetics, but recent understanding of the process of epigenetic inheritance shows that at least some aspects of Lamarck's ideas about evolution do hold true.

Lamarck will be familiar to all biology students as the author of a widely discredited theory of heredity, the "inheritance of acquired traits." However, at the time his views almost certainly influenced other biologists wrestling with the emerging field of evolution, in particular Charles Darwin (see Box 3.1). In 1861, Darwin wrote, "Lamarck was the first man whose conclusions on the subject excited much attention. This justly celebrated naturalist first published his views in 1801 he first did the eminent service of arousing attention to the probability of all changes in the organic, as well as in the inorganic world, being the result of law, and not of miraculous interposition."

Lamarck developed two laws:

1. "*In every animal which has not passed the limit of its development, a more frequent and continuous use of any organ gradually strengthens, develops and enlarges that organ, and gives it a power proportional to the length of time it has been so used; while the permanent disuse of any organ imperceptibly weakens and deteriorates it, and progressively diminishes its functional capacity, until it finally disappears.*"
2. "*All the acquisitions or losses wrought by nature on individuals, through the influence of the environment in which their race has long been placed, and hence through the influence of the predominant use or permanent disuse of any organ; all these are preserved by reproduction to the new individuals which arise, provided that the acquired modifications are common to both sexes, or at least to the individuals which produce the young.*"

Lamarck's own theory of evolution was based on the notion that an organism adapts to its environment during its own lifetime and passes on traits that have been acquired to the offspring (in modern terms the implication is that an organism will respond to events and environment by undergoing genetic alterations that can then be in some way passed on to the offspring). Offspring then adapt from where the parents left off, and evolution advances. Lamarck proposed that individuals increased specific capabilities by using them, while losing others through disuse. Lamarck believed in a teleological (goal-oriented) version of evolution, with organisms improving progressively as they evolved. Lamarck has become synonymous with pre-Darwinian ideas about evolution, now called Lamarckism.

Modern evolutionary biology accepts that the environment plays a role during natural selection, by dictating what characteristics are necessary for better reproduction opportunities. For natural selection to occur, individuals must differ somewhat genetically, in order that positive characteristics can amplify and negative ones can be deleted from the gene pool. These differences between individuals (or for that matter between cancer cells) arise from random mutations in genes – this is the mechanism underlying Darwinian evolution of individuals within a species or of cancer cells within a tumor. The environment can influence these variations (for example, radioactivity and other mutagens will damage DNA), but probably only in a random manner.

However, very recently multiple studies are indicating that we should revisit the notion that the environment may play a more direct and crucial role in evolution.

Epigenetic inheritance allows cells of differing phenotype but identical genotype to transmit their phenotype to their offspring, even when the original phenotype-inducing stimuli are no longer present. This is reasonably easy to understand with respect to somatic cells, and there is little debate any longer about the clonal evolution of cancer under the influence of somatic mutations or epigenetic changes, such as promoter methylation, and then natural selection of such changes that confer a growth advantage. However, the question still remains as to whether epigenetic inheritance plays a direct role in evolution of the organism; in this case one must somehow postulate a means whereby information not encoded in the genome of the germ cells can be transmitted to the offspring. One possible explanation for such epigenetic inheritance might be the influence of uterine environment on the developing fetus – such fetal programming has been suggested as an explanation for the observed predisposition of malnourished fetuses to develop diabetes or heart disease as adults (though this remains very contentious). Environmental factors are known to influence the emergence and reversion of epigenetic factors, allowing for the possibility that epigenetic variations at several loci and in several cells or organisms might play a role in evolution. Such an adaptive variation would be a Lamarckian form of evolution. A number of experimental studies seem to indicate that epigenetic inheritance can play a part in the evolution of complex organisms. Methylation differences between maternally and paternally inherited alleles of the mouse *H19* gene are preserved. There are also numerous reports of heritable epigenetic marks in plants.

overview). We now appreciate that the selective silencing of some genes and expression of others during development ultimately determines the phenotype of every cell, and that so-called epigenetic factors are key determinates of this. Epigenetics is one of the most exciting areas of modern biology and particularly over the last few years we have developed a much clearer understanding of how this mechanism operates, and how epigenetic factors can control gene expression. It is now clear that the genome contains information in two forms, genetic and epigenetic. The genetic information provides the blueprint for the manufacture of all the proteins necessary to create the organism, while the epigenetic information provides additional instructions on how, where, and when the genetic information should be deployed. Epigenetic information is not contained within the DNA sequence itself, but can still determine mitotic inheritance of various characteristics as surely as modifications in the DNA sequence. Thus, epigenetic factors, which include DNA methylation and histone modifications, can dictate cell fate and gene expression patterns in the progeny after cell division and are important in the normal regulation of differentiation, aging, and senescence. Epigenetic factors can even turn environmental effects into heritable changes in cell phenotypes – which at face value challenges the central dogma of genetics. However, adult patterns of methylation are erased during early embryogenesis so that in general terms cells in a new organism are believed to start life with a "clean slate" epigenetically. Subsequently, during development and adult life cells progressively acquire epigenetic modifications that they can then pass on to their progeny. Importantly, all epigenetic changes are not invariably predetermined during ontogeny and can be influenced throughout life by genetic and environmental forces. Intriguingly, such epigenetic factors may also be the explanation

for differences in phenotype observed between genetically identical twins.

The only known epigenetic modification of DNA in mammalian cells is covalent addition of a methyl group to the 5th position of cytosine within CpG dinucleotide islands, which can directly turn off gene expression ("silencing") (Fig. 11.1). For example, imprinting, the phenomenon whereby expression of a gene may be silenced depending on whether it was inherited from the mother or the father, is thought to be due to differential methylation in maternal versus paternal genes (Fig. 11.2). Around half of all genes have a CpG island in their promoter region, but most such promoter CpG islands are not normally methylated, irrespective of the expression state of the associated gene. But in areas where gene expression is silenced, such as the silenced allele of imprinted genes and the inactive X-chromosome in females, promoter-associated CpG islands are methylated. The methylation of DNA is achieved by two DNA methyltransferases (DNMTs), though exactly how these are targeted to specific DNA regions is unclear. Ultimately, methylation of a CpG island alters expression of a gene in two ways, directly by interfering with the binding of specific transcription factors to promoters and indirectly by recruiting proteins such as methyl-binding domain (MBD) proteins that associate with further enzymes called histone deacetylases (HDACs), which function to deacetylate histones and change chromatin structure. Histone deacetylation causes the condensation of chromatin, making it inaccessible to transcription factors so that the genes are therefore silenced (see Fig. 11.3).

Epigenetic changes are key factors in several diseases, and cancer in particular – the importance is underscored by the devotion of a large part of this chapter to the topic. Epigenetic regulation is often deranged in cancer cells, which typically manifest aberrant DNA methylation patterns involving an overall global decrease in DNA methylation, alongside regional hypermethylation of CpG islands in particular genes; hypermethylation may silence some tumor suppressor genes, whereas hypomethylation may activate selected proto-oncogenes. In fact, so far more than 600 genes, including tumor suppressor genes and oncogenes, are known to be regulated by epigenetic mechanisms (see also

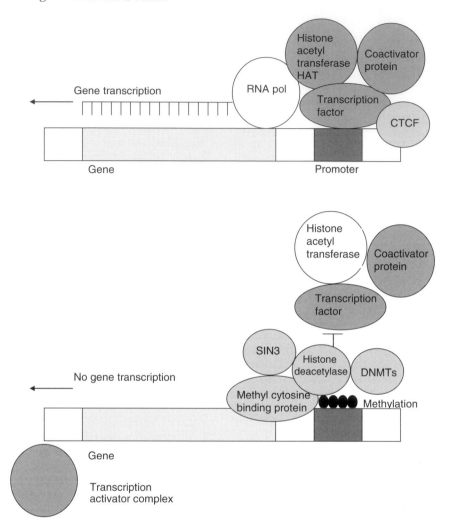

Fig. 11.1 Methylation can inhibit gene transcription. In the presence of CTCF the gene is "insulated" from methylation; in this case the gene is transcriptionally active as it will remain in the unmethylated state. Conversely, in heterochromatin CpG islands in the promoter region are methylated and these regions are transcriptionally inactive. Gene silencing following methylation is reinforced by deacetylation and interactions with repressors of transcription. In fact, DNMTs, which further mediate methylation, may be actively recruited to help maintain silencing. Methylation in turn enables the binding of a complex comprising the methylcytosine binding protein (MBP) and histone deacetylase (HDAC); some MBPs (MECP2 and MBD1 and 2) can also associate with transcriptional corepressors such as SIN3, which directly bind to HDAC, and contribute to gene silencing. HDAC promotes deacteylation of histones, which contributes to organization of nucleosomes, and also more generally repression of transcription. In order for gene expression to take place, the HDAC-SIN3 complex is displaced and a transcription activator complex (transcription factor, histone acetyl transferase-HAT- and coactivator protein) can associate with promoter elements; HAT acetylates the histones reversing effects of HDAC. In cancer many genes may be appropriately inhibited by methylation in the promoter region and this is a frequent cause of loss of tumor suppressor activity during tumorigenesis.

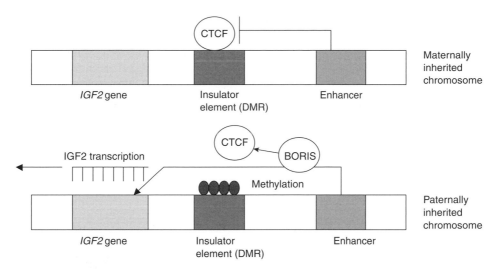

Fig. 11.2 Imprinting of the gene for IGF2. On chromosomes inherited from the female, a protein called CTCF binds to an insulator preventing interaction between the enhancer and the *IGF2* gene. *IGF2* is therefore not expressed from the maternally inherited chromosome. Because of imprinting, the insulator on the male derived chromosome is methylated; this inactivates the insulator, by blocking the binding of the CTCF protein, and allows the enhancer to activate transcription of the *IGF2* gene. It is speculated that CTCF is displaced by another protein, BORIS. The methylation patterns (imprints) on the chromosome, inherited by the zygote after fertilization, are maintained in subsequent generations by maintenance methyl transferases. After Alberts *et al.*, 2002 *Molecular Biology of the Cell*.

Table 11.1). Given the key role of epigenetic factors in silencing of tumor suppressors in cancers, novel therapeutic strategies have been developed based on the reversal of DNA methylation and the inhibition of histone deacetylation. Such therapeutic endeavors notwithstanding, there is now a major interest in new diagnostic technologies able to rapidly screen the genome for DNA methylation and histone acetylation patterns (see Chapter 18).

Although of unquestionable importance, epigenetic regulation does not involve only methylation of DNA, but as mentioned also the post-translational modification of histones. Histones may become methylated on lysine residues and this may contribute to gene silencing and imprinting; in fact, it is increasingly likely that the processes controlling DNA and histone methylation are linked and collaborate in gene silencing. Histone acetylation is a dynamic process that is regulated by histone acetyltransferases (HATs) and histone deacetylases (HDACs) – see Fig. 11.3. Deacetylation of histones is also associated with changes in chromatin structure and transcriptional repression. In fact, the cardinal features of constitutive heterochromatin (transcriptionally inactive chromatin – genes not expressed) include hypermethylation of DNA and histones and deacetylation of histones. Intriguingly, a very recent study from the group of Maria Blasco has implicated the tumor suppressor RB as a key player in the assembly of constitutive heterochromatin. As we have already seen in Chapters 4 and 8, RB is well known for its ability to repress genes required for entry into S phase and progression of the mitotic cell cycle, an action at least in part long-appreciated to involve interaction with HDAC at the promoter regions of various genes activated by E2F family transcription factors. But now this new study suggests that RB may function as a global suppressor of gene expression and that, moreover, loss of RB could result in a generalized loss of repressive chromatin and reactivation of gene expression. Importantly, epigenetic factors may be a general way by which proteins and genes communicate with one another, in particular histone acetylation is now known to contribute to the day-to-day regulation of gene expression. Thus, inhibition of gene expression by transcription factors such as c-MYC involves recruitment of

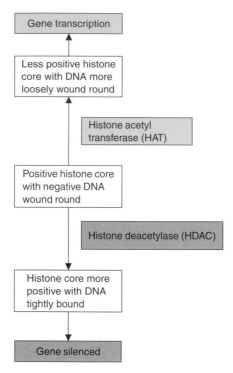

Fig. 11.3 Regulation of gene expression by histone acetylation. Gene expression is also regulated by acetylation as well as by methylation. Acetylation of histones results in a more open configuration of chromatin allowing access of transcriptional activators to the gene to be transcribed. Conversely, deacetylation results in a more closed conformation of DNA, denying access to genes. In fact, displacement of HDAC may be one way in which RB phosphorylation enables access of E2F for transcription of S-phase genes in the cell cycle.

histone-modifying corepressor complexes containing HDAC and mSIN3A (or in other such complexes, N-CoR/SMRT), which can promote deacetylation of lysines in histone H4 tails. In many cases it now appears that for transcriptional activation of multiple genes, epigenetic factors (possibly generally involving inhibitory complexes of RB and HDAC amongst others) must first be overcome. In other words, the role of epigenetic factors in controlling gene expression is extending way beyond the previous notion of such factors predominantly mediating permanent or near-permanent inactivation of genes in development. This is graphically illustrated by the means by which some genes are

inhibited and others activated during regulation of the G_1/S transition in the mitotic cell cycle, discussed in Chapter 4. To remove the brakes from the cell cycle, c-MYC, in partnership with MIZ1, inhibits expression of genes such as the CKI, *p21*^{CIP1}, by recruiting histone-modifying complexes containing mSIN3A and HDAC to the promoter regions and concurrently promotes expression of genes (cyclin-CDKs, etc.) in partnership with a different protein MAX. The net effect is hyperphosphorylation of the RB protein, which in turn displaces both RB and associated histone-modifying enzymes such as HDAC from various promoters essential in order for E2F family transcription factors to drive expression of genes needed for S phase. mSin3A is a core component of a large multiprotein corepressor complex with associated HDAC enzymatic activity. A recent computational analysis by Ron DePinho and colleagues has confirmed the wide range of genes regulated by this means. Thus, several nodal points by which mSin3A influences gene expression have been identified, including the MYC-Mad, E2F, and p53 transcriptional networks.

Some authorities have proposed the existence of a "histone code," representing the covalent modifications of histone tails that function to regulate gene transcription. According to this hypothesis, histone modifications, such as lysine methylation, that can either enhance or suppress transcription depending on the precise methylation site, are "read" by specific binding proteins that then regulate downstream events. In some ways this histone code is complementary to the DNA code and determines how and when specific genes are transcribed. In addition to methylation and acetylation, other modifications used in the "histone code" include phosphorylation, ubiquitination, sumoylation, and ADP-ribosylation within gene regulatory regions, but less is known about these in this context. Such modifications may all alter chromatin structure and accessibility of genes and function as docking sites for transcription factors or other histone-modifying enzymes.

The balance between HATs and HDACs, and thus dynamics of histone acetylation, can be altered by different exogenous agents including certain base analogs, radiation, tobacco smoke, hormones, and reactive oxygen species (see also Chapter 3). All of these could theoretically

Table 11.1 Genes known to be silenced by CpG island hypermethylation in human cancer

Gene	Normal role	Tumor	Consequences
hMLH1	DNA mismatch repair	Colon, endometrium, stomach	Frameshift mutations
BRCA1	DNA repair, transcription	Breast, ovary	Double-strand breaks
p16^INK4a	Cyclin-dependent kinase inhibitor	Multiple types	Cell cycle entry
p14^ARF	MDM2 inhibitor	Colon, stomach, kidney	Degradation of p53
p15^INK4b	Cyclin-dependent kinase inhibitor	Leukemia	Cell cycle entry
MGMT	DNA repair of O^6-alkyl-guanine	Multiple types	Mutations, chemosensitivity
GSTP1	Conjugation to glutathione	Prostate, breast, kidney	Adduct accumulation
LKB1/STK11	Serine/threonine kinase	Colon, breast, lung	Unknown
ER	Estrogen receptor	Breast	Hormone insensitivity
PR	Progesterone receptor	Breast	Hormone insensitivity
AR	Androgen receptor	Prostate	Hormone insensitivity
PRLR	Prolactin receptor	Breast	Hormone insensitivity
RARβ2	Retinoic acid receptor β2	Colon, lung, head, and neck	
VHL	Ubiquitin ligase component	Kidney, hemangioblastoma	Loss of hypoxic response
RB	Cell cycle inhibitor	Retinoblastoma	Entrance in cell cycle
THBS1	Thrombospondin-1, antiangiogenic	Glioma	Neovascularization
CDH1	E-cadherin, cell adhesion	Breast, stomach, leukemia	Dissemination
CDH13	H cadherin, cell adhesion	Breast, lung	Dissemination
FAT	Cadherin, tumor suppressor	Colon	Dissemination
APC	Inhibitor of β-catenin	Aerodigestive tract	Activation β-catenin route
SFRP1	Secreted Frizzled-related protein 1	Colon	Activation WNT signaling
COX2	Cyclo-oxygenase-2	Colon, stomach	Anti-inflammatory resistance
SOCS1	Inhibitor of JAK–STAT pathway	Liver, myeloma	JAK2 activation
SOCS3	Inhibitor of JAK–STAT pathway	Lung	JAK2 activation
GATA4	Transcription factor	Colon, stomach	Silencing of target genes
GATA5	Transcription factor	Colon, stomach	Silencing of target genes
DAPK	Proapoptotic	Lymphoma, lung, colon	Resistance to apoptosis
TMS1	Proapoptotic	Breast	Resistance to apoptosis
TPEF/HPP1	Transmembrane protein	Colon, bladder	Unknown
HOXA9	Homeobox protein	Neuroblastoma	Unknown
IGFBP3	Growth factor-binding protein	Lung, skin	Resistance to apoptosis
EXT1	Heparan sulfate synthesis	Leukemia, skin	Cellular detachment

Source: Esteller, M. (2005). Dormant hypermethylated tumor suppressor genes: questions and answers. *Journal of Pathology*, **205**(2): 172–80.

influence the phenotypes of mammalian cells epigenetically (methylation and/or acetylation, particularly of CpG islands in gene promoter regions) without changing their DNA sequence, though as can readily be appreciated most of these factors could also generate mutations.

Gene expression may also be regulated in other hitherto unexpected ways. In recent years long-suffering biologists have once again been forced to take a quantum leap in the face of new knowledge. Thus, having comparatively recently accepted that transcriptional activator and repressor proteins were only part of the story and that processing of mRNAs and the afore-mentioned epigenetic factors were also of major importance it now seems that gene expression, at least during development, is also determined by parts of the genome previously thought to be devoid of any meaningful information. In fact, we now know that some of these areas of DNA actually encode so-called "micro-RNAs" (miRNAs), which are of complementary sequence to short segments of protein-coding genes. miRNAs function similarly to small interfering RNAs (siRNAs). siRNAs have been studied for a number of years in plants and are of similar size to miRNAs, but are princip-ally involved in silencing of exogenous viral and other foreign and potentially pathogenic RNAs, which become targeted for destruction (RNA silencing). Conversely, miRNAs target endo-genous messenger RNAs thus contributing to regulation of gene expression. Broadly, these small regulatory RNAs work by forming double-stranded RNA duplex structures, which are rap-idly targeted for degradation. miRNAs are highly conserved across species, and are increasingly implicated as a major novel class of regulatory molecules operating across a wide range of devel-opmental processes in plants (where they are known to control flowering time and leaf polar-ity) and animals. Although, relatively little is yet known about the potential role of small reg-ulatory RNA molecules in cancer, they have become major tools for the regulation of gene expression for the cancer researcher. In fact, large-scale studies of cancer cell lines in which multiple genes are knocked-down by siRNA *in vitro* before phenotype analyses are currently on-going in several centers (see Chapter 18).

Aside from regulation of gene expression, the levels of oncoproteins and tumor suppressor proteins may also become abnormal through either increased or decreased degradation (Chapters 6 and 7). Lysosomes were historic-ally regarded as the principal means of protein degradation within the cytoplasm. However, recently, attention has focused on the role of ubiquitination of intracellular proteins; such ubiquitinated proteins are thereby targeted for degradation by a multiprotein complex termed the proteasome. In general it now seems that extracellular and transmembrane proteins are primarily degraded in the lysosomes, whereas intracellular proteins, including key regulators of the cell cycle and apoptosis are normally degraded by the proteasome. Self-evidently, many proteasome substrates are involved in pathways that become deregulated in cancer, and proteasome inhibitors are now entering clin-ical practice. The partitioning and localization of proteins is also an important determinant of bio-logical activity and this will also be discussed in this chapter.

CHROMATIN

DNA does not exist as naked molecules in the cell, but associates with a protein scaffold (histones) to form a complex tertiary structure referred to as chromatin [see Box 11.2]. Chro-matin was originally classed as two forms based on observational differences during interphase. These forms are now known to correlate with gene expression activity – heterochro-matin represents repressed segments of chro-mosomal DNA while euchromatin represents active segments. These two forms of DNA are interconvertible. During cell differentiation and maturation, RNA synthesis declines, and is accompanied by a corresponding conversion of euchromatin to heterochromatin – therefore, fewer genes are available for mRNA synthesis. A wide variety of cellular processes involve de-repression of previously repressed genes, and cells undergoing such gene derepression often display a reversible transformation of hetero-chromatin to euchromatin within their nuclei. A similar transformation in DNA is seen during mitosis. With the onset of prophase, the nuclear membrane dissolves and the previous euchro-matin of the interphase cell is now condensed into large chromosomal masses in preparation for the segregation and separation of the chromosomes later in metaphase. This condensation of interphase euchromatin into condensed chromosomal masses during prophase is not described as transformation to heterochromatin because these terms were applied to the appearance of chromosomes during interphase, even though it is similar at a molecular level (similar ultrastructurally and associated with a comparable decrease in the rate of RNA synthesis).

Box 11.2 Chromatin

The reader is also referred to an excellent overview of this subject in *Molecular Biology of the Cell* (Alberts *et al.*, second edition).

The length of the DNA molecule poses certain problems with respect to packing it all away in the cell nucleus, as anyone attempting the relatively trivial task of packing away a very long hosepipe will readily appreciate. Each human cell contains approximately 2 meters of DNA if stretched end-to-end; yet the nucleus of a human cell, which contains the DNA, is only about 6 μm in diameter. Imagine trying to put an 80-mile long hosepipe into your garden shed. The packaging of DNA is accomplished by a truly remarkable array of specialized proteins that bind to and fold the DNA, generating a series of contorted coils and loops that provide increasingly higher levels of organization, preventing the DNA from becoming an unmanageable tangle. Yet, somehow, despite this unbelievable complexity, the DNA remains readily accessible to various enzymes needed for replication, repair, and gene expression.

In eukaryotes, the DNA in the nucleus (genome) is divided between chromosomes, of which there are 24 different pairs in humans. Each chromosome consists of a single, very long linear DNA molecule associated with the various proteins required to fold and pack the fine DNA thread into a more compact structure. The complex of DNA and proteins is called *chromatin* (from the Greek *chroma*, "color," because of its staining properties).

Two classes of DNA-binding proteins are recognized in eukaryotic chromosomes: histones and nonhistone proteins. Histones are very abundant and maintain the first level of DNA organization,

the nucleosome. Nucleosomes are arranged roughly like beads on a string – each bead is a "nucleosome core particle" that consists of DNA wound around a protein core formed from histones. Each individual nucleosome core particle consists of a complex of eight histone proteins – two molecules each of histones H2A, H2B, H3, and H4 – and double-stranded DNA that is 146 nucleotide pairs long. On average, nucleosomes repeat at intervals of about 200 nucleotide pairs interspersed by "linker" DNA. For example, a diploid human cell with 6.4×10^9 nucleotide pairs contains approximately 30 million nucleosomes. The formation of nucleosomes converts a DNA molecule into a chromatin thread about one-third of its initial length, and this provides the first level of DNA packing.

Chromatin in a normal cell rarely adopts the extended "beads on a string" form. Instead, the nucleosomes are piled on top of one another, generating regular arrays in which the DNA is even more highly condensed – forming what is referred to as the 30-nm fiber, which is wider than chromatin in the "beads on a string" form.

As a 30-nm fiber, the typical human chromosome would still be around 100 times too big for the nucleus. Thus, a higher level of folding exists to fold the 30-nm fiber into a series of loops and coils. Each long DNA molecule in an interphase chromosome is divided into a large number of discrete domains organized as loops of chromatin, each loop comprising a folded 30-nm chromatin fiber. Interphase chromosomes are largely composed of euchromatin that is interrupted by stretches of heterochromatin, in which 30-nm fibers are subjected to additional levels of packing that usually render it resistant to gene expression (see below).

Light-microscope studies in the 1920s distinguished between two types of chromatin in the interphase nuclei of many higher eukaryotic cells: a highly condensed form and all the rest, which is less condensed. Heitz (1929) originally described that portion of the nuclear chromatin remaining condensed throughout cell interphase as heterochromatin and the rest as euchromatin. Cooper (1959) suggested that heterochromatin and euchromatin differed in their biophysical conformations and in metabolic expression of their genes but not in their basic structure of

DNA arranged within chromosomes. Since that time, increasingly detailed genetic studies have revealed that the genes within heterochromatin are repressed but can later be expressed when the heterochromatic region undergoes a transition to euchromatin. Similarly, heterochromatin displays little or no synthesis of RNA until it is converted to euchromatin.

In a typical mammalian cell, approximately 10% of the genome is packaged into heterochromatin. Although present in many locations along chromosomes, it is concentrated in specific regions, including the centromeres and telomeres. Most DNA folded into heterochromatin does not contain genes. However, those genes that are packaged into heterochromatin are not expressed, probably because heterochromatin is so compact. Some regions of heterochromatin are responsible for the proper functioning of telomeres and centromeres (which lack genes), and its formation may even help protect the genome from being overtaken by "parasitic" mobile elements of DNA. Moreover, a few genes require location in heterochromatin regions if they are to be expressed. Thus, heterochromatin should not be thought of as comprising only redundant DNA.

When a gene normally expressed in euchromatin is experimentally relocated into a region of heterochromatin, it is no longer expressed, and the gene is silenced. Such effects of location are referred to as position effects, as gene activity depends on position along a chromosome. The study of position effects has identified some intriguing properties of heterochromatin. Namely, that it is dynamic and moreover that the state of chromatin, heterochromatin or euchromatin, is inherited during cell division.

The basic unit of chromatin is the nucleosome, comprising 146 base pairs of DNA wrapped around an octamer of two molecules each of the histones H2A, H2B, H3, and H4. Adjacent nucleosomes are connected by linker DNA, and progressive coiling of nucleosomes leads to higher-order structures. DNA contained within compacted chromatin, known as heterochromatin, is not available for transcription, unless appropriate activation of remodeling processes first takes place in order to enable access of the transcription factors and transcriptional machinery to individual genes. Chromatin remodeling requires the action of two classes of proteins: those that covalently modify histones or DNA (by methylation, acetylation, etc.) and those that mobilize nucleosomes such as the SW1/SNF complex. Modification of histones has now also been linked to cancer, and interestingly the machinery involved in chromatin modification may be largely overlapping with that involved in DNA methylation in higher organisms. Certainly strong associations have been identified between the DNA methyltransferases and the histone deacetylases.

Histone acetylation, together with methylation, phosphorylation, ubiquitination, sumoylation, glycosylation, and ADP ribosylation, modulate the activity of many genes by modifying both core histones and nonhistone transcription factors. Histone acetylation is a diagnostic feature of transcriptionally active chromatin. The group of enzymes, histone acetyltransferases (HATs), involved in this crucial step of gene regulation, covalently modify the N-terminal lysine residues of histones by the addition of an acetyl group from acetyl coenzyme A. Dysfunction of these enzymes is associated with cancer.

CpG islands

These are areas of greatly increased density of the dinucleotide sequence cytosine phosphate diester–guanine, which can form regions of DNA several hundred to several thousand base pairs long. The human genome contains around 45,000 CpG islands (comprising a total of around 50 million CpG dinucleotides), mostly found at the 5' ends of genes. They are widely accepted as unmethylated in normal somatic cells except for those on the inactive X chromosome and some associated with imprinted genes. Around 50% of all human genes contain CpG islands around the promoter regions and these include housekeeping genes (essential for general cell functions) and many frequently expressed in a normal cell. As will be seen below, CpG islands are important sites for epigenetic regulation of gene expression, and are frequently aberrantly methylated in cancer cells.

EPIGENETICS

Changes to DNA and its associated proteins can alter gene expression without altering the DNA sequence. DNA is not found in isolation in the cell but is associated with proteins called histones to form a complex substance known as chromatin. Chemical modifications to the DNA or the histones alter the structure of the chromatin without changing the nucleotide sequence of the DNA. Such modifications are described as epigenetic. Changes to chromatin structure have a profound influence on gene expression: if the chromatin is condensed, the factors involved in gene expression cannot get to the DNA, and the genes will be switched off. Conversely, if the chromatin is in an "open" conformation, the genes can be expressed as needed for cellular activities. Many heritable diseases in humans are caused by DNA sequence changes (mutations) that abolish gene expression, but only relatively recently have we become aware that human diseases may also arise by inappropriate gene silencing, brought about by epigenetic modifications. Importantly, most cancers are now known to have epigenetic silencing of genes that normally act to restrain and control cell proliferation. The major forms of epigenetic modification that have been associated with transcriptional silencing in cancer cells are aberrant DNA methylation of CpG islands located in gene promoter regions (including loss of imprinting of genes) and changes in chromatin conformation involving histone deacetylation. Progress in this field has been informed by earlier studies on the mechanisms underlying inactivation of the X chromosome, where CpG islands were first functionally described.

DNA methylation is a chemical modification of the DNA molecule itself and is effected by enzymes termed DNA methyltransferases. Once methylated in or around the gene promoter regions, gene expression may be "silenced" because transcriptional activator complexes cannot bind to these promoters. Methylation may also silence gene expression by recruiting methyl-binding domain (MBD) proteins, which associate with histone deacetylases (HDACs). HDACs, as will be seen, result in deacetylation of histones and a more condensed transcriptionally inactive chromatin structure. Conversely, chromatin containing acetylated histones is more open in configuration and accessible to transcription factors. Such chromatin is potentially transcriptionally active and in the presence of appropriate transcriptional activators the gene will be expressed (note expression is still subject to regulatory factors). Histone deacetylation causes the condensation of chromatin, making it inaccessible to transcription factors and the genes are therefore silenced.

Given the key role played by gene silencing of tumor suppressor and caretaker genes by epigenetic factors in cancer it is not surprising that drugs designed to reactivate gene expression by reversing DNA methylation and inhibiting histone deacetylation have been developed. Some of these agents are already in clinical trials used alone and in combination. Moreover, the recognition that particular patterns of DNA methylation and histone acetylation may predict disease onset and outcome has fueled initiatives aiming to exploit this information to screen patient blood, urine, or biopsies in the early diagnosis of cancer and precursor states and also to individualize treatment of those with established cancers. In fact, studies of tumor-derived free DNA in the circulation (presumably resulting from tumor cell apoptosis or lysis) or from epithelial cells shed into the lumen have revealed cancer-specific methylation patterns.

Hypomethylation of genes was the first epigenetic change identified in cancer cells, and after a period of relative neglect has again become a focus of attention. In contrast to hypermethylation, hypomethylation of genes may provoke inappropriate gene activation. A notable example of this in human cancers is deregulated overexpression of c-MYC in Burkitt's lymphoma as described in Chapter 6; due to a chromosomal translocation *c-MYC* is "relocated" from a region of repressed chromatin into a region of open and active chromatin. The potential for hypomethylation to activate oncogenes in cancer cells obviously raises the possibility that drugs aiming to reactivate silenced genes could also inadvertently activate oncogenes. The risk of this remains uncertain, however, as to date administration of neither inhibitors of methylation or of deacetylation have activated oncogenes and affected growth of normal cells.

DNA methylation

The only known epigenetic modification of the DNA molecules in mammalian cells is covalent addition of a methyl group to the 5th position of cytosine within CpG dinucleotide islands, which can directly turn off gene expression ("silencing"). The methylation of DNA is achieved by two DNA methyl transferases (DNMTs), though exactly how these are targeted to specific DNA regions is unclear. The other major class of epigenetic modification involves post-translational modification of histones and chromatin remodeling. It is believed that DNA methylation may have evolved for silencing of repetitive elements, but has subsequently been adopted in order to effect transcriptional silencing in imprinting and X-chromosome inactivation. Loss of imprinting is the silencing of active imprinted genes or the activation of silent imprinted genes, and is frequently observed in many different cancers. In particular, reactivation of the normally imprinted allele of the *IGF2* gene is often seen in human cancers and is associated with resistance to apoptosis and tumor progression in animal models.

Around half of all human genes have CpG islands in their promoter region. CpG islands were previously regarded as being universally hypomethylated in normal cells unless they were in areas where gene expression is silenced, such as the silenced allele of imprinted genes and the inactive X-chromosome in females. In these cases, promoter-associated CpG islands are methylated – suggesting that these are a major site of more durable inactivation of genes. However, it is now clear that occasionally some CpG islands in the promoters of autosomal non-imprinted genes can be methylated and in fact may even rarely use this as a means of regulating gene expression. The remaining 50% of gene promoters, which are described as CpG poor (by comparison with the 50% that are CpG rich), are less equivocally thought to employ methylation in regulating transcription on a day-to-day basis. Here methylation may act to either directly block docking of transcriptional activator complexes or to recruit transcriptional repressor complexes such as those containing HDAC.

Although methylation is required for silencing of genes the actual mechanisms responsible for establishing it still remain unclear. However, some of the consequences of CpG island methylation are now known and include binding of methylated DNA-specific binding proteins to CpG islands; these then help recruit various histone-modifying enzymes responsible for restructuring the chromatin. DNA methylation by DNMTs modifies the actual DNA itself and can directly prevent gene expression by preventing transcription factors binding to promoters, but can additionally exert a more general effect by recruiting methyl-binding domain (MBD) proteins. These are associated with HDACs, which function to chemically modify histones and change chromatin structure. Chromatin containing acetylated histones is open and accessible to transcription factors, and the genes are potentially active. Histone deacetylation causes the condensation of chromatin, making it inaccessible to transcription factors and the genes are therefore silenced (see Figs 11.1 and 11.3).

Now to return to the CpG island-rich promoters. In general these CpG islands are not methylated, irrespective of the expression state of the associated gene – suggesting that methylation of these promoters is not normally involved in the day-to-day regulation of gene expression in the large majority of cases. Instead attention is directed more at the regulation of histone acetylation. Inactive genes are associated with deacetylated histones and bound complexes containing HDACs, whereas active genes have strongly acetylated histones, which presumably reconfigure the chromatin so as to allow access of the transcriptional activator complex.

Genomic methylation patterns are frequently markedly disturbed in tumor cells; which may display seemingly paradoxical region-specific hypermethylation alongside more generalized global hypomethylation. The functional significance of hypermethylation events is well-known and when these occur within the promoter of a tumor suppressor gene they can silence expression of the associated gene and provide the cell with a growth advantage as surely as inactivation by gene deletion or mutation (Fig. 11.1 and Chapter 7). Not surprisingly, there is great interest in the therapeutic targeting of aberrant or undesirable methylation patterns in cancer, and DNA methylation inhibitors are now being examined as potential anticancer agents (see Chapter 16) as these are reasonably predicted to restrain tumors by reactivating various otherwise silenced tumor suppressor genes. However, given the recent recognition of the opposing role of hypomethylation in the inappropriate activation of various oncogenes involved in tumorigenesis, metastasis, and invasion (see Chapter 12), the application of broad-spectrum treatments that affect global methylation patterns should be approached with some caution. Cancers involve both gene activation as well as silencing; inappropriate activation of oncogenes such as *c-MYC* are commonplace in most human cancers. Burkitt's lymphoma is an important historical example (see Chapter 6); c-MYC is normally found in repressed chromatin and is expressed at a low level, but as in this case, if abnormal chromosome rearrangements occurring in lymphocytes result in the relocation of this gene into a region of open and active chromatin the net effect is the overproduction of c-MYC protein, deregulated proliferation, and ultimately lymphoma. Theoretically, drugs able to reduce methylation might reactivate tumor suppressors (desirable) but could also activate oncogenes (undesirable) – the ideal scenario would be to understand how individual genes are epigenetically regulated and then selectively target therapeutically the desired gene repertoire without adversely affecting others, but this goal is some way in the future.

Finally, and as you have no doubt recognized by now, a recurring theme in much of cancer biology is the quintessential "chicken-egg conundrum"; in this case, do epigenetic factors cause cancer or do cancer cells produce epigenetic changes? Strong support for a causal role of epigenetic factors in cancer originates from the study of patients with Beckwith–Wiedemann syndrome (BWS). BWS is caused by epigenetic defects and these are specifically associated with increased cancer risk.

Acetylation and histones

As discussed earlier, histone acetylation plays an important role in the regulation of gene expression, and together with methylation can influence the binding and activity of transcriptional activator complexes (Fig. 11.1). In fact, it has recently become clear that deacetylation of histones by HDAC is a major means

by which methylation results in formation of heterochromatin and in the suppression of gene expression. Most of the human genome is packaged as transcriptionally inactive densely packed heterochromatin and this chromatin is heavily methylated. The remainder of the genome is transcriptionally active, but still subject to various stimulatory and inhibitory processes that control gene expression on a day to day basis.

Studies of heterochromatin have helped unravel the complex mechanisms by which methylation of DNA culminates in silencing of gene expression. DNA in methylated regions is packaged into dense nucleosomes, which also contain deacetylated histones such as deacetylated H3 and also H4. Histone acetylation has a direct effect on the stability of nucleosomal arrays and on chromatin structure.

Acetylation of histones is regulated by opposing enzyme activities; histone acetyltransferases (HATs) and histone deacetylases (HDACs) add or remove, respectively, acetyl groups from lysine residues on the histone N-terminal tails. Through these effects on both histone (and also non-histone proteins – see later), these enzymes play key roles in regulation of gene expression (see also section on c-MYC in Chapter 6), chromosome segregation, and development. Moreover, their deregulation has been linked to cancer.

Recent studies have identified what has become known as a "methylation mark" that may help define and separate regions of transcriptionally active chromatin from transcriptionally inactive chromatin. These marks seem to involve methylation of lysine 9 in the tail of histone H3, which marks inactive, and methylation of lysine 4 on histone H3, which marks transcriptionally active chromatin. The methylated lysine 9 appears to bind proteins required for maintaining a repressed state, but it remains to be shown how this leads to methylation of DNA. Possibilities include the facilitation of binding of DNMTs. What seems clear is that despite the complexity of epigenetic factors, methylation appears a dominant event over acetylation as in cancer inhibiting HDAC alone does not reactivate aberrantly silenced genes and hypermethylated genes, whereas these same inhibitors can if cells are first treated with demethylating drugs.

METHYLATION AND CANCER

DNA methylation is the most prominent epigenetic modification in humans, and most tumor types, including premalignant lesions, have abnormal DNA methylation patterns. Generalized loss of DNA methylation was the first epigenetic abnormality identified in cancer cells. Such loss of methylation affects non-promoter regions of DNA and structural elements but also numerous CpGs methylated in normal tissues are unmethylated in cancer cells. However, it should be borne in mind that the bulk of CpG islands are not methylated in normal cells. In fact, in concert with global hypomethylation, the most obvious aberration in cancer cells is the converse, namely, specific hypermethylation of some of the usually unmethylated CpG islands. As already mentioned, most cancer cells manifest both global hypomethylation alongside regional specific areas of hypermethylation in CpG islands. Gene silencing by hypermethylation has been frequently alluded to in this book and is well known as a major factor in inactivation of tumor suppressors in cancer. In fact, the importance of gene silencing by hypermethylation is underscored by the observation of aberrant hypermethylation in the very earliest recognizable lesions in progression of colon cancer. However, the converse, gene activation by hypomethylation, although the first recognized abnormality, is only recently being reappraised as an important player in cancer. It is now known that aberrant demethylation can activate the following genes in cancer cells: maspin (gastric cancer), *S100A4* (colon cancer), and the so-called cancer/testis genes (CT), such as *MAGE* and *CAGE*. Global demethylation might be an early mechanism favoring genomic instability (based on recent studies, exploring the role of loss of RB function might prove interesting in this context!), whereas region-specific hypermethylation might be a later event involved in progression and spread.

Inactivation of tumor suppressor genes contributes to the development of essentially all forms of human cancer (Chapter 7). This inactivation may result from epigenetic silencing associated with hypermethylation as well as from intragenic mutations. Silencing of a tumor suppressor by hypermethylation of CpG in the gene promoters was first observed for the *RB* gene, but has now been shown for numerous other gene loci in cancer cells, including *CDKN2A* (encoding ARF and p16^{INK4a}), *VHL*, *E-cadherin*, *14-3-3σ*, *BRCA1*, and *MLH1*. In fact, in some cases genes are only or predominantly silenced by methylation rather than by mutations in the DNA, as for the DNA repair gene *MGMT*, the *CDKN2B* gene (encoding the p15^{INK4b}), CKI, *HIC1* (hypermethylated in cancer 1), and *RASSF1A*. Haploinsufficient tumor suppressor genes notwithstanding, Knudson's two-hit model, discussed in Chapter 8, predicts that abnormal phenotypes arise only if both gene copies are inactivated. In those cases where one allele is mutated in the germ-line the second copy is frequently inactivated by methylation.

Some cancers appear to have multiple gene promoters methylated, which has led to the proposition of a so-called "CpG island methylator phenotype" (CIMP), by J-P. Issa and colleagues. The notion of CIMP is supported by observation of sporadic colon cancers with mismatch repair defects, though whether this model can be generally applied to other cancers remains controversial. In fact, it is still argued whether methylation is the primary cause of gene silencing in some cancers or merely contributes to the maintenance of gene silencing initiated by genetic mechanisms. Several lines of evidence confirm that maintenance of gene silencing in cancer cells requires DNA methylation, not least the findings that demethylating drugs or elimination of DNA methyltransferases can reactivate gene expression. However, it has recently been shown that histone modification and silencing of the p16^{INK4a} locus can precede promoter methylation suggesting that in at least this case methylation is not the primary cause of loss of tumor suppressor activity.

Genome-wide DNA hypomethylation occurs in many human cancers, but whether this epigenetic change is a cause or consequence of tumorigenesis is still debated. In mouse models, DNA hypomethylation may contribute to tumorigenesis possibly by promoting chromosomal instability. Mice carrying a hypomorphic DNA methyltransferase 1 (*Dnmt1*) allele, which reduces DNMT1 expression, exhibit substantial genome-wide hypomethylation in all tissues. These mice also develop aggressive T-cell lymphomas, with frequent chromosome 15 trisomy. Hypomethylation may contribute to cancer through several mechanisms including gene activation, defective DNA mismatch repair, and chromosomal instability; moreover, hypomethylation may result from dietary and carcinogen exposure linking such abnormalities to environmental factors. The causes of hypomethylation in cancer remain unclear but may include altered function of the SW1/SNF chromatin remodeling complexes. The mammalian SWI/SNF chromatin remodeling complex is composed of more than 10 protein subunits, and plays important roles in epigenetic regulation. The DNA methyltransferase, DNMT1, accounts for most methylation in mouse cells, but both *DNMT1* and *DNMT3b* genes have to be silenced in order to eliminate methyltransferase activity in human colon cancer cells and reactivate expression of the tumor suppressor p16^{INK4a} (in fact cells now lose around 95% of their total genome-wide methylation). Interestingly, all three DNMTs are moderately overexpressed in a number of cancers and overexpression of DNMT1 can also promote cellular transformation under some circumstances. It may therefore not come as a surprise to hear that DNMTs are now regarded as more complex than previously appreciated. Depending on context either inactivation or overexpression of DNMTs might contribute to tumorigenesis. Progress has been made in unraveling how these enzymes interact. DNMT3A and B are responsible for initiation of methylation, which is maintained by the action of DNMT1. However, recent studies suggest that these enzymes might silence some genes even without methylation changes either by an as yet unrecognized direct action or through interactions with other proteins such as HDACs.

CIMP and MIN and the "mutator phenotype"

As discussed in the previous chapter, genetic instability primarily may manifest as either frequent gene mutations or as more major structural rearrangements of chromosomes (CIN). Broadly two classes of genome "caretaker" systems may be identified: (1) one that normally prevents gene mutations, particularly during DNA replication or during gene expression, including the DNA mismatch repair pathway, and (2) those that detect aberrant chromosome segregation during mitosis. Any mechanisms that disrupt these pathways may cause genetic instability and thereby greatly accelerate the accumulation of further potentially cancer-causing mutations – the "mutator phenotype." The mutator phenotype hypothesis was discussed in the previous chapter and was derived largely from studies of inherited colon cancer, HNPCC. As mentioned earlier, in HNPCC mutation in one allele of a gene involved in DNA mismatch repair (e.g. *MLH1*) is inherited, but loss of the second, wild-type allele occurs with high frequency, promoting microsatellite instability (MIN). MIN is associated with mutations in numerous other genes, which contribute to progression of colon cancer.

Aberrant CpG island methylation for various individual genes has been repeatedly demonstrated in cancers, but only relatively recently has it been widely appreciated that in some tumors, groups of genes have consistently increased methylation, suggesting the existence of a more widespread defect in methylation. In sporadic colon cancer, where this has been particularly well studied, the methylation of two or more distinct genes is often strongly correlated, particularly in the subset of these cancers that also have MIN. Thus, in sporadic colon cancers, promoter methylation of *CDKN2A* (encoding the p16^{INK4a} and p14^{ARF} proteins), *THBS1* (thrombosponsin 1), and *HPP1* has been shown and most have silencing of the mismatch repair gene *MLH1* (*mutL* homolog 1).

Many genes become methylated with increasing age, and in fact many genes observed as methylated in cancers are not that different in, for example, normal aging colonic mucosa. However, a subset of genes, including several key tumor suppressors (p16^{INK4a}, MLH1, etc.) are now known to be more extensively methylated in cancers than in normal aged tissues and this phenomenon has been referred to as CIMP. One obvious possibility is that epigenetic silencing of *MLH1* might actually be a causal factor for MIN in some sporadic colon cancers, which is compatible with studies showing that methylation-inhibiting agents can reverse the mismatch repair defect in cultured colon cancer cells. Thus, parallels can readily be seen between inherited colon cancers (HNPCC) where MIN results from germline mutations in mismatch

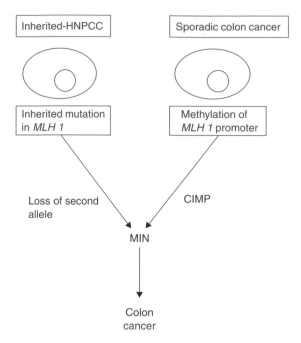

Fig. 11.4 The genetic and epigenetic route to colon cancer. Genetic instability (MIN, in particular) contributes to many colonic cancers. MIN may arise either through genetic or epigenetic mechanisms (or possibly a combination).

repair genes, and sporadic colonic tumors where MIN may result from aberrant methylation of *MLH1* (Fig. 11.4).

In fact, *MLH1* methylation could contribute to as many as 70% of sporadic colon cancers with MIN. Hypermethylation of the mismatch repair gene *MLH1* is commonly found in mismatch-repair-defective tumors. Such tumors are also more likely to manifest abnormal imprinting. Since these important observations were first made CIMP has now been demonstrated in numerous other cancers including glioblastoma, stomach, liver, pancreas, and ovary. Intriguingly, promoter hypermethylation of *MLH1* is seen in premalignant polyps even before they develop mutations.

Tumor hypomethylation has recently been strongly linked to chromosomal instability; pericentromeric satellite sequences are hypomethylated and many cancers contain unbalanced chromosomal translocations with breakpoints in pericentromeric DNA of chromosomes

1 and 16. Demethylation of satellite sequences could contribute to breakage and recombination.

These results are intriguing because they suggest that genetic instability and MIN in particular may not necessarily derive only from mutations, but also from epigenetic factors that alter gene expression. That different cancers have significantly different rates of tumor-suppressor gene silencing suggests that promoter methylation is not random and that aberrant methylation may be selected for during neoplastic transformation and might be the cause of cancer development in some cases. Although much aberrant DNA methylation may arise as a spinoff from tumorigenesis and aging, the presence of CIMP argues that at least in some circumstances DNA methylation may be a causal event.

Cancers manifesting CIMP have certain characteristics. Not surprisingly, CIMP-positive cancers are more frequent in older individuals but are also more common in women. CIMP-positive cancers also tend to have particular patterns of altered gene expression and mutations; *p53* mutations are rare whereas mutations in *KRAS* or *BRAF* are ubiquitous. In part this may be due to links between CIMP and MIN, but such aberrant features may occur in tumors without apparent MIN also.

The actual causes of aberrant methylation and CIMP are not fully understood but include nonspecific factors that might accelerate methylation such as inflammation and exposure to various carcinogens or "epimutagens" as they may be referred to in this context. In fact CIMP is more likely in colonic tumors arising on a background of inflammation such as ulcerative colitis, and particular life styles may also show correlations. One other possibility is that CIMP arises through the abnormal spreading of methylation from one gene to others; methylation spreading in cis is probably a normal feature of gene silencing, with abnormal spread contained by various proteins that operate in concert with CpG islands. This protective mechanism may become less effective with aging or potentially due to mutations in genes encoding "protective" proteins.

CIMP, as manifested by the simultaneous methylation of multiple genes, predicts a poorer prognosis in many cancers, including head and neck, lung, prostate, and acute leukemias. This observation fuels interest in employing such knowledge in the profiling of cancers (see below).

CANCER PROFILING

Much interest is developing in the concept of cancer profiling, whereby a particular observed "molecular signature" in an individual tumor might be employed to make more accurate predictions about prognosis and treatment responsiveness. The idea is that such information may enable more "individualized" or "tailored" approaches to treatment for each individual patient (the assumption made is that traditional disease classification is too broad and non-specific, based on clinical observations and rudimentary laboratory testing, and patient heterogeneity is insufficiently recognized). At a simple level further sub-classification of a given tumor type, using "molecular profiling" of the tumor or potentially of body fluids, might allow identification among individuals who traditionally are described as having the same disease (for example, ductal carcinoma *in situ*) of those most likely to develop metastases or most likely to respond to a given drug. Such molecular profiles might involve analysis of global gene or protein expression patterns in the tumor, using gene microarrays or proteomics; analyses of multiple gene alleles in the patient's genome, and potentially also epigenetic factors. Thus, the profile of CpG island hypermethylation in various malignancies might reveal an "epigenetic signature," which may be unique for each cancer "subtype."

Other than being targeted to promoters, modifications of histones, such as acetylation and methylation of lysine and arginine residues, also occur over large regions of chromatin including coding regions and nonpromoter sequences, which are referred to as global histone modifications. In a recent study, changes in such global levels of individual histone modifications were found to occur in prostate cancer and could help predict clinical outcome.

IMPRINTING AND LOSS OF IMPRINTING

Imprinting is the conditioning of the maternal and paternal genomes during gametogenesis, so that a specific allele is more abundantly or exclusively expressed in the offspring. At present around 80 genes are known to be imprinted and differential expression of the two alleles can occur in all or only some cell types or different stages of development. Imprinting is regulated by epigenetic changes such as DNA methylation, and methylation of specific localized areas, termed differential methylated regions (DMR), are characteristic of imprinted genes. It seems that the DMR contains a region, known as the imprinting control region (ICR), that is essential for controlling expression of genes lying within an imprinted domain. CCCTC-binding factor (CTCF) is an important regulator of expression of imprinted genes. By binding to the unmethylated parental allele of genes containing an ICR, CTCF allows allele-specific gene expression. CTCF effectively "reads" DNA-methylation marks and directs expression of the allele without DMR methylation. CTCF may also protect bound DMR from being methylated and, as stated in a recent review article by Keith Robertson, may even operate as a more or less general "insulator" from methylation for large regions of the genome. Loss of genomic imprinting (LOI) involves loss of the normal imprinting controls either by methylation or demethylation of the DMR or potentially by failure of CTCF function. Interestingly, with this latter in mind, cancer cells have been shown to express increased levels of a potential CTCF antagonist – BORIS. The result is either abnormal activation of the normally silent inherited allele or conversely inappropriate silencing of the normally expressed allele – examples of both scenarios have been described in human cancers.

Imprinting was first identified as disease-relevant in humans through the demonstration of paternal uniparental disomy of 11p15 in Beckwith–Wiedemann syndrome (BWS) and by maternal uniparental disomy of 15q11-q12 in Prader–Willi syndrome. BWS is characterized by multiorgan overgrowth and predisposition to embryonal tumors such as Wilms' tumor (Box 11.3), but loss of heterozygosity of 11p15 is also noted in numerous other tumors, including lung, bladder, ovarian, liver, and breast cancers. Two genes have been identified in this locus, *H19* and *IGF2*, which undergo reciprocal imprinting in the mouse and in humans, with maternal expression of *H19* and paternal expression of *IGF2*. The maternal *IGF2* gene is imprinted in normal tissues (Fig. 11.2), whereas in some Wilms' tumors and other

---

Box 11.3 Wilms' tumor and neuroblastoma

Wilms' tumor (WT) is an embryonic tumor originating from the undifferentiated renal mesenchyme. Around 2% of WT are inherited, but a genetic component is believed to also contribute to sporadic tumors. Familial WT cases generally have an earlier age of onset and an increased frequency of bilateral disease. One WT gene, *WT1* at 11p13, has been cloned, but only a minority of tumors carry detectable mutations at that locus, and it has been excluded as the predisposition gene in most WT families. Two familial WT genes have been localized, *FWT1* at 17q12-q21 and *FWT2* at 19q13.4; lack of linkage in some WT families to either of these loci implies the existence of at least one additional familial WT gene.

The Wilms' tumor 1 gene (*WT1*) plays an essential role in urogenital development and malignancies, including breast cancer and leukemia. WT1 acts to either enhance or repress transcription – which predominates depends on cell type and the context of the DNA-binding sites. This is exemplified by various reports in the literature suggesting that WT1 either inhibits or stimulates the production of the proto-oncogene c-*MYC*. Though, at least in the case of breast cancer, *WT1* may function as an oncogene in part by stimulating the expression of c-*MYC* (Han *et al.*, 2004). *WT1* is overexpressed in a number of human cancers, and is associated with a poor prognosis.

The WT1 protein is normally expressed in the developing genitourinary tract, heart, spleen, and adrenal glands and is crucial for their development; however its function at the molecular level is yet to be fully understood. Alternative splicing, RNA editing, and the use of alternative translation initiation sites generate a multitude of isoforms, which seem to have overlapping but also distinct functions during embryonic development and the maintenance of organ function. The protein is predominantly nuclear and there is evidence that the two different isoforms of WT1 (−KTS and +KTS) are involved in two different steps of gene expression control: transcription and post-transcriptional processing of RNA.

Desmoplastic small round cell tumor (DSRCT) is defined by a chimeric transcription factor, resulting from fusion of the N-terminal domain of the Ewing's sarcoma gene product *EWS* to the three

C-terminal zinc fingers of WT1. Intriguingly, this chimeric protein activates a unique set of genes as compared with the normal WT1 protein.

---

tumor types this imprinting is lost, leading to biallelic transcription of *IGF2*. Intriguingly, this mechanism contributes to Wilms' tumors primarily in Caucasian children as loss of imprinting is generally absent in Wilms' tumors in the east-Asian population. It is, therefore, likely that this epigenetic mechanism may in part explain the difference in incidence of Wilms' tumor between populations. In a very recent study it has been shown that loss of imprinting of *IGF2* in a mouse model doubles the risk of developing intestinal tumors and that this is associated with a less-differentiated phenotype of cells even in the normal colonic mucosa.

The converse scenario, whereby LOI could result in silencing of a normally expressed allele of a growth inhibitor has also recently been implicated in cancer. Thus loss of the maternally expressed allele of the gene encoding the p57^{KIP2} CKI involved in G_1/S arrest can also provide a cancer cell with a growth advantage as shown in some cases of Wilm's tumor.

INSULIN-LIKE GROWTH FACTORS (IGFs)

These regulate growth and apoptosis through interaction with the IGF1 receptor, and overexpression of the human IGF1 receptor promotes ligand-dependent neoplastic transformation in a variety of cell types. Two main subtypes of IGF proteins are known, IGF1 and IGF2; the latter appears to play a predominant role during development and is believed to be largely redundant in the normal adult. IGF1 is an important mediator of survival signaling and is involved in growth regulation in the adult and also potentially during embryogenesis, where it may compensate to some extent for inactivating mutations in the *IGF2* gene. Both proteins are ligands for the IGF1 receptor, and have been implicated in cancer.

THE ING TUMOR SUPPRESSORS

The inhibitor of growth 1 (*ING1*) gene is a member of the ING tumor suppressor family that includes at least five related genes, involved in diverse cellular processes including senescence, DNA repair, and apoptosis. ING proteins regulate gene expression by regulation of chromatin remodeling probably by acting as cofactors for HAT and HDAC. Well-known targets of ING2 are the lipid signaling molecules, phosphatidyl inositol phosphates (PIPs), also known to regulate apoptosis and motility. PIPs are also implicated in responses to DNA damage and in tumorigenesis. The PHD finger of the ING2 tumor suppressor interacts with nuclear PIPs, and may be involved in the p53-mediated apoptotic response to DNA damage. Thus, the ING family may link chromatin regulation with p53 function. However, further roles for ING independent of p53 are increasingly being appreciated.

Gliomas, are a form of primary brain tumor, in which levels of ING4 are known to be reduced. Interestingly, recent studies suggest that those tumors with the lowest levels of ING4 may be the most aggressive (highly malignant glioblastoma) and that this may in part be due to increased angiogenesis. One possible mechanism to explain this link has been suggested, namely that ING4 normally acts to suppress NF-κB, and that loss of ING4 leads to increased expression of NF-κB target genes, such as IL-8, involved in angiogenesis.

POLYCOMB GROUP PROTEINS

Polycomb group proteins (PcG) are involved in epigenetic regulation of gene expression, particularly in the determination of cell fate during development. The dysregulation of PcG genes, such as *Bmi1*, *Pc2*, *Cbx7*, and *EZH2*, has been linked with deregulated proliferation in cancer cells. PcG proteins are important for maintaining the silenced state of homeotic genes silenced during development, and together with Trithorax group proteins regulate coordinated gene activity in self-renewal of some stem cells. Biochemical and genetic studies in *Drosophila* and mammalian cells indicate that PcG proteins function in at least two distinct protein complexes: the ESC-E(Z) or EED-EZH2 complex, and the PRC1 complex. Recent work suggests that the ESC-E(Z) complex can mediate gene silencing through methylation of histone H3 on lysine 27. In addition to being involved in *Hox* gene silencing, the complex and its associated histone methyltransferase activity are important in other biological processes including X-inactivation, germline development, stem cell pluripotency, and cancer metastasis.

Overexpression of the PcG gene *Bmi1* promotes cell proliferation and induces leukemia through repression of *Ink4a/Arf* tumor suppressors. Conversely, loss of *Bmi1* leads to hematological defects and severe progressive neurological abnormalities. *Bmi1*-null mice manifest clonal expansion of granule cell precursors, which may result in development of medulloblastomas. The expression of BMI1 and patched (PTCH) are linked in many human medulloblastomas suggesting involvement of the sonic hedgehog (SHH) pathway.

DRUGS TARGETING EPIGENETIC FACTORS

It has become apparent during the last decade that reversal of aberrant epigenetic silencing might be an effective therapeutic strategy for cancer. Agents such as 5-azacytidine (5-AzaC) and 5′-deoxy-azacytidine (DAC), which inhibit the DNA methyltransferases that catalyze DNA methylation have been alluded to previously. These agents may reverse aberrant methylation of CpG islands and thereby reactivate tumor suppressor genes, which have been silenced. Inhibitors of HDAC are also of interest. HDACs and HATs catalyze the deacetylation and acetylation of lysine residues in histones thus helping to regulate chromatin structure, gene expression, and nonhistone proteins, thus influencing cell cycle progression and apoptosis. Their activity is also frequently disturbed in cancer. Several HDAC inhibitors are in clinical trials with significant activity against different hematologic and solid tumors. They have variously been shown to induce growth arrest, differentiation, apoptosis, and autophagocytic cell death of cancer cells. Early clinical results suggest a potentially useful role for HDAC inhibitors in the treatment of certain forms of lymphoma (e.g., cutaneous T-cell lymphoma) and acute leukemia. Interestingly, recent studies suggest that it is acetylation of nonhistone proteins that may be the most important means of action of these drugs and may explain lack of correlation between histone acetylation and induction of cell death. HDAC inhibitors have been shown to exert a wide range of effects on cancer cells, including disruption of corepressor complexes, induction of oxidative injury, upregulation of the expression of death receptors, generation of lipid second messengers such as ceramide, interference with the function of chaperone proteins, and modulation of the activity of NF-κB. All of these may potentially explain their benefits in cancer. Numerous patent applications describing new HDAC inhibitors have been filed. Recent studies indicate that DNA demethylating agents and HDAC inhibitors synergistically induce gene expression and apoptosis in cultured lung cancer cells, prevent lung cancer development in animals, and can induce immunogenicity and apoptosis of lung cancer cells in clinical trials.

REGULATION OF TRANSLATION

It is not intended to give an exhaustive description of how mRNA stability can influence protein synthesis, as this is covered in many excellent general molecular biology texts including *Molecular Biology of the Cell*. Many cancer relevant proteins may be regulated at this level as well as at transcription and protein stability. In fact, in a very recent study Kastan and colleagues demonstrated how important this mechanism is for p53. As described in previous chapters, p53 protein levels increase after DNA damage and it is now clear that this occurs not just through regulation of p53 degradation (MDM2 activity) but also through increased translation of p53 mRNA. In fact, translation of p53 depends on at least two proteins that bind to the 5' untranslated region (UTR) of p53; ribosomal protein L26 (RPL26) and nucleolin increase or decrease the rate of p53 translation respectively. Thus, RPL26 can induce G_1 cell-cycle arrest and augment irradiation-induced apoptosis.

RNA INTERFERENCE – SMALL REGULATORY RNA

RNA interference (RNAi) is one of the major discoveries of recent years and is predicted to revolutionize the emerging areas of systems biology and functional genomics. RNAi is an important tool for analyzing gene functions in eukaryotes and may in the future be employed for the development of therapeutic gene silencing.

RNAi is a natural post-transcriptional process through which metazoan cells suppress the expression of genes when exposed to double-stranded RNA molecules of the same sequence (sequence-specific gene silencing). RNAi occurs naturally in many organisms including nematodes, plants, fungi, and viruses and probably evolved to combat viruses and rogue genetic elements that utilize double-stranded RNA during their lifecycle (Fig. 11.5). The discovery of RNAi has profoundly altered the way we think about RNA; we now appreciate that RNA acts not only as an intermediary between genes and protein synthesis but also plays an important role in regulating gene expression. RNAi most likely serves two separate but overlapping functions:

1. To protect more complex organisms from viruses, modulate transposon activity, and eliminate aberrant transcription products, but ironically may be also employed by viruses to circumvent host defenses – in this case the regulatory RNA molecules are termed small interfering RNAs (siRNAs).
2. To regulate gene expression in development – in this case the regulatory RNAs are referred to as micro RNAs (miRNAs).

Both siRNA and miRNA are of similar size. The first evidence that dsRNA could achieve efficient gene silencing through RNAi came from studies by Fire and Mello in the nematode worm *Caenorhabditis elegans* in 1998 and subsequent studies in the fruit fly *Drosophila melanogaster*. RNAi involves three steps:

1. Long dsRNAs are cleaved by an enzyme known as Dicer into 21–23 nucleotide fragments, called siRNAs.
2. Then, siRNAs are recruited to a ribonuclease complex (RNA induced silencing complex – RISC).
3. RISC then mediates the cleavage of the target mRNA.

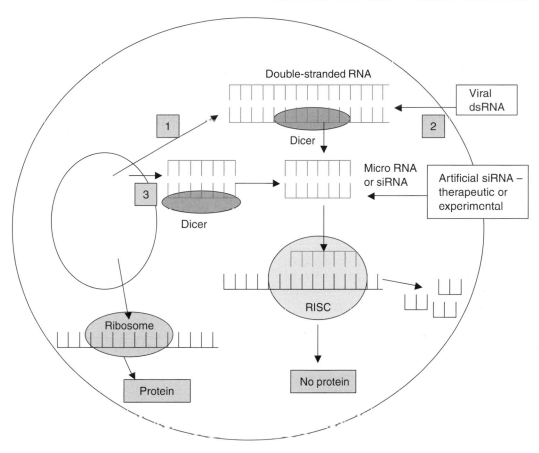

Fig. 11.5 RNAi. Cells can inhibit gene expression via small double-stranded RNA. The enzyme Dicer cuts short interfering RNAs (siRNA) from longer double-stranded RNAs deriving from (1) self-copying gene sequences, (2) viruses, or (3) regulatory RNA sequences called microRNAs. All these RNAs are cleaved by Dicer into short siRNA segments that can suppress gene expression. The short siRNA pieces unwind to form single-stranded RNAs, which combine with proteins to form a complex called RISC. The RISC captures native mRNA molecules with sequences complementary to the short siRNA sequence. If the pairing is perfect, the native mRNA is cleaved into untranslatable RNA fragments. If the pairing is not perfect then the RISC complex halts translation by preventing ribosome movement along the native mRNA.

Intriguingly, evidence is accumulating that small RNAs may, in addition to their anti-viral defense role, also act as "double agents." Recent studies have reported that viroids, small infectious particles of naked RNA, and the Epstein–Barr virus (EBV – a large DNA virus responsible for mononucleosis and other diseases, including Burkitt lymphoma), may actually employ RNA-silencing and miRNAs to inactivate host genes undesirable to the virus. This unexpected finding suggests that miRNA-mediated gene suppression might play a role in animal-virus pathogenicity and not just in host antiviral RNA activity. The possibility that miRNAs might also be involved in tumor formation must be considered.

Therapeutic and research potential of RNAi

RNAi represents a powerful new tool for the undertaking of functional genomics, which by enabling specific genes to be readily turned off enables the function of known genes to be more rapidly investigated than with earlier techniques requiring genes to be knocked out.

Although dsRNA could induce gene-specific interference in mouse embryos and oocytes, early attempts employing siRNA in mammalian systems proved difficult. These experiments probably failed because they used long dsRNAs that instead of mediating RNAi actually resulted in global decreases in mRNA and apoptosis, a response mediated by the dsRNA-dependent protein kinase. However, this nonspecific response can be avoided by employing synthetic 19–23-nucleotide dsRNAs, which elicit strong and specific suppression of gene expression in different mammalian cell lines. However, this effect is only transient due to the short half-life of synthetic RNAs. These technical problems have limited the applications of siRNAs to more complex studies, particularly *in vivo*. One potential means of overcoming this problem is to generate siRNAs within the cell with a hairpin structure called short hairpin RNAs (shRNAs). shRNAs are more stable than synthetic siRNAs, and since they are continuously expressed within the cells, this method permits long-lasting silencing of a gene of interest. Hairpin RNA is a key player in the RNA silencing pathway: When cleaved by a Dicer enzyme it generates siRNAs, which in turn guide RISC complexes to degrade target RNA.

RNAi is increasingly being explored as a means of achieving inducible cell and small organism knockouts in the laboratory, without any of the genetic disruption and more time-consuming manipulations required in traditional genetically altered knockout models. In fact, several national initiatives are currently under way to knock out multiple genes in cancer cells by use of RNAi.

RNAi in mammals, methylation, and cancer

Recent transcriptome analysis and different experimental approaches suggest that the actual number of noncoding RNAs (miRNAs, antisense transcripts and others lacking any extensive open reading frame) in eukaryotic cells is remarkably large. Mammalian cells contain multiple small non-protein-coding RNAs, including small nucleolar RNAs (snoRNAs), miRNAs, siRNAs, and small dsRNAs, which can all contribute to regulation of gene expression. They

may influence various steps in gene expression, including chromatin structure, RNA editing, RNA stability, and translation. These RNAs arise by processing of longer primary transcripts from both coding and non-coding regions of DNA. Small RNAs are now known to control a number of developmental and physiological processes, including differentiation of some cell types and, very recently, insulin secretion in mammals.

They are also likely important in diseases. Antisense transcription and production of double-stranded RNA (dsRNA) are increased in cancer cells. In a further interesting development, a potential link between RNAi and epigenetic regulation has been suggested. It has recently been shown in plants and yeast that the RNAi machinery may target genes for epigenetic silencing; dsRNA molecules targeted to gene promoter regions can induce transcriptional gene silencing in a DNA cytosine methylation-dependent manner (RNA-dependent DNA methylation). However, whether this also exists in mammalian cells is contentious still. It has been suggested that dsRNA arising from transcription of repetitive elements may be a mediator of gene silencing in mammalian cells by promoting methylation through as yet speculative interactions with DNMT or MDB. One possibility would be that in cancer cells, global low-level hypomethylation of repetitive elements in or near promoter regions could give rise to increased dsRNA homologous to the gene. The resultant activation of the RNAi machinery could then establish aberrant methylation at CpG islands. However, in a very recent study from the laboratory of Stephen Baylin, small dsRNAs targeted exclusively to one gene promoter (that for *CDH1* was chosen as this is frequently inactivated by methylation in cancer) induced transcriptional repression with chromatin changes in a manner entirely independent of DNA methylation. Moreover, they confirmed these findings in a cancer cell line modified to lack almost all capacity to methylate DNA. Thus, to date evidence for this interesting idea is lacking in mammalian cells.

A series of very recent papers published in *Nature* have begun to illustrate the potential pathogenic role of miRNAs in cancer. Altered expression of specific miRNA genes contributes to the initiation and progression of cancer. miRNAs control gene expression by targeting mRNAs and triggering either translation repression or RNA degradation. Abnormal expression of miRNAs has been observed in lung cancer, human chronic lymphocytic leukemias, lymphomas, and breast cancer, and in some of these cases the expression signatures may correlate with disease behavior, including in the latter, estrogen and progesterone receptor expression. A mechanistic role of miRNA in cancer has also been reported. The mir-17-92 polycistron is located in a region of DNA that is amplified in human B-cell lymphomas and can cooperate with c-MYC to accelerate tumor development in a mouse B-cell lymphoma model. In this case the mir-17-92 cluster may be acting to suppress apoptosis, providing one of the first functionally confirmed examples of an miRNA acting as an oncogene. Other possible examples include the inverse relationship between the let7 miRNA and RAS in lung cancer; interestingly let7 can downregulate RAS in c-elegans.

As discussed, the last five years have witnessed an explosion of research in cancer biology using RNAi. Particularly, recent developments have allowed the application of RNAi to knock down gene expression in experimental animals *in vivo*, and shRNA systems enabling tissue-specific and inducible knockdown of genes may ultimately replace more traditional mouse knockout models. In fact, RNAi technologies are currently the most widely utilized techniques in functional genomic studies. However, interest is obviously turning toward the potential therapeutic application of RNAi in diseases such as cancer, and once issues relating to siRNA stability and delivery *in vivo* are resolved this should be an exciting area in the future. Some proof-of-hypothesis studies have been completed and several Investigational New Drug applications relating to clinical trials with modified siRNA molecules have been made in the USA.

REGULATING THE PROTEINS

Protein degradation plays an equally important part in normal cellular homeostasis as protein synthesis, and in addition to removing excess enzymes or transcription factors can also supply amino acids for fresh protein synthesis. There are two key intracellular processes by which proteins are broken down: those involving lysosomes and proteasomes.

Lysosomes degrade extracellular proteins and cell-surface membrane proteins such as growth factor receptors or those employed for receptor-mediated endocytosis. Proteasomes primarily degrade endogenous proteins such as transcription factors, cyclins during cell cycle progression, and abnormal or incorrectly folded proteins arising from mutations or errors in translation. Proteasomes also produce peptides, which are presented by the major histocompatibility complex I to the immune system in order to induce antibodies. Generally the proteasome produces peptides that are six to nine amino acids in length.

The proteasome

The proteasome is an abundant multienzyme complex that represents the major route for degradation of intracellular proteins in eukaryotic cells (Fig. 11.6). The proteasome comprises the core particle (CP) and two regulatory particles, which contain recognition sites for ubiquitin. Turnover of several key cellular regulatory proteins results from targeted destruction via ubiquitination and subsequent degradation through the proteasome. The first step in proteasomal protein degradation is ubiquitination, which then targets proteins to the proteasome complex. Evidence suggests that the proteasome serves both as a disposal system for damaged cellular proteins and as a mechanism for degrading short-lived regulatory proteins that govern cellular functions such as the cell cycle, cell growth, and differentiation. Because these processes or their dysregulation are crucial steps in tumor formation, the proteasome pathway has recently become a target for therapeutic intervention.

Ubiquitin–protein ligases

The key components that regulate substrate ubiquitination are the ubiquitin–protein ligases. Ligases are a heterogeneous group comprising single proteins as well as large multiprotein complexes. Specificity of targeting is influenced by various factors including the requirement for

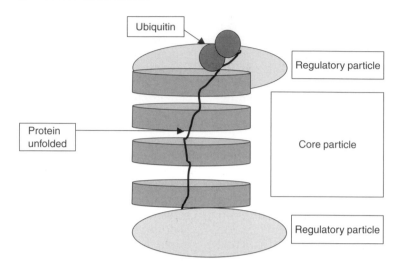

Fig. 11.6 Structure of the proteasome. The core particle comprises two copies of each of 14 different proteins, assembled in groups of 7 forming a ring. There are four rings each stacked on the other. There are two identical regulatory particles at each end of the complex, each consisting of 14 different proteins distinct from those in the core particle, 6 of which are ATPases, and some with sites that recognize ubiquitin. Proteins destined for destruction have been conjugated to a molecule of ubiquitin by ligases onto lysine residues. Additional molecules of ubiquitin bind to the first forming a chain. The complex binds to ubiquitin-recognizing site(s) on the regulatory particle and the protein unfolds, which utilizes energy from ATP. The unfolded protein translocates into the central cavity of the core particle. Several active sites on the inner surface break specific peptide bonds to form peptides averaging about eight amino acids in length. These peptides are extruded and may either be further degraded in the cytosol or incorporated in a class I histocompatibility molecule to be presented to the immune system as a potential antigen.

post-translational modifications such as phosphorylation, hydroxylation, or oxidation of the substrate and, in some cases, the ligase itself.

Many inherited diseases are now known to involve defective ligase function. Since the initial recognition that Angelman syndrome is caused by maternal deficiency of the E6-AP ubiquitin E3 ligase, several other disease-relevant E3 ligases have been identified, including autosomal recessive juvenile Parkinson disease, von Hippel–Lindau syndrome, and congenital polycythemia. Furthermore, several disorders of abnormal ubiquitin regulatory signaling are also known including at least two subtypes of Fanconi anemia and the BRCA1 and BRCA2 forms of breast and ovarian cancer susceptibility. Many other disorders affect ubiquitin pathways secondarily.

Loss of von Hippel–Lindau (VHL) protein function results in an autosomal-dominant cancer syndrome known as VHL disease, which manifests as angiomas of the retina, hemangioblastomas of the central nervous system, renal clear-cell carcinomas, and pheochromocytomas.

VHL tumor suppressor is a specific substrate-recognition component of the E3 ubiquitin complex, which regulates proteasomal degradation of the subunit of the hypoxia inducible transcription factor (HIF). Impaired VHL complex function leads to accumulation of HIF, overexpression of various HIF-induced gene products and formation of highly vascular neoplasia (Chapter 14). However, the ubiquitinating role of the VHL complex extends beyond its function in regulating HIF, as it appears to regulate the stability of other proteins that might be involved in various steps of oncogenic processes.

One major cancer-relevant E3 ubiquitin ligase is MDM2, which can antagonize the effect of the p53 protein. As discussed in Chapter 7, p53 targets genes that promote cell-cycle arrest or apoptosis in response to DNA damage and oncogene activation. MDM2 normally maintains low levels of p53 by targeting the protein for proteasome degradation, and by blocking p53 transcriptional activity. Intriguingly, as *MDM2* is also a p53 target gene this creates a negative-feedback mechanism. MDM2 activity is counteracted by

ARF in response to strong proliferation signals, such as activated oncogenes. ARF then inhibits MDM2-mediated p53 degradation. This may explain why loss of ARF may exert similar effects in tumorigenesis to loss of p53.

The F-box protein, Fbw7, a component of the SCF(Fbw7) ubiquitin ligase and a tumor suppressor, promotes proteasome-dependent degradation of c-MYC (Chapters 4 and 6), thus c-MYC activation may be a key outcome of Fbw7 loss in cancer. Fbw7 also targets cyclin E, Notch, and c-JUN, all of which may contribute to tumorigenesis if Fbw7 is lost.

Ubiquitin inhibitors act at many levels to enhance apoptosis signaling. Proteasome inhibitors can decrease Fas-like inhibitor protein (FLIP) levels in tumors activating the extrinsic pathway of apoptosis through caspase-8. The inhibitor of apoptosis proteins (IAP) E3 ligases can directly bind to caspases resulting in their degradation.

The proteasome and cancer

Various cancer-relevant proteins are degraded by the proteasome pathway, including several cell cycle regulators (see Chapter 4), including cyclins, CDC25, and some CKIs. Therefore, inhibition of proteasome-mediated degradation of these various proteins could either promote or inhibit replication depending on the balance of proteins influenced.

Multiple other proteins of importance in regulating cell growth, survival, and cancer are also regulated by the proteasome, including the tumor suppressor p53, discussed in the previous section. Most human tumors either have mutations in p53 or alterations in positive (ARF) or negative (MDM2) regulators of p53.

The NF-κB family are a group of transcription factors that are held inactive in the cytoplasm bound to the inhibitor protein IκB. Stimulation of cells by inflammation, chemotherapy, radiation, or oxidants initiates a signaling pathway culminating in the phosphorylation, ubiquitination, and degradation of IκB by the 26S proteasome; NF-κB can then translocate to the nucleus and activate transcription of various genes including growth factors, angiogenic factors, and antiapoptotic factors (such as inhibitor of apoptosis (IAP) and BCL-2).

Beta-catenin, a member of the WNT signaling pathway, is downregulated by glycogen synthase kinase-3β (GSK3β)-dependent phosphorylation of serine and threonine residues in the N-terminus of the protein, followed by ubiquitination and proteasomal degradation. In human and rodent cancers, mutations that substitute one of the critical serine and threonine residues in the GSK3β region of beta-catenin stabilize the protein and activate beta-catenin/TCF/LEF target genes.

MEN1 is a tumor suppressor gene involved in inherited tumor susceptibility in mutiple endocrine neoplasia type 1 (MEN1); it encodes a 610-amino-acid protein, called menin (see also Chapter 5). While the majority of germline mutations identified in MEN1 patients are frameshift and nonsense mutations resulting in truncation of the menin protein, various missense mutations have been identified whose effects on menin may largely be mediated via rapid degradation of mutant menin via the ubiquitin-proteasome pathway. Mutant, but not wild-type menin, may also interact with the molecular chaperone Hsp70 and with the Hsp70-associated ubiquitin ligase CHIP, which could promote ubiquitination.

PTEN encodes a major lipid phosphatase, which acts as a negative regulator of the PI3K-AKT pathway to influence G$_1$ cell cycle arrest and apoptosis (see Chapters 6 and 7). Several mechanisms of PTEN inactivation occur in primary malignancies derived from different tissues and may involve both inappropriate subcellular compartmentalization and increased/decreased proteasome degradation.

Smad4/DPC4, is a downstream mediator of TGF-beta signaling, frequently inactivated in human cancer. The ubiquitin-proteasome pathway is important for inactivating Smad4 mutants in cancer.

Therapeutic inhibition of the proteasome

Inhibition of the ubiquitin-proteasome pathway, or more specifically at the E3 ligases known to modulate apoptosis, might be a useful strategy in cancer therapy. In laboratory studies, proteasome inhibition can reduce cancer cell proliferation by preventing breakdown of proteins and transcription factors that normally restrict cell cycling. Moreover, survival of cancer cells during chemotherapy can be reduced and cancer cell death enhanced by producing cellular stress.

The recent success of the proteasome inhibitor Bortezomib, a dipeptide boronic acid analog, for treatment of relapsed, refractory myeloma demonstrates the therapeutic potential for modulating ubiquitin–protein ligase activities with synthetic agents. Bortezomib is the first such proteasome inhibitor to reach clinical trials. It has shown *in vitro* and *in vivo* activity against a variety of malignancies, including myeloma, chronic lymphocytic leukemia, prostate cancer, pancreatic cancer, and colon cancer.

The lysosomal pathway

Activation of receptor tyrosine kinase (RTK) signaling pathways is involved in tumorigenesis (Chapter 5), but it is increasingly apparent that aberrant RTK deactivation may be equally important. A major deactivation pathway, receptor downregulation, involves ligand-induced endocytosis of the RTK and subsequent degradation in lysosomes. Recent studies have shown that protein ubiquitination is key in vesicular trafficking of growth factor receptors. This degradative route is separate from the proteasome-mediated pathway, which involves polymeric chains of ubiquitin. Receptor ubiquitination takes place at the cell surface and directs internalized receptors to lumina of multivesicular bodies and the lysosome. The Cbl family of RING finger adaptors control this late sorting event, following ligand activation of receptors. Another group of E3 ubiquitin ligases, the Nedd4 family, regulates the initial sorting event, which targets receptors to clathrin-coated regions of the plasma membrane together with adaptor proteins such as epsins, which share ubiquitin-interacting motifs. Various genetic defects can prevent successful receptor downregulation through endocytosis as can certain viruses, culminating in growth factor independent signaling as receptors are channeled to default recycling pathways.

WRESTLING WITH PROTEIN TRANSIT – THE ROLE OF SUMO AND THE PML BODY

The nucleus contains a number of distinct, non-membrane bound compartments in which various different proteins are seen to concentrate.

Amongst these, promyelocytic leukemia nuclear bodies (PMLNBs) have generated a lot of interest due to their involvement in cancer. PML bodies are present in all mammalian cells, but are disrupted in acute promyelocytic leukemia, in which the PML protein becomes fused to the retinoic acid receptor. This PML–RARα hybrid protein is therapeutically activated by the ligand retinoic acid and the resultant disaggregation of the PMLNB might in part explain the benefits of such therapy. Mice that lack PML-NBs have impaired immune function, exhibit chromosome instability, and are sensitive to carcinogens. Although their direct role in nuclear activity is unclear, PMLNBs are likely involved in regulation of transcription, apoptosis, tumor suppression, and the response to viral infection. PMLNB are sites at which multisubunit complexes can form and mediate post-translational modification of proteins such as p53 in response to DNA damage. Following DNA damage, several repair factors transit through PMLNBs, supporting the view that the PMLNB may function as a dynamic sensor of cellular stress. In one model, such stresses may favor disassembly of the PMLNB enabling dispersal of DNA repair proteins.

The PML tumor suppressor protein accumulates in the PML nuclear body, but cytoplasmic PML isoforms of unknown function have also been described. Recent studies suggest that cytoplasmic PML is an important modulator of TGFβ signaling. PML-null primary cells are resistant to TGFβ-dependent growth arrest, induction of cellular senescence, and apoptosis. These cells also have impaired phosphorylation and nuclear translocation of the TGFβ signaling proteins SMAD2 and SMAD3, as well as impaired induction of TGFβ target genes. The PML–RARα oncoprotein of APL can antagonize cytoplasmic PML function, and APL cells have defects in TGFβ signaling similar to those observed in PML-null cells. PML also potentiates p53 function by regulating post-translational modifications, such as CBP-dependent acetylation and Chk2-dependent phosphorylation, in the PMLNB. PML interacts with the p53 ubiquitin-ligase MDM2 and can enhance p53 stability by sequestering MDM2 to the nucleolus. As nucleolar localization of PML may be regulated by ATR activation and phosphorylation of PML by ATR this provides

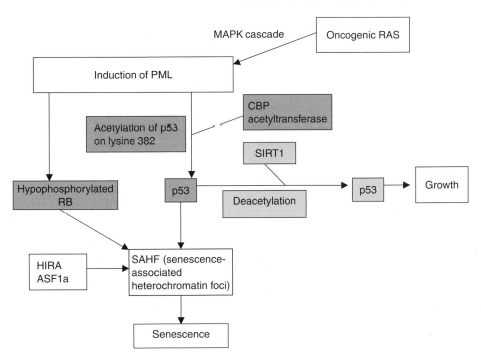

Fig. 11.7 The role of PML in senescence pathways. Upregulation of promyelocytic leukemia protein (PML) results in relocation of CBP and p53 to the PML body. CBP acetylates p53 increasing activity and expression of p53-target genes promoting senescence. An intact PML body may not be a prerequisite for this. Conversely, SIRT1 expression may antagonize PML by promoting p53 deacetylation. PML is also associated with hypophosphorylated RB and reduced E2F transcriptional activity. As cells reach senescence, a change in chromatin structure, called SAHF (senescence-associated heterochromatin foci), silences the genes that promote growth. In addition to PML, two other proteins, HIRA and ASF1a, are also involved in formation of SAHF. After Langley *et al.*, *EMBO Journal* (2002) 21, 2383—2396.

important links between DNA damage and apoptosis.

Study of the signals that may regulate PMLNB dynamics led Anne Dejean and colleagues to identify the SUMO pathway. SUMO (small ubiquitin-related modifier) family proteins are not only structurally but also mechanistically related to ubiquitin in that they are post-translationally attached to other proteins. Like ubiquitin, SUMO is covalently linked to its substrates via amide (isopeptide) bonds formed between its C-terminal glycine residue and the varepsilon-amino group of internal lysine residues. The enzymes involved in the reversible conjugation of SUMO are similar to the ligases mediating the ubiquitin conjugation. The SUMO pathway uses an E1 activation enzyme (UBA2/AOS1 heterodimer), an E2 conjugation enzyme (UBC9) and three families of E3 ligases (RANBP2, PIAS, and PC2) that are

believed to confer substrate specificity. Demodification is achived by SUMO hydrolases (SENPs), of which seven isoforms exist in mammals (Fig. 11.7).

Since its discovery in 1996, SUMO has engendered a great deal of interest because its substrates include such fundamental and important proteins as p53, c-JUN, PML, and huntingtin. SUMO modification appears to play important roles in diverse processes such as chromosome segregation and cell division, DNA replication and repair, nuclear protein import, protein targeting to and formation of certain subnuclear structures, and the regulation of a variety of processes including the inflammatory response in mammals and the regulation of flowering time in plants. Unlike ubiquitination, sumoylation does not lead to the degradation of its target proteins but rather regulates subcellular localization.

One way in which SUMO acts is to target proteins to the PMLNB (analogs to ubiquitin targeting of proteins to the proteasome discussed below). Nuclear foci containing PML bodies occur in most cells and play a role in tumor suppression. Interestingly, proteins such as c-MYC may be regulated by several processes. Stability may be influenced by ubiquitination and acetylation, and interestingly now also by targeting to the PML body. PML bodies may also control the distribution, dynamics, and function of the checkpoint protein CHFR.

Chromatin, senescence, and PML bodies

Senescent cells form so-called senescence-associated heterochromatic foci (SAHF), rich in the transcription-silencing histone H2A variant macroH2A. Heterochromatin is in general transcriptionally inactive, and SAHF contain genes associated with cell growth. Therefore, incorporation of cell cycle-related genes into heterochromatin could be a contributor to senescence. Recent studies suggest that two proteins, HIRA and ASF1a, known to be involved in chromatin structure, also contribute to formation of SAHF. As cells approach senescence, HIRA enters PML nuclear bodies, where it interacts with HP1 proteins, prior to incorporation of HP1 proteins into SAHF. Another chromatin regulator, ASF1a, is needed to complex with HIRA for the formation of SAHF and senescence. HIRA and ASF1a may drive formation of macroH2A-containing SAHF and senescence-associated cell cycle exit, via a pathway requiring the transit of heterochromatic proteins through PML bodies.

QUESTIONS REMAINING

1 Can information on the methylation status of CpG islands be employed to improve the subclassification of tumors and help predict clinical behavior and chemosensitivity?
2 To identify those enzymes directly responsible for mediating CpG island hypermethylation of tumor suppressor genes.
3 Can silenced genes be reactivated as a form of therapy in cancer cells?
4 To learn more about the role of hypomethylation in activation of oncogenes.

5 To further explore the role of protein modification by ubiquitination and sumoylation in degradation and transit of cancer-contributing proteins.
6 To further develop techniques for application of RNAi technology *in vivo*.

CONCLUSIONS AND FUTURE DIRECTIONS

Epigenetics has been described as a new frontier in biology. It is now clear that the genome contains information in two forms, genetic and epigenetic. The genetic information provides the blueprint for the manufacture of all the proteins necessary to create the organism, while the epigenetic information provides additional instructions on how, where, and when the genetic information should be used. Epigenetic information is not contained within the DNA sequence itself, but can specify mitotic inheritance of cell fate and gene expression patterns and can turn environmental effects into heritable changes in cell phenotypes. Epigenetic changes include modulation of chromatin structure, transcriptional repression, X-chromosome inactivation, and genomic imprinting. The major form of epigenetic modification in mammalian cells is DNA methylation – covalent addition of a methyl group to the 5th position of cytosine within CpG dinucleotides predominantly located in the promoter region. Recent work has revealed that DNA methylation is an important player in many processes, including DNA repair, genome instability, and regulation of chromatin structure.

Epigenetic inheritance may be mediated by two other processes aside from DNA methylation, namely histone modification and genomic imprinting.

While DNA methylation clearly enhances the ability of cells to regulate and package the genetic information, it also adds an additional level of complexity. Genomic methylation patterns are frequently altered in tumor cells, with global hypomethylation accompanying region-specific hypermethylation events. When hypermethylation events occur within the promoter of a tumor suppressor gene, this can silence expression of the associated gene and provide the cell with a growth advantage similar to that of gene deletion or mutation (Chapter 7). Not surprisingly, DNA methylation inhibitors are now being examined as potential anticancer agents (see Chapter 16). However, paradoxically, interest has again turned to the potential role of hypomethylation in the activation of genes, including those involved in malignant behaviors, metastasis, and invasion (see Chapter 12). Self-evidently this suggests that drugs which globally inhibit methylation may have both beneficial (reactivating tumor suppressors) and adverse (activating oncogenes) actions. It is important now to make progress in understanding how methylation of specific genes is regulated and how this may lead us to develop newer more selective therapeutics targeting specific genes only.

It is undoubted that RNAi technologies will continue to further research progress into the functional genomics of cancer and will ultimately yield new therapeutic agents, which will be particularly important in cases where it has proved difficult or impossible to develop suitable small molecule inhibitory agents. The next few years will likely witness startling progress in further defining the specific ligases and regulators that target individual proteins for proteasomal degradation and some of these will in time become the targets of new drugs. We are only beginning to discover how proteins are regulated by localization and partitioning, and various processes, including sumoylation, are increasingly being investigated in cancer biology. Finally, the role of chaperones such as the heat shock protein HSP 90 in regulating function and levels of numerous key proteins implicated in cancer, including p53, is being recognized and new drugs targeting HSPs are in development and some in trials.

QUESTIONS

1 Gene expression can be regulated by:
   a. Modifications in chromatin.
   b. Processing of mRNA.
   c. Small RNA molecules.
   d. Ubiquitination of transcription factor proteins.
   e. DNA methylation.
2 DNA methylation:
   a. Is only important in regulating expression of tumor suppressor genes.
   b. Is frequently altered in cancer cells.
   c. Only affects CpG islands in gene promoters.
   d. Usually alters the DNA sequence.
   e. Does not influence chromatin structure.
3 Imprinting:
   a. Is regulated by DNA methylation.
   b. Of some genes may be lost in cancer cells.
   c. Always involves inactivation of the maternally inherited allele of a gene.
   d. Is a normal feature of the *IGF2* gene.
   e. Of the *IGF2* gene is lost in all Wilms' tumors.
4 RNA interference:
   a. Does not occur naturally.
   b. May be a means of combating certain viral infections.
   c. May regulate gene expression during development.
   d. May be employed by viruses against the host cell.
   e. May be employed to manipulate gene expression in the laboratory.
5 The proteasome:
   a. Degrades proteins that have been sumoylated.
   b. May degrade proteins involved in cell cycle regulation.
   c. May be inhibited therapeutically to treat cancer.
   d. Is responsible for ubiquitinating proteins.
   e. Is a major site for degradation of misfolded proteins.

BIBLIOGRAPHY

Epigenetics

Bachman, K.E., Park, B.H., Rhee, I., Rajagopalan, H., Herman, J.G., Baylin, S.B., Kinzler, K.W. and Vogelstein, B. (2003). Histone modifications and silencing prior to DNA methylation of a tumor suppressor gene. *Cancer Cell*, **3**: 89–95.

Campos, E.I., Chin, M.Y., Kuo, W.H. and Li, G. (2004). Biological functions of the ING family tumor suppressors. *Cell Mol Life Sci.*, **61**(19–20): 2597–613.

Egger, G., Liang, G., Aparicio, A. and Jones, P.A. (2004). Epigenetics in human disease and prospects for epigenetic therapy. *Nature*, **429**: 457–463.

Feinberg, A.P. and Tycko, B. (2004). The history of cancer epigenetics. *Nat Rev Cancer*, **4**(2): 143–53.

Feinberg, A.P. and Vogelstein, B. (1983). Hypomethylation distinguishes genes of some human cancers from their normal counterparts. *Nature*, **301**(5895): 89–92.

Fukuzawa, R., Breslow, N.E., Morison, I.M., Dwyer, P., Kusafuka, T., Kobayashi, Y., Becroft, D.M., Beckwith, J.B., Perlman, E.J. and Reeve, A.E. (2004). Epigenetic differences between Wilms' tumors in white and east-Asian children. *Lancet*, **363**(9407): 446–51.

Gaudet, F., Hodgson, J.G., Eden, A., Jackson-Grusby, L., Dausman, J., Gray, J.W., cLeonhardt, H. and Jaenisch, R. (2003). Induction of tumors in mice by genomic hypomethylation. *Science*, **300**(5618): 489–92.

Issa, J.P. (2004). CpG island methylator phenotype in cancer. *Nature Reviews Cancer*, **4**: 988–993.

Kinzler, K.W. and Vogelstein, B. (1996). Lessons from hereditary colorectal cancer. *Cell*, **87**: 159–170.

Kirmizis, A., Bartley, S.M., Kuzmichev, A., Margueron, R., Reinberg, D., Green, R., and Farnham, P.J. (2004). Silencing of human polycomb target genes is associated with methylation of histone H3 Lys 27. *Genes Dev*, **18**(13): 1592–605.

Loeb, L.A. (2001). A mutator phenotype in cancer. *Cancer Res*, **61**: 3230–3239.

Rainier, S., Johnson, L.A., Dobry, C.J., Ping, A.J., Grundy, P.E. and Feinberg, A.P. (1993). Relaxation of imprinted genes in human cancer. *Nature*, **362**(6422): 747–9.

Rhee, I., Bachman, K.E., Park, B.H., Jair, K.W., Yen, R.W., Schuebel, K.E., Cui, H., Feinberg, A.P., Lengauer, C., Kinzler, K.W., Baylin, S.B. and Vogelstein, B. (2002). DNMT1 and DNMT3b cooperate to silence genes in human cancer cells. *Nature*, **416**(6880): 552–6.

Turner, B.M. (2003). Memorable transcription. *Nature Cell Biol*, **5**(5): 390–3.

Turner, B.M. (2005). Reading signals on the nucleosome with a new nomenclature for modified histones. *Nature Struct Mol Biol*, **12**(2): 110–2.

Welcker, M., Orian, A., Jin, J., Grim, J.A., Harper, J.W., Eisenman, R.N. and Clurman, B.E. (2004). The Fbw7 tumor suppressor regulates glycogen synthase kinase 3 phosphorylation-dependent c-MYC protein degradation. *Proc Natl Acad Sci U S A*, **101**(24): 9085–90.

Yamashita, K., Dai, T., Dai, Y., Yamamoto, F. and Perucho, M. (2003). Genetics supersedes epigenetics in colon cancer phenotype. *Cancer Cell*, **4**: 121–131.

Zhang, R., et al., (2005). Formation of MacroH2A-containing senescence-associated heterochromatin foci and senescence driven by ASF1a and HIRA. *Dev Cell*, **8**(1): 19–30.

Translation of mRNA

Takagi, M., Absalon, M.J., McLure, K.G. and Kastan, M.B. (2005). Regulation of p53 translation and induction after DNA damage by ribosomal protein L26 and nucleolin. *Cell*, **123**(1): 49–63.

RNA interference

Brummelkamp, T.R. *et al.* (2002). A system for stable expression of short interfering RNAs in mammalian cells. *Science*, **296**(5567): 550–3.

Elbashir, S.M. *et al.* (2001). Duplexes of 21-nucleotide RNAs mediate RNA interference in cultured mammalian cells. *Nature*, **411**(6836): 494–8.

Fire, A. *et al.* (1998). Potent and specific genetic interference by double-stranded RNA in *Caenorhabditis elegans*. *Nature*, **391**(6669): 806–11.

Matzke, M.A. and Birchler, J.A. (2005). RNAi-mediated pathways in the nucleus. *Nat Rev Genet*, **6**(1): 24–35.

Pfeffer, S. *et al.* (2004). Identification of virus-encoded microRNAs. *Science*, **304**: 734–36.

Ting, A.H., Schuebel, K.E., Herman, J.G. and Baylin, S.B. (2005). Short double-stranded RNA induces transcriptional gene silencing in human cancer cells in the absence of DNA methylation. *Nat Genet*, **37**(8): 906–10.

Wianny, F. and Zernicka-Goetz, M. (2000). Specific interference with gene function by double-stranded RNA in early mouse development. *Nat Cell Biol*, **2**(2): 70–5.

Ubiquitination and the proteasome

Adams, J. (2004). The proteasome: a suitable antineoplastic target. *Nat Rev Cancer*, **4**: 349–360

PML

de Stanchina, E., Querido, E., Narita, M., Davuluri, R.V., Pandolfi, P.P., Ferbeyre, G. and Lowe, S.W. (2004). PML is a direct p53 target that modulates p53 effector functions. *Mol Cell*, **13**(4): 523–35.

Bischof, O., Kirsh, O., Pearson, M., Itahana, K., Pelicci, P.G. and Dejean, A. (2002). Deconstructing PML-induced premature senescence. *EMBO J* **21**(13): 3358–69.

12
Cell–Matrix Adhesion, Cell–Cell Interactions, and Malignancy

Charles Streuli

KEY POINTS

- Adhesive interactions between cells and their extracellular environment occur through extracellular matrix (ECM) receptors such as integrins, and intercellular adhesion receptors such as cadherins. The sites of physical interaction are called adhesion complexes. Integrins bind matrix proteins while cadherins physically join cells together. Both organize intracellular cytostructure and transmit signals to the cell.
- Integrins containing adhesion complexes are focal centers for organizing the cytoskeleton.
- Integrins are also context-dependent checkpoints for signaling: cells monitor their local extracellular environment and only respond to growth factors properly if they are in the right place.
- Major aspects of cell physiology including cell cycle progression, suppression of apoptosis, cellular differentiation, and migration completely depend on signals being received through adhesion complexes.
- The normal spatial organization of cells within tissues is regulated by adhesive interactions. The transition from a primary tumor to a malignant one is dependent upon the loss of both the proliferative and positional controls that would otherwise maintain correct cell number and architecture. The homeostasis of a whole affected organ is disrupted in cancer.
- Malignant cells move from their site of origin and disseminate to distant organs, where they form metastases. This progression requires the acquisition of a large number of characteristics, which involve changes in cell adhesion. These include: migration; ability to invade the stroma and endothelial linings; dissemination through the circulation; survival in inappropriate tissue contexts; and environmental decoding to allow a tumor to grow in defined secondary target sites. However, the formation of metastases is inefficient and rare.
- Reduced adhesiveness at adherens junctions is a key stage in malignant progression. Loss of E-cadherin occurs through mutations as well as phosphorylation, degradation, and altered transcription. It is also lost during epithelial–mesenchymal transition.
- Invasive cells acquire the ability to migrate by altering their repertoire of integrins, activating the intracellular signaling machinery that controls the cytoskeleton and motile processes, and expressing ECM-remodeling enzymes.
- Although most of the genetic changes in carcinoma occur within epithelial cells, the supporting tissue, or stroma, has a critical role in malignant progression.
- Anoikis is a specific form of apoptosis that is induced when cells no longer receive appropriate signals from the ECM. In order for malignant cells to survive in both the stromal bed of the primary tissue that they migrate through and at secondary metastatic sites, they need to undergo activating mutations in the ECM survival pathway.

INTRODUCTION

Cellular adhesion to the ECM and to neighboring cells orchestrates tissues into functional units, and is required for cell migration (Fig. 12.1). Adhesion also regulates cell shape, via the cytoskeleton, and has a key role in determining the responses of signaling ligands to control cellular fate and differentiation, as well as survival and proliferation. Adhesion is mediated through specific classes of cell surface receptor, which join cells to each other or to the ECM, respectively. In cancer, altered adhesion influences each of the hallmarks of tumorigenesis, and is the defining characteristic of malignancy. Changes in cell–cell adhesion and cellular interactions with the ECM have severe consequences on the ability of extracellular factors to signal properly, so that cells lose their positional identity. Altered adhesion signaling results in the unscheduled migration of tumor cells, their dissemination through the circulation, and their embedment, survival, and proliferation at distant sites, thereby forming life-threatening metastases.

As discussed in other chapters, cancer is largely caused by genomic instability that arises within stem cells. It involves the escape of protective apoptosis (Chapter 8) and immune surveillance mechanisms (Chapter 13), deregulated proliferative and aging controls (Chapter 4), as well as inappropriate sampling of the environment so that tumor cells ignore normal positional cues and thereby migrate to, and survive at, distant metastatic sites (Box 12.1). This chapter is about how cells respond to their microenvironment, and how alterations in the way that cells perceive their neighbors and the ECM contribute to deregulated homeostasis.

The most common cancers arise in epithelial tissues, such as the epidermis, lungs, intestine, prostate, or breast, and are called carcinomas. Here we will consider the types of adhesive interactions that epithelial cells normally encounter, how such interactions break down in cancer, and the mechanisms that allow cells to migrate to distant sites, and survive and grow there. Many of the references are to contemporary review articles, which cover each topic in much more detail.

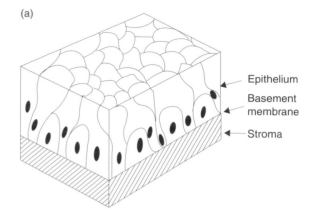

(a)

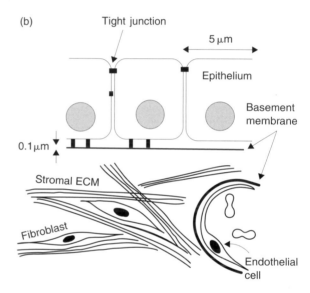

(b)

Fig. 12.1 Tissue architecture. The key components of epithelial tissues are shown as (a) a three-dimensional representation of a pseudostratified epithelium and (b) a schematic cross-section of a simple epithelium. They contain epithelial cells, separated from the stroma by a basement membrane, stromal cells such as fibroblasts and adipocytes, capillaries, neurons, and cells of the immune system. Although the main cell type to become altered genetically, and therefore neoplastic, is the epithelial cell itself, cancer progression depends on influences from other cell types. Thus, cancer is a disease where homeostasis of a whole organ is disrupted. It involves not just the cancer cells themselves, but is characterized by disturbed stromal, endothelial, and immune cells, as well as altered ECM.

Box 12.1 Benign versus malignant tumors

Benign tumor A tumor that slowly grows at its site of origin and is well circumscribed. The cells within benign tumors frequently have normal morphology and are differentiated. These tumors do not normally cause the death of an organism, but sometimes can provide the seed cells for malignant tumors. Benign tumors can be large, but they are often operable.

Malignant tumor A rapidly spreading tumor. Malignant tumors progress through a number of stages. Initially they have diffusely infiltrative margins and can show differentiated characteristics. They evolve to poorly differentiated structures that sequentially invade basement membrane, stroma, and then vasculature. It is only once a tumor has spread that a person has contracted "cancer."

Metastasis A tumor growing within another organ of the body. Although individual metastases are usually clonal in origin, they are often present in large numbers within any given organ, making them inoperable. The excessive tumor burden resulting from multiple metastases leads to the failure of essential body functions and death.

ADHESIVE INTERACTIONS WITH THE ECM

There are two main types of adhesive interactions: those with the ECM and those between cells (Fig. 12.2). Normally, parenchymal cells (this includes epithelial cells) reside on a sheet-like ECM called the basement membrane (sometimes referred to as the basal lamina). This matrix is a continuous and flexible part of the basal surface of all epithelia, which is 40–120 nm thick and forms the interaction zone with stromal ECM. Basement membranes are also associated with endothelial cells lining capillaries and with muscle and Schwann cells. The stroma provides support for epithelia and is composed of the connective tissue matrix (synonymous with stromal ECM). Stromal cells, such as fibroblasts, myofibroblasts, and adipocytes, are embedded within it, as are nerves. Blood and lymph vessels penetrate the stroma to provide nutrients and immune protection.

In cancer, epithelial cells come into contact with different types of ECM during their progression to malignancy. The basement membrane forms a boundary between epithelia and stroma both in normal tissue and premalignant lesions, such as carcinoma *in situ*. In malignancy, the cells become invasive, breaking through the basement membrane and coming into contact with stromal or other ECM. For this to happen, changes in epithelial cell-to-ECM interactions are required.

Basement membrane

The basement membrane provides critical signals for the morphogenesis of epithelial structures during development as well as for the survival, proliferation, differentiation, and cytostructural architecture of cells within them. These signals are delivered by ECM molecules themselves, and by the growth factors (GFs) they harbor. A key class of basement membrane glycoproteins are the laminins, which are cross-shaped molecules of an α–β–γ composition. There are five α, three β, and two γ laminin genes, and their protein products can assemble into at least 15 different heterotrimers. Two of the main laminin isoforms in basement membrane are laminin-1 and laminin-5. Laminin heterotrimers interact via their N′-terminal head and arm domains to form a lattice-like network of interacting proteins. They bind to transmembrane cell surface integrin and proteoglycan receptors, resulting in organization of cell shape via the cytoskeleton, and triggering of intracellular signaling pathways.

An additional basement membrane protein is collagen type-IV, which forms a self-interacting network that looks a bit like chicken-wire. Although collagen-IV does bind surface receptors, its major function is to provide rigidity to the basement membrane. The laminin and collagen networks are linked together by nidogen, stabilizing the basement membrane. The fourth major component of basement membrane is perlecan, a large proteoglycan that binds a variety of growth and morphogenesis factors, and is involved with both sequestration of factors away from cells until they are

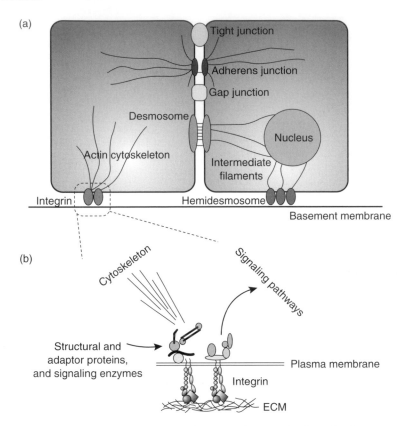

Fig. 12.2 Key adhesive interactions of epithelial cells. (a) Several types of cell–cell adhesion system bind adjacent cells together. Tight junctions form a ring around the apical surface of epithelia, and thereby ionically separate the extracellular spaces at the apical and basal surface. Adherens junctions connect cells via cadherins, and organize the microfilaments, particularly toward the apical side of polarized cells. They also sequester the transcription factor, β-catenin. Gap junctions provide ionic communication between adjacent cells. Desmosomes are large plaques that form strong intercellular bonds, again via cadherins. They are of key importance in maintaining tissue integrity under conditions of shear stress. The basal surface of epithelial cells interacts with glycoproteins within the specialized ECM, basement membrane, via both integrin receptors and transmembrane proteoglycan receptors (the latter are not shown). Integrins can either be located within multiprotein assemblies known as adhesion complexes, where they link to the actin cytoskeleton, or in very large adhesion plaques called hemidesmosomes, which link to the intermediate filament network, itself forming a bridge to the nuclear envelope. (b) In addition to binding cells to the ECM, integrin receptors regulate cell shape and polarity via the cytoskeleton, and they link to intracellular signaling pathways via structural adaptor proteins and signaling enzymes. Adherens junctions have a similar function in cytostructural regulation and signaling. Polarity: Most adhesive cells have an intracellular "direction," so that different components become spatially segregated to different sides of the cell. Epithelial cells have a basal surface, which contacts the basement membrane, and an apical surface at the opposite side of the cell. Plasma membrane lipids are different on apical and basal surfaces. Similarly, the cytoplasmic contents are different toward apical and basal poles and nuclei are usually located within the basal half.

released by proteases, as well as (in some cases) their presentation to appropriate surface receptors for signal transduction. Other minor components of the basement membrane include agrin, fibulin, and collagen types VII, XV, and XVIII.

Basement membranes are made through collaboration between epithelial and stromal cells, and are deposited to form a lamina at the interface between the two tissue compartments. Epithelia require direct signals from the basement membrane to function properly, indicating that the stroma has a critical influence on the way that epithelia behave.

Basement membrane drives polarity

The interaction of epithelial cells with the ECM at their basal surface leads to adhesion-regulated organization of the cytoskeleton and, together with a contribution from cell–cell contacts, to the establishment of cellular polarity (Fig. 12.2).

The ability of the ECM to regulate the cyto-structure of individual cells also extends to the organization of multicellular structures. For example, the functional units of mammary gland are epithelial acini, consisting of a layer of polarized cells organized around a central lumen and with their apical surfaces facing inwards (Fig. 12.14a,b). Since epithelia in this tissue secrete milk, the cells have to be topologically orientated in this way for milk to be delivered into a discrete extracellular compartment. Modeling cancer accurately in experimental tissue culture depends upon appreciating the three-dimensional nature of tissues (Fig. 12.3).

The ECM control on multicellularity and topology has important consequences for the response of cells to their local environment. For example, in a polarized epithelium, extracellular regulatory factors, or cytokines, that are supplied to one surface of an epithelium only work if their receptors are present on the same cell surface. The expression of basement membrane proteins can be reduced in malignant cancer, contributing to a loss of cell polarity and altered proliferation (Fig. 12.4).

Basement membrane orchestrates growth and survival signals

Integrin-mediated adhesion is essential for most cell types to respond to GFs. Thus, in addition to organizing the cytoskeleton and cell shape, cell–matrix interactions in epithelial cells are critical for orchestrating accurate responses to growth and survival factors. Receptors for basement membrane proteins can activate signaling pathways and are required in *trans* for many GF receptors and their downstream signaling enzymes to work properly, and thereby for normal cells to proliferate (see later and

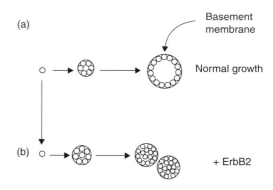

Fig. 12.3 Three-dimensional culture models for cancer. Although many aspects of cell behavior, including survival and proliferation, are frequently studied in monolayer culture, these often do not reflect the responses of cells in three-dimensional arrays *in vivo*, and extreme caution has to be taken to interpret experimental data. Fortunately, culture models are now available that mimic the three-dimensional organization of tissues, and better models are therefore being developed to study carcinoma progression. ErbB2 is a member of the EGF receptor family and is frequently upregulated in breast cancer. However, an activated ErbB2 does not transform normal breast epithelial cells, as assessed by conventional methods such as growth in soft agar. In this example, (a) breast cells plated in three-dimensional basement membrane gels grow to form hollow acini surrounded by a collagen-IV-containing basement membrane, which resemble mammary alveoli *in vivo* (Fig. 12.14b,c). Under the same culture conditions (b) the activated ErbB2 now causes the cells to proliferate so that they fill the luminal spaces and it alters their properties so that they form multiacinar structures. These resemble some forms of early breast cancer, such as comedo ductal carcinoma *in situ* (Fig. 12.14d). From Muthuswamy, S.K., Li, D., Lelievre, S., Bissell, M.J., and Brugge, J.S. (2001). ErbB2, but not ErbB1, reinitiates proliferation and induces luminal repopulation in epithelial acini. *Nature Cell Biology*, **3**: 785–92. Petersen, O.W., Ronnov-Jessen, L., Howlett, A.R., and Bissell, M.J. (1992). Interaction with basement membrane serves to rapidly distinguish growth and differentiation pattern of normal and malignant human breast epithelial cells. *Proceedings of National Academy of Sciences, U S A.* **89**: 9064–8.

Chapter 5). Integrins are also required in most normal cell types to protect them from apoptosis; death following altered or lost adhesion is a process called anoikis (see later). This effectively means that basement membrane

(a)

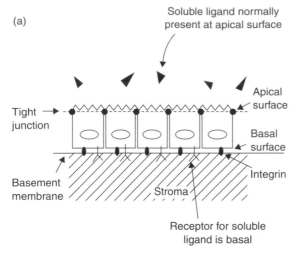

(b)

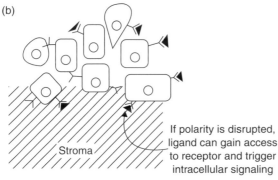

Fig. 12.4 Polarity segregates signaling ligands from their receptors and thereby controls proliferation. GF receptor signaling in simple epithelia can be regulated by polarized segregation of the ligand to a different plasma membrane compartment than the receptor. Airway epithelia express the receptor ErbB1–4 as well as one of its ligands, heregulin-α. (a) The ligand is expressed on the apical cell surface and is secreted into apical medium, while the receptor is only present basally, and cannot be activated unless heregulin-α is added ectopically to the basal cell surface or if tight junctions in intact monolayers are broken. Wounding the epithelium causes rapid heregulin-α-activated ErbB2 signaling. In this way, segregation of ligand and receptor provides an elegant mechanism for rapid GF receptor activation and tissue repair after epithelial injury. (b) Unfortunately, this has dire consequences if cellular polarity is disrupted in disease processes, because it can lead to unscheduled proliferation. For example, increased epithelial permeability in smoking-associated bronchitis can disrupt growth homeostasis and therefore contribute to tumor formation. From Vermeer, P.D., Einwalter, L.A., Moninger, T.O. *et al.* (2003). Segregation of receptor and ligand regulates activation of epithelial growth factor receptor. *Nature*, **422**: 322–6.

proteins provide essential contextual information for instructing the way that epithelial cells behave. Appropriate crosstalk between cell–ECM interactions and oncogenes is essential for tumor formation (Fig. 12.5; see also Fig. 12.10).

Stroma

The stroma has an essential supporting function for epithelia, contributing signals for epithelial morphogenesis and function, and providing a vital substratum at times of tissue repair. However, the stromal ECM has certain key properties that are important for cancer progression. First, it is composed of different proteins to the basement membrane and it harbors soluble factors and cytokines (Chapter 5). Second, it stimulates epithelial cells coming in contact with it to behave quite differently to those contacting basement membrane, for example, by changing their shape, and altering gene expression as well as proliferative and survival responses. This means that during the course of malignancy, when epithelial cancer cells invade stromal ECM, the inappropriate environment may affect cells in ways that contribute to tumor development.

Stromal ECM networks also serve to impede the passage of cells that are not specially designed to migrate through it. An important characteristic of malignant tumor cells is that they acquire mechanisms to help them migrate through the stroma, including both phenotypic changes that activate the motile machinery and inappropriate expression of proteinases that degrade collagens (see also Fig. 12.19).

Fibrillar collagen

Stromal ECM is largely composed of fibrous proteins, such as collagens and proteoglycans, which are proteins with large glycosaminoglycan (GAG) chains covalently bound to them. Collagen fibrils are made of helical bundles of collagens type I or III (in epithelial tissue stroma), to which are attached small collagens (e.g. collagen type IX) and proteoglycans (e.g. decorin). The triple-helical collagens assemble as bundles of fibrils that are organized into lattices, and through their high degree of mechanical strength

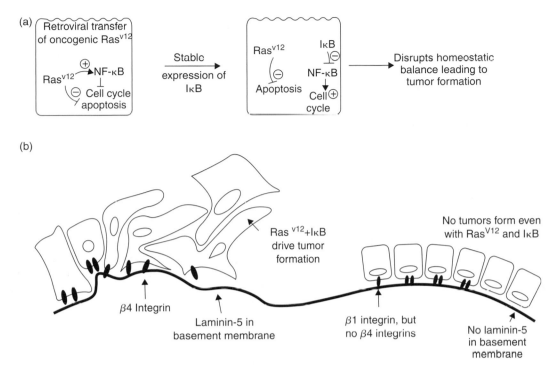

Fig. 12.5 Laminin signaling is required for oncogene-induced squamous cell carcinoma. (a) The main role of oncogenic Ras^{V12} in epithelial cells is to suppress apoptosis. However, it can also inhibit proliferation by increasing NF-κB activity. In epidermal keratinocytes NF-κB inhibits proliferation, even though it activates the cell cycle in other cell types. This brake on proliferation can be blocked by preventing NF-κB activity, which in this example is achieved by stable expression of IκB. Therefore the antiapoptotic effect of Ras^{V12} plus the proliferative response to reduced NF-κB activity cooperate in malignant tumor formation. If Ras^{V12} and IκB are expressed in human epidermal cells, which are then grafted onto mouse skin, the cells proliferate to form a tumor resembling squamous cell carcinoma. (b) However, this only occurs when the keratinocytes interact with laminin-5 in the epidermal basement membrane through β4 integrin receptors (left). If Ras^{V12} and IκB are expressed in epidermal cells from patients genetically deficient in either this form of laminin or the β4 integrin, then they are unable to drive tumor formation (right). These experiments show that there is an essential crosstalk between cell–ECM interactions and oncogenes in human tumor progression. In this case, it is likely to be because β4 integrin has a signaling role. β4 integrin is frequently upregulated in cancer, and may therefore have an important contribution in driving oncogenesis. Similar mechanisms may be important in other tumors as well. Further changes in adhesion molecules are necessary for other aspects of squamous cell carcinoma progression to full malignancy (see Fig. 12.20). From Dajee, M., Lazarov, M., Zhang, J.Y. *et al.* (2003). NF-kappaB blockade and oncogenic Ras trigger invasive human epidermal neoplasia. *Nature*, **421**: 639–43.

they provide the tensile rigidity to maintain tissue form.

Glycosaminoglycans

Glycosaminoglycans are chains of repeating disaccharides, which can be up to several hundred sugar residues long, and are negatively charged due to sulfate groups. This allows GAGs to form gels, because of the mutual repulsion of negative charges and the entrapment of large numbers of water molecules, and thereby permits the stromal ECM to withstand compressive forces. Some GAG molecules are not covalently bound to a protein core and can occupy huge spaces; a single hyaluronan molecule, for example, may consist of up to 25,000 disaccharides and can form a 300 nm cube. Because of this property, hyaluronan can separate tissue components and thereby provide a space within dense stroma for cells to migrate. Hyaluronan also activates its cell surface receptor, CD44, which is a signal-transducing receptor for promoting migration. Although important for regulated tissue morphogenesis during development, altered hyaluronan expression and its interaction with CD44 occurs in tumor cells and contributes to malignant progression.

Other stromal proteins

Other secreted glycoproteins, such as elastin and fibrillin, are also part of the stromal ECM, and contribute in their own way to the biology of the stroma; for example, the elastin fibers (made up of cross-linked networks of elastin molecules) provide resilience to tissues. Tissue fibronectin is also present throughout the stroma. This protein, which is a dimer, can provide adhesive links between collagen and cells, although cells also bind to some stromal ECM components directly. Fibronectin is one protein that cells rely on to migrate through stroma, and is therefore important in malignancy.

Tenascin is another ECM protein that has a role in cancer progression. However, rather than being adhesive, its main function appears to be antiadhesive, as it interferes with integrin-dependent cell spreading. By antagonizing stable links between cells and the ECM, tenascin can therefore promote cell migration. It is expressed during embryonic development when considerable cell movement occurs, and in the adult it is re-expressed in areas of tissue remodeling, such as at wound repair sites. Tenascin is frequently upregulated in cancer.

Altered stroma can contribute to malignancy

Although carcinomas are epithelial in origin, they usually show a strong stromal reaction, reflected in disorganized composition and cellular content of the neighboring ECM. The gene expression profile of stromal cells adjacent to primary carcinomas is widely different to normal stromal cells, and many of these changes contribute directly to tumor progression. Importantly, genomic alterations are confined to carcinoma cells and are absent from stromal cells. This

means that the epithelial tumor cells have a dominant influence over the stroma, driving it to "assist" the cancer cell's mission for dissemination. However, in some situations, the stroma can contribute directly toward the advancement of epithelial malignancy (Fig. 12.6).

Stromal ECM harbors GFs

An additional feature of the stroma is that it harbors most of the locally acting paracrine factors that control epithelial cell function: transforming growth factor-β (TGF-β) binds to latency-associated peptide (LAP); fibroblast growth factors and Wnts are sequestered by proteoglycans; insulin-like growth factors (IGFs) bind to IGF-binding proteins, which themselves are matrix bound (Chapter 5 covers GF signaling). GFs can be released from the stroma by a variety of enzymes including ECM-degrading proteinases, which are inappropriately expressed in cancer cells (see later).

Most of these proteins, and the ECM proteins themselves, are synthesized by stromal cells. Thus they have a strong controlling influence over the fate and function of epithelia. This has a key role in guiding the progression of epithelial malignancy, since cells within the stroma can be influenced by tumor cells to secrete factors that contribute to the invasive behavior of tumor cells themselves (Fig. 12.7).

Integrins

Extracellular matrix proteins bind to cells through a variety of transmembrane cell surface receptors. The most prominent are integrins, which are α–β heterodimers (Fig. 12.8). Adhesion to ECM occurs at the distal ends of the heterodimeric receptors, where the α and β subunits interact. On the cytoplasmic side, integrins bind adaptor proteins that link them to both the actin-based cytoskeleton and to enzymes that trigger signal transduction cascades. Integrins therefore integrate the extracellular anchoring elements of the ECM with intracellular proteins of the cytoskeleton. Because the interaction with ECM proteins occurs with low affinity, numerous ligand–receptor bonds are required to form stable adhesion sites for cells. The result is

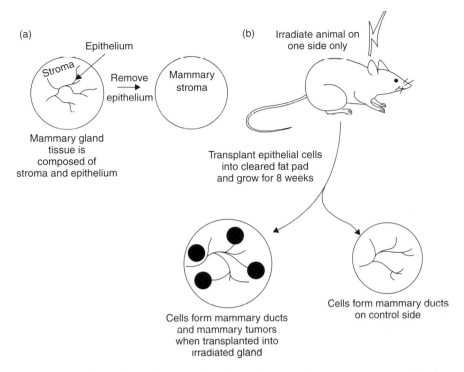

Fig. 12.6 The stroma contributes directly to epithelial malignancy. In this experiment, (a) the epithelium is removed from the mammary gland of a young 3-week-old mouse, leaving the mammary fat pads that consist of only the stromal components. (b) Eight weeks later, the mice are irradiated on one side only, and mammary epithelial cells are transplanted into both irradiated and control mammary fat pads. After a further 8 weeks, the cells in the nonirradiated glands proliferate and undergo developmental morphogenesis to form a normal-appearing ductal outgrowth. The cells in the irradiated side also form ductal outgrowths, but a large number of tumors also develop. The mammary cells injected are normal apart from harboring mutations in both p53 alleles, indicating that they are genomically unstable. Thus, ionizing radiation, which is a known carcinogen, causes changes in the stroma, which then facilitates the expression of neoplasia within the epithelial cell compartment. This type of communication between stroma and epithelium is often overlooked, but is of key importance in cancer progression. From Barcellos-Hoff, M.H. and Ravani S.A. (2000). Irradiated mammary gland stroma promotes the expression of tumorigenic potential by unirradiated epithelial cells. *Cancer Research*, **60**: 1254–60.

that macromolecular assemblies, visible in the light microscope, build up at the plasma membrane. This is rather like "velcro," where greater numbers of the same-sized bonds between two surfaces greatly increases the requirement for tensive forces to separate them. Large numbers of enzymes can be recruited to these sites, known as focal adhesions, which are effectively multiprotein signaling complexes (Fig. 12.9).

Integrin-containing adhesions exist in all cells, but some epithelia also contain a specialist type of assembly known as the hemidesmosome. These are large structures that form strong bonds between epithelial cells and the underlying interstitial ECM through a chain of molecular interactions. Instead of linking to the actin-based cytoskeleton, hemidesmosomes contain adaptor proteins that bind to intermediate filaments. Outside the cell, hemidesmosomes link to the basement membrane protein, laminin-5, via the signaling integrin, $\alpha6\beta4$ together with a transmembrane collagen called BP180, or collagen type XVII.

Integrin subunit composition frequently changes during the progression to malignancy, providing carcinoma cells with the machinery to survive and migrate in the interstitial ECM, which they do not normally come into contact

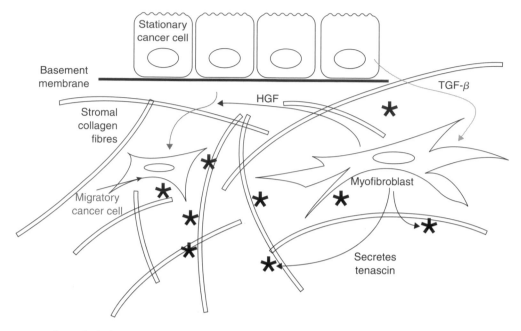

Fig. 12.7 Stromal–epithelial interactions in invasive behavior of tumor cells. Tumor cells can subvert the normal function of stromal cells so that they contribute toward malignant progression, even though they do not become altered genetically. Here, tumor cells secrete TGF-β, which recruits myofibroblasts to differentiate from fibroblasts. Myofibroblasts have two effects on carcinoma cells that encourage them to migrate through the stroma. First they secrete HGF, which causes the carcinoma cells to undergo a phenotypic transition so that they take on features of migratory mesenchymal cells. Second, myofibroblasts secrete tenascin into the stromal ECM: this is enhanced by TGF-β. As a consequence, the small GTPase Rac is activated within the tumor cells, while Rho is inhibited. These enzymes regulate the cytoskeleton so that the cells become more motile. Together, this epithelial–stromal–epithelial activation loop causes stationary carcinoma cells to become migratory. From De Wever, O., Nguyen, Q.D., Van Hoorde, L. *et al.* (2004). Tenascin-C and SF/HGF produced by myofibroblasts in vitro provide convergent pro-invasive signals to human colon cancer cells through RhoA and Rac. *FASEB Journal*, **18**: 1016–8.

with. It is well known that signaling enzymes are upregulated in many cancers, but this actually extends to adhesion-activated enzymes, such as focal adhesion kinase (FAK) and integrin-linked kinase (ILK).

Integrins integrate signaling networks

One important function of integrins is to control the activity of intracellular signaling pathways including members of the Ras family discussed in Chapters 5 and 6. These pathways in turn regulate various cellular behaviors such as proliferation, motility, polarity, differentiation, and survival. Integrins can trigger signaling pathways to some degree independently of GFs, while in the

absence of adhesion, they do not activate them efficiently (sometimes they fail to activate signaling at all). A central concept in cell–matrix adhesion is that integrins are context-dependent checkpoints for signaling: cells monitor their local extracellular environment and only permit GF/cytokine responses if they are in the right place.

Perturbations in the GF arm of this signaling network lead to oncogenesis (Chapters 5 and 6). However, mutations or epigenetic changes in the level or activity of integrin-regulated proteins also upsets the homeostatic signaling balance, thereby having a profound effect on cell growth: integrins are essential for GF responses (Fig. 12.10) but their signaling pathways can be hijacked deleteriously in

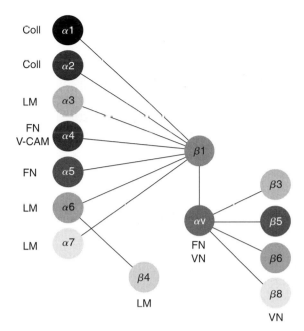

Fig. 12.8 Integrin heterodimers. There are at least 25 distinct integrin heterodimer pairs in total, made up from 18 α and 8 β subunits. Each is specific for a unique set of ligands. This diagram only shows a subset of the various ECM proteins bound by integrins. Some integrin heterodimers are promiscuous and bind several ligands, while several ECM proteins interact with cells through different integrins, depending on cell type and differentiation state. Coll, collagen I, III, IV; LM, laminins; FN, fibronectin; VN, vitronectin.

cancer (Fig. 12.5). Integrin alterations are as damaging, or even more damaging than GF receptor mutations, because they additionally affect migration, and consequently metastasis (see later).

Transmembrane proteoglycans

Some ECM proteins bind to other classes of transmembrane receptors that generate intracellular signals, and which are completely distinct from integrins. The important cell-regulatory ECM protein, laminin, binds to both a proteoglycan known as dystroglycan, as well as a lectin-like receptor β1,4-galactosyltransferase. Both of these receptors link laminin to the actin-based cytoskeleton as well as specific signaling enzymes. Dystroglycan has classically

been studied in muscle, but has subsequently been found in epithelia. Both dystroglycan and galactosyltransferase have critical roles in regulating epithelial morphogenesis during development, and they are also involved with cellular polarity and differentiation. Dystroglycan is frequently lost in carcinoma progression indicating a possible tumor-suppressor role. Syndecan and glypican are additional proteoglycan adhesion receptors now known to have a role in epithelial cell function and neoplasia.

CELL–CELL INTERACTIONS

Cell–cell interactions are also critical for many of the cell processes elaborated in earlier chapters, including the cell cycle and apoptosis. Cells interact and communicate directly with each other through a wide variety of mechanisms. Two different classes of adhesion systems are involved, multiprotein assemblies and simple receptor–ligand pairs.

Cadherins

Cell–cell adhesive junctions physically join cells together (Fig. 12.11). Two of these, adherens junctions (zonula adherens) and desmosomes, contain the most prominent cell–cell adhesion molecules, cadherins. Cadherins are transmembrane proteins that interact with each other both in *trans* (i.e. across the junctional space) and in *cis* (i.e. adjacent molecules on the same cell). They bind to each other at their distal amino-terminal ends across the intercellular space, thereby connecting adjacent cells. They also form lateral interactions between neighboring molecules on the same cell. In this way, they form large multiprotein complexes that have sufficient numbers of adhesive interactions to hold cells together. These interactions are dependent on calcium cations. Although there are many cadherin species, their binding is usually homotypic.

The prototype, epithelial (E)-cadherin, is the founder member of a small collection of "classical" cadherins, present in most epithelia in junctions that are visible in the electron

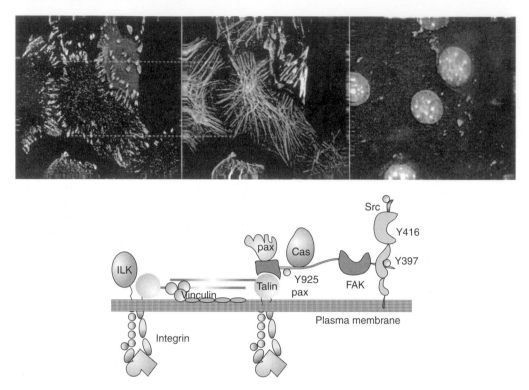

Fig. 12.9 Adhesion complexes are multiprotein assemblies at the cell–ECM interface. This fluorescence image of a cell shows the well-organized actin-based cytoskeleton (central panel) that defines the architecture of the cell, and adhesion complexes (left panel) where the cell interacts with ECM. (Please see our website for the color version of this figure.) In this case, the cell has been stained with an antibody specific for the phosphorylated, active form of focal adhesion kinase (otherwise known as pp125^{FAK} or FAK). Note that the skeletal components of the cell begin at the adhesion complexes: that is, their "feet." The dotted lines indicate coincidence of focal adhesions with the tips of cytoskeletal cables. The schematic diagram of the adhesion complex includes transmembrane integrins, which link cells to the matrix. Integrin ligation activates proximal cytoskeletal adaptor proteins, such as talin and vinculin, signaling enzymes, for example, FAK, Src, and integrin-linked kinase (ILK), and signaling adaptors such as paxillin (PAX) and 130Cas that couple to further downstream pathways. These include the Ras-Erk and PI3K-PKB axes, PKC and PLCγ, and small GTPases including Rho, Rac, and Cdc42. A large number of other proteins (not shown) are also within adhesion complexes. Adhesion-regulated signaling enzymes have key roles in controlling many aspects of cell behavior, including migration, polarity, survival, proliferation, and differentiation.

microscope. The family is now recognized to be a very large one, and includes desmosomal cadherins and protocadherins, which mediate, respectively, strong connections between epithelial cells and interneuron junctions in the CNS.

Cadherins have a key role in both developmental processes and in tissue homeostasis, but their altered expression in malignant cancer plays an important part in the dissemination of cells and the resulting metastasis (see later).

The β-catenin–Wnt connection

The cytoplasmic face of an adherens junction contains adaptor proteins that connect to actin filaments, and also to signaling molecules. One of these is β-catenin, a transcription factor that can be sequestered by E-cadherin. β-catenin is not only of importance at adherens junctions, but is also a transcription factor that is regulated by a developmental signaling cascade, the Wnt–β-catenin pathway (see Fig. 5.20).

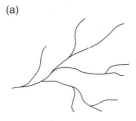

Normal mammary ductal tree

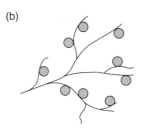

Tumors form if MT is expressed

Integrin ablation prevents tumor formation

Fig. 12.10 Integrins are essential for tumor formation. (a) The normal mammary ductal tree consists of epithelial cells that form tubular arrays. In nonpregnant animals, these ductal trees remain stable. (b) If the polyoma middle-T antigen (MT) is activated *in vivo*, multifocal tumors form within a few weeks. This is because the middle-T antigen activates numerous signal transduction pathways that are normally associated with GF receptor signaling, such as those involving c-Src, phosphatidyl inositol 3′-kinase, and the ras–Erk pathway. (c) If β1 integrin is deleted through conditional ablation of the gene, tumors do not form. This is most likely because the middle T-driven signaling pathways need active integrins in order to function properly. From White, D.E., Kurpios, N.A., Zuo, D. *et al.* (2004). Targeted disruption of beta1-integrin in a transgenic mouse model of human breast cancer reveals an essential role in mammary tumor induction. *Cancer Cell*, **6**: 159–70.

β-Catenin is not normally present as a free molecule within the cytosol because any that is not bound by E-cadherin at adherens junctions is targeted for proteasomal degradation by a complex of three proteins that include a serine-threonine kinase called glycogen synthase-3β (GSK-3β) and the tumor suppressors axin and adenomatous polyposis coli (APC; originally identified in a type of hereditary colon cancer called familial adenomatous polyposis, in which its activity is lost – see Fig. 3.5). Wnt proteins are secreted as signaling ligands, whose distribution is spatially controlled by binding to proteoglycans in the ECM (Chapters 5 and 7). They stabilize β-catenin by signaling through the cell surface receptors comprising Frizzled and LRP5/6. The resulting intracellular pathway inhibits the activity of GSK-3β, and so prevents β-catenin degradation. This leads to its nuclear translocation and interaction with a family of transcriptional repressors called Tcf/Lef. This interaction transiently switches Tcf/Lef from a transcriptional repressor to an activator, and thereby induces the transcription of numerous target genes including some involved in cell cycle regulation, apoptosis, and migration, such as c-MYC (Chapter 6).

Thus, there is a delicate relationship between extracellular signals that trigger the Wnt pathway and interepithelial interactions mediated by cadherins. Disruption of this equilibrium has a profound effect on tissue homeostasis, and leads to cancer progression. Mutations in Wnt pathway components frequently occur in cancer (Chapter 5), as do alterations in E-cadherin (see later).

Desmosomes

Cadherins also form the core of desmosomes, which provide structural integrity to cell–cell interactions in epithelia and cardiac muscle. Desmosomes are spot-welds between adjacent cells, while adherens junctions are more diffuse in nature. As with hemidesmosomes, the cytoplasmic faces of desmosomes contain adaptor proteins, such as plakoglobin and desmoplakin, that connect them to the intermediate filament network. Indeed, epithelia are networked together and to the basement membrane through contiguous intermediate

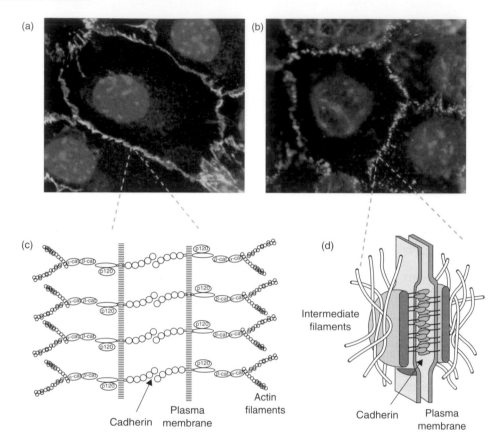

Fig. 12.11 Adhesion complexes at the cell–cell interface. (a,b) These fluorescence images of epithelial cell sheets show the localization of adhesion molecules distributed to cell–cell contact sites. (a) Cells stained for E-cadherin (grccn), which is visible in a continuous ring at the lateral junction between cells. (b) Cells stained for plakoglobin (γ-catenin), which only localizes to desmosomes and therefore has a punctate appearance. (Please see the website for the color version of these images.) Schematic diagrams of (c) an adherens junction and (d) a desmosome. Adherens junctions connect to, and organize, the actin-based cytoskeleton via the adaptor proteins α- and β-catenin. p120 is an important regulatory molecule, while β-catenin has a dual function as it is also a transcription factor. Desmosomes are strong spot-welds between cells, holding them together. They link to the intermediate filament network and thereby couple the scaffolding between hemidesmosomes and adjacent cells.

filament-to-desmosome/hemidesmosome links. Desmosomal cadherins are called desmocollins and desmogleins, and their composition differs between epithelial cell types. In complex epithelia, where more than one layer of cells lies on top of the basement membrane, the differential adhesiveness of specific desmosomal cadherins determine the cell's spatial positioning. Desmosomes can sometimes be lost during the progression of epithelial cells to malignant carcinoma, and introduction of desmosomal cadherins can suppress invasion. These cell–cell adhesion structures therefore have a tumor-suppressor function.

Other adhesive junctions

In addition to cadherin-containing structures, cells are physically cemented together by two other types of multiprotein assembly, gap junctions and tight junctions. Gap junctions form channels between cells, allowing the passage of ions and small molecules, and are thus another type of junction that networks epithelial cells. These structures contribute to cellular differentiation and they may be able to initiate signal transduction events, but it is not certain whether they have tumor-suppressing, or promoting, roles.

Tight junctions (zonula occludens) are present at the apical junction of polarized epithelia and bring the plasma membranes of adjacent cells into close apposition. They fully encircle the apical surface of polarized cells, thereby preventing the diffusion of ions and larger molecules across an epithelial layer. This is important for tissue homeostasis and if disrupted, can contribute to cancer (Fig 12.4). Tight junctions also form an intramembrane barrier, separating both proteins and phospholipids into apical and basal compartments (Fig. 12.2). Tight junctions contain a number of signal transduction molecules, including transcription factors that can be released. As with the other intercellular junctions, misregulation of tight junction components may contribute to carcinogenesis, both through loss of cell adhesiveness and polarity, and also through the unscheduled release of transcription factors, which influence cell proliferation.

A final set of contributors to cell–cell adhesion are the CAMs (cell adhesion molecules). These are a large family of transmembrane glycoproteins that contain immunoglobulin repeats within their external domains, and are calcium independent for adhesion. Several types of CAM are important for cell signaling because they bind to GF receptors in the plane of the plasma membrane, and they are therefore involved in cell signaling. The expression of many CAMs, including melanoma (M)-CAM and neural (N)-CAM, is altered in malignancy. In some cases they are required for the formation of metastases.

Cell–cell signaling junctions

A crucial set of intercellular junctions includes those that deliver direct signals through membrane-bound ligand–receptor pairs. Although these junctions are not adhesive, they can be activated when adhesive interactions between adjacent cells bring the ligand and receptors together. The Notch–Delta (see Chapter 5) and Eph–Ephrin systems are essential for patterning epithelia during development, and maintaining tissue homeostasis in mature organs. The components of these systems can be mutated or disrupted in neoplasia. Moreover, their normal function is compromised when cell–cell contacts become altered in cancer. Together this leads to abnormal signaling responses, which have knock-on effects contributing to the altered spatial disorganization and growth control of cancer cells.

Notch and Delta

Notch receptors are used widely in development and regulate cell fate decisions (also in Chapter 5). They are also involved, together with many other factors, in maintaining stem cells within their appropriate niches. These receptors, and their various ligands including Delta and Jagged, are all present on most epithelia, and emerging evidence indicates that their discrete expression patterns can be perturbed in cancer. Since stem cells are now believed to be progenitors for many types of cancer, disrupting the Notch–Delta system may have profound effects in the initial stages of tumorigenesis.

Eph and Ephrins

The Eph receptor family of receptor tyrosine kinases, the largest family of receptor tyrosine kinases, is an unusual type of system as both receptor and ligand act as signaling devices, and therefore responses can occur bidirectionally. Eph receptors provide repulsive stimuli to cells, rather than attractive ones, thereby preventing the mixing of cells between one environment and another. This leads to the formation of boundaries so that blocks of cell types become, and are maintained, separate from one another. Ephs and ephrins can also alter integrin-mediated adhesion and therefore regulate cell movement.

Ephs and ephrins are critical in the progression of certain epithelial cancers, because if they become disrupted they alter the positional identity of cells, which can lead to altered fate decisions and inappropriate proliferation. For example, in colonic epithelium, multipotent stem cells normally inhabit the lower part of the crypts of Lieberkühn, and after they are specified to become absorptive epithelial cells, they migrate slowly along the crypt–villus axis before being sloughed into the lumen of the gut when they reach the villus tip. Localized expression of Wnt causes cells at the bottom of the crypt to express Eph. As the cells migrate into the villus, they go out of range of the Wnt signal and now express an ephrin, which defines the crypt–villus boundary. However, in cancer, epithelial cells can continuously express the crypt marker and therefore cannot progress into the villus because of Eph–ephrin repulsion; they continue to proliferate in the crypt, leading to hyperplasia.

CRITICAL STEPS IN THE DISSEMINATION OF METASTASES

Benign tumors do not metastasize (Box 12.2). They form during the early stages of neoplasia and in epithelial-derived cancer they are referred to as adenomas and *in situ* carcinomas. The key characteristic of a benign tumor is that the homeostatic mechanisms controlling normal cell number are lost, through the triggering of oncogenic pathways and inactivation of tumor-suppressor genes. In order for benign tumors to grow efficiently, they need oxygen and nutrients, which are supplied through new capillaries, thus successful tumors also promote angiogenesis, often in response to hypoxia (Chapter 14).

None of these features, however, are sufficient to cause cells within primary tumors to migrate and form secondary lesions in distant parts of the body. Therefore, cell cycle dysregulation by itself does not give rise to malignant cancer. For this, further properties must be acquired (probably largely as a result of mutations or epigenetic changes), in particular, these can disable or disrupt the spatial control mechanisms that keep cells within their proper "place." As discussed in earlier chapters some of these properties may be conferred by environmental factors, not always resulting from mutations (also see Fig. 12.7). The loss of these positional controls is necessary in order to permit malignant cells to migrate away from the primary tumor, enter the circulation, and extravasate at a secondary site where a metastasis will grow provided the conditions are right. Thus, the loss of both growth

---

Box 12.2   Four characteristics of cancer cells are of key importance to enable their transition to become invasive

1. Malignant cells alter their repertoire of integrins, thereby enabling their migration through a stromal ECM and survival there (also see Fig. 12.20).
2. The composition of the stromal ECM itself changes, producing matrix proteins that promote migration; tenascin is one example of a promigratory protein that is often upregulated in the stroma adjacent to neoplasias (also see Fig. 12.7).
3. The cancer cells activate their intracellular signaling machinery that controls cytoskeleton and motile processes.
4. They also express/activate several classes of ECM-degrading proteinases and glycosidases that both remodel the ECM and release ECM-bound GFs; the latter has additional knock-on effects on the epithelial cells (also see Fig. 12.19).

Some of these changes occur as a response to epithelial-mesenchymal transitions or because the cell encounters a hypoxic, wound-repair environment. Others are achieved through acquisition of mutations.

---

and spatial control are required for cancer. Even in *Drosophila* the ability of oncogenic Ras^{V12} mutations to cause "metastases" depends on the additional alteration of genes involved in cell–cell and cell–ECM interactions, and invasion.

Fortunately, at the cellular level the metastatic process is extremely rare. In experimental models, less than 1 in 10,000 injected tumor cells are able to form metastases. The reason for this is that many changes within tumor cells need to occur in order for them to get into the circulation and survive in secondary sites. Indeed the whole environmental awareness of a cancer cell is reprogrammed during the formation of metastases, and is a consequence of acquired genomic instability that permits the sequential accumulation of mutations to form successful seed cells.

Invasion and dissemination through capillaries

Altered epithelial intercellular adhesion in malignancy is accompanied by changes in cell–matrix interactions, which together with the

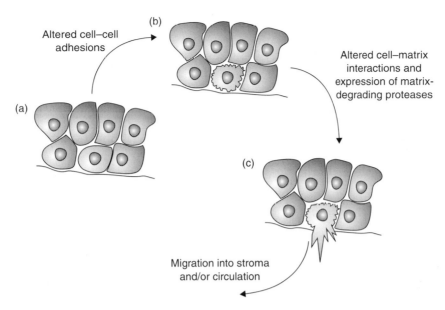

Fig. 12.12 Steps in the transition between a benign and malignant tumor. (a) Cells become DNA-repair and apoptosis defective, and acquire oncogenic properties. Their proliferation is dysregulated and normal tissue architecture is partially disrupted as a consequence. (b) Some cells may lose cell–cell adhesivity, and have ragged borders. (c) Additional migratory properties lead to exit of tumor cells from the primary neoplastic lesion. Disruption of the basement membrane is an important marker for the transition from a benign tumor to malignancy. These steps are essential to initiate tumor spread.

inappropriate activation of matrix-degrading enzymes, leads to the acquisition of invasive properties (Fig. 12.12).

In some cases, migratory cancer cells first break away from the primary tumor and enter the stroma, where they continue to proliferate as invasive lesions before encountering suitable blood vessels or lymph nodes from which to escape into the circulation. In other cases, cells that have lost adhesivity within a primary tumor enter the blood stream directly, because the capillaries that are induced to grow into the tumor by angiogenic factors provide an escape route: these cells, however, still require invasive properties in order to cross the basement membrane and endothelial cell lining of vessels.

Provided that tumor cells are strong enough to withstand the considerable hydrostatic pressures of the blood system, transit through the vasculature allows them to settle in secondary sites (Fig. 12.13).

Several factors determine the site where metastatic cells are sown successfully. One of these is vascular flow pattern. Cancer cells are frequently much larger than the small capillaries (4–8 μm diameter) that form the vascular bed of tissues, and therefore do not travel far before they become trapped. This is particularly evident in cases where tumor cells travel as clumps of cells. For example, the first vascular bed that colon tumor cells encounter is that of the liver, and many people die of colon cancer through chronic liver metastases and the resulting lack of clotting factors, failure of glucose homeostasis, and blood toxicity.

Dissemination through the lymphatics

Some cancer cells that have exited the primary tumor follow the direction of the normal interstitial fluid flow and enter the lymphatic capillaries, and thereby drain into regional lymph nodes. One mechanism facilitating this is the expression of lymphangiogenic factors such as VEGF-C. The most prominent site for metastases is the primary lymph node that the

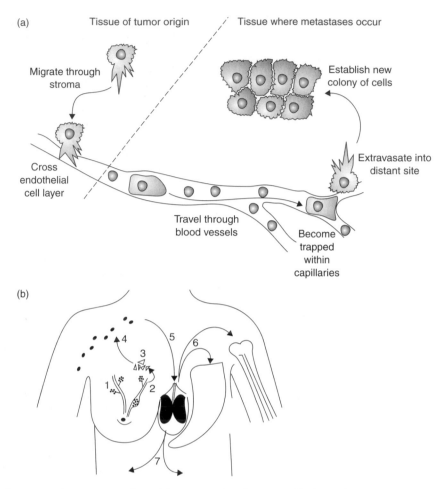

Fig. 12.13 Mechanisms of tumor spreading. (a) Passage of malignant cells from the primary tissue, where a tumor forms, to a secondary site, via the circulation. (b) Steps in the migration of malignant breast cancer cells to the organs where they colonize and form metastases. 1 – normal breast lobules are attached to ducts; 2 – formation of benign tumor, that is, ductal carcinoma *in situ* (note that cells proliferate abnormally inside the ducts); 3 – early steps in malignancy, that is, invasion into stroma; 4 – high-grade malignant tumor, that is, transfer into lymph nodes; 5 – passage of aggressive tumor cells to blood system and then heart; 6 – proximal sites of distant spread to lung and bone; 7 – metastases to other organs such as liver. The appearance of breast cancer cells in the lymph nodes is a hallmark of malignancy. If the lymph nodes of breast cancer patients have enough tumor cells to be seen histologically by a pathologist, it is most likely that these individuals will have already acquired metastases, and their chances of still being alive within 5 years are not high.

tumor cells encounter. Here the cells proliferate extensively, but can also be carried slowly through the efferent lymphatic vessels toward the thoracic duct, where the lymph is emptied into the blood. The microenvironment of a lymph node is very different to that of the stroma or primary tumor, so cells must accumulate additional changes in order to survive (these

cells survive anoikis; see later). In addition, they must become resistant to the large population of cytotoxic CD8 T cells (Chapter 13).

The lymph node is therefore a training ground for aggressiveness. Pathologists frequently determine the degree of tumor aggressiveness by examining sentinel lymph nodes (Fig. 12.14). Tumor cells travel through the

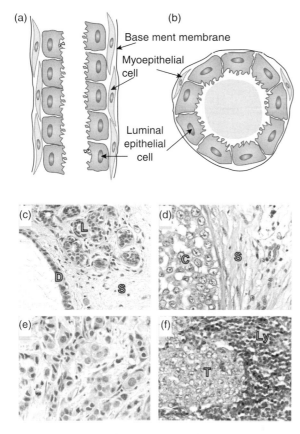

lymph vessels to the heart, and then through the arteries to other organs of the body, which they colonize.

Seed and soil

Although many metastases form at places where malignant cells become entrapped, this alone does not explain why it is that certain organs are preferred as secondary sites to others. One possible explanation for this is that tumor cells (seed) are only able to grow if they are in a conducive environment (soil), which includes the presence of appropriate proliferation and survival factors, and adhesive interactions. Thus, although the initial phases of metastasis are driven by mechanical dissemination, tumor cells need to be biochemically compatible with a foreign environment in order to be able to grow there. Such environments are tumor specific and are multifactorial, and include the expression of diverse homing and invasion molecules.

For example, transcription of chemokine receptors can be induced when the tumor environment becomes hypoxic. Chemokines are attractant cytokines that normally recruit lymphocytes to sites of infection. They encourage actin polymerization and the extension of lamellipodia, so that cells migrate along a concentration gradient that increases toward the chemokine source. Often, malignant breast cancer cells inappropriately express specific chemokine receptors (e.g. CXCR4 and CCR7), which can only be activated by the appropriate chemokines (CXCL12 and CCL21, respectively). These ligands are present in lymph nodes, thereby enticing the migration of breast cancer cells. Moreover, the CXCL12 ligand is also present at high levels in lung, liver, and bone marrow, which are organs where breast cancers metastasize to form secondary lesions, but not in other tissues, such as skin and prostate.

Thus, by subverting a homing mechanism normally used by other cell types, tumor cells of one tissue type can invade and colonize the stromal bed of another. This type of environmental reprogramming forms the molecular basis of the "seed and soil" hypothesis of metastasis, formulated in 1889.

Fig. 12.14 Tumor histology reveals aggressiveness. (a,b) Schematic diagram of a duct (a tube) and a lobule (a sphere) in mammary gland. Both are bilayered epithelia. Luminal epithelial cells (dark gray) are subtended by myoepithelial cells (light gray) and a basement membrane (thick black line). Some examples of the histology of breast tumors as they progress from benign to invasive carcinoma (please see our website for the color version of this figure). (c) Normal breast has three main compartments: lobules, ducts, and stroma. Here, lobules of epithelia (L) with central lumina are shown, together with part of a large duct (D). Note the discrete organization of lobular epithelial cells, and that the duct is lined with a simple layer of epithelium. Stroma (S) surrounds the epithelia. (d) Ductal carcinoma *in situ*. The simple bilayered epithelium of the duct has been replaced by large numbers of proliferating carcinoma cells (C). Note that the tumor has a discrete boundary with the stroma (S). (e) Advanced lobular carcinoma. This tumor arose from lobules. A large area of tissue is now taken up by proliferating cancer cells within the stroma (not visible). (f) Secondary tumor (T) metastasized to lymph node. The expansion of the tumor appears to be compressing the lymphocytes (Ly).

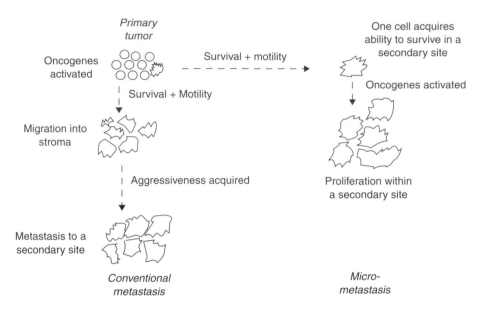

Fig. 12.15 Tumor dormancy and micrometastases. Some cells break off from the tumor prior to oncogene activation and settle in foreign sites. Although these cells might have altered adhesion mechanisms and are able to survive in an inappropriate microenvironment, they do not necessarily have a deregulated cell cycle. Thus, they have not acquired all the hallmarks of a malignant phenotype, and may remain dormant for many years. Subsequently, these cells evolve at secondary sites into metastatic lesions. An important implication of this is that an effective way to treat cancer may be to target altered proteins that occur within the metastatic lesion, rather than those that drive proliferation in the primary tumor.

Micrometastasis

Although the overt migration of mature tumor cells is a key mechanism in the formation of metastases, a new and controversial concept is that small numbers of cells may sometimes migrate to secondary sites early in the development of a cancer, and remain there in a dormant state for a long time (Fig. 12.15). These small lesions are micrometastases, which do not acquire all the hallmarks of aggressive tumor cells until often years later.

Microarray studies have identified specific sets of genes that become overexpressed in specific cancers and that are all required for successful metastases (see also Chapter 18). One such metastatic gene-expression signature for breast cells that colonize bone includes not only CXCR4, but also matrix metalloproteinases (MMPs), which facilitate invasion into the tissue, and osteopontin, an adhesive ECM protein. Importantly, this pattern

of gene expression arises within a few cells of the primary breast tumor itself, indicating that some of the genetic changes necessary to progress to a malignant tumor preexist within the benign tumor. Moreover, different subfractions of the same primary tumors have expression signatures that promote metastases to other tissues, such as the adrenal medulla, instead. Thus, subpopulations of cells within a primary tumor have heterogeneous metastatic potential. Although, these recent findings are intriguing, some caution needs to be exercised. These studies examined gene expression in groups of cells raising the possibility that specific metastatic signatures might not all be expressed by any single cell (only averaged across many); and recent advances in techniques for examining gene expression in single cells may help resolve this question.

An implication of early acquisition of "metastasis-enabling" mutations or inactivation of putative "metastasis-suppressor" genes is that the seed cells of metastases may spread early in the progression of malignant disease, sometimes prior to the detection of the primary tumor itself. Further mutations and selection may occur within such "micrometastases," leading to the evolution of a growth-competent metastatic lesion within the local environment of the distant site itself. This can occur many years after the deposition of the original micrometastasis.

Metastasis-suppressor genes

Understanding the molecular determinants that govern cancer invasiveness and metastasis will underpin development of new therapeutic strategies aimed at diagnosing or treating metastatic cancers. It has long been appreciated that mutations in oncogenes and tumor suppressors are a prerequisite for tumorigenesis, but only recently has it become clear that distinct mutations might be required, at least in some cases, for invasion and metastases. The metastasis-suppressor genes are defined as those that can suppress metastasis without affecting tumorigenicity. There is growing evidence that loss of function of metastasis suppressor genes plays an important role in cancer metastasis. The seven genes that suppress metastasis without affecting primary tumor growth that have been identified are *KAI1, CD44, mitogen activated protein kinase (MAPK) kinase 4, nm23-H1, nm23-H2, KiSS1,* and *BrMS1*. Three of these genes encode proteins (KAI1, CD44, and MAPK kinase 4) that act as metastasis suppressors of prostate cancer, while the remainder have yet to be tested in this cancer type. Loss of expression has been demonstrated for most of these genes during the clinical progression of prostate cancer to metastasis. MAPK kinase 4 and KiSS1 appear to suppress metastasis by inhibiting cancer cell growth at the secondary site. Interestingly many metastasis-suppressor genes have common roles in growth control, adhesion, and cytoskeletal reorganization, suggesting a common mechanism of metastasis suppression. Proposed candidate pathways include signaling through Src kinase and Rac GTPase.

E-CADHERIN DOWNREGULATION IN CANCER LEADS TO MIGRATION

One of the defining stages in the transition from a benign to a malignant tumor is modified cell–cell adhesion. Although in some cases this may be due to altered desmosomal function, it is primarily a consequence of reduced adhesiveness at adherens junctions. The main culprit in this aspect of carcinoma progression is loss of E-cadherin function. Once epithelial cells lose the strong interactions that normally hold them together, they can invade the stroma, provided their migration machinery is also activated. In many carcinomas, E-cadherin loss is a prerequisite for invasion and metastasis. In principle, there are two ways in which adhesion can be lost. The first is through the direct genetic or epigenetic control of the E-cadherin gene or its protein product. The second is through gross changes in the phenotype of epithelia, which can be triggered when they are induced to undergo epithelial-to-mesenchymal transitions (EMT).

Mechanisms of E-cadherin loss

There is an inverse relationship between the loss of E-cadherin in malignant tumors and patient mortality. If E-cadherin is reintroduced into cell lines derived from such tumors, they lose their ability to invade ECM in culture and to form metastases in animal models. This demonstrates the importance of E-cadherin in cancer progression, and points to its critical role as a tumor-suppressor gene (Fig. 12.16). It also indicates that understanding the mechanisms of E-cadherin loss might lead to novel therapeutic strategies for restoring its expression and thereby suppressing invasion and metastasis.

There are several examples of progressive cancer correlating with mutations in the E-cadherin gene, and this may happen at an early stage of the disease. However, this only occurs in a few specific tumor types, such as gastric cancer. The majority of mechanisms for the loss of function of the adherens junction are epigenetic (i.e. occur in ways that do not affect the genomic sequence) and can be caused in several ways.

First, cadherins (and other adhesion molecules) can be degraded from the outside by matrix-degrading MMPs. These enzymes are frequently

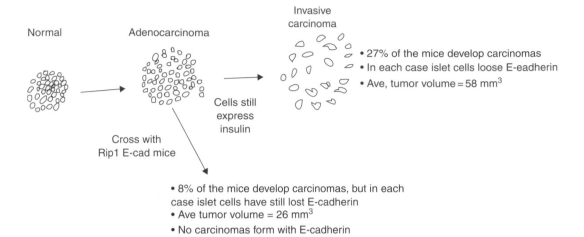

Fig. 12.16 The loss of E-cadherin is rate-limiting for tumor progression. In an experimental model of cancer progression, tumors form in the insulin-secreting islets of Langerhans in the endocrine pancreas. Transgenic mice that express the transforming SV40 T and t antigens under the control of the insulin promoter (Rip1Tag2 mice), develop cancer in a characteristic fashion that progresses through adenoma to carcinoma. The T antigens are expressed continuously from embryonic day 8.5, and by 7 weeks after birth, adenomas form. A few weeks after this, the tumors become angiogenic, and by the 11th week they become invasive. Tumorigenesis occurs because T antigen both promotes proliferation, by inactivating the retinoblastoma protein, and suppresses the proapoptotic activity of p53. E-cadherin is expressed by the normal pancreatic epithelial cells and in adenomas, but is completely lost once the tumor has progressed to the carcinoma stage. However, if the Rip1Tag2 mice are crossed with those expressing E-cadherin under the control of the same promoter (Rip1E-cad mice), the incidence of carcinoma formation is dramatically reduced. Some carcinomas do form, but these have still all lost E-cadherin expression. The experiments provide direct evidence that E-cadherin loss is required for the transition from benign adenoma to malignant carcinoma. From Perl, A.K., Wilgenbus, P., Dahl, U., Semb, H., and Christofori. G. (1998). A causal role for E-cadherin in the transition from adenoma to carcinoma. *Nature*, **392**: 190–3.

activated in malignancy (see later), resulting in the truncation of E-cadherin and loss of its adhesive function.

Second, the function of adherens junctions is acutely regulated by phosphorylation. Normally, components within the junctions such as p120-catenin are not phosphorylated. However, when tyrosine is phosphorylated by, for example, c-Src, a protein tyrosine kinase whose activity is often upregulated in cancer, they disassemble (see also Fig. 6.20). This is because phosphorylation of tyrosine can induce the ubiquitination of components in the complex via an E3 ubiquitin ligase called Hakai.

Third, the normal controls on E-cadherin gene transcription are frequently altered in cancer to cause its downregulation, even though the gene is not mutated. One mechanism is through hypermethylation, which occurs in a wide variety of carcinomas. Hypermethylation

prevents access of the transcription machinery, resulting in transcriptional inactivation.

Fourth, the levels of transcription factors that control E-cadherin expression can be altered. The E-cadherin gene is under strong transcriptional control, but many of the factors that have been identified to control its expression are repressors, rather than activators. The first cells to form during mammalian embryonic development are epithelial and they express cell adhesion molecules, including E-cadherin, to keep them together and compact them to form the morula. Development is subsequently dependent on the conversion of some of these epithelial cells to mesenchymal cells (i.e. an EMT), in order that different types of cells can form (e.g. trophectoderm and mesoderm from endoderm at the blastula stage of development) and morphogenetic cell movements can occur (e.g. during gastrulation). This requires continued suppression of the E-cadherin gene, which is carried out by several zinc-finger transcription factors that bind to E-boxes within its promoter called Snail, Slug, and SIP1. These are not expressed in normal epithelial cells, but are frequently upregulated in cancer cell lines, and in advanced cancers *in vivo*. Moreover, overexpression of SIP1 leads normal epithelial cells to invade collagen gels more easily, most likely by preventing their ability to stick to each other.

The three-dimensional structures of the interaction site between Slug, Snail, and SIP1 and the E-cadherin E-Box are currently being worked out. This knowledge will enable a hunt for small molecule inhibitors that prevent this interaction, and may therefore be used therapeutically to activate the re-expression of E-cadherin.

Recently, in a mouse model of pancreatic islet β-cell tumorigenesis (separate to that shown in Fig. 12.16), deregulated expression of the oncoprotein c-MYC (see Figs 6.9 and 6.10) was shown to trigger loss of E-cadherin expression alongside invasion of β-cells locally and into blood vessels. How c-MYC does this is the subject of ongoing studies. It is interesting to contrast this with β-cell tumors derived in Rip1Tag2 mice, in which only a small percentage of islet tumors lose E-cadherin and become invasive.

Finally, some cadherins can be upregulated in cancer; for example, neuronal (N)-cadherin is induced in invasive breast cancers. This particular cadherin is normally expressed in mesenchymal cells, so it may encourage epithelial cells to migrate into a stromal environment. Moreover, it can bind in *cis* to the fibroblast GF receptor and thereby activate signaling pathways that lead to invasion. *De novo* expression of cadherins that are not normally expressed in epithelia, but which encourage invasion, is called a "cadherin switch" and extends to several other members of the family.

EPITHELIAL–MESENCHYMAL TRANSITIONS

The mere loss of E-cadherin does not lead to cancer. Studies with mice harboring conditional-null alleles of E-cadherin demonstrate that when it is deleted in epithelial cells *in vivo*, apoptosis arises and no tumors develop. When E-cadherin

is lost, this is normally part of a larger program of alterations that occurs to change the phenotype of the cells. This program is similar to the EMT that occurs in development and converts cells from having a well-organized, stationary behavior, to ones that become separated from each other and acquire motility (Fig. 12.17). EMT is activated by factors that are secreted by stromal cells, such as fibroblasts, providing another paradigm where the stroma has a huge influence on the progression of carcinoma (Fig. 12.7). Several different types of GF are involved in EMT in cancer.

TGF-β can activate EMT

TGF-β normally inhibits proliferation of epithelial cells. At higher doses, it induces apoptosis. However, if cells acquire resistance to the growth-inhibitory and apoptotic effects of

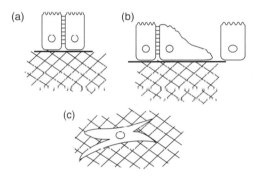

Fig. 12.17 EMTs. (a) Epithelia are normally stable, forming interactions with each other and with the ECM. These cells are polarized, and not particularly motile. (b) During EMT, signals provided by stromal cells cause epithelial cells to lose contact with one another and to alter their interactions with ECM. (c) The cells acquire a motile phenotype and take on many characteristics of fibroblasts (i.e. mesenchymal cells), such as their morphology, and expression of fibroblast markers including intermediate filaments that are characteristic of those cells. Epithelial–mesenchymal transition is a highly conserved process that is required during embryonic development in order that mesenchymal cells can be formed from the primitive epithelia of early blastocysts. This is necessary for the formation of a three-layered embryo during gastrulation. However, if the EMT program is reactivated in normal epithelial cells of adult tissues, it can contribute to malignant progression.

TGF-β, for example, if Ras becomes activated either through mutation or activation of receptor tyrosine kinases (Chapter 6), it can induce scattering and invasion. One of the ways it does this is through the expression of SIP1, which reduces cell–cell interactions. Another is that TGF-β can cause changes in the cytoskeleton through the small GTPases (see Fig. 12.7) and thereby contribute to cell migrations.

Transforming growth factor-β can collaborate with oncogenes to induce invasiveness and cancer. In healing skin wounds TGF-β becomes activated as part of the normal repair process, but if a viral oncogene such as v-*Src* is also present, for example, after infection of experimental chicks with Rous sarcoma virus, tumors develop: in fact this virus, which was the first tumor virus to be discovered (see page 200), only induces tumors at sites of wounding. Similarly, TGF-β cooperates with Ha-Ras to induce invasive spindle cell carcinomas of the skin of mice, and, moreover, the TGF-β receptor signaling pathway is altered in several human cancers.

This apparent paradox that TGF-β is tumor suppressive in normal cells, but becomes oncogenic and potentiates EMT and metastasis in later stage disease, means that therapeutic strategies to target this signaling pathway need to be designed with considerable care!

Hepatocyte growth factor drives EMT

Hepatocyte growth factor (HGF, or scatter factor) can also induce EMT by interacting with a receptor tyrosine kinase, called c-Met (Chapter 5). In monolayer-cultured epithelial cells this leads to scattering of the cells away from each other and cell migration. In cells placed in three-dimensional ECM, an experimental paradigm where the cells can aggregate to form multicellular polarized structures resembling ducts and acini (see Fig. 12.3), HGF induces branching morphogenesis. Normally this contributes to morphogenetic patterning, but if either HGF or c-Met is overexpressed or if the receptor is mutated, the signaling pathway can activate an invasion program, driving cancer. There are a wide variety of human carcinomas where alterations in either HGF or c-Met have been identified.

Activation of c-Met induces several intracellular pathways including a Ras-mediated pathway, and both phosphatidyl-3' kinase (PI3K) and phospholipase Cγ. Together, these pathways have several consequences on the cell. They promote proliferation, weaken cell–cell interactions by causing E-cadherin phosphorylation and reducing its expression, and they activate invasion by altering integrins and MMPs.

INTEGRINS, METALLOPROTEINASES, AND CELL INVASION

Epithelial cells are normally restricted to their own compartment within a tissue and do not cross basement membranes or invade stroma. However, these rules of normal behavior are broken during the progression of a benign tumor to a malignant one. Cancer cells become motile and, together with the loss of cell–cell adhesion, this allows them to become invasive (Box 12.2).

Altered integrin profiles reflect a more migratory phenotype

Cells migrate by extending plasma membrane protrusions called lamellipodia, in a polarized direction, into an ECM environment. The matrix provides mechanical support for migration. Lamellipodia engage with specific ECM molecules through integrins, which preferentially locate to the leading edge of migrating cells. This alters both the direct links with the cytoskeleton to generate traction at the front of the cell, and at the same time delivers signals to cause retraction at its rear coupled with myosin-driven movement.

Engagement is mediated by integrin activation and conformational changes within its extracellular domain so that the receptor binds with high affinity to its recognition epitopes on the ECM protein. Signals that initiate within the cytoplasm can cause the molecular reconfiguration of integrin extracellular domains, increasing their affinity for ligand. This is called "inside-out" signaling. Clustering of several integrin heterodimers within the plane of the membrane to form adhesion complexes leads to the triggering of signal cascades, involving numerous adaptor proteins and enzymes such as FAK, Src, and ILK, and thereby results in the activation of downstream effector proteins that alter the cytoskeleton (Fig. 12.9). These are "outside-in" signals. The integrin requirement for invasion is therefore both to mediate adhesion and to regulate signaling, cytoskeletal dynamics, and thus the motile machinery.

Of key importance are small GTPases of the Rho family, in particular Rac and Cdc42, which promote actin reorganization. These enzymes are switched on by guanine nucleotide exchange factors, such as αPIX, which is indirectly bound to ILK in adhesion complexes, or DOCK180, which is bound to FAK via the adaptor protein, p120^{Cas}. Once activated, Rac and Cdc42 can alter the function of cytoskeletal regulatory proteins, such as WASP and Arp2/3, to cause membrane ruffles and the extension of membrane processes. Rac and Rho, via their effector kinases, p21-activated kinase and p160^{ROCK}, phosphorylate proteins such as myosin light-chain kinase, which regulates myosin II and the motile machinery. Rac and Cdc42 have a direct effect on invasion because when activated forms of these proteins are expressed in epithelial cells they induce cellular depolarization and migration into the ECM.

Retraction at the tail of the cell occurs through a switch of integrins to a low-affinity state and dissolution of adhesion complexes. This occurs in part through a reverse inside-out signal. In addition, Erk-mediated signaling through the FAK/Src complex disassembles adhesion complexes, and together with intracellular proteases such as calpain, which cleave FAK and other proteins, this results in detachment from the ECM.

Epithelial cells tend to be fairly stationary in their normal tissue environment, but alter their expression of integrins in cancer to the profile of more migratory cells. The profile of integrins normally expressed is frequently reduced, while others, such as the αv integrins, which drive cell migration on stromal fibronectin and vitronectin, are often upregulated in malignancy, thereby contributing to invasion (Fig. 12.20). α6β4 integrin is also upregulated in several carcinomas. This integrin is usually associated with hemidesmosomes where it contributes to epithelial integrity. However, it can be released and recruited to lamellipodia during both wound healing and in malignancy, thereby increasing cell invasion. This integrin heterodimer also interacts directly with the receptor tyrosine kinases, c-ErbB2 and c-Met, the insulin signaling adaptor protein, IRS, and Shc, to promote PI3K and Ras signaling.

Although changes in integrin expression can be caused by transcription factors that are altered in cancer, they can also occur as a consequence of the cell being in a different environment. Stromal ECM, for example, strongly induces integrin expression in mammary epithelial cells. The acquisition of motile integrins can therefore be self-enhancing once an epithelial cell escapes its basement membrane constriction and enters a collagenous matrix. Many enzymes of the motile machinery, such as FAK and Rac GEFs, as well as the Ras-Erk and PI3K pathways, are upregulated in malignant cells, sometimes through mutation, which also contributes to increased migration (Chapters 5 and 6).

The altered integrin profiles of invasive cancer cells, in particular increased expression of αv integrins, informs therapeutic avenues to prevent migration in malignancy, and a number of inhibitors are currently in clinical trials (Fig. 12.18).

Serine and metalloproteinases degrade ECM to allow invasion

Activating the cellular motile machinery by itself is not sufficient for migration, particularly through a three-dimensional ECM. Cells need to relandscape their local environment in order to forge their way through the basement membrane and the dense network of fibrils such as that in the stroma *in vivo*. To do this cells require the help of ECM-degrading proteinases, of which there are several classes, including MMPs, serine proteases, and cathepsins. These enzymes are activated progressively during malignancy and promote migration by dissolving cell–cell cadherin contacts, breaking encapsulating ECM barriers, and clipping ECM proteins to expose promigratory epitopes and release migration-stimulating fragments.

Most MMPs and serine proteases are expressed by stromal cells but perform their tumor-promoting functions in the neighborhood of epithelial cells. In addition, they have an essential

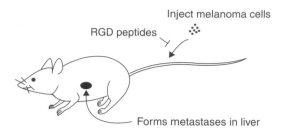

RGD peptides

Inject melanoma cells

Forms metastases in liver

Fig. 12.18 The key role of integrins in cell migration during metastasis. Melanoma cells are highly metastatic and can cause rapid patient mortality. In order for the cells to migrate and form metastases, they require integrin-mediated adhesion. ECM proteins contain specific amino acid recognition motifs to be bound by integrins, and one of these contains the sequence arginine-glycine-aspartic acid (RGD). This recognition sequence is contained within ECM molecules that promote migration such as fibronectin and vitronectin, which are bound by αv and αv integrins. In this example, melanoma cells cause experimental metastasis when injected into the tail veins of mice, but injecting an inhibitory RGD peptide at the same time can inhibit metastasis. This not only shows that integrin-mediated adhesion is necessary for metastasis to occur, it also indicates that strategies to block cell–matrix adhesion might be useful for cancer therapy. From Humphries, M.J., Olden, K., and Yamada, K.M. (1986). A synthetic peptide from fibronectin inhibits experimental metastasis of murine melanoma cells. *Science*, **233**: 467–70.

role in new blood vessel growth (angiogenesis), which is required for tumors to grow (see Chapter 14). They are normally present as inactive proenzymes, or zymogens, and become activated in the pericellular environment. This is important as it ensures that ECM remodeling does not occur throughout the stroma but instead is linked closely to aspects of cell phenotype.

For example, the serine protease, urokinase plasminogen activator (uPA), is activated at the cell surface through a balance between its receptor (uPAR) and several regulators including plasminogen activator inhibitor-1, a protease nexin, and the ECM protein vitronectin. This system directs proteolysis in the vicinity of adhesion complexes and is therefore pivotal in altering the ECM at active sites of migration. Many of the components are upregulated in malignant tumors and their levels correlate with tumor aggressiveness and poor patient prognosis.

The large family ($n = 28$) of matrix-degrading MMPs are also regulated close to the cell surface. This occurs through the removal of pro-domain, which masks the active site of the enzyme. Pro-MMP-2, for example, is activated via a complex pathway involving a transmembrane MMP (MT1-MMP or MMP-14) and the tissue inhibitor of metalloproteinase-2 (TIMP-2). MMPs are rarely altered by mutation in cancer, rather their levels are frequently increased at the transcriptional level. This can occur through *trans*-activation of MMP genes in stromal cells under the influence of cytokines secreted by the epithelial tumor cells. This is another important example of the cellular environment influencing tumor cell behavior. Occasionally, MMPs are expressed within the cancer cells themselves, via polymorphisms within the promoter region so that they are recognized by inappropriate transcription factors, or the combined activity of multiple oncogenes.

One function of ECM-degrading proteinases is to facilitate the cellular motile machinery. During migration both MMPs and serine proteases are recruited to lamellipodia, and control local ECM remodeling events that occur as the migrating cells make and break their contacts with ECM. Some ECM proteins contain cryptic domains that are recognized by integrins, and stretching the molecules or discrete MMP-catalyzed cleavage events expose them.

Another function of MMPs is to cleave the surrounding ECM molecules in order that the tumor cells can stretch out and respond to GFs. A frequent stromal host reaction to an encroaching carcinoma is excessive fibrosis, which serves to encapsulate the tumor and prevent the malignant cells from migrating. Such cells are also squeezed into a rounded geometric conformation that inhibits proliferation. Cleaving the matrix proteins will allow the tumor cells to escape their constraining mesh (Fig. 12.19).

Many MMPs are involved in cancer progression, but one, MMP-2, serves to illustrate the ways that they contribute to both the aggressiveness of tumor cells as well as the angiogenic response. MMP-2 is a broad-spectrum gelatinase that degrades many ECM proteins including laminin and collagen type IV, and mice lacking this enzyme do not support the ability of tumor cells to colonize lungs. A key function of MMP-2, together with MT1-MMP, is to

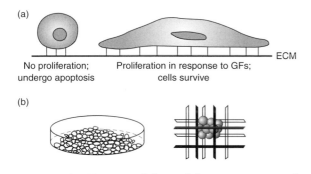

Fig. 12.19 MMPs control the proliferative response of tumor cells constrained within a three-dimensional ECM. (a) Cells that are attached to a substratum, but are forced into a rounded configuration, are unable to respond to GFs and do not proliferate. Rounded cells are, instead, prone to undergo apoptosis. This indicates that cellular geometry is critical in determining whether cells can proliferate or not. (b) When tumor cells are plated in two-dimensions on a culture dish coated with collagen (left panel) the cells are stretched out and they can proliferate well, but inside a three-dimensional collagen gel (right panel) the cells are rounded and their proliferation is severely compromised. This is because of low cyclin D3-kinase activity. However, if the cells express a surface-bound MMP, MT1-MMP, they clip the collagen fibers that enclose them, and after doing so they reorganize their cytoskeleton, stretch out, and now respond to proliferative signals. Thus, MT1-MMP allows cells that would otherwise be entrapped within a stromal or dermal matrix to proliferate, thereby providing a growth advantage. Indeed, the tumor cells expressing MT1-MMP grow much more quickly *in vivo* than those without it. This highlights the dramatic differences in growth potential of cells growing on a two-dimensional surface and the same cells embedded within a three-dimensional gel of the same matrix. As carcinoma cells exit a primary tumor and enter the stroma, they need to adapt to foreign signals and survive in the new ECM environment. They also need to proliferate, but can be prevented from doing so by the physical constraints of the stroma. The expression of MMPs provides a mechanism for carcinoma cells to proliferate at foreign sites that are frequently rich in dense networks of collagen or fibrin, and this is necessary for them to progress to full metastasis. MMP inhibitors may therefore have a dramatic therapeutic potential for cells growing in a three-dimensional environment similar to that *in vivo*, even if they have little effect on monolayer-cultured cells. From Chen, C.S., Mrksich, M., Huang, S., Whitesides, G.M., and Ingber, D.E. (1997). Geometric control of cell life and death. *Science*, **276**: 1425–1428. Hotary, K.B., Allen, E.D., Brooks, P.C., Datta, N.S., Long, M.W., and S.J. Weiss. (2003). Membrane type I matrix metalloproteinase usurps tumor growth control imposed by the three-dimensional extracellular matrix. *Cell*, **114**: 33–45.

unmask cryptic sites within specific ECM proteins; for example, these enzymes cleave the γ2 chain of laminin-5 in basement membrane to yield an armless molecule, together with some small fragments that have potent properties. One of these, DIII, is present in breast cancer, and its levels correlate with the stage of the disease. It contains an EGF-like domain and binds to the EGF receptor, triggering signal transduction and promoting motility. Other ECM proteins can be cleaved by MMPs to produce similarly active biofragments involved in cell migration and angiogenesis.

MMPs release tumor-promoting GFs

In addition to their direct effects on cell migration and indirect effects on proliferation, MMPs have other key functions in cancer progression because they release GFs that are sequestered by ECM or ECM-bound proteins. This influences angiogenesis, immune surveillance, EMT, and survival (see also Fig. 5.11). For example, IGFs are important survival and proliferation factors for epithelia but are sequestered by IGF-binding proteins. MMPs cleave IGF-BPs to release the bioactive GFs. Other classes of growth factors are secreted in an inactive state until cleaved by MMPs: TGF-α and the heparin-bound form of EGF are two examples. Similarly, TGF-β is released from its sequestering molecule, LAP, by MMPs, thereby directing the transition to a mesenchymal, migratory phenotype.

Since ECM-degrading proteinases are upregulated strongly in cancer, and they have diverse effects on migration, proliferation, survival, and EMT, they represent ideal targets for therapy. A number of strategies have been developed and are currently in clinical trials. These range from agents that inhibit their synthesis to those that block enzyme activity. As with all treatments for cancer that will be really effective in the clinic, understanding the tumor-specific contribution of specific effectors is critical in therapeutic design. We therefore need to learn much more about the individual serine proteases and MMPs that contribute to the progression of specific neoplasms. It is likely that such inhibitors will be most effective in treating early-stage disease, prior to the establishment of metastases, since this is the stage where active migration and stromal influence are most acute.

SURVIVAL IN AN INAPPROPRIATE ENVIRONMENT

Cancer cells are genetically unstable, and one might expect that the accumulation of mutations necessary to allow the growth of a secondary tumor would be rapid and frequent. One of the reasons for metastasis being rare is that, as outlined, an unfavorably large number of steps are required for tumor cells to leave the primary tumor and settle in a distant organ, so the chances of all being acquired by the same cell are small. A further driving force to prevent cells colonizing an inappropriate environment is anoikis, which is a specific form of apoptosis (Chapter 8) that is induced when cells no longer receive appropriate survival signals from the ECM. Although the trigger for this type of apoptosis may differ from that described in the earlier chapters, the downstream effector pathways are similar (see below).

Anoikis was described initially in cells that had been experimentally deprived of all contact with ECM for a few hours. This situation does not occur during the transit of malignant cells through blood vessels, since cells that travel to other tissues via the circulation only lose complete contact with ECM for a few seconds, which is not long enough to trigger anoikis. However, anoikis can be induced when cells migrate into either the lymph nodes, where there is not much ECM to contact, or into a matrix that they normally do not associate with (e.g. the stromal ECM), because any integrins that are activated there are not sufficient to maintain long-term survival. For example, normal epithelial cells from the mammary gland undergo delayed apoptosis when they come into contact with collagen I, which is the main type of collagen in the stroma. This process is not rapid, rather it takes place over several days and is due to an altered sensitivity to ECM signals, resulting in stochastic apoptosis. Anoikis is one of the key factors to maintain normal epithelial cell positioning in adult tissue homeostasis, and if this mechanism becomes deregulated, it

provides cells with the opportunity to live in a spatially inappropriate environment.

Thus cancer cells (at least those of epithelial origin), which are derived from cells that normally contact a basement membrane, will have undergone further mutations in order to survive the stromal bed of either the primary tissue they migrate through, or the secondary site to which they metastasize. Accumulation of mutations that allow cells to survive in inappropriate environments during malignant spread is as important, if not more so, than the acquisition of oncogenic mutations in proliferation genes.

Crosstalk between adhesion and GF receptors controls anoikis

Crosstalk of adhesion receptors with GF receptors and direct integrin-mediated signaling provides two separate mechanisms for environmental sensing that determines whether cells live or die. In the mammary gland, IGF is a key determinant of epithelial survival that is synthesized by stromal cells in response to growth hormone. This is an example of a stromal–epithelial communication that keeps the latter alive. However, in epithelial cells the IGF type I receptor only delivers efficient survival signals when the cells reside on a basement membrane. Signaling through the IGF-triggered PI3K and protein kinase B (PKB) pathway is inefficient in cells embedded within collagen, and they die a slow death.

Protein kinase B has been implicated in the protection of other types of cancer cells from anoikis, though it is not always activated by an integrin switch. In some tumor types, it is triggered by the unscheduled expression of a receptor tyrosine kinase, the NGF receptor, which can have dramatic effects on suppressing apoptosis and promoting metastasis.

A related crosstalk between adhesion receptors and receptor tyrosine kinases is critical for survival in other cell types too. For example, in capillaries vascular endothelial (VE)-cadherin collaborates with the vascular endothelial growth factor (VEGF) receptor to control survival and proliferation. This cell–cell adhesion molecule potentiates the PI3K–PKB signaling axis driven by VEGF, keeping the endothelial cells alive. Association with a third molecule, a phosphatase, causes specific tyrosine residues on the VEGF receptor to become dephosphorylated so that the Erk pathway is not activated, and this prevents excess proliferation in stable capillaries. When angiogenesis is occurring, the endothelial cells become motile as they invade new areas of stroma, and the cadherin junctions are dismantled, so that now VEGF can also trigger a proliferative signal. Similarly, integrins cooperate with neuregulin receptors to regulate the switch between growth and survival when oligodendrocytes mature into differentiated myelinating cells in the nervous system.

Direct integrin signals control anoikis

Integrins can also deliver survival signals directly to block anoikis by inactivating the apoptosis machinery. For example, in fibroblasts, decreased adhesion results in Jnk signaling leading to p53 activation, while in detached endothelial cells the reduced activation of Erk leads to lower levels of the Fas antagonist, c-FLIP, and apoptosis triggered through the Fas pathway. In some cases, altered integrin expression within cancer cells can contribute directly to the suppression of anoikis (Fig. 12.20).

In mammary epithelial cells, a slightly different apoptotic pathway is activated when integrin signaling is abrogated or altered. The proapoptotic protein, Bax, is maintained in the cytosol by integrin signaling, but the loss of adhesion to ECM leads to rapid and synchronous translocation of Bax to mitochondria, thereby inducing apoptosis (Fig. 12.21).

Many of the components linking adhesion to Bax in mammary epithelia (e.g. FAK) are altered in breast cancer, or their activities are enhanced. This provides cancer cells that have migrated into a stromal environment with the ability to survive when they would otherwise have undergone apoptosis, and thereby gather further mutations that contribute to the final stages of malignancy.

ECM and alteration of p53/ATM responsiveness

In several cancer cells, the accumulation of mutations is exacerbated by inappropriate ECM environment because the apoptotic response to DNA damage is not fully functional. If DNA damage occurs through chemical attack or radiation, a normal cell responds either by activating a repair pathway or, in the case of heavily damaged DNA, by triggering a p53-dependent apoptosis response (Chapter 7). The levels of p53 are kept in check by a balance between Mdm2, which targets p53 to the proteasome, and p19^{Arf}, which inhibits Mdm2. As outlined in Chapters 7 and 10, DNA damage normally results in phosphorylation of p53 so that Mdm2 cannot bind p53, and an increase in p19^{Arf} that sequesters Mdm2. Together this leads to stabilized p53 and transcription of Bax, Noxa, and Puma – all proteins involved in apoptosis and discussed in Chapter 8. However, in some cancer cells (e.g. melanoma and sarcoma), integrins are necessary for this response. Lost or reduced adhesion to ECM leads to lower p19^{Arf} levels, and the cells fail to stabilize p53 after treatment with DNA-damaging agents. As a consequence, less apoptosis occurs and the cells accumulate more chromosomal rearrangements.

Moreover, a separate pathway for driving apoptosis in response to damage, mediated by c-Abl, is also compromised by loss of ECM adhesion. c-Abl is a protein tyrosine kinase that is present in adhesion complexes and is regulated by integrins. Its activity is increased by DNA damage, but only if cells interact with ECM and the integrins are active. Certain cancer cells therefore escape the c-Abl-mediated damage response if they are not in an appropriate ECM environment.

Thus, conventional therapies for cancer that kill cells by inducing apoptosis, are not likely to be very effective in some types of malignant cells that are displaced from their normal ECM, growing either in the stroma, lymph nodes, or secondary sites, because the coupling between drug and apoptosis is less efficient in these microenvironments. In other words, cancer cells may be more resistant to apoptosis. Moreover, designer drugs such as those that target c-Abl (e.g. Glivec or STI571) reduce the apoptotic DNA damage response even further. However, treatments that activate integrins may be a positive benefit and enhance the usefulness of chemo- and radiation therapy. Since integrins are also involved in cell motility, this strategy may need

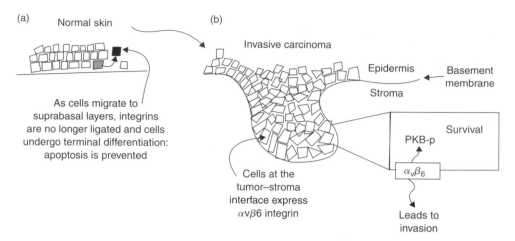

Fig. 12.20 An integrin switch can suppress anoikis and participate in the progression of skin cancer. (a) Keratinocytes normally express α2β1, α3β1, and α6β4 integrins, and have low levels of the αv integrins, usually αvβ5. β1 integrins are essential for stem cell maintenance in the basal layer, but as keratinocytes differentiate, the cells migrate away from the basement membrane into the suprabasal stratifying layers. Here, they are able to survive in the absence of integrin ligation and undergo terminal differentiation. Keratinocytes are uniquely different to other epithelia in this respect. (b) Under conditions where integrins are not ligated, cancerous keratinocytes expressing αvβ5 integrin undergo apoptosis. The mechanism for this appears to be through a suppression of PKB, or Akt via the integrin β5 subunit, though the details of how the integrin links to the PKB pathway are not yet clear. It probably acts as a fail-safe device to delete keratinocytes that are unable to differentiate properly. During the subsequent transition to squamous cell carcinoma, there is an integrin switch that results in elevated αvβ6 integrin and a corresponding decrease in the αvβ5 heterodimer. This leads to reactivation of PKB and suppression of anoikis. This is a unique mechanism to allow growth of tumor cells as they progress through the stroma. αvβ6 integrin also contributes to invasion of squamous cell carinoma, because it promotes increased expression of MMPs. Some integrins may therefore have multiple roles in cancer progression, and in the case of squamous epithelium, altered integrin profiles can both protect cells from apoptosis and drive invasion. From Janes, S.M. and Watt, F.M. (2004). Switch from alphavbeta5 to alphavbeta6 integrin expression protects squamous cell carcinomas from anoikis. *Journal of Cell Biology*, **166**: 419–31. Thomas, G.J., Lewis, M.P., Hart, I.R., Marshall, J.F. and Speight, P.M. (2001). AlphaVbeta6 integrin promotes invasion of squamous carcinoma cells through up-regulation of matrix metalloproteinase-9. *International Journal of Cancer*, **92**: 641–50.

to be tempered with other approaches to block migration. This is discussed in greater detail in Chapters 5 and 16.

QUESTIONS REMAINING

1 What are the detailed mechanisms that cause epithelial cells to lose the controls on their positional identity?
2 What is the precise contribution that specific ECM-degrading proteinases have in tumor spread?
3 Do the signaling pathways leading to anoikis resistance really have a role in driving malignancy?
4 Do micrometastases make a major contribution to human cancer?
5 Which effective therapies can be devised to target each of the processes that are altered during the progression from a benign to a malignant tumor?

6 What strategies can be developed to prevent EMT and to inhibit the contribution of stromal cells to malignancy?

CONCLUSIONS AND FUTURE DIRECTIONS

Cancer is the disease process whereby cells lose the normal controls not only on growth but also on their spatial location. Altered adhesion therefore has a critical role in cancer, and is the main characteristic of cells that become malignant and progress to form metastases. Genetic selection, or epigenetic changes affecting adhesion (e.g. EMT), reduce cell–cell adhesion, increase migration through the stroma, and activate ECM signaling pathways. Epithelial cells are normally under strong homeostatic

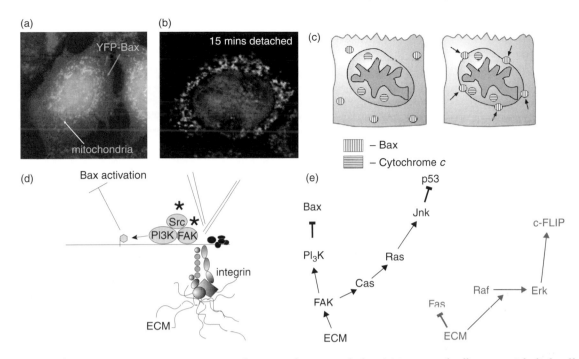

Fig. 12.21 The proapoptotic protein, Bax, regulates anoikis in epithelia. (a) In normal adherent epithelial cells, Bax is located within the cytosol (revealed by transfected YFP-Bax). The mitochondria are discrete. (b) When the same cells are detached from the ECM so that integrins are unligated and therefore unable to send survival signals, all the cytosolic Bax translocates rapidly to mitochondria. This can be seen by codistribution of YFP-Bax with the mitochondrial marker HSP70 (red). [Please see our website for our color versions of (a) and (b).] (c) This redistribution of Bax is schematized, showing relocalization of Bax to mitochondria. In an experimental situation, this process occurs homogeneously throughout a cell population. Cytochrome *c* remains within mitochondria for a long time after Bax translocation and eventually is released stochastically into the cytosol. Cells commit to apoptosis immediately after cytochrome *c* release. (d) Gene transfer experiments with plasmids expressing a dominant-negative version of FAK followed by rescue with constitutively active forms of Src, PI3K identify an integrin-triggered pathway that regulates the subcellular distribution of Bax. (e) Different cell types activate different intracellular signaling pathways downstream of integrins. FAK-mediated pathways suppress Bax in epithelia and p53 in fibroblasts, while an integrin pathway has effects on both Fas and c-FLIP in endothelial cells. From Gilmore, A.P., Metcalfe, A.D., Romer, L.H., and Streuli, C.H. (2000). Integrin-mediated survival signals regulate the apoptotic function of Bax through its conformation and subcellular localization. *Journal of Cell Biology*, **149**: 431–46. Ilic, D., Almeida, E.A., Schlaepfer, D.D., Dazin, P., Aizawa, S., and Damsky, C.H. (1998). Extracellular matrix survival signals transduced by focal adhesion kinase suppress p53-mediated apoptosis. *Journal of Cell Biology*, **143**: 547–60. Aoudjit, F. and Vuori, K. (2001). Matrix attachment regulates Fas-induced apoptosis in endothelial cells: a role for c-flip and implications for anoikis. *Journal of Cell Biology*, **152**: 633–43.

control, and mechanisms have evolved to prevent both the unscheduled accumulation of cells and inappropriate spatial positioning within tissues. However, disruption of homeostasis causes malignant cells to survive in, and migrate through, diverse ECM environments, thereby forming life-threatening metastases.

Although the concepts underpinning mechanisms of tumor spread are becoming established,

the details of how this occurs are still very uncertain. Much work has been done on a small set of cancer cell lines that have been acclimatized for efficient growth in tissue culture, but very little is known about the specific mechanisms of tumor invasion and survival *in vivo*. Still less is understood about the differences that diverse tumor types employ to metastasize.

Since metastasis is the major cause of cancer deaths, the key to keeping this disease under control (and therefore for keeping people alive for longer) is to understand the detailed mechanisms of dissemination. Effective therapies therefore need to be devised for targeting both migratory carcinoma cells (early-stage disease) and those that survive in distant sites (late-stage), but which avoid normal cells. Given the large array of events that contribute to malignant progression, the future is bright for the introduction of designer drugs for adhesion-specific processes. One always has to remember, though, that unwanted consequences may sometimes occur through surprising mechanisms, and to guard against this.

ACKNOWLEDGMENTS

Sincere thanks to Dr Keith Brennan and Professor David Garrod for helpful and critical reviews of the manuscript.

BIBLIOGRAPHY

Adhesive interactions with the ECM

Bissell, M.J. and Radisky, D. (2001). Putting tumors in context. *Nature Reviews: Cancer*, **1**: 46–54.

Colognato, H. and Yurchenco, P.D. (2000). Form and function: the laminin family of heterotrimers. *Developmental Dynamics*, **218**: 213–34.

Cell–cell interactions

Cavallaro, U. and Christofori, G. (2004). Cell adhesion and signaling by cadherins and Ig-CAMs in cancer. *Nature Reviews: Cancer*, **4**: 118–32.

Steps in the dissemination of metastases

Chambers, A.F., Groom, A.C., and MacDonald, I.C. (2002). Dissemination and growth of cancer cells in metastatic sites. *Nature Reviews: Cancer*, **2**: 563–72.

Fidler, I.J. (2003). The pathogenesis of cancer metastasis: the "seed and soil" hypothesis revisited. *Nature Reviews: Cancer*, **3**: 453–8.

Epithelial–mesenchymal transitions

Thiery, J.P. (2002). Epithelial–mesenchymal transitions in tumor progression. *Nature Reviews: Cancer*, **2**: 442–54.

Integrins, MMP, and cell invasions

Coussens, L.M., Fingleton, B., and Matrisian, L.M. (2002). Matrix metalloproteinase inhibitors and cancer: trials and tribulations. *Science*, **295**: 2387–92.

Giancotti, F.G. and Tarone, G. (2003). Positional control of cell fate through joint integrin/receptor protein kinase signaling. *Annual Review of Cell and Developmental Biology*, **19**: 173–206.

Hood, J.D. and Cheresh, D.A. (2002). Role of integrins in cell invasion and migration. *Nature Reviews: Cancer*, **2**: 91–100.

Ridley, A.J., Schwartz, M.A., Burridge, K. *et al.* (2003). Cell migration: integrating signals from front to back. *Science*, **302**: 1704–9.

Schatzmann, F., Marlow, R., and Streuli, C.H. (2003). Integrin signaling and mammary cell function. *Journal of Mammary Gland Biology and Neoplasia*, **8**: 395–408.

Trusolino, L. and Comoglio, P.M. (2002). Scatter-factor and semaphorin receptors: cell signalling for invasive growth. *Nature Reviews: Cancer*, **2**: 289–300.

Werb, Z. (1997). ECM and cell surface proteolysis: regulating cellular ecology. *Cell*, **91**: 439–442.

Survival in an inappropriate environment

Douma, S., Van Laar, T., Zevenhoven, J. *et al.* (2004). Suppression of anoikis and induction of metastasis by the neurotrophic receptor TrkB. *Nature*, **430**: 1034–9.

Green, K.A. and Streuli, C.H. (2004). Apoptosis regulation in the mammary gland. *Cellular and Molecular Life Sciences*, **61**: 1867–1883.

Lewis, J.M., Truong, T.N., and Schwartz, M.A. (2002). Integrins regulate the apoptotic response to DNA damage through modulation of p53. *Proceedings of the National Academy of Sciences USA*, **99**: 3627–32.

13

Tumor Immunity and Immunotherapy

Cassian Yee

Everything should be made as simple as possible, but not simpler.   Albert Einstein (1879–1955)

KEY POINTS

- The immune system developed as a means for an organism to distinguish foreign invaders from normal host tissues.
- The tumor cell evolves from normal tissues that do not usually provoke an immune response. But through the expression of mutated products or breaching of normal tissue barriers leading to an inflammatory milieu the tumor cell may elicit an immune response.
- The immune response to a tumor cell initially activates early immune cells (the innate response) and then immune cells involved in antigen-specific recognition (the adaptive response).
- However, tumor cells can evolve and evade the immune response leading to the clinical appearance of cancer.
- By dissecting the manner in which the immune response is activated, how immune cells recognize tumor cells, and the factors involved in augmenting or suppressing a tumor-specific immune response, effective immunotherapeutic strategies for cancer can be developed.

INTRODUCTION

The hypothesis that the immune system plays a role in the antitumor response and can be manipulated for the treatment of cancer was advanced as early as the 1890s, when William Coley used bacterial extracts to treat patients with sarcoma. It was thought that the ensuing inflammatory response successfully induced regression of large tumors by activating the immune system and inducing immune cells to attack the tumor. Since then, nonspecific forms of immunotherapy

have been used with varying degrees of success for the treatment of a limited number of malignancies – (i) BCG adjuvant for superficial bladder cancer and high-dose IL-2 for metastatic melanoma (Atkins *et al.* 1999), and (ii) donor lymphocyte infusions for leukemic relapse after allogeneic stem cell transplant (Kolb *et al.* 1990). However, these strategies were often accompanied by serious and potentially life-threatening toxicities. Advances in immunology, in the understanding of the requirements for T-cell activation and tolerance, and the development of novel technologies to analyze and augment immune response, now provide tumor immunologists with the opportunity to translate more broadly applicable principles in immunology to the practice of treating patients with cancer in a more specific and effective manner.

This chapter on tumor immunology begins with a description of the endogenous immune response, followed by a discussion of the components involved, [including T cells, dendritic cells (DCs), tumor antigens], and finally a summary of immunotherapeutic strategies arising from an understanding of the antitumor immune response.

Endogenous Immune Response

The endogenous immune response comprises two phases – the *innate* response that provides the initial line of defense against tumors, and *adaptive* response that provides longer lasting, antigen-specific immunity. These events are usually precipitated in response to a "danger signal" in the host that can be characterized by the introduction of foreign antigens or disruption of the normal microenvironment by invading tumor (see Box 13.1). As a first step, natural killer (NK) cells, representing the "rapid response" component of innate immunity, release pro-inflammatory cytokines such as interferon-γ leading to a cascade of soluble factors that include chemokines to recruit and activate macrophages and antigen presenting cells (APCs) such as DCs (Fig. 13.1). Macrophages kill tumor directly through the release of lysosomal enzymes, reactive oxygen intermediates, and nitric oxide. DCs collect tumor fragments and tumor proteins and migrate to draining lymph nodes, where they process and present these proteins to antigen-specific CD4 and CD8 T cells, leading to initiation of an adaptive immune response (Fig. 13.2). NK cells that can also mediate direct tumor cytotoxicity provide some degree of initial protection, but it is the adaptive response characterized by expanding populations of antigen-specific T cells and their differentiation into memory cells that provides the greatest potential for long-term immunoprotection against cancer. Naïve T cells, when they first encounter antigen (also known as priming), undergo a process of differentiation that leads to

---

Box 13.1   The "danger signal" in immune response

Activation of the innate and subsequently adaptive responses is believed to be precipitated by molecules that are released by cells undergoing stress or abnormal cell death and have often been described as "danger signals" (Gallucci & Matzinger 2001). In the case of immunity to foreign pathogens, cell products from infectious organisms, for example, lipopolysaccharides from gram-negative bacteria, double-stranded RNA from viruses, or zymosan from Fungi can all lead to a "danger signal" and a productive immune response. Empirical attempts by scientists and clinicians to activate an immune response by recapitulating endogenous immunity have been largely achieved through the use of adjuvants – which are an ill-defined mixture of bacterial components or irritants likely to induce inflammation and which, in the presence of a vaccine for example, can lead to the desired antigen-specific immune response. Only recently has the molecular basis for this initiating response been defined. A family of receptors expressed by cells of the innate response – NK cells and APCs, in particular DCs, allow these cells to perceive these "danger signals" and are known as pattern recognition receptors. These include the Toll-like receptors (TLRs) of which there are 10 varieties in mammals, each TLR recognizing specific ligands. What is more interesting is that these ligands are not limited to foreign pathogens, but may also be activated by host products, such as heat-shock proteins. Other recognition receptors, not included within the family of TLR, include Fcγ receptors that recognize opsonized (antibody-bound) antigens, scavenger receptors, NOD (nucleotide-oligmerization domain family) receptors that bind bacterial peptidoglycans and "stress" activated molecules, such as MICA which appear on virally infected or aberrant, that is, transformed cells and are believed to play an important role in antitumor immunity through the activation of stimulatory ligands on NK cells and CD8+ T cells. It is believed that transformed tumor cells by upregulation of surface MICA or the release of HSP following cell death provide a "danger signal" to DCs, activating DCs, inducing the release of cytokines and chemokines, and drawing effector cells to the site of attack, resulting in the cascade of events that lead to an adaptive T-cell response.

Normal

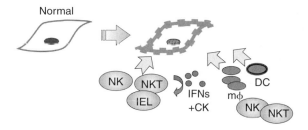

Fig. 13.1 Innate antitumor immunity. Development of tumor cells results in a proinflammatory milieu and activation of NK, NKT, and IEL (intraepithelial) cells. Release of cytokines and chemokines leads to recruitment of macrophages and additional innate effectors that provide an initial line of defense, and DCs that initiate the adaptive immune response.

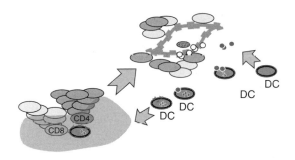

Fig. 13.2 Adaptive anti-tumor immunity. DCs take up tumor antigen, and process and present antigen to CD4 and CD8 T cells in draining lymph nodes. Presentation of tumor antigen leads to CD4 and CD8 T-cell activation (cf. Fig. 13.3 for details). Activated CD4 and CD8 T cells traffic to tumor sites where they mediate antigen-specific effector response through the release of cytotoxic granules, Fas–FasL interaction, and recruitment of secondary effectors (cf. Fig. 13.4 for details).

an effector population mediating tumor killing (see Box 13.2). A fraction of these cells persist as memory T cells. Memory T cells are T cells "at-the-ready," requiring a much shorter time to become fully activated and expand than naïve T cells. This population of memory antigen-specific T cells provides for a more robust and rapid response when antigens reappear.

Cancer immunosurveillance

That these defense mechanisms provide an immunologic basis for suppression of tumor cells by the endogenous immune response was hypothesized more than 40 years ago and has recently been supported by a number of murine models and clinical observations. In 1957, Burnet postulated that *"small accumulations of tumour cells may develop ... and provoke an effective immunological reaction with regression of the tumour and no clinical hint of its existence"* (Burnet 1957).

However, initial studies evaluating the incidence of chemically induced or spontaneous tumors in immunosuppressed mice treated by thymectomy or antilymphocyte sera led to conflicting results. Malignancies that did develop in immunodeficient mice were often virally induced or represented by lymphomas attributed to chronic antigenic stimulation from infectious agents. When athymic nude mice were evaluated for the development of spontaneous or chemically induced tumors, there was no increase in tumor incidence when compared with immunocompetent mice. Although it was not known at the time, athymic nude mice do retain a functional immune system comprising a small population of T cells and thymic-independent populations of NK cells. In the 1990s, mice were engineered with deficiencies in lymphocyte-specific recombinase that was responsible for antigen receptor rearrangement (RAG-2-/- mice). Such mice that possessed no NK, T, or B cells were found to develop sarcomas more frequently and with shorter latency period than wild-type mice.

Interferon-γ was also shown to be protective against the growth of spontaneous and chemically induced tumors and pivotal experiments using mice deficient for the interferon-γ receptor, or its signaling molecule, STAT1, demonstrated an increased incidence of chemically-induced tumors that was over 10 times greater than wild-type controls (Kaplan *et al.* 1998). Mice deficient in both RAG-2 and STAT1 exhibited, in addition to inducible sarcomas, spontaneous intestinal and mammary cancers (Kaplan *et al.* 1998). Further, mice lacking perforin, which is released within cytolytic granules upon lymphocyte stimulation to effect tumor cell killing, were found to be more susceptible to sarcomas and spontaneous lymphomas. Interestingly, chemically induced tumors developing in immunodeficient (RAG-2-/-) mice, when injected into wild-type,

---

Box 13.2   T-cell priming and differentiation / T-cell memory response

During development, the pluripotent stem cell gives rise to lymphoid progenitors that mature into T cells in the thymus. Mature T cells leaving the thymus into the circulation are each endowed with a singular antigen specificity determined by the TCR they bear on the surface. Together, these T cells form the repertoire of all possible specificities the organism can respond to and are represented by "naïve" T cells on the basis of not having previously encountered antigen. Presentation of antigen, for example, by DCs, selects for and activates a preexisting naïve T cell ❶, leading to its differentiation into an effector T cell and clonal expansion. Effector cells possess the ability to release cytokines, toxic granules (perforin, granzyme B), and mediate tumor cell killing ❷. Following an effector response, the majority of antigen-specific T cells undergo apoptosis ❸, but a small fraction develop into memory T cells ❹ that can survive in a quiescent state for many years after the antigen has been eliminated and are found to reside in lymphoid and mucosal tissues ❺. Upon antigen reexposure ①, these memory cells are quickly called into action, having the capacity to be rapidly activated and expand ② and kill tumor ③.

---

immunocompetent mice, were found to be more readily rejected than those tumors developing in immunocompetent mice. These results suggested that the innate immune response in immunologically intact hosts may be instrumental in sculpting the immunogenic phenotype of tumors and may, over time, result in immunoselection that leads to tumor immune escape.

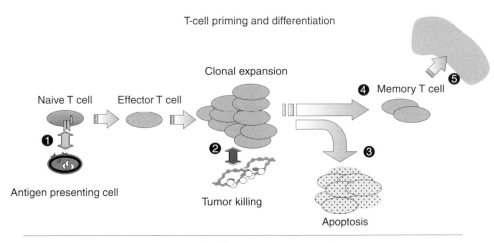

T-cell priming and differentiation

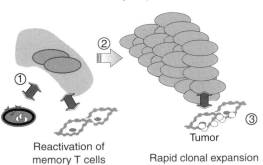

T-cell memory response

Cancer and immunosuppression

The role of the immune system in suppressing the development and progression of human malignancies was suggested initially in observational studies evaluating infiltrating or intratumoral lymphocytes in melanoma biopsy specimens. A correlation between the extent of mononuclear infiltration and prognosis has been extended for breast, colon, and most recently ovarian cancer in which patients whose tumors contained T cells experienced a six- to eightfold greater 5-year overall survival rate than those whose tumors contained no T cells. Conversely, patients who are chronically and/or severely immunosuppressed, as a result of stem cell ablation, HIV, or induced immunosuppression following transplantation, experience a higher overall incidence of malignancy than the general population. Although melanomas and occasionally colon, lung, and bladder cancers are observed, most tumors that develop are virally related, for example, EBV-post-transplant lymphoproliferative disease progressing to lymphoma, Kaposi's sarcoma, and HPV-associated malignancies. This may be a consequence of a shorter latency period for viral-associated malignancies and the shorter lifespan of immunosuppressed patients, which preclude the appearance of longer-developing malignancies. In some cases, reversal of immunosuppression has resulted in dramatic responses.

EFFECTOR CELLS IN TUMOR IMMUNITY

Effectors of adaptive immunity can be ascribed to a "humoral" and a "cellular" arm represented, respectively, by B cells that mediate effects through production of antibodies, and T cells that interact directly with target cells through the T-cell receptor (TCR). In humoral immunity, antibodies binding surface proteins on tumor cells can kill via complement activation or by bridging targets with cytotoxic cells through a process known as ADCC (antibody-dependent cell-mediated cytotoxicity). In this process, the Fc portion of antibody couples with receptors on macrophages or NK cells that then effect cell killing. Although antibodies are highly effective

in vitro, convincing evidence that antibody responses elicited *in vivo* play a critical role in antitumor immunity is weak. However, the significance of humoral responses with respect to tumor immunity has been supported by the identification of serum antibodies to potentially immunogenic tumor antigens (see the section on "SEREX", pp. 398–9) and the successful therapy of patients using monoclonal antibodies (see the section on "Antibody Therapy", pp. 405–6).

Dendritic cells

Dendritic cells are specialized or "professional" APCs that are activated during the innate immune response and are uniquely equipped to take up and present antigen to effector cells of the adaptive immune system – the antigen-specific CD4 and CD8 T cells. DCs are so-named because of pseudopods or "dendrites" which are processes that extend from the cell to facilitate antigen presentation. *In vivo*, the induction of an antitumor immune response may occur by tumor cells presenting antigen directly to T cells, but it is believed that a more common and robust pathway for tumor-specific T-cell activation *in vivo* is by cross-presentation. This is a process by which antigens released by necrotic, dying, or apoptotic tumors are taken up by DCs and re-presented to T cells under more favorable stimulatory conditions in the tumor-draining lymph node.

Dendritic cells can be characterized in their immature or mature forms based on contrasting surface and functional phenotype (Table 13.1). In their immature form, DCs are well equipped to capture antigens through surface receptors such as the C-type lectins (e.g. DEC-205, mannose receptors), $\alpha v\beta 5$ integrins, or CD36 for internal processing and presentation. DC activation via "danger signals" mediated through some of these abovementioned receptors and other surface receptors (e.g. the TLRs), leads to DC maturation (Gallucci & Matzinger 2001) in the presence of bacterial or viral products, TNF-α, or prostaglandins. It is also thought that in addition to DC activation by these receptors, cooperation of CD4 helper T cells is required to "license" DCs with the capacity to stimulate CD8 T cells through interaction of the CD40 ligand on CD4 T cells with CD40 on DCs. Upon

Table 13.1 Immature versus mature DCs – surface and functional phenotypic differences

	Immature DC	Mature DC	Activation signals for DC maturation
Morphology		Increased "veil" and dendrite appearance	Bacterial products:
Phagocytosis	+++	–	LPS (lipopolysaccharide)
Costimulatory molecules			Teichoic acid
HLA-DR (*MHC CLASS II)			CpG DNA
CD40	+/–	+++	Viral products:
CD80 (B7-1)			ds RNA
CD83			
CD86 (B7-2)			CD40 ligand
Chemokine receptors			
CCR7	–	++	TNFα, PGE2
CCR2	++	+	
CCR6			

maturation, further antigen uptake by DCs is downregulated, and, in preparation for optimizing T-cell activation, DCs upregulate surface expression of the T-cell costimulatory molecules (CD80, CD83, and CD86), and secrete cytokines such as IL-7 and IL-12 that facilitate full T-cell activation. In the case of tumor immunity, DCs circulate through the blood and accumulate at tissue sites in response to chemokines arising from the site of tumor necrosis or inflammation. As immature DCs, antigen is collected and processed for presentation on the surface in the context of MHC molecules. Following maturation and upregulation of surface costimulatory molecules, lymphokines, and the chemokine receptor CCR7, DCs traffic to lymph nodes where T-cell activation can occur.

T cells

Antigen presentation and T-cell stimulation

In contrast to B cells that provide "humoral" immunity through the production of soluble antibodies, T cells mediate "cellular" immunity by interacting directly with their target cells. T cells achieve specificity for cells expressing the target antigen through the surface TCR that recognizes fragments of antigen (usually peptide fragments) presented by the major histocompatibility complex (MHC) through either a Class I or Class II processing pathway (Fig. 13.3). Since proper antigen processing and presentation by the MHC is critical to antigen-specific immunity, a brief description of the MHC complex is presented.

The MHC is encoded by highly polymorphic genes that cluster on chromosome 6 in humans and are codominantly expressed. Human MHC molecules are called human leukocyte antigens (HLAs) and are divided into Class I and Class II HLA or MHC that present peptides to CD8 and CD4 T cells, respectively. For the most part, Class I MHC molecules are represented in humans by HLA-A, -B, and -C family of alleles, and Class II MHC by HLA-DR, -DQ, and -DP families. The Class I MHC complex comprises three parts: the MHC-encoded heavy α chain, a non-MHC-encoded β2-microglobulin chain, and a 8–11-mer peptide sitting in a groove formed by the polymorphic region of the α chain. The Class II MHC complex comprises three parts, the polymorphic α and β chains and a 10–30+ mer peptide sitting in a groove formed by the two chains. These parts are assembled in cytosolic compartments together with peptides derived from tumor proteins that have undergone proteosome-mediated cleavage to peptides of the appropriate length. The peptide–MHC

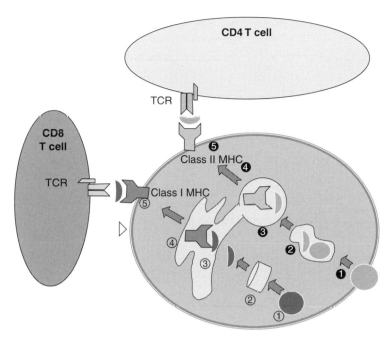

Fig. 13.3 Antigen presentation. In the Class I pathway (①–⑤), cytosolic proteins are processed by proteasomes ② into peptide fragments, transported through the endoplasmic reticulum (ER, ③), complexed with Class I MHC ④ and β2-microglobulin, and brought to the surface ⑤ where they are presented to CD8 T cells. In the Class II pathway, extracellular protein antigens are endocytosed ❶, degraded into peptide fragments ❷, combined with MHC ❸, and presented to the surface ❹ where they are presented to CD4 T cells ❺.

complex is then transported to the surface where the immunogenic peptide sitting in the MHC groove is presented to the TCR. The T cell recognizes the peptide *only* in the context of the MHC complex; therefore, mutations affecting any component of the antigen-presenting machinery can abrogate specific T-cell recognition and killing of tumor cells.

Costimulatory and inhibitory T-cell signals

In addition to the interaction of the TCR with the peptide–MHC complex, the activation of T cells can be modulated by the engagement of surface costimulatory or accessory molecules by their respective ligands on antigen-presenting cells (see Box 13.4). The most prominent of these are the signals provided by CD28 upon binding to B7-1 (CD80) or B7-2 (CD86) on APC. B7–CD28 interaction mediates signals that can fully activate an antigen-driven T-cell response, enhance T-cell survival by upregulation of antiapoptotic proteins such as BCL-x_L and drive proliferation. Absence of B7 has been associated with T-cell anergy while engineered expression of B7 in potentially immunogenic tumor cells can induce tumor rejection in murine models. While B7–CD28 interaction appears to be critical to the generation or priming of an effective antitumor

response, it does not influence the effector or killing phase of T cells. Hence, T cells generated with a B7-transduced tumor vaccine can eradicate B7-negative tumor. Other costimulatory molecules that deliver a positive signal to T cells include ICOS (inducible costimulator), OX40, 4-1BB, and other B7 family members (e.g. B7-H3). Accessory/adhesion molecules, such as ICAM-1 and LFA-1, are also critical to T-cell recognition. These molecules converge in and reinforce the TCR–peptide–MHC synapse, by forming in aggregate with other molecules, a supramolecular activation complex to facilitate delivery of a longer-lasting, more potent T-cell signal.

Cytotoxic T lymphocyte associated antigen-4 (CTLA-4) delivers a negative regulatory signal to activated T cells and competes with CD28 for binding to B7 on target cells (see Box 13.3). CTLA-4 is an inducible receptor with a greater affinity for B7 than CD28; however, in contrast to CD28, its surface expression is nonconstitutive and relatively short lived. CTLA-4 is believed to provide an immunologic "brake" to prevent overly robust and potentially damaging overstimulation by suppressing T-cell proliferation through IL 2 inhibition and downregulation of cell cycle activity. CTLA-4-deficient mice develop splenomegaly and a lymphoproliferative

---

Box 13.3   T-cell–APC interaction

Tumor-derived peptides are processed within the APC and presented by the peptide MHC complex to their cognate TCR.

For CD4 T cells, optimal T-cell activation requires the ligation of costimulatory molecules such as CD28 with B7. Upregulation of CD40 ligand following TCR engagement delivers a "licensing" signal to APC that results in increased B7 expression and production of T-cell modulatory cytokines such as IL-12. T-cell activation is downregulated by the inducible inhibitory receptor, CTLA-4, which blocks CD28 mediated signals and competes for B7 binding.

For CD8 T cells, activation following TCR engagement with the Class I MHC complex may be enhanced through costimulatory signals delivered by 4-1BB and other counterreceptors. IL-2 and other cytokines produced by activated CD4 T cells provide growth signals to cognate CD8 T cells.

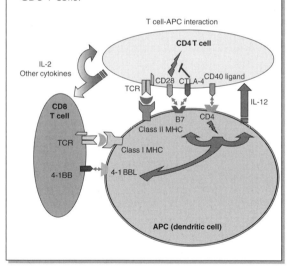

pathology. Since many tumor target antigens are also normal self-proteins, eliminating CTLA-4 inhibition may provide a means of breaking tolerance to self-antigens and augment an otherwise muted T-cell response to tumor. Administration of anti-CTLA-4 antibody in some murine models results in organ autoimmunity but can also lead to rejection of previously nonimmunogenic tumors. In clinical trials, administration of anti-CTLA-4 antibody has produced signs of autoimmune toxicity as well as tumor regression in individuals receiving a tumor-specific vaccine.

T lymphocytes   T cells can generally be divided into helper CD4 T cells and cytotoxic or killer CD8 T cells. Helper CD4 T cells recognize antigen in the context of MHC Class II and can be further differentiated into Th1 and Th2 subsets on the basis of distinct cytokine and receptor profiles. Th1 CD4 T cells produce IL-2 and interferon-γ, express IL-12 and IL-18 receptors, and regulate T-cell immunity, while Th2 T cells produce IL-14, IL-15, and IL-13, and regulate B cell immunity. It is believed that a Th1-type response would be beneficial in antitumor immunity since it mobilizes a T-cell-mediated response.

Cytotoxic CD8 T cells recognize antigen in the context of MHC Class I and, when activated, release perforin and toxic granules that mediate direct cell killing by punching holes in the cell membrane to facilitate entry of enzymatic packets (granzymes A and B). Although most studies have weighed in on a greater role for the cytotoxic CD8+ T lymphocyte (CTL) in tumor eradication, the helper CD4 T lymphocyte has also been shown to be a vital component in the induction and maintenance of a competent antitumor immune response. Not only have tumor antigen-specific responses been identified for CD4 T cells but the presence of CD4 T cells may be required for CTL responsiveness. Acting in concert, both CD4 and CD8 T cells provide for synergistic mechanisms of tumor killing. CD8 T cells kill tumor cells through the release of perforin and granzymes A and B, or through engagement of the death receptor, Fas, through Fas ligand (FasL) expressed on activated T cells. FasL–Fas interaction leads to a form of cell death known as apoptosis. In contrast to necrosis or death due to cell injury, apoptosis or programmed cell death involves a stepwise cascade of events initiated by receptor engagement at the cell surface (in this case, Fas), leading to DNA fragmentation. CD4 T cells can kill tumor cells directly by FasL–Fas engagement, as well as through indirect mechanisms that involve the recruitment of nonspecific effectors, such as macrophages and eosinophils, that can act even on MHC-negative tumors (Fig. 13.4).

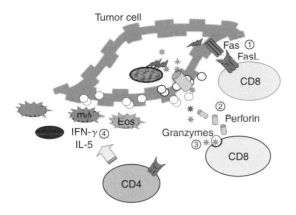

Fig. 13.4 Effector mechanisms of T cells. Activated CD8 T cells deliver a "death" signal to tumor cells through FasL–Fas interaction ①. CD8 T cells may also kill tumor cells directly through perforin and granzymes released upon engagement of the TCR②. Perforin exocytosed in CTL granules forms pores in the tumor cell membrane. Granzymes enter tumor cells through pores and induce tumor cell death③. CD4 T cells can mediate tumor death through Fas interaction. Activated CD4 T cells may also mediate cytotoxicity indirectly through the release of interferon-γ and IL-5 to recruit tumoricidal macrophages (mφ) and eosinophils (Eos) ④.

NK cells

Natural Killer cells are activated during the innate response by the inflammatory milieu that is established by invading tumor cells. These effector cells are not antigen specific and do not express a TCR but do kill tumor through killer *activating* receptors (KARs) expressed on their surface. Engagement of KARs with tumor-derived ligands, such as MICA and MICB, which are upregulated in infected or "stressed" cells, such as tumor cells, leads to NK cell activation and tumor cell death. NK cells also engage self-MHC Class I molecules on target cells through *inhibitory* receptors (killer inhibitory receptors–KIRs), perhaps, as a means of preventing autoreactivity. The loss of MHC expression on tumor cells, a process that can develop during carcinogenesis and immunoselection lends itself to preferential NK cell activation. The contribution of NK cells to the endogenous antitumor response *in vivo* may be best exemplified in athymic nude mice that have no T cells, but retain a population of functional NK cells that

appears to be sufficient to mediate tumor resistance. In humans, NK-type cells can be expanded *in vitro* with high doses of IL-2 for adoptive transfer; however, in this setting, their efficacy is less well defined and treatment is often accompanied by serious toxicity. The *in vivo* augmentation of NK-type cells may also be one mechanism by which high-dose IL-2 therapy has shown some clinical effect in the treatment of patients with metastatic melanoma or renal cancer.

Regulatory T cells

A population of T cells with regulatory properties that control autoimmune and antitumor responses was postulated as early as 1975; however, convincing evidence for their existence has been elusive. Recently, a population of CD4+, CD25+ T cells that possess immunosuppressive function has been identified. This discovery has led to a renewed understanding of the role of regulatory cells. CD4+ regulatory T cells (Tregs) are represented by two subsets – naturally occurring Tregs representing 5–10% of peripheral T cells and induced Tregs that develop from conventional CD4+ CD25- T cells. Naturally occurring Tregs mediate their suppressive properties through cell-to-cell contact by an unknown mechanism. Although activation is dependent on TCR engagement, their suppressive effects are nonspecific. They are known to express glucocorticoid-induced TNF receptor (GITR) and CTLA-4, a known T-cell inhibitor of T-cell costimulation. However, the role of CTLA-4 and GITR in mediating the suppressive effects of naturally occurring Tregs is not well defined. Induced or adaptive Tregs can be generated from conventional CD4+ CD25 T cells following *in vitro* exposure to antigen and IL-10 and the induced Treg cells themselves appear to mediate their inhibitory properties through the production of IL-10 and TGF-β. Tregs have been found to be fundamental for the control of autoimmune responses in several murine models, such as inflammatory bowel disease, and depletion of CD25+ T cells has been shown to mediate immune rejection of various murine tumors *in vivo*, presumably through the release of suppressive effects on T cells targeting shared self-tumor antigens (Shimizu *et al.* 1999). Elevated frequencies of CD4 Treg cells have also been described in cancer patients, leading to the design of clinical trials involving the administration of anti-CD25 antibodies to augment an endogenous antitumor immune response.

TUMOR ANTIGENS

Methods for identifying immune targets

T-cell-defined antigens Although it had been possible for some time to isolate tumor-antigen-specific T cells from the peripheral blood or tumor-infiltrating populations of lymphocytes by iterative *in vitro* restimulation with autologous tumor cells, it was not until 1991 that the first T-cell-defined human tumor antigen, MAGE-1 (for Melanoma Antigen-1) , was discovered (van der Bruggen *et al.* 1991). In this strategy, a tumor-specific CD8+ CTL line was first generated *in vitro* following repeated stimulation of lymphocytes with autologous tumor cells. Target cells engineered to express the restricting allele was then transfected with the cDNA library of the target tumor and the autologous tumor-specific CD8+ CTL line was used to screen pooled samples of the transfected target cells (Fig. 13.5). The cDNA cells lysed by tumor-specific CTL was extracted, sequenced, and compared to a gene database to identify the targeted antigen. Based on this strategy, many other investigators since have successfully isolated additional tumor-associated antigens in

this T-cell-defined manner (Table 13.2). These studies have also been extended to the identification of Class-II-restricted antigens using CD4 T cells to screen for cDNA library modified to facilitate presentation through the Class II MHC alleles. Interestingly, a number of identified Class-I-restricted antigens are shared among patients with similar tumors and, in some cases, even among different tumor types. Furthermore, several of these antigens are represented by normal nonmutated self-proteins such as tyrosinase, a melanosomal protein found in pigmented cells such as normal melanocytes.

SEREX Serological Recombinant Expression cloning (SEREX) has been used to identify potentially immunogenic tumor antigens from a wide variety of human cancers. SEREX is a simple, robust immunoscreening technique in which auto-antibodies present in patient serum samples are used to identify tumor proteins within tumor-derived cDNA expression libraries (Tureci *et al.* 1997). Currently, over 2000 SEREX clones for several human cancers have been identified using this procedure. SEREX immunoscreening identifies tumor antigens on the basis of a T-cell-dependent IgG antibody response. The induction of an IgG response suggests that a cellular immune response was operative and, indeed, both antibody and T-cell responses to several human tumor antigens, such as NY-ESO-1, HER2/neu, PSA,

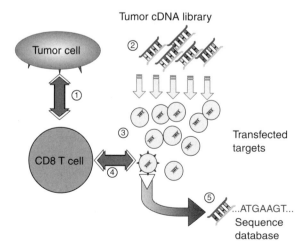

Fig. 13.5 Identification of T-cell-defined tumor antigens. CD8+ T cells are stimulated *in vitro* with autologous tumor cells to generate tumor-specific CD8 T-cell clones ①. To identify the tumor antigen recognized by these T-cell clones, a cDNA library is first prepared from the tumor cell line ② and used to transfect HLA-matched nontumor target cells (in this case, COS cells) ③. These cells are screened using the tumor-specific T-cell clone ④ and those recognized by the T cell are isolated and their cDNA extracted, sequenced, and compared with a genetic database ⑤ to identify the antigen targeted by the tumor-specific T cell.

Table 13.2   Tumor antigens

Antigen	Tissue expression	
	Tumors	Normal tissues
Self antigens associated with normal differentiation		
Tyrosinase	Melanoma	Melanocytes
MART1/MelanA	Melanoma	Melanocytes
gp100	Melanoma	Melanocytes
Prostate-specific antigens	Prostate cancer	Prostate
CD20, idiotype	B-cell malignancies	B cells
Her-2-neu	Breast, ovarian cancer	
Self-antigens associated with tumorigenic phenotype		
Survivin	Most tumors	
Telomerase	Most tumors	
p53	Most tumors	
WT1	Leukemias, lung, other solid tumors	
Cancer-testis /oncofetal antigens		
MAGE-1, MAGE-3	Ovarian, breast, GI, melanoma	Testis, placenta
GAGE and others	H&N, melanoma	Testis, placenta
NY-ESO-1	Ovarian, GI, breast, melanoma, lung, others	Testis, placenta
CEA	GI, breast	GI, breast
Mutated Antigens		
CASP-8	Head and neck cancer	–
CDK4-kinase, MUM-3	Melanoma	
β-catenin	Melanoma, lung	–
Viral Antigens		
HPV	Cervical cancer	–
EBV	Hodgkin's lymphoma, H&N, PTLD	–

and p53, have been detected within the same cancer patient. The majority of one class of tumor-associated antigens, the Cancer-Testis (CT) antigens, have been identified in this manner.

Gene expression profiling   The use of cDNA microarrays and more sophisticated methods including Serial Analysis of Gene Expression (SAGE) has been used to identify over-expressed genes among tumor cells compared with normal tissues. Combining this technology with immunomagnetic selection or laser capture microdissection to enrich for tumor cells represents a potentially powerful method for antigen mining. Once overexpressed or uniquely expressed genes have been identified, determining if such genes are in fact antigenically relevant targets is critical. This can be resolved by using algorithms to identify potential epitopes within the gene sequence that are predicted to bind given MHC alleles with high affinity or are more likely to be processed and presented by antigen presenting machinery. These predicted epitope sequences would then be validated empirically by assessing their ability to generate antigen-specific T-cell responses *in vitro*. Alternatively, APCs can be engineered to express the target gene (by viral transduction, nucleotide transfection, or other means) and then validated through *in vitro* testing for T-cell reactivity.

Classification of tumor antigens

In the context of antibody therapy, tumor antigens may be represented by any overexpressed proteins on the surface of tumor cells, and antibody therapy has led by far immunotherapy advances into the clinical arena. However, such surface proteins may not necessarily be immunogenic, that is, capable of inducing an immune response *in vivo*. For the purposes of vaccine studies or adoptive cellular therapy, immunization or *in vitro* generation of antigen-specific effectors can only be readily achieved against targets for which an immune repertoire exists and against which an effective immune response can be elicited. Table 13.2 lists tumor antigens that are known to elicit a cellular or humoral response and represent potential targets for cellular immunotherapy.

Tumor antigens can be classified on the basis of tumor specificity and function. Surprisingly, many tumor antigens initially described represented an increasingly important paradigm: the identification of normal self-proteins as tumor-associated antigens. Most of these differentiation type antigens were described in melanoma and play a role in melanin production. The reason for the preponderance of antigens that were initially identified in melanoma is due, perhaps, to the fact that melanoma is known to be amenable to immune manipulation (i.e. the use of nonspecific immunomodulators such as IL-2 and IFN can result in clinical responses), and that tumor lines and melanoma-specific T-cell lines for construction and screening of cDNA libraries were more commonly available. For melanoma, the appearance of vitiligo (depigmentation) in patients is predictive of a clinical response and suggests that such shared antigens are in fact therapeutically relevant. Targeting self proteins is accompanied by the risk of autoimmune toxicity. In the cases where the disease organ is surgically removed (e.g. prostate cancer), or toxicities are acceptable to the patient (e.g. vitiligo), targeting self-proteins may be a feasible strategy.

Those normal self-antigens that are overexpressed in tumors and are associated with a tumorigenic or proliferative phenotype provide the added benefit that immune escape through loss of antigen expression is less likely to occur. Among these include survivin (an antiapoptotic protein), telomerase (the ribonucleoprotein responsible for telomere synthesis), and WT-1 (a transcription factor overexpressed in leukemia and other cancers). However, the safety and efficacy of targeting such self-antigens are presently unknown and are being evaluated in a number of current trials in immunotherapy.

Truly tumor-specific antigens, such as the mutated cell cycle regulator – CDK4 kinase, or virus-associated proteins – HPV E6/E7 and EBV LMP proteins, represent favorable targets for immunotherapy, but are often limited to a small number of cancer types or may be associated with inherent immune escape mechanisms, such as altered antigen presentation in the case of the E6 and E7 proteins associated with HPV+ cervical cancer.

Cancer-testis antigens represent a distinct family of tumor-associated antigens with normal expression limited to the adult testis and fetal placenta. These antigens are variably expressed in a wide variety of tumors, including melanoma, breast, lung, and ovarian cancers, with generally increased expression in metastatic sites. Although their function is unknown, these genes are regulated by methylation and agents that induce hypomethylation and/or histone acetylation can lead to increased CT antigen expression in tumors. Because of their relative immunogenicity and potential for inducible expression, CT antigens are being evaluated in a number of clinical trials as immunotherapy targets.

T-cell repertoire and affinity for tumor antigens

The repertoire of T cells in the adult immune system that can recognize the vast array of antigens and their peptide fragments is represented by the naïve T-cell population that is shaped during embryonic development by events in the thymus. During thymic development, potentially autoreactive T cells are deleted while potentially beneficial T cells are retained. This process of thymic selection involves presentation of autoantigens to developing T cells and is dependent on the affinity of the TCR for its ligand such that high-affinity interactions lead to T-cell deletion, low-affinity interactions are ignored, and moderate-affinity interactions retained. The resulting repertoire consists of a precursor or naïve T-cell population that is predominantly capable of responding to foreign antigens and a lesser population of lower-affinity potentially self-reactive T cells.

Therefore, one overriding concern when targeting tumor antigens is whether T cells of sufficient affinity can be generated from a normal T-cell repertoire that not only harbors a very low precursor frequency of self-reactive T cells, but whose high affinity responders may have been deleted as a consequence of thymic development. For differentiation antigens in melanoma, T cells responding to MART-1 or tyrosinase exhibit at least one log lower affinity for cognate targets than virus-specific T cells in the same individual. This is further compounded when peptides are used in vaccine therapy such that supraphysiologic peptide concentrations are more likely to elicit low-affinity responses and delete high-affinity T cells in general.

T-cell avidity is largely dependent on the TCR affinity for its cognate peptide-MHC complex and is primarily a function of, and inversely related to, the dissociation constant (K_d). Altering epitope peptide sequences to enhance peptide binding to MHC or altering peptides that influence TCR contact with the peptide–MHC complex may elicit T cells with improved TCR affinity. Indeed, intentional mutations in the CDR3 (peptide–MHC binding) region of tumor–antigen specific TCR and amino acid substitutions in the TCR contact residues of the presenting epitope have been shown to lead to more robust higher-affinity T-cell responses. Engineering T cells to express such high-affinity receptors represents one rational strategy to the development of effective T-cells responses in adoptive cellular therapy.

ANTIGEN-SPECIFIC THERAPY OF CANCER

In general, there are two basic strategies to augment the antigen-specific immune response to tumor cells – vaccine therapy and adoptive therapy (see Box 13.4). Vaccine therapy involves the use of an immune stimulator – be it DNA, protein, or cell-based – that is administered to induce a tumor-specific response in the host,

often through repeated injections. It is anticipated that the immunogenic properties associated with the vaccine reagent will elicit an immune response that will be amplified *in vivo* with each boost. Adoptive therapy involves the *ex vivo* isolation and expansion of immune effector cells followed by their adoptive transfer into the host to augment the antitumor response. Generally, vaccination strategies are less labor intensive, especially in cases where the reagent is readily available (e.g. peptide-based vaccines). However, the magnitude and quality of the response may be limited by *in vivo* constraints. Adoptive therapy provides the opportunity to manipulate immune effectors *ex vivo* and thus more rigorous control over the intended immune response.

Vaccine therapy

Tumor vaccines represent an attractive modality for the treatment of patients with cancer because of their potential expediency and universal applicability. These vaccines can be tumor-cell based vaccines, derived from previously whole cells that have been digested, irradiated, or transfected and often combined with adjuvant to potentiate the immune response, or antigen-specific vaccine formulations that target known tumor associated antigens.

Tumor cell-based vaccines

Tumor cell lines Vaccines generated from established cell lines can provide multivalent antitumor responses against antigens that may share expression with patient tumor. This eliminates the requirement for "tailor-made" vaccines using autologous tumor cells, but may be less effective for any given individual. Allogeneic cell-based vaccines that have been fully developed include a whole cell vaccine and a tumor cell lysate vaccine admixed with adjuvant that are currently undergoing Phase III testing in clinical studies in patients with melanoma.

Autologous tumor Autologous tumor cell-based vaccines are often more labor intensive to produce, requiring the isolation and short-term *in vitro* propagation of tumor cells from a biopsy sample. However, they offer the advantage that

Box 13.4 Antigen-specific immunotherapy

Augmentation of tumor antigen-specific cellular response can be achieved by vaccination or adoptive cellular therapy strategies. For vaccines, protein, peptides, or antigen-charged DCs are manufactured *ex vivo* and used as immunogens to stimulate the immune response *in vivo*. An increased frequency of antigen-specific T cells is elicited *in vivo* after repeated immunizations. For adoptive therapy, T cells of defined specificity and phenotype are isolated *ex vivo* by iterative stimulations. Tumor-specific T cells isolated are expanded *ex vivo* to large numbers and infused into patients to augment the *in vivo* immune response.

A source of stimulator or APCs is required for either vaccine or adoptive therapy (① ❶)

For vaccine therapy (①–④):

② The vaccine reagent:
- Whole tumor cells, cell lysates, or tumor cell lines genetically modified, for example, with GM-CSF to augment immunogenicity can be used as vaccines.
- Peptides or DNA encoding the antigen of interest may be used directly as vaccine reagents.
- DCs charged with peptide or protein, or transfected with recombinant vectors encoding the antigen of interest, represent a potentially greater immunogen.
- Adjuvant (e.g. BCG) or adjuvant cytokines (e.g. IL-12) may be added to enhance immunoreactivity

③ The route of administration may be:
- intradermal
- intramusclar (e.g. plasmid DNA)
- intralymphatic.

④ Augmentation of antigen-specific immunity may require several booster immunizations after initial priming; maximal levels may not be achieved until several weeks after vaccination.

For adoptive therapy (❶–❺):

❷ Leukapheresis is a procedure that collects white blood cells from patients and provides peripheral blood mononuclear cells as a source of DCs and T cells for *in vitro* stimulation cultures.

❸ Isolation of antigen-specific T cells *in vitro* requires a source of stimulator and responder cells. Stimulator cells can be DCs pulsed with peptide or engineered to express antigen following transfection or viral infection with recombinant vectors encoding the target gene of interest. Responder cells can be PBMC enriched for CD4 or CD8

T cells. Iterative restimulation cycles are performed to augment the fraction of antigen-specific T cells *in vitro*.

❹ T cells demonstrating specificity for the target antigen are identified and expanded in flasks or culture bags to several billion using a combination of feeder cells, TCR stimulation and cytokines.

❺ Expanded T cells are adoptively transferred into patients; high frequencies of *in vitro*–generated antigen-specific T cells may be achieved *in vivo* with multiple infusions.

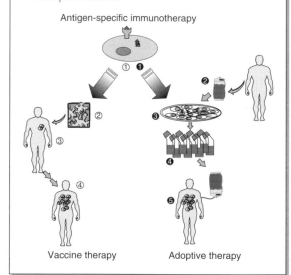

Antigen-specific immunotherapy

Vaccine therapy   Adoptive therapy

that they are more likely to express antigens associated with tumor cells for that particular individual. These tumor vaccines may be represented by (i) whole tumor cells that are admixed with adjuvant or genetically modified to enhance immunogenicity; (ii) cellular lysates produced by sonication, freeze-thaw, or other disruptive preparations; or (iii) apoptotic bodies derived from tumor cells exposed to UV or γ-irradiation, antibody, or pharmacologic treatment (e.g. betulinic acid in the case of melanoma). Vaccines composed of cellular lysates or apoptotic products take advantage of the antigen uptake capacity of autologous DCs *in vivo* followed by representation of tumor antigens to stimulate an immune response. Whole tumor cells cultured for short term *in vitro* can be engineered to express adjuvant-type cytokines or costimulatory molecules to enhance the tumor-specific immune response. In a murine model of melanoma testing the efficacy of syngeneic tumor transduced with various cytokines, GM-CSF and IL-12 were found to be the most potent in eradicating tumor – IL-12 through the augmentation of TH1 responses, and GM-CSF through the activation of APCs, such as DCs. Phase I clinical studies utilizing surgically excised tumor cells engineered to express GM-CSF, have been performed in patients with metastatic melanoma, renal cancer, prostate cancer, and advanced lung cancer (Soiffer *et al.* 2003). Clinical responses were observed in several patients accompanied by the presence of activated/mature DCs and granulocyes at vaccination sites, and a brisk infiltration of CD4 and CD8 T cells at tumor sites associated with tumor necrosis. Since efficient viral-mediated transfection of primary tumor lines can be problematic, methods to enhance local production of adjuvant cytokines have been addressed by peritumoral injection of recombinant virus or DNA encoding GM-CSF (for example) or by coadministration of irradiated autologous tumor cells with a cell line (e.g. K562) that stably expresses the cytokine of interest. Clinical studies utilizing these approaches are currently underway and may represent a more easily applicable vaccination strategy utilizing autologous tumor with cytokine adjuvants

Peptide-based vaccines Clinical grade peptides, usually 9–10 amino acids long, corresponding to Class-I-restricted epitopes of tumor-associated antigens, were one of the first reagents used in clinical trials to elicit antigen-specific antitumor immune responses. More recently, peptides of greater length (>14 up to 22 amino acids), selected on the basis of known Class-II-restricted epitopes or representing consensus sequences predicted to bind Class II MHC, have been evaluated for their T-helper function. Clinical results from studies using peptide reagents, however, have often been inconsistent even when similar peptides were used and were confounded by a lack of correlation between measured CTL frequency and clinical responses. This may be due to both the imprecise tools for immunologic monitoring that were used in these earlier studies and a requirement for more robust adjuvants and means of antigen presentation – addressed in future studies with the use of DCs, GM-CSF, CD40 ligand, and other reagents.

DC vaccines Since DCs provide optimal antigen presentation and T-cell stimulation, they represent ideal vaccine reagents if a source of such cells can be identified and engineered to express target antigens of interest. Human DC can be procured by density-based elutriation, dedifferentiation of CD14+ monocytes, or treatment of CD34+ precursors. Density-based methods rely on the differential buoyancy of DCs and are limited by the relatively low (<1%) frequency of DC among peripheral blood mononuclear cells. CD34+ cells enriched from the peripheral blood by immunomagnetic bead separation, for example, and treated with GM-CSF, IL-4, and TNF provides a purified population of human DCs. The use of monocyte-derived DCs perhaps represents the most easily achievable source of human DC. Adherent or CD14-selected PBMC when cultivated in the presence of GM-CSF and IL-4 yields an enriched population of DCs after 7 days. Subsequent addition of maturation factors (monocyte-conditioned media, TNF, LPS, CD40L, or a cocktail of IL 1, IL-6, TNF, and prostaglandin) results in upregulation of adhesion and costimulatory molecules on the DC surface. With any of these approaches, the addition of FLT3-ligand, a hematopoietic cell growth factor, *in vivo* or *in vitro* can lead to further expansion of DCs.

Engineering DCs to present the target antigen of interest can be accomplished by a variety of methods. DCs can be charged with whole or partially processed protein. For example, PAP (prostatic acid phosphatase) for the treatment of prostate cancer or immunoglobulin idiotype for the treatment of B-cell malignancies. Antigens for which the immunogenic epitope and restricting MHC allele are known can be targeted by pulsing autologous DCs with peptides corresponding to these epitopes as in the case of melanoma for which several HLA-A2-restricted epitopes have been characterized. Since clinical grade peptides can be readily obtained, this represents the most widely used approach in current clinical trials. In cases where the epitope has not been identified, DC may be engineered to express the target antigen of interest following infection with recombinant viral (e.g. adenoviral) vectors, transfection with plasmid DNA introduced by lipid-based carriers or, when coated with gold pellets, impelled into cells by a "gene gun" and RNA in a "naked" form, or combined with specialized transfection reagents and/or electroporation. These strategies provide a highly efficient means of engineering antigen expression in DCs with minimal cytotoxicity. Whole tumor-derived RNA has even been used to transfect DCs and generate multivalent T-cell responses. Finally, custom vaccines can be generated using DCs fused with autologous tumor cells to form tumor/DC hybrids. In this way, antigens known to be associated with a given patient's tumor can be optimally presented by autologous DCs.

As with any vaccine reagent, the source (CD34-selected, monocyte-derived, or Flt3L–expanded), dose, schedule, route of administration (intralymphatic, intradermal, or intramuscular), and method of antigen delivery remain to be established. However, DC vaccines currently represent one of the most promising translational applications of tumor immunology.

Plasmid-DNA-based vaccines

Plasmid-DNA-based vaccines   The surprising finding that plasmid DNA encoding the target antigen, when introduced intramuscularly, can elicit an antigen-specific immune response has been exploited in preclinical and clinical studies of tumor immunotherapy. DNA may be delivered by intramuscular injection or, when coated on inert beads, by a helium blast from a "gene gun." It is hypothesized that localized expression of injected DNA in myocytes or neighboring cells, together with the "danger signals" provided by local trauma, facilitates DC accumulation, antigen uptake, and cross-presentation of priming of T cells by DC in draining lymph nodes. To facilitate immunogenicity, adjuvants such as IL-12 or GM-CSF may be coadministered either in protein form or as DNA encoding these genes. Incorporation of immunostimulatory sequences, represented by methylated, palindromic CpG-oligonucleotide sequences, into the target DNA may also provide an adjuvant signal through TLRs expressed on DCs. Methods to improve the efficacy of DNA vaccines include the use of cationic lipids to increase delivery as well as the use of this approach in conjunction with other antigen delivery methods in a prime-boost design that involves DNA immunization followed by peptide, viral, or protein-based immunization as a boost to initial priming.

Adoptive therapy

Nonspecific cellular therapy   The *ex vivo* expansion and adoptive transfer of immune effector cells was pioneered in large part by work in the 1970s, first in murine models and then in clinical trials. Exposure of peripheral blood mononuclear cells to supraphysiologic doses of IL-2 mediated the expansion of nonspecific effectors known as lymphokine activated killer (LAK) cells that included a population of NK and NK-like cells. LAK infusions provided only a modest response and significant toxicities, related in part to the effector cells as well as to a vascular leak syndrome associated with the high doses of IL-2 that were needed to sustain LAK cells *in vivo*. Efforts to improve efficacy included the use of tumor-infiltrating lymphocytes (TIL) culled from tumor biopsy samples. It was hypothesized that these cells would be more effective than nonspecific LAK cells in eradicating tumors and indeed this was demonstrated in murine models. For clinical studies, TIL cells were expanded *in vitro* with high-dose IL-2 and infused in numbers as high as 10^{11} accompanied by high-dose IL-2 administration (Rosenberg *et al.* 1994). Although initial results were promising, TIL therapy was also associated with significant toxicity and it was not clear if beneficial effects were any greater than with HD IL-2 alone in patients with melanoma.

For patients who develop lymphoproliferative malignancies due to severe immunosuppression or who relapse with certain leukemias following allogeneic stem cell transplant, peripheral blood lymphocytes from the donor (donor lymphocyte infusions, DLIs) have been successfully used to treat patients with >75% response rate. However, such results were often accompanied by severe GVHD and myelosuppression.

Antigen-specific T-cell therapy   Broader application of an adoptive T-cell therapy strategy with decreased toxicities would be desirable. With the advent of tumor antigen identification and newer methods of T-cell expansion, it became feasible to enrich for a population of antigen-specific effectors *in vitro* and expand these to a desired magnitude for infusion. Current methods for eliciting tumor antigen-specific T cells *in vitro* use, as stimulator cells, DCs charged with protein, peptide, or transfected to express the target antigen of interest. The use of a TCR ligand, such as anti-CD3, either in combination with irradiated feeder cells or crosslinked to reagents coexpressing T-cell costimulatory ligands, such as anti-CD28 and 4-1BB, provides an optimized T-cell signal for activation. When cocultivated with T-cell lymphokines, such as IL-2 or IL-15, a high rate of proliferation can be achieved *in vitro* to obtain several billion antigen-specific T cells from an enriched cell line or single T-cell clone. Initial studies using this strategy, although labor intensive, provided more rigorous control over the specificity, phenotype, and magnitude of the intended immune response. Transferred T cells can be enumerated from the peripheral blood or tumor site, and their duration of *in vivo* persistence determined using immunologic methods (Yee & Greenberg 2002). Adoptively transferred T cells that had been isolated and expanded *in vitro* were found to persist *in vivo*, traffic to tumor sites, and mediate an antigen specific immune response that in some cases led to tumor regression (Dudley *et al.* 2002; Yee *et al.* 2002). Methods to enhance the efficacy of adoptive therapy include the coadministration of lymphokines known to promote T-cell proliferation and survival, such as IL-15 or IL-7, the combined use of both T helper CD4 T cells as well as cytolytic CD8 T cells and their use in the post-lymphoablative setting where homeostatic response to lymphopenia may facilitate "engraftment" and *in vivo* expansion of transferred effector T cells. As methods are established for rapid isolation by flow sorting or tetramer-specific isolation, for example, and more efficient *in vitro* expansion, adoptive cellular therapy as a treatment strategy may become more broadly applied.

Antibody therapy of cancer

Antibody therapy represents one of the success stories of bench to bedside technology. The development of monoclonal technology led to the practical application of antibody-targeted therapy as an approved treatment for patients with cancer. Although the question of long-term immunoprotection that is not predicted with antibody therapy remains, this modality has proven to be highly effective in a number of malignancies and greater use of this modality may be predicted on the heels of studies that use antibodies to target more fundamental mechanisms of tumorigenesis such as angiogenesis.

In contrast to T cells, antibodies can only target surface proteins. Upon binding, antibodies mediate tumor cell killing by various mechanisms (Figure 13.6). Engagement of the Fc portion of the antibody by NK cells and macrophage can direct killing of antibody-coated tumor cell. Activation of complement by the Fc portion of the antibody leads to complement-mediated lysis. Binding of antibody to tumor antigen can initiate the apoptotic cascade leading to tumor

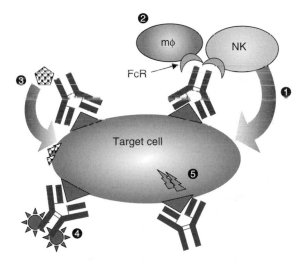

Fig. 13.6 Mechanisms of antibody-mediated killing. NK cells ❶ and macrophages ❷ docking through their Fc receptors to the Fc segment of antibodies bound to tumor cell surface antigen mediate cell lysis, a process known as ADCC. Alternatively, complement activation through the Fc portion of antibodies leads to a cascade of events resulting in a cell lytic complex that inserts itself into the cell membrane leading to tumor cell death ❸. Antibodies bound to radionuclides or toxins mediate cell death following internalization ❹. Blockade of growth factor receptors and/or activation of downstream signals lead to apoptosis and cell death ❺.

cell death or induce cell cycle arrest. Conjugation of antibody to a radionuclide such as yttrium-90 and iodine-131 forms radioimmunoconjugates that selectively deliver radiation to tumor. Conjugation to biological toxins such as ricin or diphtheria toxin and cytotoxic drugs such as doxorubicin or calicheamicin mediate cell death following antibody internalization. Antibody binding to growth factor receptors leads to receptor blockade and inhibition of downstream signaling events.

For B-cell lymphomas that are comprised of a monoclonal tumor cell population, tumor-specific idiotype immunoglobulins or B-cell specific markers (e.g. CD20) expressed on the surface represent suitable antibody targets. The tumor-specific anti-idiotype strategy pioneered by Levy *et al.* provided a proof of principle for the safety and efficacy of such a strategy, but each reagent that was produced was, by definition, patient specific (Miller *et al.* 1982). The B-cell-specific antibody, rituximab, was the first anticancer monoclonal antibody licensed by the FDA in 1997 and was used for the treatment of patients with low grade B-cell lymphoma. It is now an integral drug in the treatment of CD20-positive non-Hodgkin's lymphoma. Antibodies targeting other surface markers on hematopoietic cell surface, such as CD22 for B-cell lymphomas, CD52 for chronic lymphocytic leukemia, and CD33 and CD45 for acute leukemia have also been licensed for clinical use or are undergoing clinical studies.

Trastuzumab, an antibody targeting the HER2 tyrosine-kinase receptor overexpressed in breast cancer, represents the first FDA-approved antibody for use against solid tumors. Mechanisms of action of trastuzumab include inhibition of cell proliferation by downregulation of HER2 receptor expression, cell cycle arrest, and ADCC and CDC. Other antibodies targeting solid tumors include cetuximab, an antiepidermal growth factor receptor, for the treatment of colon cancer and head and neck cancer and Bevacizumab, an antivascular endothelial growth-factor receptor that has been FDA-approved for metastatic colon cancer but because of its broader mechanism of action (antiangiogenesis) can be considered for the treatment of other vascularized solid tumors including renal cell cancer and lung cancer.

CYTOKINE THERAPY OF CANCER

Cytokine therapy represents a "nonspecific" approach to immunotherapy that does not target any particular tumor antigen. However, as a strategy to augment preexistent effector responses, when used alone, or as an adjuvant to antigen-specific approaches described above, the use of cytokines may play a greater role in optimizing the antitumor response.

IL-2

IL-2 is the prototypic antitumor cytokine. Its discovery in 1976 as a growth promoting factor for T cells and subsequently NK cells led to its eventual use in clinical trials and FDA approval for the treatment of renal cell carcinoma and metastatic melanoma.

IL-2 affects many immune effectors; it can augment the cytolytic activity of NK cells, induce LAK activity, and enhance the direct cytotoxicity of monocytes against tumor targets. IL-2 plays a major role in expanding naïve and antigen-activated CD4 and CD8 T cells during the adaptive immune response; however, in later stages, it may also promote activation-induced cell death (AICD) presumably as a means of limiting T-cell expansion. IL-2 is produced primarily by CD4 T cells following TCR engagement, and its expression is upregulated by crosslinking of the costimulatory molecule CD28. Engagement of the CD28 coreceptor by antibody or its natural ligand, B7, results in tumor rejection in animal models, in large part due to its ability to enhance IL-2 production and upregulate anti-apoptotic signals (i.e. Bcl-x$_L$) leading to enhanced T-cell survival. The IL-2 receptor is composed of three subunits, IL-2Rα, β, and γ. The $\gamma-\beta$ heterodimeric complex alone can bind IL-2 with intermediate affinity and is capable of mediating an intracellular signal responsible for the downstream effects of IL-2 receptor engagement. Antigen activation upregulates expression of the α chain; the resulting heterotrimeric IL-2R$\alpha\beta\gamma$ complex engages in higher affinity interactions during the amplification process of T-cell expansion. In tumor animal models, IL-2 is instrumental in tumor eradication and in some cases can supplant the use of CD4 T cells as a helper component when cytolytic CD8 T cells are used.

IL-7 and IL-15

The proliferative signals induced by IL-2 are believed to be mediated only when the IL 2R β and γ chain heterodimerize. It is not surprising then that the "common" γ chain is shared among the several cytokines that influence T-cell growth and function: IL-4, 7, 9, 15, and 21. Among these, IL-15 and IL-7 have a unique role in maintaining T-cell proliferation that is distinct from IL-2. Although IL-15 shares both the IL-2Rβ and γ subunits, its α chain is distinct. IL-15 receptor engagement promotes expansion and survival of memory T cells *in vitro* and *in vivo*, in contrast to the AICD effect induced in activated T cells with prolonged exposure to IL-2. The IL-15 receptor is expressed on activated and memory T cells but not naïve T cells and appears to play an important role in the maintenance of T memory cell homeostasis. IL-7 also shares the common γ chain receptor with IL-2 and IL-15. This cytokine plays a role in lymphopoiesis, that is, during the development of precursor T cells and is important in the survival of naïve T cells in the periphery. In concert with IL-15, IL 7 also supports the survival of the memory T cell-population. It appears that in the lymphopenic (lymphocyte-depleted host, e.g. following radiation), IL-15 and IL-7 are required for homeostasis of memory T cells while in the normal (lymphocyte-replete) host, IL-15 has a predominant effect on this population. Preclinical studies have shown that IL-15 and IL-7 alone has a robust antitumor effect mediated by the *in vivo* expansion of T cells.

IL-21

In addition to a receptor that shares the common γ chain, IL-21 is also structurally related to IL-2, 4, and 15. Its role in the immune response is related to both the innate and adaptive arms of immunity. IL-21 supports the expansion, effector function, and maturation of NK cells and augments the antigen-specific T-cell response *in vitro* and *in vivo*. Murine tumor models demonstrate a robust IL-21-mediated T-cell response that is tumor specific and superior

in effectiveness compared with other γ chain receptor cytokines, IL-2 or IL-15. Its safety and efficacy in the clinical setting remain to be seen.

IL-12

IL 12, although not a member of the common γ chain family, plays a vital role in NK-cell activation, and in the differentiation of CD4 T helper (Th) precursor cells into Th1 cells. IL-12 is produced by macrophages and DCs. Its receptor is a heterodimeric molecule composed of an α chain and β chain expressed on activated T and NK cells. It is a proinflammatory cytokine known to activate NK cells resulting in secretion of IFN-γ. IFN-γ in turn can activate macrophages (e.g. to produce nitric oxide), mediate adaptive immunity through Th1 differentiation, and induce CD8 T-cell cytotoxicity. A role for antiangiogenesis has also been ascribed to IL-12. Animal models have demonstrated a robust antitumor effect of IL-12 that is dependent on NK cells alone, CD8 T cells alone, or both CD4 and CD8 T cells. Clinical studies of IL-12 have been conducted alone or in combination with vaccines with modest success and occasional toxicities.

Interferons

Interferons, initially discovered as viral inhibitors, have proven to be important modulators of the immune response and therapeutic reagents for the treatment of specific viral and malignant diseases. Interferon-α, which is produced by macrophages *in vivo*, plays a role in the activation of NK cells during an innate immune response but has other effects including antiangiogenesis and upregulation of expression of MHC and adhesion molecules. In the clinical setting, interferon-α has demonstrated efficacy against chronic myeloid leukemia and was a mainstay of treatment until the advent of newer agents such as the tyrosine kinase inhibitors (i.e. imitanib or STI-571). The use of IFN-α for the treatment of metastatic melanoma has been studied in well-designed randomized clinical trials, which demonstrate a significant increase in response rates and survival among patients with nonmetastatic high risk disease when used alone (Kirkwood 2002). Due to its immunomodulatory properties, IFN-α has also been combined with chemotherapy (bio-chemotherapy) to induce responses in patients with metastatic melanoma. Other malignancies amenable to interferon therapy include Hairy cell leukemia, Kaposi's sarcoma, and non-Hodgkin's lymphoma.

The initial enthusiasm in the use of cytokines such as IL-2 and interferon in patients with advanced cancer has been mitigated by dose-limiting toxicities, modest responses, and lack of a durable effect. More recently cloned cytokines such as IL-12, IL-7, IL-15, and IL-21 have demonstrated potent antitumor efficacy and low toxicity in preclinical studies. Their use alone, or in combination with vaccine or adoptive cellular therapy, to augment the frequency and *in vivo* persistence of antigen-specific T cells represents a cause for renewed interest.

TUMOR IMMUNE EVASION

The process of immunosurveillance suggests that immunoselective pressure will eventually result in the outgrowth of tumors specially equipped to evade the immune response. This is known as tumor immune evasion or immune privilege. For example, cells among a heterogeneous tumor population that have lost expression of the target antigen, epitopes, or the machinery to present such antigen will escape immune detection. The outgrowth of such tumor cells under endogenous immunoselection or following specific immunotherapy was first reported in murine models, but has since been observed in patients with cancer. It has also been suggested that primary tumors may evade detection by a process of "immunologic ignorance" in which tumors are sequestered from immune recognition by a stromal barrier. Even if recognition is achieved by antigen-specific T cells, however, tumor cells may subvert the effector response.

In general, tumor immune escape mechanisms can be divided into "intrinsic" and "extrinsic" factors. Intrinsic factors are (i) alterations in tumor antigen presentation, including loss of antigen expression, mutations affecting the immunogenic epitope, and defects in the antigen-presenting machinery that preclude effective presentation of the epitope to T cells;

(ii) counterattack or expression of molecules that inhibit T-cell effector function or viability such as IL-10 and TGF-β (cytokines that inhibit T-cell function), galectin-1 (carbohydrate binding proteins that induce apoptosis and inhibit effector function), kynurenines (tryptophan metabolites resulting from indoleamine dioxygenase upregulation that lead to T-cell apoptosis), and negative costimulatory molecules (B7-H1 or PD-L1 that induce IL-10 production and T-cell apoptosis); and (iii) inhibitors of apoptosis that are immunoprotective mechanisms limiting T-cell-mediated lysis of tumor cells and include the family of IAP (inhibitors of apoptosis proteins) such as survivin, c-FLIP (protects from Fas-induced apoptosis), and PI9 (a serpin protease inhibitor that inhibits granzyme B activity).

Extrinsic factors represent the influence of noncancer cells in the T-cell antitumor response. These include regulatory T cells previously described (see the section on "Regulatory T cells", p. 397). By direct T-cell contact, in the production of TGF-β, indirectly through CTLA-4-mediated upregulation of indoleamine dioxygenase or some other unknown mechanisms, these regulatory cells have been shown to have a profound T-cell suppressive effect and have been found at tumor sites in proportionally greater numbers. "Suppressor"-type DCs have also been described in playing a modulatory role in the T-cell immune response and may act through upregulation of IDO activity leading to the production of tryptophan metabolites. Suppressor myeloid cells, defined phenotypically as CD11b+, Gr-1+, are also believed to play a role in mediating T-cell apoptosis through production of TGF-β and depletion of arginine (required for T-cell signaling).

This list of potential mechanisms is by no means comprehensive and remains an area of intense interest for tumor immunologists. As more mechanisms are discovered, several questions arise: what is the interplay among these varied actors in attenuating an otherwise effective antitumor response? What draws these suppressor or regulatory cells to the tumor site and activates their suppressive function? Do these mechanisms exist *in vivo* and what is their relevance in subverting a clinical response?

Studies designed to address these mechanisms of immune escape include the use of multivalent vaccines or the transfer of effector cells targeting multiple antigens. This, together with the targeting of antigens representing proteins essential for tumorigenesis or tumor survival, may undermine the outgrowth of antigen-loss tumor variants as one mechanism of immune escape. The adjunctive use of chemotherapy or radiation therapy to reduce the tumor burden and presensitize tumor cells to apoptosis may facilitate the effector response of immunotherapeutic strategies. Methods for selectively eliminating the impact of regulatory or suppressor cells are also now finding their way into clinical trials. As we gain a greater understanding of the molecular machinery responsible for tumor-induced immune escape, it may be possible to address these hurdles in a systematic and comprehensive fashion.

CONCLUSIONS AND FUTURE DIRECTIONS

The immune response to tumor cells involves a complex interplay of antigen-presenting cells, effector cells, cytokines, and chemokines that evolves over time and space. An understanding of basic immunologic principles can lead to insights into the reasons for failure of the endogenous antitumor immune response and an opportunity to manipulate components of the immune system to augment an antigen-specific effect. The identification of tumor antigens capable of eliciting immunity was one of the first steps toward achieving this goal. Now, an understanding of the mechanisms of T-cell recognition and costimulation has led to the possibility of vaccinating patients using DCs and the development of methods to isolate and expand antigen-specific CD4 and CD8 T cells *ex vivo* for adoptive transfer. The role of cytokines and chemokines in bringing together many of these effectors of innate and adaptive immunity yields yet another opportunity to augment the antitumor immune response. The use of monoclonal antibodies, which are already in clinical use, foreshadows the evolution of immunotherapy as a more broadly applied modality for the treatment of patients with cancer. The potential synergy of combining immunotherapy with chemotherapy, cytokines and chemokines with a tumor-specific vaccine, or antiangiogenic antibodies with adoptive T-cell therapy, may provide additional weapons in the anticancer armamentarium. The application of immunotherapy at earlier stages of malignancy may provide the opportunity for more complete and durable response in patients for whom more conventional therapy would be ineffective. However, many questions remain unanswered. For example, what is the significance of regulatory T cells in tumor immunity? What is the best strategy for optimal vaccination? What phenotypic qualities are desired of effector cells for adoptive therapy? How can cytokines and chemokines be integrated into the use of vaccines in delivering immunogens to the site of activation and augmenting the ensuing response? What triggers cells down the path of immunologic memory to ensure long-term immunoprotection? How can we identify and address obstacles of immune escape? With more precise immunologic tools and preclinical models at our disposal, it is hoped that many of these questions can be answered.

QUESTIONS

1 One of the following cell types is not commonly associated with an adaptive immune response:
   a. DCs.
   b. CD8 T cells.
   c. NK cells.
   d. CD4 T cells.

2 Proteins associated with the following viral pathogens are a potential immunotherapeutic target for cervical cancer:
   a. EBV.
   b. CMV.
   c. HSV.
   d. HPV.

3 Mature DCs exhibit the following properties:
   a. Enhanced phagocytosis.
   b. Increased expression of costimulatory molecules.
   c. (a) and (b).
   d. Neither (a) nor (b).

4 The following are all examples of a γ chain receptor cytokine *except*:
   a. IL-2.
   b. IL-7.
   c. IL-12.
   d. IL-15.

5 Which of the following statements regarding CTLA-4 is false:

a. CTLA-4 is an example of a positive costimulatory receptor on T cells.
b. CTLA-4 can compete for the same ligand as CD28.
c. Treatment with anti-CTLA-4 antibody can lead to autoimmune toxicity.
d. Anti-CTLA-4 antibody is being used in clinical trials for the treatment of patients with cancer.

BIBLIOGRAPHY

Endogenous immune response

Atkins, M.B., Lotze, M.T., Dutcher, J.P. et al. (1999). High-dose recombinant interleukin 2 therapy for patients with metastatic melanoma: analysis of 270 patients treated between 1985 and 1993. *Journal of Clinical Oncology*, **17**: 2105–16.
Burnet, M. (1957). Cancer–a biological approach. *British Medical Journal*, **1**: 841–7.
Kolb, H.J., Mittermuller, J., Clemm, C. et al. (1990). Donor leukocyte transfusions for treatment of recurrent chronic myelogenous leukemia in marrow transplant patients. *Blood*, **76**: 2462–5.

Effector cells in tumor immunity

Gallucci, S. and Matzinger, P. (2001). Danger signals: SOS to the immune system. *Current Opinion in Immunology*, **13**: 114–19.
Kaplan, D.H., Shankaran, V., Dighe, A.S. et al. (1998). Demonstration of an interferon gamma-dependent tumor surveillance system in immunocompetent mice. *Proceedings of the National Academy of Sciences, USA*, **95**: 7556–61.
Shimizu, J., Yamazaki, S., and Sakaguchi, S. (1999). Induction of tumor immunity by removing CD25+CD4+ T cells: a common basis between tumor immunity and autoimmunity. *Journal of Immunology*, **163**: 5211–18.

Tumor antigens

van der Bruggen, P., Traversari, C., Chomez, P. et al. (1991). A gene encoding an antigen recognized by cytolytic T lymphocytes on a human melanoma. *Science*, **254**: 1643–7.

Tureci, O., Sahin, U., and Pfreundschuh, M. (1997). Serological analysis of human tumor antigens–molecular definition and implications [review]. *Molecular Medicine Today*, **3**: 342–9.

Antigen-specific therapy of cancer

Dudley, M.E., Wunderlich, J.R., Robbins, P.F. et al. (2002). Cancer regression and autoimmunity in patients after clonal repopulation with antitumor lymphocytes. *Science*, **298**: 850–4.
Miller, R.A., Maloney, D.G., Warnke, R., and Levy, R., (1982). Treatment of B-cell lymphoma with monoclonal anti-idiotype antibody. *New England Journal of Medicine*, **306**: 517–22.
Rosenberg, S.A., Yannelli, J.R., Yang, J.C. et al. (1994). Treatment of patients with metastatic melanoma with autologous tumor-infiltrating lymphocytes and interleukin 2. *Journal of National Cancer Institute*, **86**: 1159–66.
Soiffer, R., Hodi, F.S., Haluska, F. et al. (2003). Vaccination with irradiated, autologous melanoma cells engineered to secrete granulocyte-macrophage colony-stimulating factor by adenoviral-mediated gene transfer augments antitumor immunity in patients with metastatic melanoma. *Journal of Clinical Oncology*, **21**: 3343–50.
Yee, C. and P. Greenberg. (2002). Modulating T-cell immunity to tumors: new strategies for monitoring T-cell responses. *Nature Reviews: Cancer*, **2**: 409–19.
Yee, C., Thompson, J.A., Byrd, D. et al. (2002). Adoptive T cell therapy using antigen-specific CD8+ T cell clones for the treatment of patients with metastatic melanoma: *in vivo* persistence, migration, and antitumor effect of transferred T cells. *Proceedings of the National Academy of Sciences, USA*, **99**: 16168–73.

Cytokine therapy of cancer

Kirkwood, J. (2002). Cancer immunotherapy: the interferon-alpha experience. *Seminar in Oncology*, **29**: 18–26.

14

Angiogenesis

David T. Shima and Christiana Ruhrberg

KEY POINTS

- New blood vessel growth (angiogenesis) plays a central role during the onset and progression of cancer, and the inhibition of vessel growth can therefore be exploited to control tumor growth and metastasis.
- The model of the angiogenic switch postulates that the transition from a normally quiescent blood vessel to a proliferative and invasive endothelial cell sprout during tumor growth is due to an imbalance between stimulators and inhibitors of angiogenesis.
- The most potent known stimulators of tumor angiogenesis are the secreted growth factors VEGF and FGF; they are likely to cooperate with other signaling molecules such as angiopoietins, members of the EPH/EPHRIN families, and NOTCH ligands to control the patterning of vascular networks and vessel structure.
- Endogenous inhibitors of angiogenesis act either locally to modify vessel patterning or systemically to suppress new vessel growth.
- Endostatin and angiostatin are two examples of endogenous inhibitors that are secreted by the primary tumor to suppress tumor metastasis.
- Several endogenous and engineered angiogenesis inhibitors have entered clinical trials for cancer therapy; amongst these, an antibody blocking VEGF function has already proven beneficial for the treatment of colon cancer.

INTRODUCTION

New blood vessel growth plays a central role during the onset and progression of cancer. Accordingly, inhibiting vessel growth has become an important concept in the design of more efficient cancer therapies. In this chapter, we review our current understanding of the molecular mechanisms that govern vascular growth and differentiation and highlight their significance for tumor expansion. As it is becoming increasingly evident that pathological blood vessel growth reactivates numerous signaling pathways that control vascularization of the embryo, we will relate our current knowledge of vascular development to specific observations

derived from studies of tumor vascularization. In addition, we will consider the notion that pathological vascularization may depend on additional mechanisms that do not operate in the embryo.

The vasculature is a flexible conduit for the delivery and exchange of nutrients, wastes, hormones, and immune cells. During embryogenesis, blood vessels and the heart are the first recognizable organ system to develop. In the adult, the continued maturation and maintenance of the blood vessel network is critical for tissue metabolism and homeostasis as well as for repair processes such as inflammation and wound healing. Excessive vascular growth plays a central role during the onset and progression of a number of adult diseases such as diabetic retinopathy and cancer.

Blood vessels are comprised of two main cellular components, endothelial and mural cells (Fig. 14.1). The endothelium is a continuous, cylindrical epithelial sheet that creates the vessel lumen. The endothelial cells directly interface with blood and are responsible for maintaining a nonthrombogenic surface. A basal lamina separates the endothelium from the second component of vessels, the contractile mural cells. These cells surround the endothelial tube and are primarily responsible for the modulation of vascular tone. In large caliber vessels such as arteries, veins, arterioles, and venules the mural cells are called smooth muscles and are organized into multiple cell layers. At the level of the microvasculature, including capillaries and post-capillary venules, a single discontinuous layer of cells, called pericytes, surrounds the endothelium. Homotypic and heterotypic communication between the endothelium and mural cells contributes to the control of vascular physiology and growth.

GENERAL PRINCIPLES OF NEW VESSEL GROWTH

In the adult, the vasculature normally represents an extreme example of a stable, quiescent population of cells, with an estimated turnover time

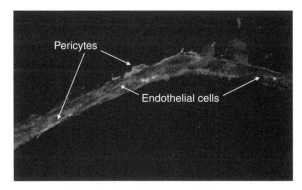

Fig. 14.1 Blood vessels are comprised of endothelial and mural cells. Double label immunofluoresence reveals that capillaries are formed by an inner layer of PECAM-positive endothelial cells and an outer layer of smooth muscle actin-positive pericytes.

of an adult capillary endothelial cell of approximately 1000 days. In contrast, the average turnover for a highly regenerative tissue, such as the gut epithelium, is 2–3 days. The growth of new blood vessels occurs in the adult normally only under the influence of specific physiological stimuli, such as hormonal fluctuations during the female reproductive cycle or chronic increases in tissue metabolism, to stimulate new vessel growth.

In contrast to adult life, blood vessel growth occurs rapidly and in almost all tissues of the developing embryo. During embryogenesis, two well-recognized mechanisms of blood vessel growth operate, which are referred to as vasculogenesis and angiogenesis (Box 14.1; Fig. 14.2). Vasculogenesis entails the *de novo* formation of blood vessels from endothelial cell precursors, or angioblasts, which are derived from the primitive mesenchyme. In the embryo, vasculogenesis is the mode of development for the major axial vessels such as the aorta and for the vascular plexus of some endoderm-derived organs

---

Box 14.1 The main mechanisms of blood vessel growth

Vasculogenesis: Formation of new blood vessels from mesenchymally-derived endothelial cell precursors in the embryo.

Angiogenesis: Formation of new blood vessels from preexisting vessels by sprouting growth.

Incorporation of circulating endothelial cell progenitors into preexisting vessels: The bone marrow of adults contains progenitor cells, which have the capacity to differentiate into mature endothelial cells and have therefore been termed endothelial progenitor cells (EPCs). In response to growth factor signals, EPCs have been found to exit the bone marrow, travel within the vascular system to sites of neovascularization or tissue damage, and incorporate there into nascent vessels.

---

such as the spleen. Angiogenesis involves the formation of new vessels by sprouting growth from preexisting vessels (Fig. 14.3). This is the main mechanism not only for vascularization of neural and mesodermal tissues in the embryo, for example the brain and limb bud, but also the main mechanism for the neovascularization of adult tissues. Extensive remodeling of blood

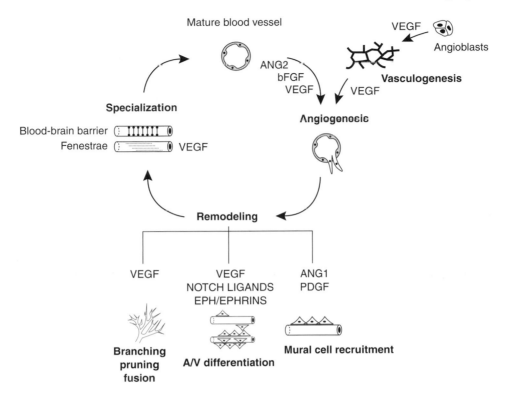

Fig. 14.2 The classes of secreted or cell-surface-associated molecules that cooperate to stimulate vascular growth and differentiation. The life cycle of endothelium may begin with the assembly of angioblasts into a primitive vascular plexus (vasculogenesis), or with the reactivation of quiescent vessel endothelium. Through the process of sprouting growth (angiogenesis), the vessel network begins to expand. New vessels acquire mural cells and adopt an arterial or venous identity; concomitantly, the vascular network continues to branch and remodel into a hierarchical vascular tree. Lastly, vessels specialize according to local physiological needs, for example to form the blood-brain barrier or the fenestrated sieve plates found in the kidney glomerulus. With the notable exception of blood-brain barrier function, most of the processes illustrated are stimulated by VEGF; several other factors act upstream of VEGF or cooperate with VEGF to affect specific aspects of vessel growth and specialization.

vessels occurs subsequent to or in parallel with network formation by angiogenic sprouting, to modify vessel caliber by fusion or change network properties through anastamosis or selective regression of vessel segments. In addition, remodeling involves the recruitment of mural cells, the acquisition of venous or arterial properties, and an ill-understood increase in vessel stability and resilience that may be linked to changes in basement membrane composition and reciprocal signaling between mural and endothelial cells. Taken together, these mechanisms of vessel maturation appear to make the adult vasculature more refractory to angiogenic activation than is the case for embryonic vessels.

In this context, it is interesting to note that some angiogenesis inhibitors regulate vascular growth in the adult, but are not essential for the development of the embryonic vasculature (see below). Finally, endothelial cells within the remodeling blood vessels undergo tissue-specific specializations, for example the acquisition of specialized cell-to-cell adhesions, termed tight-junctions, or transcellular channels, termed fenestrae (Fig. 14.2).

A third mode of vascular growth has recently been described, which involves the incorporation of mobile endothelial progenitors into the existing adult vasculature (Box 14.1). This mode of vascular growth may bear some mechanistic

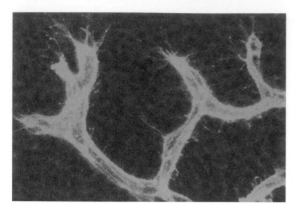

Fig. 14.3 Angiogenic sprouting of blood vessels in the developing brain. Isolectin B4-positive blood vessels extend filopodia-studded sprouts to initiate new branch formation during brain vascularization in the mid-gestation mouse embryo.

parallels to embryonic vasculogenesis, but its general significance is not clear as yet. It is thought that mobile endothelial progenitors derived from the bone marrow can be recruited from the circulation into existing vasculature during tumor angiogenesis. Alternatively, bone-marrow-derived cells may contribute to tumor angiogenesis by adding to the mural complement of the vasculature.

PATHOLOGICAL NEOVASCULARIZATION: TUMOR VESSELS

In stark contrast to physiological neovascular responses, which consist of well-governed bursts of vessel growth limited in both time and space, blood vessel growth that accompanies diseases such as diabetes, psoriasis, and cancer is often characterized by chaotic growth, patterning, and dysfunction. During tumorigenesis, endothelial cell proliferation rapidly increases, with turnover time decreasing from approximately 1000 days in quiescent vessel beds to 50–60 hours. Nascent tumor vessels are also morphologically distinct, as they can be tortuous and of highly variable shape and size; moreover, they often lack a normal complement and juxtaposition of mural cells. In addition to assuming a chaotic structure, the vessels in

tumors form disorganized networks bereft of the intricate pattern observed in normal tissues, and there is rarely a clear organization into arterial and venous compartments. Enhanced vascular permeability often accompanies tumor angiogenesis and contributes to high interstitial pressures in the tumor microenvironment and chaotic blood flow. It has been suggested that both the tortuous nature of vessels and the high interstitial pressure inside the tumor impair delivery of conventional chemotherapeutics into the tumor tissue.

Several different factors likely contribute to the formation of the chaotic tumor vasculature. First, host tissues influence the patterning of growing vessels by providing local cues, for example during development, to control branching morphogenesis, the copatterning of vessels and nerves, and the positioning and connectivity of the arterio-venous compartments. Such a well-orchestrated elaboration of vessel pattern is most likely not recapitulated during neoplasia, because a disorganized, heterogeneous, and chaotically expanding tumor mass could only provide similarly disorganized angiogenic cues. Second, the genetic alterations associated with tumor cell transformation and malignancy may trigger a gross imbalance in the levels of tumor-derived stimulators and inhibitors of angiogenesis (see next section). Finally, an increase in activated immune cells and local tumor-induced changes in the stroma influence tumor vessel morphogenesis and permeability, in part simply by adding even more players to the local cytokine and growth factor "soup."

BASIC CONCEPTS IN TUMOR ANGIOGENESIS: THE ANGIOGENIC SWITCH

Pathological blood vessel growth is found in tissues with elevated metabolic demands or reduced blood flow as a means to increase the capacity of the microvasculature. For example, the ischemic retina elaborates angiogenic factors to stimulate neovascular growth in diabetic retinopathy. In the case of cancer, an increased metabolic load and reduced oxygen provision to the tumor core are believed to be the primary trigger for angiogenesis. The original concept

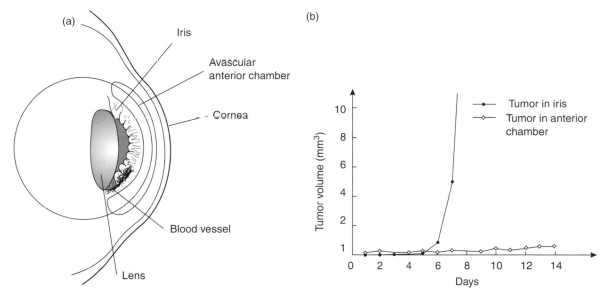

Fig. 14.4 An experimental proof of principle – the importance of tumor angiogenesis for malignant growth. In 1972, Judah Folkman's group performed seminal experiments to prove the importance of neoangiogenesis for malignant tumor growth in the eye (Gimbrone *et al.*, 1972): When small tumor fragments were implanted under the cornea, malignant tumors formed rapidly if the transplantation site was located adjacent to highly vascularized iris tissue. In contrast, when such tumor fragments were placed under the cornea, but within the avascular anterior chamber, tumor cells continued to proliferate for a little while, but these tumors switched to a dormant state once they reached 1 mm in diameter. The modification of this original transplantation approach led to the development of the first widely used model system for the identification of tumor-derived angiogenic factors, the rabbit cornea pocket assay.

defining the importance of tumor angiogenesis postulates that the expansion of solid tumors beyond 1–2 mm^3 depends on the growth of new capillaries (Fig. 14.4). Importantly, the intimate relationship between tumor cells and blood vessels does not just promote the growth of the primary tumor, but can also contribute to metastatic spread. Moreover, neoangiogenesis also appears to play a role in the establishment of "liquid tumors", that is, hematological malignancies; for example, leukemia is associated with neoangiogenesis in the bone marrow, and leukemic cells can be seen clustered on the new vessels.

The predominant model explaining the transition from a quiescent capillary to a proliferative and invasive endothelial cell sprout during tumorigenesis is based upon the idea that the maintenance of vascular quiescence is due to a precise balance between stimulators and inhibitors of angiogenesis. According to this model, which was first proposed by Folkman and colleagues, a rise in the concentration of stimulators initiates an angiogenic response, whereas the return to quiescence is promoted by a decrease of stimulators. Vice versa, a reduction in the usual level of inhibitors would make endothelial cells more susceptible to stimulatory cues. Such changes in the levels of stimulators and inhibitors can be induced by a combination of genetic alterations within the tumor cells and physiological demands of the growing tumor mass, which together trigger the "angiogenic switch" of a slowly growing tumor. This angiogenic switch requires that the tumor either produces its own stimulators of angiogenesis or induces its surrounding cells to secrete such factors. In parallel, tumor cells growing within an avascular environment, for example the epidermis tissue, need to induce the breakdown of the basement membrane to allow invasion of blood vessels. In some instances, for example during metastasis, the switch to an angiogenic phenotype requires that tumor cells exit their host vessel and induce new vessel sprouts from neighboring vessels (cooption).

Importantly, the term "switch" does not necessarily imply that a specific gene mutation is required – the "switch" may be thrown early as an inevitable accompaniment of the growth-promoting activity of various oncogenes, whose main role may be in driving cellular proliferation.

The idea that tumor growth and spread are intimately linked to an angiogenic switch, which therefore might be targeted to control cancer, catalyzed an intensive hunt for factors that are elaborated by tumor cells to stimulate angiogenesis. In the wake of this quest, the past three decades of research have shed much light on the complex molecular interplay between the endothelium and its environment both during tumor growth and embryonic development. This effort has given rise to groundbreaking new treatments for cancer that, in turn, validate the antiangiogenesis hypothesis first formalized by Folkman.

VASCULAR GROWTH AND DIFFERENTIATION FACTORS: STIMULATORS OF THE ANGIOGENIC SWITCH

Numerous stimulators of capillary endothelial cell growth, termed angiogenic factors, have been identified since the search began 30 years ago. Traditionally, assignment of angiogenic activity to biomolecules was dependent on their ability to increase endothelial cell proliferation and/or to stimulate new capillary growth using *in vivo* assays, such as the chick chorioallantoic membrane (CAM) or the rodent cornea pocket (Box 14.2). Using these assays, various laboratories have described the purification of angiogenic factors not only from capillary-rich sources, such as the brain and retina, but also from various untransformed and tumor-derived cell lines. With the advent of reverse mouse genetics, we are now able to test the role of these candidates in vascular development and determine their functional requirements for the elaboration of a functional vascular tree. In addition, the careful analysis of mouse knockouts has revealed unforeseen roles for several matrix and signaling molecules in

---

Box 14.2 Angiogenesis assays

The CAM assay

The chick chorioallantoic membrane (CAM) assay is perhaps the most widely used assay for screening purposes, because it is a relatively simple and inexpensive method to identify angiogenic factors and is suitable for large-scale screening. Grafts of tissue and polymers or sponges soaked in putative angiogenic factors are placed on the CAM of developing chick embryos through a window in the eggshell. If an angiogenic factor is released, an increase of vessels is observed around the graft within 4 days after implantation. Blood vessels entering the graft can be visualized and counted under a stereomicroscope. Because young chick embryos lack a mature immune system, this assay is particularly useful to study angiogenesis elicited by tumor tissues. The major disadvantage of this assay is that the CAM contains a well-developed vascular network at the time of implantation, which can make it difficult to discriminate between new capillaries and existing ones.

The rabbit cornea pocket assay

The rabbit cornea is normally avascular. However, if sponges, polymers, or tissues containing angiogenic substances are implanted into pockets in the corneal stroma, vessels can grow out of the limbal region into the cornea. In albino rabbits, the vascular response can be readily quantified by stereomicroscopy and image analysis after perfusion of the cornea with Indian ink. This method is very reliable, but technically more demanding and more expensive than the CAM assay, and it is therefore not an ideal screening assay. Moreover, where suitable, the CAM assay may be preferred for ethical reasons.

---

vascular growth and differentiation. As a result, it has become clear that no signaling pathway alone controls both vascular growth and organization. Rather, a number of different pathways cooperate to build, branch, and mature the growing vessel network.

Below, we review findings on several key proteins implicated in the creation of new blood vessels during embryonic development and highlight their relevance to tumor biology. The factors that act as negative regulators of vascular development and their relevance to tumor angiogenesis are less well understood. We will therefore discuss this topic only briefly. However, we will draw attention to the concept of endogenous angiogenesis inhibitors in the adult, because these factors, despite being non-essential for embryogenesis, can be specifically exploited in the clinic to block tumor angiogenesis.

Vascular endothelial growth factor and its receptors

The most potent and versatile angiogenic factor described to date is vascular endothelial growth factor (VEGF) (Fig. 14.2). VEGF is a secreted, homodimeric glycoprotein with endothelial cell-specific mitogenic activity and the ability to stimulate angiogenesis *in vivo*. VEGF also stimulates vascular permeability, with an effect 50,000 times greater than that of the vasoactive substance histamine. In fact, the observation that tumor growth is associated with increased microvascular permeability provided the basis for VEGF's original discovery and, accordingly, it was first named vascular permeability factor (VPF). In addition to the founding member, the VEGF gene family now includes at least five new members that have been implicated in various aspects of cardiovascular growth and function.

VEGF and its tyrosine kinase linked receptors, KDR/FLK1 and FLT1, are expressed in regions of blood vessel growth during embryogenesis. KDR/FLK1 is expressed in regions of the early mesoderm that are presumed to give rise to angioblasts, and it is the earliest known molecular marker for the endothelial cell lineage. During later stages of embryogenesis, FLT1 and KDR/FLK1 mRNA are both restricted to hematopoietic lineages and the endothelium of vascular cords and blood islands, with VEGF mRNA expressed in adjacent embryonic tissues. The mRNA levels for VEGF and its receptors decrease significantly postnatally, but expression is upregulated in the endothelium of tissues with ongoing angiogenesis, such as tumors, and is also elevated at sites proximal to fenestrated endothelium. The observation that VEGF expression is regulated by hypoxia and glucose explains why VEGF levels are elevated in the ischemic regions of tumors and the retina and provides a molecular link between alterations in local tissue metabolism and growth factor control of angiogenesis.

Experimental data provide strong evidence for a necessary role of VEGF in developmental, adult physiological, and pathological blood vessel growth. In the mouse embryo, targeted disruption of VEGF or its receptors leads to early embryonic lethality due to severe defects in vasculogenesis. The postnatal blockade of VEGF function inhibits physiological neovascularization during bone growth and various aspects of the reproductive cycle. Consistent with a necessary role for VEGF in ischemia, neutralization of VEGF protein reduces the normal course of angiogenesis and disease severity in several experimental model systems.

VEGF's role as an angiogenic factor during diabetes, cancer, and chronic inflammation has made it a key target in the inhibition of angiogenesis and angiogenesis-dependent pathologies. In contrast, the major clinical problem of rescuing the viability of tissues starved of a proper blood supply due to obstruction or circulatory disorders demands clinical methodology for stimulating controlled blood vessel growth, for example, to alleviate coronary heart or diabetic disease in humans. Takeshita, Isner, and colleagues first demonstrated the ability of VEGF to stimulate collateralization in experimental models of limb ischemia, and now major research efforts are focused on delivering VEGF protein, cDNA, or VEGF-expressing virus to human patients suffering from peripheral and coronary ischemia.

A key feature of VEGF, which is becoming increasingly relevant to its use in human gene therapy, is that it is not a single protein but refers to a group of at least seven isoforms in humans; the predominant forms are secreted proteins of 121, 165, and 189 amino acids (VEGF121, VEGF165, and VEGF189; Fig. 14.5). These isoforms are generated by alternative mRNA splicing, which determines the presence or absence of heparin binding domains; the splicing pattern seems to be subject to complex temporal and spatial regulation during development. The biological significance of the VEGF isoforms is still under investigation. Two general ideas, by no means mutually exclusive, can be put forward to explain the function of the VEGF isoforms. The first model suggests that each isoform elicits a different signal within

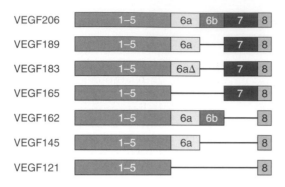

Fig. 14.5 Schematic illustration of the human VEGF-A isoforms. The VEGF isoforms are generated by alternative splicing from a single copy gene, termed *VEGF-A*. The isoforms differ by the absence or presence of domains that confer binding to heparin sulfate proteoglycans and neuropilin receptors and are encoded by exons 6 (6a or 6b) and 7. The VEGF183 isoform contains a deletion within exon 6a. The VEGF121 isoform does not contain any exon 6 or exon 7 domain. The murine orthologs of VEGF189, VEGF165, VEGF145 and VEGF121 have been described; they are referred to as VEGF188, VEGF164, VEGF144 and VEGF120, because they are shorter by one amino acid.

the responding endothelial cell, perhaps through isoform-specific cell surface receptors. The interaction of VEGF165, but not VEGF121, with a cell surface receptor termed neuropilin (NP-1) supports this model. The presence of NP-1 amplifies KDR/*FLK1*-mediated chemotaxis and mitogenesis in cultured endothelial cells, either by directly increasing the affinity of VEGF165 for KDR/*FLK1* or by enhancing receptor clustering. The second model suggests that differing affinities of the isoforms for the heparin-rich matrix and cell surface results in their distinct localization within the extracellular milieu, thus providing a mechanism to regulate VEGF availability to target cells.

While precise mechanistic details remain to be resolved, recent *in vivo* findings underscore the need to discern the relevance of VEGF isoform function to therapies aimed at pro- and antiangiogenesis in humans. Mice with targeted mutations that restrict VEGF expression to solely the 120 amino acid isoform, which lacks both NP-1 and heparin binding capacity, have recently been described. VEGF120 mice are born with a grossly normal vascular tree, but suffer from severe defects in postnatal blood vessel

growth and symptoms of ischemic cardiomyopathy. These observations suggest that VEGF isoforms exhibit a functional specificity that might be exploited in the design of proangiogenic therapies for ischemic cardiac and peripheral tissue disease.

The expression of VEGF isoforms has been evaluated in various types of human cancer to determine their significance as prognostic factors for tumor growth and metastasis and to correlate isoform expression profiles with the properties of the tumor vasculature. However, a unifying theory of VEGF isoform function during tumor vascularization has not yet emerged. In contrast, in mouse models of tumor angiogenesis differential VEGF isoform expression correlates with vascular network properties, as found in the embryo. In both situations, VEGF120 or VEGF188 alone do not support the formation of a robust vascular network. In contrast, VEGF164 alone is sufficient for the formation of a functional vasculature, presumably because of its ability to bind heparin while remaining partially mobile. However, it is the combination of soluble and heparin-binding VEGF isoforms that maximizes tumor growth, presumably because full vessel network functionality is achieved only when all isoforms cooperate.

In addition to their diverse properties with respect to vascular patterning, the VEGF isoforms also differ in their potency to promote inflammation, as VEGF164 is more proinflammatory than the other isoforms. The heightened inflammatory properties of VEGF164 may contribute to the onset and progression of some forms of pathological neovascularization. However, the relevance of this observation to tumor angiogenesis has not yet been fully addressed.

Fibroblast growth factors

The fibroblast growth factor (FGF) family of polypeptides was the first class of angiogenic factors described. Acidic FGF (aFGF or FGF1) and basic FGF (bFGF or FGF2) are potent mitogens and trophic factors for cultured endothelial cells, and both have potent angiogenic properties in the CAM and cornea assays (Box 14.2). The presence of FGFs, particularly bFGF, in tissues associated with new capillary growth made it an early favorite for a general angiogenic factor *in vivo* (Fig. 14.2).

Molecular cloning and biochemical analysis has led to the unexpected discovery that both aFGF and bFGF lack a signal sequence for secretion via exocytosis, therefore suggesting they normally reside inside the cell. Yet, in support of a role for these FGFs in extracellular signaling, a family of high-affinity tyrosine kinase cell surface receptors for FGFs has been described. This apparent contradiction might in part be explained by the finding that bFGF is released from cultured cells and animal tissues *in vivo* upon mechanical injury. These data have led investigators to suggest that the FGFs may exert angiogenic activities in situations associated with cellular damage, thus acting as a "wound hormone." In situations where tissue injury does not play a role, the nonsecreted FGFs may act as intracrine signaling factors that translocate to the nucleus to exert their activity. Consistent with this idea, both aFGF and bFGF possess a nuclear localization sequence and aFGF has been found in the nucleus of cultured vascular endothelial and smooth muscle cells.

Despite the fact that aFGF and bFGF are potent stimulators of angiogenesis in *in vitro* models, the targeted disruption of both aFGF and bFGF does not appear to impair vascular development and results only in mild defects in hematopoesis and wound healing. It therefore appears that, in contrast to VEGF, neither one of these FGFs is required for vascular growth. However, these FGFs may prime adult vessels to mount a VEGF response when they are present at unusually high levels in certain pathological situations. In support of the latter idea, FGFs can induce VEGF expression in the cornea assay and in cultured vascular smooth muscle cells and VEGF and FGFs synergize to promote neoangiogenesis *in vitro* and *in vivo*.

Over the past decade, the FGF family has rapidly expanded in size and now includes 20 members, many of which are predicted to be secreted proteins. FGF3 and FGF4 can induce angiogenesis in the CAM assay and FGF7 induces neovascularization in the cornea assay (Box 14.3). FGF5 and FGF7 have also been implicated in angiogenesis. However, we will have to await the analysis of the relevant knockout mouse models to assess whether any of the more novel FGFs play an essential role in vascular growth and morphogenesis.

Angiopoietins and the TIE2 receptor tyrosine kinase

The angiopoietins are ligands for the endothelial-specific receptor tyrosine kinase TIE2 and are thought to act in a complementary and coordinated fashion with VEGF (Fig. 14.2). Four angiopoietin ligands termed ANG1 to ANG4 are known, and together they provide the first example of a growth factor family in vertebrates that consists of both receptor activators and receptor blockers: while ANG1 and ANG4 act as TIE2 agonists, ANG2 and ANG3 behave as antagonists.

Angiopoietins are likely to play a later role than VEGF in vascular development, since both ANG1 and TIE2 deficient mice form a normal primitive vascular plexus and therefore develop further than VEGF and KDR/FLK1 mutants. However, failure to stabilize and remodel the primary vascular plexus subsequently causes embryonic lethality of ANG1 deficient mice around the time of mid-gestation. Ultrastructural studies suggest that these problems are due to a disruption of ANG1-mediated interactions between the endothelium and its supporting cells, that is smooth muscle cells. In support of the idea that ANG1 signaling is important for vessel remodeling, mutation of its receptor TIE2 is responsible for a heritable human disease characterized by thin-walled blood vessels with markedly reduced smooth muscle layers. Overexpression of ANG1 in the embryo results in hypervascularization, presumably by decreasing the normal amount of vessel regression that accompanies development. In contrast, overexpression of ANG2 results in embryonic lethality with phenotypes similar to those of the ANG1 and TIE2 knockouts, presumably because ANG2 blocks ANG1 activity by competing for TIE2.

In support of the notion that angiopoietins also control the stability of mature vessels, *ANG1* is widely expressed in adult tissues, while expression of *ANG2* is present at sites of active physiological angiogenesis, such as the female reproductive tract and the placenta. Detailed analysis of *ANG2* expression in the ovary has revealed that vessel growth and sprouting occurs at sites where *ANG2* is coexpressed with VEGF. In contrast, expression of *ANG2* in the absence of VEGF results in vessel regression. Taken together, these observations have led to the following model: ANG1 expression in the mesenchyme normally activates endothelial TIE2 to promote mural cell recruitment. Since ANG2 can compete for binding to TIE2, ANG1-dependent TIE2 activation is blocked in the presence of ANG2, and this leads to blood vessel destabilization. The exposed endothelium of destabilized vessels either degenerates, causing vessel regression, or, in the presence of VEGF, proliferates to yield net vessel growth. In support of the model of synergistic VEGF and angiopoietin signaling, their balance controls vascular permeability and tumor vascularization. It was shown that the initial vascularization of tumors is accompanied by high levels of ANG2, but these early vessels regress, causing necrosis in the tumor center. VEGF upregulation at the ischemic tumor margin then results in coexpression with ANG2 and a second wave of more stable vascularization is induced. Thus, sites of ANG2 expression correlate with vessel plasticity, and the outcome of vessel growth versus regression is decided by the presence of VEGF (Fig. 14.2).

Similar to ANG1, platelet-derived growth factor (PDGF-B) has been implicated in mural cell recruitment to blood vessels, but it remains to be elucidated how PDGF-B and angiopoietin signaling synergize to ensure mural cell investment of growing vessels (Fig. 14.2).

EPH/EPHRIN signaling

The transmembrane EPH receptors and their membrane-bound ephrin ligands comprise a signaling pathway with well-characterized functions during the development of several different organ systems, including the vasculature. EPHs and EPHRINs are often reciprocally expressed at tissue compartment boundaries and are best known for their roles in neural crest cell migration and axon guidance; however, it was attempts to better understand their role in neural patterning by Wang, Anderson and colleagues that revealed their role in vascular development. *EphrinB2* is widely expressed in the embryo, but within the vasculature it is restricted to arterial endothelium. Significantly, expression of its receptor *EphB4* predominates in the endothelium of veins. These findings provided the first molecular distinction between arteries and

veins. Targeted inactivation of *EphB4* or *ephrinB2* leads to failure in the remodeling of the primitive vascular plexus and subsequent lethality at mid-gestation.

Other EPH/EPHRIN family members are expressed in the vasculature, but expression is not always complementary between arteries and veins (e.g. ephrin A1 and B1 are expressed in both, and EphB3 is present in veins and the aortic arches). Instead, ligand/receptor expression is often reciprocal between blood vessels and their surrounding tissues, suggesting that paracrine signaling may occur between the endothelium and the mesenchyme. Based on the findings to date, it seems reasonable to suggest that members of the EPH/EPHRIN family are acting in the vasculature much in the same way as they do in the nervous system, where the complementary ligand/receptor expression patterns provide guidance cues by defining spatial boundaries in the developing embryo.

A number of different EPHRINs and EPH receptors are overexpressed in a wide variety of human cancers. For example, EPHRIN A1 and its receptor EPHA2 have been found on tumor vessels and tumor cells and may therefore contribute to both tumor-induced angiogenesis and tumor growth/spread, possibly by stimulating an autocrine loop. EPH/EPHRIN expression by tumor cells might also attract or organize their vascular supply, in analogy to the role of this signaling system during embryogenesis. In support of this idea, it has recently been shown that xenografted EPHB4-expressing tumor cells can attract EPHRIN B2-positive blood vessels to increase their blood supply. However, it is presently not known whether disregulated EPH/EPHRIN expression normally contributes to the vascularization of tumors or the chaotic organization of their vessel networks; further research will be necessary before we can conclude whether EPH/EPHRIN signaling might provide a valid target for antiangiogenic therapy in cancer.

NOTCH signaling

Like the EPH/EPHRIN system, the NOTCH signaling pathway has been studied extensively in embryos, where it regulates the specification of cell fate through local cell interactions. The NOTCH proteins are transmembrane receptors that are activated by membrane-spanning ligands of the DELTA, SERRATE and JAGGED families. Both NOTCH receptors and ligands are expressed in specific vascular compartments, that is, arterial or venous endothelium or vascular smooth muscle cells/pericytes.

During vascular patterning, VEGF may act upstream of NOTCH to promote arterialization, but it also promotes other aspects of angiogenesis: loss of *Notch4* and/or *Notch1* impairs vascular morphogenesis in the embryo, and loss of function for *Notch3* or *Jagged 1* causes hereditary vascular diseases in humans. The contribution of NOTCH signaling to tumor angiogenesis is not yet clear. However, recent genetic studies in the mouse and human provide compelling evidence for a role of NOTCH signaling in the adult vasculature, raising the possibility that this pathway may present a target for tumor angiogenesis. For example, DELTA-4 is an endothelial-specific NOTCH ligand that is upregulated in tumor vasculature (see Chapter 5 for a more detailed discussion of NOTCH).

ROLE OF INHIBITORS IN ANGIOGENESIS

The paradigm of the "angiogenic switch" postulates that tumor angiogenesis may be stopped in its tracks if angiogenesis inhibitors are administered at levels that exceed those of angiogenesis promoters. Accordingly, the identification of angiogenesis inhibitors has become a subject of much research activity.

Embryonic vessel branching and physiological neovascularization both demand the existence of angiogenesis inhibitors that act at a short range to limit and refine vascular morphogenesis. Such factors would act in synergy with local angiogenic stimuli to control the balance between endothelial cell proliferation, apoptosis, and migration within growing tissues. The need for such factors is demonstrated in an analogs example of branching growth, that of lung branching morphogenesis, where FGFs promote budding from the epithelium, while BMPs restrict budding to certain regions of the epithelium. The recent description of model systems to study vascular branching patterns should allow us to validate several matrix and signaling molecules in candidate antiangiogenic factors with local patterning capacity. Such candidate genes include the mammalian sprouty proteins, as the overexpression of some family members inhibits endothelial cell proliferation *in vitro* and leads to defective vessel network formation in the mouse embryo.

In addition to short-range inhibitors of angiogenic growth, several naturally occurring compounds operate systemically in adults and have been termed endogenous inhibitors of angiogenesis. Some of these endogenous inhibitors have been discovered, because they are produced by primary tumors and inhibit the growth of secondary tumors, that is, metastases. Because these inhibitors of angiogenesis are likely to be more stable than most angiogenic proteins, they are thought to remain in the circulation at relatively higher levels. Only a drop in endogenous angiogenesis inhibitors following removal of the primary tumor would increase relative levels of angiogenesis stimulators and thereby activate neoangiogenesis in previously dormant metastases. Accordingly, the systemic administration of such inhibitors should prevent metastatic growth after surgical removal of the primary tumor. Two such tumor-derived factors that have entered clinical trials for antiangiogenic therapy of cancer are angiostatin, an internal fragment of plasminogen, and endostatin, an internal fragment of collagen XVIII. Interestingly, mice lacking plasminogen or collagen XVIII develop a normal embryonic circulation, suggesting that postnatal angiogenesis is in part controlled by mechanisms that do not operate in the embryo. Consistent with this idea, mice lacking the endogenous angiogenesis inhibitor thrombospondin 1 also develop normally; however, they are less sensitive to oxygen-induced vessel obliteration in the eye, and in the presence of activated *Neu/ErbB2* oncogene, they are more susceptible to develop highly vascularized breast tumors.

The specific role of thrombospondin 1 in controlling postnatal, but not developmental, vascularization may be linked to its involvement in the maintenance of a quiescent and differentiated endothelial phenotype. At the cellular level, thrombospondin 1 inhibits endothelial cell migration and growth factor mobilization and promotes endothelial cell apoptosis. In contrast, VEGF is able to stimulate proliferation and is antiapoptotic, and it has therefore been suggested that the relative balance of VEGF and thrombospondin 1 determines the outcome of oxygen-induced neovascularization in the eye. In an analogs mechanism, tumor angiogenesis may result when increased VEGF expression is accompanied by or is due to the derepression of thrombospondin 1 following the oncogenic activation of genes such as *RAS* and *MYC*. Like angiostatin and endostatin, thrombospondin 1 has now entered the first clinical trials to prevent tumor angiogenesis.

CONCLUSIONS AND FUTURE DIRECTIONS

The growing list of experimentally validated stimulators and inhibitors of angiogenesis suggests that there are many opportunities for the development of new therapeutics to complement traditional cancer therapies. Accordingly, the number of antiangiogenic compounds in clinical trials has grown steadily in the past 15 years (http://www.cancer.gov/clinicaltrials; http://www.cancer.gov/clinicaltrials/developments/anti-angio-table). These promising therapeutics fall into several different classes, including endogenous inhibitors of angiogenesis and engineered drugs such as those that diminish VEGF signaling; others that directly target and debilitate rapidly growing endothelium; or a third class that prevents extracellular matrix degradation and remodeling. The clinical success of an inhibitory antibody directed to VEGF in patients with colorectal cancer provides the first unequivocal evidence for the hypothesis that attacking their blood supply will inhibit the growth of human tumors. The exciting outcome of this antiangiogenesis approach has also encouraged clinical efforts aimed at inhibiting blood vessel growth in organ-specific pathologies such as the treatment of age-related macular degeneration. Finally, these promising clinical experiences have, in turn, identified new areas for future angiogenesis research in the basic science laboratory. For example, it will be important to elucidate the molecular mechanisms that mediate tumor-specific responses to antiangiogenics and look for novel and more specific targets present only on the tumor vasculature of adults. A better molecular understanding of tumor angiogenesis will in part be driven by the exchange of novel insight amongst scientists working on cancer, embryonic vascular development, eye disease, and other related disciplines.

QUESTIONS

1   What phrase describes a shift in the balance of angiogenesis stimulators and inhibitors, leading to enhanced vascular and tumor growth?
    a. Neovascularization.
    b. Angiogenic switch.
    c. Arteriolization.
    d. Neoplasia.
2   VEGF refers to a collection of protein isoforms that differ in their ability to bind to?
    a. Angiopoietins.
    b. Glucose.
    c. Heparin.
    d. Veins.
3   Which of the following would represent the best inhibitor of angiogenesis?
    a. VEGF.
    b. Mural cells.
    c. Thrombospondin 1.
    d. EPHRINs.
    e. Endothelial progenitors.

BIBLIOGRAPHY

Mechanisms of angiogenesis

Engerman, R.L., Pfaffenbach, D., and Davis, M.D. (1967). Cell turnover of capillaries. *Laboratory Investigation*, **17**: 738–43.
Rhodin, J.A. (1967). The ultrastructure of mammalian arterioles and precapillary sphincters. *Journal of Ultrastructure Research*, **18**: 181–223.
Risau, W. (1997). Mechanisms of angiogenesis. *Nature*, **386**: 671–4.
Ruhrberg, C. (2003). Growing and shaping the vascular tree: multiple roles for VEGF. *Bioessays*, **25**: 1052–60.

Molecular control of angiogenesis

Adams, R.H., Wilkinson, G.A., Weiss, C. *et al.* (1999). Roles of ephrinB ligands and EphB receptors in cardiovascular development: demarcation of arterial/venous domains, vascular morphogenesis, and sprouting angiogenesis. *Genes and Development*, **13**: 295–306.

Betsholtz, C. and Raines, E.W. (1997). Platelet-derived growth factor: a key regulator of connective tissue cells in embryogenesis and pathogenesis. *Kidney International*, **51**: 1361–9.

Davis, S. and Yancopoulos, G.D. (1999). The angiopoietins: Yin and Yang in angiogenesis. *Current Topics in Microbiology and Immunology*, **237**: 173–85.

Krebs, L.T., Xue, Y., Norton, C.R. *et al.* (2000). Notch signaling is essential for vascular morphogenesis in mice. *Genes and Development*, **14**: 1343–52.

Powers, C.J., McLeskey, S.W., and Wellstein, A. (2000). Fibroblast growth factors, their receptors and signaling. *Endocrinology Related Cancer*, **7**: 165–97.

Robinson, C.J. and Stringer, S.E. (2001). The splice variants of vascular endothelial growth factor (VEGF) and their receptors. *Journal of Cell Science*, **114**: 853–65.

Takeshita, S., Zheng, L.P., Brogi, E. *et al.* (1994). Therapeutic angiogenesis. A single intraarterial bolus of vascular endothelial growth factor augments revascularization in a rabbit ischemic hind limb model. *Journal of Clinical Investigation*, **93**: 662–70.

Wang, H.U., Chen, Z.F., and Anderson, D.J. (1998). Molecular distinction and angiogenic interaction between embryonic arteries and veins revealed by ephrin-B2 and its receptor Eph-B4s. *Cell*, **93**: 741–53.

Tumor angiogenesis

Folkman, J. (1992). The role of angiogenesis in tumor growth. *Semin Cancer Biol*, **3**: 65–71.

Hanahan, D. and Folkman, J. (1996). Patterns and emerging mechanisms of the angiogenic switch during tumorigenesis. *Cell*, **86**: 353–64.

Holash, J., Maisonpierre, P.C., Compton, D. *et al.* (1999). Vessel cooption, regression, and growth in tumors mediated by angiopoietins and VEGF. *Science*, **284**: 1994–8.

Knighton, D., Ausprunk, D., Tapper, D., and Folkman, J. (1977). Avascular and vascular phases of tumor growth in the chick embryo. *British Journal of Cancer*, **35**: 347–56.

Lyden, D., Hattori, K., Dias, S. *et al.* (2001). Impaired recruitment of bone-marrow-derived endothelial and hematopoietic precursor cells blocks tumor angiogenesis and growth. *Nature Medicine*, **7**: 1194–201.

Senger, D.R., Galli, S.J., Dvorak, A.M., Perruzzi, C.A., Harvey, V.S., and Dvorak, H.F. (1983). Tumor cells secrete a vascular permeability factor that promotes accumulation of ascites fluid. *Science*, **219**: 983–5.

Anti-angiogenesis: concepts and target molecules

Folkman, J. (1971). Tumor angiogenesis: therapeutic implications. *New England Journal of Medicine*, **285**: 1182–6.

Gimbrone, M.A., Jr., Leapman, S.B., Cotran, R.S., and Folkman, J. (1972). Tumor dormancy *in vivo* by prevention of neovascularization. *Journal of Experimental Medicine*, **136**: 261–76.

Kerbel, R. and Folkman, J. (2002). Clinical translation of angiogenesis inhibitors. *Nature Reviews: Cancer*, **2**: 727–39.

Ruhrberg, C. (2001). Endogenous inhibitors of angiogenesis. *Journal of Cell Science*, **114**: 3215–16.

Angiogenesis assays

Ribatti, D., Vacca, A., Roncali, L., and Dammacco, F. (1996). The chick embryo chorioallantoic membrane as a model for in vivo research on angiogenesis. *International Journal of Developmental Biology*, **40**: 1189–97.

Gimbrone, M.A., Cotran, R.S., Leapman, S.B., and Folkman, J. (1974). Tumor growth and neovascularization: an experimental model using the rabbit cornea. *Journal of National Cancer Institute*, **52**: 413–27.

15
Diagnosis of Cancer

Anne Thomas and William Steward

New ideas pass through three periods:
It can't be done.
It probably can be done, but it's not worth doing.
I knew it was a good idea all along!

Arthur C. Clarke

KEY POINTS

- Even with all the advances in the diagnosis of cancer, the initial history taking of the patient remains one of the most vital "investigations."
- Improvements in genetic testing and immunohistocytochemistry ensure that we can make the diagnosis of cancer more reliably and on ever smaller specimens. Moreover some of the tests are indicators of prognosis as well as diagnosis.
- Although there are screening programs available for some tumors, we need to roll this out to the other common tumor types such as non-small-cell lung cancer.
- We are now entering a new era of imaging in the diagnosis of cancer with the availability of PET/CT and even more advanced techniques.

INTRODUCTION

This chapter discusses the presentations of some of the most common cancers. The range of diagnostic tests are reviewed and the advantages and disadvantages of individual techniques are summarized. Future developments and potential new diagnostic techniques are reviewed (Box 15.1).

The diagnosis of cancer can be remarkably difficult to make. This is because patients may have nonspecific symptoms that mirror those found in common benign conditions. For example, a patient with colon cancer may have similar symptoms to another with a benign irritable bowel syndrome. In addition, some patients may

hide their symptoms due to fear of the possible diagnosis of cancer and its likely treatment. Sadly, this means that many patients present late and already have advanced disease at the time of diagnosis.

It is essential to confirm that there is an underlying malignancy that is the cause of a patient's symptoms and this ultimately relies on obtaining a representative sample of tissue in most instances. This will also provide information on the type of tumor so that the appropriate treatment can be chosen. Unfortunately, it is frequently difficult to establish the histological diagnosis to confirm a suspected cancer. Tumors often contain areas that vary markedly

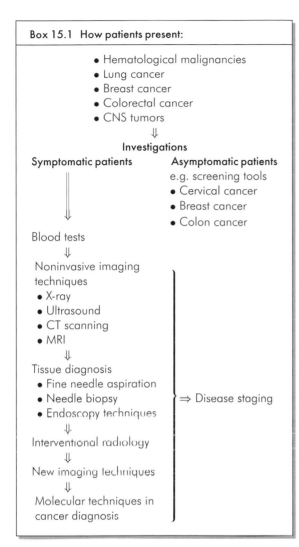

Box 15.1 How patients present:

- Hematological malignancies
- Lung cancer
- Breast cancer
- Colorectal cancer
- CNS tumors

⇓

Investigations

Symptomatic patients

Asymptomatic patients
e.g. screening tools
- Cervical cancer
- Breast cancer
- Colon cancer

Blood tests
⇓
Noninvasive imaging techniques
- X-ray
- Ultrasound
- CT scanning
- MRI
⇓
Tissue diagnosis
- Fine needle aspiration
- Needle biopsy
- Endoscopy techniques   } ⇒ Disease staging
⇓
Interventional radiology
⇓
New imaging techniques
⇓
Molecular techniques in cancer diagnosis

Many of the investigative procedures used in oncology are uncomfortable and inconvenient for the patient. They are often difficult to interpret, with false positives and negatives. It is also important not to submit patients to unnecessary tests, which can cause morbidity (even mortality), huge costs to the healthcare system, and inappropriate distress. There should always be a logical sequence of investigations and for many conditions there are now protocols to follow that should facilitate the ability to make an accurate diagnosis in the shortest time possible.

Fortunately in oncology there has been a steady development of investigative procedures that are more reliable and less invasive. In the future, the hope is that not only can a diagnosis be made quickly with minimal discomfort to the patient, but also that more molecular information about the tumor can be obtained with the techniques available. This could enable the grading of the biological aggressiveness of a tumor and the provision of information that would allow the choice of individualized treatment using targeted therapies, increasing efficacy, and reducing toxicity.

CLINICAL MANIFESTATIONS

Patients with cancer present in a variety of ways, depending on the local effect of the primary tumor, the effects of any sites of metastatic spread, or from an indirect metabolic abnormality such as hypercalcemia. There may also be nonspecific symptoms resulting from production of a variety of cytokines by the malignant cells; symptoms include anorexia, weight loss, and lethargy. Finally, a variety of malignancies may produce symptoms and signs distant from the primary site or its metastases – termed paraneoplastic syndromes – and these are often related to the central nervous system. The clinical manifestations of the hematological malignancies (lymphoma and leukemia) and solid tumors involving lung, breast, and bowel have been discussed in Chapter 2, and tumors of the nervous system are discussed in more detail below. Further information on presenting symptoms in other tumors can be found in general texts.

in their concentration of malignant cells and frequently have large islands of necrotic and acellular tissue. As a result, the site from which a biopsy is taken may not actually contain tumor cells. A classic example of this arises when trying to make a diagnosis of pancreatic cancer; a lesion in the head of the pancreas can cause a desmoplastic reaction (a local inflammatory process), which, when biopsied, reveals reactive rather than neoplastic cells. Further attempts at biopsy may fail and the clinician is left with the difficult decision of whether to make the diagnosis on clinical grounds alone, with all the implications this has for the prognosis that is given to the family, and the potential for toxicity of the treatment that will be offered.

Tumors arising from the nervous system

Tumors of the nervous system can be divided into those that arise centrally and those in the periphery. Patients with brain tumors may present with symptoms of raised intracranial pressure such as headache, often worse in the morning, drowsiness, and nausea and vomiting. The actual site of the tumor may cause specific symptoms. For example, a tumor in the motor cortex can cause paralysis, tumors in the temporal region are often associated with epilepsy, and tumors in the cerebellum typically cause ataxia, nystagmus, and double vision. There may be endocrine effects of central nervous tumors that are particularly prominent with those originating from the pituitary gland. The pituitary is usually responsible for the production of several peptide hormones that in turn regulate the production of almost all other hormone-secreting tissues in the body, including the thyroid, adrenal cortex, and gonads. Pituitary tumors are usually reasonably well-differentiated adenomas that continue to produce and secrete their usual hormone though unfortunately they lose the ability to appropriately regulate the output. Even pituitary tumors that are in themselves nonsecretory may promote the secretion of hormones by other pituitary cells, most notably of the hormone prolactin, by compressing part of the gland and preventing inhibitors of hormone secretion from reaching their target cells. Moreover, the pituitary also secretes growth hormone, which can drive the production of the pleiotropic growth factor, IGF1. Sometimes, pituitary tumors originating in one cell type may compromise the function of other hormone-secreting cell types, causing pituitary failure (the effects depend on the hormones no longer secreted and include: loss of sexual characteristics, fertility, and libido, due to defective secretion of LH and FSH; adrenal failure, due to loss of ACTH; hypothyroidism, due to loss of TSH; and a still poorly understood complex of problems associated with loss of growth hormone in adults though in children this can cause a severe problem in growth). Pituitary tumors may also extend out from their usual location (the sella turcica or Turkish saddle) into a part of the brain containing the optic pathways to cause visual field defects (such as tunnel vision) or into the hypothalamus to cause a number of problems with salt and water balance. Sometimes the tumors may bleed and the patient will notice a sudden deterioration in vision associated with headache and signs of hypopituitarism.

Tumors involving the spinal cord cause symptoms depending on the site at which they arise. Patients may experience pain at that level and, if there is resultant compression of the spinal cord, weakness may occur below the lesion, with associated constipation, bladder dysfunction, and changes in sensation. This is a medical emergency requiring rapid intervention to reduce the risk of permanent paralysis.

INVESTIGATIONS IN ONCOLOGICAL PRACTICE

Interpretation of hematological and biochemical tests

In general, routine blood tests do not make the diagnosis of cancer but may indicate that there is a serious underlying illness (Table 15.1). The most common changes that can accompany a malignancy are anemia (too few red blood cells), polycythemia (too many red blood cells due to excess production of erythropoietin), neutropenia (too few white blood cells), leucocytosis (too many white blood cells), thrombocytosis or thrombocytopenia (too many or too few platelets), elevated liver enzymes, and reduced renal function. The erythrocyte sedimentation rate may rise in patients with cancer, but it is a nonspecific test that measures increased "thickening" of the blood that often results from increased production of inflammatory proteins by the liver, and is also elevated in infections and several other conditions. In leukemias, the peripheral blood film is essential as there may be a normocytic, normochromic anemia (a reduction in red blood cells without alteration in size or hemoglobin content of the remaining cells, and thus can be distinguished from another common cause of anemia – iron deficiency), though this again occurs in many other generalized diseases, but occasionally leukemic blast cells are present, which is specific. There may be an excess of white cells and a low platelet count. If

Table 15.1   Interpretation of blood results in the diagnosis of cancer

Abnormal blood test	Possible interpretation
Low hemoglobin	Anemia due to bleeding from tumor
	Anemia due to bone marrow involvement by tumor
Elevated white cell count	Infection
	Bone marrow involvement by tumor
Low white cell count	Neutropenia due to chemotherapy
	Bone marrow failure
Elevated urea and creatinine	Renal impairment, which may be due to the kidneys being obstructed by the tumor, dehydration, or damage by chemotherapy
Elevated urea with anemia	Gastrointestinal hemorrhage
Hyponatremia and hyperkalemia	Addison's syndrome due to tumor involving the adrenal glands
Hyponatremia	Syndrome of inappropriate ADH secretion in small-cell lung cancer
Hypercalcemia	Indirect effect of metastatic cancer or bone metastases
Hypokalemia, hypomagnesemia	Renal tubular damage due to chemotherapy or diarrhea
Elevated c-reactive protein	Nonspecific indicator of infection or inflammation
Hypoalbuminemia	Liver impairment or malnutrition/cachexia
Elevated alkaline phosphatase	Bone metastases or liver metastases
Elevated alkaline phosphatase with elevated bilirubin	Biliary obstruction
Elevated hepatic transaminases (AST, ALT)	Liver damage, which could be due to chemotherapy or metastases

the tumor has spread to the bone marrow in solid malignancies, a leukoerythroblastic picture will be seen with circulating primitive cells. In some solid tumors the routine full blood count will show normocytic, normochromic anemia and, in those tumors where there has been blood loss, an iron deficient anemia occurs.

Routine biochemistry may indicate varying degrees of renal failure if the tumor has obstructed the renal outlet. If the tumor has blocked the biliary tract, an obstructive picture in the liver enzymes will be seen with an elevated bilirubin and alkaline phosphatase. Bone metastases and some lung cancers will also cause a rise in alkaline phosphatase. The liver transaminases, aspartate, and alanine amino transferase, will be increased in patients with liver injury that may be from alcohol, infection, or cancer. Lactic dehydrogenase (LDH) is an enzyme associated with tissue injury and is often increased in patients with a large tumor burden, lymphoma, or liver metastases. An elevated corrected calcium level indicates hypercalcemia, which may be due to bone metastases, or be a metabolic consequence of an underlining tumor such as lymphoma. Low sodium (hyponatremia) may represent the syndrome of inappropriate ADH secretion, which is often associated with an underlying small-cell lung cancer. A low albumin may indicate liver involvement with tumor or be a general indicator of the poor nutritional state of the patient.

Tumor markers

The first tumor marker to be detected was in 1848, when Bence Jones discovered that the secreted light chain proteins of myeloma were detectable in the urine. Since that time it has become evident that there is a range of proteins and glycoproteins that are produced by tumors and which can be detected in the blood using immunoassays. It is now clear that biomarker

Table 15.2 Tumor markers and their uses

Name	Type	Cancer	Use
Carcinoembryonic antigen (CEA)	Oncofetal protein	Colorectal, breast	M
Alpha-fetoprotein (αFP)	Oncofetal protein	Hepatocellular	S, D, P, M
		Germ cell tumors	D, P, M
CA 19-9	Cancer-related antigen	Pancreas	D, M
CA15-3	Cancer-related antigen	Breast	D, M
CA 125	Cancer-related antigen	Ovary	D, M
Prostate-specific antigen (PSA)	Cancer-related antigen	Prostate	S, D, M
Beta choriogonadotropin (βHCG)	Hormone	Germ cell tumors	D, P, M
		Choriocarcinoma	D, P, M
Calcitonin	Hormone	Medullary carcinoma of thyroid	D, M
5-Hydroxy indoleacetic acid (5-HIAA)	Hormone	Carcinoid	D, M
Lactate dehydrogenase	Enzyme	Germ cell tumors	P, M
		Lymphomas	

D, diagnosis; P, prognosis; S, screening; M, monitoring of treatment

production may result either from increased expression of a normal gene/protein (because of increasing cell mass) or because of cancer-related alterations in gene/protein expression. Unfortunately, such markers may not be tumor specific and a degree of cross-reactivity with antigens of normal tissues is frequently seen. Therefore, the value of different tumor markers varies; theoretically, biomarkers could be useful in diagnosis, predicting prognosis, or response to treatment and aiding follow-up of a patient, but despite increasing research in this area, surprisingly few biomarkers are clinically useful at present. Tumor markers can be broadly divided into three groups, oncofetal proteins, cancer-related antigens, and hormones (Table 15.2). Here we focus on tumor markers that are measurable in the serum (i.e. can be readily measured in large numbers of patients relatively easily and noninvasively). However, biomarkers of cancer may also be detected by analyses of tumor samples though in this case testing is self-evidently restricted to patients with an accessible tumor. With the hope that some cancer biomarkers will also be cancer-important proteins, at least some of these will become the target of antibody-directed therapies.

An example of an oncofetal protein tumor marker is carcinoembryonic antigen (CEA). Although this marker may be produced in up to 80% of people with advanced bowel cancer, it will also be elevated in patients with inflammatory bowel disease and other benign conditions. Therefore CEA is usually not helpful in diagnosis but can be useful in measuring response to therapy or identifying the presence of recurrent disease at follow-up.

Cancer related antigen tumor markers include prostate-specific antigen (PSA), CA125 and CA19-9. PSA is elevated in both inflammatory and malignant prostate disease. This means that it can support but not guarantee a diagnosis of cancer and can be very useful in monitoring the response to treatment. It has also been proposed as a marker to be used for population screening but its utility as a screening tool remains controversial. Currently in the United States most middle-aged men will know their PSA level, is in stark contrast to the United Kingdom where a national prostate screening program using PSA does not exist.

CA125 is elevated in approximately 80% of patients with nonmucinous ovarian cancer.

Again it can be extremely useful in monitoring response to treatment and detecting early relapse during follow-up. In many patients, the serum CA125 may become elevated before there is radiologically confirmed disease relapse and in this scenario the timing of further therapy is a management dilemma – should retreatment be instituted at the time of biochemical or symptomatic/clinical relapse? Good models now exist to help interpret CA125 rises and they provide guidance of their optimal use as response criteria in clinical trials. There are no data to recommend CA125 as a screening tool for ovarian cancer; this is because CA125 can be elevated in benign conditions such as endometriosis and cirrhosis.

CA19-9 is a tumor marker that is elevated in approximately 70% of patients with pancreatic malignancy. It is often a useful diagnostic tool, in combination with biopsy, to establish the etiology of pancreatic lesions. It can be particularly valuable to support a diagnosis in the significant number of patients for whom histological confirmation proves impossible.

Examples of hormones that are also tumor markers include human chorionic gonadotropin (HCG) and calcitonin. Calcitonin is elevated in patients with medullary carcinoma of the thyroid and is used in detection, diagnosis, and follow-up after treatment. HCG is worth discussing in some detail, as it is a tumor marker that has almost perfect characteristics. It consists of two subunits, α and β, and is produced normally by the placental trophoblast. The β unit is elevated in early pregnancy and provides the basis for most pregnancy tests. Therefore, as long as early pregnancy is excluded, an elevated HCG is pathognomonic of choriocarcinoma. In addition, the level of HCG reflects the volume of tumor and as such it is a valuable prognostic marker.

Both HCG and alpha-fetoprotein (AFP) are highly specific for testicular germ cell tumors. Again, the level of these markers is proportional to the tumor volume and they are valuable as prognostic indicators in patients with teratoma. They play a central role in monitoring the response to treatment. Experience is key here, as we now know that in some cases AFP can become elevated when chemotherapy is initiated. The level may remain high despite successful treatment and does not necessarily indicate residual active disease. The cause of this phenomenon is not clear but AFP is elevated in patients with liver disease and particularly hepatocellular carcinoma. Its persistent elevation may be due to the chemotherapeutic agents causing some damage to the liver parenchyma.

There are now published recommendations for the use of tumor markers in clinical practice. In the future, it is hoped that these markers will have greater specificity and sensitivity. Already there are research tools available that allow circulating tumor cells to be identified by polymerase chain reaction (PCR) in the blood of patients with breast and prostate cancer. In other tumor types, such as bladder and colorectal, tumor cells are being detected in excreted specimens using markers of oncogenes or oncogene products.

Genetic tests

Genetic abnormalities occur frequently in patients with leukemia and cytogenetic analysis is carried out in all patients presenting with this disease. Leukemic cells have somatically acquired cytogenetic abnormalities, which can be useful for both diagnosis and prognosis (Table 15.3). The Philadelphia chromosome was the first described chromosomal abnormality to be associated with a malignancy. It results from a reciprocal translocation between the long arm of chromosomes 22 and 9 and is present in 97% of cases of chronic myeloid leukemia (see Chapter 6). The consequence of this genetic abnormality is the gene *BCR-ABL*, which encodes a fusion protein that is critical in the development of chronic myeloid leukemia. This is the target of the new agent, imatinib, which is discussed further in Chapter 18.

Diagnostic DNA testing is now available for a number of rare hereditary cancers. Examples of these familial phenotypic cancers are shown in Table 15.4 (see Chapter 3 also). The majority are inherited in an autosomal dominant manner. To detect inherited mutations present in the germ line (and therefore theoretically in all cells in the organism) DNA is usually extracted from white blood cells (can be done via a simple blood test) and then profiled for cancer predisposition genes and mutations. In monogenic inherited

Table 15.3 Chromosomal analysis in leukemia

Leukemia	Abnormality	Implication
CML	t(9;22)(q34.1;q11.2)	Diagnostic
CLL	Trisomy 12	Poor prognosis
	Deletion 13q14	Worst prognosis
ALL	t(9;22)(q34.1;q11.2)	Poor prognosis
	t(11;14(p13;q11)	Poor prognosis
	t(8;14)(q24;q11)	Poor prognosis
AML		
M2	t(8;21)(q22;q21)	Good prognosis in young adults
M3	t(15;17)(q22;q11)	Best prognosis of all AML
M4	Inversion of 16	Good prognosis
M5	11q23 abnormalities	Poor prognosis

Table 15.4 Familial cancer syndromes

Autosomal dominant	Defective gene	Chromosome location	Malignancy
Retinoblastoma	*RB1*	13q	Eye
Wilms' tumor	*WT1*	11p	Kidney
Li–Fraumeni	*P53*	17p	Sarcoma/CNS/leukemia
Neurofibromatosis 1	*NF1*	17q	Neurofibromas
Von Hippel–Lindau	*VHL*	3p	Hemangioblastomas, renal cell
Familial adenomatous polyposis	*APC*	5q	Colon
Hereditary nonpolyposis colon cancer	*MLH1* and *MSH2*	3p, 2p	Colon/endometrium
Breast ovary families	*BRCA1* and *BRCA2*	17q	Breast/ ovary
Multiple endocrine neoplasia 1	*MEN-1*	11q	Pancreatic islet cell/adrenal cortex/thyroid
Multiple endocrine neoplasia 2a	*MEN2/RET*	10q	Thyroid/pheochromocytoma

cancer syndromes, single gene defects are usually responsible and a specific gene mutation can be identified by various molecular genetics techniques (Chapter 18). In some cases such as in familial adenomatous polyposis coli, a mutated *APC* gene can be looked for in relatives of an affected patient to determine if they also carry the same specific mutation (as the index case in your clinic) conferring a high risk of developing colon cancer themselves. (See also Chapters 1, 2 and 3 where aspects of genetic testing are discussed.) Genetic testing can therefore provide some individuals with their personal lifetime risk of developing a cancer. These presymptomatic subjects can then enter intensive screening programs, or elect to have prophylactic surgery performed such as a total colectomy.

The vast majority of individuals however want predictive genetic testing because they have a strong family history of cancer. Examples here would be those families where there appear to be high numbers of breast, bowel, and ovary cancers. It is known that the *BRCA1* and *BRCA2* genes account for the vast majority of cases of hereditary breast cancer and 50% of cases of hereditary ovarian cancer. Mutations of the *BRCA* genes lead to accumulation of mutations in tumor-suppressor genes and oncogenes and the subsequent development of malignancy (Chapters 2 and 10). However, this familial cause of breast cancer only accounts for a minority of the total number of cases, most of the cases being sporadic. Currently we only recognize some of the responsible *BRCA* mutations. This means that where there is a large family with a known mutation we can screen for that using DNA from the blood of living relatives. We can then confidently confirm whether a patient carries that defect and offer prophylactic treatment. However, where there are no living relatives it can be hard or impossible to perform the linkage or necessary mutation analysis. Moreover for breast cancer there is seldom just a single faulty gene responsible and with such heterogeneous defects it is not possible to 100% accurately what someone's lifetime risk of cancer is. In this scenario, it is a very difficult decision to make as to whether to undergo bilateral mastectomy with or without bilateral salpingo-oophrectomy. In oncological practice now in the United States and Europe these families are referred directly to a family history clinic where there are specialist geneticists and counselors.

It is estimated that at least 1 in 10 of all cancers is due to an inherited genetic predisposition. This therefore means that while DNA testing is currently available, it is only actually helpful in a small number of people, representing a minority of patients with cancer. The sequencing of the human genome has focused huge amounts of research in this area and in the future the value of genetic testing is likely to increase for the whole population.

Screening tests

The objective of screening a population for cancer is to detect tumors at the earliest possible stage while they are potentially treatable and thus curable. This straightforward aim is actually a complex process. In an ideal screening program, the natural history of the disease should be well understood with a recognizable early stage that is treatable. Both the sensitivity (the chance that someone with the disease will test positive) and the specificity (the chance that someone without the condition will test negative) should be high. The test should be acceptable to the public and the cost should be balanced against the benefit it provides. Unfortunately, most proposed screening tests have failed to meet all of these criteria. There are wide differences in the screening available in the United States and United Kingdom. To date in the United Kingdom, national screening programs are only available for breast and cervical cancer. In the United States many patients are also screened for prostate and colorectal cancer.

Cervical cytology

The screening test used in the early detection of cervical cancer is the Papanicolaou smear (Pap smear). This involves taking a scraping of cells from the transformation zone of the cervix (the area where columnar epithelium changes to squamous morphology). The precancerous lesion in the cervix is cervical intraepithelial neoplasia (CIN). Cytological studies can grade the CIN from I to III, where CIN III, is the most severe grade (approximately 30% of these precancerous lesions will develop into invasive carcinomas if left untreated). By treating patients with CIN changes, it is thought that the incidence of invasive cancers could be reduced by up to 90% if the screening test was offered to women every 3 years. There has also been interest in screening patients for the human papillomavirus (HPV) which is present in the majority of women with cervical malignancies. It is likely that over the next decade the development of vaccines to HPV will be successful.

In practice, there have been problems with the cervical cytology-screening program both in the United States and United Kingdom. Some of these difficulties have been related to the accurate reporting of CIN I to III. Quality assurance and audit have, to a large extent, rectified the inter observer variability that was a concern in early years. However, those most at risk of cervical cancer are women who have been sexually active with multiple partners at a young age, who are generally of poor socioeconomic background. These are not typically the individuals who come forward for screening and efforts are still ongoing to encourage these women to attend these programs.

Mammography

Mammography involves an X-ray examination of the breast. There have been numerous randomized controlled studies investigating the effectiveness of mammography for detecting breast cancers at an early stage. At the outset, the use of mammography as a screening test for breast cancer held much promise as this is one of the most common malignancies, and mammography can detect tumors before patients are themselves aware of them. However, to date the overall results from screening programs worldwide have been the subject of much debate with doubt being raised over whether real reductions in mortality have been seen. There does seem to be an increased rate in mammographic detection of early cancer, but this does not necessarily correlate with a survival advantage. The false positive rate is considerable, which makes the cost of screening high, both financially and in terms of the anxiety to women.

Currently in the United Kingdom, women are invited for screening on a 3 yearly basis, although the number of interval cancers, that is, those cancers that develop in between the appointments, remains high. It is also contentious as to what age women should commence screening. In the United Kingdom, the Forrest report suggested screening of women between the ages of 50 and 64. This has now been widened to include younger women.

Colonoscopy and fecal occult blood testing

Colorectal cancer should be an ideal tumor to screen. It is increasing in frequency in the Western world, and there is a clear precancerous lesion, the adenomatous polyp, which may form years before transformation to invasive malignancy. The genetic stages involved in the adenoma to carcinoma sequence have been well described. In the general population, there is a lifetime risk of about 20% of developing an adenoma and approximately 2–5% of these will then become an invasive cancer. There have been large studies looking at the usefulness of fecal occult blood testing, flexible sigmoidoscopy, and colonoscopy as screening tools for colorectal cancer.

Colonoscopy involves passing a fiber-optic scope through the bowel, allowing both detection and treatment of abnormalities. Patients have to take strong laxatives to ensure that the bowel is clear and the endoscope is not obscured by feces. It is then possible to examine the left side of the large bowel (flexible sigmoidoscopy) or large bowel from the rectum to the cecum and terminal ileum (colonoscopy). One of the major advantages of this test is that abnormal areas can be photographed and biopsied so that a definitive diagnosis may be made (Fig. 15.1). With the advance in technology in this area it is also possible to perform endoscopic resections of small polyps such as adenomas in the bowel. This means that a potentially neoplastic lesion can be either biopsied or removed at the time of the investigation. The downside of the procedure

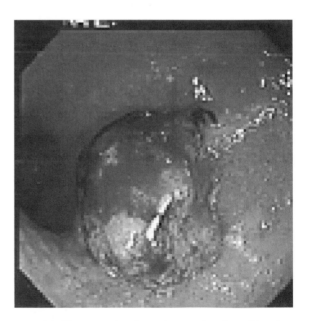

Fig. 15.1 Photograph of carcinoma in sigmoid colon seen through colonoscope.

is that it is uncomfortable, expensive, and relies on appropriately trained staff. As a result of these disadvantages, fecal occult blood testing is often used as a simple noninvasive test with only those who have positive results proceeding to colonoscopy.

A positive fecal occult blood test suggests that a tumor may be present in the colon. However, in the past there have been high false positive and negative results. With the advent of more sensitive fecal occult blood techniques, it does appear that this is a useful screening tool. Once there has been a positive fecal occult blood test the patient needs to undergo flexible sigmoidoscopy or colonoscopy. A large randomized trial of patients aged 45 to 74 years was carried out offering people either 2 yearly fecal occult blood testing or routine medical management. Those with positive results underwent colonoscopy and a 15% reduction in mortality was seen in the screened population. In the United States, flexible sigmoidoscopy is used as the screening test for colorectal cancer in patients over the age of 50 years with claims being made for 40% reductions in the incidence of malignancy in the population who attend.

Noninvasive imaging techniques

Plain film X-ray

The information provided by plain radiographs in the detection of cancer is limited when compared with other available imaging techniques. Plain radiographs of the skeleton can be used to detect lytic or sclerotic bone metastases. A skeletal survey, which includes plain views of the skull, thoracic spine, lumbar spine, and pelvis can be useful to determine the extent of multiple myeloma. A follow-up chest radiograph in a patient who has experienced a chest infection may show an obstructing pulmonary tumor.

Ultrasound

There have been major advances in ultrasonography over the past 10 years. The technique involves interpreting the different echo patterns achieved when sound is passed through an organ from an ultrasound transducer. These echoes are then displayed as a two-dimensional image. There are several advantages to this technique as it is cheap, quick, and noninvasive and does not subject patients to radiation. The disadvantages are that it is only useful for investigating certain organs, it is operator dependent, and the images cannot be used as a reference to be followed for tumor response in clinical trials. Areas where ultrasound is particularly useful include visualization of liver metastases and determining the nature of testicular masses. Ultrasound is often used to distinguish solid from cystic lesions and is therefore useful to differentiate between a benign and a malignant lesion in the kidney, the thyroid, and the ovary.

When a person presents with abdominal symptoms, an abdominal ultrasound is frequently the first screening test performed. While it is often effective at assessing the etiology of conditions such as obstructive jaundice, it may be less sensitive in imaging the pancreas due to overlying bowel gas. Advances in this technique have recently occurred that reduce such limitations and include the development of endoscopic ultrasound. This is of increasing importance in the staging of esophageal and pelvic tumors. There is also considerable research into novel contrast agents, which may increase the sensitivity and resolution of ultrasonography.

Computerized tomography

Computerized tomography (CT) scanning (Box 15.2, Fig. 15.2) currently provides the cornerstone of imaging in cancer medicine. Its accuracy is now extremely high and the advances in spiral (helical) and, more recently, multislice techniques have increased the quality and reduced the time taken to acquire the necessary images (Fig. 15.3). In addition, new techniques in three dimensions (3D imaging and virtual CT imaging) have paved the way for virtual bronchoscopy and colonoscopy. The advantages of these techniques are that they are relatively noninvasive with a high level of acceptability to patients. They can also guide the radiologist to a site of interest when undertaking a biopsy. CT scanning is therefore useful in the diagnosis of cancer, in determining the anatomical extent of disease (stage), and also as the follow-up imaging technique when monitoring the tumor response to treatment. It is now common practice when performing CT to include oral contrast in the examination of the upper GI tract and rectal contrast in the imaging of the lower gastrointestinal (GI) tract. In addition, IV contrast can be administered at the

---

Box 15.2 CT and MRI acquisition

- *Computerized tomography (CT)* A collimated X-ray beam moves synchronously with its detectors across a slice of the body part of interest. A computer processes the transmitted X-irradiation for each element, or pixel within the slice. A value is generated (Hounsfield number after the inventor) depending on the density of the structure (bone is 1000 units, water is 0 and air is −1000). The difference in density (X-ray attenuation) between the different structures in the slice makes it possible to differentiate areas of normal and abnormal tissue on the computer-generated image.
- *Magnetic resonance imaging (MRI)* The hydrogen nucleus is a proton whose electrical charge creates a local electrical field. Therefore it is possible to align protons with a strong magnetic field. When a radio frequency wave is applied at 90° to the alignment of the protons, they resonate and spin before returning to their previous alignment. During this process, images are made in the different phases of relaxation of the protons, for example, T1, T2 sequences. By comparing these different sequences it is possible to distinguish between normal and abnormal tissues.

---

time of the imaging to give detailed information on the vascularity of the region of interest.

Magnetic resonance imaging Magnetic resonance imaging (MRI) (Box 15.2) is an extremely valuable technique that provides excellent anatomic definition of structures and avoids the use of ionizing radiation. There are, however, some contraindications including the presence of cardiac pacemakers or other metallic foreign bodies. In addition, the procedure can be claustrophobic for some people and there is, on average, a 10% refusal rate. It is of particular value in imaging the brain (Fig. 15.4) and spinal cord and can show areas of nerve compression and cerebral tumors with great accuracy. It is also possible to differentiate between normal and malignant tissues by altering the sequencing on the MRI. The contrast agent, gadolinium, is used to help differentiate tumor from surrounding normal tissue (dynamic contrast enhanced MRI).

Contrast enhanced MRI is the most sensitive imaging technique. It allows detailed imaging of

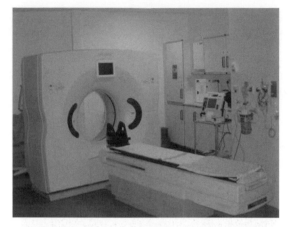

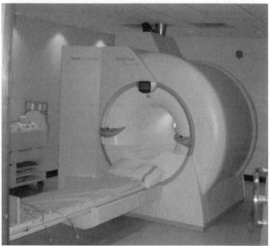

Fig. 15.2 CT scanner (top) and MRI scanner. Note much longer imaging tube in which patient lies in the MRI scanner.

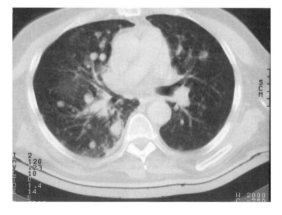

Fig. 15.3 CT image of thorax showing heart in center with surrounding lung in which circular metastases are clearly seen.

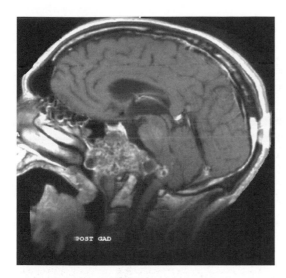

Fig. 15.4 MRI image of brain. Note the high anatomical detail demonstrated. There is a large tumor in the pituitary region.

the tumor in relation to the surrounding vasculature and other organs. Preoperatively this enables surgeons to assess whether the tumor is technically resectable in its entirety or whether other treatment is necessary first. For example, MRI of the rectum can evaluate the extent to which the tumor invades surrounding structures such as the vagina or seminal vesicles. Studies are ongoing to evaluate the possibility of using MRI as a screening tool for breast cancer. MRI can also be very useful in establishing the nature of liver lesions. There are new contrast agents available containing super paramagnetic iron oxide particles, which can help distinguish hepatic cysts from metastases. These particles are taken up by the Kupffer cells within the liver, and cause a "drop out" in signal in normal liver. Metastatic lesions do not contain these Kupffer cells and therefore do not take up the particles and appear as high intensity signals. Many new contrast agents are in advanced stages of development and should further increase the value of MRI in the diagnosis, staging, and follow-up of malignancies.

Tissue diagnosis

Although imaging and tumor markers can provide strong evidence to support a diagnosis of malignancy, it is of paramount importance that tissue samples are obtained from abnormal areas within the body so that they can be sent to the pathology laboratory where a definitive diagnosis can be made. In addition, it is now possible to use an increasing number of immunohistochemistry markers to understand the biological nature of tumors (e.g. to provide an assessment of how aggressive they are likely to be). The ease of obtaining a biopsy depends on the site of the abnormality. There are some tumor deposits that are easy to biopsy by direct excision (e.g. subcutaneous deposits). However, the abnormal structure may be quite deep and as such poorly accessible. There are a number of techniques available to obtain adequate tissue samples without repeated attempts being necessary; for instance concurrent imaging (usually with ultrasound) guides the biopsy needle with great accuracy to the appropriate site and minimizes the risk of damaging other organs. Methods of obtaining tissue or cells for analysis include cytology, fine needle aspiration, needle biopsy, incision and excision biopsy, and the use of endoscopy.

Cytological examination of body fluids For some tumors it may be appropriate for a small amount of body fluid (e.g. sputum, ascitic fluid, cerebral spinal fluid, pleural fluid, or urine) to be collected and spun down using a "CytoSpin". Any cells present are then stained and examined under the microscope immediately after collection. It is often possible to establish whether cancer cells are present and give a rapid diagnosis. The main disadvantage is the frequency of false negative results, normally because the specimen taken is not large enough. An advantage of this technique is that it is a quick procedure that can be performed as an outpatient with minimal discomfort.

Fine needle aspiration A fine needle aspiration (FNA) involves the passage of a hypodermic needle into the abnormal area under investigation. Ultrasound or CT can be used to guide the needle to the suspected tumor if it is not possible to palpate it. Cells are then aspirated and put on a microscope slide and sent to the cytology laboratory. This technique is particularly useful for assessing breast lumps in the outpatient setting.

It can discriminate between reactive and malignant cells and may sometimes provide further information about tumor type if specific staining techniques are employed.

Needle biopsy   This technique is more invasive than FNA as a larger biopsy needle (trucut) is used under a local anesthetic. This means that an actual core of tissue can be sampled and sent to the pathology laboratory for fixing and mounting in wax. As this is a larger specimen, more information is potentially available. For example, an FNA of a breast lump can determine whether the lump is merely a benign fibroadenoma or is a malignant tumor. If a trucut needle biopsy is used, it is possible to carry out immunohistochemical analysis and determine whether the breast cancer is expressing estrogen or progesterone receptors and to what grade the tumor belongs.

Incision and excision biopsy   These techniques involve removing a sample of tissue under local anesthetic. For example, a punch biopsy using a small circular blade can be used to take a sample from a skin lesion. In practice, it is often preferable to excise an abnormal area totally. This is known as an excision biopsy and in some cases the biopsy itself may be curative. When this form of biopsy has occurred, it is important to examine the periphery of the lesion to see whether the tumor has involved the microscopic margins. If this is the case, then further excision would usually be required.

With more advanced ultrasound and CT techniques, it is now often unnecessary that a patient should undergo a surgical procedure, such as laparotomy, to determine the nature of abnormal tissue. In most instances, a guided biopsy will provide sufficient tissue to make diagnosis.

Bronchoscopy   When a patient presents with an abnormal area on a chest radiograph, the initial investigation is almost always a bronchoscopy. This involves having a fiber-optic instrument passed through the nose or mouth into the pulmonary tree. It gives extremely good access to lobar and proximal bronchi but is not usually helpful in assessing more peripheral lesions. When a bronchoscopy is performed, the abnormal area can be visualized (Fig. 15.5) and either

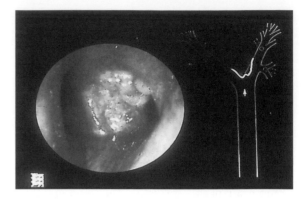

Fig. 15.5 Photograph of carcinoma of bronchus as visualized down bronchoscope. The site of the tumor is illustrated on the diagram to the right.

washed out (and the washings sent for cytological examination), or biopsied when a sample of tissue can be sent to the pathology laboratory. In some cases, it is possible to actually establish, at the time of the bronchoscopy, whether a tumor is likely to be operable or not (e.g. tumors within 2 cm of the main carina would be deemed inoperable). Sometimes a rigid bronchoscope has to be used and this is performed by a thoracic surgeon under general anesthetic. The main indication for the latter procedure is to visualize more difficult tumors and to have more control over blood loss at the time of biopsy. The rigid bronchoscope can also be used to study the mediastinum and this again helps determine the operability of a tumor.

Endoscopy of the upper GI tract   Upper GI endoscopy is used to assess the esophagus, stomach, and first part of the duodenum. It involves the passage of fiber-optic instruments through the esophagus into the stomach. Abnormal areas can be visualized and biopsies taken. When undertaken by a skilled operator, a positive biopsy is available in about 90% of patients. The areas where false negative biopsies occur are usually when submucosal gastric tumors are present. In this situation, a double biopsy may be necessary. Obviously this carries with it a higher risk of perforation and this complication always has to be balanced against the potential value of the information to be gained.

Endoscopy is extremely useful to identify the source of hemorrhage. If this is from an ulcerated area where a bleeding vessel is noted, the vessel can be sclerosed using an injection (sclerotherapy). It is also possible to use laser techniques to ablate obstructing tumors. This is a simple method that can be carried out under light sedation and rapidly relieves symptoms of dysphagia.

Endoscopic retrograde cholangiopancreatography (ERCP) is a variation of standard endoscopy. The scope used has a side arm that allows cannulation of the pancreatic duct. Dye can then be injected to delineate the anatomy of the pancreatic and biliary system. Malignant strictures can be identified and brushings taken to produce cells for cytological examination. It may also be possible to alleviate an obstruction causing jaundice by placing a stent through the stricture.

Additional techniques used in diagnosis

There are some patients for whom a tissue or cytological diagnosis cannot be made. This may occur when the site of abnormality cannot be accessed or when biopsies fail to contain tumor. Other techniques may be employed to support a diagnosis of malignancy.

Interventional radiology   Interventional radiology usually involves the insertion of cannulae into the vascular system to obtain blood samples or inject local contrast media. With the advent of superior CT and MRI techniques these invasive diagnostic procedures have a diminishing role. Portal venous sampling is occasionally used to localize pancreatic endocrine tumors such as gastrinomas. These are "functioning" tumors and produce gastrin. It is possible to perform percutaneous transhepatic portal venous sampling and measure the hormone levels in the superior mesenteric and portal veins. The sensitivity of this test is over 95% and the venous sampling gives precise anatomical mapping of the tumor so that radical resection can be planned.

Angiography   Angiography is used to accurately delineate the vasculature of the organ under study. It is most commonly used in the diagnostic evaluation of liver, renal, or pancreatic tumors. Again with the advent of more sophisticated imaging techniques such as MRI, angiography is being used less frequently. It may sometimes be used to image the vascular blood supply around a pancreatic tumor and thereby allow surgeons to determine whether the tumor is resectable. Another situation is the use of angiography to determine the blood supply of liver lesions. It is now possible to administer intraarterial chemotherapy or, more recently, radiolabeled microspheres, for the treatment of liver metastases.

The technique of transarterial chemoembolization is being used more widely now to treat a number of tumor types including primary liver carcinomas. With this technique, high dose chemotherapeutic agent is injected locally into the tumor bed and the blood supply to the tumor is also compromised. Usually this is a palliative procedure, but with technical advances this could be used with curative intent in the future.

Radionucleotide imaging   Many different radionucleotide studies are now used to detect primary and metastatic tumors. Most are specific for a particular tissue rather than a tumor. They can be divided into general scans and tumor-specific scans. The most common of the general scans images the bones. Radiolabeled bisphosphonates are administered intravenously to the patient who is then scanned with a gamma camera. A positive bone scan involves the presence of "hot spots" or areas of high signal intensity. They usually occur due to increased levels of blood flow and osteoblastic activity (Fig. 15.6). The bone scan is therefore extremely useful in the detection of sclerotic bone metastases. It is now possible to use radioactive strontium to image and treat bone metastases as the radioactivity is localized to tumors and emits sufficient energy to kill neoplastic cells.

A highly tumor-specific radiolabeled scan targets thyroid tissue. [131]Iodine-labeled sodium iodide is taken up to a high level by the organ whereas minimal activity is found in other sites. This can be used to stage a thyroid cancer or at a therapeutic level to treat thyroid malignancy. Another tumor-specific scan is the adrenal scan

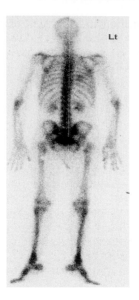

Fig. 15.6 Radioisotope bone scan. Note the increased uptake of radioactive tracer producing "hotspots" in the scapulae and limbs.

using ^{131}iodine-labeled metaiodobenzylguanidine (MiBG). This agent is valuable in assessing any metastatic deposits of pheochromocytomas. Octreotide is a synthetic somatostatin analog that is used in the treatment of neuroendocrine tumors. ^{111}In-pentetreotide is used in an "octrescan" and utilizes the characteristic of the agent to be taken up by tumor cells which express somatostatin receptors. The octreoscan is valuable to stage, and evaluate whether the neuroendocrine tumor will respond to somatostatin as a therapeutic maneuver.

New imaging techniques

Positron emission tomography Positron emission tomography (PET) scanning is a technique that measures physiological function by imaging blood flow, and cell metabolism, using radiolabeled agents. It is the most sensitive functional imaging technique currently available in clinical practice. It can be used to support a diagnosis of cancer by use of the radiolabeled glucose analog, 18-fluorodeoxyglucose (^{18}FDG). Radionuclides chosen for this technique usually have a short half-life, which means that the cyclotron (used to produce them) needs to be geographically close to the imager. It is therefore a very expensive technique that is currently available in only a limited number of cancer centers.

Malignant tumors metabolize glucose at a faster rate than benign lesions or normal tissue. As this scanning technique measures the rate of consumption of glucose, it can be used to distinguish between benign and malignant tumors. The anatomical resolution of PET (Plate 15.1) is inferior to current MRI and CT scans and as such, PET often needs to be used in collaboration with CT or MRI. An example of the value of PET is in the assessment of response to treatment in lymphomas (especially Hodgkin's disease). It is relatively common that lymph node masses fail to return to a normal size after chemotherapy and radiotherapy. It is likely that the majority of these residual abnormalities represent scar tissue and the PET scan can be performed to establish whether that is indeed the case or if there is any functionally active Hodgkin's disease remaining.

In the brain, PET has been used to differentiate between high and low grade tumors and therefore remains extremely useful in predicting the prognosis for patients. It also has a large role to play in the preoperative assessment of the extent of disease in lung cancer, enabling much greater accuracy in predicting which patients are suitable for resection surgery. There are likely to be a number of future developments with the production of many selective, high affinity radiopharmaceuticals that will allow characterization of tissues as well as the assessment of drug distribution and other parameters of pharmacodynamics. The technique is also being incorporated with CT image acquisition to provide functional and high quality anatomical information about an area that can be obtained during the same examination.

Functional magnetic resonance imaging Another technique that is becoming available is functional MR imaging. This technique complements PET and it has some advantages, particularly the fact that there is no need for an onsite cyclotron as it does not depend on short lived radiotracers. It also provides much greater anatomical detail. At the moment, these MR-based techniques are still research tools but it is likely they will be accepted into routine practice in the next few years. The molecule that has been most widely used to date is ^{31}Phosphate. This has been assessed for the follow-up of responses to treatment in patients with breast cancer. Another molecule being studied following treatment with 5-fluorouracil is ^{19}fluoride.

THE ROLE OF MOLECULAR PATHOLOGY IN CANCER DIAGNOSIS

In the past, the pathologist used light microscopy with hematoxylin and eosin staining to assess morphological features of a tumor specimen. The arrival of electron microscopy provided more information relating to subcellular features. In recent years, molecular pathology has assumed an increasing and often fundamental role in establishing a definitive diagnosis. With more sensitive immunohistochemical techniques becoming available, the biological behavior of the tumor can be predicted. In addition, new molecular genetic assays can provide predictive information as to which treatment option would be superior. For example, the tumor can be stained to see if proteins coding for mismatch repair genes are present, as these would suggest resistance to platinum chemotherapy agents. It is now also possible to use sensitive techniques that enable the detection of cancer using very small tumor specimens.

One of the hurdles in developing new molecular techniques has been that the majority of tumor specimens are fixed in formalin and then embedded in paraffin. It is only recently with the development of techniques that allow the extraction of DNA from these specimens that molecular biology can be more routinely utilized. When DNA is extracted, it can be directly sequenced by southern blotting and restriction fragment polymorphism. With the sequencing of the human genome, it is now possible to identify point mutations, translocations, amplifications, deletions, microsatellite length instability, and changes in methylation. If a tumor is known to harbor a specific mutation in a given gene then the altered sequence can serve as a marker for clonality.

Flow cytometry

Flow cytometry is a useful tool to analyze the cellular DNA content in tumor specimens. The DNA index (abnormal cell DNA content relative to normal cells) may indicate a duplication (polyploidy) or loss (aneuploidy) of chromosomes. It is also possible to assess the DNA content of tumor cells and determine which phase of the cell cycle they are in. This can provide information regarding the prognosis of tumors. Numerous different fluorescent dyes are available and can be used to label cells and count the numbers with certain characteristics (e.g. proportion dying from apoptosis), which can support a diagnosis and could help with the choice of therapy.

Fluorescent *in situ* hybridization

Fluorescent *in situ* hybridization (FISH) allows specific staining of any given region of the genome. It is possible to use a fluorescent technique to identify chromatin within tumor sections. The DNA probes used for this are a mixture of DNA fragments containing chromosome-specific repeat sequences. FISH is therefore able to identify changes in the genome (for example, deletions in regions of the chromosomes) and provides quantitative data.

Microchips have also greatly enhanced molecular pathological techniques. DNA microarrays consist of numerous cDNA probes that detect the expression of many thousands of genes in one experiment. This means that one sample can be screened to define which genes are up- or downregulated or alternatively large numbers of tissues can be examined for the expression of selected genes. These tissue microarrays could be of major importance not only in the diagnosis of disease but also in predicting the behavior of a tumor.

DISEASE STAGING

The accurate staging of a tumor provides information on its anatomical distribution and is essential to ascertain its most appropriate treatment. In general, tumors are staged using the Tumor, Node, Metastasis, (TNM) classification (Box 15.3). This standardization of staging

---

Box 15.3   Key significant event for this chapter

There is no doubt that the development of the TNM classification of malignant tumors has had a huge impact on the successful treatment of cancer. Up until that time, there was no internationally accepted, uniform system to stage tumors accurately. This meant that it was difficult to compare the outcome of different treatments in a specific disease type as there was no clear idea of how heterogeneous the patient population treated was. With the advent of new treatments and clinical studies it became paramount to have one recognized staging system that would accurately record the anatomical extent of a tumor, thus ensuring physicians were comparing "like with like." In addition, it means that patients can be given more accurate prognostic information about their disease. With the advances in radiological techniques, the TNM classification still is central to the interpretation of the radiological reports (Hermanek *et al.*, 2002).

---

is recorded in a consensus publication by the American Joint Committee on Cancer (AJCC), and the Union Internationale Contre le Cancer (UICC). It is now in its sixth edition. In general, the more advanced the stage of disease, the worse the prognosis and greater the likelihood that systemic therapy will be necessary.

While anatomical staging gives important information regarding the extent of the cancer, it is also necessary to know the degree of differentiation (grade) of the tumor and other factors such as whether there is vascular or lymphatic invasion by malignant cells. Many of these factors are now being incorporated into more sophisticated staging systems to try to refine the appropriate treatment decisions so that only those most likely to benefit will be given specific treatments.

There are now a number of molecular markers that can also be examined in tumors to help define prognosis. For example, in breast cancer the presence of the oncogene *c-erbB2* is associated with a worse outcome. In this situation, the use of the monoclonal antibody, trastuzumab (herceptin), which recognizes this oncogene, can improve progression-free survival in the affected patient. In oncology, we are now entering the era of targeted therapies. In an ideal

world, an oncologist would know the genetic make-up of an individual's cancer so that therapy could be individualized with drugs that were most likely to be effective. This would prevent unnecessary toxicity from an ineffective drug. As staging techniques become ever more sophisticated they will incorporate increasing information about the biology of individual cancers and play an even more pivotal role in treatment decisions.

CIRCULATING TUMOR CELLS

Clinically, relatively large numbers of circulating tumor cells (CTCs) are found in the blood of some cancer patients, yet surprisingly relatively small numbers of metastases result (see Chapter 12). Importantly, in clinical cancer, exemplified by breast cancer, such CTCs are readily detected and their presence may correlate with prognosis. It is likely that only a subset of CTCs are capable of establishing metastases (in breast cancer probably those that are CD44+/CD24−). Survival of CTCs in the circulation is efficient (often the majority survive) and does not necessarily correlate with the degree of malignancy of the cells, and might therefore limit the usefulness of measuring CTCs in clinical practice at this time. Moreover, even detection of small numbers of cancer cells in secondary organs (micrometasases) may also not be simply predictive of their ability to form significant secondary tumors.

QUESTIONS REMAINING

1 There is a need for a reliable screening tool to detect nonsmall-cell lung cancer at such an early stage that curative treatments are possible.
2 Now that we are able to detect very small nodules (less than 5 mm) on multislice CT, we need to advance our technique so that we can characterize them into small metastases or benign disease.
3 We already have used our knowledge in immunohistochemistry to identify genes that confer a worse prognosis for patients, for example *HER2* in breast cancer. We need to develop this understanding especially for the more common cancers, so that we can identify the patients "at risk" to direct our

more aggressive treatments at them, thus reducing the treatment necessary for the "low risk" patients.

4 More effort must be made to improve our education programs so that patients seek help early for symptoms so that all cancers can be diagnosed at an early stage when curative treatments are available.

5 In this era of biological targeted therapies we are aware that conventional imaging techniques measuring tumor volume may not be helpful since the agents may be cytostatic rather than cytotoxic. Therefore the development of novel imaging techniques to determine the biological effectiveness of these agents is of paramount importance.

CONCLUSIONS AND FUTURE DIRECTIONS

Making a diagnosis of cancer involves a process of first having a high index of suspicion based on the clinical history, which relies on a knowledge of typical patterns of presentation, and then proceeding to a series of investigations aimed at confirming or refuting that suspicion. The investigations should be undertaken in a logical sequence, minimizing the number of tests, discomfort, and cost. Results should be available as rapidly as possible to allow early intervention when necessary and to reduce anxiety. Simple biochemistry and hematology measurements taken on blood specimens can provide an indication of malignancy, particularly if blood loss has occurred or if metastases have interfered with liver or renal function. Serum tumor markers can be valuable if elevated as they will support a diagnosis of malignancy and may provide information on prognosis, response, and relapse. Imaging is fundamental to the diagnosis, staging, and follow-up of most malignancies. Plain radiographs have a limited role but can be helpful to show bronchial neoplasms or bone metastases. Cross-sectional imaging with CT or MRI has revolutionized the anatomical resolution obtained and can highlight the likely site of a primary tumor and any metastases. The addition of contrast medium provides further information that can help distinguish abnormalities, which indicate a greater likelihood of malignancy. Despite all these techniques, the final unequivocal diagnosis depends on pathological examination of a specimen (aspirated cells

or biopsy) obtained from the area of suspicion. This is often facilitated by directing the biopsy needle with ultrasound, CT, or MRI.

Many new imaging techniques are being developed. The ability to do three-dimensional reconstructions on CT images has already led to the development of virtual bronchoscopy and colonoscopy. New contrast agents are available allowing more clear-cut delineation of structures and greater differentiation between benign and malignant lesions. There has been much controversy about the role of sentinel lymph node mapping. This is a technique where the injection of dye and radiolabeled tracer is injected to identify the first node in the regional lymphatics into which the primary tumor drains. It allows for extensive lymph node dissection where this node is positive and allows for a patient with a negative sentinel lymph node to have more conservative surgery. There is now convincing data to support the use of this technique particularly in patients with breast cancer.

Advances in diagnostic techniques, often based on increased understanding of the molecular changes that occur in malignancy, are being developed. These are providing a more confident means of diagnosis on smaller specimens of tissue or numbers of cells and give more information on prognostic features and the potential to gain benefit from new targeted therapies. Increasingly, treatment decisions are being driven by knowledge of the biology of each cancer and information on this is obtained at the time of diagnosis.

QUESTIONS

1 Patients with lymphoma and B symptoms typically complain of:
a. Liver pain.
b. Night sweats.
c. Pain on drinking alcohol.
d. Shortness of breath.
e. Weight loss of more than 10% of their body weight.

2 The tumor marker bHCG in patients with testicular tumors is useful in:
a. Diagnosis.
b. Monitoring the response to therapy.
c. Prognosis.
d. Screening.
e. Therapeutic strategies.

3 A validated screening program exists for:
  a. Breast cancer.
  b. Cervical cancer.
  c. Colorectal cancer.
  d. Gastric cancer.
  e. Nonsmall-cell lung cancer.

BIBLIOGRAPHY

General

Devita, V.T., Hellman, S., and Rosenberg, S.A. (2001). *Cancer Principles and Practice of Oncology*, 6th edn. Philadelphia, PA: Lippincott Williams & Wilkins.

Hermanek, P., Sobin, L.H., and Wittekind, C. (2002). *UICC TNM Classification of Malignant Tumors*. New York: John Wiley & Sons.

Ring, A., Smith, I.E., and Dowsett, M. (2004). Circulating tumor cells in breast cancer. *Lancet Oncology* **5**: 79–88.

Roskell, D.E. and Buley, I.D. (2004). Fine needle aspiration cytology in cancer diagnosis. *British Medical Journal* **329**: 244–5.

Toghill, P.J. (1997). *Essential Medical Procedures*, 1st edn. London: Arnold.

Varella-Garcia, M. (2003). Molecular cytogenetics in solid tumors: laboratory tool for diagnosis, prognosis and therapy. *The Oncologist* **8**(1): 45–58.

Zilva, J.F., Pannall, P.R., and Mayne, P.D. (1987). *Clinical Chemistry in Diagnosis and Treatment*, 5th edn. London: Edward Arnold, pp. 421–5.

Tumor markers

Bast, R.C., Ravdin, P., Hayes, D.F., *et al*. (2001). 2000 update of recommendations for the use of tumor markers in breast and colorectal cancer: clinical practice guidelines of the American Society of Clinical Oncology. *Journal of Clinical Oncology* **19**(6): 1865–8.

Rustin, G.J., Bast, C.R., Kelloff, G.J., *et al*. (2004). Use of CA-125 in clinical trial evaluation of new therapeutic drugs for ovarian cancer. *Clinical Cancer Research* **10**: 3919–26.

Screening

Brawley, O. and Kramer, B.S. (2005). Cancer screening in theory and practice. *Journal of Clinical Oncology* **23**(2): 293–300.

Crum, C.P., Abbott, D.A., and Quade, B.J. (2003). Cervical screening from the Papanicolaou smear to the vaccine era. *Journal of Clinical Oncology* **21**(10): 224–30.

Forrest Report (1986). *Breast Cancer Screening*. London: Department of Health and Social Security.

Frankel, S., Smith, G.D., Donovan, J., *et al*. (2003). Screening for prostate cancer. *Lancet* **361**: 1122–8.

Gotzsche, P.P. and Olsen, O. (2000). Is screening for breast cancer with mammography justifiable? *Lancet* **355**(9198): 129–34.

Hardcastle, J.D., Chamberlain, J.O., Robinson, M.H., *et al*. (1996). Randomized controlled trial of faecal-occult-blood screening for colorectal cancer. *Lancet* **348**: 1472–7.

Ransohoff, D.F. and Sandler, R.S. (2002). Screening for colorectal cancer. *New England Journal of Medicine* **346**: 40–4.

Tyczynski, J.E., Bray, F., and Parkin, D.M. (2003). Lung cancer in Europe in 2000: epidemiology, prevention, and early diagnosis. *Lancet Oncology* **4**(1): 45–55.

Genetics

Garber, J.E. and Offit, K. (2005). Hereditary cancer predisposition syndromes. *Journal of Oncology* **23**(2): 276–92.

Imaging

Eary, J.F. (1999). Nuclear medicine in cancer diagnosis. *Lancet* **354**: 853–7.

Gray, B.N., Van Hazel, G., Anderson, J., *et al*. (2001). Randomized trial of SIR-spheres and chemotherapy versus chemotherapy alone for treating patients with liver metastases from primary large bowel cancer. *Annals of Oncology* **12**: 1711–20.

New imaging techniques

Berger, A. (2003). Positron emission tomography. *British Medical Journal* **326**: 1449.

Bomanji, J.B., Costa, D.C., and Ell, P.J. (2001). Clinical role of positron emission tomography in oncology. *Lancet Oncology* **2**(3): 157–64.

Friedberg, J.W. (2003). PET scans in the staging of lymphoma: current status. *The Oncologist* **8**(5): 438–47.

Jakub, J.W., Pendas, S., and Reintgen, D.S. (2003). Current status of sentinel lymph node mapping and biopsy: facts and controversies. *The Oncologist* 8(1): 59–68.

Vansteenkiste, J., Fischer, B.M., Dooms, C., and Mortensen, J. (2004). Positron-emission tomography in prognostic and therapeutic assessment of lung cancer: systematic review. *Lancet Oncology* 5(9): 531–40.

Van Tinteren, H., Hoekstra, O.S., Smit, E.F., *et al.* (2002). Effectiveness of positive-emission tomography in the preoperative assessment of patients with suspected non-small cell lung cancer. *Lancet* 359: 1388–92.

16

Treatment of Cancer: Chemotherapy and Radiotherapy

A.L. Thomas, R.A. Sharma, and W.P. Steward

The art of medicine consists in amusing the patient while nature cures the disease.
Voltaire, French author, humanist, rationalist, and satirist (1694–1778)

KEY POINTS

- Major advances have been made in the delivery of radiotherapy to minimize the damage to normal tissues.
- The incorporation of chemoradiation regimens in the treatment of solid tumors such as cervical cancer has improved outcomes.
- By exploiting the genetic differences between normal and tumor cells, new targeted therapies have been discovered such as imatinib in CML and herceptin in breast cancer.
- The range of hormone therapies to treat breast cancer has improved so now effective second- and third-line treatments are available.
- The support of patients undergoing cytotoxic chemotherapy is much improved so that the dose intensity of regimens has been optimized and the quality of life for patients undergoing chemotherapy is better.

INTRODUCTION

Chemotherapeutic agents exert their effect by killing cells that are rapidly dividing. The agents are therefore not tumor-cell specific and cause their toxic effect by killing normal cells that are dividing, for example, hair follicle cells and gastrointestinal mucosa. In recent years, our understanding of the molecular pathways controlling the growth of both normal and tumor cells has improved significantly. By exploiting the differences between normal and malignant cells, we can target pathways and receptors unique to the cancer cells, thus avoiding the indiscriminate universal killing of dividing cells of conventional cytotoxics. We are therefore entering the era of biologically targeted therapies in cancer treatment.

Radiotherapy is the application of *ionizing radiation* to treat disease. Ionizing radiation is electromagnetic radiation and elementary particles, which deposit energy in materials through the processes of excitation and ionization events. The forms of ionizing radiation in common use are photon beams (X-rays and gamma rays) and electrons (β-particles). Radiotherapy has been in use for over 100 years, and currently over 60% of patients diagnosed with cancer will receive it at some time during their illness. There have been major technical advances in delivery of radiotherapy in recent years.

444

Hormones are thought to be etiological agents in at least half of the cancers diagnosed in developed countries. Common cancers, such as breast and prostate, often express receptors for sex hormones (e.g. female hormones, such as estradiol, or male hormones, such as dihydrotestosterone). The growth of such cancers is at least in part dependent on hormonal stimulation. The aim of hormonal therapy is to *withdraw the growth stimulus*, either by reducing hormone production or by interfering with the receptor–ligand interaction or signaling associated with agonist activity. Subclasses of estrogen and progesterone receptors are of particular interest with regard to breast cancer, and androgen receptors are relevant to the treatment of prostate cancer. The outline of the chapter is summarized in Figure 16.1.

AIM OF RADIOTHERAPY

X-rays were discovered in 1895 by Wilhelm Röntgen. The principle of X-ray production is the same as that discovered by Röntgen – the sudden deceleration of high-speed electrons results in a release of part of the energy as X-rays. Radiation therapy involves delivery of a precisely measured dose of irradiation to a defined tumor volume while inflicting minimal damage to surrounding healthy tissue. The therapeutic objectives need to be carefully considered, particularly as both the efficacy and the toxicity of radiotherapy are dose dependent. In *radical radiotherapy*, there is a realistic prospect of cure, so significant short-term morbidity and a risk (usually less than 5%) of long-term morbidity is accepted. Whereas, in *palliative radiotherapy*, where the aim is improved symptoms and quality of life, side effects are undesirable.

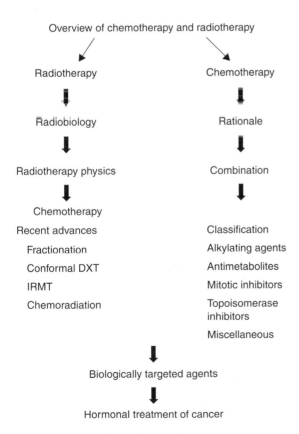

Fig. 16.1 Concept figure for the chapter.

RADIOTHERAPY PHYSICS

The biological effects of radiation are a result of absorption of energy from the radiation. Since absorbed dose is energy divided by mass, radiotherapy is prescribed in units of Grays, where 1 Gray (GY) = 1 joule of energy absorbed per kilogram of mass.

When photons pass through matter, they interact with that matter through a number of different physical processes. Low-energy photons (e.g. those used in diagnostic X-rays) interact with electrons that are bound within atoms, resulting in the *photoelectric effect*. This effect is highly dependent on atomic number, and therefore precisely differentiates tissues such as bone from other tissues such as water. At higher photon energies, the photoelectric effect becomes less important, and photons tend to interact with unbound electrons within the matter. This physical process is called *Compton scatter*. For photon energy produced by linear accelerators (≥ 4 MeV), the effect of photoelectric absorption is negligible and other physical processes predominate.

External beam radiotherapy can be delivered using a number of different modalities.

X-ray tubes. These produce X-rays of energy in the range of 80–300 kV (superficial and ortho-voltage radiotherapy) and are used for the radical treatment of skin cancer and the palliative treatment of skin and bone metastases.

Cobalt gamma ray machines. These release gamma rays equivalent to 2 MV X-rays, and are cheap to operate and service. However, due to safety concerns, because of the low energy produced and the less sharp edges of the beam (called the penumbra), they have been largely replaced by linear accelerators.

Linear accelerators. These use electromagnetic waves to accelerate electrons, which bombard a tungsten target resulting in the production of X-rays with energies of 4–25 MV. These machines enable administration of high doses of radiation to deep-seated tumors and minimize the dose to normal tissues by careful treatment planning (see below). By removal of a tungsten target, they also permit treatment with electrons.

Electrons. They penetrate a short distance, dependent on their energy. They are used to treat skin and lip lesions, particularly areas overlying sensitive structures. They may also be used to treat the whole body (e.g. mycosis fungoides).

Brachytherapy. It involves a radioactive source, usually caesium-137, iodine-125, or iridium-192, placed close to the tumor. An example of brachytherapy is the use of intracavity therapy treatment of cervical cancer. The doses depend on the radioisotope used, and the dose received decreases with the square of the distance from the source (the *inverse square law*).

Systemic therapy. It involves administration of *unsealed sources*. This is exemplified by the treatment of thyrotoxicosis and thyroid cancers, where oral dosing with iodine-131 has been successfully employed for more than half a century. Radioiodine therapy is remarkably safe and is not known to increase risk of other cancers, the only common side effect being hypothyroidism.

RADIOBIOLOGY

Ionizing radiation causes extensive cellular damage, principally via the formation of free radicals. Cells have a considerable capacity for

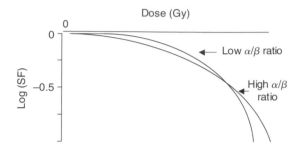

Fig. 16.2 Survival curves for cells with different α/β ratios subjected to irradiation. The more steeply curving survival curve has a lower α/β ratio when fitted to the linear quadratic equation. Note log scale. SF = surviving fraction following irradiation.

repair and cell survival depends on a number of factors (see Chapters 8 and 10). In an average cell, 1 Gy causes damage to over 1000 bases in DNA, approximately 1000 single-strand breaks in DNA and approximately 40 double-strand breaks. *Double-strand DNA breaks* are considered the most significant types of molecular damage, and correlate with the probability of cell survival. Typically, survival curves are continuously bending, with a slope that steepens as the dose increases (see Fig. 16.2). Mathematically, a continuously bending curve is most simply described by a linear quadratic equation of the form: surviving fraction = exponential $(-\alpha D - \beta D^2)$, where D is the given single dose and α and β are parameters characteristic of the cells concerned. Since the *ratio of α/β* gives the relative importance of the linear dose term and the quadratic dose term for those cells, this ratio is used to characterize the radiosensitivity of a particular tissue type.

To improve the *therapeutic ratio* (differential effect on malignant tissue compared with normal tissue), the total dose of radiotherapy can be divided into small daily amounts – *fractionation*. Fractionated radiotherapy inflicts more damage on tumor cells than on normal cells for a given dose of radiation. Moreover, cell and tissue kinetics usually favor recovery and repopulation of damaged normal tissue over tumor tissue (see Box 16.1). Although tumors are very diverse, the radiobiological properties of most tumors are similar to those of acute responding tissues, that is, a high α/β ratio results in moderate sensitivity to changes in fraction size and some dependence

---

Box 16.1 Radiobiological factors that influence response to fractionated radiotherapy (the five Rs)

1. Radiosensitivity. The intrinsic radiosensitivity of different tumor types determines the likelihood of survival following irradiation. Classically, lymphoma, myeloma, and small cell lung cancer are considered more radiosensitive than most other tumor types.
2. Repair. Cells differ in their capacity to repair DNA damage, particularly single- and double-strand breaks in DNA. Repair is considered more effective in nonproliferating cells, but it is now known that repair processes take at least 6 hours to be performed effectively.
3. Repopulation. Surviving tumor cells may proliferate more rapidly than those in normal tissues. The overall treatment time influences the repopulation of the tumor cells and the acute toxicity in acute responding normal tissues.
4. Reoxygenation. Hypoxic cells are more resistant to radiation than oxygenated cells. Giving tumor cells time to reoxygenate may improve their radiosensitivity.
5. Redistribution. Cells in certain phases of the proliferative cycle (e.g. G_2 and M phases) are more radiosensitive than cells in other phases of the cell cycle. Given time, cells can redistribute themselves throughout the phases of the cells.

Box 16.2 Acute side effects of radiotherapy

1. Mucosa of the gastrointestinal tract. Common side effects include mucositis (including loss of taste), and esophagitis (heartburn). Good oral hygiene and suitable analgesics are important during radiotherapy in order for adequate oral intake. A soft diet is advised, but if the patient is unlikely to tolerate this, a percutaneous gastrostomy tube should be inserted before the commencement of radical radiotherapy. This tube can then be used for the administration of feeds, liquids, and drugs.
2. Brain. Nausea, vomiting, alopecia, and transient worsening of neurological symptoms are common side effects of radiotherapy. Antiemetics and corticosteroids are common treatments used to palliate these symptoms.
3. Abdomen. Nausea and vomiting are common side effects of radiotherapy, usually requiring strong antiemetics such as serotonin receptor antagonists.
4. Pelvis. Diarrhea, dysuria, and urinary frequency are common side effects of radiotherapy to the pelvis. Antidiarrheal agents are often prescribed, and midstream urine samples should be used to check for opportunistic bacterial or fungal infections.
5. Skin. Dry desquamation can advance to moist desquamation, a result of radical radiotherapy. Clear good skin and the avoidance of friction are important. Water-based moisturizers and corticosteroid creams can be applied topically.

---

on the total treatment times. It has been shown that extending total treatment time or allowing significant gaps during treatment can decrease the efficacy of the radiotherapy.

The timing and severity of injury to normal tissue depends on the rate of turnover of the mature cells in the tissue. Epithelial and hematopoietic tissues have rapid turnover and are therefore susceptible to *acute effects* over days or weeks (Box 16.2). Tissues with slower turnover are susceptible to late effects, with a timescale of months or years. In fact, risk of major late effects to spinal cord, lung, liver, and kidney are often the dose-limiting factors in radiotherapy. Since stem cells of late responding tissues have more curved survival curve (i.e. a lower α/β ratio) than cells of acute responding tissues, the late responding tissues are particularly sensitive to changes in fraction size. The probability of *long-term toxicity* can therefore be diminished significantly by using small fractions of radiotherapy rather than larger fractions, but this

consideration must be balanced against the effect of the overall treatment time on the tumor tissue.

TREATMENT PLANNING

Management plans are highly individualized and based on the nature of the cancer, the general health of the patient and the goal of therapy. Treatment planning involves a number of steps, as shown in Figure 16.3. The *gross tumor volume* consists of all known microscopic diseases, including any abnormal regional lymph nodes. The *clinical target volume* encompasses the gross tumor volume plus regions considered to harbor potential microscopic disease. The *planning tumor volume* provides a margin around the clinical target volume to allow for internal target motion, other anatomical motion

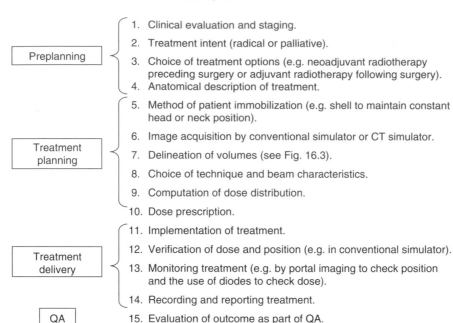

Preplanning
1. Clinical evaluation and staging.
2. Treatment intent (radical or palliative).
3. Choice of treatment options (e.g. neoadjuvant radiotherapy preceding surgery or adjuvant radiotherapy following surgery).
4. Anatomical description of treatment.

Treatment planning
5. Method of patient immobilization (e.g. shell to maintain constant head or neck position).
6. Image acquisition by conventional simulator or CT simulator.
7. Delineation of volumes (see Fig. 16.3).
8. Choice of technique and beam characteristics.
9. Computation of dose distribution.
10. Dose prescription.

Treatment delivery
11. Implementation of treatment.
12. Verification of dose and position (e.g. in conventional simulator).
13. Monitoring treatment (e.g. by portal imaging to check position and the use of diodes to check dose).
14. Recording and reporting treatment.

QA
15. Evaluation of outcome as part of QA.

Fig. 16.3 The radiotherapy planning process. The most common staging system used is the tumor node metastasis (TNM) system.

during treatment (e.g. respiration) and variations in treatment setup. Finally, a margin must be added to account for radiation physical characteristics, such as the penumbra. These target volumes are shown in Figure 16.4.

Patient immobilization is used when highly sensitive structures are close to target, for example, spinal cord or brain. A thermoplastic shell is often used to immobilize the head and keep the neck position constant with the advantages that treatment fields can be drawn on the shell rather than the patient and the shell can be fixed to the treatment couch.

Recent advances

Fractionation. In continuous hyperfractionated accelerated radiotherapy (CHART), radiotherapy is given three times per day and completed in 12 days. In the first large randomized clinical trial in the United Kingdom, the use of CHART in the treatment of nonsmall cell lung cancer appears to have improved survival compared with conventional fractionation. (*Hyperfractionation* = smaller doses per fraction; *acceleration* = shortening the overall treatment time.)

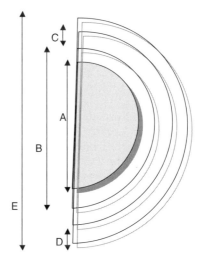

Fig. 16.4 Schematic representation of "volumes" in radiotherapy. The treatment volume includes tumor volume, potential areas of local microscopic disease around tumor, and margin of surrounding normal tissue and a margin added on account of physical considerations. In the figure, A shows gross tumor volume, B shows clinical target volume, C shows internal margin (which cannot be altered significantly), D shows setup margin (a physical concept with some scope for alteration), and E shows planning target volume.

Conformal radiotherapy. Advances in tumor "simulation" using CT scanning and special software to recreate physical movements enable unprecedented reconstruction of tumor anatomy in three dimensions. When this is coupled with delivery devices such as multileaf collimators, where beam shape can be varied in three dimensions, this further limits the exposure of normal tissue to damaging radiation. The contemporary treatment of prostate cancer with radical radiotherapy is an example of this technique and initial results suggest that a higher dose can be delivered to the prostate with a lower chance of late effects in the rectum.

Stereotactic radiotherapy. The treatment of small malignant lesions in the brain and benign arteriovenous malformations has been revolutionized by the use of sterotactic radiotherapy. More than 200 beams of gamma rays may be focused on one point, allowing tightly conformed doses of radiotherapy to be delivered as a single dose or as fractionated doses. Lesions must meet stringent criteria in order to be considered amenable to stereotactic radiotherapy.

Chemoradiation. The concomitant use of chemotherapy and radiotherapy is a curative modality in the treatment of anal, head and neck, and cervical cancer, and may be considered as an alternative to surgery. Although the concomitant use of chemotherapy with radiotherapy increases acute and late damage to normal tissues, it is an established standard of care for carcinoma of the cervix and anus. In squamous head and neck cancer, meta-analyses have suggested a small but consistent survival benefit from the addition of chemotherapy to radical radiotherapy. The expanding use of chemoradiation is currently an exciting area of research, particularly the combination of newer chemotherapy drugs and biological therapies with radiotherapy in the treatment of gastrointestinal cancers.

Rationale of chemotherapy

To understand the rationale of cytotoxic chemotherapy, it is important to recognize the features of tumor growth (see Chapter 5). The factors responsible for determining the growth of a tumor include the cell cycle time, growth fraction, that is, the number of cells that are undergoing cell division, and the number of cells. Also the more cancer cells there are, the more likely that some will become resistant to chemotherapy and prevent cure. The asymmetric sigmoidal growth curve ("Gompertzian" growth curve) describes the natural history of tumor growth. By the time they are clinically apparent (around 10^9 cells or more) most tumors are in the relatively slow phase of growth (near the plateau of the Gompertzian growth curve). Unfortunately, this is the time that chemotherapeutic agents are least likely to prove effective. However, if the tumor can first be resected then subsequent chemotherapy should be more effective as remaining cancer cells are now likely to be in logarithmic growth.

Surgery

Historically, surgery was the only treatment option for cancer, and in the modern era often still offers the best chance of a cure. There are six main areas where surgery plays a fundamental role in the treatment of cancer (Box 16.3).

Recently, laparoscopic or "key hole" surgery has become a more attractive proposition in cancer patients. Alongside reduced morbidity and hospital stay, recent data from the United States suggest that this technique can be as effective as removal at laparotomy, at least for some cancers.

Combination chemotherapy

In modern oncological practice, chemotherapeutic agents are used in combination rather than as sequential single therapies. For a particular tumor type, only drugs that are known to be effective as single agents should be incorporated and ideally these drugs should be used at their optimal dose and schedule. It is imperative that the combination of drugs chosen should have a minimal overlap in toxicity to achieve this goal. The treatment should be delivered on an intermittent basis, with the shortest possible time between treatments that allows the recovery of the most sensitive normal tissue (e.g. the bone marrow or gut). This intermittent scheduling takes advantage of the observation that tumor cells recover more slowly from cytotoxic damage than do their normal counterparts. Thus, with each sequential cycle, the tumor population should be increasingly

---

Box 16.3 Role of surgery in oncology

Surgery for diagnosis and staging. Historically, surgery was important in both the diagnosis of patients with cancer and also their staging. Before the advent of accurate CT scanning, surgery was necessary for staging, for example, the staging of lymphoma. This indication for surgery is now declining although accurate surgical staging of patients with ovarian cancer remains of fundamental importance. With ultrasound and CT-guided techniques, it is now possible to biopsy areas of the body, which previously were inaccessible. Again, the need for surgery still exists, for example, in the excision biopsy of lymph nodes potentially involved by lymphoma.

Surgery as treatment of the primary tumor. When a tumor is confined to the anatomic site of origin, surgery usually is the main form of treatment. Obviously, the extent of that surgery is important to ensure that the tumor has been completely excised locally. The intention of curative surgery is to remove the entire solid tumor "en bloc." There are really only very limited indications now to perform debulking surgery. One of these indications would be in the treatment of extensive ovarian cancer.

Surgery to resect metastases. Nowadays, surgery is used not only to resect primary tumors but also to perform metastasectomies. An example of this would be to remove isolated liver metastasis in the treatment of colorectal cancer or pulmonary metastasis in the treatment of soft tissue sarcoma. Studies have shown extremely good survival rates for patients with liver involvement from colorectal cancer where chemotherapy has been used to down-stage metastasis prior to surgical resection.

Surgery as preventative treatment. For a patient who is at high risk of developing a malignancy, surgery can be used as a preventative technique. The rationale here is to remove the organ at risk prior to the development of cancer. An example of this would be to perform a panproctocolectomy in patients in their teens that have polyposis coli. The issue of prophylactic surgery in women that carry the *BRCA1* and *BRCA2* genes is more complicated. To fully reduce their risk of malignancies, prophylactic bilateral mastectomies need to be performed in combination with salpingo-oopherectomy. Clearly, this is extremely major surgery and the patient needs to have undergone genetic counseling before considering this.

Reconstructive surgery. Once a patient has undergone radical excision of their primary tumor, surgery may have a role in reconstructing the organ removed. Women are now offered a number of reconstruction options following a radical mastectomy. The cosmetic results are excellent and part of this has been due to the development of better microsurgical techniques. In patients that undergo panproctocolectomy, pouches can be made to produce a false rectum so that patients do not need to have a permanent stoma bag. The success of this type of bowel surgery has been greatly aided by the development of highly accurate stapling devices that can reliably perform anatomizes and reduce the leak rate.

Palliative surgery. Finally, there still remains a palliative role for surgery. This could be in performing a toilet mastectomy for a patient with a large ulcerating breast cancer or a gastric bypass procedure for a patient with a distal gastric cancer causing gastric outlet obstruction. With the development of endoscopy techniques, sometimes it is possible to place a stent through the scope. It is likely that we will see more endoscopic treatments being performed in the next few years.

---

depleted leaving the normal population relatively unaffected. Wherever possible, it is preferable to use drugs with known synergistic killing effects, for example, the combination of oxaliplatin and 5-fluorouracil (5-FU). Another possibility is to use a combination of drugs that can kill cancer cells at different stages of the cell cycle (cell cycle dependent drugs).

Some regimens use alternating cycles of different drug combinations. One of the main principles behind this practice is the Goldie–Coldman hypothesis. The model predicted that the mutation of tumor cells to drug resistance occurred at a rate proportional to the genetic instability of a particular tumor. Therefore, the probability that a tumor at diagnosis would contain resistant clones would be proportional to both the tumor size and the inherent mutation rate. Theoretically, tumors should initially respond to treatment but recur as resistant clones expand. Therefore, the Goldie–Coldman hypothesis predicted that drug resistance could be present even with small tumors and the maximum chance of cure would occur if all available effective drugs were used simultaneously. This was the basis behind the development of using combinations of effective, noncross-resistant drugs in *alternating* cycles.

In the 1980s, Norton and Day reviewed the Goldie–Coldman hypothesis. In their overview, they predicted that if two agents were used and if the less effective was given first, the overall outcome would be superior to initial administration of the more effective agent. This has been the rationale for using sequential therapies in the treatment of some tumor types, for example, breast cancer.

Indications for chemotherapy

Chemotherapy is the systemic treatment most commonly used to treat advanced cancer. In a limited number of situations, it may be used as the primary or sole modality of treatment (e.g. chemosensitive cancers, such as leukemia or teratoma). For the majority of the common solid tumors, chemotherapy is frequently used to reduce the volume of disease and palliate symptoms caused by the cancer. A further indication for chemotherapy is to use it as an adjuvant after the primary tumor has been controlled by either surgery or radiotherapy. Here the rationale is to eradicate subclinical micrometastatic disease and reduce the risk of recurrence. An example where adjuvant chemotherapy is extremely useful is in patients with breast cancer who have nodal involvement.

Neoadjuvant chemotherapy is being used increasingly to debulk or downstage primary tumors prior to the definitive treatment, which could be either surgery or radiotherapy. An example of this would be to use neoadjuvant chemotherapy to downstage a large breast primary tumor, so that conservative breast surgery could be offered rather than radical mastectomy.

In the majority of cases, chemotherapy is used systemically either intravenously or orally. It may, however, be administered regionally, for example, intrathecally in the treatment of hematological malignancies with a high likelihood of central nervous system involvement. It can also be infused directly into the blood supply of liver metastases from colorectal cancer, or into the peritoneum in the treatment of advanced ovarian cancer.

CLASSIFICATION OF CYTOTOXIC DRUGS

Historically, anticancer agents have been classified according to the phase of the cell cycle during which they work. With the increasing number and complexity of chemotherapeutic agents, classification according to their mode of action is far more relevant.

Alkylating agents

These were the first drugs to successfully treat cancer, and comprise a diverse group of highly reactive molecules that react with DNA via the alkyl chemical group ($R–CH_2$) to promote single- or double-strand DNA breaks and DNA crosslinking. These effects result in apoptosis or growth arrest (Chapter 10).

Some alkylating agents have two available alkyl groups that form crosslinks and are known as bifunctional. These are more cytotoxic than alkylating agents with only one available alkyl group. The most commonly used alkylating agents bind in the major groove of DNA at the N_7 position of guanine. Minor groove binders are currently in development. In addition, the alkylating agents also bind to RNA and other cell proteins. Exactly how much these interactions contribute to the overall toxic effect remains the subject of further research.

Types of alkylating agents include:

- Nitrogen mustards, for example, melphalan, chlorambucil, cyclophosphamide, ifosfamide.
- Alkyl sulfonates, for example, busulfan.
- Aziridines, for example, thiotepa.
- Tetrazines, for example, dacarbazine.
- Nitrosoureas, for example, BCNU, CCNU.
- Metal salts, for example, cisplatin, carboplatin, oxaliplatin.

Melphalan is a nitrogen-mustard derivative of the amino acid phenylalanine, mainly used intravenously in the treatment of myeloma. Chlorambucil is a nitrogen mustard administered orally to patients with lymphoma. The nitrosoureas are a group of compounds that were developed in the 1960s by the US National Cancer Institute Screening Program. They alkylate DNA through formation of chloroethyl diazonium hydroxide and isocyanate. BCNU and CCNU are particularly lipophilic and as such are commonly used to treat brain tumors. The tetrazines, dacarbazine and temozolomide, are small molecules that release a reactive diazonium ion that alkylates DNA. Dacarbazine is used to treat malignant melanomas, sarcomas, and Hodgkin's disease, and temozolomide is a new agent active in malignant gliomas.

Platinum compounds

Cisplatin, carboplatin, and oxaliplatin (1,2-diaminocyclohexane platinum), are widely used in the treatment of malignancies. They all inhibit DNA synthesis through the formation of intrastrand crosslinks in the DNA and formation of DNA adducts. Serendipity played a major part in the discovery of cisplatin in 1969. At that time, ongoing studies were looking at the effect of electric currents on the growth of bacteria. Bacterial growth was reduced and it transpired that platinum from the electrodes was responsible. Cisplatin is very effective in the treatment of testicular tumors, but use can be limited by kidney damage (nephrotoxicity), though this can in part be reduced by adequate hydration protocols. Carboplatin is an analog of cisplatin and has less nephrotoxicity and neurotoxicity. It has many applications, particularly in the management of lung and ovarian cancer.

Oxaliplatin is a third-generation platinum analog, and differs from the other platinum drugs by having a large 1, 2-diaminocyclohexane (DACH) ligand in its structure. It is thought that the presence of the DACH group renders the molecule nonrecognizable by the mismatch repair process. This means that the DNA repair mechanisms responsible for platinum resistance are not so effective in repairing oxaliplatin-induced DNA damage. Oxaliplatin is therefore useful in tumors where cisplatin has failed, and also in colorectal cancer, a tumor type not classically treated by platinum drugs. Oxaliplatin does have neurotoxicity, although this differs from that typically seen with cisplatin. Patients develop a peripheral neuropathy with cumulative drug exposure but, in addition, they also develop an acute neuropathy that is characterized by cold induced tingling (paraesthesia) in the fingertips and toes after infusion. This can also affect the throat after cold drinks and the patient develops a feeling of choking. Studies have demonstrated that oxaliplatin has a synergistic affect when administered concomitantly with 5FU chemotherapy, and therefore this is the most common oxaliplatin regimen used.

Antimetabolites

They structurally resemble the naturally occurring purines and pyrimidines involved in nucleic acid synthesis. Since they are mistaken by the cell for normal metabolites, they have two modes of action. The first is to inhibit key enzymes required for DNA synthesis, and the second is to become incorporated into the DNA and RNA and therefore produce incorrect codes and cause strand breaks or premature chain terminations. The antimetabolites are S-phase specific drugs (Chapter 4), because their primary effect is on DNA synthesis. They can be divided into four groups.

Folate antagonists   They are exemplified by Methotrexate, which inhibits dihydrofolate reductase (DHFR), an essential enzyme in folate metabolism. Dihydrofolate reductase maintains intracellular folate in the fully reduced tetrahydrofolate form required for synthesis of thymidine monophosphate and other purine nucleotides. Methotrexate is structurally related to folic but has a greater affinity for dihydrofolate reductase, which it binds and inactivates. It is possible to overcome the metabolic block and the activity of dihydrofolate reductase by administering folinic acid. This is a tetrahydrofolate and therefore provides an alternative source for the synthesis of nucleic acids. This has importance, particularly when high dose methotrexate is used, as folinic acid can be administered to the patient as a "rescue" from the toxic effect of the treatment. Clinically, the drug can be given orally, intravenously, or at a low dose intrathecally. It is excreted and changed in the urine and widely distributed in body fluids. The acute side effects of methotrexate are similar to those of the other S-phase agents, namely bone marrow suppression, mucositis, and diarrhea. Methotrexate is most commonly used in the treatment of non-Hodgkin's lymphoma, acute lymphoblastic leukemia, choriocarcinoma, and breast cancer.

Over the last 10 years, new antifolates have been developed including raltitrexed and pemetrexed. Raltitrexed is a specific inhibitor of thymidylate synthase (TS), thus reducing the nucleotides necessary for DNA synthesis. Due to toxicity, it is generally reserved for patients with colorectal cancer who have coexistent cardiac disease. Pemetrexed is a multitargeted pyrrolopyrimidine-based antifolate. When polyglutamated, it targets a number of folate dependent enzymes, for example, DHFR and TS. It has activity in a number of tumor types including mesothelioma.

Arabinosides Cytosine arabinoside is an analog of deoxycytidine. It is a competitive inhibitor of the enzyme DNA polymerase. Its principal action is as a false nucleotide competing for the enzymes, which are responsible for converting cytidine to deoxycytidine and for incorporating deoxycytidine into DNA. It is used mainly in the treatment of acute leukemia. Gemcitabine is a second-generation pyrimidine analog. It is a difluoronated analog of cytosine arabinoside and is metabolized by nucleotide kinase intracellularly to the active diphosphate and triphosphate. The triphosphate derivative of gemcitabine inhibits the enzyme deoxycytidine diaminase, which is responsible for deactivating gemcitabine. This increases the intracellular half-life of gemcitabine. It is administered intravenously and is the standard therapy for pancreatic tumors and is also used in combination with platinum in the treatment of nonsmall cell lung cancer.

Fludarabine is another analog of cytarabine. It is an adenosine analog, which is incorporated into DNA and inhibits enzymes important in DNA synthesis, such as ribonucleotide reductase and DNA polymerase. It is a highly effective chain terminator. Fludarabine is also incorporated into RNA and its proteins, although the cytotoxic effect of this is not known. It is the most active single agent in the treatment of chronic lymphatic leukemia.

Antipyrimidines 5-Fluorouracil is a fluoropyrimidine that inhibits DNA synthesis by inhibiting the main enzymes involved in the synthesis of cytosine and thymine. 5FU enters the cell via the uracil transport mechanism. It is converted to fluorouridine diphosphate (FudR) by the enzyme thymidine phosphorylase and then to its active metabolites 5-fluoro-2 deoxyuridine monophosphate (FdUMP) by the enzyme thymidine kinase. In the presence of its cofactor 5,10-methylene tetrahydrofolate, FdUMP forms covalent bonds with TS, creating a complex, which inhibits the formation of thymidine from deoxyuridine monophosphate (dUMP), thus inhibiting DNA synthesis. In clinical practice, it is usual to administer folinic acid in combination with 5FU therapy as this has been shown to increase the effectiveness of the drug presumably by stabilizing the binding of FdUMP to TS. 5FU is also erroneously incorporated into RNA instead of uracil and thus inhibits RNA synthesis. One cause of chemoresistance to 5FU is that increase of dUMP may overcome the inhibition of TS by FdUMP. In an attempt to overcome this, specific TS inhibitors are being developed.

5-Fluorouracil is a prime example of a chemotherapy agent whose activity and toxicity depends upon its scheduling. Up until recently, it was only possible to give 5FU intravenously due to its unpredictable oral bioavailability. There are now oral 5FU analogs available: capecitabine is activated to 5FU in a three step enzymatic reaction with tumor-selective generation by exploiting the higher level of thymidine phosphorylase in tumor compared with normal tissue. The pharmacokinetics of orally administered capecitabine essentially mimic a continuous infusion, rather than bolus 5FU. Clearly, it is a significant advantage to be able to administer tablets to patients, rather than use infusional chemotherapy with central venous catheters and all their potential complications. The second oral fluoropyrimidine is UFT, a combination of ftorafur and uracil. It needs to be given in combination with folinic acid. S1 is a combination of ftorafur and potassium oxonate. The ftorafur inhibits dihydropyrimidine deaminase and the oxonate accumulates in the gastrointestinal tract and inhibits the breakdown of 5FU, which theoretically should increase the availability of the active

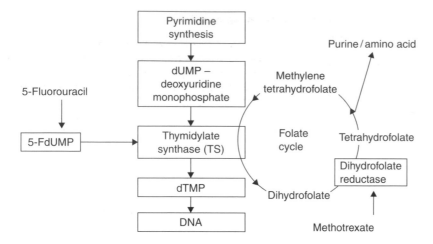

Fig. 16.5 Mechanisms of action of the antimetabolites methotrexate and 5-fluorouracil. These agents both interfere with the folic acid (folate) cycle, which particularly compromises DNA synthesis in replicating cells.

metabolites of 5FU in the tumor. Antimetabolite pathways are simplified in Figure 16.5.

Antipurines There are two purine antagonists: 6-mercaptopurine and 6-thioguanine. The latter is an analog of guanine and is now rarely used in clinical practice. 6-mercaptopurine inhibits purine synthesis by being erroneously incorporated into DNA and thus interfering with DNA replication. It is normally given orally and used in the treatment of leukemia.

Mitotic inhibitors

These include the vinca alkaloids, vincristine, vinblastine, and vinorelbine, and the taxanes. The vinca alkaloids are either naturally occurring (from the periwinkle plant), or semisynthetic. They bind to tubulin and prevent assembly of the mitotic spindle (Chapter 4) required for chromosome segregation in mitosis. Vincristine is used in the treatment of lymphomas, leukemias, and breast cancer, and its major dose limiting toxicity is peripheral neuropathy. Vinblastine is used in lymphoma and nonsmall cell lung cancer, but may cause deficiency in platelets (thrombocytopenia). Vinorelbine is the most recently discovered vinca alkaloid and tends to cause less neuropathy than the older agents. The taxanes are more recent agents that in contrast to the vinca alkaloids promote the assembly of the mitotic spindle but instead inhibit its disassembly. Paclitaxel is an extract from the

bark of the pacific yew and initially its development was restricted due to limited supplies and also its solubility properties. Docetaxel was synthesized from the needles of the tree and also a more commonly found yew species. Both are now made semisynthetically and are used in treating ovarian and breast cancer. These drugs have created great interest due to their perceived efficacy and also controversy where their use has been restricted due to cost. Numerous studies are ongoing and should resolve some of these issues. Bone marrow suppression is one of the major limiting side effects, though paclitaxel may cause neuropathy. Hypersensitivity reactions are well recognized and it is now standard practice to premedicate patients with steroids and H_2 inhibitors to prevent this. Docetaxel is characterized by a fluid retention syndrome with the development of edema and weight gain. The etiology of this is unclear but again premedication usually prevents the overall fluid retention effect.

More recently, other tubulin polymerizing agents have entered clinical trials. In particular, epothilone B shows promise as it appears to be active in tumor types that are taxane resistant.

Topoisomerase inhibitors

Topoisomerases are enzymes that control the coiling and uncoiling of DNA and are important during DNA replication. Topoisomerase 1 binds to double-stranded DNA and causes a single-strand break allowing super coiled regions of DNA to unravel. Following religation the DNA is able to complete replication. Interestingly, levels of topoisomerase 1 are known to be higher in malignant compared with normal tissues, especially in the colon and ovaries. Inhibitors of topoisomerase 1, such as irinotecan and topotecan, interfere with the topoisomerase–DNA bond and prevent the religation of the single-strand DNA break. Thus during replication, the replication fork religation interacts with the single-strand break in the DNA forming an irreversible double-strand break in the DNA. The majority of cells will then die by apoptosis or growth arrest in G_2 phase. The topoisomerase 2 enzyme binds to complementary double-strand DNA causing both strands to cleave and move apart allowing a second segment of double-strand DNA to pass through the cleavage site. This site then religates. Topoisomerase 2 inhibitors prevent the religation and therefore maintain double-strand DNA breaks that impede DNA replication.

Camptothecin was the first topoisomerase 1 inhibitor to be discovered. Its further development was hindered by toxicity and studies are ongoing using a pegylated formulation. Irinotecan is now widely used in the treatment of colon cancer. Following intravenous administration, it is activated by carboxylesterase to 27-ethyl-10-hydroxycamptothecin (SN-38). The major route of excretion of SN38 is through the liver and therefore care has to be taken when prescribing irinotecan for patients with abnormal liver function tests. Another side effect of irinotecan therapy is diarrhea, but this responds well to prophylactic administration of atropine and if treated promptly will respond to constipating agents, such as loperamide and codeine. Failure to introduce these agents quickly can result in extreme diarrhea that, particularly if it coincides with the myelosuppression, can be life threatening.

Etoposide is the most widely used topoisomerase 2 inhibitor being used in Hodgkin's disease, non-Hodgkin's lymphoma, small cell lung cancer, and testicular teratoma. It is another example of a drug that is schedule dependent and is more effective given over 5 days than on 1 day. It can be given both orally and intravenously and the major dose limiting side effect is myelosuppression. In addition, alopecia and mucositis are common.

Miscellaneous

There are a group of chemotherapy agents that were initially developed as antibiotics. These are a group of related antimicrobial molecules produced by *Streptomyces* species in culture. They were found to have significant cytotoxicity and therefore are now classified as antitumor antibiotics. They include the anthracyclines, doxorubicin, daunorubicin, epirubicin, and idarubicin, and the nonanthracycline antibiotics, bleomycin and actinomycin D. They act partly as alkylating agents but also intercalate DNA and inhibit topoisomerase 2 function. In addition, they can generate free radicals, which cause oxidative damage to cellular proteins, thus inhibiting their action.

Doxorubicin is the oldest anthracycline and remains very widely used in clinical practice. Doxorubicin, daunorubicin, and epirubicin are all given intravenously. Idarubicin is an oral synthetic analog of daunorubicin. The main toxicity after anthracycline treatment is myelosuppression. Daunorubicin and doxorubicin and, to a lesser extent, epirubicin, cause cumulative cardiotoxicity and there is a maximum recommended total dose. Consequently, anthracyclines should be avoided in patients with a significant history of cardiac disease. Dexrazoxane is an agent that reduces the incidence of cardiac toxicity in patients with breast cancer receiving doxorubicin by preventing the formation of free radicals.

Actinomycin D is a DNA intercalator and blocks the transcription of DNA by intercalating base pairs and preventing RNA from recognizing and binding to the DNA. It is a very toxic drug and is usually given intravenously as a single injection, mainly in the treatment of childhood cancers.

Bleomycin is another antibiotic and causes single- and double-strand DNA breaks through the production of free radicals. It is most active on cells during G_2 phase and causes cell cycle arrest in G_2. In may be administered either intramuscularly or intravenously. It is renally excreted and appears to concentrate mainly in the skin and lungs. This is important since the main side effect is pulmonary fibrosis, particularly after high flow oxygen therapy. Mucocutaneous side effects are also seen and these may vary from hyperpigmentation to severe mucosal ulceration. Bleomycin plays a central role in the combination therapy of testicular tumors and lymphomas. It causes very little myelosuppression and is thus an ideal agent to be used in combination regimens.

Enzymes

Asparaginase is an example of this class of cytotoxic drug. It is an enzyme that degrades asparagine, an essential amino acid for protein and nucleic acid synthesis. Tumor cells, unlike their normal counterparts, have little or no asparaginase synthetase, the enzyme necessary for making asparagine. They therefore rely on exogenous sources of asparagine. Following treatment with asparaginase, the lack of this essential amino acid inhibits protein and nucleic acid synthesis. It is used most commonly in the treatment of acute leukemias and the main side effect is that of hypersensitivity reaction.

Hydroxyurea blocks the action of ribonucleoside diphosphate reductase, and interferes with the synthesis of DNA. It causes leukopenia and therefore is used in the treatment of myeloproliferative disorders, such as polycythemia.

Biologically targeted agents

Following the recent spectacular progress in understanding the fundamental molecular basis of cancer, we are now entering the era of biologically targeted therapies. By exploiting the well-described differences between normal and malignant cells, new agents have been specifically developed to target gene expression and signaling pathways deregulated in the cancer cells. Some of these agents will be cytostatic rather that cytotoxic (see Chapter 5, and "Differentiation Therapy") and therefore new measures of treatment success will be needed in clinical trials alongside more conventional measures of tumor volume. In developing signal transduction inhibitors, most efforts have been directed against receptor tyrosine kinase and G protein receptors.

These newer agents have been discussed in detail in Chapter 5 (Box 16.4), but a brief description of processes targeted is provided here.

Epidermal growth factor receptor tyrosine kinases (EGFR TK)

These activate numerous signaling pathways involved in tumorigenesis (Fig. 16.6), including

---

Box 16.4 Key significant event for this chapter

One of the most disabling side effects of conventional chemotherapy is nausea and vomiting. The discovery of the $5HT_3$ receptor and then development of the $5HT_3$ receptor antagonists has revolutionized the delivery of chemotherapy. Prior to the availability of these drugs, the antiemetics were not very effective. For the majority of the time, patients were given the best drugs available and were sedated to sleep through the profound nausea or were given high dose metoclopramide. For some, this caused significant acute dystonic reactions that were worse than the vomiting itself. Now with the $5HT_3$ receptor antagonists, chemotherapy can be given to an outpatient as a matter of routine and the quality of life for patients is significantly better.

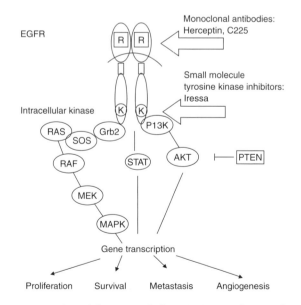

Fig. 16.6 Simplification of the EGFR signaling pathway showing the site of action of the monoclonal antibodies and the small molecule tyrosine kinase inhibitors.

the RAS–MAPK cascade and the PI3K pathway (Chapters 5 and 6), and may thus be targeted in a number of ways therapeutically – examples are shown in Table 16.1.

Signaling through RAS–MAPK and PI3K

RAS and downstream signaling occupy central roles in cancer (Chapter 6) and may be targeted by inhibiting agents at various levels including: farnesyl transferase (Table 16.2), Raf, and MAPK pathway. Inhibitors of mTOR (mammalian target of rapamycin), such as rapamycin, may antagonize some of the growth promoting signals mediated through the PI3K–AKT signal transduction pathways. Other mTOR inhibitors are

under evaluation including rapamycin analogs such as CCI-779.

Matrix metalloproteinase inhibitors

There are a number of enzymes involved in proteolysis of the extracellular matrix, which allows tumor cells to invade (Chapter 12). The family of matrix metalloproteinases (MMPs), which include gelatinases, stromalysins, and collagenases, are the major group of enzymes involved. Their activity is ultimately terminated by tissue inhibitors of metalloprotinases (TIMPs). Synthetic metalloproteinase inhibitors have been developed. Of these, marimastat has been the most widely studied. Initially, early

Table 16.1   EGFR inhibitors

Drug	Target	Company	Tumor	Clinical status
Anti-EGFR monoclonal antibodies				
Trastuzumab herceptin	HER-2	Genentech	Breast	Licensed
Cetuximab, Erbitux C225	EGFR/HER-1	Imclone/Merck	Colorectal	Licensed
ABX-EGF	EGFR/HER-1	Abgenex	Solid tumors	Phase I
2C4, Pertuzumab, Omnitarg	HER-2	Genentech	Solid tumors	Phase I
EMD72000	EGFR/HER-1	Merck	Solid tumors	Phase II
H-R3	EGFR/HER1	NIO	Solid tumors	Phase I

Drug	Description	Tumor	Company	Clinical status
Anti-EGFR TK inhibitors				
ZD1839, Gefitinib Iressa,	Quinazoline, EGFR PO	NSCLC	Astrazeneca	Licensed
OSI-774 Tarceva, Erlotinib	Quinazoline EGFR, HER-2	Breast	Genentech, Roche	Phase III
CI-1033	Quinazoline PAN-HER	Breast	Pfizer	Phase II
GW2016	Quinazoline EGFR, ErbB-2	Solid tumors	GlaxoSmithKline	Phase I
EKB-569	Quinazoline EGFR, HER-2	Solid tumors	Wyeth	Phase I

Table 16.2 Farnesyl transferase inhibitors

Drug	Company	Tumor types	Phase of development
SCH66336 Lonafarnib, Sarasar	Schering-Plough	NSCLC	Phase III
R115777 Tipifarnib, Zarnestra	Orthobiotech, Janssen	NSCLC	Phase III
BMS214662	Bristol Myers Squibb	NSCLC	Phase II

Table 16.3 Angiogenesis inhibitors

Drug	Company	Target	Status
Monoclonal antibody			
Bevacizumab Avastin	Genentech	VEGF-A	Licensed in colorectal cancer
Soluble receptor chimeric protein			
VEGF-trap		VEGF-A	Phase I
Small molecule TK inhibitors			
SU6668	Sugen/Pfizer	VEGFR-1,2, Flt-3 PDGFR, FGFR	Phase III
SU11248	Sugen/Pfizer	VEGFR-1,2,3, PDGFR, c-Kit	Phase III
PTK787/ZK	Novartis	VEGFR-1,2,3, c-Kit, PDGFR	Phase III
ZD6474	AstraZeneca	VEGFR-2, EGFR	Phase I
Vascular targeted agents			
Combretastatin A	Bristol Myers Squibb	Endothelial tubulin	Phase II
Inhibitors of endothelial cell proliferation			
Angiostatin			Phase I
Thalidomide		TNF-α	Approved in US

clinical studies showed these agents to have promising activity. As yet their potential has not been realized in randomized phase III controlled studies.

Angiogenesis inhibitors

These are discussed in detail in Chapter 14 (Table 16.3).

Proteasome inhibitors

Discussed in chapter 11, these drugs, exemplified by Velcade (bortezomib, formerly known as PS341), the first selective proteasome inhibitor with proven preclinical activity in several tumor models and clinical efficacy in patients with refractory or relapsed multiple myeloma, all prevent degradation of various proteins in the proteasome. They show different degrees of selectivity for the proteins whose degradation is affectd suggesting that further advances may enable the targeting of smaller numbers of desired predetermined proteins without impeding degradation of proteins where increasd stability would be detrimental. Velcade is believed to be effective in myeloma by interfering with normal function of the NFκB signaling pathway, thus reducing tumor cell replication and angiogenesis and fostering apoptosis.

Gene therapy

Dysfunctional tumor suppressor genes are the most common genetic lesions identified in human cancers. Working versions of tumor suppressor genes can be restored by introducing them into cancer cells by gene transfer using adenoviral vectors. This approach has been well-studied with respect to direct inoculation of a replication-defective adenovirus that expresses p53. This treatment can induce growth arrest and apoptosis in cancer cells and have very low toxicity, but has as yet to prove its worth in large clinical studies.

Illustration of potential application of targeted therapy in cancer

Recent studies have confirmed the effectiveness of herceptin in advanced stages of breast cancer and, more recently, also in early disease. In fact, testing for the presence of the HER2 mutation and thus selecting patients for herceptin treatment is rapidly becoming common practice, at least in health systems where the costs of treatment can be accommodated. Gastrointestinal stromal tumors (GISTs) are treatment resistant and have a poor prognosis, with high rates of recurrence after surgery. The identification of the central role of c-KIT mutations in progression of GISTs has been exploited by employing imatinib, a small-molecule inhibitor of tyrosine kinases, including BCR-ABL and the PDGFr and also KIT. Imatinib had already revolutionized treatment of CML and now has had a major impact on tumor response rates, survival, and quality of life in GISTs also. Resistance is frequently a problem with single agents in cancer therapy and several new targeted agents are being tested in GISTs, which are or become resistant to imatinib therapy.

To realize the potential of targeted therapy in cancer, it is important to address the emergence of drug resistance in treated patients. Mutant forms of BCR-ABL, KIT, and the EGF receptor (EGFR) have been found that confer resistance to the drugs imatinib, gefitnib, and erlotinib respectively. This will most likely need to be addressed by combination therapy to reduce risk of resistance cells emerging and by developing further agents which can still effectively inhibit growth of these resistant cells. The future potential for targeted therapy in cancers is illustrated by renal cell carcinoma (RCC), which is highly resistant to traditional treatments. New targets for treatment have been identified, many deregulated by mutations in the von Hippel-Lindau (*VHL*) gene, including proteins required for angiogenesis and RCC progression, such as VEGF and various GFs. As discussed in Chapter 14, several agents in development target VEGF (bevacizumab), the VEGF receptor (PTK787, SU11248). Drugs can target various growth factor RTKs such as the PDGF receptor (SU11248 and BAY 43-9006), or the EGF receptor (gefitinib and Cetuximab). Signaling downstream of the RTKs can be targeted, thus the RAF-MAPK pathway can be targeted at the level of RAF (BAY 43-9006, ISIS 5132) or MEK (CI-1040 and PD184352). The PI3K/AKT/mTOR pathway is disrupted by CCI-779. In metastatic colorectal cancer first-line treatments include 5-flurouracil, irinotecan, and oxaliplatin, but other combination therapies are currently under investigation. In particular the potential of using agents that target VEGF and EGFR are being explored.

Hormonal therapy for breast cancer

In premenopausal women, biologically active estrogens are synthesized within the ovarian follicles under the influence of the *pituitary hormones*, namely follicle stimulating hormone (FSH) and luteinizing hormone (LH). Thus, estrogen levels can be lowered either by oophorectomy (i.e. surgical removal of the ovaries), radiotherapy of the ovaries (artificial radiation menopause), or by administration of drugs that suppress the secretion of LH and FSH by the pituitary (chemical castration).

Chemical castration is utilized in the treatment of endometriosis, breast cancer, and prostate cancer and is usually achieved using *gonadotrophin-releasing hormone* (GnRH) agonists (seemingly paradoxically). This works because continuous stimulation by GnRH results in desensitization of the response. However, prior to the onset of desensitization, the GnRH agonists may cause an initial surge in pituitary levels of LH and FSH that can be detrimental in the treatment of malignancy since it may result in a "tumor flare."

The *estrogen receptor status* of breast cancer may predict the usefulness of hormonal manipulation. In metastatic disease, more than 50% of patients with estrogen receptor positive (ER +ve) tumors experience regression of the cancer with hormonal therapy, compared with less that 10% of those with estrogen receptor negative (ER −ve) tumors.

Tamoxifen is a *selective estrogen receptor modulator* (SERM). It acts as a competitive antagonist for estrogen receptors, although it can have partial agonist activity for subclasses of estrogen receptors. Tamoxifen is capable of causing tumor regression in endocrine sensitive tumors in both premenopausal and postmenopausal women. Both tamoxifen and its metabolites are highly protein bound, and it takes 8 weeks for serum levels of the principal metabolite to reach steady state. In addition to the treatment of metastatic breast cancer, tamoxifen is used as adjuvant therapy in women who have undergone surgery with or without radiotherapy as the primary treatment for their cancer. In the adjuvant setting, tamoxifen is currently administered for 5 years. The principal toxicities of tamoxifen are similar to those of the menopause, such as hot flushes, weight gain, nausea, and vaginal discharge. Although rare, patients should be warned about the possibility of thromboembolic events, cataracts and a slightly increased risk of endometrial cancer following long-term usage.

In an attempt to improve efficacy and reduce the effect on bone and endometrium, more selective SERMs have been developed. These include those that are tamoxifen-like including toremifene, fixed ring compounds, such as raloxifene, and selective estrogen receptor downregulators (SERDs), such as fulvestrant. All these compounds are currently in clinical development both as chemoprevention agents, and as therapy for early and advanced breast cancer.

In postmenopausal women, androgens from the adrenal glands act as the principal source of estrogens by conversion in peripheral tissues, particularly subcutaneous fat. The aromatase enzyme system is the principal class of enzymes responsible for this conversion. *Aromatase inhibitors* are in common use in postmenopausal women with breast cancer. Aminoglutethimide was the first to be developed, but its use was associated with lethargy, hypotension, nausea, hypothyroidism, reversible agranulocytosis, and rash. More selective aromatase inhibitors have emerged over the past decade, which can be subdivided into nonsteroidal and steroidal compounds. Common toxicities of these drugs include nausea, headache, hot flushes, and weight gain. Unlike tamoxifen, aromatase inhibitors may decrease bone density and predispose to osteoporosis.

The overall response rate to hormonal manipulation in metastatic breast cancer is approximately 30%. Complete clinical remission is uncommon, and tumor regression may take 2–3 months. Partial responses and stable disease can be observed for many years in some patients. The median duration of response to first hormonal treatment is 18 months, and therapy is generally continued until disease progression. Second, third, and even fourth line hormonal manipulation may be considered in these patients, either alone or in combination with palliative chemotherapy or radiotherapy. Combinations of hormonal therapies are not generally used since no impact on survival superior to monotherapy has been shown.

Hormonal treatment of prostate cancer

The growth of benign and malignant prostate cells is controlled by androgens. The most potent androgen is dihydrotestosterone, produced in the testis by the enzyme 5-α reductase.

Androgen ablation is the primary treatment for men with metastatic prostate cancer. This can be achieved by orchidectomy or the use of GnRH analogs. Approximately 85% of metastatic prostate cancers are androgen sensitive at diagnosis, and response to androgen ablation is often dramatic. Androgen deprivation can also be used in localized prostate cancer to decrease the size of the prostate prior to definitive treatment with radiotherapy or surgery.

Since GnRH analogs are receptor agonists and can cause a transient stimulation of pituitary hormone production, it is important that androgen receptor antagonists are coadministered initially to prevent symptoms from "tumor flare." Androgen receptor antagonists in common use include flutamide and cyproterone acetate. These drugs act at the androgen receptor by inhibiting the agonist activity of dihydrotestosterone. Common side effects include decreased libido, hot flushes, breast tenderness, and impotence. Treatment is continued until disease progression and the median duration of response to first-line hormonal therapy is about 18 months. The prognosis is poor once the tumor develops androgen resistance, although androgen deprivation is generally continued beyond this point to reduce the rate of tumor progression. Diethylstilbestrol (stilbestrol) is sometimes used to treat metastatic prostate cancer which is androgen resistant and the antifungal drug, ketoconazole, which blocks testicular and adrenal steroidogenesis, is currently in clinical trials in this setting.

Androgen deprivation with GnRH analogs is also used in the *adjuvant setting* following radical radiotherapy for locally advanced disease in the hope of reducing the chances of tumor recurrence. The optimum duration of such therapy is unclear, and treatment is currently continued for 2–3 years in most countries. The side effects of GnRH analog treatment include impotence, decreased libido, mild anemia, and decreased bone density. Following short-term therapy, the effects are reversible.

QUESTIONS REMAINING

1 Identify genes that will predict what toxicity a patient will experience with different chemotherapy drugs. For example, we know that patients with dihydropyrimidine dehydrogenase (DPD) deficiency will have significantly more side effects when treated with 5FU. We need to identify more genes that will allow us to individualize chemotherapy so patients can avoid toxicity.
2 Further genomic work needs to be carried out so we can identify predictive genes for response. This again will allow us to "tailor" therapy for patients to ensure that they are receiving the most effective drugs.

3 We have growth factors available that will reliably increase white cell levels and red cell levels postchemotherapy. There is still a need for a reliable and effective growth factor to stimulate platelet recovery after chemotherapy.
4 We need to develop more oral chemotherapy options since patients much prefer to take chemotherapy tablets in the comfort of their homes rather than be subjected to numerous injections.
5 Although striking advances have been made in the delivery of radiotherapy, we need to understand more about intensity-modulated radiotherapy and also develop further drugs that are activated in hypoxic conditions to improve radiosensitivity.

CONCLUSIONS AND FUTURE DIRECTIONS

Options for the treatment of cancer are expanding at a great rate. Technological advances are facilitating the development of new radiotherapy techniques that enable the delivery of higher doses to the total tumor volume, sparing surrounding tissues. As a result, greater local tumor control appears possible and the limiting toxicity of radiotherapy – damage to adjacent normal cells – is minimized. With greater tolerability and efficacy, radiotherapy may take an increasing role in combination with surgery and chemotherapy to reduce the risk of local recurrence of a variety of tumors.

With the ever increasing understanding of the biology of cancer, and the massive expansion of interest in oncology by the pharmaceutical industry, new systemic therapies are being developed at an unprecedented rate. Analogs of existing cytotoxics are in development and have the potential to reduce the toxicity of chemotherapy while retaining or increasing efficacy. New targets, including the cyclin-dependent kinases, provide options for developing agents that have greater efficacy or that could have additive or synergistic activity with existing drugs. Novel strategies aimed at developing drugs that target receptors or components of the cell signaling pathways are currently providing the largest number of new molecules entering clinical trials. Some of these have already entered clinical practice (e.g. imatinib in chronic myeloid leukemia, herceptin in breast cancer, bevacizumab in colon cancer) and provide efficacy with minimal toxicity. Over the next decade, research will focus on developing new targeted drugs and also on determining predictors of response using microarray assays, proteomics and genomics. We may thus be able to tailor therapies to the individual depending on aspects of the biology of their cancer.

As many of the new therapies are cytostatic or do not target the cancer cell (e.g. angiogenesis inhibitors), we may also shift our paradigm of care from one that aims to eradicate all tumor cells into one that aims to stop progression and allow the patient to live with prolonged stable disease. The way in which we conduct clinical trials will need addressing to ensure that we use appropriate objectives. It is not adequate to just use a reduction in tumor volume as an efficacy outcome; for the targeted therapies we need to develop surrogate endpoints to confirm we are "hitting" the intended target. For some new agents, moving their indication from the metastatic to the adjuvant setting may take many years, with huge randomized studies required to establish improvement in overall survival. We need to accelerate this process so that patients are given the most active drugs as soon as possible; this may mean using progression free survival as the primary endpoint of studies rather than overall survival.

It is hoped that with further development of biological agents, the side-effect profile of treatment will improve. Nevertheless, it will be important to develop even better agents to treat adverse events such as chemotherapy induced emesis and thrombocytopenia. Greater understanding of chemoresistance will allow us to understand why some cancers, such as pancreas cancer, remains so difficult to treat. There is the real prospect that we can make a major impact on survival rates for many of the common cancers if these new approaches fulfill their potential.

QUESTIONS

1 Chemoradiation techniques have improved survival in the following tumor types:
   a. Anal cancer.
   b. Cervical cancer.
   c. Ovarian cancer.
   d. Sarcoma.
   e. Small cell lung cancer.

2 The following cytotoxic drugs are alkylating agents:
   a. 5-Fluorouracil.
   b. Doxorubicin.
   c. Irinotecan.
   d. Oxaliplatin.
   e. Vinblastine.

3 Hormone therapy in breast cancer is commonly associated with the following side effect:
   a. Alopecia.
   b. Hot flushes.
   c. Myelosuppression.
   d. Vaginal dryness.
   e. Weight gain.

4 The following are monoclonal antbodies targeting the EGFR:
   a. Bevacizumab.
   b. Cetuximab.
   c. Gefitinib.
   d. Rituximab.
   e. Trastuzumab.

BIBLIOGRAPHY

General

Devita, V.T., Hellman, S., and Rosenberg, S.A. (2001). *Cancer Principles and Practice of Oncology*, 6th edn. Philadelphia, Lippincott-Raven. pp. 363–520.

Nottage, M., and Siu, L.L. (2002). Principles of clinical trial design. *Journal of Clinical Oncology*, **20**(18s), 42s–46s.

Perry, M.C. (2001). *The Chemotherapy Source Book*, 3rd edn. Philadelphia, Lippincott, Williams & Wilkins.

Weiner, L. M. (2000). *The Cancer Protocol Guide*, 1st edn. Philadelphia, Lippincott, Williams & Wilkins.

Radiotherapy

Green, J.A., Kirwan, J.J., Tierney, J., *et al.* (2001). Systematic review and meta-analysis of randomized trials of concomitant chemotherapy and radiotherapy for cancer of the uterine cervix: better survival and improved distant recurrence rates. *Lancet*, **358**: 781–6.

Hanks, G.E. (2000). Conformed radiotherapy for prostate cancer. *Annals Medicine*, **32**: 57–63.

Nutting, C., Dearnaley, D.F., and Webb, S. (2000). Intensity modulated radiation therapy: a clinical review. *British Journal of Radiology*, **73**: 459–69.

Saunders, M., Dische, S., Barrett, A., *et al.* (1999). Continuous hyperfractionated, accelerated radiotherapy (CHART) versus conventional radiotherapy in non-small cell lung cancer: mature data from the randomized multicentre trial. *Radiotherapy and Oncology*, **52**(2): 137–48.

Sims, E., Doughty, D., Macauley, E., *et al.* (1999). Stereotactically delivered cranial radiation therapy: a ten year experience of linac-based radiosurgery in the UK. *Clinical Oncology*, **11**: 303–20.

Steel, G.G. (ed.) (1993). *Basic Clinical Radiobiology*. London: Arnold.

Specific chemotherapy drugs

Galmarini, C.M., Mackey, J.R., and Dumontet, C. (2002). Nucleoside analogues and nucleobases in cancer treatment. *Lancet Oncology*, **3**: 415–24.

Gligorov, J., and Lotz, J.P. (2004). Mictrotubule stabilization in the treatment of solid tumors: Role of the Taxanes. *The Oncologist*, **9** (S2): 3–8.

Hazarika, M., White, R.M., Johnson, J.R., *et al.* (2004). FDA Drug Approval Summaries: Pemetrexed (Alimata). *The Oncologist*, **9**: 482–7.

Ibrahim, A., Hirschfield, S., Cohen, M.H., *et al.* (2004). FDA Drug Approval Summaries: Oxaliplatin. *The Oncologist*, **9**: 8–32.

Pizzolato, J.F., and Saltz, L. (2003). The camptothecins. *Lancet*, **361**: 2235–42.

Hormonal therapy

Campos, S.M. (2004). Aromatase inhibitors for breast cancer in postmenopausal women. *The Oncologist*, **9**: 126–36.

Howell, S.J., Johnston, S.R., and Howell, A. (2004). The use of the selective estrogen receptor modulators and selective estrogen receptor down-regulators in breast cancer. *Best Practice and Research Clinical Endocrinology & Metabolism*, **18**(1): 47–66.

Loblaw, D.A., Mendelson, D., Talcott, J.A. *et al.*, (2004). American Society of Clinical Oncology Recommendations for the initial hormonal management of androgen-sensitive metastatic, recurrent or progressive prostate cancer. *Journal of Clinical Oncology*, **22** (14): 2927–41.

Targeted agents

Ahmed, S.I., Thomas, A.L., and Steward, W.P. (2004). Vascular Endothelial Growth Factor (VEGF) inhibition by small molecules. *Journal of Chemotherapy*, **16**: 8–12.

Eskens, F.A.L.M. (2004). Angiogenesis inhibitors in clinical development; where are we now and where are we going? *British Journal of Cancer*, **90**: 1–7.

Ferrara, N. (2004). Vascular Endothelial Growth Factor as a target for anticancer therapy. *The Oncologist*, **9** (S1): 2–10.

Gschwind, A., Fischer, O. M., and Ullrich, A. (2004). The discovery of receptor tyrosine kinases: targets for cancer therapy. *Nature Reviews*, **4**(5): 361–70.

Janmaat, M.L., and Giaccone, G. (2003) Small-molecule epidermal growth factor receptor tyrosine kinase inhibitors. *The Oncologist*, **8**: 576–86.

Laskin, J.J., and Sandler, A.B. (2003). Epidermal growth factor receptor: a promising target in solid tumors. *Cancer Treatment Reviews*, **30**: 1–17.

Mauro, M.J., O'Dwyer, M., Heinrich, M.C., *et al.* (2002). STI571: A paradigm of new agents for cancer therapeutics. *Journal of Clinical Oncology*, **20**(1): 325–34.

Sebti, S.M. (2003). Blocked pathways: FTIs shut down oncogene signals. *The Oncologist*, **8**(S3): 30–8.

Swanton, C. (2004). Cell-cycle targeted therapies. *Lancet Oncology*, **5**: 27–36.

17

Caring for the Cancer Patient

Nicky Rudd and Esther Waterhouse

It is better to light a candle than to curse the darkness.
A man's dying is more the survivors' affair than his own.

Chinese proverb
Thomas Mann

KEY POINTS

- Not all patients with cancer will require specialist palliative care.
- Care can be shared between oncologist and palliative medicine physician.
- Patients may need intermittent palliative medicine input from diagnosis.
- Much can be done by those trained in principles of general palliative care.

INTRODUCTION

The growth of palliative medicine and palliative care services (PCS) has been rapid in the last 10 years. The expansion of these services has led to increased specialization of palliative care. Not all patients with cancer will require specialist palliative care as their symptoms and problems can be dealt with by staff, such as oncology ward nurses, who are able to give general supportive care. Hospices and palliative care units are increasingly used as tertiary centers prioritizing those patients with complex physical or psychological problems. Oncology patients may need interventions from palliative medicine at any time during their illness and in the United Kingdom, they may be under the care of both the oncologist and palliative medicine physician simultaneously.

In the United States, there is an artificial boundary driven by the reimbursement system of Medicare and Medicare hospice benefit. This results in patients being transferred to the hospice system only when life-prolonging treatments are ineffective and death is imminent. This is beginning to change with the National Consensus Project Clinical Practice guidelines from the United States stating that "the effort to integrate palliative care into all health care for persons with debilitating and life-threatening illnesses should help to ensure that pain and symptom control, psychological distress, spiritual issues and practical needs are addressed with patient and family throughout the continuum of care."

KEY CONCEPTS

The model shown in Figure 17.1 illustrates the continuum of palliative care, which should be accessed in acute hospitals, hospices, or the community, and differentiates between palliative care and what is called terminal care in the United Kingdom and hospice care in the United States.

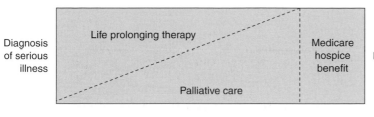

Fig. 17.1 Palliative care's place in the course of illness (National Consensus guidelines).

COMMUNICATION WITH THE CANCER PATIENT

Excellent care is difficult to achieve without good communication. By finding out what your patients are thinking, and tailoring information to what they want to know, communication can be markedly improved. Current training emphasizes key tasks in communication, which are outlined below:

- Elicit the patient's main problems, the patient's perceptions of these, and the physical, emotional, and social impact of these problems on the patient and family;
- tailor information to what the patient wants to know, checking his or her understanding;
- determine how much the patient wants to participate in decision making (when treatment options are available);
- discuss treatment options so that the patient understands the implications;
- maximize the chance that the patient will follow agreed decisions about treatment and advice about changes in lifestyle.

In order to be able to do all these tasks, oncologists need to be able to communicate clearly and effectively.

It is now widely acknowledged that many doctors struggle with effective communication. They feel pressure of time and worry that if they explore distress during a consultation they will be faced with a situation that they cannot handle. Consequently, many doctors use "blocking strategies" to prevent further disclosure.

Blocking strategies

- Offering "premature advice or reassurance"

  Patient: *"I am so worried..."* (unsaid – "about my wife")
  Doctor: (assuming she knows what the patient is worried about) *"the chemotherapy will work well and you'll be feeling better before you know it"*

- Explaining distress as normal

  Patient: *"I'm so frightened"*
  Doctor: *"everybody I see feels frightened at first but that soon goes"*

- Attending to physical aspects only

  Patient: *"I'm very worried..."*
  Doctor: *"I see. How are your bowels?"*

- Jollying patients along

  Patient: *"oh I am upset about my cancer"*
  Doctor: *"but we should be able to cure it"*

- Switching topic

  Patient: *"my wife is so upset at the moment"*
  Doctor: *"now have you got your blood test forms for the next cycle of chemo?"*

Doctors use strategies like these in the mistaken belief that they help, both the patient, by preventing them from getting upset, and the doctor, by minimizing the distress that they are exposed to. In fact the opposite is true, they prevent patients from disclosing their anxieties and problems and lead to increased distress for the patient and the doctor. Exploring these concerns can lead to better joint management of problems. It can also aid compliance with drugs and treatment programs.

WHEN IS PALLIATIVE CARE APPROPRIATE FOR CANCER PATIENTS?

Many oncologists will look after patients until their death. The increasing use of palliative chemotherapy has prolonged the palliative phase in many patients. However, this has also raised new problems, including when to stop palliative chemotherapy and how to ensure patient's agreement with the decision.

Palliative care may be required at any time during the patient's treatment. Referral to the palliative care team (PCT) should also be considered as chemotherapy treatment is discontinued.

PALLIATIVE CARE ASSESSMENT

The assessment of patients needs to be specific, detailed, and encompass psychological, social, and existential issues as well as physical problems. Only by paying close attention to detail will the physician be able to identify the cause of problems and treat them effectively. For instance, a patient's pain may seem nonopioid responsive not because the morphine is ineffective but because the patient is fearful of addiction and has not been taking the tablets.

SYMPTOM CONTROL

A retrospective case study of 400 patients referred to PCTs showed that 64% had pain, 34% anorexia, 32% constipation, 32% weakness, and 31% dyspnea. The most common symptoms are covered in this chapter. For other problems, the reader may consult the reference list at the end of the chapter.

Pain   This may result from the tumor itself, indirectly from tissues related to the cancer, or from other unrelated causes. In a retrospective study in 1995, 2074 carers were asked about the patient's experiences in the last year of life. They reported that 88% of patients were in pain at some time, 47% of patients' pain was either only partially or not controlled at all by general practitioners (GPs), and 35% had pain partially or not at all controlled by hospital doctors.

A study of 200 cancer patients referred to a pain clinic showed that 158 had pain caused by tumor growth, visceral involvement was the cause in 74 cases, bone secondaries in 68 cases, soft tissue invasion in 56 cases, and nerve compression/infiltration in 39 cases. Many patients had more than one type of pain.

Principles of pain control

The analgesic ladder (Fig. 17.2) is a simple scheme which emphasizes the stepwise approach

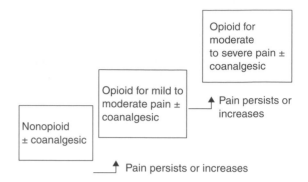

Fig. 17.2 WHO analgesic ladder.

to pain due to cancer and the need to take regular analgesics.

Step 1 – simple nonopioids, e.g. paracetamol/ acetaminophen, nonsteroidal anti-inflammatory drugs (NSAIDs)
Step 2 – opioids for mild to moderate pain e.g. co-codamol, codeine, dihydrocodeine
Step 3 – opioids for moderate to severe pain, e.g. morphine

Move up the ladder if the current step is ineffective. All the medication has to be given regularly and orally unless unable to do so.

Some patients will go directly from Step 1 to Step 3. Regular nonsteroidals or regular paracetamol can be used as opiate sparing drugs.

Opioids

There are a variety of long and short-acting opioids available that can be used for patients requiring step 3 analgesics, however not all opioids are available in all countries (Table 17.1). Morphine remains the most frequently used opioid. Others can be used either for a different route of administration or in case of side-effects.

Common side-effects of opioids are drowsiness, nausea, constipation, xerostomia, vivid dreams, and itching. Opiate toxicity is characterized by pinpoint pupils, visual hallucinations, twitching, agitation and confusion, and respiratory depression. If patients are experiencing side-effects from one formulation, it is worth trying another drug.

Patients' needs for opioids will vary markedly. It is therefore logical to ascertain a patient's

Table 17.1 Commonly used opioids

Morphine	4 h, 12 h and 24 h preparations (can be used parenterally)
Oxycodone	4 or 12 h preparation (parenteral preparation available)
Hydromorphone	4 or 12 h preparation
Methadone	Useful for patients with neuropathic pain and partially opiate-sensitive pain.
Fentanyl	Transdermal patch. Change every 3 days. Reservoir of drugs for 12–24 h after patch removed. Only use in stable pain. Oral transmucosal preparations are also available for buccal use. Intravenous analogs are becoming available.
Buprenorphine	Low dose transdermal patch and sublingual preparation
Diamorphine	High solubility, can be given subcutaneously either as a stat dose or via a syringe driver (not available worldwide)
Levorphanol	Orally or parenterally
Oxymorphone	Suppository or parenteral preparation

needs before commencing long-acting preparations. The usual method is to start a patient on a regular 4-hourly dose of short-acting opioid judging the dose on age, renal function, and previous drug requirement. They are also allowed "breakthrough doses" for pain. After 48 hours, it should be possible to calculate their 24 hour opiate requirement and convert to a long-acting preparation. Whatever be the long-acting preparation, there should be a short-acting opiate given for breakthrough pain.

Cancer pain syndromes

Bone pain. It is a very common cause of cancer pain. Many patients will achieve good pain control with radiotherapy and bisphosphonates. NSAIDs may be effective or analgesics as per the analgesic ladder may be required.

Incident pain. It is most commonly seen in patients with spinal bone metastases and may be one of the most difficult pain syndromes to treat effectively. The patient will generally be pain-free at rest but experience severe and sometimes incapacitating pain when weight-bearing. If the pain is treated with sufficient analgesia to cover the periods of activity, they will be sedated and overdosed at rest and if the dose is only sufficient for rest pain, they will stay immobile.

Management includes optimum treatment for the cause, including chemotherapy, radiotherapy, and bisphosphonates. Opiates should be used as previously described with the addition of coanalgesics for bone pain and neuropathic pain. It may be necessary to change to methadone or add corticosteroids. Epidural or intrathecal analgesia may be required. Physiotherapists and occupational therapists may be useful in moderating either activity or the environment in order to reduce episodes of pain.

Liver capsule pain. It is due to the distention of the capsule and may be associated with diaphragmatic irritation. The pain may respond to NSAIDs or steroids, otherwise opiates may be required.

Bowel pain. Intra-abdominal cancer may present with poorly localized pain due to peritoneal inflammation, vascular or lymphatic blockage, or bowel spasm secondary to serosal or intraluminal disease. Constipation may be an additional problem due to drugs, bowel dysfunction or a combination of both.

Nerve compression/plexopathy. Patients with pelvic tumors, pancreatic cancer, spinal secondaries, apical lung tumors, and head and neck tumors may have severe neuropathic pain (Table 17.2). Patients will typically describe their pain as "burning" or "stabbing" and may have associated loss of sensation or reflexes. The pain may be opiate-sensitive in approximately one third of patients, others will need the addition of drugs specifically for neuropathic pain such as gabapentin.

Patients with difficult pain should be referred to the palliative care or pain team.

Coanalgesics

These are drugs that are used, often in combination with opioids, for a variety of pains, often neuropathic in origin. They include

Table 17.2   Descriptors in neuropathic pain

Allodynia	Increased sensitivity such that stimuli that are normally innocuous produce pain
Hyperalgesia	Increased sensitivity to a painful stimulus
Hyperpathia	Prolonged response to a stimulus even after its removal

Table 17.3   Commonly used coanalgesics

Nonsteroidal anti-inflammatories (NSAIDs)	Bone pain / other pain particularly if inflammatory component potentiates opiates
Tricyclic anti-depressant, e.g. amitriptyline	Neuropathic pain. Initial drowsiness may subside as analgesic effect becomes effective after 5 days
Antiepileptics, e.g. car-bamazepine, gabapentin	Neuropathic pain. Newer drugs have less interactions and side-effects
Steroids	Useful for short periods for pain due to nerve compression
Ketamine	Effective for neuropathic pain. Suppress hallucinations with haloperidol or diazepam
Benzodiazepines	For muscle spasm and myofascial pain
Paracetamol/ acet-aminophen	Used as a coanalgesic to potentiate the effects of opiates

antidepressants and antiepileptic agents such as gabapentin, though the mechanism of benefit is not fully understood (Table 17.3).

It may be necessary to give patients a combination of these drugs.

Anesthetic techniques in pain control

These techniques are used to either block the pain pathway or deliver analgesia through another route (e.g. spinally).

Nerve blocks. They are usually carried out in patients who have pain localized to a nerve distribution and do not respond to analgesia. They are usually carried out with local anesthetic in the first instance to assess response before the nerve is ablated with phenol, cryotherapy, or radio frequency.

Epidural and intrathecal catheters. They may be used to place opiates, local anesthetic, and steroids spinally for patients with back or pelvic pain.

RESPIRATORY SYMPTOMS

Breathlessness is a common symptom and occurs in many cancers. It is thought that toward the end of life respiratory muscle weakness plays a large part but earlier on in the illness there are many treatable causes. Thus, breathlessness may result from the cancer itself (pleural effusion, lung metastases, or massive ascites), but also from preexisting conditions or anticancer treatment (anemia).

There is evidence in lung cancer that non-drug treatments can be of help. Examples of such treatments include teaching breathing exercises and energy conservation techniques and explaining the cause of the breathlessness. Additional drug treatments, including β_2 agonists should be used if required.

A recent systematic review has shown benefit of opiates orally or subcutaneously for the treatment of cancer-related breathlessness. Benzodiazepines may improve panic associated with breathlessness. Some patients who are hypoxic may benefit from supplemental oxygen.

Terminal breathlessness

Patients are often very frightened of suffocating to death. They and their family members need lots of explanation and reassurance. It is often easier to treat breathlessness at the terminal stage because sedation is less of an anxiety and patients and families often accept some drowsiness if symptom control is good. Paradoxically some patients become more alert once their breathlessness has been adequately treated.

Generally a syringe driver containing an opiate and an anxiolytic is used; many centers use diamorphine and midazolam. Extra doses should be given if needed by the patient.

Nausea and vomiting

Nausea and vomiting are common symptoms in cancer patients and, in the experience of the authors, often poorly treated by nonspecialists. This seems to be because of a lack of understanding of the mechanisms causing the symptoms, and the different treatments available.

Assessing nausea and vomiting is essential. The history should include the time course of the problem, precipitating and relieving factors (including previously tried drug and nondrug treatments), the relationship between the nausea and the vomiting, and the amount of vomitus produced at each vomit. Knowing whether vomiting relieves the nausea can be very helpful. Examination and selected use of tests such as blood biochemistry are also helpful.

Some common syndromes and suggested treatments are outlined below.

Delayed gastric emptying

This encompasses a range of problems, from mechanical gastric outlet obstruction to opiate-induced reduced gastric motility. Patients often have little nausea, but may feel themselves "filling up" before having a large vomit. If they are managing to eat, the vomit will contain undigested food. The vomit may be effortless and often leaves patients feeling much better.

Dopamine antagonists, such as metoclopramide and domperidone, are the treatment of choice. An open-label study suggested efficacy of metoclopramide although it was not placebo controlled. Metoclopramide speeds up gastric emptying at gastric level (acting on D_2 and $5HT_4$ receptors) as well as centrally, whereas domperidone acts mainly on D_2 receptors in the stomach. Therefore most palliative medicine physicians will use metoclopramide. Patients with a mechanical outlet obstruction may benefit from steroids. Drugs should be given parenterally to start with, as oral absorption will be poor.

Chemically induced nausea

Common causes in patients with advanced cancer are uremia and hypercalcemia. If a patient who has been stable on opiates develops nausea, particularly with signs of opioid toxicity, it may be that their renal function is impaired causing opioid accumulation and toxicity. Patients with chemically induced nausea complain of a constant nausea with intermittent retching, which usually fails to relieve the nausea. They rarely manage to eat much.

The most potent dopamine blocker is haloperidol, and this is usually the treatment of choice. Side-effects are rare at low doses. If this is ineffective, when given parenterally, then levomepromazine should be used.

Emesis caused by chemotherapy and radiotherapy is usually treated with $5HT_3$ antagonists, according to written protocols. Despite revolutionizing chemotherapy induced nausea and vomiting, $5HT_3$ antagonists have not found a clear place in the management of nausea and vomiting from other causes.

Hypercalcemia is the commonest life-threatening metabolic disorder related to malignancy (Table 17.4). It is seen in 10% of patients with malignancy, most frequently in patients with myeloma, breast, lung, and renal cancers. Up to 20% of cases occur in the absence of bone metastases. It is an indicator of a poor prognosis, with a median survival of 3–4 months. The underlying metabolic problems include increased osteoclastic bone resorption, decreased renal clearance of calcium, and enhanced absorption from the gut.

Table 17.4 Humoral factors and hypercalcemia

Parathyroid hormone-related protein (PTH r P)
    Activates PTH receptors in tissue
    Stimulates osteoclastic bone resorption and
      renal tubular absorption
Transforming growth factors (TGF)
    TGF are functionally and structurally similar
      to epidermal growth factors
    Increases osteoblastic activity
    TGFβ increases production of PTH r P in
      breast cancer cells
Osteoclast-activating factors
    Tumor necrosis factor (TNFα)
    Lymphotoxin (TFNβ) produced by activated T
      lymphocytes and myeloma cells
Interleukin 6 (IL6)
Interleukin 1 (IL1)

Some cancers produce parathyroid hormone-related protein, transforming growth factors, or osteoclast activating factors.

Symptoms are often vague with most patients developing malaise and lethargy followed by thirst, nausea and constipation, and then confusion. It is important to look for hypercalcemia as treatment is often effective at resolving symptoms.

Treatment involves rehydration with intravenous IV fluids followed by IV bisphosphonates. Resolution of symptoms often takes several days. Some patients require repeated infusions at intervals to maintain normocalcemia; oral bisphosphonates, although effective in this context, have poor bioavailability and tolerability and are therefore not often used in the palliative setting. Other treatments such as calcitonin may be necessary in intractable cases.

Stretch and distortion of gastrointestinal (GI) tract. Can be due to a host of causes, such as intraabdominal tumors, constipation, bowel obstruction, and retroperitoneal lymph nodes. It is usually necessary to treat the underlying cause; steroids can also be helpful in reducing peri-tumor edema; they may also have a direct antiemetic effect.

Cranial causes. Raised intracranial pressure can cause nausea. High-dose dexamethasone is usually indicated and palliative radiotherapy may be used. An antiemetic acting at the vomiting center, such as cyclizine or levomepromazine, would be the symptomatic treatment of choice.

Anxiety. Can cause nausea and may cause vomiting.

Nondrug treatments. Should be remembered when assessing patients with nausea and vomiting. Sometimes simple changes to the environment, such as controlling odor or changing portion sizes, can be enough to significantly reduce symptoms. Remember that patients' tastes often change such that previously palatable food is no longer tolerated.

Intractable nausea. Not all nausea and vomiting is amenable to treatment despite best efforts. In these cases it is important to be realistic with the patients, most of whom will accept control of their nausea with intermittent vomits if that is the best that can be achieved.

BOWEL OBSTRUCTION

Malignant bowel obstruction (MBO) is a common problem in advanced abdominal and pelvic cancers. Estimates of frequency amongst patients with these cancers are difficult to ascertain. Some studies have quoted up to 51% of patients with gynecological malignancies and 28% of patients with GI malignancies developing bowel obstruction; the true figure may be higher.

Obstruction can be complete or incomplete and the degree of obstruction, and therefore of symptoms, may change over time. Patients present with a mixture of distended abdomen, altered bowel habit (which may be constipation), colicky abdominal pain, nausea, and vomiting. They also often have a constant background abdominal pain.

The pattern of symptoms varies according to the level of obstruction:

High (stomach)	Large volume vomits
	Persistent nausea, may be relieved by vomiting
	Little distension
	Constipation, develops late
Small bowel	Minimal distension
	Abdominal colic
Large bowel	Abdominal distension
	Colic
	Constipation

Managing bowel obstruction

Assess the cause of the obstruction. Is it amenable to surgery? Is chemotherapy appropriate? Would a stent be helpful? Stents can be useful for gastric outlet obstruction and, occasionally, for large bowel obstruction. Is decompression with a naso-gastric tube likely to help? In a few patients it is the only way to maintain control of vomiting.

In advanced MBO, it is rarely possible to stop the patient vomiting and most palliative physicians will aim for control of nausea and reduction of vomiting to once every 1 or 2 days. It is important that all involved agree on the aims of the treatment.

Colicky abdominal pain may benefit from an antispasmodic such as hyoscine butylbromide (buscopan) in a syringe driver.

Most patients with MBO will be able to eat small amounts, although these may be vomited again later. Most palliative care units manage MBO without long-term parenteral hydration, although some patients find symptoms of thirst and nausea ease with small amounts of fluid.

CONSTIPATION

63% of elderly people in hospital are constipated. Causes include debility, immobility, depression, inadequate fluid intake and drugs, particularly opioids, tricyclic antidepressants, diuretics, and ferrous sulfate. Tumor effects include spinal cord compression and direct compression of bowel by tumor. Other concurrent problems may exist, e.g. diabetes, local anal pathology.

It is best to prevent constipation before it becomes established, by ensuring adequate fluid intake and use of laxatives for all those on weak and strong opioids.

FATIGUE

Fatigue is a commonly reported problem in advanced cancer but is rarely identified and treated. It can be caused by the cancer or by the treatment. Treatable causes of fatigue include:

- Anemia – it is not clear how much of the fatigue related to advanced disease responds to treatment of anemia;
- metabolic and endocrine disorders;
- psychological issues including depression;
- anticancer treatment.

Dexamethasone can help cancer-related fatigue but the effect is often short-lived and side-effects are common. Psychostimulants can be helpful and in the United States they are the treatment of choice for depression in the terminally ill, as they have a rapid onset of action. However, they can cause agitation and insomnia, which limit compliance. Trial evidence for the use of psychostimulants in cancer-related fatigue is minimal.

Table 17.5 Common problems causing anorexia or cachexia

Problem	Treatment
Mouth problems	
Dry mouth	• Consider side-effects
Intraoral infections	• Treat infections promptly
Sore mouth	• Mouth washes
	• Analgesics, e.g. diamorphine mouthwash, NSAIDs, adcortyl in orabase, gelclair
Dysphagia	
Tumor	• Consider stent or palliative radiotherapy
Infection	• Treat infections promptly
Nausea	• See Section on Nausea
Malabsorption	
Enzyme deficiency	• Pancreatic supplements
Chronic diarrhea	• Treat diarrhea
Anxiety and depression	• See Section on Anxiety and depression
Altered taste	• Increase spiciness of food
	• Add lemon juice

CACHEXIA AND ANOREXIA

Cachexia occurs in over 50% of patients with solid tumors. It is characterized by loss of both fat and muscle. The anorexia–cachexia syndrome (Table 17.5) is driven by production of cytokines, particularly tumor necrosis factor (TNF). There are also numerous changes in the humoral regulation of appetite, energy use, carbohydrate, and protein metabolism that cause loss of muscle as well as fat.

Drug treatment

- Metoclopramide may be helpful if gastric stasis is a contributing factor.

- Progesterone, e.g. megestrol acetate 160–800 mg (this may take one to two weeks to work). Long-term use of progesterones may induce adrenal suppression.
- Corticosteroids – use as short courses, which can be repeated if necessary.

Other drugs such as thalidomide (TNF inhibitor) or creatine may be useful in the future.

PSYCHOLOGICAL PROBLEMS

The three main psychological problems causing morbidity in patients with cancer are depression, anxiety, and adjustment disorders.

Approximately 35–50% of patients with cancer will have a psychiatric disorder, the prevalence of which is approximately 22% in the general population. The skill for the oncologist and other health professionals involved in the cancer patient's care is to distinguish between sadness associated with a life-threatening illness and a psychiatric disorder.

Depression

Studies have found the prevalence of depression in cancer patients to vary from 3.2 to 42%. A more realistic figure is 5–15% fulfilling the criteria for major depression. Risk factors are previous psychiatric problems, young age, lack of social support, decreasing physical function, and the presence of pain.

Depression can be difficult to diagnose in cancer patients. A commonly used screening tool is the Hospital Anxiety and Depression Tool.

The UK National Institute for Clinical Excellence (NICE) guidance for palliative and supportive care states that "the psychological needs of all patients and carers are assessed on a continual basis throughout the patient pathway, with particular attention to points that are particularly challenging, such as the time prior to diagnosis and after bereavement."

This guidance mandates regular screening of all cancer patients, using some type of screening tool incorporated into routine cancer care.

Drug treatment of depression is as effective in this population as those who are physically well. Selection of drugs may be determined by other symptoms, for instance a tricyclic antidepressant may be useful for those with neuropathic pain or insomnia, and selective serotonin reuptake inhibitors (SSRIs) should be used with care in those with preexisting nausea.

Anxiety

This is common in patients with cancer. However, some patients will present with debilitating symptoms. Much can be done with good communication and clear information. Some patients will require benzodiazepines, which are generally well tolerated. Patients should be reassured that these drugs are being used appropriately as even dying patients may require reassurance over addiction.

Adjustment disorders

These occur when patients are unable to adapt to a severe life event and cope inflexibly in response to their illness. The definition of an adjustment disorder is rather nebulous; it is a disorder characterized by a variety of clinically significant behavioral or emotional symptoms that occur as a result of some triggering event or stressor. It occurs in approximately 25–30% of patients with cancer but it is often overlooked. There may be overlap with depression but it can only be diagnosed if depression and anxiety disorders are excluded. Brief psychotherapy has been proven to be effective in this disorder.

THE DYING PATIENT

Ongoing communication with the dying patient and their family remains essential. The physician needs to consider discontinuing ineffective treatments. In the United States, a high proportion of oncology patients die in intensive care. In the United Kingdom, the drive is to ascertain patients' choice in their place of death.

Relatives may have some difficulty in agreeing to discontinuing treatments such as IV fluids or antibiotics as they may perceive it as helping the patient to die. They need to be reassured that the patient is dying with or without active treatment and that most patients are more comfortable with good nursing care rather than invasive procedures.

All these decisions can be made much more easily if end-of-life issues have been discussed with the patient and they have made their preferences known. Living wills can be very helpful both to the medical team and the family. Discussing death may seem a daunting task but many patients welcome a discussion on what is, after all, at the forefront of their minds. They will also be reassured that you are not just interested in their tumor response but also what happens to them when treatment is no longer effective.

Symptom control in the dying patient

Important principles in care of the dying patient are the attention to individual symptoms and anticipation of problems such as agitation and inability to swallow. If the patient's symptoms have been previously under good control, they may not experience worsening of symptoms as they die. Problems occurring frequently are listed below.

Inability to swallow. This should be anticipated. Syringe drivers delivering drugs in a 24 hour subcutaneous dose are commonly used to replace necessary oral medications.

Agitation. This is common in the dying patient and has a variety of etiologies. It may be a manifestation of distress and inability to communicate but it is important to exclude and treat causes that are reversible, e.g. urinary retention, constipation, and pain. Discuss with the relatives what they think may be the cause and consider, with their agreement, sedation with benzodiazepines, e.g. midazolam, or levomepromazine, if appropriate. These can be given via the syringe driver.

Retained secretions. As patients become weaker, they are unable to clear oral secretions or cough effectively. This leads to accumulation of saliva at the back of the throat – "the death rattle." The patient may be relatively unaware of this but it often causes distress to relatives sitting by the bedside. Hyoscine or glycopyrronium can be used subcutaneously, otherwise tipping the bed, or in the last resort, suctioning, can

be used. Other symptoms can be treated as previously described.

Palliative care services. These have expanded exponentially over the last 15 years in the United Kingdom particularly. Services prior to this were sporadic with much of the drive for development being from local groups, resulting in diverse services across the country due to a lack of regional planning. There have been recent efforts to remedy this and most regions will have services both in the acute sector and the community.

SUPPORTIVE CARE

"Supportive care" is a relatively new phrase to describe an age-old concept, that of caring for patients and those close to them during difficult times. This concept has now been formalized in NICE guidelines for supportive and palliative care. The definition from the National Council for Hospices and Specialist Palliative Care Services is that supportive care:

helps the patient and their family to cope with cancer and treatment of it from pre-diagnosis, through the process of diagnosis and treatment, to cure, continuing illness or death and into bereavement. It helps the patient to maximize the benefits of treatment and to live as well as possible with the effects of the disease. It is given equal priority alongside diagnosis and treatment.

Provision of supportive care is therefore not dependent on the stage or type of cancer, or indeed on the predicted treatment outcome. Supportive care can be seen as an umbrella term, covering all services, generalist and specialist, which may be needed to help patients and families.

AN EXAMPLE OF THE CARE OF A CANCER PATIENT

Mr A has a chest X-ray suspicious of lung cancer. He attends outpatients and the lung cancer nurse is present when the doctor is seeing him. After discussing the probable diagnosis, she gives him an information package. She then coordinates his diagnostic process, making sure that he is aware of what is going on and that he understands the need for further tests (information giving).

After the diagnosis is confirmed she makes contact with him by phone – due to high levels of distress she visits him and his family at home (psychological support). He agrees to attend the local lung cancer support group (user involvement, self help, and support). As Mr A is the main carer for his wife who has severe multiple sclerosis (MS), the nurse contacts other professionals to arrange for her to go into respite care while he is having treatment. She also refers him for specialist benefits advice (social support). Following his treatment he remains breathless and is referred to the local breathlessness program (rehabilitation).

He finds reflexology particularly helpful (complementary therapies). Later he develops significant psychological distress and she refers him to the community specialist PCTs for further assessment (palliative care). His distress seems to be due to a loss of faith in God and he is referred to the hospice chaplain (spiritual support). During the next few weeks he requires ongoing palliative care involvement to control his worsening pain.

His terminal illness is managed at home with support from the primary care and specialist palliative care teams, and this is where he dies (end-of-life-care). His wife requires a lot of support from the primary care team after his death (bereavement care).

QUESTIONS REMAINING

1 Increased understanding of the underlying mechanisms of the process of dying.
2 More large multicenter randomized controlled trials of analgesics for cancer pain; this is problematic for many reasons including:
   a. many patients in the early stages of their disease are participating in trials of anticancer treatment;
   b. there are numerous practical and ethical issues around research on patients in the last few weeks or months of life.
3 A better understanding of which services are most effective for this population.

CONCLUSIONS AND FUTURE DIRECTIONS

This chapter has reviewed the common problems experienced by cancer patients and described a palliative approach.

Good cancer care, including palliative care, relies on the input from a diverse range of health professionals, access to whom should be on a need basis rather than on stage of disease. Effective communication between professionals as well as with the patient is paramount.

Palliative medicine works most effectively if integrated into oncological units and not limited to end-of-life care.

Work continues worldwide not only to increase recognition of palliative medicine and its achievements but also to educate patients and health professionals about available treatments. A first and important step would be worldwide availability of opiates for those in pain.

QUESTIONS

1 An 84-year-old man with metastatic carcinoma of the prostate presents with the following symptoms. Match them with the likely underlying problem:
 (i) Acute confusion and vomiting.
 (ii) Urinary retention and fecal incontinence.
 (iii) Back pain, constipation, and urinary retention.
 (iv) Pain left knee and reduced mobility.
 (v) Confusion and fever.

Answer items

 (a) Bone metastases left hip.
 (b) Renal failure.
 (c) Urinary tract infection.
 (d) Cauda equina compression.
 (e) Constipation.

Comment

1. Renal failure can occur secondary to urinary outflow obstruction. Beware of iatrogenic causes, e.g. NSAIDs.
2. Constipation is very common in patients with cancer especially if they are on morphine and can cause urinary retention. "Diarrhea" may be due to fecal overflow.

3. Cauda equina compression can present gradually and with vague symptoms of weakness, numbness, urinary/fecal incontinence, a numb sacrum, and legs just feeling "odd."

2 A 65-year old patient with malignant pain is about to start morphine. Which of the following statements are correct?
(a) Patients should be observed for respiratory depression.
(b) All patients should start laxatives with opiates unless contraindicated.
(c) There is no ceiling dose of morphine.
(d) The majority of patients will experience nausea or vomiting.
(e) Addiction can be a problem for some patients.
(f) Doses can be limited by drowsiness.
(g) The dose will inevitably increase with time.
(h) The prescription of morphine will shorten the patient's life.

Comment These are all the worries expressed by patients and doctors. Approximately one third will experience nausea and vomiting. Tolerance can occur but others will remain on a stable dose for some time. There is no evidence that morphine per se will shorten life.

Drowsiness can be a problem for some patients. Most patients require laxatives unless malabsorbing, ileostomy, etc.

3 For each cause of nausea and vomiting, select the most useful drug:
(i) Initial opiate-induced nausea.
(ii) Gastric stasis.
(iii) Vestibular dysfunction.
(iv) Chemotherapy-induced nausea.
(v) Associated with bowel obstruction.

Drugs
(a) Hyoscine hydrobromide.
(b) Metoclopramide.
(c) Ondansetron.
(d) Haloperidol.
(e) Cyclizine.

BIBLIOGRAPHY

Service provision

Morrison, R.S. and Meier, D.E. (2004). Palliative care. *The New England Journal of Medicine*, **350**: 2582–90.

National Consensus Guidelines (2004). National Concerns Project for Quality Palliative Care. *Clinical Practice Guidelines*, April 2004; 57–60.

Communication

Maguire, P. and Pitceathly, C. (2002). Key communication skills and how to acquire them. *British Medical Journal*, **325**:697–700.

General symptoms

Potter, J. (2003). Symptoms in 400 patients referred to palliative care services: prevalence and patterns. *Palliative Medicine*, **17**:310–14.

Addington-Hall, J.M. and McCarthy, M. (1995). Dying from cancer: results of a national population-based investigation. *Palliative Medicine*, **9**:295–305.

Pain

Mercadante, S., Radbruch, L., Caraceni, A. *et al.* (2002). Episodic (breakthrough) pain: consensus conference of an expert working group of the European Association for Palliative Care. *Cancer*, **94**(3):832–9.

Sindrup, S.H. and Jensen, T.S. (1999). Efficacy of pharmacological treatments of neuropathic pain: an update and effect related to mechanism of drug action. *Pain*, **83**:389–400.

World Health Organization (1986). Cancer Pain Relief. WHO, Geneva.

Foley, K.M. (1985). The treatment of cancer pain. *The New England Journal of Medicine*, **313**:84.

Breathlessness

Bredin, M., Corner, J. *et al.* (1999). Multicentre randomized trial of nursing intervention for breathlessness in patients with lung cancer. *British Medical Journal*, **318**:901.

Jennings, A.l, Davies, A.N. *et al.* (2004). Opioids for the palliation of breathlessness in terminal illness (Cochrane Review). The Cochrane Library Issue 4.

Abernethy, A.P. *et al.* (2003). Randomized, double-blind placebo-controlled crossover trial of sustained release morphine for the management of refractory dyspnoea. *British Medical Journal*, **327**:523–8.

Booth, S., Anderson, H. *et al.* (2004). The use of oxygen in the palliation of breathlessness. A report of the expert working group of the Scientific Committee of the Association for *Palliat Medicine/Respiratory Medicine*, **92**(1):66–77.

Bower, M. and Cox, S. (2004). Endocrine and metabolic complications of advanced cancer. In Doyle, D. *et al.* (eds), *Oxford Textbook of Palliative Medicine*, 3rd edn. Oxford: Oxford University Press, pp 687–702.

Bruera, E., Sweeney, C. *et al.* (2003). A randomized controlled trial of supplemental oxygen versus air in cancer patients with dyspnoea. *Palliative Medicine*, 17(3):659–63.

Gastrointestinal symptoms

Wilson, J. *et al.* (2002). Long-term safety and clinical effectiveness of controlled-release metoclopramide in cancer associated dyspepsia syndrome – a multicentre evaluation. *Journal of Palliative Care*, 18(2):84–91.

Skinner, J. and Skinner, (1999). Levomepromazine for nausea and vomiting in advanced cancer. *Hospital Medicine*, 60(8):568–70.

Feuer, D.J. and Broadley, K.E. (2003). Corticosteroids for the resolution of malignant bowel obstruction in advanced gynaecological and gastrointestinal cancer (Cochrane Review). In: *The Cochrane Library*, Issue 4, 2003. Chichester, UK: John Wiley & Sons.

Complementary therapies

Pan, C.X. *et al.* (2000). Complementary and alternative medicine in the management of pain, dyspnoea and nausea and vomiting near the end of life. A systematic review. *Journal of Pain and Symptom Management*, 20(5):374–87.

Cachexia and asthenia

Dein, S. and George, R. (2002). A place for psychostimulants in palliative care? *Journal of Palliative Care*, 18(3):196–9.

Knobel, H. *et al.* (2003), The validity of, E.O.RTC QLQ-C30 fatigue scale in advanced cancer patients and cancer survivors. *Palliative Medicine*, 17(8):664–72.

Strasser, F. (2004) Pathophysiology of the anorexia/cachexia syndrome. In Doyle, D. *et al.* (eds), *Oxford Textbook of Palliative Medicine*, 3rd edn. Oxford: Oxford University Press, pp 520–33.

Jatoi, A. Jr. and Loprinzi, C.L. (2001). Current management of cancer-associated anorexia and weight loss. *Oncology*; 15(4):497–502.

Loprinzi, C.L. *et al.* (1993). Phase, I.I.I evaluation of four doses of megestrol acetate as therapy for patients with cancer anorexia and/or cachexia. *Journal of Clinical Oncology*, 11(4): 762–7.

Loprinzi, C.L. *et al.* (1999). Randomized comparison of megestrol acetate versus dexamethasone versus fluoxymesterone for the treatment of cancer anorexia/cachexia. *Journal of Clinical Oncology*, 17(10):3299–306.

Loprinzi, C.L., Fonseca, R., Jensen, M.D. (1996). Induction of adrenal suppression by megestrol acetate. *Annals of Internal Medicine*, 124(6):613.

Depression and anxiety

Zabora, J., Brintzenhofenszoc, K., Curbow, B., Hooker, C., Piantadosi, S. (2001). The prevalence of psychological distress by cancer site. *Psychooncology*, 10:19–28.

Alexander, P.J., Dinesh, N., Vidyasagar, M.S. (1993). Psychiatric morbidity among cancer patients and its relationship with awareness of illness and expectations about treatment outcome. *Acta Oncologica*, 32:623–6.

Breitbart, W. *et al.* (2000). Depression, hopelessness and desire for hastened death in terminally ill patients with cancer. *The Journal of American Medical Association*, 284:2907–11.

Minagawa, H., Uchitomi, Y., Yamawaki, S., Ishitani, K. (1996). Psychiatric morbidity in terminally ill cancer patients – a prospective study. *Cancer*, 78:1131–7.

Endicott. J. (1984). Measurement of depression in patients with cancer. *Cancer*, 53:2243–8.

Zigmond, A.S., Snaith, R.T. (1983). The Hospital Anxiety and Depression Scale. *Acta Psychiatrica Scandinavica*, 67:361–70.

SUGGESTED READING

Simpson, K., Budd, K. (eds). *Cancer Pain Management*. Oxford: Oxford University Press, 2000.

Faull, C., Carter, Y., Woof, R. (eds). *Handbook of Palliative Care*. Oxford: Blackwell Science, 1998.

Lloyd-Williams. M. (ed). *Psychological Issues in Palliative Care*. Oxford: Oxford University Press, 2003.

Doyle, D., Hanks, G., Cherney, N.I., Calman, K. (eds). *Oxford Text of Palliative Medicine*. Oxford: Oxford University Press, 2004.

Twycross, R., Wilcock, A. *Symptom Management in Advanced Cancer* (2nd edn). Abingdon: Radcliffe Publishing, 1997.

Twycross, R.G. *et al.* (eds). *Palliative Care Formulary*, 2nd edn. Abingdon: Radcliffe Publishing.

18
Genomics and Proteomics in Cancer Research and Diagnosis

Norbert C.J. de Wit

Do not read, think!
The best way to get a good idea is to get a lot of ideas.

Arthur Schopenhauer
Linus Pauling

KEY POINTS

- There are orders of magnitude more functional proteins than genes, due to various post-translational modifications.
- Although, much can be learned from analysis of gene expression, it is the proteins that directly regulate cellular phenotype and must ultimately be the target of study in cancer biology.
- High throughput techniques that produce vast amounts of gene and protein expression data in single studies can only be fully analyzed and understood by the involvement of mathematicians, statisticians, and informaticians – Systems biology.
- Techniques that average expression of proteins/genes across thousands or millions of cells will not reveal the important role of cell–cell interactions and heterogeneity.
- Particularly in cancer where stromal and other interactions in the cancer cell's microenvironment are so crucial new techniques able to localize expression changes to single cells without disruption of the tissue will be essential.

INTRODUCTION

The sequencing of the human genome and model organisms such as the rat, mouse, *Caenorhabditis elegans* (nematode worm), and *Drosophila melanogaster* (fruit fly) has revolutionized biology. Although it has only started, the revolution in biology could be compared to the twentieth century revolution in physics. The impact of this revolution will not only be felt in the biological sciences but will ultimately transform clinical medicine. Noninvasive diagnosis and treatment, disease prediction, and personalized medicine are all expected developments resulting from the genomics revolution. Biology has become an information science dealing with enormous amounts of data. The interaction between chemistry, biology, bioinformatics, mathematics, statistics, and physics is no longer an advantage but a sheer necessity to deal with the complicated biological systems and the large data quantities acquired with high throughput techniques.

A quote by Paul Nurse, Nobel laureate, illustrates this best, "perhaps a proper understanding of the complex regulatory networks making up cellular systems like the cell cycle will require a

... shift from common sense thinking. We might need to move into a strange more abstract world, more readily analyzable in terms of mathematics than our present imaginings of cells operating as a microcosm of our everyday world."

INFORMATION FLOW IN THE CELL

Human life starts as a single cell, a fertilized egg, which develops into a complex organism that contains trillions of cells and thousands of cell types. Ironically cancer too starts as a single cell including Darwinian tendencies. All information that is needed for the creation and functioning of a human being and also other organisms is encoded in the genome. In essence, a genome is nothing more than the encrypted encyclopedia of life albeit a very difficult one.

An illuminating quote from Sir John Sulston, Nobel laureate, who as director of the Sanger center in Cambridge, was a central figure in the recent sequencing of the human genome, refers to his own efforts and others in this venture – "The genome is a hieroglyph and all we have done is dig it up and brush the sand off. We still have to understand most of it."

Genomes of the mouse, human, and *Drosophila* are remarkably similar and most genes are conserved between species. It is not so much the genes in the genome but the regulatory networks that are largely responsible for the differences between species. This information is also encoded in the genome. The high homology between species explains why, for example, the mouse or the fruit fly is a useful "model organism" to unravel human physiology and disease.

The genome is transcribed into mRNA and the mRNA is subsequently translated into proteins. The human genome codes for at least 30,000 genes (the number of functional genes is still a matter of much debate), which are transcribed into many more mRNAs that code for over a suggested 100,000 proteins. At each level, many variations occur making the system very complex (Fig. 18.1). Individuals differ on the genome level by approximately 10 million DNA single nucleotide polymorphisms also known as SNPs, but this amounts to only 0.1% of the genome. Small changes in the DNA can have fatal effects. Cystic fibrosis for example is an endocrine gland disorder that is caused by mutations in the cystic

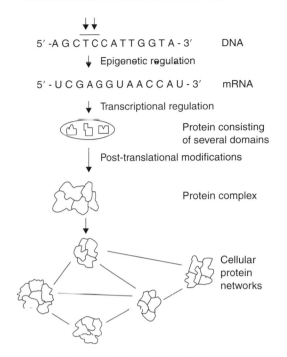

Fig. 18.1 Information flow in the cell – from genome to proteome to cellular networks. On each level, variation has an effect on cell function and abnormal variation can underlie cancer.

fibrosis transmembrane conductance regulator (*CFTR*) gene and affects several organs. The CFTR-protein functions as an ATP-regulated chloride channel and deletion of three base pairs in the *CFTR*-gene results in deletion of the amino acid phenylalanine. The mutated chloride channel is dysfunctional, which causes the production of abnormally viscous mucus. In lungs, the viscous mucus is an ideal growth medium for bacteria and the continuous bacterial growth results in chronic respiratory infections. Ultimately, the permanent tissue damage leads to a range of complications resulting in death.

Breast cancer, among other cancers, is also associated with specific mutations in the genome, but its relationship to a single gene mutation is not as clear as in cystic fibrosis. Many mutations in *BRCA1* and *BRCA2* genes have been linked to increased chances of developing breast cancer, but these mutations do not predict all cases of breast cancer. This highlights our limited understanding of gene mutations and genetic variation between individuals (including SNPs and other polymorphisms). Many diseases are multi-factorial, different gene mutations conspire with environmental factors (e.g. lifestyle) to influence disease development and outcome. Attributing the role of SNPs and gene mutation to numerous associated diseases is a field of much research.

The next level of system regulation is in the structure of DNA. Expression of genes is not just regulated by transcription factors. Chemical modifications of DNA can alter chromatin structure and therefore gene expression without changing the DNA sequence. The enzyme DNA methyltransferase methylates DNA in CG-nucleotide rich areas called CpG-islands, and alters chromatin structure. Methylated DNA forms condensed chromatin, which is not accessible to transcription factors and leads to immediate gene silencing. Epigenetic regulation involves DNA methylation, chromatin remodeling, methyl-binding domain proteins, and histone deacetylation (see also Chapter 11).

When chromatin is accessible to transcription factors, a gene can be transcribed into mRNA. Each gene can potentially give rise to a number of mRNAs, which in turn can be post-transcriptionally regulated. Post-transcriptional regulation usually means splicing of the mRNA transcript, but the translation of these transcripts into protein is also regulated. Transcripts are generally translated into proteins but there is still much debate surrounding how this is related to protein levels. Synthesized proteins are often post-translationally modified. The original protein product can be modified after synthesis by, for example, the attachment of a phosphate group (phosphorylation), sugar moiety (glycosylation), or ubiquitin group (ubiquitination). There are hundreds of protein modifications possible that can be either permanent or temporary. Post-translational modifications have a multitude of effects and control many cellular processes.

For example, synthesized proteins are tagged with a modification to ensure that the protein is transported to the correct cellular compartment. The modification is akin to a postal code on a letter and guarantees punctual and correct delivery of the protein. As with postal services this is not flawless. These modifications, ubiquitinylation and sumoylation in particular, can also determine when and how rapidly a protein is degraded, and thus help control levels in the cell.

Alternatively, proteins can also be modified to perform a specific action in a signal transduction pathway. Protein kinases play a crucial role in activating signal transduction pathways by phosphorylating tyrosine, threonine, or serine residues in proteins. These post-translational modifications are transient and activate the protein for a short period of time. Termination of this phosphorylation induced cellular signal often occurs through dephosphorylation by protein phosphatases. Each post-translationally modified protein is a different protein because it has a different function. Hence, the roughly 30,000 genes probably encode for over a million functionally different proteins.

However, proteins rarely function independently in a cell (proteins are "sociable") and usually form complexes that function as little cellular motors. Proteins are the individual parts of these motors and the protein complexes interact with each other forming protein networks. It is these networks that are essential for cellular communication.

One of the largest cellular motors is the well-known spliceosome. The spliceosome consists of approximately 145 proteins and 5 small RNA molecules. The spliceosome complex is dynamic and undergoes multiple assembly stages and conformational changes during the splicing process.

Identifying and characterizing protein complexes, and unraveling pathways and cellular networks are of critical importance to understand cell biology.

Figure 18.1 illustrates the complexity of cells and organisms. The integrated study of a biological model is termed "systems biology." Although systems are often seen as cells, organs, or even ecosystems, a system approach can also focus on a cellular pathway or a tumor.

Biological systems are highly dynamic and organisms adapt and change constantly to respond to external influences. A sudden large amount of exercise for example leads to muscle ache due to muscle adaptation, or the immune system is activated to deal with bacterial infections. This all leads to changes at various levels in an organism and a biological system is therefore never static.

Studying dynamic systems such as cells, cancers, or organs is extremely challenging and requires many expertises. Figure 18.1 shows the informational hierarchy and the flow of information in biological systems (central dogma of biology). This chapter roughly follows this flow of information and will focus on new techniques that play an important role in cancer biology research. Important discoveries and developments are discussed and, where possible, they are linked to previous chapters. Many applications of technical advances in life science research are pioneered in cancer research. Moreover, cancer research plays a crucial role in the translation of laboratory techniques into clinical applications and key examples are discussed. The end of the chapter has a number of internet resources that are given as additional information to this chapter.

MODEL ORGANISMS AND CANCER MODELS

Yeast, fruit fly (*D. melanogaster*), zebra fish, puffer fish, mouse, rat, and the nematode worm *C. elegans* are all model organisms. These organisms have been extensively used to study and understand biology. Much information is available on these organisms and our understanding of human biology has increased by studying them. Each organism is used for different reasons and an organism is often either a genetic, experimental, or genomic model. The mouse is a particularly interesting model because its genome is organized similar to the human genome. Much has been learnt about normal and cancer physiology by using mouse models.

Two premises underlie the use of various animal models in cancer biology. First is that such models can contribute to our knowledge of cancer mechanisms, particularly when informed by knowledge derived from study of human cancers. Second, such models can be crucial test-beds for evaluating therapeutic targets and strategies, aimed at treating, curing, or preventing cancer. In fact, recent advances in regulatable genetic systems allow the proof of hypothesis testing of inactivating specific cancer genes/proteins even before a suitable drug has been developed (see Appendix 18.1).

Mice have been employed in cancer research since 1894. Initially, studies focused on same-species tumor transplantations and drug treatment studies. In the 1920s, inbred strains that had been shown to be prone to cancer were first disseminated among cancer researchers. Many more strains of mice were identified following the founding of the Jackson Laboratory in Bar Harbor, Maine, now a preeminent source of mouse cancer models for the research community. In 1962, the identification of the "nude" mutant mouse (that has lowered immunity and shows minimal rejection) enabled studies of transplantation of human tumors, which proved a major advance for cancer research. In the 1980s, breakthroughs in molecular biology resulted in the development of genetically altered mice. Oncogenes, or other genes that cause cancer, can now be investigated in much greater detail. Engineering of genetically altered mice has made it possible to create models to address specific processes.

It has been suggested that transformation of a normal cell into a cancer cell requires an estimated four to seven rate-limiting mutational events, that confer specific properties on the developing cancer cell:

- Loss of proliferative control,
- the failure to undergo programmed cell death (apoptosis),
- angiogenesis,
- tissue remodeling,
- invasion of tumor cells into surrounding tissue and, finally,
- metastatic dissemination of tumor cells to distant organs.

In man, the molecular analysis of these multiple steps is rarely possible due to the limited availability of tumor specimens obtained prospectively from all tumor stages. In contrast, mouse models of tumorigenesis allow the reproducible isolation of all tumor stages, including normal tissue, which are then amenable to pathological, genetic, and biochemical analyses.

The list of specific mouse models available is continuously expanding and includes those with various checkpoint and tumor-suppressor proteins knocked out, or mutated, with oncogenes overexpressed or with alterations in caretaker genes. Moreover, increasingly, modern systems enable the control of such genetic alterations in a tissue and time-dependent fashion (see Chapters 6 and 7 for discussion of cancer models in the study of oncogenes and tumor suppressors respectively).

Mice have proved invaluable as cancer models in the following areas in particular:

- The many well-characterized inbred strains allow the mapping and identification of cancer susceptible genes.
- They are a rich source of transforming viruses carrying oncogenes.
- Transgenic mice carrying candidate oncogenes provided direct evidence for the oncogenic potential of these genes.
- Knockout mice of tumor-suppressor genes have given insight into the normal (developmental) function of these genes and permitted assessment of the oncogenic potential, when mutated.
- Cross-breeding of different genetically altered mice allows the identification of consecutive steps in the transformation process (oncogene collaboration).

During the last decade, much effort has been directed at improving mouse models and making them resemble human cancer more closely. In particular, we are increasingly paying attention to the introduction of similar mutations in a suitable genetic background, and also to achieve appropriate spatio-temporal control of gene expression and the extent to which a particular oncogene is overexpressed (or tumor-suppressor gene inactivated) in a particular cell or tissue.

These newer systems overcome some of the limitations of traditional transgenic and KO as models of sporadic cancers as follows:

- Most cancer-causing genes are not mutated throughout the ontogeny of the organism but instead arise as somatic mutations in the adult.
- The widespread expression of the gene mutation creates a different microenvironment than during initiation of sporadic cancers in which a few tumor cells are surrounded by many normal cells (role of cell–cell contact or paracrine signaling).
- Many tumor-suppressor gene knockouts show severe developmental defects or early embryonic lethality precluding study of their role in cancer.
- Many knockouts show a very distinct tumor predisposition pattern not necessarily corresponding to the tumor spectrum seen in man. This is exemplified by many earlier p53 knockouts (but may have been overcome by a very recent development from the laboratory of Tyler Jacks).
- One cannot test whether the genetic alteration is required for maintenance as well as initiation of the cancer. This requires the capacity to switch genes on and off.
- One needs to be able to follow the tumorigenic process *in vivo*. Therefore, imaging techniques

permitting monitoring of tumor growth are useful – including use of luciferase or other light-emitting reporters that can be monitored in the living animal. This improves the utility of spontaneous mouse models for conducting therapeutic intervention trials.

- It is useful to be able to track cells during tumorigenesis *in vivo* in order to identify the originating cell for cancer formation and to address the requirement for stem cells in this process. Lineage pulse-chase tracking tools have been developed to achieve this.

To identify multiple polymorphic regulators of cancer development, the effects of gene alterations can be examined in mouse strains of different backgrounds.

In the ideal model, one should have a defined strain background (or several for comparison), and should be able to switch genes on and off, ideally in a small subset of cells in a tissue in a time-dependent manner. Combining the use of Cre/Lox P techniques with hormone-inducible chimeric proteins (see Appendix 18.1 – techniques for generating genetically-altered mice) may allow this to be achieved, alongside new developments for gene knockdown (RNAi) that may soon be more readily employed *in vivo*.

ARRAY BASED TECHNOLOGIES

The first array experiment was performed in 1995, and measured mRNA expression levels in the small flowering plant *Arabidopsis thaliana*. This array contained 48 probes to measure differences in mRNA levels between wild-type plants and a transgenic line overexpressing the transcription factor HAT4. Current technology allows measurement of mRNA levels covering complete genomes in a fraction of the time the first experiment took. The most established array types are the DNA microarray chip for measuring mRNA expression levels and SNP microarray chips (SNP-chips) to identify polymorphic variation in the population (see Box 18.1). However, other array types and applications have been developed such as comparative genomic hybridization arrays (CGH-arrays), CpG island microarrays (CGI arrays), and the ChIP-on-Chip approach (see Box 18.2). These

Box 18.1 Microarrays

DNA microarrays, GeneChips, Arrays, and SNP-chips are all synonyms for the same technique. Microarray technology is based on the process of hybridization. Watson and Crick discovered that adenine (A) binds to thymine (T) and cytosine (C) binds to guanine (G). This biological phenomenon underlies hybridization and as a result two complementary DNA sequences will hybridize to form double stranded DNA. Equally, mRNA will hybridize to a DNA or mRNA strand as long as the strands are complementary. The hybridization process is so specific that a single strand of 25 nucleotides can pick out its matching partner from a mixture containing millions of different sequences. Microarrays take advantage of this process.

In essence, microarrays are a large number of parallel hybridizations on a chip. The chip surface (substrate) can be quartz wafer (silicon), glass, nylon, or nitrocellulose. Probes are usually cDNA or oligonucleotide strands that are attached to the chip surface. Microarrays can be produced by either spotting of the probes or by *in-situ* synthesis of the probes on the substrate. The *in-situ* synthesis of probes combines photolithography and combinatorial chemistry. The latest microarrays contain more than 45,000 probes on a single chip covering whole genomes and the probe density is still rising. Before hybridization to the chip, amplified mRNA or DNA (target) is labeled with a fluorochrome. After hybridization, each probe on the chip is scanned and the signal intensity is a semiquantitative measure for the amount of mRNA or DNA hybridized to each probe.

There are two different microarray systems available, which are either one color or two color technology. If, for example, we are interested in breast cancer we could identify differences between breast cancer cells and normal cells. This would give us insight into cancer pathology. In a two color experiment mRNA extracts are labeled with either green fluorescent Cy3 (healthy cells) or red fluorescent Cy5 (cancerous cells) label after mRNA amplification. Equally mixed Cy3/Cy5 labeled mRNA is hybridized to the microarray and microarray chips are scanned. Color changes indicate differences in mRNA levels between healthy cells and cancerous cells. Yellow (equal intensity of red and green dye) means no

change in expression and green or red gradients indicate a change in expression.

One color experiments (Affymetrix technology) use a single fluorescent label. Labeled mRNA of healthy cells or cancerous cells is hybridized to separate microarray chips. Changes in mRNA levels are calculated by comparing the signal intensities of the separate microarrays.

Expression analysis is the most common application of microarrays. mRNA expression levels are measured in cells or tissue and an array provides a snapshot of the cell's dynamic transcriptome. Changes between cell states can give information on pathway activation and other cellular processes occurring. Application of this technique ranges from fundamental cellular biology (e.g. cell cycle) to clinical applications (class prediction).

SNP-chips are similar to the microarrays used for expression analysis, but are mainly used to screen for SNPs in the genome to link genotype to phenotype. The starting material is not mRNA but genomic DNA, which is also amplified and labeled before hybridization. SNP-chips have been used for different studies such as loss of heterozygosity in cancer.

array systems are mostly custom built and their use is not as widespread as microarrays for mRNA expression analysis.

Single nucleotide polymorphisms, the HapMap, and identifying cancer genes

Linking a monogenic (single gene) variation to a phenotype, that is, disease is relatively straightforward, and numerous examples have been described already in Chapters 3 and 7. However, diseases are seldom the result of either single gene defects or of a single polymorphism. In fact, most common diseases are multigenic (genetically complex traits) and the process of identifying these complex traits is difficult and involves several steps: (i) linkage or association, (ii) fine mapping of the genetic changes, (iii) sequence analysis, and (iv) functional tests of the candidate genes. Identifying genes in the genome that underlie disease starts with either linkage or association. Linkage or association studies

Box 18.2 CGH arrays, CpG island microarrays, and ChIP-on-Chip

Comparative Genomic Hybridization arrays (CGH arrays) are a chip-based version of the classical cytogenetic technique that allows for rapid detection and mapping of genome-wide changes in DNA sequence copy numbers between normal and abnormal genomes. CGH arrays detect relative DNA copy number changes that provides information on chromosomal deletion, duplication, and unbalanced translocation often seen in cancer. The advantage of CGH arrays over normal CGH is the much higher resolution and speed. CHG arrays have up to 1,000 times higher resolution and are much faster to generate information with.

Conventional cDNA microarrays have been used for comparative genomic hybridization but only abnormalities in known genes (cDNA present on the chip) are detected and not in other regions of the genome. The use of normal microarrays is therefore limited. Current approaches use amplified genomic clones (pieces of human genome inserted into bacteria for replication) as probes on the array surface covering all nonredundant sequence of the genome instead of metaphase chromosomes used in conventional CGH.

CpG island microarrays (CGI arrays) are used to identify DNA methylation sites in a genome. Methylation occurs in CG-rich areas (CpG sites) of the genome and these methylated regions are termed CpG islands. CpG islands are common in exons and promoter regions but are rare in other parts of the genome. CGI arrays have probes that are selected on their potential methylation sites and are therefore CpG-island enriched.

For genome methylation analysis, genomic DNA is purified and selectively amplified for methylated regions. The amplified DNA and control DNA are labeled with CyDyes and hybridized to the chip. CGI arrays are two color technology and, labeled test (Cy5) and control samples (Cy3) are always mixed before hybridization.

ChIP-on-Chip research tool combines chromatin immunoprecipitation (ChIP) and DNA arrays or CGI arrays. The ChIP-on-Chip approach is used to identify new promoter regions in genomes, but can equally be used to identify differences in transcriptional silencing of genes by methyl-CpG binding proteins in cancer or healthy cells.

Chromatin is immunoprecipitated using specific antibodies raised against transcription factors or methyl-CpG binding proteins. If, for example, binding sites of the E2Fα transcription factor are determined, chromatin fragments that bind an E2Fα transcription factor are precipitated using an E2Fα antibody. DNA is enriched for chromatin fragments that are bound to E2Fα and these fragments are amplified and then hybridized on a chip against a control sample (not enriched DNA). This allows for the detection of E2Fα transcription factor DNA binding sites.

This methodology for binding sites of methyl-CpG binding proteins is identical and chromatin fragments are precipitated with a methyl CpG binding protein specific antibody.

identify regions, not a gene, on a chromosome – not necessarily on a single chromosome – that are "linked or associated" to disease or disease susceptibility.

Linking studies are generally successful in familial monogenic disorders, which are often high-risk but rare mutations. The selection of families with a rare disease from the population serves as a filter and selects for the specific mutated allele. Genomic variation of the diseased family members is compared to members of the family without the specific disease. However, linkage studies are not very powerful for identifying common disease alleles, such as cardiovascular disease, most cancers, and diabetes. Association studies are more useful in finding candidate disease genes in common diseases.

Genomic regions identified by linkage or association analysis, called quantitative trait loci, are still large and can correspond to 200–300 genes in a DNA sequence. Fine mapping is required to reduce this large number of candidate genes that might be associated with the disease and to zoom into the region of interest. This reduction is needed to make functional studies feasible. Studying 300 genes extensively in relation to a disease with current technology would simply take too much time and resources. This fine mapping results in identification of "minimal quantitative trait loci" covering several genes and numerous polymorphic variants.

DNA sequence analysis is needed to identify these candidate variations. Each nucleotide variant and combinations of nucleotide variants in a single gene and between genes have to be identified, ranked, and ultimately tested in a population to confirm the association.

All these steps can be undertaken by microarray analysis and specific microarrays are commercially available. Due to the variation and the large number of polymorphisms in any genome, custom arrays are popular. The stepwise approach works as a funnel reducing a full genome to a small number of genes and nucleotide variants. Identification of candidate genes and nucleotide variants is in itself not conclusive evidence and functional tests are needed. This is perhaps the most difficult step because methods to demonstrate the link between phenotype and nucleotide variant are limited. The best conclusive experiment is by experimentally changing phenotypes. Phenotype "swaps" are only successful if the correct nucleotide variant or correct combination of variants is changed.

Association studies are an emerging field and much effort has been directed at improving techniques and resources for researchers active in this area. The international HapMap consortium was set up to determine common patterns of DNA sequence variation in the human genome, the HapMap, to facilitate the identification of disease genes and mutations. Discovering genetic variation will contribute to advances in understanding pathogenesis, diagnosis, treatment, drug response, and metabolism. Identifying all possible SNPs in the genome at once would be an impossible task. Taking advantage of our understanding of genetics, a genomic map with genetic markers can be constructed to facilitate discovery of genetic factors underlying disease. These genetic markers are SNPs (tagSNPs) and a small number of tagSNPs identify a haplotype.

Haplotypes are stretches of DNA sequence that are inherited as an uninterrupted block of sequence during meiosis. These stretches of sequence contain many SNPs and a combination of a small number of these SNPs (tagSNPs) can uniquely identify a haplotype (Fig. 18.2). An understanding of meiosis is crucial to appreciate how haplotypes and therefore the HapMap can be used for discovery of genetic factors underlying disease.

Meiosis is the recombination of the paternal and maternal chromosomes during sex cell formation. This recombination event results in a new hybrid genome combining alleles of parental genomes, but the merging of the chromosomes does not happen from gene to gene. The recombination occurs in genomic blocks, that is, large stretches of DNA coding for many genes, and SNPs in that block travel together. The combination of a number of these SNPs uniquely identifies a genomic block, a haplotype. Due to the relatively short existence of human species the number of haplotypes is limited per chromosomal region. New haplotypes can be formed by recombination events and new additional mutations.

Different haplotypes can be characterized by identifying which tagSNPs are present in the particular genomic region. By haplotype mapping peoples' genome on the basis of tagSNPs we can detect if certain haplotypes occur more frequently in a diseased population compared to a normal population. If one haplotype is predominantly present in the diseased population, the disease associated polymorphism or mutation is on that specific haplotype. Identifying the haplotype and therefore the region of the genome that is associated with the disease, a more focused study can be undertaken to identify the specific SNP or combination of SNPs that underlie the disease.

As discussed in the introduction, not all breast cancer cases can be explained by SNPs or mutations in the known hereditary breast cancer genes (*BRCA1*, *BRCA2*, *TP53*, *CHK2*, and *ATM*). Yet, chance alone cannot account for the many cases of breast cancer that occur in some families. However, although these strands of evidence suggest that they should exist, the search for new genes that account for the many cases of hereditary breast cancer, not explained by known mutations has been largely unsuccessful. The HapMap will make association studies easier and will help to identify new genes that underlie common cancers such as breast and others. Microarray chips will be used containing all the tagSNPs selected by the HapMap project to identify the haplotypes. Subsequent microarray experiments will be undertaken to identify the SNPs or mutations in the haplotype region causing disease and these experiments will include fine-mapping, sequence

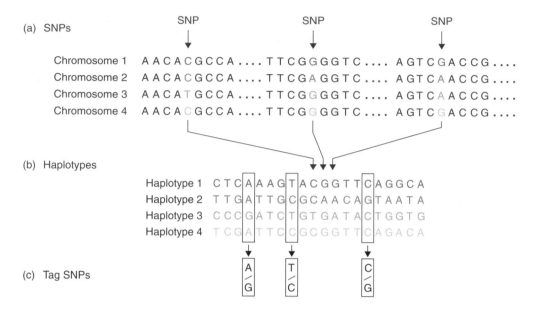

(a) SNPs

(b) Haplotypes

(c) Tag SNPs

SNPs, haplotypes and tag SNPs. a, SNPs. Shown is a short stretch of DNA from four versions of the same chromosome region in different people. Most of the DNA sequence is identical in these chromosomes, but three bases are shown where variation occurs. Each SNP has two possible alleles; the first SNP in panel a has the alleles C and T. b, Haplotypes. A haplotype is made up of a particular combination of alleles at nearby SNPs. Shown here are the observed genotypes or 20 SNPs that extend across 6,000 bases of DNA. Only the variable bases are shown, including the three SNPs that are shown in panel a. For this region, most of the chromosomes in a population survey turn out to have haplotypes 1-4. c, Tag SNPs. Genotyping just the three tag SNPs out of the 20 SNPs is sufficient to identify these four haplotypes uniquely. For instance, if a particular chromosome has the pattern A-T-C at these three tag SNPs, this pattern matches the pattern determined for haplotype 1. Note that many chromosomes carry the common haplotypes in the population.

Fig. 18.2 HapMap-Consortium (2003). The International HapMap Project. *Nature.* **426**: 789–96. Reproduced by permission of *Nature.*

analysis, and functional testing of candidate genes.

In the future, a patient's genome will be screened for all disease SNPs by SNP-array and this will allow for personalized medicine. Most diseases are a combination of genetic predisposition and environmental factors such as lifestyle. If patients are aware of disease susceptibility they can act upon it by, for example, taking medication or changing their lifestyle. This already happens for breast cancer; women with a familial history of breast cancer are screened for *BRCA1* and *BRCA2* mutations. In the event of increased risk (i.e. mutated genes) preventative mastectomy is often carried out, which markedly reduces the chance of developing breast cancer.

Cancer mRNA expression analysis

Before the development of microarray technology, cellular mRNA levels were studied comparing expression levels between, for example, normal and diseased cells after amplifying the mRNA by polymerase chain reaction (PCR).

Identifying gene transcription levels for many genes that might play a role in a cellular pathway or disease is very time consuming, especially if no prior knowledge is available. Which gene should be chosen?

The use of microarrays to study expression levels has revolutionized biology, but has come with many challenges. The advantage of microarrays is that all genes in a genome and, in future, all splice variants can be screened for expression. This gives a complete overview of the cellular state. Most genes in the genome are known but a number are merely bioinformatically predicted genes. Less is known about splice variants which are often only predicted from DNA sequence. This will change in the future as more experiments will have been undertaken.

Expression analysis studies have confirmed many findings in cancer research. More importantly, these studies have resulted in many new insights in cancer biology and mRNA expression analysis is proving to be a very useful tool for cancer classification, cancer diagnosis, and disease outcome prediction.

Challenges in mRNA expression analysis and other microarray-based techniques are extensive. The large datasets require new statistical methods to analyze the data (see Box 18.3), extract useful data, remove random variation or

---

Box 18.3 Microarray data analysis

The large amounts of data acquired by microarray experiments make data interpretation difficult. Dealing with such large datasets requires specialized expertise and a very good understanding of statistics and perhaps even mathematics. A successful experiment starts with good experimental design and ends with correct interpretation of results. When using high-throughput methods like microarrays the principle of "garbage in, garbage out" applies and statistical support is essential for getting useful data and results. The British geneticist and statistician Ronald Fisher already pointed out the need for good statistical support in research in 1938. He wrote: "to call in the statistician after the experiment is done may be no more than asking him to perform a post-mortem examination; he may be able to say what the experiment died of."

There are many software tools available for microarray data analysis, both commercial and public domain tools. There is software for mRNA expression analysis and genetic variation analysis (e.g. SNP analysis), but there are numerous tools that focus on a single aspects of microarray experiments. There is much debate on how to correctly analyze microarray expression data. New algorithms and analysis techniques are rapidly developed and there is still no consensus on the best methods. However, most biologists use software packages that roughly contain similar tools such as data normalization, statistical analysis, principal component analysis, and clustering. The main function of these software tools is visualization of the data and reducing the complexity of the datasets. When more than 30,000 transcripts are screened in different samples (e.g. cancer and normal cells) making sense of such large amounts of data is difficult and a powerful software is required.

In mRNA expression analysis experiments, normalization of datasets is important as it corrects for variation in (i) experimental preparation of the mRNA, (ii) fluorescent labeling, and (iii) chip hybridization. Removal of this systematic bias is needed to ensure correct interpretation of the data.

Statistical analysis of the data is undertaken to filter out the interesting genes that show significant changes in expression as a result of the biological process studied. The statistical tools usually include parametric and nonparametric statistical tests often incorporating multiple testing corrections.

Clustering algorithms are applied to group genes that have a similar expression pattern. Genes that show similar expression patterns are thought to have a role in the same biological process or pathway. Clustered genes are therefore in theory biologically related. Clustering can only be done in large microarray experiments where, for example, there are several time points. The better known clustering methods are hierarchical clustering, k-means clustering, and QT-clustering.

---

bias, and confirm findings. The biggest challenge perhaps is to make biological sense of all the information available. The data acquired are only a snapshot of the cell state and as cellular systems are highly dynamic much information is lost. This can be partly addressed by undertaking time course experiments, but the full dynamics of cells are never captured.

Research is therefore increasingly directed by in-silico experimentation using many experimentally obtained datasets, which are analyzed and modeled. Mathematical models are built using the empirical data and prior knowledge of the biological system. Such models describe the biological system and biologically relevant predictions or conclusions are then confirmed or explored using conventional laboratory techniques.

Comparing mRNA expression patterns from cancer tissue or cell lines with a control will identify cellular changes associated with cancer. Changes in cellular growth pathways can be detected where growth signals increase and growth inhibitory signals decrease (antigrowth). Changes in gene expression related to evasion of programmed cell death (apoptosis), angiogenesis, and tissue invasion pathways, genes involved in the extracellular matrix and increased activity of proliferation genes can be detected, for example. These changes can be identified by increased or decreased levels of single gene transcripts, but often several genes are affected that play a role in the same biological process or pathway. Besides cellular pathways, cell types can be distinguished in microarray data when tissue biopsies are analyzed. Human tissue is heterogeneous and cell types present can be identified by clustering cell type specific expressed genes.

Many microarray experiments have been undertaken to identify differences in mRNA expression between normal and cancerous tissue. In most cases, molecular signatures can be deduced that are specific for a cancer and are based on mRNA expression patterns. Analyzing all these datasets together, a meta-analysis, allows for identification of common transcriptional changes in cancers that describe neoplastic transformation. Moreover, common differences between differentiated and undifferentiated cancers have been identified. Such differences in mRNA expression underlie the clinical outcome of cancer. Undifferentiated cancers are more aggressive than differentiated cancers and chances of survival are significantly smaller for undifferentiated cancers.

Microarray experiments have highlighted what the molecular origin of this phenomenon is. Undifferentiated cancers maintain a more disordered state than differentiated ones and an increased level of cellular proliferation and invasion is maintained. This chaotic cell state results in a greater tumor aggressiveness ultimately leading to poor patient outcome. Three genes play a particularly important role in maintaining the undifferentiated phenotype. *EZH2*, *H2AFX*, and *H2AFZ* are overexpressed in undifferentiated cancers and the corresponding proteins are involved in chromatin remodeling and broad-spectrum transcriptional regulation. The protein EZH2 is important in transcriptional memory and both H2AFX and H2AFZ are histone variant proteins that are known to control euchromatin–heterochromatin transition. These three genes modulate the expression of potentially hundreds of genes and are likely to play a role in keeping the undifferentiated cancer phenotype. Plate 18.1 shows the list of genes that play a

role in undifferentiated cancers and neoplastic transformation in many cancer types.

Microarrays can be used for diagnostic purposes and are ideally suited for molecular classification of cancers. Breast cancer, for example, can have very different clinical outcomes and although breast cancers cannot always be separated by morphology using histopathological techniques, these cancers of the breast are likely to be very different diseases. Using microarrays in combination with advanced statistical analysis techniques has improved cancer diagnosis. Microarray analysis not only allows for differentiation between cancer and normal tissue or cells, but can also subclassify cancers. There are ample examples that demonstrate the usefulness of microarrays in cancer diagnosis (Perou *et al.* 2000; Shipp *et al.* 2002; Sorlie *et al.* 2003) and for breast cancer this subclassification has also been correlated to clinical outcome. The gene-expression signature of a cancer tissue biopsy can be used as a predictor of survival in breast cancer. Breast tumor gene expression profiles proved very useful for overall survival prediction and the probability of remaining metastasis-free. The main advance of microarray approaches is the better prediction of outcome compared to other criteria for classifying high and low risk cancer and survival. NIH and St Gallen criteria misclassified a clinically significant number of patients compared to the tumor-gene-expression signature leading to under- and overtreatment. Moreover, microarray data indicate that the ability of a tumor to metastasize is probably an inherent genetic property of (breast) cancer and not an acquired potential in later stages of tumorigenesis.

However, this highlights another problem with many microarray studies, namely that of cellular heterogeneity. Studies, which examine gene expression in tissue (even when areas of that tissue may have been selected by techniques such as laser capture microdissection), will inevitably average gene expression across multiple cells. Particularly, in the context of cancers, which may contain multiple subclones of cells of increasing genetic instability, the demonstration of a catalog of mutant genes must not be assumed to mean that every cancer cell carries all such mutations. In other words, the presence of metastases-related gene mutations in the primary tumor prior to overt metastatic spread should be interpreted with caution until it can be confirmed that single cells in the original tumor carry all the same mutations as those in the early metastasis. This is an important point as we need to also account for the well-documented presence of large number of cancer cells in the blood of patients (circulating tumor cells – CTCs) with no secondary tumors and often it takes some considerable amount of time before they develop. If the mutations required for metastases were already present then why are metastases not formed?

This not withstanding, data from "whole tumor" microarray studies will likely have a major impact on future therapies. Thus, treatment of cancer will have to be "tailored" and a more aggressive systemic treatment for the more malign forms of cancer would be required compared to less aggressive localized tumors.

In conclusion, microarray analysis can facilitate cancer diagnosis and can classify cancers by their molecular-gene-expression profile. Tumor-gene-expression profiles allow for tumor class discovery leading to tumor class and survival outcome prediction.

The appropriate classification of cancer is essential for successful treatment. Microarray data have been used effectively for class prediction, but tumor class prediction by microarray is expensive, time consuming, and requires specialized expertise. Microarray-based diagnostic and prognostic testing will become more widespread in the future as costs come down and supporting analysis software becomes available. In the meantime translational research is undertaken to identify the genes that have the highest predictive value. The objective is to develop a simple multigene assay for routine clinical use. This can be done by, for example, quantitative PCR to measure gene-expression levels in the tumor for a set of marker genes. The diagnostic and/or predictive test will give identical answers to microarrays but is more cost-effective and faster. Such a predictive test has been developed for diffuse large-B-cell lymphoma.

Molecular tests for cancer diagnosis are used to add further predictive power to standard classification and outcome prediction protocols. In the future, molecular data might supercede other nongenomic information. However, for all molecular tests tumor biopsies are required, which remains mostly an invasive procedure.

CGH arrays, CpG island microarrays, and ChIP-on-Chip

CGH arrays, CpG arrays, and ChIP-on-Chip approaches are not as established as expression or DNA microarrays and certainly have not reached the stage of routine clinical application (see Box 18.2). However, they have enormous potential to improve our insights in cancer biology.

Comparative genomic hybridization arrays are promising to be very useful in identying new cancer related genes and chromosomal abnormalities. Genomic mutations, unbalanced translocations, deletions, and duplications can be detected throughout the entire genome using CGH arrays. All these biological events can underline cancer development and previous methodology was not as advanced in detecting such genomic changes as CGH arrays. The main limitation has been insufficient genomic resolution to detect small genomic changes.

CGH can detect DNA copy number aberrations that are indicative of loss or gain changes in the genome. Moreover, due to the array setup the exact location on the genome is known and the breaking point and the genes that are involved are identified relatively easily.

CGH arrays have already resulted in new insights in breast, ovarian, and gastric cancer. Additionally, CGH arrays could become a very useful diagnostic and prognostic tool when the underlying chromosomal changes are correlated with clinical outcome. Unlike mRNA levels, genomic changes are less dynamic and are not influenced by stress, medication, and other factors that affect mRNA expression levels. Moreover, chromosomal changes often lead to changes in gene or protein expression necessary for cancer to arise. The further development of CGH arrays, in particular increased resolution, will contribute to the understanding of genomic aberrations that will lead to advances in research diagnosis.

Functional protein changes can not only be the result of quantitative changes but also can result from chromosomal abnormalities. The Bcr-Abl fusion protein is formed as a result of a chromosomal translocation. The first exon of *ABL1* gene is replaced by sequences of the *BCR* gene, which leads to the expression of the Bcr-Abl fusion protein. The expression of the Bcr-Abl fusion protein is linked to different forms of leukemia.

Depending on the cell type Bcr-Abl expression results in morphological transformation, enhanced proliferation, or abrogation of growth factor/adhesion dependence. All these biological changes are the result of abnormal constitutive tyrosine kinase activity of the fusion protein Bcr-Abl.

In ovarian and breast cancer, DNA copy numbers and mRNA levels of *RAB25*-gene, a small GTPase, are predictive of the clinical outcome. Increased DNA copy number and mRNA levels of *RAB25* in ovarian and breast cancer are prognostic for a poor survival chance as *RAB25* underlies the aggressiveness of epithelial cancers such as ovarian and breast cancer.

Identification of chromosomal breakpoints, translocations, and loss or gain events are key to further advances in cancer diagnosis and treatment. CGH arrays is the best method currently available to identify cancer related chromosomal changes.

CpG island arrays (CGI arrays) have been developed to map epigenetic changes in the genome. Epigenetics is a field that is evolving rapidly and identification of methylated CpG sites in the human genome is essential to understand chromatin remodeling and gene expression changes underlying disease. It is known that Prader–Willi syndrome, Angelman's syndrome, and Beckwith–Wiedemann syndrome are caused by gene-methylation abnormalities.

CpG islands are often found in gene promoter regions and are usually 0.5–4.0 kb in length. Hypermethylation of these promoter regions is the best categorized epigenetic change to occur in cancer and hypermethylation is found in almost every type of human neoplasm. If epigenetic changes, for example, affect genes regulating cell proliferation, uncontrolled cell division can occur resulting in cancer. Epigenetic abnormalities underlying cancer are described in Chapter 11.

ChIP-on-Chip arrays are a variation on CGI arrays. The array-based technology is identical to CGI arrays, but the preparation of DNA is different. Chromatin is precipitated with specific antibodies to chromatin binding proteins such as methyl CpG-binding proteins. The antibody used can be changed according to the chromatin binding protein. Differences in the chromatin precipitated provides information on chromatin remodeling and therefore potential differences in transcriptional activity.

ChIP-on-Chip approaches have demonstrated that methylation associated transcriptional silencing exists in cancer and target genes have been identified. In breast cancer, homeobox gene *PAX6* and the prolactin hormone receptor gene show variation in transcriptional activity.

Both CGI array and ChIP-on-Chip methodology provide information on gene transcription and in particular, how DNA sequence and chromatin structure play a role in transcriptional regulation. There remains much unknown about transcriptional regulation in cancers but both approaches will contribute to our understanding.

PROTEOMICS

Proteins are the final product of cellular synthesis and DNA transcription leads in most cases to protein production. Since Watson and Crick published their landmark paper on DNA our understanding of cellular processes has improved, but much is still a mystery. Studying the end product of cellular synthesis, proteins, is therefore key to understanding cellular processes.

Proteomics is a relatively new research discipline and the proteome is often defined as the protein complement of a cell, organ, that is, brain or heart, or genome. Tools often used in proteomics are mass spectrometry (MS), high performance liquid chromatography (HPLC) and two-dimensional gel (2D-gel) electrophoresis (see Box 18.4).

Box 18.4 Tools in proteomics

The main tools in proteomics research are two-dimensional polyacrylamide gel electrophoresis (2D-gels), liquid chromatography (LC) and mass spectrometry (MS). 2D-gels and nano-flow high performance liquid chromatography (HPLC) are protein separation techniques and MS is an analytical technique to identify and characterize proteins.

Cells contain thousands of proteins and to identify these proteins powerful separation techniques are needed. 2D-gels and LC are mostly used to separate individual proteins in complex protein mixtures.

Box 18.4 (continued)

Two dimensional gel electrophoresis is a polyacrylamide-gel-based technique that allows for separation of proteins in two-dimensions. Proteins are separated in the first dimension on isoelectric point using pH gradient strips. The pH gradient strips are then transferred to a polyacrylamide gel and the proteins are separated based on molecular weight in the second dimension. Proteins are visualized by staining the gel akin to developing a photo. The end result is a large gel with up to several thousand small spots that are all different proteins (Fig. 18.3). Before mass spectrometric analysis, proteins are digested by a protease into peptides. Proteases, such as trypsin, cut proteins into peptides at specific sites of a protein sequence. There are several reasons why proteins are digested into peptides, but in a few words it makes the identification process much easier. An alternative to 2D-gels is LC that can also be used to reduce the complexity of a protein mixture by separating the (digested) proteins based on certain characteristics like hydrophobicity or charge. The LC-system is usually directly connected to a mass spectrometer (LC-MS).

Mass spectrometry is an essential technique to identify proteins and is used in any proteomics experiment. The three main parts of a mass spectrometer are the ionization source, the mass analyzer, and the detector. The two main ionization sources are electrospray and MALDI (matrix assisted laser desorption ionization). The ionization source transfers nonvolatile compounds (peptides and proteins) into the gas phase. After ionization, the peptides are guided into the mass spectrometer to the mass analyzer. The mass analyzer separates the peptides or proteins based on their mass before hitting the detector. The time a peptide or protein takes to travel through the mass analyzer before it hits the detector can be used to calculate the mass of the protein or peptide. The main mass analyzers in mass spectrometers are the time of flight (ToF) mass analyzer, the quadrupole mass analyzer, and the quadrupole ion trap. Mass spectrometers are used to obtain either (i) sequence information of a single peptide that allows for unambiguous identification of a protein or (ii) a mass fingerprint of the peptides from a digested protein that allows protein identification without sequencing the

individual peptides. A peptide mass fingerprint is like a human fingerprint – it allows for unique identification of a protein.

An excellent basic introduction on peptide sequencing, protein identification, and mass spectrometers is written by Steen and Mann (2004).

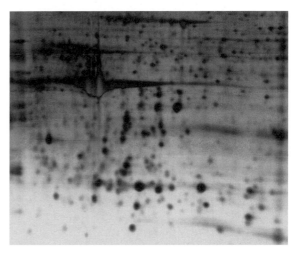

Fig. 18.3 A section of a 2D-gel from smooth muscle protein extract. Y-axis is separated on molecular weight. X-axis is separated on isoelectric point.

Proteomics started off by identifying as many proteins in a cell or tissue as possible. Cataloging proteins provided an insight into the complexity of the proteome. However, the proteome is highly dynamic and changes with respect to its environment, cell state, disease, and so on. Identifying proteins in itself does not provide a sufficient understanding of the cellular dynamics and a shift towards quantitative proteomics occurred. Relative increases and decreases of proteins in tissue as a result of, for example, cancer are studied. Moreover, higher order structures such a protein interaction dynamics and inter- and intracellular communication networks have not been (fully) identified. This section deals with the various proteomic approaches that have increased our understanding of the proteome. Cancer is a biological system and proteomic techniques have been applied to understand cancer cell dynamics on the protein level. Examples of proteome research will be discussed to illustrate the use of these techniques. Cancer models vary from mouse models to yeast and this has been discussed in an earlier part of the chapter and book.

Quantitative proteomics

Studying relative changes in protein quantities in a cell or tissue can provide useful information about proteins that are involved in, for example, cancer. It is difficult to find out if changes are a consequence of, or responsible for, the disease state. The main methods to identify protein changes are 2D-gel electrophoresis, chemical labeling of proteins like isotope coded affinity tagging (ICAT), and metabolic labeling (see Box 18.5).

Box 18.5   Quantitative proteomics

Identifying all proteins in a cell is of limited importance as we know the majority of expressed genes through the genome sequence. Identifying changes in protein quantity or post-translational modification – which also leads to a quantitative change of a functional protein – is far more interesting. These proteome changes result from altered normal cellular processes or disease and identifying these changes help us understand cellular processes.

Identifying quantitative changes in the proteome can be done by combining (i) 2D-gel electrophoresis with MS or (ii) HPLC techniques with MS (LC-MS). MS is not a quantitative technique and is therefore combined with 2D-gel electrophoresis or protein labeling strategies such as ICAT or metabolic labeling.

After separating proteins on a 2D-gel the proteins are visualized with silver stain or fluorescent dyes. These staining techniques are semiquantitative and allow for identification of quantitative protein changes between proteins from, for example, cancerous and healthy breast tissue. Comparing 2D-gels from diseased and normal tissue with specialized analysis software can lead to identification of proteins that play an important role in cancer. 2D-gel electrophoresis technology is still improving. DiGE (difference in-gel electrophoresis) is one of these technological improvements and is the proteomics variation on the two color microarray experiment also using Cy-Dyes. Proteins are labeled with different fluorochromes (Cy3 or Cy5) before two-dimensional separation and relative protein quantities can be measured.

The advantage of LC-MS methods over 2D-gel electrophoresis is the increased sensitivity and the speed of acquisition, but MS is not a quantitative technique. Using LC-MS methods for identifying proteome changes requires additional methodology for quantifying protein changes. This is done by differentially labeling proteins with a tag akin to a two-color microarray experiment. However, the proteins are not labeled with a different fluorochrome but with a different mass tag as mass spectrometers separate on differences in mass. Labeled proteins are digested into peptides by a protease before mass spectrometric analysis. A mass spectrometer can analyze only a small number of peptides simultaneously and a powerful separation technique (HPLC) is needed before a complex mixture of peptides is analyzed.

ICAT (isotope-coded affinity tag) is a chemical labeling method that uses heavy and light labels that react specifically with the thiol group of the amino acid cysteine in all proteins. The labels are identical except for the carbon (C) atoms, which in the heavy mass tag are replaced by ^{13}C atoms – the heavy isotopic form of ^{12}C. By replacing nine ^{12}C atoms by ^{13}C atoms a mass gain of 9 Da is achieved, which can be measured by MS. Differentially labeled protein samples (e.g. from cancerous cells and normal cells) are mixed in equal amounts. After protease digestion, identical peptides with different mass tags can be separated by their mass in the mass spectrometer but will still coelute from the LC-column. The success of this approach depends on the existence of cysteine in a protein and the efficiency of the labeling. If a protein does not contain a cysteine or if the peptide is not labeled successfully, the protein cannot be quantified. Two labeled peptides for each protein are needed to identify and quantify protein changes by statistical means.

Because chemical labeling is never 100% efficient, new methods have been developed for *in vivo* labeling or metabolic labeling. SILAC (stable isotope labeling by amino acids in cell culture) is an *in vivo* labeling method for cell culture.

Box 18.5 (continued)

Essential amino acids needed for cell growth are substituted by heavy ^{13}C forms in the cell culture media. Essential amino acids used are ^{13}C-arginine and ^{13}C-leucine, which have a mass difference of 6 Da with the normal ^{12}C form. The cells are "forced" to consume these amino acids and incorporate them in the proteins that are synthesized. After a couple of days and several cell divisions later all ^{12}C-arginines are replaced by ^{13}C-arginine in every single protein in the cell. By mixing normal ^{12}C-arginine cells with ^{13}C-arginine labeled cells (induced) proteome changes can be measured by LC-MS approaches.

Not only cells but also yeast, the nematode worm *C. elegans*, and *D melanogaster* have been labeled with heavy isotopes (^{15}N and ^{13}C). Completely heavy-isotope labeled mice and rats are only a matter of time.

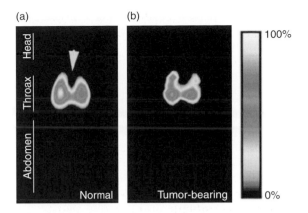

Fig. 18.4 Gamma-scintigraphic images of normal and tumor bearing rats. *In vivo* images were obtained after ^{125}I-APP (^{125}I-aminopeptidase-P) monoclonal antibodies were injected intravenously. A difference in lung shape indicates tumor growth. From Oh, P., Li, Y., Yu, J., *et al.* (2004). Subtractive proteomic mapping of the endothelial surface in lung and solid tumors for tissue-specific therapy. *Nature*, **429**, 629–35. Reproduced by permission of *Nature*.

Two-dimensional gel electrophoresis was the first method that allowed for quantitative proteomics and was initially developed by O'Farrell in 1975 (Fig. 18.3; 2D-gel). Since then the technique has been improved, developed further, and standardized, including much needed analysis software. 2D-gel electrophoresis has become the default method in the last 10 years for quantitative proteomics despite many limitations. 2D-gels resolve mainly the high abundant proteins in a cell, which are usually of less interest. Moreover, 2D-gel electrophoresis has limited resolving power resulting in loss of proteins that are either very small, very large, basic, or membrane proteins. However a focused application of 2D-gels can be very successful. For example, resolving differently post-translationally modified proteins on 2D-gels can be very successful or separating protein components of a purified protein complex can be undertaken by 2D-gel. In essence, 2D-gels have limitations in resolving power and sensitivity, but can still be a useful technique. Some interesting literature examples on 2D-gel-based proteomics can be found in the reference section (Lewis *et al.* 2000; van Anken *et al.* 2003).

Proteomic research has shown that organ microenvironments are dynamic and change as a result of tumor growth. Tumor cell membranes, endothelial cell membranes, and stromal proteome, in particular stromal enzymatic activity, have been found to alter. The chaperone protein HSP90α plays an important role in cancer cell invasion in several implanted cancer cell lines and is likely to be of importance in other tumors. Aminopeptidase-P and annexin A1 have been found by a 2D-gel proteomic approach to be differently expressed at the blood-tissue interface of lung tumors. By gamma-scintigraphic imaging with labeled antibodies, differences in lung shape between normal and tumor-bearing rats could be detected (see Fig. 18.4).

Isotope coded affinity tagging was one of the first chemical peptide labeling techniques that allowed for high-throughput proteomics by LC-MS. The main disadvantage of ICAT is that the labeling is dependent on a chemical reaction that has to be optimal for successful labeling of cysteine. However, ICAT can be used on any biological system whether it is cell lines, (human) tissue, yeast, or bacteria. An example of the application of ICAT is the quantitative proteomic analysis of Myc oncoprotein function in mammalian cells (Shiio *et al.* 2002).

Stable isotope labeling by amino acids in cell culture (SILAC) and heavy nitrogen (^{15}N) labeling are the main types of metabolic labeling. Metabolic labeling is a very efficient method of proteome-wide labeling of proteins. Unlike chemical labeling like ICAT, the cell incorporates the heavy isotope form (see Box 18.5). Currently, the most promising quantitative proteomics approach is metabolic labeling due to the efficient incorporation of the label. However, SILAC can only be used for cell-culture-based proteomic experiments. In contrast to SILAC, ICAT, and 2D-gel electrophoresis can be used on both tumor biopsies and cell-culture-based experiments. The appropriate application of proteomic techniques is an important part of the experimental design.

Metabolically labeled cells or organisms do not require a complicated two-dimensional gel electrophoresis step to separate and semiquantify proteins prior to LC-MS analysis. Generally, SILAC or ^{15}N labeled protein extracts are mixed with a control in a 1 : 1 ratio and then separated by one-dimensional gel electrophoresis before LC-MS analysis. The mass spectra are processed to identify and quantify proteins. Although SILAC has a large potential not much research has been published yet, but an example will be discussed in the section on post-translational modifications.

As mentioned in the introduction of this chapter, proteins can also be altered by post-translational modifications, which results in a new protein variant. Post-translational modifications of proteins is a subject of much research as they underlie many cellular processes and this will be discussed in the next section.

Post-translational modifications

Oxidation, sumoylation, methylation, acetylation, ubiquitination, phosphorylation, and glycosylation of proteins are all post-translational modifications that influence protein function, cellular destination, and more. A change of function effectively corresponds to formation of a new protein. Knowledge of the modification site, kinetics, regulation, and functional effects of a modification is of importance to understand cellular consequences of a post-translational modification.

All post-translational modifications are of great importance in a cell even if they do not occur as frequently as, for example, glycosylation and phosphorylation. Discussing all post-translational modifications or even the modifications named in this book is beyond the scope of this chapter. Many post-translational modifications can be found elsewhere in this textbook, in research papers, or on the internet. Because glycosylation and phosphorylation are well known post-translational modifications they will be briefly discussed here.

Glycosylation is the most complex post-translational modification in eukaryotes. It plays a role in for example intracellular regulation, cell–cell communication, cell–substrate recognition, and secretory pathways. The sugars that form building blocks for glycosylation of proteins are fucose (Fuc), galactose (Gal), N-acetylgalactosamine (GalNac), glucose (Glc), mannose (Man), N-acetylneuroaminate (NeuNAc), sialic acid (Sia), and N-acetylglucosamine (GlcNAc). Although few sugars make up a glycan, the glycans that are attached to a protein as a post-translational modification are highly complex due to the large number of sugar combinations formed.

There are two sites in a protein were glycosylation occurs: N-glycosylation and O-glycosylation. N-linkage occurs on the amide nitrogen atom in the side chain of asparagine and O-linkage occurs on the oxygen atom in the side chain of serine or threonine. Potential glycosylation sites are detected in the amino acid sequence and are very specific sites. Mass spectrometry is a powerful technique to investigate glycosylation of proteins.

Phosphorylation of proteins is regulated through kinases and phosphatases on specific sites of the protein. Serine, threonine, and tyrosine are the three sites of phosporylation. Phosphorylation results in activation or deactivation of a protein, which in turn often switches cellular pathways off or on. These activities are highly regulated and (de)phosphorylation leads to amplification of the signal. Overactivity of certain kinases underlies cancer. The Bcr-Abl fusion protein, as discussed previously in this chapter, is an example of kinase overactivity that underlies cancer.

Temporal phosphotyrosine signaling of the EGF receptor pathway has been studied by SILAC. Multiple SILAC experiments were undertaken to study the phosphorylation

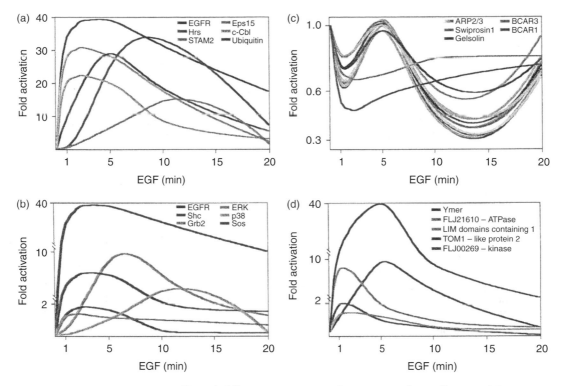

Fig. 18.5 Continues activiation profiles of different catogories of EGFR signaling effectors. (a) Key proteins involved in receptor internalization and endosomal trafficking. (b) Proteins from the Ras-MAP kinase pathways. (c) Proteins involved in actin remodeling. (d) Novel proteins found to be activated by the EGFR signaling pathway From Blagoev, B., Ong, S.E., Kratchmarova, I. and Mann, M. (2004). Temporal analysis of phosphotyrosine-dependent signaling networks by quantitative proteomics. *Nature Biotechnology*, **22**, 1139–45. Reproduced by permission of *Nature*.

dynamics of the EGFR pathway over a time course of 20 minutes. Figure 18.5 shows how the phosphorylation status changes over time for functionally clustered proteins. These temporal changes demonstrate how elaborate and complicated signal transduction pathways are, bearing in mind that this only represents phosphotyrosine-dependent signaling networks for the EFG receptor in a single cell system.

Protein complexes and cellular networks

Proteins rarely function alone and form complexes with other proteins to undertake a certain action in the cell. Proteins can be activated by a post-translational modification (see Box 18.6) and can travel between cellular compartments to interact with other proteins to perform a specific

Box 18.6 Protein–protein interactions and networks

Protein–protein interactions are the basis of cellular communication. All proteins have interaction domains, which are mostly conserved between species. This explains why the findings in yeast can explain much about human intracellular signaling. Protein domains can be constitutionally active or can be activated by a post-translational modification. Kinases, for example, can activate protein domains by phosphorylation that leads to binding of the activated domain with a recognition domain. Phosphotyrosine sites in proteins for example bind to Src homology-2 domain (the SH2 recognition domain). Identifying post-translational modifications and protein–protein interactions is needed to unravel interaction networks. Not only

Box 18.6 (continued)

identification of protein complexes, which are essentially multiple protein–protein interaction, is of interest but also the temporal kinetics of these biological processes.

Identification of post-translational modifications can be done by MS. Modifications of a protein or peptide lead to a change in mass and the modification site can be identified by sequencing the protein by mass spectrometry.

To identify cellular protein complexes, molecular techniques need to be used to "fish" the protein complexes out of the cell. This is done by inserting a specific affinity tag at the 3'-end of the gene in the genome. When the protein is expressed it is expressed as a fusion protein with an affinity tag attached to the C-terminal end of the protein. This affinity tag allows for a very selective purification of that particular protein out of the complete cellular extract and interacting proteins are purified at the same time. This tandem affinity purification (TAP) method is a two step purification technique and uses specific affinity columns to purify the protein complex (Fig. 18.6). The TAP-tag consists of two binding domains of protein-A and a calmodulin binding peptide with protease cleavage sites in between. Protein-A binds very strongly to immunoglobulin-G (IgG) and the tagged protein including interacting proteins binds to the IgG affinity column. The complex is eluted from the column by cleaving off a part of the TAP-tag with a protease. The eluent is purified further by a second affinity column to which the calmodulin binding peptide binds. The protein complex is eluted off the column after several wash steps and purified proteins can be identified by MS. The identified proteins are part of a functional protein complex.

There are other tagging approaches known but the principles are identical.

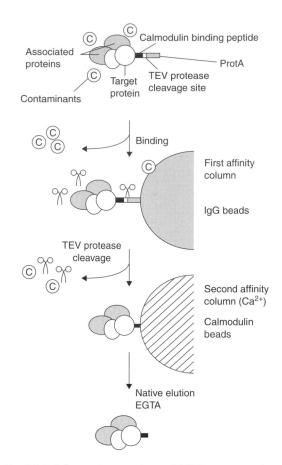

Fig. 18.6 Schematic overview of TAP tag procedure. A protein of interest is expressed in the cell with a TAP tag. Cultured cells are lysed and the complex is purified with the first affinity column (IgG-beads). A specific protease, TEV-protease, then cleaves the complex off the affinity column. The complex is eluted from the column onto the second affinity column and further cleaned up. Finally, the cleaned up complex is eluted off the calmodulin beads by EGTA and can be analyzed by MS. All interacting proteins can be identified. From Puig, O., Caspary, F., Rigaut, G., Rutz, B., Bouveret, E., Bragado-Nilsson, E., Wilm, M. and Seraphin, B. (2001). The tandem affinity purification (TAP) method: a general procedure of protein complex purification. Reprinted from *Methods*. **24**, 218–29. Copyright (2001), with permission from Elsevier.

task. Even ATP-driven conformational changes are of importance in complex formation. Hence the importance of three-dimensional structural studies.

Identification of protein complexes is key to elucidating and understanding intracellular protein networks. Protein complexes are ordered but dynamic structures that assemble, store, and pass on biological information.

Tandem affinity purification strategies have been used to purify protein complexes from yeast, *Drosophila*, and human cell lines. Identified protein complexes were used to unravel cellular communication networks using mathematics and bioinformatics. A connectivity map

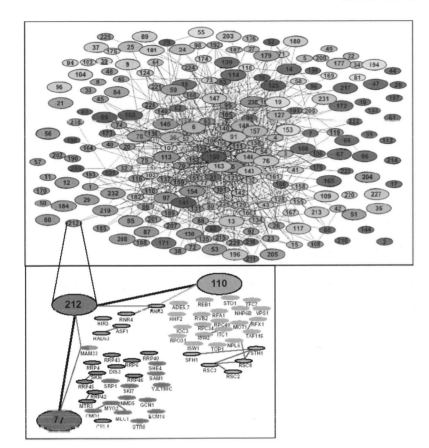

Fig. 18.7 Connectivity map of the yeast proteome established through TAP-tagging approaches. More than 80% of the proteome is highly connected. Links were established between complexes sharing at least one protein. Individual complexes are color coded. Red is cell cycle; dark green is signaling; dark blue is transcription, chromatin structure and DNA maintenance; pink is protein and RNA transport; orange is RNA metabolism; light green is protein synthesis and turnover; brown is cell polarity and structure; violet is intermediate and energy metabolism; light blue is membrane biogenesis and traffic. (Please see our website for the color version of this figure.) From Gavin, A.C., Bosche, M., Krause, R. *et al.* (2002). Functional organization of the yeast proteome by systematic analysis of protein complexes. *Nature*, **415**, 141–7. Reproduced by permission of Nature.

of yeast protein network is shown in Fig. 18.7. More than 80% of the yeast proteome is highly connected through sharing proteins in different complexes. Thus certain proteins are highly promiscuous and are essential for cellular communication structures.

Most research on protein complexes is done on yeast, but by screening human protein orthologs our understanding of human proteins is increased. Orthologs are proteins from a common ancestor and have an identical function in different species. Similar tag approaches have been used to label proteins and localize them in the various cellular compartments of yeast or cells. By labeling proteins with a green fluorescent protein tag (GFP-tag) the subcellular compartment in which the protein is localized can be detected, which provides information on its function. For yeast this was done as a large screen that provided a wealth of information about thousands of proteins of the (yeast) proteome.

The majority of research, such as protein network elucidation, has been undertaken in yeast. This is mainly because yeast is a eukaryotic organism that can be very easily manipulated compared to mammalian cell lines. The acquired information is still very useful for research into the human proteome, due to evolutionary conservation.

One of the few exceptions on protein interaction work is the research undertaken on the TNF-α/NF-κB pathway in a TNF-α responsive HEK293 mammalian cell line. Thirty-two proteins known to be part of the TNF-α/NF-κB pathway were TAP tagged. Tagged proteins were copurified with interacting proteins and 680 proteins including 241 interactions were identified. Twenty-eight proteins were selected for validation and 10 new functional modulators of the pathway were verified using a functional assay combined with RNA-interference (RNAi).

TRAF7 and TBKBP1 had no previous functional annotation and were among the 10 proteins verified by RNAi to be actively involved in the TNF-α/NF-κB pathway. TRAF7 formed a complex with Hsp 70, PFDN2, MARK2, MEK5b, and C20orf126. TBKBP1 interacts with TBKBP2 to form a complex with TBK1. TBKBP is an abbreviation for TBK binding protein. More information on TAP-tagging of TNF-α/NF-κB pathway can be found in Bouwmeester *et al.* (2004).

Unraveling of a pathway allows for target selection to modulate pathway activation in disease by pharmaceutical compounds. Proteomics is therefore one of the key drives behind target selection for drug development.

All this information about proteins from mammalian cells and yeast can be cataloged in databases and provides a valuable resource for researchers. Sequence information, structure, post-translation modifications, interacting proteins, cellular function, and location can be found in databases. Some links to databases can be found at the end of the chapter.

Clinical applications of proteomics

Proteomic methodology can be used for identifying drug targets and development of diagnostic tests. In particular, unraveling of pathways and identification of diagnostic or prognostic markers has much potential.

A promising new approach in proteomics research is the use of plasma for identifying new diagnostic targets. Plasma is an important body fluid that can be acquired in a noninvasive way.

For diagnostic purposes, there has been much interest in serum protein fingerprinting using proteomic techniques. The hypothesis is that cancer and other diseases (e.g. inflammation) release protein waste into the blood stream. When plasma is analyzed by mass spectrometric techniques a protein fingerprint is obtained. This fingerprint can be specific for a disease state and would allow for diagnosis using specialized software. Such proteomic-based diagnosis does not require any invasive procedures like taking tumor biopsies and has a quick turnaround. Development of such proteomic-based diagnostic tools is at an early stage and validation of proteomic-based tests is required before its introduction into the clinic.

CONCLUSIONS AND FUTURE DIRECTIONS

As already discussed earlier for gene expression studies, one very important area in cancer research is that relating to cellular heterogeneity within a tumor. Thus, within most tumors there are a large number of different cell types other than the cancer cells themselves, including inflammatory and immune cells recruited to the tumor, tissue stroma (now known to play a crucial role in tumorigenesis), vascular endothelial cells, and others. Moreover, even the cancer cells themselves are not genetically identical and may vary quite significantly from one another (due to clonal evolution and genetic instability). Even seemingly identical cells may vary in gene expression due to interactions between neighboring cells (lateral inhibition effects through Wnt signaling, recent exciting reports of Myc activated cells affecting growth in neighboring identical cells, etc). All these effects can create major headaches for the investigator bent upon molecular and proteomic profiling. Profiling any combination of cells, no matter how seemingly homogeneous, will inevitably result in the averaging of gene expression across several cells. Results obtained must therefore be interpreted with caution, for example, finding mutations in *Ras*, *p53*, *TERT*, and *Myc* in a tumor would have very different ramifications if all these were present in single cells as distinct from different combinations of these being in different individual cells. One way of circumventing this issue is to profile single cells isolated from the tumor or from the circulating blood. Unfortunately, despite tools available to achieve this, such as LCMD, FACS, and "immunopanning," the problem of analyzing gene/protein expression in single cells is still not fully resolved.

Microarray and proteomic techniques will undoubtedly develop further in the coming years and will become even more powerful research tools. Yet, the large amounts of data acquired with these high-throughput methods demand new approaches for confirming results. Currently, only the most interesting genes or proteins are verified by standard laboratory techniques.

Completion of the HapMap, sequencing of new genomes, identification of the human transcriptome, and proteome for different organs will further increase our understanding of (human) biology. The ultimate goal is to have mathematical models of cellular systems where one can navigate through the biological system from genome to protein network as depicted in Figure 18.1.

All the developments discussed will contribute to identification of new drug targets and diagnostic biomarkers. Personalized medicine will become a reality due to genotyping, disease outcome prediction, and predictive diagnostic tests. Additionally, cancer treatment is currently very general and more targeted treatment will be the outcome of all the developments discussed in this chapter. To achieve improved cancer treatment, a better understanding of cancer as a disease and as a biological system is needed.

At the end of this chapter and book, it is difficult not to be amazed by the complexity of cancer biology. It is at times worthwhile to step back and appreciate how controlled biological systems or organisms are. More importantly how little actually goes wrong considering the complexity of biology.

QUESTIONS

1 What is the most common array type used in research?
   a. CGH arrays.
   b. mRNA expression array.
   c. Antibody arrays.
   d. ChIP-on-Chip.
   e. CGI arrays.
2 What is the correct flow of information in the cell?
   a. DNA, mRNA, protein, post-translational modification, protein complexes.
   b. mRNA, DNA, protein complexes, post-translational modification, protein.
   c. DNA, protein, mRNA, post-translational modification, protein complexes.
   d. Protein complexes, post-translational modification, protein, mRNA, DNA.
3 What is the most commonly used method to separate out a cellular protein extract?
   a. ICAT.
   b. Liquid chromatography.
   c. TAP-tagging.
   d. Two-dimensional gel electrophoresis.

Appendix 18.1 Techniques for the generation of genetically-altered mouse models of cancer

Gene knockouts (targeting) and transgenesis represent the two most direct and powerful approaches for analyzing gene function in higher organisms. These techniques allow the study of inactivation or activation of any given genes within the context of the whole organism during its life. This is often essential as processes such as cellular growth, replication, and survival (central to diseases such as cancer) are critically dependent on networks of regulatory factors that are only accurately modeled *in vivo*. Thus, although wherever possible researchers strive to conduct studies in isolated cells or tissues sometimes it is unavoidable to examine the consequences (phenotype) of a given gene in the animal. Both approaches have been successfully employed to derive strains of animals for the study of the molecular basis of cancer. Transgenesis defines any process that involves the transfer of a gene from one species to another. However, the term is often used in order to describe the insertion in the mouse genome of a modified mouse, or human genes, in order to dissect the gene function.

Gene knockouts The most widely known method for generating an abnormality in a mouse gene to establish its function is to make a "knockout," that is, using molecular biological techniques to make mice without a working copy of the gene. Despite the undoubted power of this technique it is not without potential problems. Obviously, this approach can only be used when there is preliminary evidence that a particular known gene might be relevant to a particular known disease process. Moreover, functional redundancy (i.e. where the function of a gene can be replaced by the action of another gene) may prevent any obvious changes in phenotype when the gene under study is knocked out. This is particularly likely if the gene under study is knocked out throughout the ontogeny of the organism – where the knockout may even prove lethal if no "salvage" pathways can be invoked. Obviously, this may result in erroneous conclusions as to the role of a given gene in the adult. Similar problems may also pertain to transgenic mice, and for this reason considerable efforts

Appendix 18.1 (continued)

are being devoted to refining mouse mutant technologies, such as developing regulatable systems, which enable a gene (or its protein product) to be activated or knocked out in both tissue-specific and time-dependent manners.

Transgenesis

Transgenes can be directed to the germ line by direct microinjection of the DNA encoding the gene under study in the pronucleus of a fertilized egg or by using embryonic stem (ES) cell technology.

A wide variety of vectors have been employed for transgenic studies including:

1 Regulatory sequences (e.g. promoter enhancer elements) of the gene under study are used to direct the expression of a reporter gene, for example, the gene encoding bacterial β-galactosidase for which simple and relatively sensitive histochemical detection exist. These studies can establish directly *in vivo* the stages and the tissue domains in which the gene under study is expressed.
2 Using several well-characterized promoters in order to drive the expression of the gene under study and alter the physiological pattern of gene expression. Typically these experiments lead to over and/or misexpression of the gene and they enable the analysis of the ensuing phenotype. In a significant development of the latter approach, inducible or lineage-restricted promoters can be used in order to achieve further control of the pattern of expression of the gene.

The pronuclei of fertilized mouse eggs can be most easily injected at the one-cell stage when they are at their maximum size. For most strains of mice this is between 24 and 28 hours after mating. Females used for the generation of fertilized eggs for pronuclear injection are typically induced to "superovulate," that is, to ovulate a larger number of eggs that normal [similar to techniques used for *in vitro* fertilization (IVF) in man]. Superovulation is achieved by sequential administration of the hormones follicle stimulating hormone (FSH) and then leutinizing hormone (LH) (or suitable analogs).

For microinjection, eggs are recovered from the oviduct and treated with hyaluronidase in order to free the egg from the cumulus cells and are transferred to a petri dish where they are injected.

Either of the two pronuclei can be injected. However, the male pronucleus is larger and is injected in preference. After injection, eggs are either transferred directly into the oviduct of a 0.5 dpc foster mother or cultured *in vitro* to the two-cell stage (overnight) and transferred the following morning.

The generation of targeted mutation in mammalian genomes that can be transmitted through the germ line (and the effect of which can be assessed at the level of the whole organism) has been facilitated by the development of techniques for use of ES cells. ES cells were first established in culture in 1981, and within a decade had been employed for germ-line transmission of retroviral sequences and for germ-line transmission of ES cell mutations at various loci. Soon researchers had succeeded in achieving targeted mutations by homologous recombination in ES cells at various loci including the loci for HPRT and β2-microglobulin. To date, several hundred mouse genes have been targeted in ES cells providing a unique tool for the dissection of gene function in the intact organism *in vivo*. Most of the ES cell lines currently in use are derived from the 129 mouse strain (129/Sv), which is homozygous, wild-type at the agouti locus. In order to generate chimeric mice, the 129 cells are injected in nonagouti blastocysts, for example, from the C57BL/6 strain. In this way, the agouti and nonagouti offspring can be distinguished visually at approximately 1 week.

ES cells are derived from blastocysts collected from the uterine horns and are cultured on "feeder" layers of immature fibroblasts. These feeder cells are prevented from growing by irradiation or treatment with mitomycin C (an inhibitor of DNA synthesis). After 1–2 days in culture, the cells from the inner cell mass start to divide and after another day or two the clump of cells derived from the inner cell mass is disaggregated with trypsin-EDTA and transferred to a new dish. Colonies are inspected and those with stem-cell-like morphology are recovered and propagated with regular subculturing in media that minimize the differentiation of the cells. In most instances, gene targeting in ES cells by homologous recombination relies on some knowledge of the gene under study and gene targeting is achieved by introducing a targeting vector in the form of naked DNA in the ES cells by electroporation.

The simplest targeting vectors consist of two gene fragments (of at least several hundred, typically of several thousand nucleotides) separated by an unrelated sequence that may replace the corresponding gene sequence through a double crossover event. It is essential that the unrelated sequence contains a genetic marker (e.g. the gene encoding neomycin phosphotransferase) which, once expressed, confers resistance to neomycin or neomycin analogs in order to select for vector integration (whether random or homologous).

For blastocyst injection with targeted ES cells, normal 3.5 dpc blastocysts are collected and cultured briefly to promote expansion. For microinjection, individual blastocysts are collected on a holding pipette and positioned such that the inner mass is located away from the injection needle. ES cells (10–15) are injected in the blastocel cavity prior to reimplantation.

Regulatable systems

Controlled or inducible misexpression of genes is a new and powerful tool for analyzing gene function. Various systems are now available for regulating gene expression in a temporally as well as spatially controlled manner. Genetically engineered mice are invaluable biological tools in biomedical research. The development of both transgenic and targeted mutant mice has allowed researchers to study gene function in the context of a whole mammalian organism. These studies, however, have limitations. For example, the deletion of a gene required during embryonic development often results in embryonic or perinatal lethality, making it impossible to study the effects of the gene ablation at later developmental ages. In addition, unregulated overexpression of transgenic gene products may have unwanted physiological or toxic effects. The development of inducible expression systems has allowed researchers to overcome some of the problems associated with transgenic and targeted mutagenesis studies. The two best-developed systems for this are the tetracycline-regulated systems developed by Hermann Bujard (University of Heidelberg) and the ER^{TAM} system first described *in vitro* and more recently adapted to mouse transgenesis *in vivo*.

Tet-regulatable gene expression

The Tet-Off and Tet-On expression systems are binary transgenic systems in which expression from a target transgene is dependent on the activity of an inducible transcriptional activator. In both the Tet-Off and Tet-On systems, expression of the transcriptional activator can be regulated both reversibly and quantitatively by exposing the transgenic animals to varying concentrations of tetracycline or derivatives such as doxycycline (Dox).

The Tet-Off and Tet-On systems are complementary, and the decision to choose one over the other depends on the particular experimental strategy. In the Tet-Off system, transcription is inactive in the presence of doxycycline; conversely, in the Tet-On system, transcription is active in the presence of doxycycline.

In the Tet-Off expression system, a tetracycline-controlled transactivator protein (tTA), which is composed of the Tet repressor DNA binding protein (TetR) from the Tc resistance operon of *Escherichia coli*, transposon Tn10, fused to the strong transactivating domain of VP16 from Herpes simplex virus, regulates expression of a target gene that is under transcriptional control of a tetracycline-responsive promoter element (TRE). The TRE is made up of Tet operator (tetO) sequences fused to a minimal promoter (commonly the minimal promoter sequence derived from the human cytomegalovirus (hCMV) immediate-early promoter). In the absence of doxycycline, tTA binds to the TRE and activates transcription of the target gene. However, in the presence of doxycycline, tTA cannot bind to the TRE, and expression from the target gene remains inactive.

The Tet-On system is based on a reverse tetracycline-controlled transactivator, rtTA. Like tTA, rtTA is a fusion protein comprised of the TetR repressor and the VP16 transactivation domain; however, a four amino acid change in the tetR DNA binding moiety alters rtTA's binding characteristics, such that it can only recognize the tetO sequences in the TRE of the target transgene in the presence of doxycycline. Thus, in the Tet-On system, transcription of the TRE-regulated target gene is stimulated by rtTA only in the presence of doxycycline.

Appendix 18.1 (continued)

The two-vector nature of the Tet-Off and Tet-On systems allows tissue-specific promoters to drive *tTA* or *rtTA* expression, resulting in tissue-specific expression of the TRE-regulated target transgene. Further, the ability to strictly regulate the level of *rtTA* and *tTA* activity allows the investigator to regulate activation of the target gene both quantitatively and temporally. Successful application of both the repressible and activatible systems has been achieved in mice and have provided important information on tumor formation and regression in particular. Problems have been reported including mosaic induction, background leakiness (particularly with the *rtTA* system), or no detectable expression of the transactivator. Many of the issues may be revolved by use of optimum promoters to drive expression of the *tTA* and characterization of suitable responder lines. An additional problem is the lengthy lag phase in the *in vivo* response of the system following drug administration or removal.

Transgenes encoding regulatable chimeric proteins

The general concept of utilizing the well-studied ligand-binding properties of the steroid hormone receptors in the design of inducible systems has proved highly successful. This approach offers several advantages over the "Tet" system, notably a much speedier regulation of 'activity' of the transgenic protein.

Several groups have generated transgenic mice in which a transgene encodes a fusion (chimeric) protein between the ligand-binding domain of a modified estrogen receptor (now only responsive to synthetic estrogenic compounds such as tamoxifen and 4-hydroxy tamoxifen, but not to endogenous estrogens) and various proteins, which the researcher may wish to regulate. To date, the most prominent successes have been in regulating activity of c-Myc in various cancer models and in regulating the activity of a Cre recombinase. In the latter case, this potentially allows targeted gene knockouts in a time-dependent as well as tissue-dependent manner. In general, these fusion proteins are inactive except in the presence of the ligand, most likely because the ligand-binding domain is complexed with Hsp90 proteins in the cytoplasm (or nucleus), preventing c-Myc or Cre recombinase from reaching their targets.

Regulatable mouse models of tumorigenesis using a "switchable" form of the c-Myc protein – MYCER^TAM

The c-MYCER^TAM construct encodes a chimeric protein comprised of human c-MYC fused at its C terminus to the hormone-binding domain of a mutant murine estrogen receptor. Directed expression of the chimeric protein to the suprabasal epidermis or pancreatic islet β-cells was achieved by placing c-MYCER^TAM cDNA under the control of an involucrin promoter or an insulin promoter, respectively. Although c-MYCER^TAM protein is expressed constitutively in these tissue compartments, in the absence of the specific ligand, 4-hydroxy tamoxifen (4-OHT), c-MYCER^TAM remains inactive through its association with heat shock protein 90 (HSP90). 4-OHT administration leads to HSP90 dissociation and activation of c-MYCER^TAM as the protein is now able to bind its partner Max and become transcriptionally active.

Hit and run

Many mouse models of cancer have been accused of not accurately recapitulating the kinds of spontaneous mutations that characterize most human cancers – neither embryonic insertion nor tissue-specific transgenic mice can usually address this. What is required is a system whereby individual cells will acquire an oncogenic mutation, but are otherwise surrounded by normal cells.

One strategy for achieving this was used by Tyler Jacks and colleagues, who designed what they have termed a genetic "Trojan horse." Simply, they introduced latent *K-ras* genes into mice. The genes were inactivated because they had duplicated segments of DNA that prevented the genes from activating themselves. However, as the mice developed, individual cells undergo rare spontaneous, sporadic recombination events that deleted one copy of the duplicated sequence, such that the *K-ras* gene was activated and began to initiate tumor development. This technique is a variant of a "hit-and-run" gene-targeting technology developed by HHMI investigator Allan Bradley at

Baylor College of Medicine. The two-part technique consists of "hitting" cells with an inserted mutated gene and then allowing the recombination event to "run", activating the inserted gene.

INTERNET RESOURCES

General background

http://www.wellcome.ac.uk/en/genome/index.html

This is an outstanding website with up-to-date news, interviews with scientists, and background information regarding the genome and proteome. The main sections are the genome, genes and body, genes and society, and tackling disease.

http://www.dnai.org/

This is a beautiful website providing all the background information on DNA, both historical and biological. All key scientists are discussed and placed in context in an interactive way. The DNA-interactive website is useful for both student and lecturer.

http://www.ygyh.org/

The "your genes your health" website is a component of the DNA-interactive website. A number of genetic disorders, including cystic fibrosis, are discussed and illustrated.

http://www.ncbi.nlm.nih.gov/About/primer/

This up-to-date website gives an extensive primer on science and includes bioinformatics, genome mapping, microarrays, SNPs, genetics, and much more. This website gives further background to this chapter.

Research resources

http://www.geneontology.org/

This database allows for searching protein function of most known proteins. The gene ontology database archives protein information according to function. The database can be searched on protein name, for example, *RAB25* gene that is involved in ovarian and breast cancer. Useful for all genes discussed in this chapter and book.

http://www.expasy.ch/

This website provides similar information on proteins as the gene ontology database, but focuses on the proteome, for example, searching for RAB25 gives three hits in different species. RAB25 human variant gives detailed information on this protein.

http://www.genome.jp/kegg/pathway.html

This is a wonderful site that contains most known pathways ranging from metabolism to cellular processes. By selecting one pathway, for example, glycolysis, the complete pathway is shown including all the proteins that are a part of that pathway.

http://kinasedb.ontology.ims.u-tokyo.ac.jp:8081/

This is a database that contains all information on human, mouse, and rat kinases. Interactions and pathway can be found for specific kinases.

http://www.abrf.org./index.cfm/dm.home

This website lists all known post-translational modifications including their weight. By typing in "All" in the box all post-translational modifications are listed.

Research institutes and organizations

http://www.hapmap.org/

This is the official website of the HapMap Consortium. News and background information can be found on this website.

http://www.epigenome.org/

This website provides some background on the epigenome project between the Sanger Institute, Epigenomics AG, and CNG.

http://cgap.nci.nih.gov/

This website provides information on cancer research and diagnosis. The goal of this community is to determine gene expression profiles of normal, precancer, and cancer cells, leading eventually to improved detection, diagnosis, and treatment for the patient. The website is a repository for scientists interested in tools for cancer research.

http://www.sanger.ac.uk/

This is the official website of the Sanger Institute, one of the major contributors to genome research. A third of the human genome was sequenced at the Sanger Institute and a lot of background information on previous and current research can be found on this website.

BIBLIOGRAPHY

Information flow in the cell

Balmain, A., Gray, J. and Ponder, B. (2003). The genetics and genomics of cancer. *Nature Genetics*, **33**, 238–244.

Bell, J. (2004). Predicting disease using genomics. *Nature*, **429**, 453–456.

Brown, D. and Superti-Furga, G. (2003). Rediscovering the sweet spot in drug discovery. *Drug Discovery Today*, **8**, 1067–1077.

Evans, W.E. and Relling, M.V. (2004). Moving towards individualized medicine with pharmacogenomics. *Nature*, **429**, 464–468.

Hood, L. and Galas, D. (2003). The digital code of DNA. *Nature*, **421**, 444–448.

Jones, P.A. and Baylin, S.B. (2002). The fundamental role of epigenetic events in cancer. *Nature Reviews. Genetics*, **3**, 415–428.

Narod, S.A. and Foulkes, W.D. (2004). BRCA1 and BRCA2: 1994 and beyond. *Nature Reviews. Cancer*, **4**, 665–676.

Ponder, B.A. (2001). Cancer genetics. *Nature*, **411**, 336–341.

Model organisms and cancer models

Jain, M., Arvanitis, C., Chu, K., Dewey, W., Leonhardt, E., Trinh, M., Sundberg, C.D., Bishop, J.M. and Felsher, D.W. (2002). Sustained loss of a neoplastic phenotype by brief inactivation of MYC. *Science*, **297**, 102–104.

Chin, L., Tam, A., Pomerantz, J., Wong, M., Holash, J., Bardeesy, N., Shen, Q., O'Hagan, R., Pantginis, J., Zhou, H., Horner, J.W., 2nd, Cordon-Cardo, C., Yancopoulos, G.D. and DePinho, R.A. (1999). Essential role for oncogenic Ras in tumor maintenance. *Nature*, **400**, 468–472.

Moody, S.E., Sarkisian, C.J., Hahn, K.T., Gunther, E.J., Pickup, S., Dugan, K.D., Innocent, N., Cardiff, R.D., Schnall, M.D. and Chodosh, L.A. (2002). Conditional activation of Neu in the mammary epithelium of transgenic mice results in reversible pulmonary metastasis. *Cancer Cell*, **2**, 451–461.

Pelengaris, S., Littlewood, T., Khan, M., Elia, G. and Evan, G. (1999). Reversible activation of c-Myc in skin: induction of a complex neoplastic phenotype by a single oncogenic lesion. *Molecular Cell*, **3**, 565–577.

Pelengaris, S., Khan, M. and Evan, G.I. (2002). Suppression of Myc-induced apoptosis in beta cells exposes multiple oncogenic properties of Myc and triggers carcinogenic progression. *Cell*, **109**, 321–334.

Pelengaris, S., Abouna, S., Cheung, L., Ifandi, V., Zervou, S. and Khan, M. (2004). Brief inactivation of c-Myc is not sufficient for sustained regression of c-Myc-induced tumors of pancreatic islets and skin epidermis. *BMC Biology*, **2**, 26.

Shachaf, C.M., Kopelman, A.M., Arvanitis, C., Karlsson, A., Beer, S., Mandl, S., Bachmann, M.H., Borowsky, A.D., Ruebner, B., Cardiff, R.D., Yang, Q., Bishop, J.M., Contag, C.H. and Felsher, D.W. (2004). MYC inactivation uncovers pluripotent differentiation and tumor dormancy in hepatocellular cancer. *Nature*, **431**, 1112–1117.

SNPs, HapMap, and identifying cancer genes

Carlson, C.S., Eberle, M.A., Kruglyak, L. and Nickerson, D.A. (2004). Mapping complex disease loci in whole-genome association studies. *Nature*, **429**, 446–452.

Daly, M.J., Rioux, J.D., Schaffner, S.F., Hudson, T.J. and Lander, E.S. (2001). High-resolution haplotype structure in the human genome. *Nature Genetics*, **29**, 229–232.

Futreal, P.A., Coin, L., Marshall, M., Down, T., Hubbard, T., Wooster, R., Rahman, N. and Stratton, M.R. (2004). A census of human cancer genes. *Nature reviews. Cancer*, **4**, 177–183.

Glazier, A.M., Nadeau, J.H. and Aitman, T.J. (2002). Finding genes that underlie complex traits. *Science*, **298**, 2345–2349.

Goldstein, D.B. and Weale, M.E. (2001). Population genomics: linkage disequilibrium holds the key. *Current Biology*, **11**, R576–579.

HapMap-Consortium (2003). The International HapMap Project. *Nature*, **426**, 789–796.

Zondervan, K.T. and Cardon, L.R. (2004). The complex interplay among factors that influence allelic association. *Nature Review. Genetics*, **5**, 89–100.

Cancer mRNA expression analysis

Alizadeh, A.A., Eisen, M.B., Davis, R.E., Ma, C., Lossos, I.S., Rosenwald, A., Boldrick, J.C., Sabet, H., Tran, T., Yu, X., Powell, J.I., Yang, L., Marti, G.E., Moore, T., Hudson, J., Jr., Lu, L., Lewis, D.B., Tibshirani, R., Sherlock, G., Chan, W.C., Greiner, T.C., Weisenburger, D.D., Armitage, J.O., Warnke, R., Levy, R., Wilson, W., Grever, M.R., Byrd, J.C., Botstein, D., Brown, P.O. and Staudt, L.M. (2000). Distinct types of diffuse large B-cell lymphoma identified by gene expression profiling. *Nature*, **403**, 503–511.

Creighton, C., Kuick, R., Misek, D.E., Rickman, D.S., Brichory, F.M., Rouillard, J.M., Omenn, G.S. and Hanash, S. (2003). Profiling of pathway-specific changes in gene expression following growth of human cancer cell lines transplanted into mice. *Genome Biology*, **4**, R46.

Lossos, I.S., Czerwinski, D.K., Alizadeh, A.A., Wechser, M.A., Tibshirani, R., Botstein, D. and Levy, R. (2004). Prediction of survival in diffuse large-B-cell lymphoma based on the expression of six genes. *New England Journal of Medicine*, **350**, 1828–1837.

Ma, X.J., Salunga, R., Tuggle, J.T., Gaudet, J., Enright, E., McQuary, P., Payette, T., Pistone, M., Stecker, K., Zhang, B.M., Zhou, Y.X., Varnholt, H., Smith, B., Gadd, M., Chatfield, E., Kessler, J., Baer, T.M., Erlander, M.G. and Sgroi, D.C. (2003). Gene expression profiles of human breast cancer progression. *Proceedings of the National Academy of Sciences of the United States of America*, **100**, 5974–5979.

Perou, C.M., Sorlie, T., Eisen, M.B., van de Rijn, M., Jeffrey, S.S., Rees, C.A., Pollack, J.R., Ross, D.T., Johnsen, H., Akslen, L.A., Fluge, O., Pergamenschikov, A., Williams, C., Zhu, S.X., Lonning, P.E., Borresen-Dale, A.L., Brown, P.O. and Botstein, D. (2000). Molecular portraits of human breast tumors. *Nature*, **406**, 747–752.

Pittman, J., Huang, E., Dressman, H., Horng, C.F., Cheng, S.H., Tsou, M.H., Chen, C.M., Bild, A., Iversen, E.S., Huang, A.T., Nevins, J.R. and West, M. (2004). Integrated modeling of clinical and gene expression information for personalized prediction of disease outcomes. *Proceedings of the National Academy of Sciences of the United States of America*, **101**, 8431–8436.

Rhodes, D.R., Yu, J., Shanker, K., Deshpande, N., Varambally, R., Ghosh, D., Barrette, T., Pandey, A. and Chinnaiyan, A.M. (2004). Large-scale meta-analysis of cancer microarray data identifies common transcriptional profiles of neoplastic transformation and progression. *Proceedings of the National Academy of Sciences of the United States of America*, **101**, 9309–9314.

Ross, D.T., Scherf, U., Eisen, M.B., Perou, C.M., Rees, C., Spellman, P., Iyer, V., Jeffrey, S.S., Van de Rijn, M., Waltham, M., Pergamenschikov, A., Lee, J.C., Lashkari, D., Shalon, D., Myers, T.G., Weinstein, J.N., Botstein, D. and Brown, P.O. (2000). Systematic variation in gene expression patterns in human cancer cell lines. *Nature Genetics*, **24**, 227–235.

Shipp, M.A., Ross, K.N., Tamayo, P., Weng, A.P., Kutok, J.L., Aguiar, R.C., Gaasenbeek, M., Angelo, M., Reich, M., Pinkus, G.S., Ray, T.S., Koval, M.A., Last, K.W., Norton, A., Lister, T.A., Mesirov, J., Neuberg, D.S., Lander, E.S., Aster, J.C. and Golub, T.R. (2002). Diffuse large B-cell lymphoma outcome prediction by gene-expression profiling and supervised machine learning. *Nature Medicine*, **8**, 68–74.

Sorlie, T., Tibshirani, R., Parker, J., Hastie, T., Marron, J.S., Nobel, A., Deng, S., Johnsen, H., Pesich, R., Geisler, S., Demeter, J., Perou, C.M., Lonning, P.E., Brown, P.O., Borresen-Dale, A.L. and Botstein, D. (2003). Repeated observation of breast tumor subtypes in independent gene expression data sets. *Proceedings of the National Academy of Sciences of the United States of America*, **100**, 8418–8423.

van de Vijver, M.J., He, Y.D., van't Veer, L.J., Dai, H., Hart, A.A., Voskuil, D.W., Schreiber, G.J., Peterse, J.L., Roberts, C., Marton, M.J., Parrish, M., Atsma, D., Witteveen, A., Glas, A., Delahaye, L., van der Velde, T., Bartelink, H., Rodenhuis, S., Rutgers, E.T., Friend, S.H. and Bernards, R. (2002). A gene-expression signature as a predictor of survival in breast cancer. *New England Journal of Medicine*, **347**, 1999–2009.

CGH arrays, CpG island microarrays, and ChIP-on-Chip

Albertson, D.G. (2003). Profiling breast cancer by array CGH. *Breast Cancer Research and Treatment*, **78**, 289–298.

Ballestar, E., Paz, M.F., Valle, L., Wei, S., Fraga, M.F., Espada, J., Cigudosa, J.C., Huang, T.H. and Estcller, M. (2003). Methyl-CpG binding proteins identify novel sites of epigenetic inactivation in human cancer. *The EMBO Journal*, **22**, 6335–6345.

Balog, R.P., de Souza, Y.E., Tang, H.M., DeMasellis, G.M., Gao, B., Avila, A., Gaban, D.J., Mittelman, D., Minna, J.D., Luebke, K.J. and Garner, H.R. (2002). Parallel assessment of CpG methylation by two-color hybridization with oligonucleotide arrays. *Analytical Biochemistry*, **309**, 301–310.

Bird, A. (2002). DNA methylation patterns and epigenetic memory. *Genes Development*, **16**, 6–21.

Chen, C.M., Chen, H.L., Hsiau, T.H., Hsiau, A.H., Shi, H., Brock, G.J., Wei, S.H., Caldwell, C.W., Yan, P.S. and Huang, T.H. (2003). Methylation target array for rapid analysis of CpG island hypermethylation in multiple tissue genomes. *American Journal Pathology*, **163**, 37–45.

Egger, G., Liang, G., Aparicio, A. and Jones, P.A. (2004). Epigenetics in human disease and prospects for epigenetic therapy. *Nature*, **429**, 457–463.

Gitan, R.S., Shi, H., Chen, C.M., Yan, P.S. and Huang, T.H. (2002). Methylation-specific oligonucleotide microarray: a new potential for high-throughput methylation analysis. *Genome Research*, **12**, 158–164.

Mantripragada, K.K., Buckley, P.G., de Stahl, T.D. and Dumanski, J.P. (2004). Genomic microarrays in the spotlight. *Trends in Genetics*, **20**, 87–94.

Oostlander, A.E., Meijer, G.A. and Ylstra, B. (2004). Microarray-based comparative genomic hybridization and its applications in human genetics. *Clinical Genetics*, **66**, 488–495.

Weinmann, A.S., Yan, P.S., Oberley, M.J., Huang, T.H. and Farnham, P.J. (2002). Isolating human transcription factor targets by coupling chromatin immunoprecipitation and CpG island microarray analysis. *Genes Development*, **16**, 235–244.

Yan, P.S., Perry, M.R., Laux, D.E., Asare, A.L., Caldwell, C.W. and Huang, T.H. (2000). CpG island arrays: an application toward deciphering epigenetic signatures of breast cancer. *Clinical Cancer Research*, **6**, 1432–1438.

Yan, P.S., Chen, C.M., Shi, H., Rahmatpanah, F., Wei, S.H. and Huang, T.H. (2002). Applications of CpG island microarrays for high-throughput analysis of DNA methylation. *Journal of Nutrition*, **132**, 2430S–2434S.

Proteomics and quantative proteomics

Jessani, N., Humphrey, M., McDonald, W.H., Niessen, S., Masuda, K., Gangadharan, B., Yates, J.R., 3rd, Mueller, B.M. and Cravatt, B.F. (2004). Carcinoma and stromal enzyme activity profiles associated with breast tumor growth *in vivo*. *Proceedings of the National Academy of Sciences of the United States of America*, **101**, 13756–13761.

Krijgsveld, J., Ketting, R.F., Mahmoudi, T., Johansen, J., Artal-Sanz, M., Verrijzer, C.P., Plasterk, R.H. and Heck, A.J. (2003). Metabolic labeling of C. elegans and D. melanogaster for quantitative proteomics. *Nature Biotechnology*, **21**, 927–931.

Lewis, T.S., Hunt, J.B., Aveline, L.D., Jonscher, K.R., Louie, D.F., Yeh, J.M., Nahreini, T.S., Resing, K.A. and Ahn, N.G. (2000). Identification of novel MAP kinase pathway signaling targets by functional proteomics and mass spectrometry. *Molecular Cell*, **6**, 1343–1354.

Li, J., Steen, H. and Gygi, S.P. (2003). Protein profiling with cleavable isotope-coded affinity tag (cICAT) reagents: the yeast salinity stress response. *Molecular Cell Proteomics*, **2**, 1198–1204.

Ong, S.E., Foster, L.J. and Mann, M. (2003a). Mass spectrometric-based approaches in quantitative proteomics. *Methods*, **29**, 124–130.

Ong, S.E., Kratchmarova, I. and Mann, M. (2003b). Properties of 13C-substituted arginine in stable isotope labeling by amino acids in cell culture (SILAC). *Journal of Proteome Research*, **2**, 173–181.

Shiio, Y., Donohoe, S., Yi, E.C., Goodlett, D.R., Aebersold, R. and Eisenman, R.N. (2002). Quantitative proteomic analysis of Myc oncoprotein function. *The EMBO Journal*, **21**, 5088–5096.

Steen, H. and Mann, M. (2004). The ABC's (and XYZ's) of peptide sequencing. *Nature Reviews. Molecular Cell Biology*, **5**, 699–711.

van Anken, E., Romijn, E.P., Maggioni, C., Mezghrani, A., Sitia, R., Braakman, I. and Heck, A.J. (2003). Sequential waves of functionally related proteins are expressed when B cells prepare for antibody secretion. *Immunity*, **18**, 243–253.

Post-translational modifications

Blagoev, B., Ong, S.E., Kratchmarova, I. and Mann, M. (2004). Temporal analysis of phosphotyrosine-dependent signaling networks by quantitative proteomics. *Nature Biotechnology*, **22**, 1139–1145.

Larsen, M.R. and Roepstorff, P. (2000). Mass spectrometric identification of proteins and characterization of their post-translational modifications in proteome analysis. *Fresenius Journal of Analytical Chemistry*, **366**, 677–690.

Mann, M. and Jensen, O.N. (2003). Proteomic analysis of post-translational modifications. *Nature Biotechnology*, **21**, 255–261.

Protein complexes and cellular networks

Aloy, P., Bottcher, B., Ceulemans, H., Leutwein, C., Mellwig, C., Fischer, S., Gavin, A.C., Bork, P., Superti-Furga, G., Serrano, L. and Russell, R.B. (2004). Structure-based assembly of protein complexes in yeast. *Science*, **303**, 2026–2029.

Blagoev, B., Kratchmarova, I., Ong, S.E., Nielsen, M., Foster, L.J. and Mann, M. (2003). A proteomics strategy to elucidate functional protein-protein interactions applied to EGF signaling. *Nature Biotechnology*, **21**, 315–318.

Bouwmeester, T., Bauch, A., Ruffner, H., Angrand, P.O., Bergamini, G., Croughton, K., Cruciat, C., Eberhard, D., Gagneur, J., Ghidelli, S., Hopf, C., Huhse, B., Mangano, R., Michon, A.M., Schirle, M., Schlegl, J., Schwab, M., Stein, M.A., Bauer, A., Casari, G., Drewes, G., Gavin, A.C., Jackson, D.B., Joberty, G., Neubauer, G., Rick, J., Kuster, B. and Superti-Furga, G. (2004). A physical and functional map of the human TNF-alpha/NF-kappa B signal transduction pathway. *Nature Cell Biology*, **6**, 97–105.

Gavin, A.C., Bosche, M., Krause, R., Grandi, P., Marzioch, M., Bauer, A., Schultz, J., Rick, J.M., Michon, A.M., Cruciat, C.M., Remor, M., Hofert, C., Schelder, M., Brajenovic, M., Ruffner, H., Merino, A., Klein, K., Hudak, M., Dickson, D., Rudi, T., Gnau, V., Bauch, A., Bastuck, S., Huhse, B., Leutwein, C., Heurtier, M.A., Copley, R.R., Edelmann, A., Querfurth, E., Rybin, V., Drewes, G., Raida, M., Bouwmeester, T., Bork, P.,Seraphin, B., Kuster, B., Neubauer, G. and Superti-Furga, G. (2002). Functional organization of the yeast proteome by systematic analysis of protein complexes. *Nature*, **415**, 141–147.

Pawson, T. and Nash, P. (2003). Assembly of cell regulatory systems through protein interaction domains. *Science*, **300**, 445–452.

Puig, O., Caspary, F., Rigaut, G., Rutz, B., Bouveret, E., Bragado-Nilsson, E., Wilm, M. and Seraphin, B. (2001). The tandem affinity purification (TAP) method: a general procedure of protein complex purification. *Methods*, **24**, 218–229.

Clinical applications of proteomics

Petricoin, E.F., Ardekani, A.M., Hitt, B.A., Levine, P.J., Fusaro, V.A., Steinberg, S.M., Mills, G.B., Simone, C., Fishman, D.A., Kohn, E.C. and Liotta, L.A. (2002). Use of proteomic patterns in serum to identify ovarian cancer. *Lancet*, **359**, 572–577.

Villanueva, J., Philip, J., Entenberg, D., Chaparro, C.A., Tanwar, M.K., Holland, E.C. and Tempst, P. (2004). Serum peptide profiling by magnetic particle-assisted, automated sample processing and MALDI-TOF mass spectrometry. *Analytical Chemistry*, **76**, 1560–1570.

Glossary

5-FU: 5-fluorouracil.

5HIAA: 5-hydroxyindoleacetic acid.

5HT: 5-hydroxytryptamine.

ADH: Antidiuretic hormone.

AFP: α-fetoprotein.

AKT: Protein kinase B, mediates survival signals downstream of RAS-PI3K.

Allele: One of a set of alternative forms of a gene in a diploid cell one will be on each of the 2 homologous chromosomes.

Anaphase: Stage in mitosis when duplicated pairs of chromosomes separate and move apart.

Angioblast: A primordial mesenchymal cell type from which embryonic blood cells and vascular endothelium differentiate.

Angiogenesis: Formation of new blood vessels from preexisting vessels by sprouting from existing blood vessels.

Antigen-induced cell death: Excess antigen density can lead to effector cell death following TCR engagement in the absence of an adequate costimulatory (i.e. CD28) signal.

APC: Adenomatous polyposis coli – Tumor suppressor that inactivates beta-catenin.

APC/C: Anaphase-promoting complex/cyclosome. Ubiquitin ligase complex needed for degradation of cyclin B and securin for anaphase entry and exit from mitosis.

Apoptosis: Programmed cell death or cell suicide.

ARF: Tumor suppressor encoded by a transcript that is read from a same exon than the one encoding INK4a but in an Alternative Reading Frame.

Athymic nude mice: A mutant strain of mice with a genetic defect that affects the development of the thymus and hair follicles. These mice lack a thymus and therefore generally fail to develop T cells. These cells, however, retain a population of NK cells.

ATM: Ataxia telangiectasia mutated.

ATR: ATM- and Rad3-related.

Autocrine: Extracellular signals (for example those regulating replication or growth) released by the same cell on which those signals then elicit a response. For example the production of IGF by tumor cells then acting on those same tumor cells to support survival or growth.

BCG: Bacillus Calmette-Guerin is a proinflammatory mixture of killed mycobacteria (the organism responsible for tuberculosis) that acts as an adjuvant. When given to patients by injection, it promotes T-cell and macrophage activation.

BCNU: *Bis*-colorethyl nitrosourea.

CCNU: *Cis*-chloroethyl nitrosourea.

Cdk – cyclin-dependent kinases: Kinases that phosphorylate protein substrates on serine and threonine residues that form complexes and are activated by binding to cyclins.

CEA: Carcinoembryonic antigen.

Cell cycle: The mitotic cell cycle is composed of four phases: two gap phases: G_1 when the cells respond to the environment and make the decision to proceed through mitosis and G_2, a period between the DNA synthesis phase S and mitosis M, where cells check the integrity of their DNA before proceeding through division (M).

Checkpoint: A point in the cell cycle during which cells pause and wait until the conditions are favorable for the cells to proceed through the next phase of the cell cycle.

Chemokine: Chemokines are chemotactic factors that influence the migration of immune cells through interaction with specific receptors. They recruit immune cells to sites of inflammation and regulate traffic through lymphoid tissues. CCL-19 that binds to the chemokine receptor CCR7 is responsible for lymphocyte and dendritic cell migration to lymph nodes. The chemokine CXCL8 or IL-8 recruits neutrophils to sites of inflammation through interaction with the CXCR1 chemokine receptor.

CIN: Cervical intraepithelial neoplasia.

Circulating endothelial cell progenitor cells: The bone marrow of adults contains progenitor cells that have the capacity to differentiate into mature endothelial cells.

CKI-(CDKI also used) cyclin kinase inhibitor: Small polypeptide proteins that bind to and inhibit cyclin-cdk holoenzymes. They can inhibit cell cycle entry and progression in response to various checkpoints. They include two main families, the INK4 and CIP/KIP.

INK4 family of CKI: All of these proteins are inhibitors of CDK4/6. INK4a or p16^{INK4a} (p16^{Ink4a} nonhuman): p15^{INK4b} (p15^{Ink4b} nonhuman): are CKI protein encoded by the CDKN2B and Ink4b genes in man and nonhuman respectively. p18^{INK4c} (p18^{Ink4c} nonhuman): are CKI proteins encoded by the CDKN2C (human) and Ink4c (nonhuman) genes respectively. p19^{INK4d}: not to be confused with p19ARF ARF-(p14^{ARF} in man and p19Arf rodent): tumor suppressor proteins encoded by a transcript that is read from the same gene exon as the one encoding INK4a (CDKN2A) but in an Alternative Reading Frame. They are involved in activating another very important tumor suppressor, p53. Thus, ARF can sequester MDM2 (a protein which otherwise binds and targets p53 for degradation) thus stabilising the p53 protein in response to "oncogenic stress".

CIP/KIP family of CKI: Negative regulators of cyclin E- and cyclin A-CDK2 and of cyclin B-CDK1. p21^{CIP1} also referred to as p21^{CIP1}/WAF1 (p21^{Cip1} nonhuman): are CKI proteins encoded by the CIP1(WAF1) and Cip1 genes respectively. p21^{CIP1} is one of the major transcriptional targets of p53. p27^{KIP1} (p27^{Kip1} nonhuman): are CKI proteins encoded by the KIP1 (human) and Kip1 (nonhuman) genes respectively. p27^{KIP1} is a key inhibitor of the G1/S transition and acts downstream of various inhibitory growth factors such as TGF-β. p57^{KIP2} (p57^{Kip2} nonhuman): are CKI proteins encoded by the KIP2 (human) or Kip2 (nonhuman) genes respectively.

CT: Computerized tomography.

Cyclin dependent kinases (CDK): Together with their cyclin partners represent the cell cycle engine.

Cyclins: Proteins that exhibit expression cycles throughout the mitotic cell cycle, hence their name. Together with cyclin dependent kinases with which they form active holoenzyme complexes, they drive many of the key processes of the cell cycle. Cyclin D1 in man is encoded by the CCND1 gene, cyclin D2 by CCND2.

Cytokine: Cytokines are soluble factors that generally influence the growth and function of immune cells through interaction with specific cytokine receptors on the cell surface. Cytokines, such as interleukin -2 (IL-2), IL-7, or IL-15 can induce cellular proliferation. Other cytokines modulate T-cell development, for example, IL-12 and Interferon -γ promote the differentiation of T cells into becoming T-helper -1 (Th1) – cells that are characterized by the ability to release IL-2 upon activation. Those cytokines that act specifically on lymphocytes are called lymphokines.

Differentiation: Process by which cells become irreversibly committed to a specialized cell type.

DISC: Death-inducing signaling complex (DISC). DISC recruits and activates the initiator procaspase-8 in the extrinsic pathway of apoptosis.

DNA – deoxyribonucleic acid: Formed by covalently linked deoxyribonucleotides that forms the genetic code; carries the genetic information.

Donor lymphocyte infusions: To treat for leukemic relapse in the recipient of an allogeneic stem cell transplant, lymphocytes are collected from the original stem cell donor (by a process called leukapheresis) and infused into the recipient to induce a graft versus leukemia response. In a graft versus leukemia response, the donor lymphocytes kill host leukemic cells following recognition of minor antigens on the host leukemic cells.

DSB: Double strand break in DNA.

DTIC: Dimethyltriazenoimidazole carboxamide (Dacarbazine).

EBV Post-transplant lymphoproliferative disease: Post-transplant lymphoproliferative disease (PTLD) often arises following stem or organ transplantation in heavily immunosuppressed patients as a B-cell neoplastic disease that may be causally related to EBV (Epstein Barr Virus) reactivation in the absence of a functional T-cell response.

EBV–LMP: The latent membrane proteins of the Epstein Barr virus are associated with Hodgkin's disease and nasopharyngeal carcinoma. Some of these proteins (i.e. LMP-1) are believed to play a role in carcinogenesis due to its transforming properties in cell lines.

EGF: Epidermal growth factor.

EGFR: Epidermal growth factor receptor.

Endocrine: Extracellular signals (for example those regulating replication or growth) which operate at a distance usually via the circulation. Such factors are termed hormones and include insulin produced by the pancreatic beta cell acting on distant muscle or liver cells.

Endothelial cell: Epithelial-type cells that line the cavities of blood vessels, lymph vessels and the heart.

ER: Endoplasmic reticulum.

ERCP: Endoscopic retrograde cholangiopancreatography.

ER^{TAM}: Modified estrogen receptor.

ESR: Erythrocyte sedimentation rate.

FBC: Full blood count.

FdUMP: 5-Fluoro-2-deoxyuridine monophosphate.

FGF: Fibroblast growth factor.

FISH: Fluorescent *in situ* hybridization.

Flk1: Fetal liver kinase 1, a receptor for vascular endothelial growth factor; also known as KDR or VEGFR2.

Flt1: Fms-like tyrosine kinase 1, a receptor for vascular endothelial growth factor; also known as VEGFR1.

FNA: Fine needle aspiration.

FOB: Fecal occult blood.

FPC: Familial polyposis coli.

FSH: Follicle-stimulating hormone.

5-FU: 5-fluorouracil.

G1: Period in the cell cycle after mitosis during which cells sense their environment and decide to enter a second cell division cycle, exit cycle or die.

G1/S transition: Once a cell passes this point in the mitotic cell cycle the cell will usually go on to replicate even in the absence of growth-promoting stimuli unless a problem is detected such as DNA damage or improper chromosome segregation in which case checkpoints can arrest the cycle or kill the cell. This key transition is mediated by hyperphosphorylation of the RB protein and is promoted by growth factors via Cyclin D and Cyclin E- CDK complexes and inhibited by various CKIs including p15^{Ink4}, p21^{Cip1} and p57^{Kip2}, some of which are triggered by inhibitory growth factors such as TGFβ.

G-CSF: Granulocyte colony stimulating factors.

GDNF-Glial cell line-derived neurotrophic factor (GDNF): a ligand for the RET tyrosine kinase.

GH: Growth hormone.

Growth factor (GF): Is often used as a catch-all term for proteins encompassing all of the following functional categories – *Mitogens*: promote cell division largely by allowing G1/S transition in the cell cycle. *Growth factors*: promote an increase in cell mass (cell growth), by enhancing protein synthesis. *Survival factors*: that promote cell survival by suppressing apoptosis.

GSK3β – Glycogen synthase kinase 3β. Originally identified as a modulator of glycogen metabolism, GSK3β is now known to play a key regulatory role in a variety of cellular processes including initiation of protein synthesis, growth and proliferation, differentiation, apoptosis, and is essential for embryonic development as a component of the Wnt signaling cascade. GSK3ß interacts with ubiquitin E3 ligases to target various proteins for degradation through the ubiquitin-proteasome pathway. These include key growth and cell cycle regulators such as c-MYC, c-JUN, cyclin D1, Converesly, GSK3β activity is inhibited by signaling through the RAS-PI3K-AKT pathway.

HCG: Human chorionic gonadotrophin.

5HIAA: 5-hydroxyindoleacetic acid.

Homeostatic regulation: Immune responses are tuned to return to a basal resting state in a process called homeostasis. Following antigen-specific proliferation, most effector cells undergo apoptosis or are inhibited by regulatory mechanisms as the antigen is eliminated and the immune response wanes. Conversely, in the lymphopenic state, induced by lymphodepletion or in the case of RAG knockout mice, homeostatic mechanisms involving the upregulation of lymphokines, such as IL-15 and IL-7, are induced to expand the pool of lymphocytes to a basal level.

HPV associated malignancies: HPV or human papilloma virus is believed to be the strongly associated with cervical cancer, the most common being the HPV subtype 16. HPV is also less commonly associated with head and neck cancer. It is believed that the HPV oncoproteins E6 and E7 may play a role in tumorigenesis.

HtrA2/Omi: Reaper-related serine protease released during MOMP, encoded by the Prss25 gene.

5HT: 5-hydroxytryptamine.

IAP: Inhibitor of apoptosis protein.

IFN: Interferon.

IGF: Insulin like growth factor. Two types are known, IGF-1 and IGF-2.

Immune privilege: A term which has sometimes been applied to tumor immune evasion is "immune privilege." Immune privileged sites were initally used to describe normal tissues critical to the survival of the organism (e.g. reproductive organs) or tissues where an inflammatory response would be detrimental (e.g. central nervous system) and where the immune response is normally suppressed. In many instances, it has been found that mechanisms operative at such sites may also exist in tumors.

Immunologic synapse: The region of interaction between the TCR and the peptide–MHC complex. In the mature synapse, which forms when cognate TCR–peptide–MHC interaction occurs, the immunologic synapse is reinforced by concentric outer ring of adhesion molecules and polarization of the microtubule organizing complex to the region of the synapse. This leads to downstream signals responsible for full T-cell activation and cell division.

INK4: Inhibitory protein of cyclin-dependent kinase 4. Small polypeptide proteins that bind to and inhibit cyclin/cdk holoenzymes (see cyclin and cdk).

INK4a/ARF (CDKN2a): Gene locus encodes two proteins that regulate two of the most important tumor suppressor pathways represented by p53 and RB.

Laser capture microdissection (LCM): Tissue is viewed under a microscope and those areas containing the chosen cells, based on histologic appearance or specialized staining are targeted by a near infrared laser. A transparent transfer film immediately overlying the chosen cells is transiently heated by the laser, fuses with the cell(s) underneath, and is then lifted off the specimen so that the captured cells can be subjected to further analysis, In this way, homogenous cell populations can be selected away from other "contaminating" cells.

LH: Luteinizing hormone.

LHRH: Luteinizing hormone releasing hormone.

MAPK: Mitogen activated protein kinase. One of the most important signaling pathways activated by mitogens involves the RAS/MAP Kinase cascade.

MEN: Multiple endocrine neoplasia.

Mitosis: Process of cell division during which one cell gives rise to two identical daughter cells.

MOMP: Mitochondrial outer membrane permeability, key process involved in the mitochondrial pathway of apoptosis. Leads to release of key proteins such as cytochrome C, endonuclease G, AIF, Smac/DIABLO and Htra2/Omi. These can engage both caspase-dependent and –independent pathways of apoptosis.

MRI: Magnetic resonance imaging.

Myc protein: A transcription factor protein initially isolated from avian acute leukemia viruses that cause Myelocytomatosis in chickens.

Neoplasia = Cancer = Tumor: Result of inappropriate proliferation and loss of control.

Neovascularization: Vascularization of adult tissue in excess of developmentally predetermined vascularization.

NER: Nucleotide excision repair of DNA damage.

NHEJ: Nonhomologous end joining.

P53: A transcription factor protein that, when activated by stress signals, induces the transcription of many proteins to enforce growth arrest or apoptosis; 53 represents the molecular weight of the protein.

Paracrine: Extracellular signals (for example those regulating replication or growth) which operate at a local level between adjacent cells for example the production of angiogenic factors by tumor cells acting on adjacent endothelial cells.

PCR: Polymerase chain reaction.

PDGF: Platelet derived growth factor.

Pericyte: A connective tissue-type cell containing smooth muscle actin, which is found in the outer layer of small blood vessels.

PET: Position emission tomography.

Phase III clinical trial: Clinical trials are conducted in stages from Phase I to Phase III or Phase IV. Phase I studies address the safety, or maximally tolerated dose of a new treatment agent. Phase II studies seek to determine the potential efficacy of new treatments. Phase III studies are randomized trials, that randomly allocate a portion of patients to a "control" arm (usually the best available treatment at the time) and another portion to a treatment or experimental arm or arms. In this way, the superiority (or lack thereof) of the new treatment over standard treatment can be evaluated. Phase IV studies usually examine novel applications of existing previously approved treatments. Phase III studies are often a final step before federal regulatory approval for a new treatment is considered.

PLAP: Placental alkaline phosphates.

PR: Progesterone receptor.

PSA: Prostate specific antigen.

PTEN: Phosphatase and TENsin homolog on human chromosome TEN. Phosphatase that removes the phosphates groups from phosphatidyl-inositol 3, 4, 5-triphosphate (PIP3) to generate a PI (3,4)P2 and counteract the action of the phosphatidyl inositol 3 kinase (PI3K) that converts PIP2 to PIP3.

PTH: Parathyroid hormone.

Quiescence: A cellular state where cells are in a reversible growth arrest.

RAG: Recombination-activating gene. The B-cell receptor is composed of immunoglobulin (Ig) chains responsible for binding antigen. The T-cell receptor is composed of a heterodimer of two polypeptide chains containing Ig-like domains. The *RAG-1* and *RAG-2* genes encode for lymphocyte-specific recombinases responsible for DNA recombination events required to form functional Ig and TCR genes. The absence of RAG-1 or -2 function leads to the absence of mature B and T cells.

Ras protein: A protein that binds GTP in the active state and GDP in the inactive state and relays signals from receptors at the surface of the cell to the nucleus. Was first identified in Rat sarcomas, hence its name.

Retinoblastoma: Tumor of the retina. Name of the tumor suppressor that when mutated or deleted induces this tumor type.

RTK: Receptor tyrosine kinase.

S phase: The phase in the cell cycle during which cells synthesize DNA and double their chromosome content from $2N$ to $4N$.

SAGE (Serial analysis of gene expression): SAGE provides quantitative expression profiling without a requirement for previous knowledge of genes that are to be detected. All genes, including novel genes, are tagged by a short 14–21 bp sequence (SAGE tags). These SAGE tags, which are uniquely associated with a given gene and can be used to identify it, are joined serially to allow efficient sequencing, amplified by polymerase chain reactions (PCR) and then digitally analyzed for levels of expression. This information forms a library that can then be used to analyze differential expression among cells that have been subjected to SAGE analysis.

SCF: Ubiquitin ligase complex named after its three main protein subunits- SKP1/CUL1/F-box protein.

Senescence: An irreversible state reached by the cells after a certain number of cell divisions.

Characterized by cell's flat morphology, irreversible growth arrest, and expression of the β-galactosidase protein.

Smac/DIABLO: Smac (second mitochondria derived activator of caspase).

S phase: The phase in the mitotic cell cycle when DNA is duplicated.

STAT: Signal transducers and activators of transcription. A family of signaling molecules associated with cytokine receptors. Cytokine receptor binding leads to phosphorylation of STAT proteins that migrate to the nucleus and bind to promoter regions of various genes.

SVCO: Superior vena caval obstruction.

TK: Tyrosine kinase.

TNFα: Tumor necrosis factor type alpha.

TGFβ: Transforming growth factor type beta.

TNM: Tumor node metastatic staging classification.

TS: Thymidylate synthase.

TSH: Thyroid-stimulating hormone.

Tumor suppressor: Protein present in all cells at low levels but induced in response to stress to counter inappropriate cell proliferation by inducing growth arrest or apoptosis. Its loss predisposes cells to cancer.

Vascular leak syndrome: *In vivo* administration of lymphokines, such as IL-2, and immunotoxins can lead to a condition involving damage to vascular endothelial cells, leakage of intravascular fluid into tissues, and organ failure, known as vascular leak syndrome or VLS. VLS represents a dose-limiting adverse effect in patients receiving high-dose IL-2. It is believed that a three amino acid consensus sequence in IL-2 is responsible for this initiating damage.

Vasculogenesis: Formation of new blood vessels from angioblasts in the embryo.

VEGF: Vascular endothelial growth factor; also referred to as VEGF-A.

WBC: White blood count.

Answers to Questions

Chapter 2

1 a. False
b. True
c. True
d. False

2 a. False
b. True
c. False
d. True

3 a. False
b. False
c. True
d. True

Chapter 3

1 a. False
b. True
c. False
d. True
e. True

2 a. False
b. True
c. False
d. True
e. False

3 a. True
b. True
c. True
d. True
e. True

4 a. True
b. True
c. True
d. False
e. True

5 a. True
b. False
c. True
d. True
e. False

Chapter 4

1 a. False
b. True
c. True
d. True
e. False

2 a. False
b. True
c. True
d. True
e. False

3 a. True
b. False
c. True
d. True
e. False

4 a. False
b. False
c. False
d. False
e. False

5 a. False
b. True
c. True
d. True
e. False

Chapter 5

1 a. True
   b. False
   c. False
   d. True
   e. False

2 a. False
   b. True
   c. False
   d. False
   e. True

3 a. True
   b. True
   c. False
   d. True
   e. False

4 a. False
   b. True
   c. False
   d. True
   e. True

5 a. True
   b. False
   c. False
   d. True
   e. False

Chapter 6

1 a. True
   b. False
   c. False
   d. True
   e. False

2 a. False
   b. True
   c. False
   d. True
   e. True

3 a. False
   b. True
   c. False
   d. True
   e. True

4 a. False
   b. True
   c. True
   d. True
   e. False

5 a. True
   b. False
   c. False
   d. True
   e. True

Chapter 7

1 a. True
   b. True
   c. True
   d. True
   c. True

2 a. False
   b. True

3 a. True
   b. True
   c. True
   d. True
   e. True

4 a. False
   b. True

5 a. True
   b. False
   c. False

Chapter 8

1 a. False
   b. True
   c. True
   d. False
   e. True

2 a. True
   b. False
   c. True
   d. False
   e. True

3 a. True
   b. False
   c. True
   d. True
   e. False

4 a. True
   b. True
   c. True
   d. False
   e. False

5 a. True
   b. True
   c. True
   d. False
   e. False

Chapter 9

1 a. True
   b. False
   c. True
   d. True
   e. False

2 a. True
   b. False
   c. False
   d. True
   e. False

3 a. True
   b. True
   c. False
   d. False
   e. True

4 a. False
   b. False
   c. True
   d. True
   e. False

5 a. False
   b. True
   c. True
   d. False
   e. True

Chapter 10

1 a. False
   b. False
   c. True
   d. True
   e. True

2 a. False
   b. False
   c. True
   d. False
   e. False

3 a. False
   b. True
   c. True
   d. True
   e. False

4 a. True
   b. False
   c. False
   d. True
   e. False

5 a. True
   b. False
   c. True
   d. True
   e. False

Chapter 11

1 a. True
   b. True
   c. True
   d. True
   e. True

2 a. False
   b. True
   c. False
   d. False
   e. False

3 a. True
   b. True
   c. False
   d. True
   e. False

4 a. False
 b. True
 c. True
 d. True
 e. True

5 a. False
 b. True
 c. True
 d. False
 e. True

Chapter 13

1 c
2 d
3 b
4 c
5 a

Chapter 14

1 b
2 c
3 c

Chapter 15

1 a. False
 b. True
 c. False
 d. False
 e. True

2 a. True
 b. True
 c. True
 d. False
 e. False

3 a. True
 b. True
 c. True
 d. False
 e. False

Chapter 16

1 a. True
 b. True
 c. False
 d. False
 e. False

2 a. False
 b. True
 c. False
 d. True
 e. False

3 a. False
 b. True
 c. False
 d. True
 e. True

4 a. False
 b. True
 c. False
 d. False
 e. True

Chapter 17

1 (i) b; (ii) e; (iii) d; (iv) a; (v) c
2 a. False; b. True; c. True; d. False;
 e. False; f. True; g. False; h. False
3 (i) d; (ii) b; (iii) a; (iv) c; (v) e

Chapter 18

1 b
2 a
3 d

Index

Page numbers in *italics* refer to figures and those in **bold** to tables